STANLEY E. GUNSTREAM
Pasadena City College

ANATOMY & PHYSIOLOGY

with Integrated Study Guide

second edition

Boston Burr Ridge, IL Dubuque, IA Madison, WI New York San Francisco St. Louis
Bangkok Bogotá Caracas Lisbon London Madrid
Mexico City Milan New Delhi Seoul Singapore Sydney Taipei Toronto

McGraw-Hill Higher Education

A Division of The ***McGraw-Hill*** *Companies*

ANATOMY & PHYSIOLOGY WITH INTEGRATED STUDY GUIDE,
SECOND EDITION

This book is printed on recycled, acid-free paper containing 10% postconsumer waste.

3 4 5 6 7 8 9 0 QPD/QPD 0 9 8 7 6 5 4 3 2

ISBN 0–697–16022–X

Vice president and editorial director: *Kevin T. Kane*
Publisher: *Colin H. Wheatley*
Sponsoring editor: *Kristine Tibbetts*
Developmental editor: *Patrick F. Anglin*
Marketing manager: *Heather K. Wagner*
Project manager: *Joyce M. Berendes*
Senior production supervisor: *Mary E. Haas*
Coordinator of freelance design: *Michelle D. Whitaker*
Photo research coordinator: *John C. Leland*
Supplement coordinator: *Brenda A. Ernzen*
Compositor: *Carlisle Communications, Ltd.*
Typeface: *10/12 Melior*
Printer: *Quebecor Printing Book Group/Dubuque, IA*

Freelance cover/interior designer: *Kaye Farmer*
Cover illustration: *Brian Evans*

The credits section for this book begins on page 525 and is considered an extension of the copyright page.

Anatomy and Physiology Laboratory Textbooks

Benson, Gunstream, Talaro, Talaro	Complete Version—Cat, 7th edition Intermediate Version—Cat, 5th edition Intermediate Version—Fetal Pig, 4th edition Short Version—7th edition
Gunstream, Benson, Talaro, Talaro	Essentials Version—2nd edition
Benson, Talaro	Human Anatomy—6th edition

Some of the laboratory experiments included in this text may be hazardous if materials are handled improperly or if procedures are conducted incorrectly. Safety precautions are necessary when you are working with chemicals, glass test tubes, hot water baths, sharp instruments, and the like, or for any procedures that generally require caution. Your school may have set regulations regarding safety procedures that your instructor will explain to you. Should you have any problems with materials or procedures, please ask your instructor for help.

www.mhhe.com

CONTENTS

PREFACE

Anatomy and Physiology with Integrated Study Guide, second edition, is designed for students who are enrolled in a one-semester course in human anatomy and physiology. The scope, organization, writing style, depth of presentation, and pedagogical aspects of the text have been tailored to meet the needs of students preparing for a career in one of the allied health professions.

These students usually have diverse backgrounds, including limited exposure to biology and chemistry, and this presents a formidable challenge to the instructor. To help meet this challenge, this text is written in clear, concise English and simplifies the complexities of anatomy and physiology in ways that enhance understanding without diluting the essential subject matter.

Themes

There are two unifying themes in this presentation of normal human anatomy and physiology: (1) the relationships between structure and function of body parts and (2) the mechanisms of homeostasis. In addition, interrelationships of the organ systems are noted where appropriate and useful.

Organization

The sequence of chapters progresses from simple to complex. The simple-to-complex progression also is used within each chapter. Chapters covering an organ system begin with anatomy to ensure that students are well prepared to understand the physiology that follows. Each chapter concludes with a brief consideration of common disorders that the student may encounter in the clinical setting. An integrated study guide, unique among anatomy and physiology texts, is located between the text proper and the appendices.

Study Guide

The *Study Guide* is a proven mechanism for enhancing learning by students and now features full-color line art. There is a study guide of four to eight pages for each chapter. Students demonstrate their understanding of the chapter by labeling diagrams and answering completion, matching, and true/false questions. The completion questions "force" students to write and spell correctly the technical terms that they must know. Each chapter study guide concludes with a few critical-thinking, short-answer essay questions where students apply their knowledge to clinical situations.

Answers to the *Study Guide* are included in the *Instructor's Manual* to allow the instructor flexibility: (1) answers may be posted so students can check their own responses, or (2) they may be graded to assess student progress. Either way, prompt feedback to students is most effective in maximizing learning.

Other Learning Aids

A variety of additional learning aids are incorporated to facilitate the learning process and to stimulate interest in the subject.

Chapter Preview and Learning Objectives

Each chapter begins with a list of major topics discussed in the chapter, and under each topic the learning objectives are noted. This informs students of the major topics to be covered and their minimal learning responsibilities.

Key Terms

Several features have been incorporated to assist students in learning the necessary technical terms that often are troublesome for beginning students.

1. A list of *Selected Key Terms* with definitions, and including derivations where helpful, is provided at the beginning of the chapter to inform students of some of the key terms to watch for in the chapter.
2. Throughout the text, key terms are in bold or italic type for easy recognition, and they are defined at the time of first usage. A *phonetic pronunciation* follows where students need help in pronouncing the term. Experience has shown that students learn only terms that they can pronounce.

3. *Keys to Medical Terminology* in appendix A explains how technical terms are structured and provides a list of prefixes, suffixes, and root words to further aid an understanding of medical terminology.
4. At the end of each chapter, a section titled *Building Your Vocabulary* provides a list of *Selected New Terms* that allows students to review their understanding of the new terms. Also, a few *Related Clinical Terms* are defined with phonetic pronunciations to help students to start building a clinical vocabulary.

Figures and Tables

Over 300 new, high quality, full-color illustrations are coordinated with the text to help students visualize anatomical features and physiological concepts. Tables are used throughout to summarize information in a way that is more easily learned by students.

Clinical Boxes

Numerous boxes containing related clinical information are strategically placed throughout the text. They serve to provide interesting and useful information related to the topic at hand. The clinical boxes are identified by a *stethoscope icon* for easy recognition.

In-Text Review Questions

Review questions at the end of major sections challenge students to assess their understanding before proceeding.

Chapter Summary

Each chapter summary consists of a numerical listing of key concepts for each major heading of the chapter. The summary provides a quick review of each chapter for the student.

Study Activities

In this section, students are directed to complete the chapter study guide and the chapter learning objectives. Both activities aid and reinforce learning and allow students to recognize those topics that they need to study further.

Check Your Understanding

Each chapter concludes with a brief quiz, composed of completion questions, that allows students to evaluate their understanding of chapter topics. Answers are provided in appendix B for immediate feedback.

Changes in the Second Edition

Every chapter has been improved to make this edition even more suitable for allied health students.

1. Most chapters have been substantially rewritten to update the subject matter and to simplify the narrative.
2. Nearly all figures are new, and they present the subject matter with even greater clarity.
3. Many new tables have been added that summarize information in ways that enhance student learning.
4. In-text review questions have been inserted at the end of major sections to challenge students to assess their understanding before proceeding.
5. Tables of normal values for common blood and urine tests and a list of common medical abbreviations have been added to the appendices.

Multimedia Correlations

This second edition introduces the Dynamic Human, Version 2.0, 3-D Visual Guide to Anatomy and Physiology CD-ROM, which interactively illustrates the complex relationships between anatomical structures and their functions in the human body. The program covers each body system, demonstrating clinical concepts, histology, and physiology. The Dynamic Human icon appears in appropriate figure legends to alert the reader to the corresponding information. A list of correlating figures to specific sections of The Dynamic Human, Version 2.0, follows this preface.

A set of five videotapes contains nearly 53 animations of physiological processes integral to the study of human anatomy and physiology. Entitled "WCB's Life Science Animation (LSA) Videotape Series," these videotapes cover such topics as cell division, genetics, and reproduction. The LSA 3-D Videotape with 42 key biological processes is included in these correlations. Videotape icons appear in appropriate figure legends to alert the reader to these animations. A list of the figures that relate to the animations follows this preface.

Ancillaries

Instructor's Manual

This helpful manual provides the following for each chapter:

1. lecture outline;
2. instructional suggestions;
3. list of related films and videos;
4. answers to the *Study Guide;*
5. a test-item file of 40 to 50 multiple-choice questions with answers.

Laboratory Manual

Anatomy and Physiology Laboratory Textbook, Essentials Version by Stanley E. Gunstream, Harold J. Benson, Arthur Talaro, and Kathleen P. Talaro, all of Pasadena City College. This excellent lab text presents the fundamentals of human anatomy and physiology in an easy-to-read manner that is appropriate for students in allied health programs. It is designed especially for the one-semester course; it features a simple, concise writing style, 37 self-directing exercises, full-color photomicrographs in the Histology Atlas, and numerous illustrations in each exercise. The lab text is accompanied by an Instructor's Handbook and Slides.

Also Available from McGraw-Hill . . . .

- *The Dynamic Human CD-ROM, Version 2.0* (0-697-38935-9) consists of 3-D and other visualizations of relationships between human structure and function.
- *The Dynamic Human Videodisc* (0-697-38937-5) contains all of the CD-ROM animations, with a bar code directory.
- *Virtual Physiology Lab CD-ROM* (0-697-37994-9) has 10 simulations of animal-based experiments common in the physiology component of a laboratory course; allows students to repeat experiments for improved mastery.
- *WCB Anatomy and Physiology Videodisc* (0-697-27716-X) has more than 30 physiological animations, line art, and photomicrographs, with a bar code directory.
- *WCB's Life Science Animations (LSA)* contains 53 animations on VHS videocassettes; Chemistry, The Cell, and Energetics (0-697-25068-7); Cell Division, Heredity, Genetics, Reproduction, and Development (0-697-25069-5); Animal Biology No. 1 (0-697-25070-9); Animal Biology No. 2 (0-697-25071-7); and Plant Biology, Evolution, and Ecology (0-697-26600-1). Another available videotape is Physiological Concepts of Life Science (0-697-21512-1). A new 3-D videotape (0-07-290652-9) is also available with 42 key biological processes all narrated and animated in vibrant color with dynamic three-dimensional graphics.
- *WCB Anatomy and Physiology Videotape Series* consists of four videotapes, free to qualified adopters, including Blood Cell Counting, Identification and Grouping (0-697-11629-8); Introduction to the Human Cadaver and Prosection (0-697-11177-6); Introduction to Cat Dissection: Cat Musculature (0-697-11630-1); and Internal Organs and Circulatory System of the Cat (0-697-13922-0).
- *Human Anatomy and Physiology Study Cards,* third edition (00697-26447-5) by Kent Van De Graaff, Ward Rhees, and Christopher Creek is a boxed set of 300 illustrated cards (3×5 in.), each of which concisely summarizes a concept of structure or function, defines a term, and provides a concise table of related information.
- *Coloring Guide to Anatomy and Physiology* (0-697-17109-4) by Robert and Judith Stone consists of outline drawings and text that emphasize learning through color association. Students retain information through a meditative exercise in color-coding structures and correlated labels. This can be an especially effective aid for students who more easily remember visual concepts than verbal ones.
- *An Atlas to Human Anatomy* (0-697-38793-3) by Dennis Strete and Christopher Creek is a new full-color atlas that contains over 200 full-color photographs and over 150 black-and-white illustrations that accompany and portray the necessary detail of human anatomy.
- *Atlas of the Skeletal Muscles,* third edition (0-07-290332-5) by Robert and Judith Stone illustrates each skeletal muscle in a diagram that the student can color, and provides a concise table of the origin, insertion, action, and innervation of each muscle.
- *Laboratory Atlas of Anatomy and Physiology,* second edition (0-697-39480-8) by Douglas Eder et al. is a full-color atlas containing histology, human skeletal anatomy, human muscular anatomy, dissections, and reference tables.

Acknowledgments

The development and production of this second edition has been the result of a team effort. My dedicated and creative teammates at McGraw-Hill have contributed greatly to the finished product. I gratefully acknowledge and applaud their efforts. It has been a pleasure to work with these gifted professionals at each step in the process. I am especially appreciative of the continued support given by Colin Wheatley and Kris Tibbetts. Pat Anglin and Laura Beaudoin played key roles in polishing the manuscript. Michelle Whitaker created an eye-catching and functional design, and Joyce Berendes skillfully guided the production process. All have made significant contributions.

The following instructors have served as critical reviewers:

Tracey A. Bergeron
McIntosh College

Lu Anne Clark
Lansing Community College

Joan Ebersole
Greater Altoona Career and Technology Center

David J. Pavlat
Central College

Mary M. Ryan
Central School of Practical Nursing

Marcia L. Scott
Anderson School of Practical Nursing

Halcyon Watkins
Prairie View A & M University

Their suggestions have been very helpful, and I am grateful for their input.

I especially thank Margie for her patience and support.

S.E.G.

LIFE SCIENCE 3D ANIMATIONS CORRELATION GUIDE

Chapter 2

2.1	Module 1	Atomic Structure and Covalent and Ionic Bonding
2.2	Module 1	Atomic Structure and Covalent and Ionic Bonding
2.3	Module 1	Atomic Structure and Covalent and Ionic Bonding
2.12	Module 7	Enzyme Action
2.14	Module 13	Structure of DNA

Chapter 3

3.5	Module 3	Cellular Secretion
3.10	Module 4	Diffusion
3.11	Module 5	Osmosis
3.18	Module 13	Structure of DNA
3.19	Module 18	Transcription
	Module 19	Translation
3.21	Module 14	DNA Replication

Chapter 7

7.5	Module 40	Muscle Contraction

Chapter 10

10.2	Module 41	Hormone Action

Chapter 13

13.9	Module 34	How T Lymphocytes Work

Chapter 14

14.13	Module 37	Gas Exchange
14.14	Module 37	Gas Exchange
14.15	Module 37	Gas Exchange

Chapter 15

15.17	Module 36	Digestion Overview
15.20	Module 36	Digestion Overview
15.22	Module 36	Digestion Overview

Chapter 16

16.6	Module 38	Kidney Function
16.8	Module 38	Kidney Function
16.9	Module 38	Kidney Function
16.10	Module 38	Kidney Function

LIFE SCIENCE ANIMATIONS CORRELATION GUIDE

Chapter 2

2.2	Tape 1	Concept 1	Formation of an Ionic Bond
2.12	Tape 6	Concept 1	Lock and Key Model of Enzyme Action
2.15	Tape 1	Concept 11	ATP as an Energy Carrier

Chapter 3

3.1	Tape 1	Concept 2	Journey into a Cell
3.5	Tape 1	Concept 4	Cellular Secretion
	Tape 6	Concept 4	Cellular Secretion
3.11	Tape 6	Concept 2	Osmosis
3.14	Tape 6	Concept 3	Active Transport
3.15	Tape 1	Concept 3	Endocytosis
3.16	Tape 1	Concept 5	Glycolysis
	Tape 1	Concept 6	Oxidative Phosphorylation
	Tape 1	Concept 7	Electron Transport Chain and the Production of ATP
	Tape 6	Concept 6	Electron Transport Chain and Oxidative Phosphorylation
3.17	Tape 1	Concept 5	Glycolysis
	Tape 1	Concept 6	Oxidative Phosphorylation
	Tape 1	Concept 7	Electron Transport Chain and the Production of ATP
	Tape 6	Concept 6	Electron Transport Chain and Oxidative Phosphorylation
3.21	Tape 2	Concept 15	DNA Replication

Chapter 7

7.1	Tape 3	Concept 29	Levels of Muscle Structure
7.5	Tape 3	Concept 30	Sliding Filament Model of Muscle Contraction
7.7	Tape 1	Concept 11	ATP as an Energy Carrier
7.8	Tape 1	Concept 5	Glycolysis
	Tape 1	Concept 6	Oxidative Phosphorylation
	Tape 1	Concept 7	Electron Transport Chain and the Production of ATP
	Tape 6	Concept 6	Electron Transport Chain and Oxidative Phosphorylation

Chapter 8

8.3	Tape 3	Concept 22	Formation of Myelin Sheath
8.5	Tape 6	Concept 6	Conduction of Nerve Impulses
	Tape 6	Concept 23	Saltatory Nerve Conduction
8.6	Tape 6	Concept 6	Conduction of Nerve Impulses
	Tape 6	Concept 23	Saltatory Nerve Conduction
8.7	Tape 6	Concept 8	Synaptic Transmission
18.18	Tape 3	Concept 25	Reflex Arcs

Chapter 9

9.8	Tape 3	Concept 27	Organ of Corti
9.10	Tape 3	Concept 26	Organ of Static Equilibrium
9.17	Tape 6	Concept 9	Visual Accommodation

Chapter 10

10.2	Tape 6	Concept 10	Action of Steroid Hormone on Target Cells
10.3	Tape 3	Concept 28	Peptide Hormone Action (cAMP)
	Tape 6	Concept 11	Action of T_3 in Target Cells
	Tape 6	Concept 12	Cyclic AMP Action

Chapter 11

11.3	Tape 4	Concept 40	A, B, O Blood Types

Chapter 12

12.1	Tape 4	Concept 37	Blood Circulation
12.7	Tape 4	Concept 32	Cardiac Cycle and Production of Heart Sounds
12.8	Tape 4	Concept 37	Blood Circulation
12.10	Tape 4	Concept 38	Production of Electrocardiogram
12.11	Tape 4	Concept 38	Production of Electrocardiogram

Chapter 13

13.9	Tape 4	Concept 43	Types of T Cells
13.10	Tape 4	Concept 41	B Cell Immune Responses
	Tape 4	Concept 42	Structure and Function of Antibodies

Chapter 15

15.17	Tape 4	Concept 36	Digestion of Lipids
15.20	Tape 4	Concept 34	Digestion of Carbohydrates
15.22	Tape 4	Concept 35	Digestion of Proteins

Chapter 17

17.3	Tape 2	Concept 19	Spermatogenesis
17.4	Tape 2	Concept 19	Spermatogenesis
17.8	Tape 2	Concept 20	Oogenesis
17.9	Tape 2	Concept 20	Oogenesis

Chapter 18

18.3	Tape 2	Concept 21	Human Embryonic Development
18.8	Tape 2	Concept 21	Human Embryonic Development

DYNAMIC HUMAN 2.0 CORRELATION GUIDE

Chapter 1

1.2	Cardiovascular/Anatomy Digestive/Anatomy Endocrine/Anatomy Immune & Lymphatic/Anatomy Muscular/Anatomy Nervous/Anatomy Reproductive/Anatomy Respiratory/Anatomy Skeletal/Anatomy Urinary/Anatomy
1.4	Human Body/Explorations/Anatomical Orientation/Planes
1.5	Human Body/Explorations/Anatomical Orientation/Planes
1.6	Human Body/Explorations/Anatomical Orientation/Planes
TA1.1	Human Body/Clinical Concepts/Clinical Imaging/CT
1.9	Human Body/Explorations/Visible Human/Male/Thorax Human Body/Explorations/Visible Human/Female/Thorax
1.10	Human Body/Explorations/Visible Human/Male/Abdomen Human Body/Explorations/Visible Human/Female/Abdomen

Chapter 2

TA2.1 Human Body/Clinical Concepts/Clinical Imaging/PETT

Chapter 3

3.1 Human Body/Anatomy/Cell Components

Chapter 4

4.1 Human Body/Histology/Simple Squamous Epithelium
4.3 Human Body/Histology/Simple Columnar Epithelium
4.4 Human Body/Histology/Pseudostratified Ciliated Columnar Epithelium
4.5 Human Body/Histology/Stratified Squamous Epithelium
4.6 Human Body/Histology/Transitional
4.10 Human Body/Histology/Hyaline Cartilage
4.11 Human Body/Histology/Elastic Cartilage
4.12 Human Body/Histology/Fibrocartilage
4.13 Skeletal/Histology/Compact Bone
4.15 Muscular/Histology/Skeletal Muscle (cross section)
Muscular/Histology/Skeletal Muscle (longitudinal)
4.16 Muscular/Histology/Cardiac Muscle
4.17 Muscular/Histology/Smooth Muscle
4.18 Nervous/Histology/Dorsal Root Ganglion Neurons

Chapter 6

6.1 Skeletal/Explorations/Cross Section of a Long Bone
6.2 Skeletal/Explorations/Cross Section of a Long Bone
6.5 Skeletal/Anatomy/Gross Anatomy
6.6 Skeletal/Anatomy/Gross Anatomy/Axial Skeleton/Skull
Skeletal/Anatomy/3D Viewer: Cranial Anatomy
6.7 Skeletal/Anatomy/Gross Anatomy/Axial Skeleton/Skull
Skeletal/Anatomy/3D Viewer: Cranial Anatomy
6.8 Skeletal/Anatomy/Gross Anatomy/Axial Skeleton/Skull
Skeletal/Anatomy/3D Viewer: Cranial Anatomy
6.9 Skeletal/Anatomy/Gross Anatomy/Axial Skeleton/Skull
Skeletal/Anatomy/3D Viewer: Cranial Anatomy
6.10 Skeletal/Anatomy/Gross Anatomy/Axial Skeleton/Skull
Skeletal/Anatomy/3D Viewer: Cranial Anatomy
6.12 Skeletal/Anatomy/Gross Anatomy/Axial Skeleton/Vertebral Column
6.13 Skeletal/Anatomy/Gross Anatomy/Axial Skeleton/Vertebral Column
6.14 Skeletal/Anatomy/Gross Anatomy/Axial Skeleton/Vertebral Column
6.15 Skeletal/Anatomy/Gross Anatomy/Axial Skeleton/Vertebral Column
6.16 Skeletal/Anatomy/Gross Anatomy/Axial Skeleton/Thoracic Cage
Skeletal/Anatomy/3D Viewer: Thoracic Anatomy
6.17 Skeletal/Anatomy/Gross Anatomy/Appendicular Skeleton/Pectoral Girdle
6.18 Skeletal/Anatomy/Gross Anatomy/Appendicular Skeleton/Pectoral Girdle
6.19 Skeletal/Anatomy/Gross Anatomy/Appendicular Skeleton/Upper Limb
6.20 Skeletal/Anatomy/Gross Anatomy/Appendicular Skeleton/Upper Limb
6.21 Skeletal/Anatomy/Gross Anatomy/Appendicular Skeleton/Upper Limb
6.22 Skeletal/Anatomy/Gross Anatomy/Appendicular Skeleton/Pelvic Girdle
6.23 Skeletal/Anatomy/Gross Anatomy/Appendicular Skeleton/Pelvic Girdle
6.25 Skeletal/Anatomy/Gross Anatomy/Appendicular Skeleton/Lower Limb
6.26 Skeletal/Anatomy/Gross Anatomy/Appendicular Skeleton/Lower Limb
6.27 Skeletal/Anatomy/Gross Anatomy/Appendicular Skeleton/Lower Limb
6.29 Skeletal/Explorations/Synovial Joints
6.30 Skeletal/Explorations/Synovial Joints

Chapter 7

7.1 Muscular/Anatomy/Skeletal Muscle
7.2 Muscular/Histology/Skeletal Muscle (longitudinal)
Muscular/Histology/Skeletal Muscle (cross-sectional)
7.4 Muscular/Explorations/Neuromuscular Junction
7.5 Muscular/Explorations/Sliding Filament Theory
7.11 Muscular/Explorations/Muscle Action Around Joints
7.14 Muscular/Anatomy/Body Regions/Head and Neck
7.15 Muscular/Anatomy/Body Regions/Abdomen and Back
7.16 Muscular/Anatomy/Body Regions/Abdomen and Back
7.17 Muscular/Anatomy/Body Regions/Pectoral Girdle and Upper Arm
7.18 Muscular/Anatomy/Body Regions/Pectoral Girdle and Upper Arm
7.19 Muscular/Anatomy/Body Regions/Forearm
7.20 Muscular/Anatomy/Body Regions/Forearm
7.21 Muscular/Anatomy/Body Regions/Thigh
7.22 Muscular/Anatomy/Body Regions/Thigh
7.23 Muscular/Anatomy/Body Regions/Thigh
7.24 Muscular/Anatomy/Body Regions/Lower Leg
7.25 Muscular/Anatomy/Body Regions/Lower Leg
7.26 Muscular/Anatomy/Body Regions/Lower Leg

Chapter 8

8.1 Nervous/Histology/Spinal Neuron
8.8 Nervous/Anatomy/3D Viewer: Cranial Anatomy
8.9 Nervous/Anatomy/Spinal Cord Anatomy
8.10 Nervous/Anatomy/Gross Anatomy of the Brain
8.12 Nervous/Anatomy/Gross Anatomy of the Brain
8.15 Nervous/Anatomy/Spinal Cord
Nervous/Histology/Spinal Cord
8.18 Nervous/Histology/Spinal Neurons
Nervous/Explorations/Reflex Arc

Chapter 9

9.3 Nervous/Histology/Vallate Papilla
9.4 Nervous/Explorations/Taste
Nervous/Explorations/Zones of Taste
9.5 Nervous/Explorations/Olfaction
9.6 Nervous/Explorations/Hearing
9.8 Nervous/Histology/Organ of Corti
9.9 Nervous/Explorations/Hearing
9.10 Nervous/Explorations/Static Equilibrium
9.13 Nervous/Explorations/Vision
Nervous/Histology/Eye
9.16 Nervous/Histology/Eye
9.18 Nervous/Histology/Retina
9.21 Nervous/Explorations/Nearsightedness vs. Farsightedness

Chapter 10

10.1 Endocrine/Anatomy/Gross Anatomy
10.6 Endocrine/Explorations/Endocrine Function
10.7 Endocrine/Anatomy/Gross Anatomy/Thyroid Gland
Endocrine/Histology/Thyroid Gland
10.10 Endocrine/Anatomy/Gross Anatomy/Adrenal Glands
10.11 Endocrine/Anatomy/Gross Anatomy/Pancreas
Endocrine/Histology/Pancreas

Chapter 11

11.13 Immune & Lymphatic/Clinical Concepts/Blood Type

Chapter 12

12.1 Cardiovascular/Explorations/Heart Dynamics/Blood Flow
12.2 Cardiovascular/Anatomy/3D Viewer: Thoracic Anatomy
12.3 Cardiovascular/Anatomy/Gross Anatomy of the Heart
12.4 Cardiovascular/Anatomy/Gross Anatomy of the Heart
12.5 Cardiovascular/Anatomy/Gross Anatomy of the Heart
12.7 Cardiovascular/Explorations/Heart Dynamics/Cardiac Cycle

12.8	Cardiovascular/Explorations/Heart Dynamics/Blood Flow
12.9	Cardiovascular/Explorations/Heart Dynamics/Conduction System
12.10	Cardiovascular/Explorations/Heart Dynamics/Electrocardiogram
12.11	Cardiovascular/Explorations/Heart Dynamics/Electrocardiogram
12.13	Cardiovascular/Histology/Vasculature Cardiovascular/Explorations/Generic Vasculature
12.14	Cardiovascular/Explorations/Generic Vasculature/Arteriole
12.15	Cardiovascular/Explorations/Generic Vasculature/Capillary
12.25	Cardiovascular/Explorations/Generic Portal System

Chapter 13

13.1	Immune & Lymphatic/Anatomy/Gross Anatomy Immune & Lymphatic/Explorations/Lymph Formation and Movement
13.3	Immune & Lymphatic/Anatomy/Gross Anatomy Immune & Lymphatic/Explorations/Lymph Formation and Movement
13.5	Immune & Lympathic/Histology/Lymph Node
13.6	Immune & Lymphatic/Histology/Lymph Node
13.7	Immune & Lymphatic/Anatomy/Gross Anatomy
13.8	Immune & Lymphatic/Anatomy/Microscopic Components Immune & Lymphatic/Explorations/Specific Immunity
13.10	Immune & Lymphatic/Explorations/Specific Immunity

Chapter 14

14.1	Respiratory/Anatomy/Gross Anatomy
14.3	Respiratory/Histology/Nasal Cavity
14.6	Respiratory/Anatomy/Gross Anatomy
14.7	Respiratory/Histology/Alveoli Respiratory/Histology/Bronchiole
14.8	Respiratory/Explorations/Mechanics of Breathing
14.10	Respiratory/Clinical Concepts/Spirometry
14.12	Respiratory/Histology/Alveoli
14.13	Respiratory/Explorations/Gas Exchange
14.14	Respiratory/Explorations/Gas Exchange
14.15	Respiratory/Explorations/Gas Exchange

Chapter 15

15.1	Digestive/Anatomy/Gross Anatomy Digestive/Anatomy/3D Viewer: Digestive Anatomy
15.2	Digestive/Histology/Duodenum Digestive/Histology/Duodenal Villi
15.3	Digestive/Explorations/Oral Cavity
15.4	Digestive/Anatomy/Gross Anatomy
15.5	Digestive/Explorations/Oral Cavity
15.6	Digestive/Histology/Tooth
15.7	Digestive/Histology/Submandibular Gland
15.8	Digestive/Histology/Fundic Stomach
15.13	Digestive/Anatomy/Gross Anatomy
15.15	Digestive/Anatomy/Gross Anatomy
15.16	Digestive/Histology/Duodenal Villi
15.17	Digestive/Explorations/Digestion
15.18	Digestive/Anatomy/Gross Anatomy
15.20	Digestive/Explorations/Digestion
15.22	Digestive/Explorations/Digestion

Chapter 16

16.1	Urinary/Anatomy/Gross Anatomy
16.2	Urinary/Anatomy/Kidney Anatomy Urinary/Anatomy/Nephron Anatomy Urinary/Anatomy/3D Viewer: Nephron
16.3	Urinary/Anatomy/Nephron Anatomy Urinary/Anatomy/3D Viewer: Nephron
16.4	Urinary/Anatomy/Kidney Anatomy
16.6	Urinary/Explorations/Urine Formation
16.8	Urinary/Explorations/Urine Formation
16.9	Urinary/Explorations/Urine Formation
16.10	Urinary/Explorations/Urine Formation
16.13	Urinary/Histology/Bladder Urinary/Clinical Concepts/Ultrasound of Bladder

Chapter 17

17.1	Reproductive/Anatomy/3D Viewer: Male Reproductive Anatomy Reproductive/Anatomy/Male Reproductive Anatomy
17.2	Reproductive/Histology/Testis
17.3	Reproductive/Explorations/Male: Spermatogenesis
17.4	Reproductive/Explorations/Male: Spermatogenesis
17.7	Reproductive/Anatomy/3D Viewer: Female Reproductive Anatomy Reproductive/Anatomy/Female Reproductive Anatomy
17.8	Reproductive/Explorations/Female: Oogenesis
17.9	Reproductive/Histology/Ovary Reproductive/Explorations/Female: Oogenesis
17.10	Reproductive/Histology/Ovarian Follicle
17.11	Reproductive/Anatomy/3D Viewer: Female Reproductive Anatomy Reproductive/Anatomy/Female Reproductive Anatomy
17.13	Reproductive/Explorations/Female: Ovarian Cycle Reproductive/Explorations/Female: Menstrual Cycle

Chapter 18

18.20	Reproductive/Explorations/Female: Amniocentesis

NOTE TO STUDENTS

You are starting a fascinating and challenging study of human anatomy and physiology. The course will be rigorous, but you can improve your chances of success by taking advantage of the learning aids found in this text. The suggestions noted below have been helpful to many students. Consider incorporating them into your study habits.

1. You can best profit from your instructor's lecture if you have prepared for it by studying the reading assignment *before* you attend the lecture. If you do this, you will know the new terms that will be encountered and at least the general aspects of the subject matter. This will enable you to better understand the lecture and take better lecture notes. It will save you study time later on.
2. Before you read a chapter, you need to do two things. First, examine the *Chapter Preview and Learning Objectives.* This will inform you of the sequence of major topics to be covered and your minimal learning responsibilities. Second, study the list of *Selected Key Terms* to learn their meanings. After you have done these two things, you will be prepared to learn effectively as you read the chapter
3. As you read a chapter, watch for new terms and be sure to learn their pronunciation and meaning before going on. Use the *Glossary* and *Keys to Medical Terminology* as necessary to assist you. Key terms are in bold or italic print so that you can easily recognize them. Where pronunciation help is needed, a *phonentic pronunciation* immediately follows in parentheses to help you pronounce the term correctly. Say the term out loud a few times to be sure that you know the correct pronunciation. This will help you remember the term.

   The phonetic pronunciation breaks terms into syllables and includes marks denoting long vowel sounds and major accents. Vowels marked with a line above the letter are pronounced with a long sound—the same sound as when saying the letter. Some examples follow:

   - ā as in take
   - ē as in be
   - ī as in time
   - ō as in hole

   Vowels without these marks are pronounced with short sounds as in the following examples:

   - a as in above
   - e as in pet
   - o as in pot
   - u as in mud
   - i as in hip

   The accent mark indicates the major accent in the term, such as in terminology (ter-min-ol′-ō-jē) and anatomy (a-nat′-ō-mē).

   You can best learn new terms and their meanings by preparing *flash cards* from 3 × 5 index cards. Place the term on one side and the definition on the other. *Place only one term on each card.*
4. Use a highlighter pen liberally to mark the key statements in each paragraph as you read. This will help you identify the key points to study later on. You may want to make flash cards to help you learn these key points.
5. Review questions are strategically located in the text to allow you to check your understanding of each major section. It is important to be able to answer these questions correctly before proceeding.
6. After reading the chapter, review the *Chapter Summary* to be sure that you understood the key points of the chapter. This section will give you a quick review of the chapter later on as well.
7. Learning new terms is one of the more difficult aspects of a course in anatomy and physiology. After reading a chapter, use *Building Your Vocabulary* to help you master the terminology. Review the list of *Selected New Terms* to be sure that you know their meanings. If not, look them up in the chapter or in the glossary. Then learn the *Related Clinical Terms* to start building your clinical vocabulary.
8. After you think that you understand the topics pretty well, complete the *Study Activities* to reinforce your understanding and bring to your attention those topics that you need to study a bit more.

   - Complete the *Study Guide* for the chapter you are studying. It covers the major points of the chapter, and it includes a few critical thinking questions that apply your knowledge to clinical situations. You may find it easier to remove the chapter study guide from the text before completing it. You can keep the study guides in a loose-leaf notebook. Your instructor has the answer key so you can check your responses.
   - Complete the *Learning Objectives* listed on the first page of each chapter. If you can explain the concepts and mechanisms correctly in your own words, you know the topics well.
9. Finally, complete the *Check Your Understanding* self-test. The answers to the questions are located in appendix B for immediate feedback. If you have thoughtfully completed all of these study aides and have answered the questions correctly, you may be confident of your understanding.

CHAPTER ONE

Introduction to the Human Body

Chapter Preview & Learning Objectives

Anatomy and Physiology
- Contrast anatomy and physiology.

Levels of Organization
- List and describe the levels of organization in the human body.
- List the major organs and functions for each organ system.

Directional Terms
- Use directional terms to describe the location of body parts.

Body Regions
- Locate the major body regions on a chart or manikin.

Body Planes and Sections
- Name and describe the four planes used in making sections of the body or body parts.

Body Cavities
- Name the two major body cavities, their subdivisions and membranes, and locate them on a chart or manikin.
- Name the organs located in each body cavity.

Abdominopelvic Subdivisions
- Name the abdominopelvic quadrants and nine regions, locate them on a chart or manikin, and list the major internal organs found in each.

Maintenance of Life
- Describe the general nature of metabolism.
- List the five basic needs essential for human life.
- Define homeostasis and explain its relationship to both normal body functions and disorders.

Chapter Summary
Building Your Vocabulary
Study Activities
Check Your Understanding

I

SELECTED KEY TERMS

Anatomy (ana = apart; tom = to cut) The study of the structure of living organisms.
Anterior (ante = before, in front of) The abdominal or ventral side of the body.
Appendicular (append = to hang) Pertaining to the extremities (arms and legs).
Axial (ax = axis) Pertaining to the longitudinal axis of the body.
Cephalic (cephal = head) Pertaining to the head.
Cervical (cervic = neck) Pertaining to the neck.
Homeostasis (homeo = same; sta = make stand or stop) Maintenance of a relatively stable internal environment.
Meninges (mening = membrane) Membranes covering the brain and spinal cord.
Metabolism (metabole = change) The sum of the chemical reactions (changes) in the body.
Parietal (paries = wall) Pertaining to the wall of a body cavity.
Pericardium (peri = around; cardi = heart) The membrane surrounding the heart.
Peritoneum (peri = around; ton = to stretch) The membrane lining the abdominal cavity and covering the abdominal organs.
Physiology (physio = nature; logy = study of) The study of the functioning of living organisms.
Pleura (pleura = rib) The membrane lining the thoracic cavity.
Posterior (post = after, behind) The dorsal or backside of the body.
Visceral (viscus = internal organ) Pertaining to organs in a body cavity.

You are beginning a fascinating and challenging study—the study of the human body. As you progress through this text, you will begin to understand the complexity of organization and function found in the human organism. Keep in mind that the goals of your study are to learn (1) how the body is structured and organized and (2) how the body functions.

This first chapter provides an overview of the human body to build a foundation of knowledge that is necessary for your continued study. Like the chapters that follow, this chapter introduces a number of new terms that must be learned. It is important that you start to build a vocabulary of technical terms and continue to develop it throughout your study.

Anatomy and Physiology

Knowledge of the human organism is obtained primarily from two scientific disciplines—anatomy and physiology—and each consists of a number of subdisciplines.

Human **anatomy** (ah-nat′-ō-mē) is the study of the structure and organization of the body and its parts. There are two subdivisions of anatomy. *Gross anatomy* involves the dissection and examination of various parts of the body without magnifying lenses. *Microanatomy* consists of the microscopic examination of tissues and cells.

Human **physiology** (fiz-ē-ol′-ō-jē) is the study of the function of the body and its parts. Physiology involves observation and experimentation, and it usually requires the use of specialized equipment and materials.

In your study of the human body, you will see that there is always a definite relationship between the anatomy and physiology of the body and body parts.

Levels of Organization

The human body is complex, so it is not surprising that there are several levels of structural organization, as shown in figure 1.1.

Chemical Level

At the simplest level, the body is composed of chemical substances that are formed of atoms and molecules. *Atoms* are the fundamental building blocks of chemicals, and atoms combine in specific ways to form *molecules.* Some molecules are very small, such as water molecules, but others may be very large, such as the macromolecules of proteins.

Cellular Level

Various combinations of chemical substances form the trillions of cells composing the body. **Cells** are the basic structural and functional units of the body because all of the processes of life occur within cells. The body is composed of many different types of cells. Muscle cells, blood cells, and nerve cells are examples.

Cells contain subunits called *organelles* that carry out specific functions within the cells. Organelles are formed of both small and large molecules.

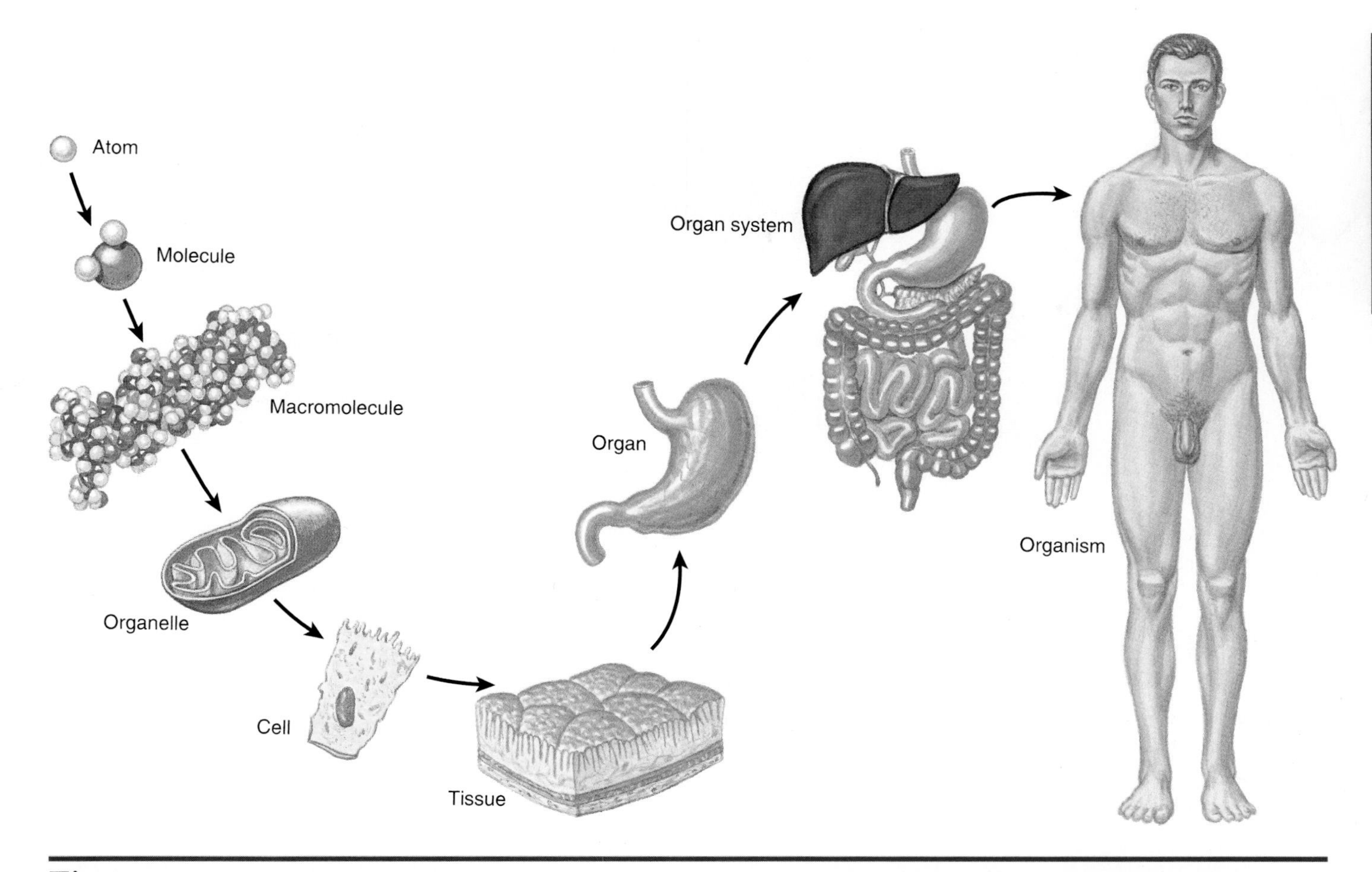

Figure 1.1 Levels of organization. Atoms form the lowest, or simplest, level. Each higher level increases in complexity.

Tissue Level

Similar types of cells are usually grouped together in the body to form a tissue. Each body **tissue** consists of an aggregation of similar cells that perform similar functions. There are several different types of tissues in the body. Muscle and nerve tissues are examples.

Organ Level

Each **organ** of the body is composed of two or more tissues that work together, enabling the organ to perform its specific functions. The body contains numerous organs, and each has a definite form and function. The stomach, heart, and brain are examples of organs.

Organ System Level

The organs of the body are arranged in functional groups so that their independent functions are coordinated to perform specific system functions. These coordinated, functional groups are called **organ systems.** The digestive and nervous systems are examples of organ systems. Most organs belong to a single organ system, but a few organs are assigned to more than one organ system. For example, the ovaries and testes belong to both the reproductive and endocrine systems.

Figure 1.2 illustrates the organ systems of the body and lists the major components and functions for each system.

Effective communication among health-care personnel depends upon the use of anatomical and physiological terms to provide precise meaning. These technical terms usually have Latin roots and endings. To change such terms from singular to plural requires a change in the ending, but not the addition of an *s*. Instead, terms ending in *us* are changed to *i*, those ending in *a* are changed to *ae*, and those ending in *um* are converted to *a*.

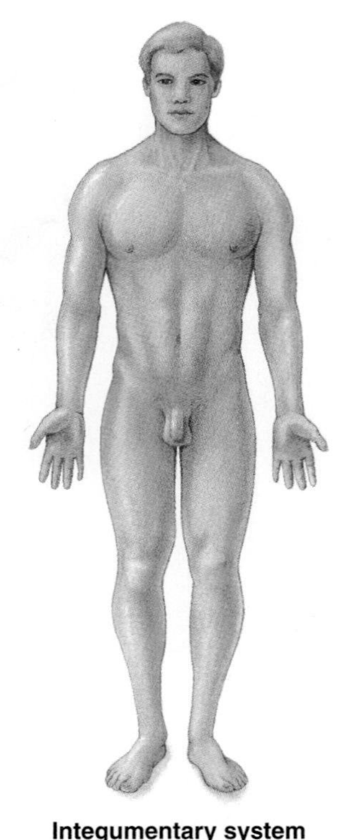

Integumentary system

Components: skin, hair, nails, and associated glands
Functions: protects underlying tissues and helps regulate body temperature

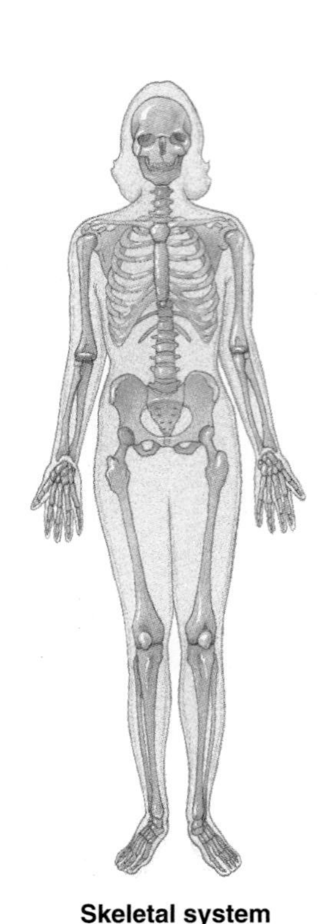

Skeletal system

Components: bones, ligaments, and associated cartilages
Functions: supports the body, protects vital organs, stores minerals, and forms blood cells

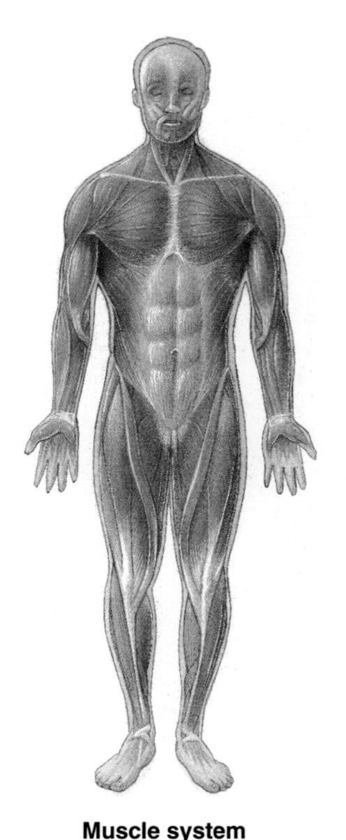

Muscle system

Components: skeletal muscles, cardiac muscle of the heart, and smooth muscle in walls of internal organs
Functions: moves the body and body parts and produces heat

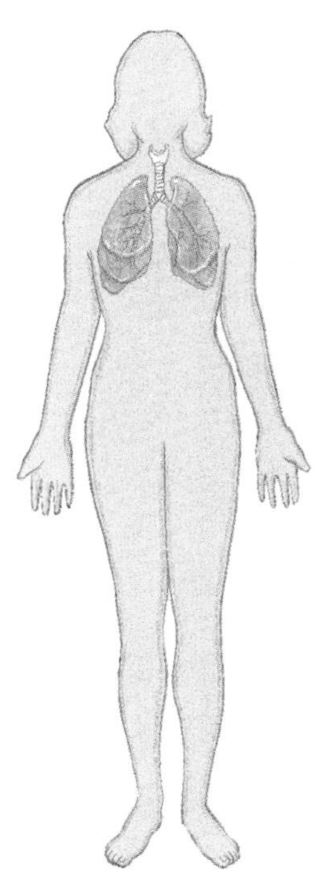

Respiratory system

Components: nasal cavity, pharynx, larynx, trachea, bronchi, and lungs
Functions: exchanges O_2 and CO_2 between air and blood in the lungs

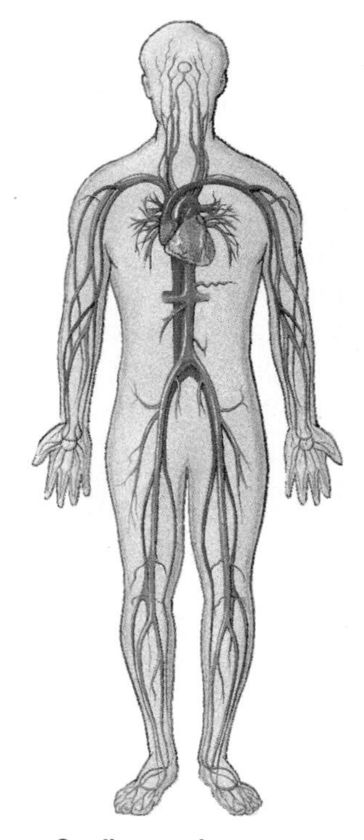

Cardiovascular system

Components: blood, heart, arteries, veins, and capillaries
Functions: transports materials to and from the body cells

Figure 1.2 The organ systems of the body.

Organismic Level

The highest organizational level is the *organismic level,* the human organism as a whole. All of the organizational levels from chemicals to organ systems contribute to the functioning of the entire body.

> ***What are the organizational levels of the human body?***
> ***What are the major organs and general functions of each organ system?***

Directional Terms

Directional terms are used to describe the relative position of a body part. The use of these terms conveys a precise meaning enabling the listener or reader to locate the body part of interest. It is always assumed that the body is in a standard position, the *anatomical position,* in

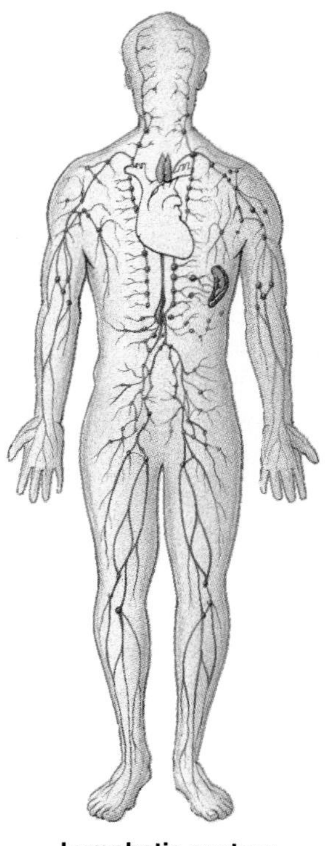

Lymphatic system

Components: lymph, lymphatic vessels, and lymphatic tissue
Functions: collects, cleanses, and returns interstitial fluid to the blood; provides immunity

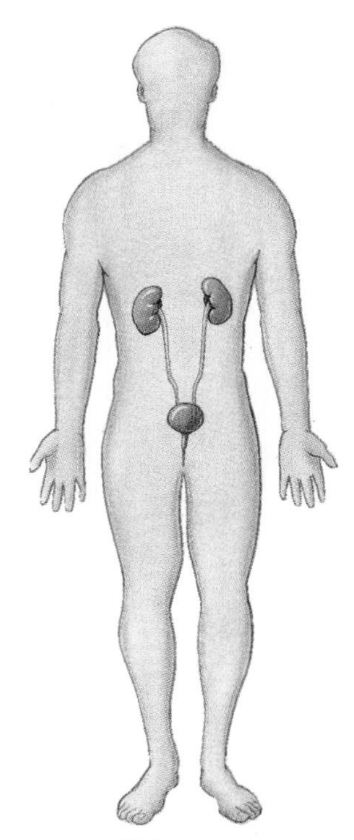

Urinary system

Components: kidneys, ureters, urinary bladder, and urethra
Functions: regulates volume and composition of blood by forming and excreting urine

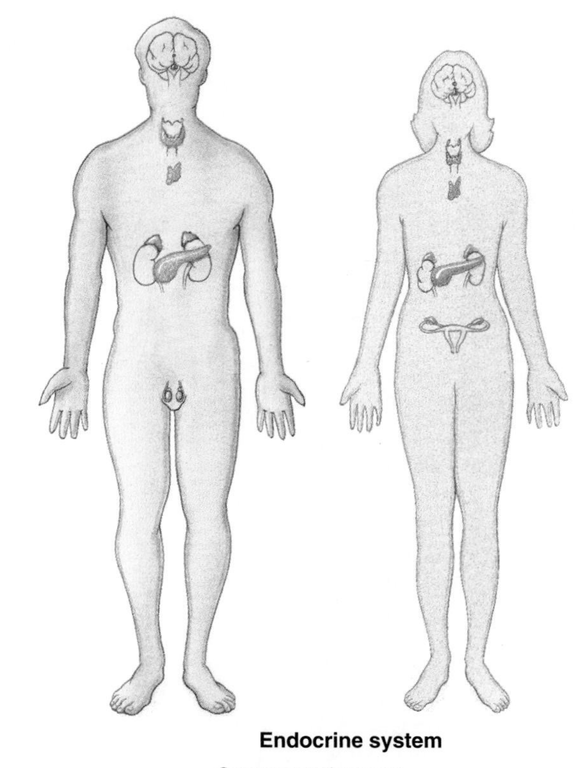

Endocrine system

Components: hormone-producing glands
Functions: secretes hormones that regulate body functions

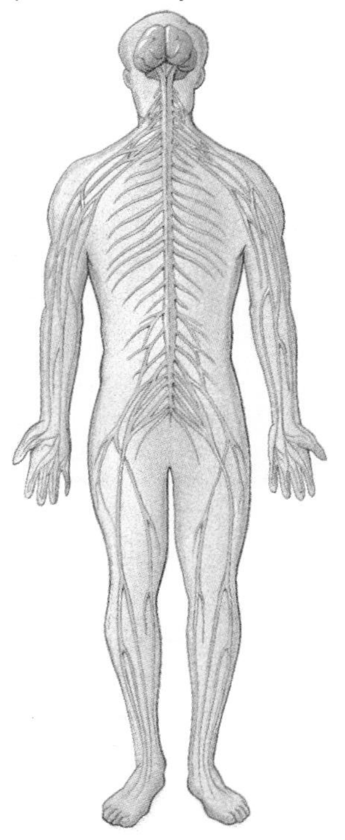

Nervous system

Components: brain, spinal cord, nerves, and sensory receptors
Functions: rapidly coordinates body functions and enables learning and memory

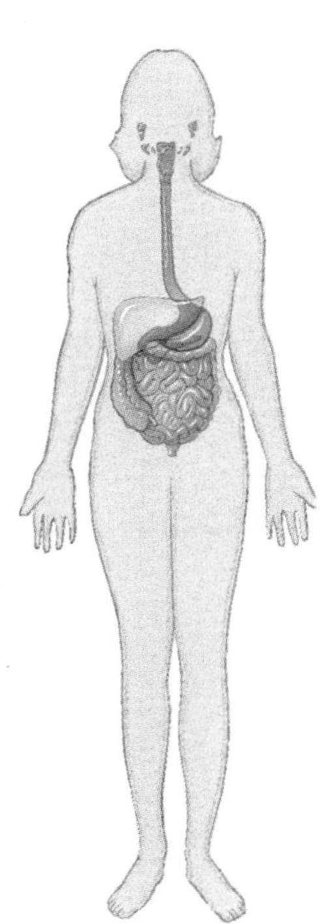

Digestive system

Components: mouth, pharynx, esophagus, stomach, intestines, anus, liver, pancreas, and associated structures
Functions: digests food and absorbs nutrients

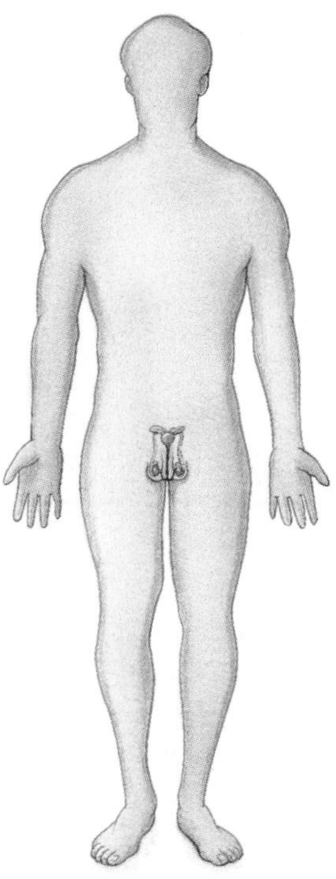

Male reproductive system

Components: testes, vasa deferentia, prostate gland, bulbourethral gland, and penis
Functions: produces sperm and transmits them into the female vagina during sexual intercourse

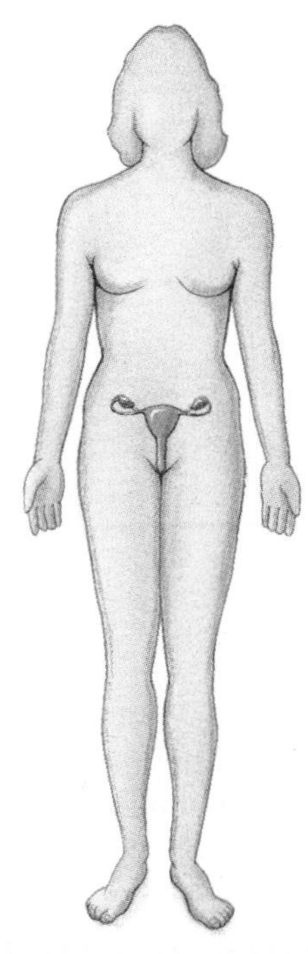

Female reproductive system

Components: ovaries, uterine tubes, vagina, and vulva
Functions: produces ova, receives sperm, provides intrauterine development of offspring, and enables birth of baby

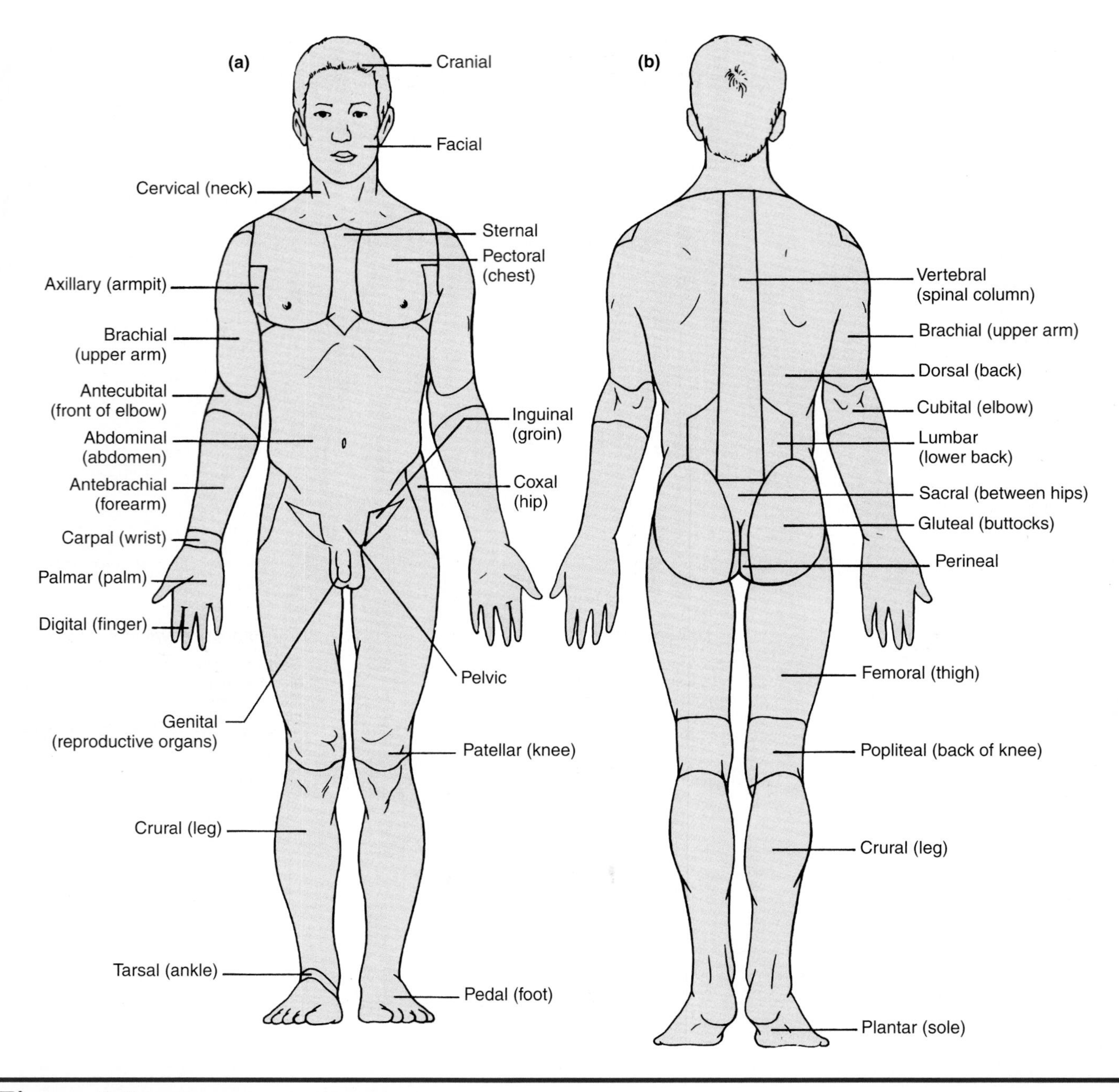

Figure 1.3 Major regions of the body.

which the body is standing upright with arms at the sides and palms of the hands facing forward, as in figure 1.3. Directional terms occur in pairs, and the members of each pair have opposite meanings, as noted in table 1.1.

Body Regions

The human body consists of an **axial** (ak′-sē-al) **portion,** the head, neck and trunk, and an **appendicular** (ap-pen-dik′-ū-lar) **portion,** the upper and lower extremities (arms and legs). Each of these major portions of the body is divided into regions with specialized names to facilitate communication and to aid in the location of body components.

The major body regions are listed in tables 1.2 and 1.3 to allow easy correlation with figure 1.3, which shows the locations of the major regions of the body. Take time to study the names, pronunciations, and locations of the body regions.

Body Planes and Sections

In studying the body or organs, you often will be observing the flat surface of a section that has been

Table 1.1 Directional Terms

Term	Meaning
Anterior (ventral)	Toward the front or abdominal surface of the body
Posterior (dorsal)	Toward the back of the body
Superior (cephalad)	Toward the head
Inferior (caudad)	Away from the head
Medial	Toward the midline of the body
Lateral	Away from the midline of the body
Parietal	Pertaining to the outer boundary of body cavities
Visceral	Pertaining to the internal organs
External (superficial)	Toward or on the body surface
Internal (deep)	Away from the body surface
Proximal	Part of an extremity that is nearer the point of attachment of the extremity to the trunk
Distal	Part of an extremity that is farther from the point of attachment of the extremity to the trunk
Central	At or near the center of the body or organ
Peripheral	External to or away from the center of the body or organ

Table 1.2 Major Regions of the Head, Neck, and Trunk

Region	Location
Head and Neck	
Cephalic (se-fal′-ik)	The head; composed of cranial and facial regions
Cervical (ser′-vi-kal)	The neck
Cranial (krā′-nē-al)	The part of the head containing the brain
Facial (fā′-shal)	The face
Anterior Trunk	
Abdominal (ab-dom′-i-nal)	The region between the lowest ribs and the pelvis (hip bones)
Abdominopelvic (ab-dom-i-nō-pel′-vik)	Composed of abdominal and pelvic regions
Inguinal (ing′-gwi-nal)	The groin; depressed regions at the junction of the thighs to the anterior trunk
Pectoral (pek′-tōr-al)	The chest
Pelvic (pel′-vik)	The region enclosed by the pelvic bones
Sternal (ster′-nal)	The region over the breastbone and between the two pectoral regions
Posterior Trunk	
Dorsum (dor′-sum)	The posterior surface of the thorax
Gluteal (glu′-tē-al)	The buttocks
Lumbar (lum′-bar)	The lower back region between the lowest ribs and the pelvis
Sacral (sāk′-ral)	The region over the sacrum and between the buttocks
Vertebral (ver-tē′-bral)	The region over the vertebral column or backbone
Lateral Trunk	
Axillary (ak′-sil-lary)	The armpits
Coxal (kok′-sal)	The hips
Inferior Trunk	
Genital (jen′-i-tal)	The external reproductive organs
Perineal (per-i-nē′-al)	The small region between the anus and external reproductive organs

Table 1.3 Major Regions of the Extremities

Region	Location
Upper Extremity	
Antebrachial (an-tē-brā′-kē-al)	The forearm
Antecubital (an-tē-kū′-bi-tal)	The anterior portion of the elbow joint
Brachial (brā′-kē-al)	The upper arm
Carpal (kar′-pal)	The wrist
Cubital (kū′-bi-tal)	The posterior portion of the elbow joint
Digital (di′-ji-tal)	The fingers
Palmar (pal′-mar)	The palm of the hand
Lower Extremity	
Crural (krū′-ral)	The leg
Digital (di′-ji-tal)	The toes
Femoral (fem′-ōr-al)	The thigh
Patellar (pa-tel′-lar)	The anterior portion of the knee joint
Pedal (pe′-dal)	The foot
Plantar (plan′-tar)	The sole of the foot
Popliteal (pop-li-tē′-al)	The posterior portion of the knee joint

produced by a cut through the body or a body part. Such sections are made along specific *planes.* These well-defined planes—transverse, sagittal, and coronal planes—lie at right angles to each other as shown in figure 1.4. It is important to understand the nature of the plane along which a section was made in order to understand the three-dimensional structure of an object being observed. Locate the planes described on figure 1.4.

Transverse, or **horizontal, planes** divide the body into superior and inferior portions, and they are perpendicular to sagittal planes and the longitudinal axis of the body. Sections made along transverse planes are often called *cross sections.*

Sagittal planes divide the body into right and left portions and are parallel to the longitudinal axis of the body. Sections made along sagittal planes are often called longitudinal sections. A **midsagittal,** or **medial, plane** passes through the midline of the body and divides the body into equal left and right halves.

Coronal, or **frontal, planes** divide the body into anterior and posterior portions. These planes are perpendicular to sagittal planes and parallel to the longitudinal axis of the body.

How do sagittal, transverse, and coronal planes differ from one another?

Body Cavities

There are two major cavities of the body that contain internal organs: the dorsal (posterior) and ventral (anterior) cavities. Note the locations and subdivisions of these cavities in figures 1.5 and 1.6.

The **dorsal cavity** is subdivided into the **cranial cavity,** which houses the brain, and the **spinal cavity,** which contains the spinal cord. Note in figure 1.5 how the cranial bones and the vertebral column form the walls of the dorsal cavity and provide protection for these delicate organs.

The **ventral cavity** is divided by the *diaphragm,* a thin dome-shaped sheet of muscle, into a superior **thoracic cavity** and an inferior **abdominopelvic cavity.** The thoracic cavity is protected by the *rib cage* and contains the heart and lungs. The abdominopelvic cavity is subdivided into a superior **abdominal cavity** and an inferior **pelvic cavity,** but there is no structural separation between them. The abdominal cavity contains the stomach, intestines, liver, gallbladder, pancreas, spleen, and kidneys. The pelvic cavity contains the urinary bladder, sigmoid colon, rectum, and internal reproductive organs.

What organs are located in each subdivision of the dorsal cavity?
What organs are located in each subdivision of the ventral cavity?

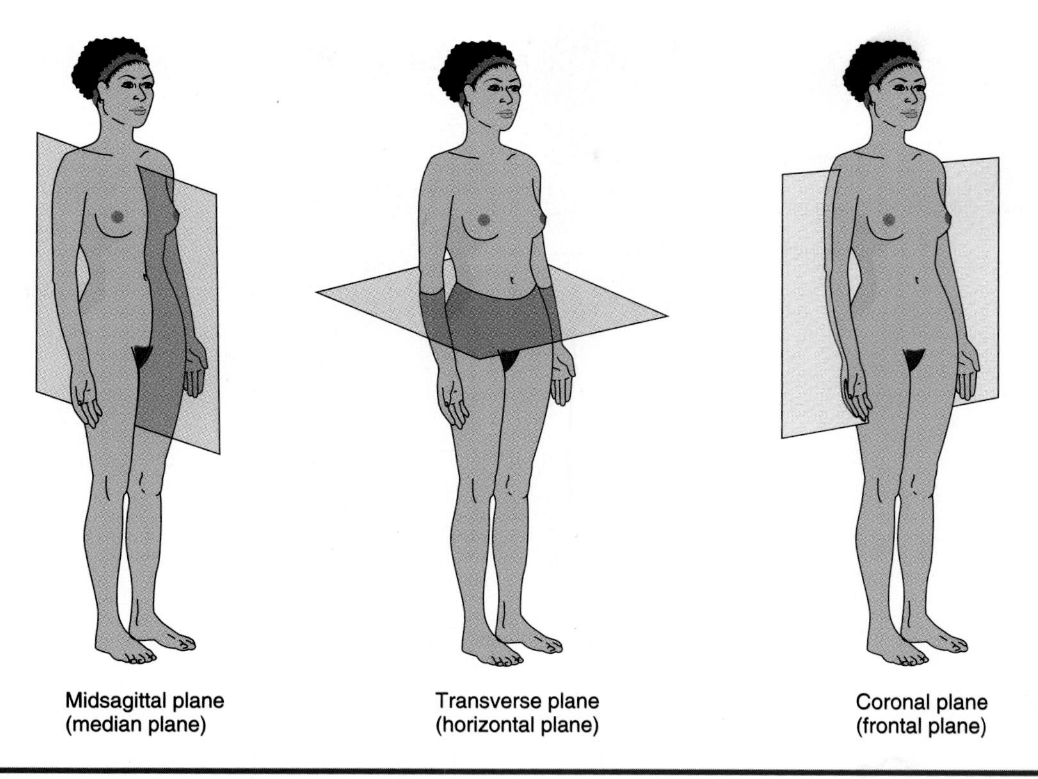

Figure 1.4 The planes of the body.

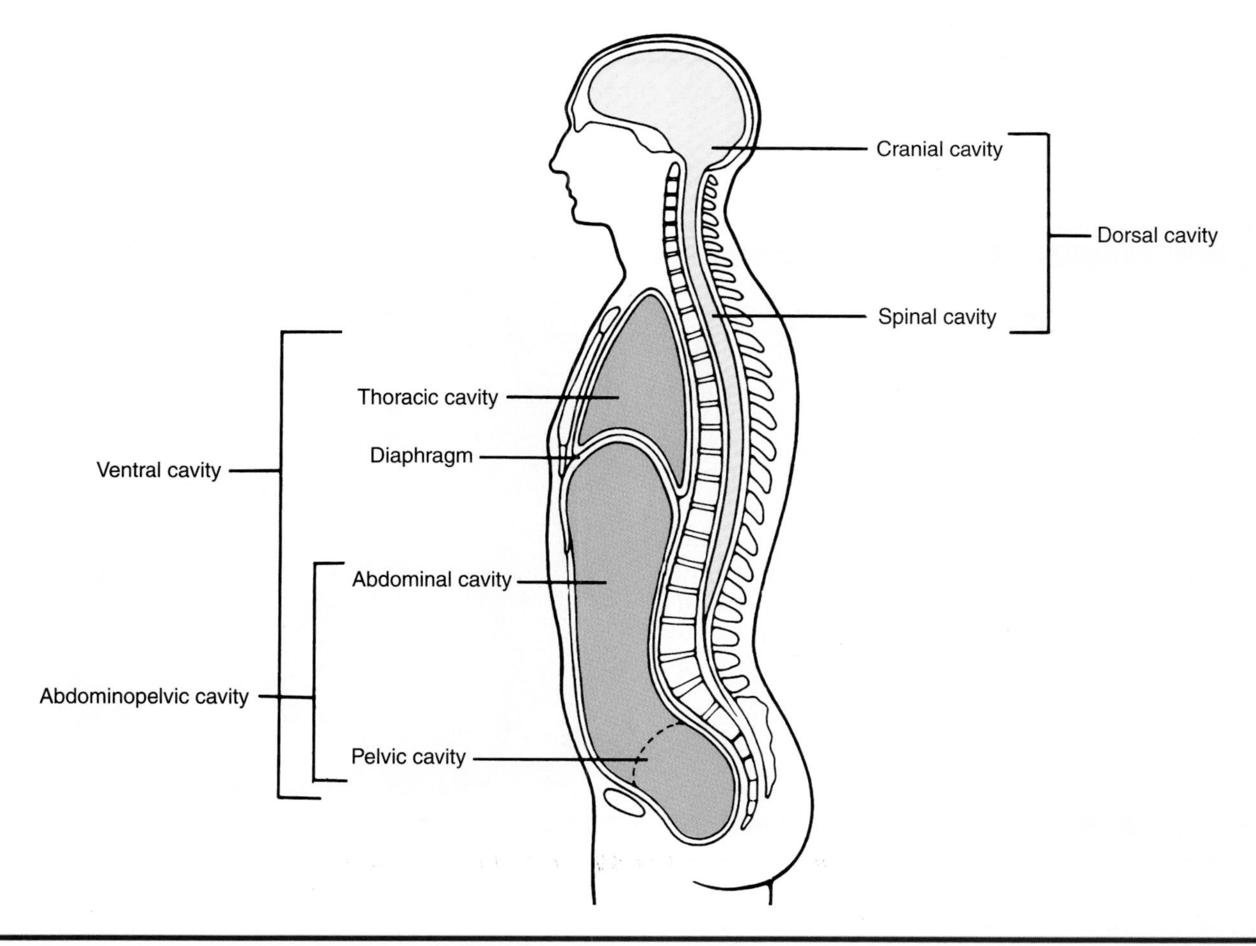

Figure 1.5 A midsagittal section of the body showing the dorsal and ventral cavities and their subdivisions.

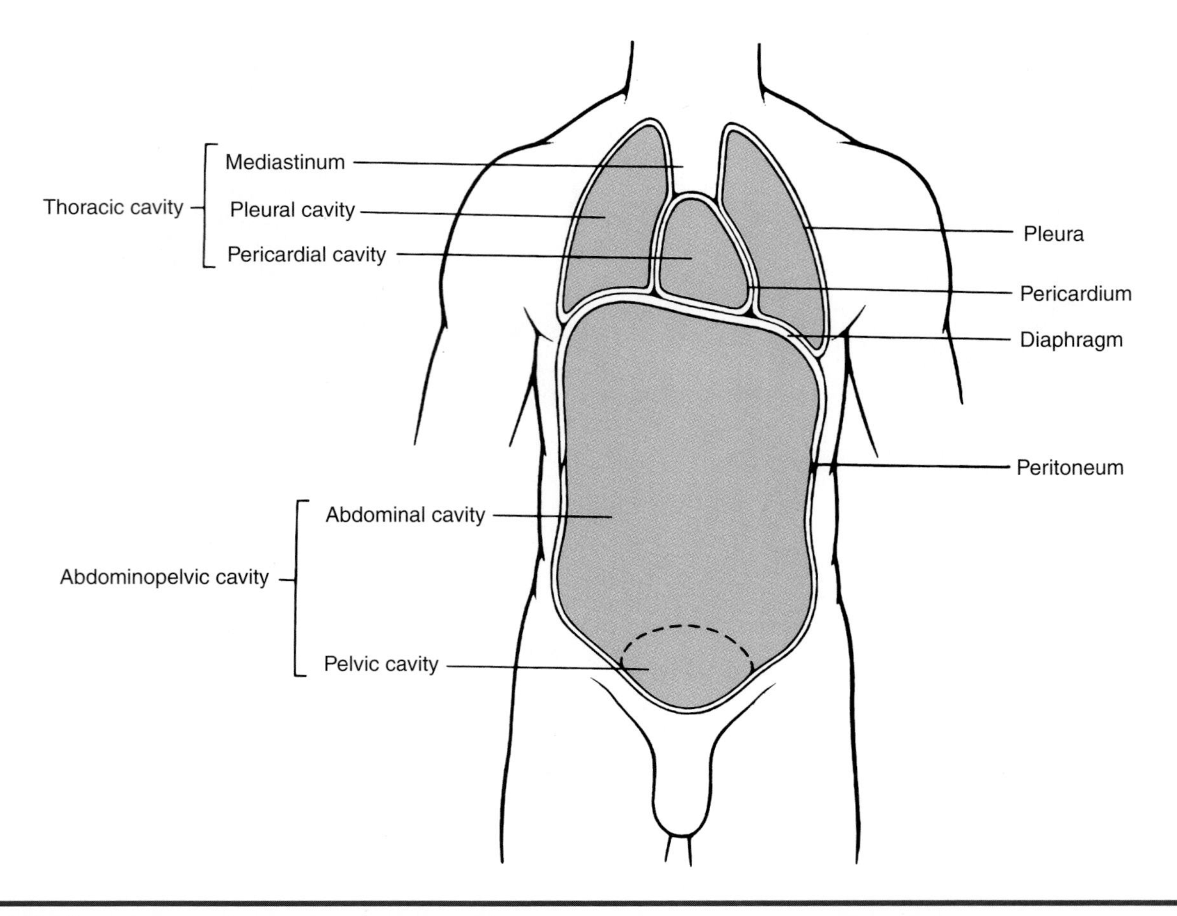

Figure 1.6 A frontal (coronal) section of the body showing the ventral cavity and its subdivisions.

Physicians use certain types of diagnostic imaging systems, for example *computerized tomography (CT)* and *magnetic resonance imaging (MRI),* to produce images of sections of the body to help them diagnose disorders. In computerized tomography, an X-ray emitter and an X-ray detector rotate around the patient so that the X-ray beam passes through the body from hundreds of different angles. X rays collected by the detector are then processed by a computer to produce sectional images on a screen for viewing by a radiologist. A good understanding of sectional anatomy is required to interpret CT scans.

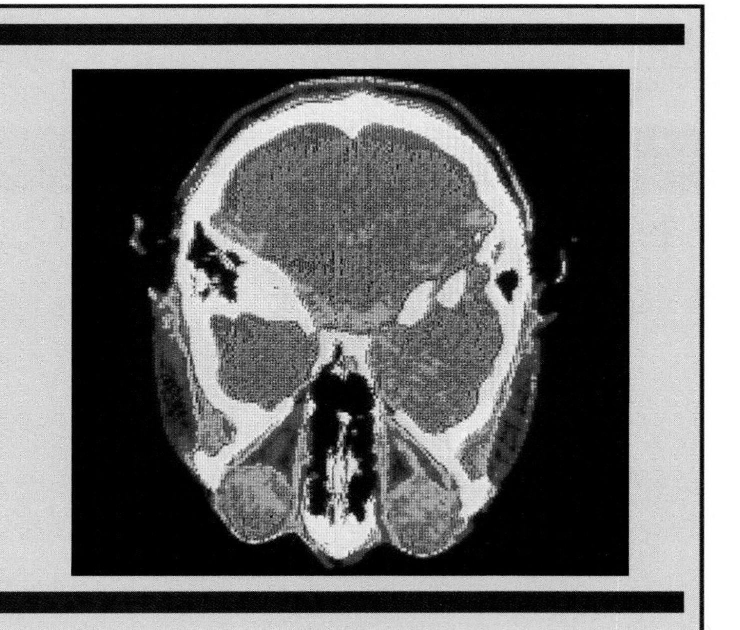

Membranes of Body Cavities

The membranes lining body cavities support and protect the internal organs in the cavities.

Dorsal Cavity Membranes

The dorsal cavity is lined by three protective membranes that are collectively called the **meninges** (me-nin′-jēz). The outer membrane is attached to the wall of the dorsal cavity while the inner membrane tightly envelops the brain and spinal cord.

Ventral Cavity Membranes

The ventral cavity is lined with *serous membranes* that secrete a watery fluid on their surfaces. Serous membranes provide smooth, moist, friction-reducing surfaces for the enclosed internal organs.

The thoracic cavity is lined by the **pleurae,** or **pleural membranes.** The walls of the left and right portions of the thoracic cavity are lined by the *parietal* (pah-rī′-e-tal) *pleurae.* The outer surfaces of the lungs are covered by the *visceral* (vis′-er-al) *pleurae.* The parietal and visceral pleurae are separated by a thin film of *serous fluid,* a watery fluid secreted by the pleurae. Serous fluid reduces friction as the pleurae rub against each other as the lungs expand and contract during breathing. The potential space (not an actual space) between the parietal and visceral pleurae is known as the **pleural cavity.**

The left and right portions of the thoracic cavity are divided by a membranous partition, the *mediastinum* (mē-dē-a-stī′-num). Organs located within the mediastinum include the heart, thymus gland, esophagus, and trachea.

The heart is enveloped by the **pericardium** (per-i-kar′-dē-um), which is formed of membranes of the mediastinum. The thin *visceral pericardium* is tightly adhered to the surface of the heart, but the thicker *parietal pericardium,* a double membrane, forms a loosely fitting sac around the heart. The potential space between the visceral and parietal pericardia is the **pericardial cavity,** and it contains serous fluid to reduce friction as the heart contracts and relaxes.

The walls of the abdominal cavity and the surfaces of abdominal organs are lined with the **peritoneum** (per-i-tō-nē′-um). The *parietal peritoneum* lines the walls of the abdominal cavity but not the pelvic cavity. It descends only to cover the upper portion of the urinary bladder. Since the kidneys are located posterior to the parietal peritoneum, only a portion of their surfaces are covered by it. The *visceral peritoneum,* an extension of the parietal peritoneum, covers the surface of the abdominal organs. Double-layered folds of the visceral peritoneum, the *mesenteries* (mes′-en-ter″-ēs), extend between the abdominal organs and provide support for them. The potential space between the parietal and visceral peritoneal membranes is called the **peritoneal cavity.**

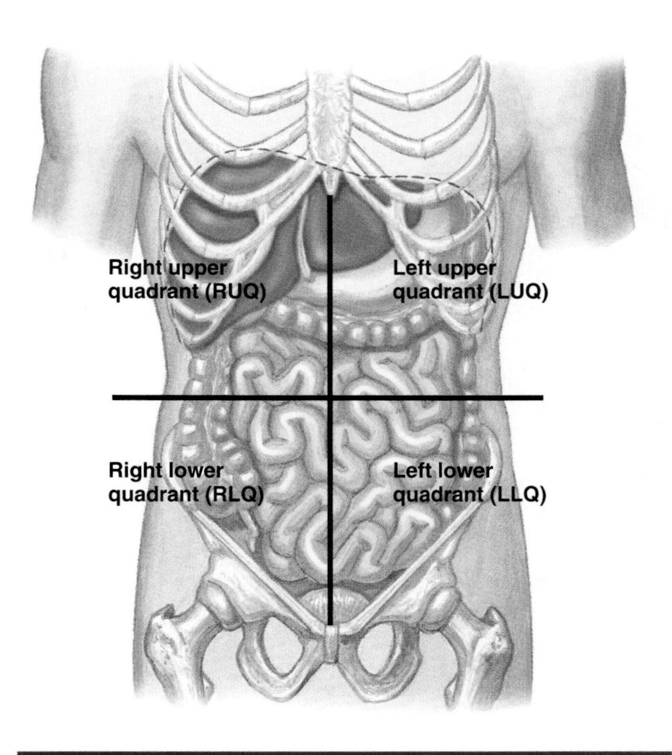

Figure 1.7 The four quadrants of the abdominopelvic area.

> ***What membranes line the dorsal and ventral cavities?***

Abdominopelvic Subdivisions

The abdominopelvic surface is subdivided into either four quadrants or nine regions to aid physicians and nurses in locating underlying organs in the abdominopelvic cavity. By knowing the internal organs that are located within each quadrant or region, organs that may be responsible for a patient's complaint of abdominopelvic pain may be identified.

The four quadrants are formed by two planes that intersect at the umbilicus (navel), as shown in figure 1.7. Note the organs within each quadrant.

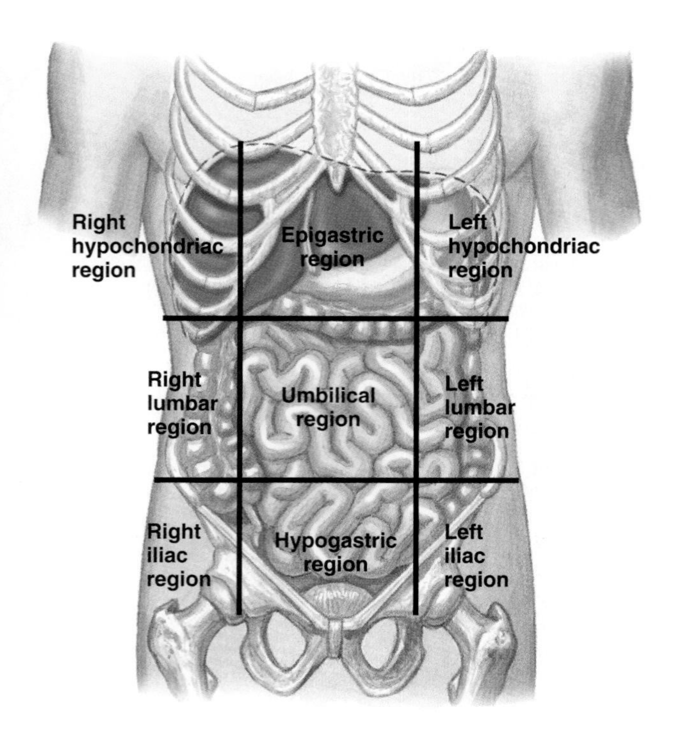

Figure 1.8 The abdominopelvic area is subdivided into nine regions.

The nine regions are formed by the intersection of two sagittal and two transverse planes as shown in figure 1.8. The sagittal planes extend downward from the midpoints of the collarbones. The superior transverse plane lies just inferior to the lowest ribs, and the inferior transverse plane lies just below the hip bones.

Identify the organs located within each quadrant and region by comparing figure 1.7 to table 1.4, and figure 1.8 to table 1.5.

Study figures 1.9 and 1.10 to increase your understanding of the locations of the internal organs and associated membranes.

Now examine the colorplates that follow this chapter. They show an anterior view of the body in progressive stages of dissection that reveals major muscles, blood vessels, and internal organs. Study these plates to learn the normal locations of the organs of the ventral cavity. Also, check your understanding of the organs within each abdominal quadrant and region.

What are the four quadrants and nine regions of the abdominopelvic region?

Table 1.4 Organs in Abdominopelvic Quadrants

Quadrant	Organs
Right upper	Gallbladder; most of liver and duodenum; right kidney; parts of pancreas, small intestine, ascending and transverse colon
Right lower	Appendix; cecum; parts of ascending colon, small intestine, right ureter, urinary bladder, rectum
Left upper	Stomach; spleen; parts of duodenum and pancreas; left kidney; parts of left ureter, small intestine, transverse and descending colon
Left lower	Parts of small intestine, descending and sigmoid colon, rectum, left ureter, urinary bladder

Table 1.5 Organs in Abdominopelvic Regions

Region	Organs
Right hypochondriac	Gallbladder; parts of liver, transverse colon, and right kidney
Right lumbar	Ascending colon; parts of small intestine and right kidney
Right iliac	Appendix; cecum; parts of small intestine
Epigastric	Parts of liver, stomach, pancreas, duodenum, transverse colon
Umbilical	Parts of duodenum, small intestine, kidneys, ureters
Hypogastric	Urinary bladder; rectum; parts of ureters, small intestine, and sigmoid colon
Left hypochondriac	Spleen; parts of stomach, transverse colon, left kidney
Left lumbar	Descending colon; parts of left kidney and small intestine
Left iliac	Parts of small intestine, descending and sigmoid colon

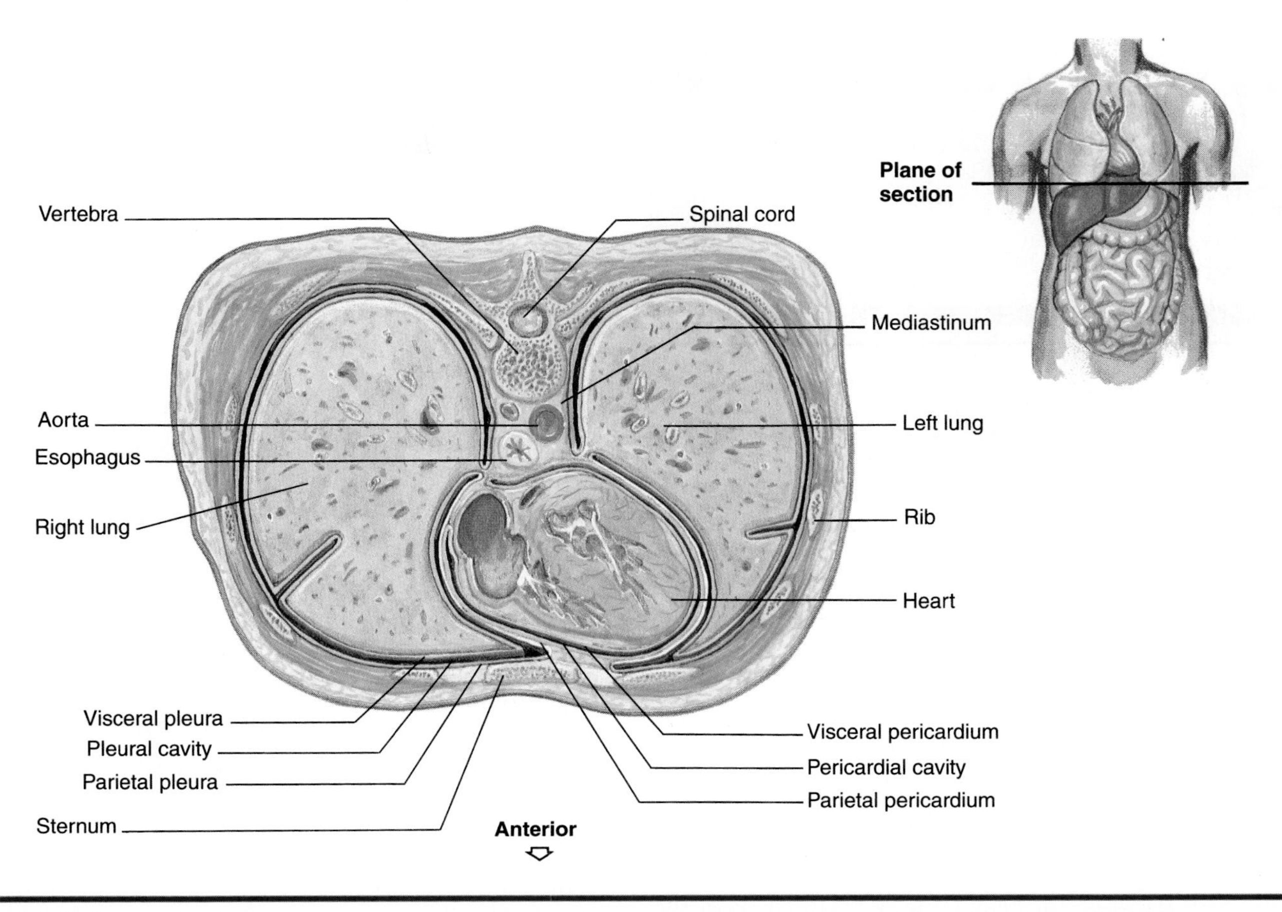

Figure 1.9 A transverse section through the thorax reveals the serous membranes associated with the heart and lungs (*superior view*).

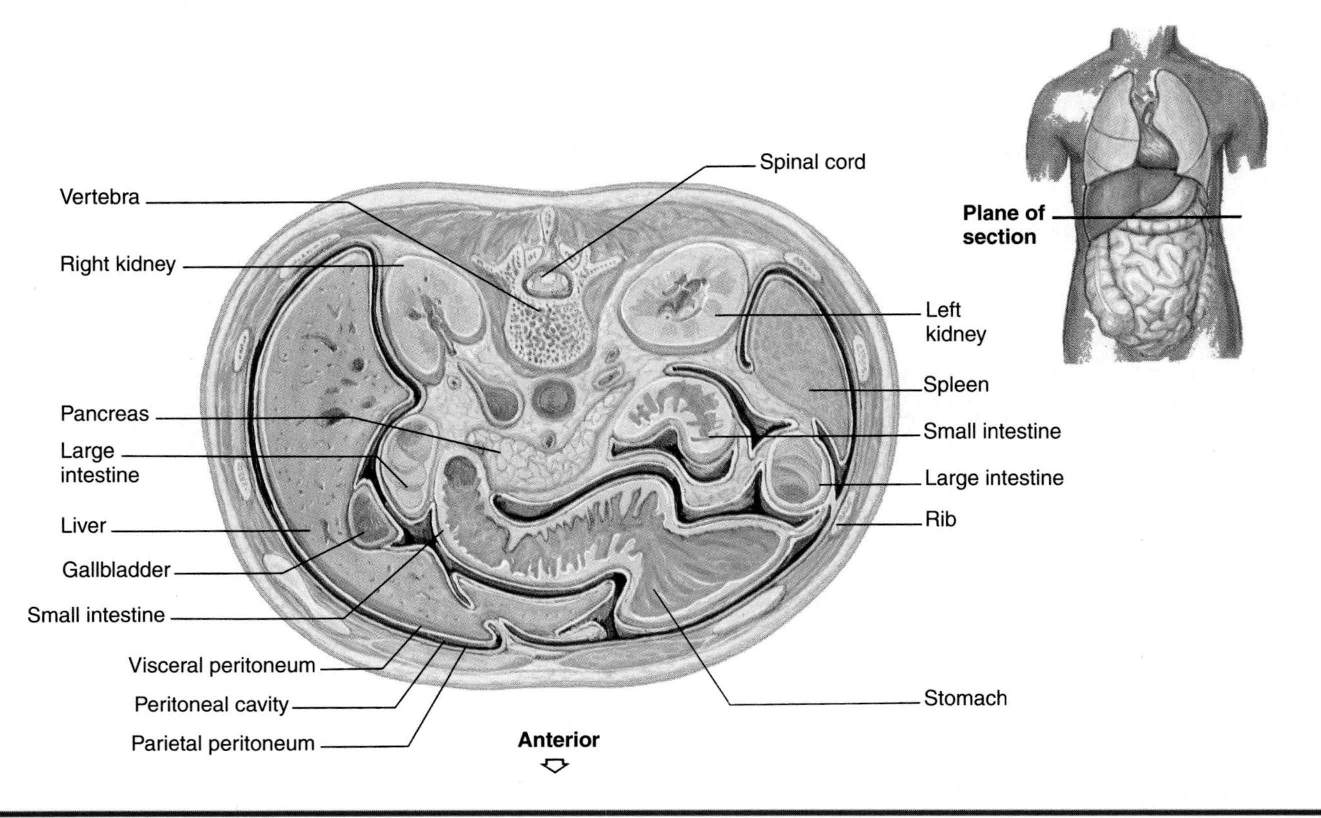

Figure 1.10 Transverse section through the abdomen (*superior view*).

When a patient complains about abdominopelvic pain, a physician is able to identify organs that are possibly involved by knowing the organs located in each abdominopelvic quadrant or region.

Maintenance of Life

Humans, like all living organisms, exhibit the fundamental processes of life. **Metabolism** (me-tab′ō-lizm) is the term that collectively refers to all of the life processes, that is, all of the chemical reactions that collectively are known as life.

There are two phases of metabolism: anabolism and catabolism. *Anabolism* (ah-nab′-ō-lizm) refers to processes that use energy and nutrients to build the complex organic molecules that compose the body. *Catabolism* (kah-tab′-ō-lizm) refers to processes that break down complex molecules into simpler molecules.

Life is fragile. It depends upon the normal functioning of trillions of body cells, which, in turn, depends upon factors needed for survival and the ability of the body to maintain relatively stable internal conditions.

Survival Needs

There are five basic needs that are essential to human life:

1. **Food** provides chemicals that serve as a source of energy and raw materials to maintain cells of the body.
2. **Water** provides the environment in which the chemical reactions of life occur.
3. **Oxygen** is required to convert organic nutrients into energy, which powers life processes.
4. **Body temperature** must be maintained close to 37°C (98.6°F) to allow the chemical reactions of human metabolism to occur.
5. **Atmospheric pressure** is required for breathing to occur.

Homeostasis

Body cells function best when the internal environment of the body—blood and intercellular fluid—is relatively stable. Therefore, the composition and concentration of blood and intercellular fluid need to be kept within rather narrow limits by physiological processes. The maintenance of a relatively stable internal environment is known as **homeostasis.**

Homeostasis maintains a relatively stable internal environment in spite of the fact that materials are continuously entering and exiting the blood and intercellular fluid. Thus, the internal environment is kept in a dynamic, rather than a static, balance.

The maintenance of a normal concentration of nutrients in the blood is an example of homeostasis. Nutrients in food we eat are absorbed into the blood. Excess nutrients are stored in the liver and in fat cells. As body cells take up and use nutrients, nutrient concentration in the blood declines, but stored nutrients are released to bring the nutrient concentration back to normal levels. When the nutrient supply in blood declines to a critical point, certain brain cells trigger a hunger response, which stimulates us to eat food to replenish the supply of nutrients.

What is homeostasis and why is it important?

CHAPTER SUMMARY

Anatomy and Physiology

1. Human anatomy is the study of body structure and organization.
2. Human physiology is the study of body functions.

Levels of Organization

1. The body consists of several levels of organization of increasing complexity.
2. From simple to complex, the organizational levels are chemical, cellular, tissue, organ, organ system, and organismic.
3. The organs of the body are arranged in coordinated groups called organ systems.
4. The organ systems of the body are:

integumentary	cardiovascular
skeletal	lymphatic
muscle	respiratory
nervous	urinary
endocrine	reproductive
digestive	

Directional Terms

1. Directional terms are used to describe the relative positions of body parts.
2. Directional terms occur in pairs, with the members of a pair having opposite meanings.

anterior—posterior	proximal—distal
superior—inferior	external—internal
medial—lateral	parietal—visceral
central—peripheral	

Body Regions

1. The body is divided into two major portions: the axial portion and the appendicular portion.
2. The axial portion is subdivided into the head, neck, and trunk.
3. The head and neck contain cervical, cranial, and facial regions. The cranial and facial regions combine to form the cephalic region.
4. The trunk consists of anterior, posterior, lateral, and inferior areas.
5. Anterior trunk regions include the abdominal, inguinal, pectoral, pelvic, and sternal regions. The abdominal and pelvic regions combine to form the abdominopelvic region.
6. Posterior trunk regions include the dorsal, gluteal, lumbar, sacral, and vertebral regions.
7. Lateral trunk regions are the axillary and coxal regions.
8. Inferior trunk regions are the genital and perineal regions.
9. The appendicular portion of the body consists of the upper and lower extremities.
10. The upper extremity is attached to the trunk at the shoulder. Regions of the upper extremity are the antebrachial, antecubital, brachial, carpus, digital, and palmar regions.
11. The lower extremity is attached to the trunk at the hip. Regions of the lower extremity are the crural, femoral, patellar, pedal, popliteal, tarsal, and plantar regions.

Body Planes and Sections

1. Well-defined planes are used to guide sectioning of the body or organs.
2. The common planes are transverse, sagittal, and coronal.

Body Cavities

1. There are two major body cavities: dorsal and ventral.
2. The dorsal cavity consists of the cranial and spinal cavities.
3. The ventral cavity consists of the thoracic and abdominopelvic cavities.
4. The thoracic cavity lies superior to the diaphragm. It consists of two lateral pleural cavities that are separated by the mediastinum. The pericardial cavity lies within the mediastinum.
5. The abdominopelvic cavity lies inferior to the diaphragm. It consists of a superior abdominal cavity and an inferior pelvic cavity.
6. The body cavities are lined with protective and supportive membranes.
7. The meninges consist of three membranes that line the dorsal cavity and enclose the brain and spinal cord.
8. The parietal pleurae line the inner walls of the pleural cavities, and the visceral pleurae cover the external surfaces of the lungs.
9. The parietal pericardium is a saclike membrane in the mediastinum that surrounds the heart. The visceral pericardium is attached to the surface of the heart.
10. The parietal peritoneum lines the walls of the abdominal cavity but does not extend into the pelvic cavity. The visceral peritoneum covers the surface of internal abdominal organs.
11. The mesenteries are double-layered folds of the visceral peritoneum that support internal organs.

Abdominopelvic Subdivisions

1. The abdominopelvic region is subdivided into either four quadrants or nine regions as an aid in locating underlying organs.
2. The four quadrants are:

right upper	left upper
right lower	left lower

3. The nine regions are:

epigastric	right lumbar
left hypochondriac	hypogastric (pubic)
right hypochondriac	left iliac (inguinal)
umbilical	right iliac
left lumbar	

Maintenance of Life

1. Metabolism is the sum of all life processes, and it consists of anabolism, the synthesis of body chemicals, and catabolism, the breakdown of body chemicals.
2. The basic needs of the body are food, water, oxygen, body temperature, and atmospheric pressure.
3. Homeostasis is the maintenance of a relatively stable internal environment.

BUILDING YOUR VOCABULARY

1. **Selected New Terms**
   abdominal cavity, p. 8
   cranial cavity, p. 8
   distal, p. 7
   dorsal cavity, p. 8
   frontal plane, p. 8
   inferior, p. 7
   lateral, p. 7
   medial, p. 7
   midsagittal plane, p. 8
   organ, p. 3
   organ system, p. 3
   pelvic cavity, p. 8
   pericardial cavity, p. 11
   pleural cavity, p. 11
   proximal, p. 7
   sagittal plane, p. 8
   spinal cavity, p. 8
   superior, p. 7
   thoracic cavity, p. 8
   transverse plane, p. 8
   ventral cavity, p. 8

2. **Related Clinical Terms**
   meningitis (men-in-jī′-tis) Inflammation of the meninges.
   pathology (path-ol′-ō-jē) Medical specialty studying anatomical and physiological changes caused by disease.
   peritonitis (per-i-tō-nī′-tis) Inflammation of the peritoneum.
   podiatry (pō-dī′-ah-tre) Medical specialty treating foot disorders.

STUDY ACTIVITIES

1. Complete the **Chapter 1 Study Guide,** which begins on page 397.
2. Complete the **Learning Objectives** listed on the first page of this chapter.

CHECK YOUR UNDERSTANDING

(Answers are located in appendix B.)

1. A study of body functions is called _____ .
2. Blood, the heart, and blood vessels compose the _____ system.
3. Rapid coordination of body functions is the function of the _____ system.
4. The fingers are located _____ to the wrist.
5. The upper and lower extremities compose the _____ portion of the body.
6. The posterior surface of the knee is known as the _____ region.
7. The thigh is known as the _____ region.
8. The _____ body cavity is divided into the cranial and _____ cavities.
9. The gallbladder is located in the _____ quadrant and the _____ abdominopelvic region.
10. The _____ separates the left and right portions of the thoracic cavity.
11. The abdominal cavity is lined by the _____.
12. The maintenance of a relatively stable internal environment is called _____.

COLORPLATES OF THE HUMAN BODY

The seven colorplates that follow show the basic structure of the human body. The first plate shows the anterior body surface and the superficial anterior muscles of a female. Succeeding plates show the internal structure as revealed by progressively deeper dissections.

Refer to these plates often as you study this text in order to become familiar with the relative locations of the body organs

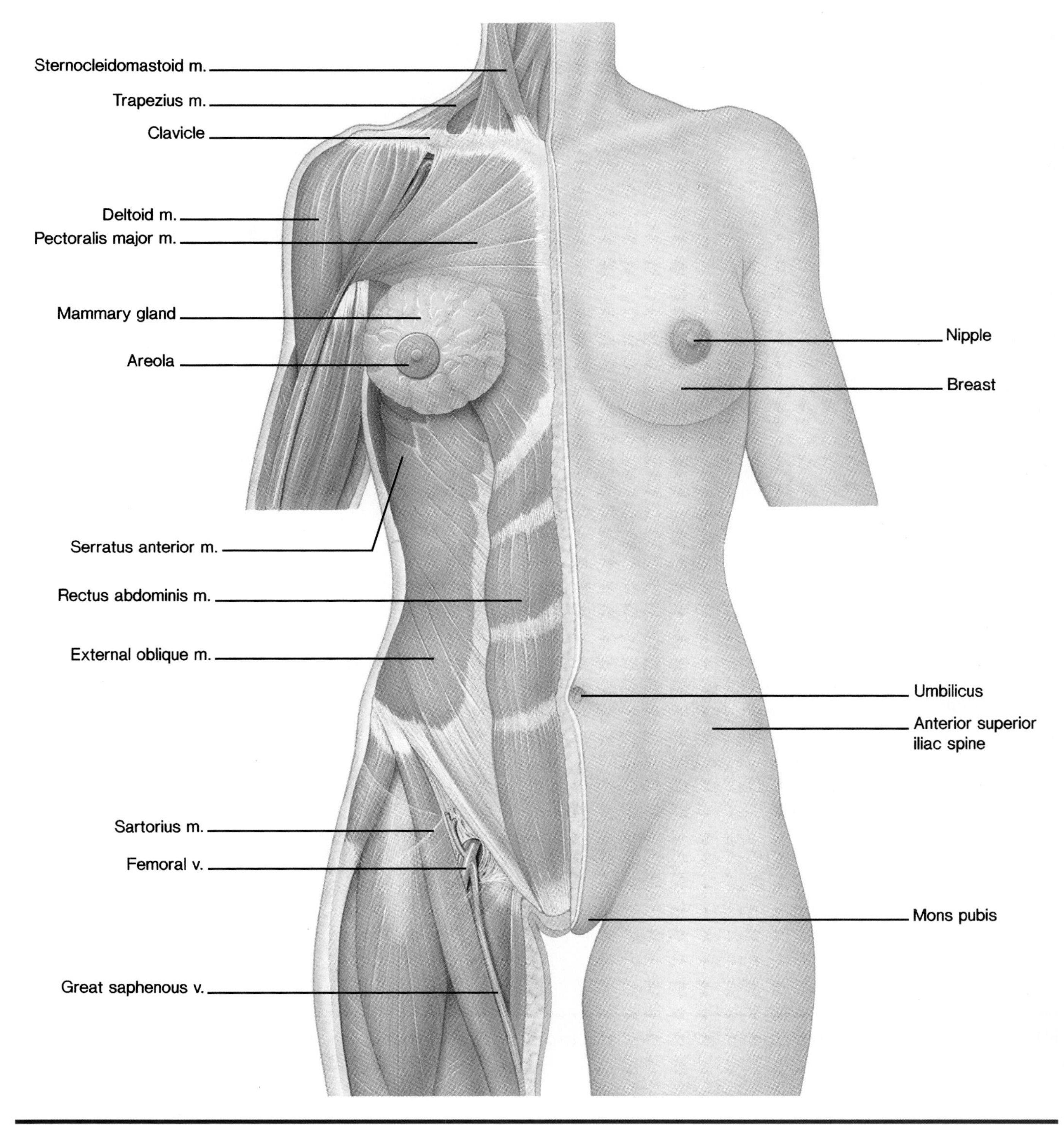

Plate 1 Female torso with skin removed to expose superficial muscles. (m. = muscle, v. = vein)

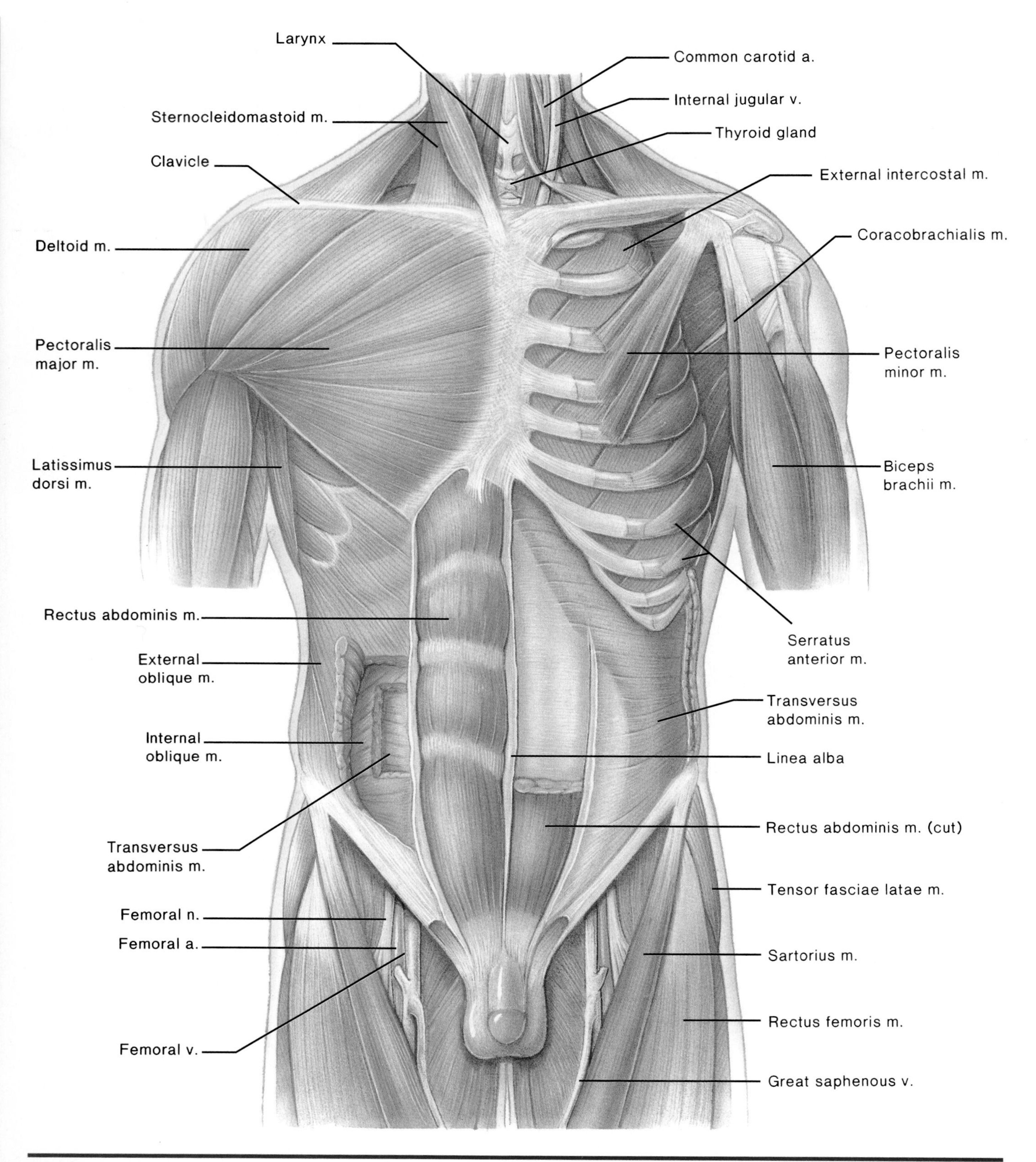

Plate 2 Male torso with deeper muscles exposed. (m. = muscle, v. = vein, n. = nerve, a. = artery)

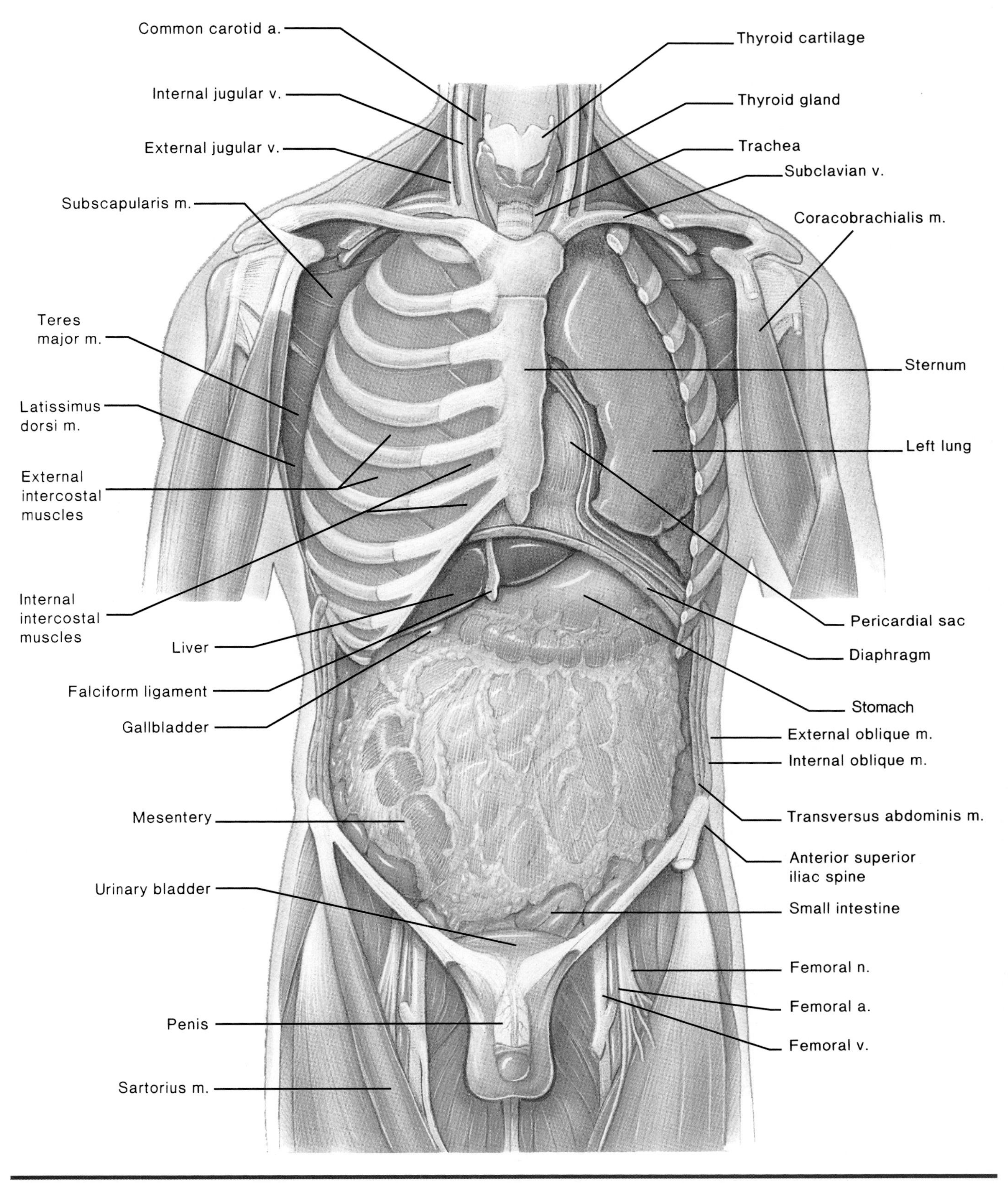

Plate 3 Male torso with some internal organs exposed. (m. = muscle, v. = vein, a. = artery)

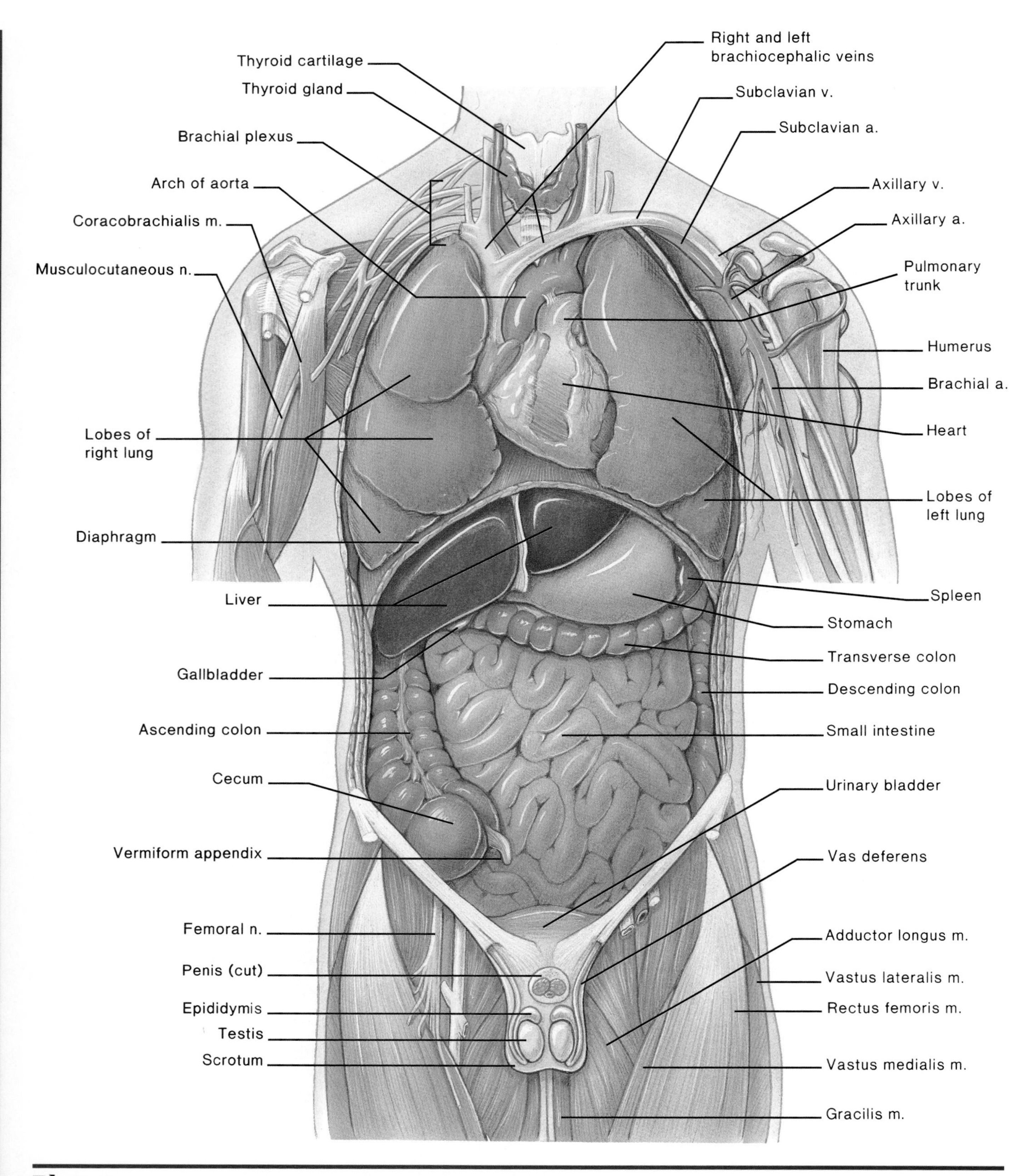

Plate 4 Male torso with thoracic and abdominopelvic organs exposed (m. = muscle, v. = vein, a. = artery).

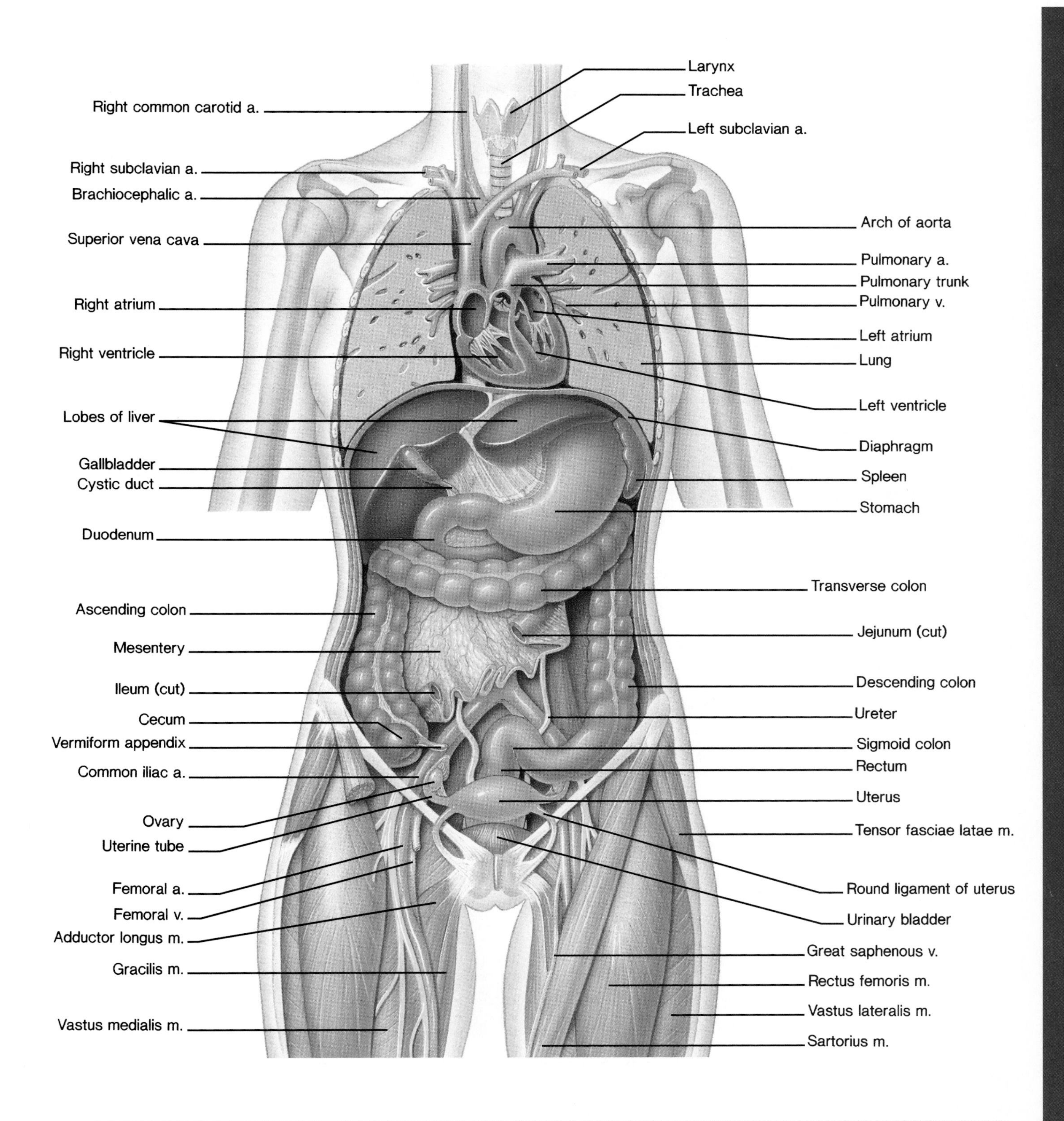

Plate 5 Female torso with lungs and heart in coronal section and small intestine removed. (v. = vein, a. = artery)

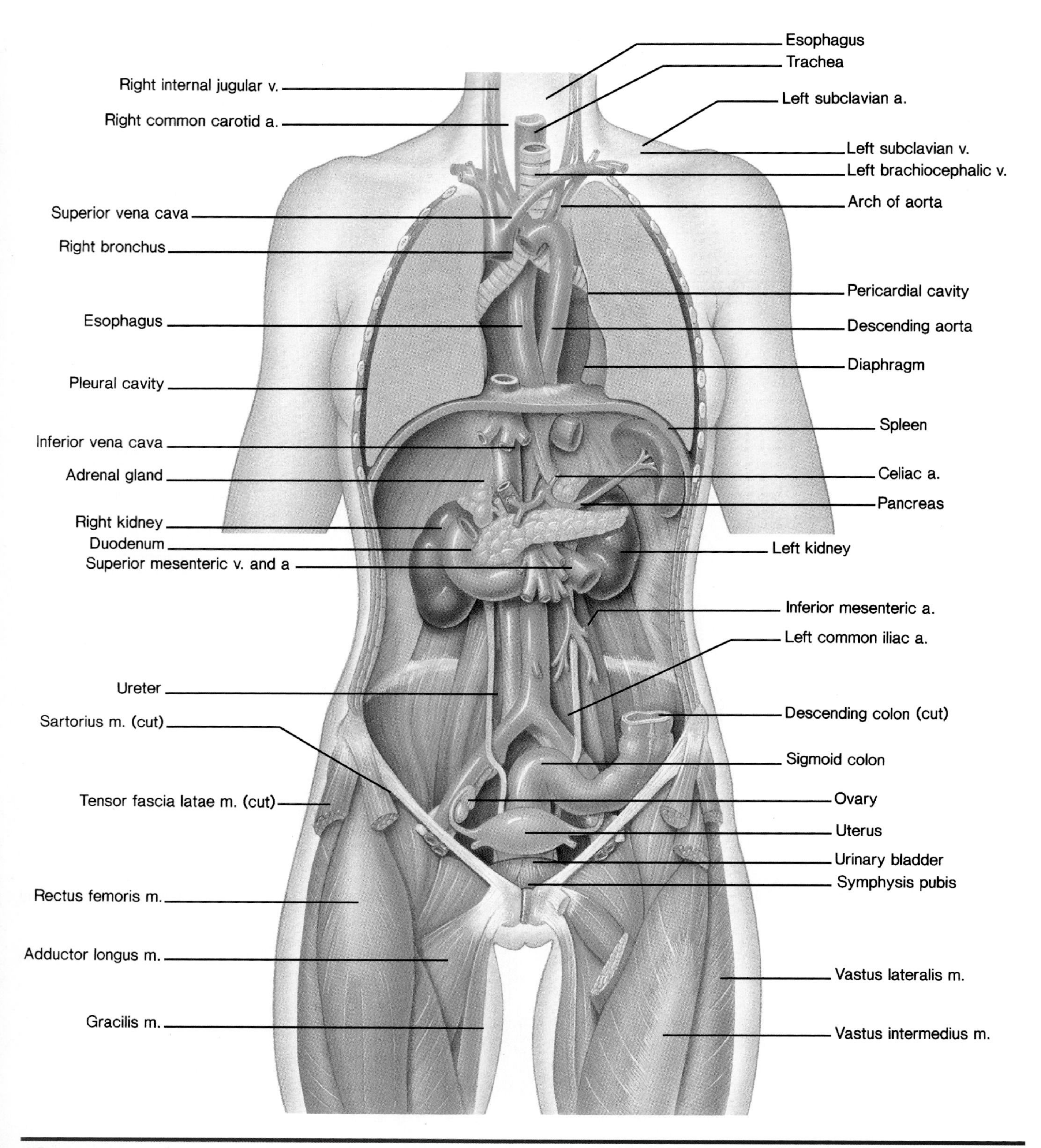

Plate 6 Female torso with some organs removed to expose esophagus, deep vessels, kidneys, and ureters. (v. = vein, a. = artery)

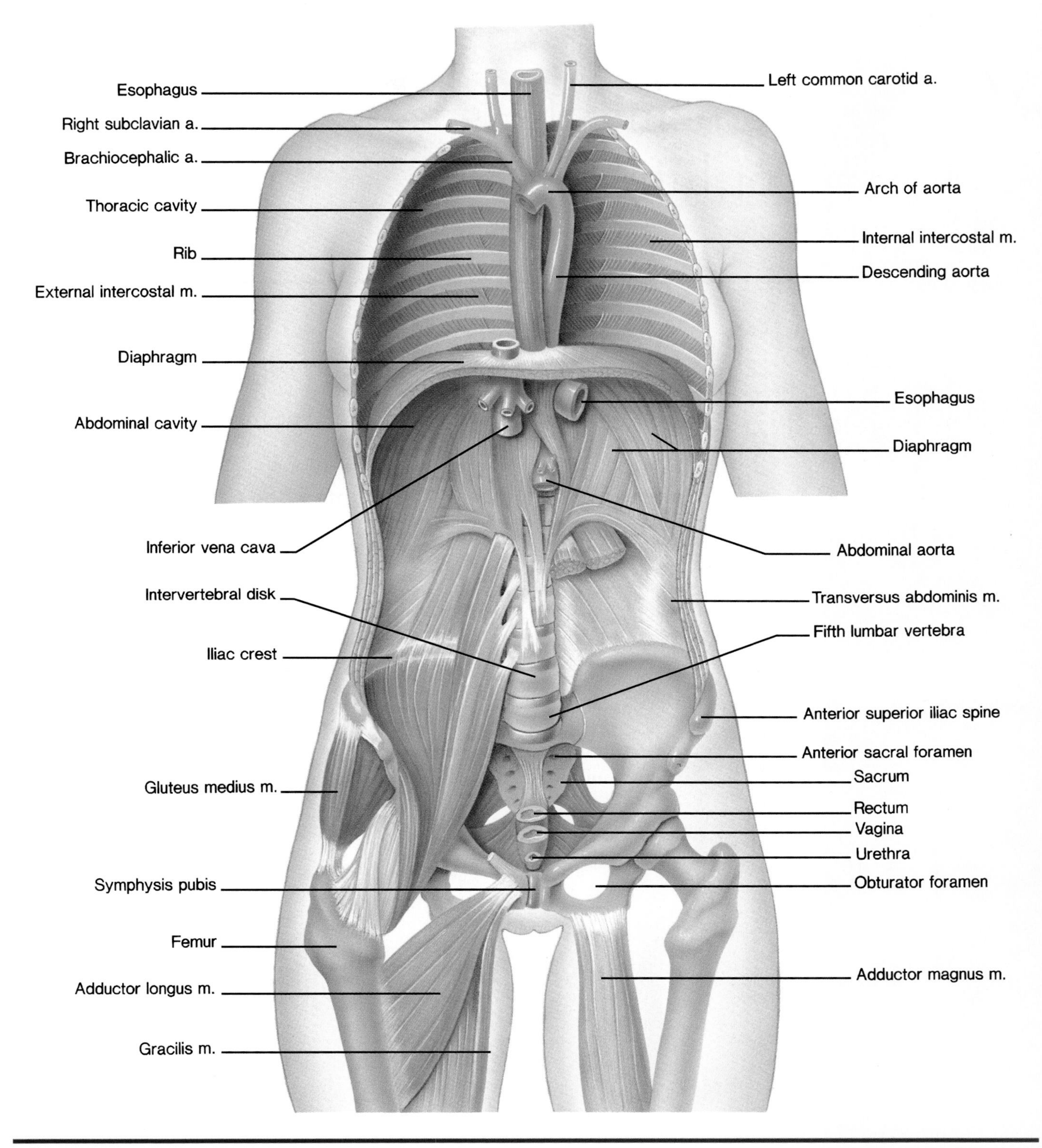

Plate 7 Female torso with viscera removed to expose posterior walls of thoracic and abdominopelvic cavities. (m. = muscle, v. = vein)

TWO

2

C H A P T E R

Chemical Aspects of Life

Chapter Preview & Learning Objectives

Atoms and Elements
- Describe the basic structure of an atom.
- Distinguish between atoms, isotopes, and radioisotopes.

Molecules and Compounds
- Explain the meaning of a chemical formula.
- Compare ionic, covalent, and hydrogen bonds.
- Compare synthesis and decomposition reactions.

Compounds Composing the Human Body
- Distinguish between inorganic and organic compounds.
- Explain the importance of water and its locations in the body.
- Compare acids and bases.
- Explain the use of the pH scale.
- Explain the importance of mineral salts.
- Distinguish between carbohydrates, lipids, proteins, and nucleic acids and their roles in the body.
- Explain the role of enzymes.
- Describe the composition and role of ATP.

Chapter Summary
Building Your Vocabulary
Study Activities
Check Your Understanding

SELECTED KEY TERMS

Atom (atomos = indivisible) The smallest unit of an element.
Carbohydrate (carbo = carbon; hydr = water) A group of organic compounds composed of C, H, and O usually in a 1:2:1 ratio; mono-, di-, and polysaccharides.
Compound (componere = to place together) A substance formed of two or more elements combined in a fixed ratio.
Decomposition (de = down from) The breakdown of complex molecules into simpler molecules.
Element A substance that cannot be broken down into simpler substances by chemical means.
Enzyme (en = in; zym = ferment) A protein that catalyzes chemical reactions.
Isotope (iso = equal) An atom of an element having the same number of protons and electrons but a different number of neutrons than most atoms of the element.
Lipid (lip = fat) A group of organic compounds containing fats.
Molecule (molecula = little mass) The smallest unit of a compound; composed of two or more atoms chemically combined.
Monosaccharide (mono = one; sacchar = sugar) A simple sugar; a carbohydrate molecule composed of a single saccharide unit.
Nucleic acid (nucle = kernel) A complex organic molecule composed of nucleotides.
Protein A group of nitrogen-containing organic compounds formed of amino acids.
Synthesis (syn = together; thesis = to put) Formation of a molecule by combining atoms or simpler molecules.

Chemistry is the scientific study of substances, especially the interaction of substances. A basic knowledge of chemistry is necessary for health-care professionals because the human body is composed of chemicals, and the processes of life are chemical interactions.

Atoms and Elements

Anything that has weight and occupies space is **matter.** The entire physical universe, both living and nonliving, is composed of matter. All matter is composed of **chemical elements,** substances that cannot be broken down by chemical means into simpler substances. Carbon, hydrogen, and nitrogen are examples of chemical elements.

There are 92 naturally occurring chemical elements, and 26 elements are found in the human body. About 96% of the body weight consists of only four elements: hydrogen, oxygen, carbon, and nitrogen. Many of the elements in the body are called *trace elements* because they occur in exceedingly small amounts. Table 2.1 lists the elements composing 99.9% of the human body.

Table 2.1 The Most Abundant Elements Composing the Human Body

Name	Symbol	% by Weight	Importance
Oxygen	O	65	Component of most organic molecules and water
Carbon	C	18.5	Forms "backbone" of all organic molecules
Hydrogen	H	9.5	Component of organic molecules, including all foods; part of water molecule
Nitrogen	N	3.2	Component of proteins and nucleic acids
Calcium	Ca	1.5	Component of teeth and bones; essential for blood clotting, muscle contraction, and nerve function
Phosphorus	P	1.0	Component of teeth, bones, nucleic acids, ATP, and many proteins
Potassium	K	0.4	Essential for nerve function and muscle contraction
Sulfur	S	0.3	Component of many proteins
Sodium	Na	0.2	Essential for nerve function; helps maintain water balance in body fluids
Chlorine	Cl	0.2	Helps maintain water balance in body fluids
Magnesium	Mg	0.1	Component of many proteins

Atomic Structure

An **atom** (a′-tom) is the smallest unit of an element that participates in chemical reactions. Atoms of a given element are similar to each other, and they are different from atoms of all other elements. Atoms of different elements differ in size, weight, and how they interact with other atoms.

Observe the diagrammatic representation of an atom in figure 2.1. The **nucleus** (nū′klē-us) of an atom is centrally located, and it consists of two kinds of subatomic particles: protons and neutrons. Each **proton** (prō′-ton) has a positive electrical charge but each **neutron** (nū′-tron) has no charge. Since protons are located in the nucleus, an atom's nucleus is always positively charged.

Electrons (ē-lek′-tronz) rotate at high speed around the nucleus in *energy levels,* or *electron shells.* Each electron has a negative electrical charge. Since an atom contains the same number of electrons as protons, an atom has no charge and is electrically neutral. Table 2.2 summarizes the characteristics of these subatomic particles.

The atoms of each element are characterized by a specific atomic number, symbol, and atomic weight. These characteristics are used to identify the element. The *atomic number* indicates the number of protons and electrons in each atom. The *symbol* is a shorthand way of referring to an element or to an atom of the element. The *atomic weight* is the sum of the number of protons and neutrons in each atom. For example, the element carbon has an atomic number of 6, a symbol of C, and an atomic weight of 12. From this information, you know that an atom of carbon has six protons, six electrons, and six neutrons. Examine table 2.3, which lists the atomic characteristics for some common elements in the human body.

Isotopes

All atoms of an element have the same number of protons and electrons, but some atoms may have a

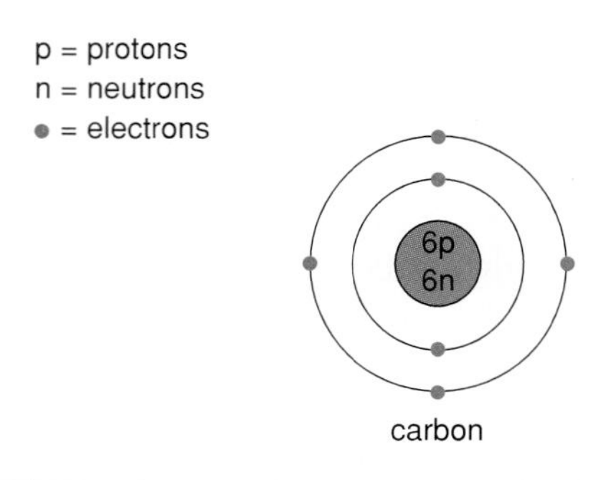

Figure 2.1 Diagram of a carbon atom. The number of protons and electrons are equal, which makes the atom electrically neutral.

Table 2.2 Characteristics of Subatomic Particles

Particle	Electrical Charge	Location
Proton	Positive (+)	Nucleus
Neutron	0	Nucleus
Electron	Negative (−)	Electron shells

Table 2.3 Atomic Characteristics of the Common Elements in the Body

Element	Symbol	Protons	Neutrons	Atomic Weight*	Electrons in Valence Shell
Hydrogen	H	1	0	1	1
Carbon	C	6	6	12	4
Nitrogen	N	7	7	14	5
Oxygen	O	8	8	16	6
Sodium	Na	11	12	23	1
Magnesium	Mg	12	12	24	2
Phosphorus	P	15	16	31	5
Sulfur	S	16	16	32	6
Chlorine	Cl	17	18	35	7
Potassium	K	19	20	39	1
Calcium	Ca	20	20	40	2

*Isotopes not included.

different number of neutrons. An atom of an element with a different number of neutrons is called an **isotope** (ī′-so-tōp). For example, carbon has three isotopes: ^{12}C, ^{13}C, ^{14}C. All isotopes of an element have the same chemical properties because they have the same number of protons and electrons.

Certain isotopes of some elements have an unstable nucleus that emits high-energy radiation as it breaks down to form a more stable nucleus. Such isotopes are called *radioisotopes.* Certain radioisotopes are used in the diagnosis of disorders and in the treatment of cancer. See the clinical box on page 28.

What is the relationship between matter, elements, and atoms?
What is the basic structure of an atom?

Molecules and Compounds

Atoms combine chemically in specific ways to form **molecules** (mol′-e-kūlz). For example, two atoms of oxygen may combine to form a molecule of gaseous oxygen (O_2), and an atom of sodium may combine with an atom of chlorine to form a molecule of sodium chloride (NaCl). Sodium chloride is a **chemical compound** because it is composed of two elements chemically combined in a *fixed ratio.* A molecule is the smallest unit of a compound that exhibits the properties of the compound. A compound can be broken down into its component elements by chemical means.

Chemical Formulas

A *chemical formula* expresses the chemical composition of a molecule. Chemical symbols indicate the elements of the atoms involved, and subscripts identify the number of atoms of each element in the molecule. A chemical formula expresses both the composition of a single molecule and the composition of a compound. For example, the chemical formula for water is H_2O, which indicates that two atoms of hydrogen combine with one atom of oxygen to form a water molecule.

Chemical Bonds

It is electrons that are involved in the formation of **chemical bonds,** which join atoms together to form a molecule. A chemical bond is a force of attraction between two atoms. It is not a structure. An atom combines with another atom in order to fill its **valence shell,** its outermost shell of electrons. To do this, atoms either (1) donate or receive electrons or (2) share electrons.

The first shell of electrons, the shell closest to the nucleus, can hold a maximum of two electrons even if it is the only electron shell. An atom with two or more electron shells reacts with other atoms to fill its valence shell with eight electrons. Note the number of electrons in the valence shell of atoms of the elements in table 2.3.

Ionic Bonds

Atoms that form ionic bonds do so by donating or receiving electrons. Atoms with one, two, or three electrons in the valence shell tend to donate electrons. Atoms with five, six, or seven electrons in the valence shell tend to receive electrons.

Consider the interaction of sodium and chlorine in the formation of sodium chloride (table salt), as shown in figure 2.2. Sodium has a single electron in its valence shell while chlorine has seven electrons in its valence shell. The transfer of an electron from sodium

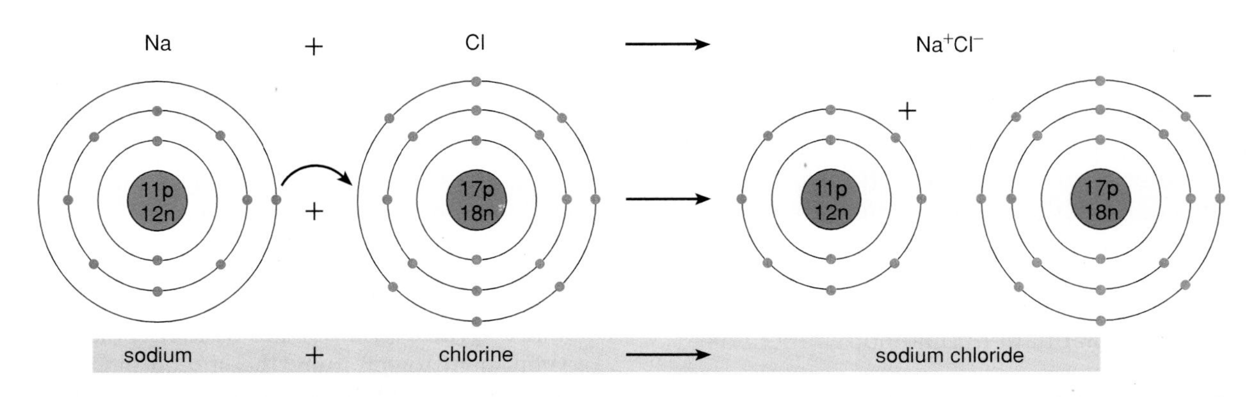

Figure 2.2 The formation of sodium chloride by an ionic reaction. The transfer of an electron from sodium to chlorine converts sodium to a positive ion and chlorine to a negative ion. The attraction between these oppositely charged ions is an ionic bond.

Nuclear medicine is the medical specialty that uses radioisotopes in the diagnosis and treatment of disease. Very small amounts of weak radioisotopes may be used to tag biological molecules in order to trace the movement or metabolism of these molecules in the body. Special instruments can detect the radiation emitted by the radioisotopes and identify the location of the tagged molecules.

In *nuclear imaging,* the emitted radiation creates an image on a special photographic plate. In this way, it is possible to obtain an image of various organs or parts of organs where the radioisotopes accumulate. A new computerized imaging technique called positron emission tomography (PET) uses certain radioisotopes that emit positrons (positively charged electrons), and it enables precise imaging similar to computerized tomography (CT) scans. PET can be used to measure processes, such as blood flow, rate of metabolism of selected substances, and effects of drugs on body functions. It is a promising new technique for both the diagnosis of disease and the study of normal physiological processes.

Another form of nuclear medicine involves the use of radioisotopes to kill tumor cells. Certain radioisotopes may be attached to specific biological molecules and injected into the blood. When these molecules accumulate in cancerous tissue, the emitted radiation kills the cancerous cells. A similar effect is obtained by implanting pellets of radioactive isotopes directly in cancerous tissue.

A PET scan of the brain.

to chlorine causes the sodium atom to have a net electrical charge of +1 and the chlorine atom to have a net electrical charge of −1.

Atoms with a net electrical charge, either positive or negative, are called **ions.** Thus the transfer of an electron from sodium to chlorine has (1) resulted in the valence shell of each atom being filled with electrons and (2) produced a sodium ion (Na^+) and a chlorine ion (Cl^-). The force of attraction that holds these ions together is an **ionic bond.** Ionic bonds always are formed between ions of unlike electrical charges. Positively charged ions, such as Na^+, are called **cations.** Negatively charged ions, such as Cl^-, are called **anions.**

When ionic compounds are dissolved in water, their molecules tend to **dissociate (ionize)** releasing ions. Such compounds are called **electrolytes** (ē-lek′-trō-lītz) because when dissolved they can conduct an electrical current. The composition and concentration of electrolytes in the body must be kept within narrow limits to maintain normal body functions.

Covalent Bonds

Atoms that form molecules by sharing electrons are joined by **covalent bonds.** The shared electrons orbit around each atom for part of the time so that they may be counted in the outer shell of each atom. Thus, the valence shell of each atom is filled. The structural components of living cells consist of molecules formed by covalent bonds.

A water molecule is a simple example of covalent bonding (figure 2.3). The oxygen atom has only six electrons in its outer shell, and each hydrogen atom has a single electron. A covalent bond is formed when a hydrogen atom shares its electron with an oxygen atom and the oxygen atom shares one of its electrons with the hydrogen atom. It takes two shared electrons, one from each atom being bonded, to form a covalent bond. In this way, the two hydrogen atoms are bonded covalently to a single oxygen atom to form a water molecule (H_2O). The shared electrons complete the outer shell of each atom.

Hydrogen Bonds

When a hydrogen atom is covalently bonded to an oxygen or nitrogen atom, the shared electrons spend less time in the hydrogen atom and more time in the oxygen or nitrogen atom. Therefore, the hydrogen atom is slightly positively charged, and the oxygen or nitrogen atom is slightly negatively charged. This sets the stage for the formation of a **hydrogen bond**—a

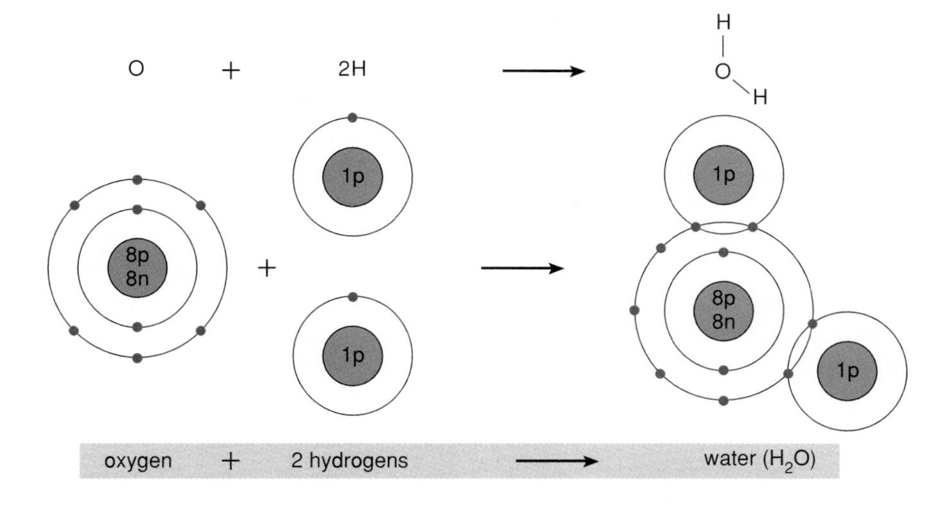

Figure 2.3 The formation of a water molecule by a covalent reaction. Two hydrogen atoms and one oxygen atom share electrons to complete their outer electron shells. Two shared electrons compose one covalent bond.

weak attractive force between a slightly positive hydrogen atom and a slightly negative oxygen or nitrogen atom at a different site in the same molecule or in a different molecule.

For example, hydrogen bonds between water molecules cause water molecules to cling together in the liquid state (figure 2.4). Boiling water breaks the hydrogen bonds, and individual water molecules are released as steam.

Hydrogen bonds do not join atoms together to form molecules, but they play vital roles in living organisms, as you will see later.

> ***How do ionic and covalent bonds join atoms to form molecules?***
> ***What are hydrogen bonds?***

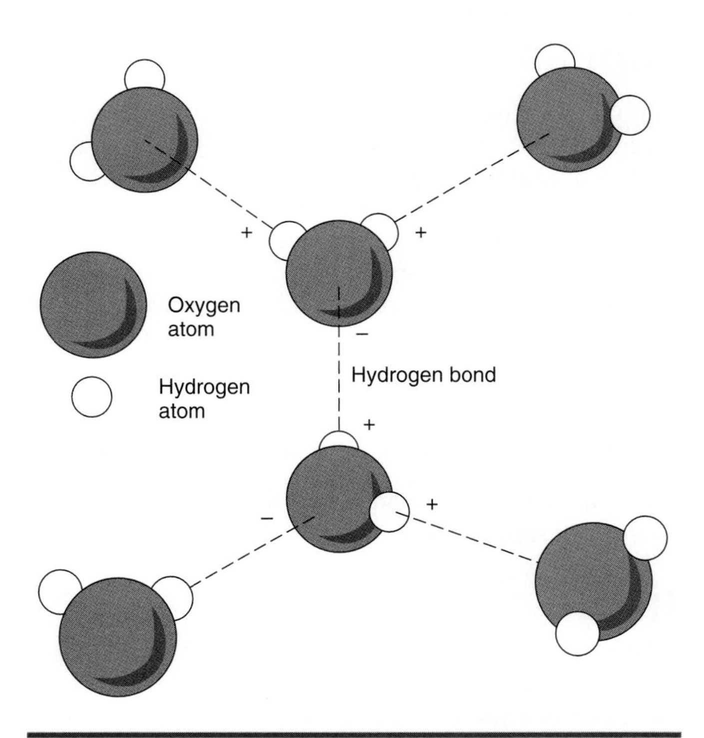

Figure 2.4 Hydrogen bonds form between adjacent water molecules. A hydrogen bond forms between a slightly positive hydrogen atom of one molecule and a slightly negative oxygen atom of an adjacent molecule.

Chemical Reactions

In chemical reactions, bonds between atoms are formed or broken, and the result is a new combination of atoms. There are two basic types of chemical reactions: synthesis and decomposition reactions.

In a **synthesis reaction,** simple substances (atoms or molecules) combine to form a more complex substance. The reaction of hydrogen and oxygen (*reactants*) to form water (*product*) is an example. Synthesis reactions produce complex molecules used in the growth and repair of body parts. Synthesis reactions may be generalized as:

$$A + B \rightarrow AB$$

A **decomposition reaction** is just the reverse. A complex substance is broken down into simpler substances. For example, water can decompose to form hydrogen and oxygen. Decomposition reactions are used to break down food molecules to form nutrients usable by body cells. Decomposition reactions may be generalized as:

$$AB \rightarrow A + B$$

Many chemical reactions are reversible, meaning that the product(s), on the right side of the arrow, can change back into the reactants, on the left side of the arrow. Several factors determine the reversibility of a reaction, such as energy available and relative abundance of reactants and products. Reversible reactions are indicated by a double arrow:

$$A + B \rightleftharpoons AB$$

How do synthesis and decomposition reactions differ?

Compounds Composing the Human Body

Compounds composing the human body include both inorganic and organic substances. Molecules of **inorganic compounds** may contain either carbon or hydrogen in the same molecule, *but not both.* Molecules of **organic compounds** always contain *both* carbon and hydrogen, and they usually also contain oxygen. The carbon atoms form the "backbone" of organic molecules. Table 2.4 compares the major differences between inorganic and organic compounds.

Major Inorganic Compounds

The major inorganic compounds in the body are water, most acids and bases, and mineral salts.

Water

Water is a very important compound. It composes about two-thirds of the body weight, which makes it the most abundant compound in the body. Water is able to absorb or release large amounts of heat without much change in its own temperature. Thus, it helps to distribute heat throughout the body and prevents sudden fluctuations in body temperature.

Water is an excellent **solvent** and dissolves many substances **(solutes),** especially ionic compounds. Body fluids are aqueous (water) solutions in which numerous solutes are dissolved. Chemical reactions that occur in the body take place in an aqueous solution.

Water serves as a lubricant, reducing friction where internal body parts rub together, and it aids the movement of food through the digestive tract.

The specific locations where water is found in the body are called **water compartments.** They are:

Intracellular fluid (ICF): water within cells; about 65% of the total body water.
Extracellular fluid (ECF): all water not in cells; about 35% of the total body water.
- **Tissue fluid (interstitial fluid):** water in tiny spaces between cells.
- **Plasma:** fluid portion of blood.
- **Lymph:** fluid in lymphatic vessels.
- **Specialized fluids:** water in more limited locations, such as serous fluid (fluid secreted by membranes lining the ventral body cavity).

Acids and Bases

Inorganic acids and bases are ionic compounds that **dissociate** and release ions when dissolved in water.

An **acid** is a compound that dissociates to release **hydrogen ions (H^+)** in an aqueous solution. The strength of an acid depends on the degree of dissociation, that is, the degree to which hydrogen ions are released. The stronger the acid, the greater is the degree of dissociation, which results in a greater concentration of H^+. Hydrochloric acid (HCl) is a strong acid. It dissociates into hydrogen (H^+) and chloride (Cl^-) ions.

$$HCl \rightarrow H^+ + Cl^-$$

A **base** is a compound that dissociates to release **hydroxyl ions (OH^-)** in an aqueous solution. The stronger the base, the greater is the degree of dissocia-

Table 2.4 Comparison of Inorganic and Organic Compounds

Inorganic Compounds	Organic Compounds
Molecules contain relatively few atoms	Molecules usually contain many atoms
Molecules do *not* contain both carbon and hydrogen	Molecules *always* contain both carbon and hydrogen
Molecules usually formed by ionic bonding	Molecules always formed by covalent bonding

tion, which results in a greater concentration of OH^-. Sodium hydroxide (NaOH) is a strong base. In an aqueous solution, it dissociates to form sodium (Na^+) and hydroxyl (OH^-) ions.

$$NaOH \rightarrow Na^+ + OH^-$$

pH: Acid-Base Balance Chemists have developed a pH scale that is used to indicate the acidity or alkalinity of a solution, that is, the relative concentrations of hydrogen ions (H^+) and hydroxyl ions (OH^-) in a solution. The **pH scale** ranges from 0 to 14. Figure 2.5 shows the relative concentration of H^+ and OH^- at each pH value. The concentration of H^+ decreases and the concentration of OH^- increases as the pH values increase. These ions are equal in concentration at pH 7, so a solution with a pH of 7 is neither an acid nor a base. For example, pure water has a pH of 7.

Solutions with a pH less than 7 are acids, and those with a pH greater than 7 are bases. The lower the pH value below 7, the more acidic is the solution, and the higher the pH value above 7, the more basic is the solution. There is a 10-fold difference in the concentrations of H^+ and OH^- when the pH changes by one unit. For example, an acid with a pH of 4 has a concentration of H^+ that is 10 times greater than that of an acid with a pH of 5. Table 2.5 lists the pH values of a few common substances.

Buffers Cells of the body are especially sensitive to pH changes. Even slight changes can be harmful. The hydrogen ion concentration (pH) of blood and other body fluids is maintained within narrow limits by the lungs, kidneys, and buffers in body fluids. A **buffer** is a chemical or a combination of chemicals that either picks up excess H^+ or releases H^+ to keep the pH of a solution rather constant.

Buffers are extremely important in maintaining the normal pH of body fluids, but they can be overwhelmed by a disruption of homeostasis. The normal pH of the blood is 7.35 to 7.45. In *acidosis,* the pH ranges from 7.0 to 7.3, and the patient feels tired and disoriented. In *alkalosis,* the pH range is 7.5 to 7.8, and the patient feels agitated and dizzy. Greater variations may be fatal.

Salts

Like inorganic acids and bases, **mineral salts** are ionic compounds that dissociate in an aqueous solution, but they do not produce hydrogen and hydroxyl ions. The most important salts in the body are sodium, potassium, and calcium salts. Calcium phosphate is the most abundant salt because it is a main component of bones and teeth. For example, sodium chloride (NaCl), a common salt in body fluids, dissociates into sodium (Na^+) and chloride (Cl^-) ions.

$$NaCl \rightarrow Na^+ + Cl^-$$

Mineral salts provide ions that are essential for normal body functioning. Physiological processes in which ions play an essential role include blood clotting, muscle and nerve functions, and pH and water balance (table 2.6).

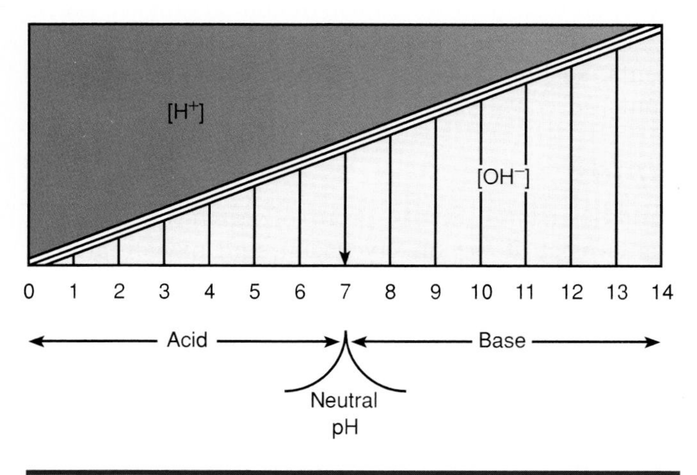

Figure 2.5 The pH scale. The relative concentration of hydrogen ions (H^+) and hydroxyl ions (OH^-) at each pH is indicated by the diagonal line. Acids have a pH less than 7, and bases have a pH greater than 7.

Table 2.5 pH Values of Selected Substances

Substance	pH Value
Lemon juice	2.3
Apple cider	3.0
Tomato juice	4.2
Coffee	5.0
Urine	5.0–8.0
Saliva	6.5–3.5
Pure water	7.0
Blood	7.35–7.45
Egg white	8.0
Borax	9.2
Milk of magnesia	10.5
Household ammonia	11.0
Limewater	12.3

Table 2.6 Important Inorganic Ions

Ion	Symbol	Functions
Bicarbonate	HCO_3^-	Helps maintain acid-base balance
Calcium	Ca^{+2}	Major component of bones and teeth; required for muscle contraction and blood clotting
Carbonate	CO_3^{-2}	Major component of bones and teeth; helps maintain acid-base balance
Chloride	Cl^-	Helps maintain water balance
Hydrogen	H^+	Helps maintain acid-base balance
Hydroxyl	OH^-	Helps maintain acid-base balance
Phosphate	PO_4^{-3}	Major component of bones and teeth; required for energy transfer
Potassium	K^+	Required for muscle and nerve function
Sodium	Na^+	Required for muscle and nerve function; helps maintain water balance

Where is water located in the body, and why is it so important?
What is the relationship between acids, bases, and pH?
What are mineral salts, and why are they important?

Major Organic Compounds

The major organic compounds of the body are carbohydrates, lipids, proteins, and nucleic acids (table 2.7). Another compound, adenosine triphosphate, is also considered here because it plays such a vital role in the transfer of energy within cells.

Carbohydrates

Carbohydrates (kar″-bo-hī′-drātz) are formed of carbon, hydrogen, and oxygen. In each carbohydrate molecule, there are two hydrogen atoms for every oxygen atom. Carbohydrates are the primary source of nutrient energy for cells of the body. Carbohydrates are classified according to molecular size as monosaccharides, disaccharides, or polysaccharides (figure 2.6).

The simplest carbohydrates are **monosaccharides** (mon-ō-sak′-ah-rīds), or simple sugars. For example, **glucose** ($C_6H_{12}O_6$) is a six-carbon simple sugar that is the major carbohydrate fuel for cells. It is often called blood sugar since it is the form in which carbohydrates are transported to body cells. Fructose and galactose are other six-carbon simple sugars found in foods. Six-carbon monosaccharides are chemically combined to produce more complex carbohydrates.

The chemical combination of two monosaccharides forms a **disaccharide** (dī-sak′-ah-rīd), a double sugar. The common disaccharides in foods are **maltose,** or malt sugar (glucose + glucose), **sucrose,** or table sugar (glucose + fructose) and **lactose,** or milk sugar (glucose + galactose).

A **polysaccharide** (pol-ē-sak′-ah-rīd) is formed by the chemical combination of many monosaccharide units. Two polysaccharides are important to our study: glycogen and starch. Both are formed of many glucose units.

Glycogen is the storage form of carbohydrates in the body. Some of the excess glucose in blood is converted into glycogen and stored primarily in the liver, but small amounts are stored in muscle cells. Glycogen serves as a reserve energy supply that can be quickly converted into glucose. For example, whenever the level of blood glucose declines, the liver converts glycogen into glucose to raise the blood glucose level.

Starch is the storage form of carbohydrates in plants, so it is present in many foods derived from plants.

Lipids

Lipids are a large, diverse group of organic compounds that consist of carbon, hydrogen, and oxygen atoms. Carbon atoms form the backbone of the molecules, and there are many times more hydrogen atoms than oxygen atoms. The most abundant lipids in the body are triglycerides (fats), phospholipids, and steroids.

Molecules of **triglycerides** (trī-glys′-er-īds), or **fats,** consist of one **glycerol** (glys′-er-ol) molecule and three **fatty acid** molecules joined together (figure 2.7). Triglycerides are the most concentrated energy source found in the body. Excess nutrients (energy reserves) are stored as triglycerides in the fat cells of the body, primarily around internal organs and under the skin. Fats are not soluble in water.

Table 2.7 Important Organic Compounds

Compound	Building Units	Examples	Functions
Carbohydrates	Monosaccharides (simple sugars)	Glucose	Primary energy source for cells
		Starch	Storage form in plants; common in plant foods
		Glycogen	Storage form in animals; stored in liver and muscles
Lipids	Glycerol, fatty acids, phosphate	Triglycerides (fats)	Energy source and storage
		Phospholipids	Cell structure
		Steroids	A variety of functions (e.g., sex hormones promote sexual development)
Proteins	Amino acids	Structural proteins	Cell structure
		Functional proteins	Nonstructural functions (e.g., enzymes catalyze chemical reactions)
Nucleic acids	Nucleotides	DNA	Directs cell functions by controlling protein synthesis
		RNA	Helps in protein synthesis

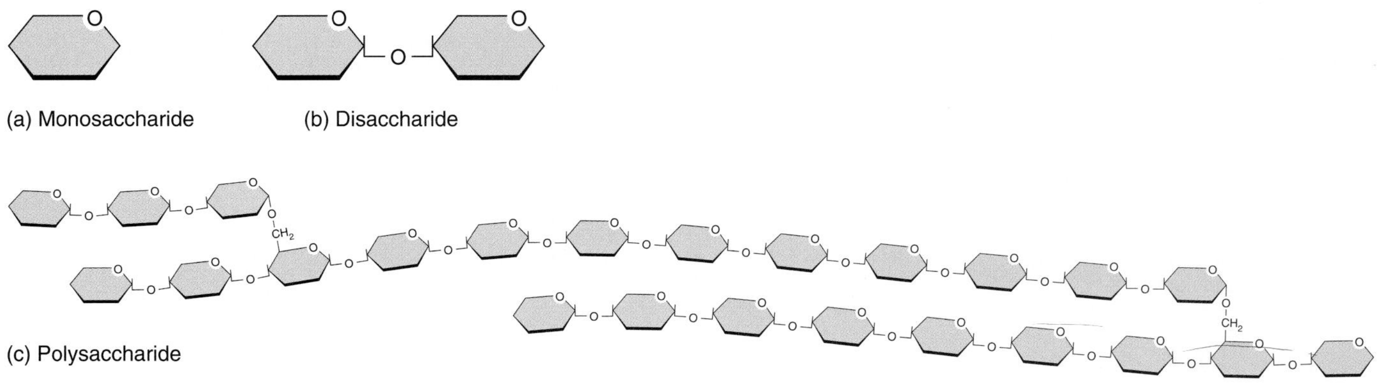

(a) Monosaccharide

(b) Disaccharide

(c) Polysaccharide

Figure 2.6 (*a*) The building units of carbohydrates are monosaccharides. (*b*) Two monosaccharides combine to form a disaccharide. (*c*) The combination of many monosaccharide units form a polysaccharide.

Triglycerides may be classified as either saturated or unsaturated fats. In *saturated fats* (animal fats), the bonds of the carbon atoms in the fatty acids are saturated (filled) by hydrogen atoms so that the carbon–carbon bonds are all single bonds. Saturated, or "hydrogenated," fats are solid at room temperature. Examples are butter, lard, and margarine. Excessive saturated fats in the diet are associated with an increased risk of heart disease.

In *unsaturated fats* (plant fats), all carbon bonds in the fatty acids are not filled with hydrogen atoms, and one or more double carbon–carbon bonds are present. Monounsaturated fats have a single double bond; polyunsaturated fats contain two or more double bonds. Unsaturated fats occur as oils, for example olive oil and corn oil. Figure 2.8 illustrates saturated and unsaturated fatty acids.

Phospholipids (fos″-fō-lip′-idz) have molecules similar to triglycerides. The basic difference is that one of the fatty acids is replaced with a phosphate-containing group. Unlike fats, phospholipids are soluble in water. They are major components of cell membranes. Figure 2.9 shows the basic structure of a phospholipid molecule.

Steroids constitute another group of lipids, and their molecules characteristically contain four carbon rings. Cholesterol and sex hormones are examples of steroids. Figure 2.10 shows the structure of cholesterol.

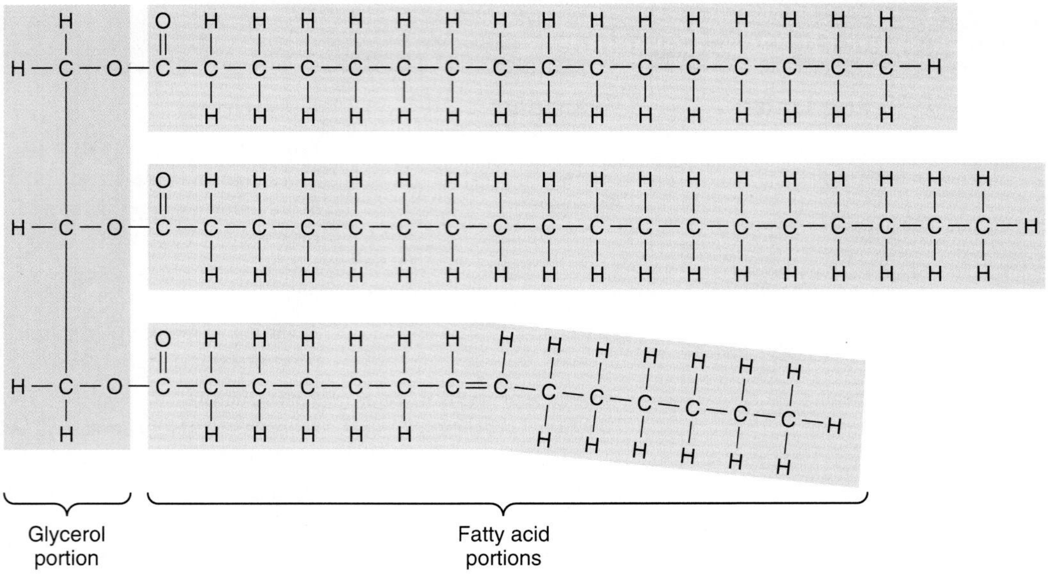

Glycerol portion

Fatty acid portions

Figure 2.7 A triglyceride molecule consists of three fatty acids joined to a glycerol molecule.

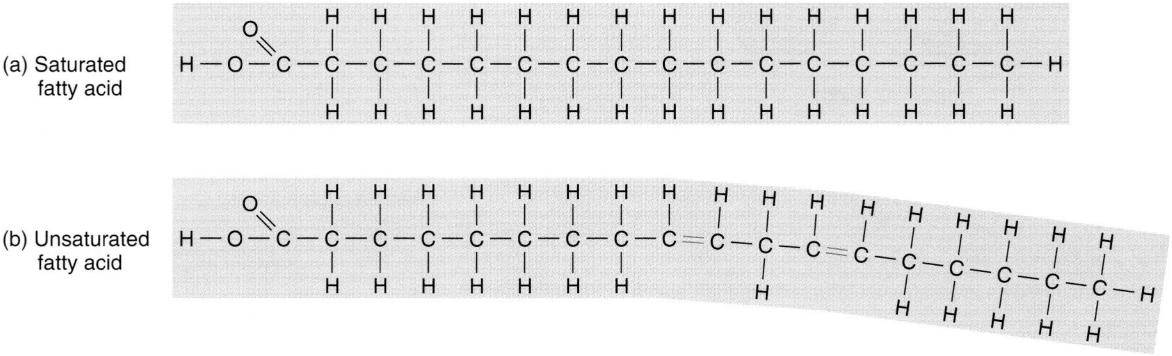

(a) Saturated fatty acid

(b) Unsaturated fatty acid

Figure 2.8 Molecular structures of (*a*) a saturated fatty acid and (*b*) an unsaturated fatty acid. Note that the double bonds (*red*) in the unsaturated fatty acid cause the molecule to bend.

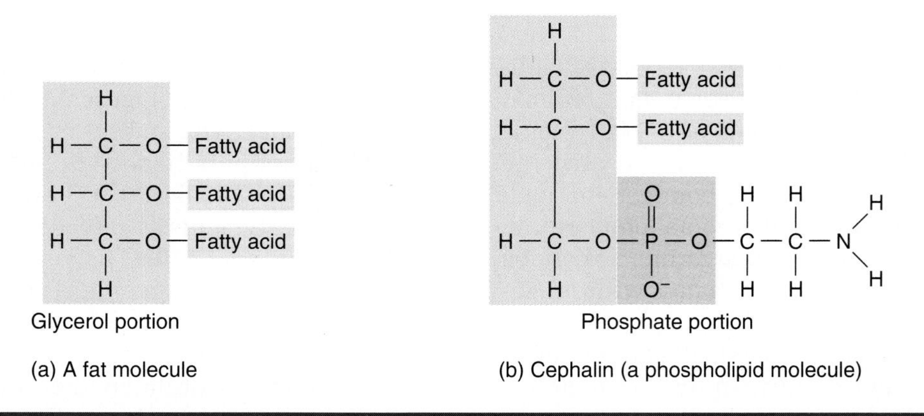

Figure 2.9 Comparison of triglyceride (fat) and phospholipid molecules. (*a*) A triglyceride consists of three fatty acids attached to a glycerol. (*b*) In a phospholipid, one fatty acid is replaced by a phosphate-containing group.

Cholesterol (kō-les′-ter-ols) is an essential component of body cells and serves as the raw material for the synthesis of other steroid molecules. But cholesterol can be hazardous when present in the diet in excessive amounts. See the clinical box on this page.

Proteins

Proteins (pro′-te-ins or pro′-tēns) are large, complex molecules composed of smaller molecules (building units) called **amino acids.** There are 20 different kinds of amino acids, and each is composed of carbon, hydrogen, oxygen, and nitrogen. As the name implies, each amino acid consists of an *amine group* (—NH_2) and an *acid group* (—COOH), but it is the *R group,* the remainder of the molecule, that distinguishes the different kinds of amino acids. Figure 2.11 depicts the basic structure of an amino acid.

Amino acids are joined by **peptide bonds.** Two amino acids bonded together form a **dipeptide.** A chain up to 50 amino acids long forms a **polypeptide.** A chain of 50 to thousands of amino acids forms a protein. Hydrogen bonds between various parts of a polypeptide or protein molecule cause the molecule to twist and fold. The result is that each specific polypeptide or protein has a unique three-dimensional shape.

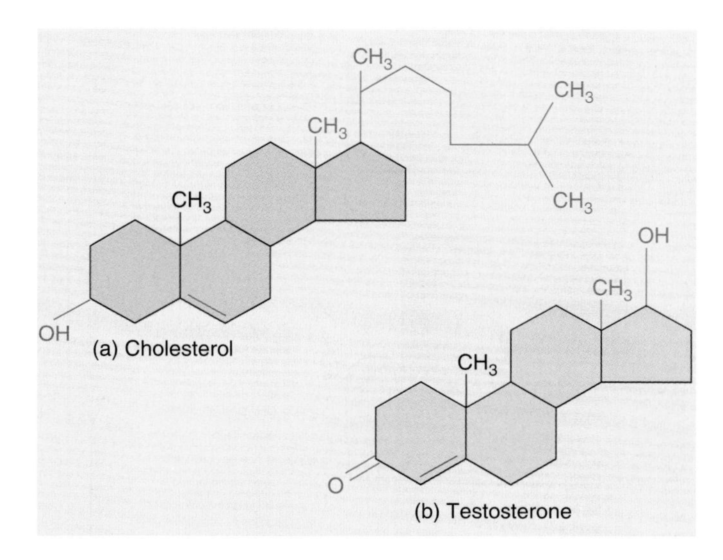

Figure 2.10 Steroids include cholesterol and testosterone (male sex hormone). Note the four carbon rings, which are characteristic of steroids.

Considerable evidence exists linking high concentrations of cholesterol in the blood with heart disease, which is caused by a decreased blood flow to the heart muscle. Cholesterol deposits in the walls of the coronary arteries restrict the blood flow, depriving the heart of needed nutrients and oxygen. Since the primary dietary source of cholesterol is red meats, egg yolks, and milk products, the American Heart Association (AHA) recommends that adults reduce the amounts of these foods in the diet. Red meats and milk products also contain large amounts of saturated fats, and saturated fats are readily converted into cholesterol by the liver. Unsaturated fats in the diet tend to decrease the concentration of cholesterol in the blood. Therefore, the AHA recommends a low-fat diet with a ratio of at least 1.5 unsaturated fats to 1.0 saturated fats. The "typical American diet" usually contains far more cholesterol and saturated fats than are recommended by the AHA.

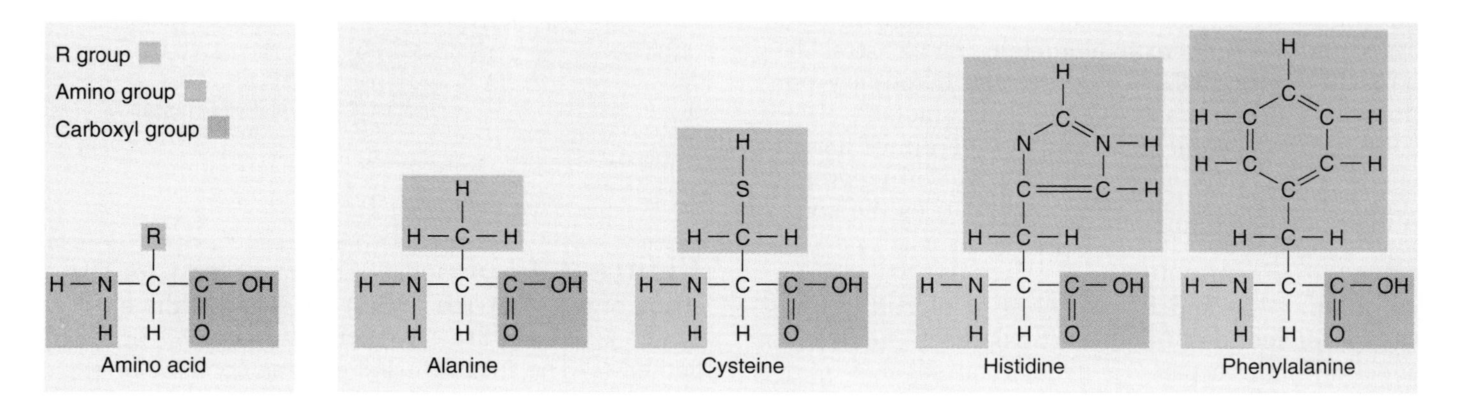

Figure 2.11 Amino acids. (*a*) The basic structure of an amino acid. (*b*) Representative amino acids. Note that the amino group and carboxyl (acid) group are the same in all amino acids. It is only the R group that is different among different amino acids.

Proteins may be classified as either structural or functional proteins. *Structural proteins* compose parts of body cells and tissues, where they provide support and strength in binding parts together. Ligaments, tendons, and contractile fibers in muscles are composed of structural proteins. *Functional proteins* perform a variety of different functions in the body. *Antibodies,* which provide immunity, and *enzymes,* which speed up chemical reactions, are examples of functional proteins.

Enzymes Without **enzymes,** the body's chemical reactions would occur too slowly to maintain life. Body cells contain thousands of enzymes, and each enzyme *catalyzes* (speeds up) a particular chemical reaction. An enzyme catalyzes a reaction by binding to a substrate (reactant) molecule and weakening a chemical bond so it breaks faster and more easily.

The enzyme must have just the right shape to fit onto the substrate, somewhat like a piece of a puzzle or like a key fits a lock (figure 2.12). After the reaction is complete, the products and enzyme separate. Since the enzyme is not altered in the reaction, it may be recycled and used again. A single enzyme can catalyze thousands of reactions.

Like other proteins, the three-dimensional shape of an enzyme is determined by hydrogen bonds. And hydrogen bonds are easily broken by several factors, such as temperature and pH changes, poisons, and radiation. If an enzyme's hydrogen bonds are altered, its shape is changed, and the enzyme is inactivated (denatured). Since it cannot bind to the substrate, the reaction it catalyzes will not occur. If the reaction is vital, the result can be fatal.

Nucleic Acids

Nucleic (nū-klā′-ic) **acids** are composed of very large molecules. Two types of nucleic acids occur in cells: DNA and RNA. **Deoxyribonucleic** (dē-ok″-se-rī″-bō-nū-klā′-ic) **acid (DNA)** composes the hereditary portion of chromosomes in the cell nucleus. DNA contains the *genetic code,* which determines hereditary traits and cellular functions by controlling protein synthesis. **Ribonucleic acid (RNA)** carries the coded instructions from DNA to the cellular machinery involved in protein synthesis.

Both DNA and RNA consist of repeating units called **nucleotides** (nū′-klē-ō-tīds). Each nucleotide consists of three parts: a five-carbon sugar, a phosphate group, and an organic base. Figure 2.13 shows four nucleotides that form part of a strand of a nucleic acid. Note that the five-carbon sugars and phosphate groups form the backbone of the strand and that the organic (nitrogen-containing) bases project from the sugar molecules.

DNA consists of two coiled strands that are joined by hydrogen bonds (figure 2.14). The five-carbon sugars in DNA are *deoxyribose* sugars. In contrast, RNA consists of a single strand, and its five-carbon sugars are *ribose* sugars.

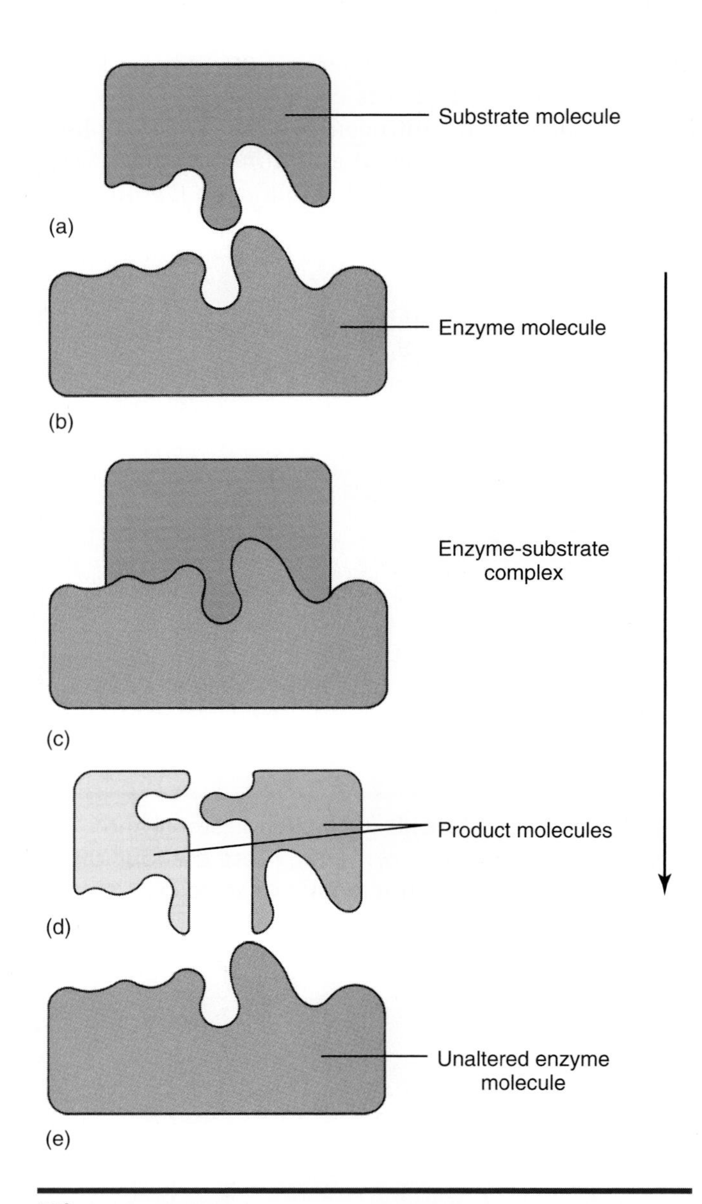

Figure 2.12 A model of enzyme action. (*a*) The shape of the substrate and (*b*) the shape of the enzyme's active site are complementary and will fit together. (*c*) The substrate and enzyme combine briefly, enabling a chemical reation that splits the substrate molecule. (*d*) The product molecules separate from the enzyme, and (*e*) the unaltered enzyme may be recycled.

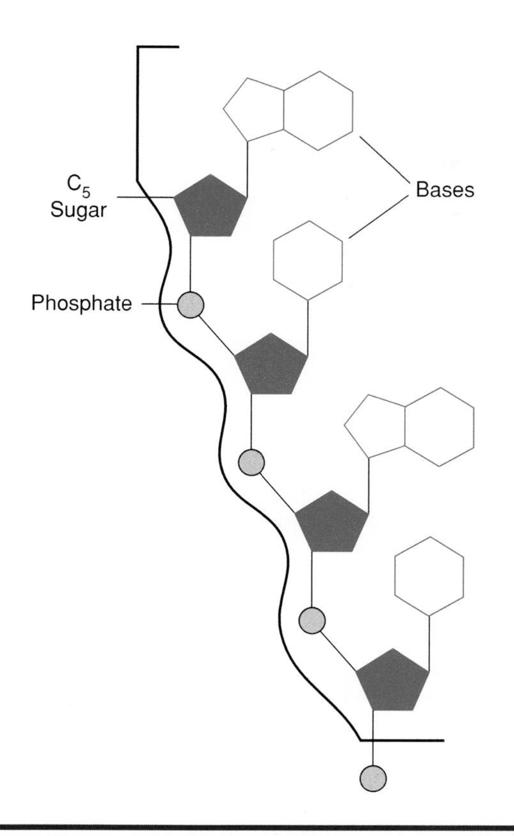

Figure 2.13 Four nucleotides of a nucleic acid molecule. Each nucleotide consists of a 5-carbon sugar, a phosphate group, and an organic base. The phosphate groups join the sugars of the nucleotides together to form the strand, while the organic bases project from the sugar molecules.

Adenosine Triphosphate (ATP)

ATP, or **adenosine** (ah-den′-ō-sēn) **triphosphate** (trī-fos′-fāt), is a modified nucleotide that consists of adenosine and three phosphate groups. The last two phosphate groups are joined to the molecule by special bonds called *high-energy phosphate bonds.* In figure 2.15, these bonds are represented by wavy lines. Energy in these bonds is released to power chemical reactions within a cell. In this way, ATP provides immediate energy to keep cellular processes operating. It is the *only* molecule that performs this role.

Energy extracted from nutrients by cells is temporarily held in ATP and then released to power chemical reactions. When the terminal high-energy phosphate bond of ATP is broken and energy is transferred, ATP is broken down into **adenosine diphosphate (ADP)** and a low-energy phosphate group (—P). The addition of a high-energy phosphate group to ADP re-forms ATP. ATP is continuously broken down into ADP to release energy, and it is re-formed as energy is made available from nutrients.

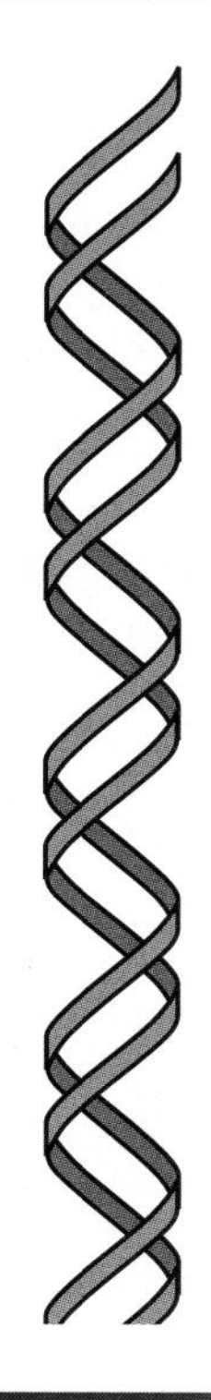

Figure 2.14 DNA consists of two coiled strands of nucleotides joined by hydrogen bonds.

What distinguishes the chemical structure and functions of carbohydrates, lipids, proteins, and nucleic acids?
What is the role of enzymes in body cells?
What is ATP, and what is its role in body cells?

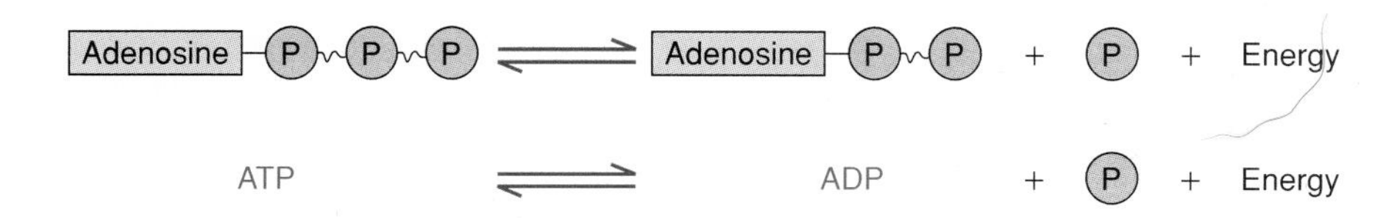

Figure 2.15 The breakdown of ATP forms ADP and a phosphate group and releases energy to power cellular reactions. ATP is synthesized from ADP, phosphate, and energy extracted from nutrients.

CHAPTER SUMMARY

Atoms and Elements

1. Matter is composed of elements, substances that cannot be broken down into simpler substances by chemical means.
2. Oxygen, carbon, hydrogen, and nitrogen form 96% of the human body by weight.
3. An atom is the smallest unit of an element.
4. An atom consists of a nucleus formed of protons (+1) and neutrons (0) and electrons (−1) that orbit around the nucleus.
5. Elements are characterized by their atomic numbers, symbols, and atomic weights.
6. Isotopes of an element have differing numbers of neutrons.
7. Radioisotopes emit radiation.

Molecules and Compounds

1. A molecule is formed of two or more atoms joined by chemical bonds; it is the smallest unit of a compound.
2. A compound is formed of two or more elements combined in a fixed ratio.
3. A chemical formula indicates the number of atoms of each element in the molecule.
4. Chemical bonds join atoms to form molecules.
5. An ionic bond is the force of attraction between two ions with opposite electrical charges. It results from one atom donating one or more electrons to another atom.
6. A covalent bond is formed between two atoms by the sharing of electrons in the valence shell.
7. A hydrogen bond is a weak force of attraction between a slightly positive H atom and a slightly negative atom either in the same molecule or in different molecules.
8. Synthesis reactions combine simpler substances to produce more complex substances.
9. Decomposition reactions break down more complex substances into simpler substances.

Compounds Composing the Human Body

1. Inorganic compounds do *not* contain both carbon and hydrogen. Organic compounds contain *both* carbon and hydrogen.
2. Water (H_2O) is the most abundant inorganic compound in the body, and it is the solvent of living systems.
3. There are two major water compartments: intracellular fluid (65% of body water) and extracellular fluid (35% of body water).
4. An acid releases H^+ in an aqueous solution, and a base releases OH^- in an aqueous solution.
5. A salt releases positively and negatively charged ions in an aqueous solution, but they are neither H^+ nor OH^-.
6. pH is a measure of the relative concentrations of H^+ and OH^- in a solution.
7. A buffer keeps the pH of a solution relatively constant by picking up or releasing H^+.
8. Carbohydrates are composed of C, H, and O with a 2:1 ratio between H and O.
9. Monosaccharides, or simple sugars, are the building units of carbohydrates. Disaccharides consist of two simple sugars. Polysaccharides are formed of many simple sugars.
10. Lipids are a diverse group of organic compounds that include triglycerides, phospholipids, and steroids.
11. Triglycerides (fats) consist of three fatty acids bonded to glycerol. Unsaturated fats differ from saturated fats by having one or more double carbon–carbon bonds in their fatty acids. Excess nutrients are stored as fats.
12. Phospholipids consist of two fatty acids and a phosphate-containing group bonded to glycerol.
13. Steroids are an important group of lipids that includes sex hormones and cholesterol.
14. Proteins are large molecules formed of many amino acids. The 20 different kinds of amino acids are distinguished by their R groups.
15. Amino acids are joined by dipeptide bonds.
16. Structural proteins form parts of cells and tissues. Functional proteins include enzymes and antibodies.
17. Enzymes catalyze chemical reactions.
18. Nucleic acids are very large molecules formed of many nucleotides.
19. A nucleotide consists of a five-carbon sugar (ribose in RNA and deoxyribose in DNA), a phosphate group, and an organic base.
20. There are two types of nucleic acids: deoxyribonucleic acid (DNA) and ribonucleic acid (RNA). DNA determines hereditary traits and controls cellular functions. RNA works with DNA in the synthesis of proteins.
21. Adenosine triphosphate (ATP) is a modified nucleotide that temporarily holds energy in high-energy phosphate bonds and releases that energy to power chemical reactions in a cell.

BUILDING YOUR VOCABULARY

1. **Selected New Terms**
   acid, p. 30
   amino acid, p. 35
   ATP, p. 37
   base, p. 30
   buffer, p. 31
   covalent bond, p. 28
   dipeptide, p. 35
   disaccharide, p. 32
   DNA, p. 36
   electron, p. 26
   electrolyte, p. 28
   hydrogen bond, p. 28
   ionic bond, p. 28
   mineral salt, p. 31
   neutron, p. 26
   pH, p. 31
   phospholipid, p. 33
   polypeptide, p. 35
   polysaccharide, p. 32
   proton, p. 26
   RNA, p. 36
   solute, p. 30
   solvent, p. 30
   steroid, p. 33
   triglyceride, p. 32

2. **Related Clinical Terms**
   acidosis (as″-id-ō′-sis) Abnormally low blood pH (< 7.3).
   alkalosis (al″-kah-lō′-sis) Abnormally high blood pH (> 7.5).
   dehydration (dē-hī-drā′-shun) A significant decrease in the water content of the body.
   hypovolemia (hī-pō-vol-ēm′-ia) Abnormally low volume of blood plasma.

STUDY ACTIVITIES

1. Complete the **Chapter 2 Study Guide,** which begins on page 403.
2. Complete each of the **Learning Objectives** listed on the first page of this chapter.

CHECK YOUR UNDERSTANDING

(Answers are located in appendix B.)

1. An atom contains the same number of electrons as ______.
2. An ______ is an atom with a net electrical charge.
3. A ______ is the smallest unit of a compound.
4. When atoms share electrons a ______ bond is formed.
5. The most abundant compound in the body is ______.
6. An acid has a pH ______ than 7.
7. In a ______ reaction, smaller molecules are combined to form a larger molecule.
8. The building units of a carbohydrate are ______.
9. The building units of proteins are ______.
10. The building units of nucleic acids are ______.
11. Glycerol and fatty acids combine to form ______.
12. ______ is the energy carrier that releases energy to power cellular processes.

THREE

CHAPTER 3
The Cell

Chapter Preview & Learning Objectives

SELECTED KEY TERMS

Active transport Movement of substances through a plasma membrane requiring the expenditure of energy by the cell.
Cell (cella = room, cell) Structural and functional unit of living organisms.
Cellular respiration Breakdown of organic nutrients, which releases energy to form ATP.
Centrioles (centr = center) Paired cylindrical organelles adjacent to the nucleus that form the spindle during cell division.
Chromosome (chrom = color; soma = body) A threadlike or rodlike structure in cell nuclei that is composed of DNA and protein.
Cytoplasm (cyt = cell; plasma = molded) The semifluid substance located between the cell nucleus and the plasma membrane.
Diffusion Passive movement of molecules from an area of higher concentration to an area of lower concentration.
Endoplasmic reticulum System of membranes extending throughout the cytoplasm.
Filtration The forcing of small molecules through a membrane by hydrostatic pressure.
Mitochondrion Organelle where aerobic cellular respiration occurs.
Mitosis (mit = thread; -sis = condition) Separation and distribution of chromosomes to daughter cells during mitotic cell division.
Nucleus (nucle = kernel) Spherical organelle containing chromosomes and controlling cellular functions.
Organelle (-elle = little) A specific part of a cell that performs a specific function.
Osmosis The diffusion of water through a selectively permeable membrane.
Passive transport Movement of substances through a plasma membrane without expenditure of energy by the cell.
Phagocytosis (phag = to eat; cyt = cell) Engulfing of particles by the cell.
Pinocytosis (pino = to drink) Engulfing of liquid droplets by the cell.
Ribosome (ribo = ribose; soma = body) Tiny cytoplasmic organelle, composed of RNA and protein, that is the site of protein synthesis.
Selective permeability The characteristic of a plasma membrane that permits some, but not all, substances to pass through it.

Like all living organisms, humans are composed of **cells.** In fact, there are about 75 trillion cells in the human body. Many different kinds of cells compose the body, and each type is specialized to perform specific functions. Although these cells vary in size, shape, and function, they exhibit many structural and functional similarities.

Human cells are very small and are visible only with a microscope. Knowledge of cell structure is based largely on the examination of cells with an electron microscope, a type of microscope that provides magnifications up to 200,000× or more.

Cell Structure

Although human cells are small, they are amazingly complex with many specialized parts. The composite cell in figure 3.1 illustrates the major structures known to compose human cells. These structures are shown as they appear in electron microscope images. Most, but not all, of these structures are found in each human cell. As each part of a cell is discussed, note its structure and relationship to other structures in figure 3.1.

The Plasma Membrane

The **plasma membrane** forms the outer boundary of a cell. It consists of two layers of phospholipid molecules, aligned back-to-back, with their fatty acid tails forming the interior of the membrane (figure 3.2). A few cholesterol molecules are scattered among the phospholipids. These lipids provide a barrier between water-soluble materials inside and outside the cell. Protein molecules are attached to or embedded in the phospholipid layers. Globular proteins that extend through the membrane serve as passageways for transport of substances into and out of cells. Some membrane proteins are receptors for enzymes and hormones, special molecules that influence cell function. Some proteins in combination with carbohydrates serve as identification markers, enabling cells to recognize each other.

All materials that enter or exit a cell must pass through the plasma membrane. Since it allows only certain molecules to enter or exit the cell, it is said to be *selectively permeable.* Whether or not molecules can pass through the membrane is determined by a number of factors that include molecular size, solubility, ionic charges, and attachment to carrier molecules.

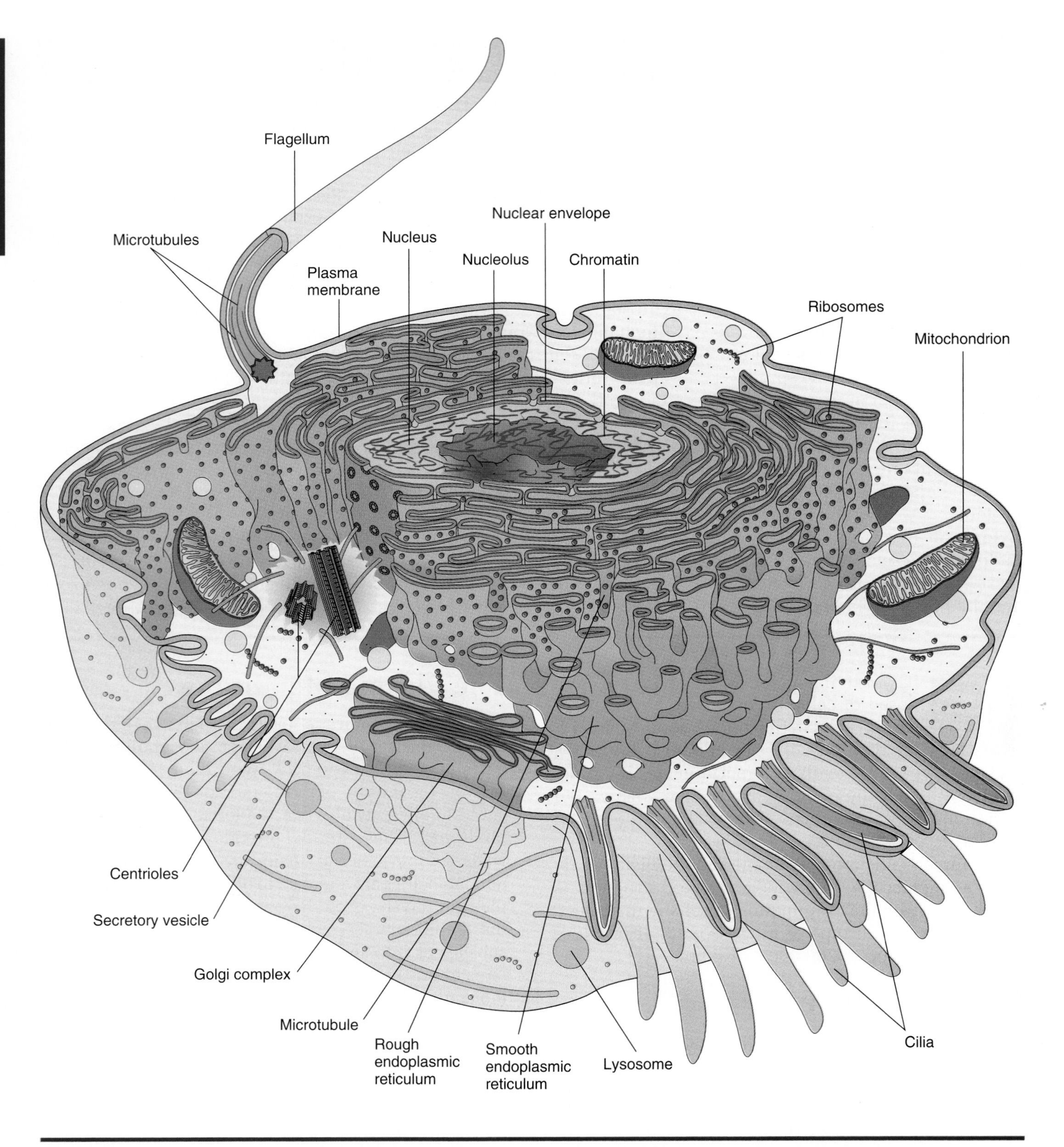

Figure 3.1 A composite human cell shows the major structures composing cells. (The organelles are not drawn to scale.)

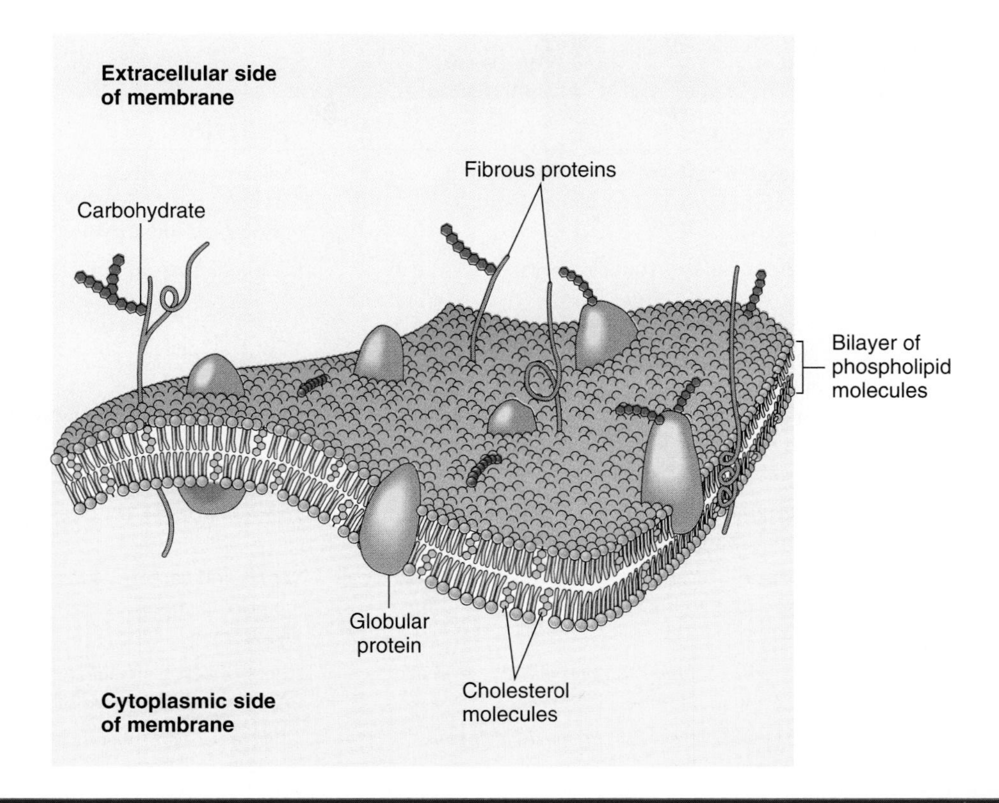

Figure 3.2 The cell membrane is composed primarily of phospholipids, with proteins scattered throughout the lipid bilayer.

Cytoplasm

The interior of the cell is filled with the fluid or gel-like **cytoplasm** (sī′-tō-plazm). It contains 75% to 90% water plus an assortment of organic and inorganic compounds. It is semitransparent and usually slightly thicker than water, but at times it may have almost a gel-like consistency. It is supported by (1) intracellular membranes and (2) the cytoskeleton.

Organelles

A variety of **organelles** (or-gah-nel′z), or tiny organs, are surrounded by the cytoplasmic fluid. Each type of organelle performs a specific role in the cell. Organelles are distinguished by size, shape, structure, and function. Table 3.1 summarizes the structure and functions of the major parts of a cell.

Nucleus

The largest organelle is the **nucleus** (nū′klē-us), a spherical or egg-shaped structure that is slightly more dense than the surrounding cytoplasm. It is separated from the cytoplasm by a double-layered **nuclear envelope** containing numerous pores that aid the movement of materials between the nucleus and cytoplasm.

Chromosomes (krō′-mō-sōms), the most important structures within the nucleus, consist of DNA and protein. The DNA of chromosomes contains coded instructions that control the functions of the cell. When a cell is not dividing, chromosomes are extended to form thin threads that appear as *chromatin* (krō′-mah-tin) *granules* when viewed microscopically, as in figure 3.3. During cell division, the chromosomes coil and shorten and become rod-shaped. Each human body cell contains 23 pairs of chromosomes, with a total of 46 in all.

One or more dense spherical bodies, the **nucleoli** (nū-klē-ō-lē), are also present in the nucleus. A nucleolus consists of RNA and protein.

Ribosomes

Ribosomes are tiny organelles that appear as granules within the cytoplasm even in electron photomicrographs. They are composed of ribosomal RNA (rRNA) and protein, which are pre-formed in a nucleolus before migrating from the nucleus into the cytoplasm. Ribosomes are the sites of protein synthesis in cells. They may occur singly or in small clusters.

Table 3.1 Summary of Cells Parts

Component	Structure	Function
Plasma membrane	Phospholipid bilayer with proteins and cholesterol molecules embedded in it	Selectively controls movement of materials into and out of the cell; has receptors for hormones and enzymes
Cytoplasm	Gel-like fluid surrounding organelles	Site of numerous chemical reactions
Organelles		
Nucleus	Largest organelle; contains chromosomes and nucleoli	Controls cellular functions
Endoplasmic reticulum (ER)	System of membranes extending through the cytoplasm; RER has ribosomes on the membrane; SER does not	Serves as sites of chemical reactions; channels for material transport within cell
Ribosomes	Tiny granules of rRNA and protein either associated with RER or free in cytoplasm	Sites of protein synthesis
Golgi complex	Series of stacked membranes near nucleus; associated with ER	Sorts and packages substances in vesicles for export from cell or use within cell; forms lysosomes
Mitochondria	Contain a folded membrane within a smaller exterior membrane	Sites of aerobic cellular respiration that form ATP from breakdown of nutrients
Lysosomes	Small vesicles containing strong digestive enzymes	Digest foreign substances or worn-out parts of cells
Microfilaments	Thin rods of protein dispersed in cytoplasm	Provide support for cell; contraction gives cell movement
Microtubules	Thin tubules dispersed in cytoplasm	Provide support for cell, cilia, and flagella; form spindle during cell division
Centrioles	Two short cylinders formed of microtubules; located near nucleus	Form spindle fibers during cell division
Cilia	Numerous short, hairlike projections from certain cells	Move materials along cell surface
Flagella	Long whiplike projections from sperm cells	Enable movement of sperm

Endoplasmic Reticulum

The numerous membranes that extend from the nucleus throughout the cytoplasm are collectively called the **endoplasmic reticulum** (en″-dō-plas′-mik rē-tik′-ū-lum), or **ER** for short. These membranes provide some support for the cytoplasm and form a network of channels that facilitate the movement of materials within the cell.

There are two types of ER: rough ER and smooth ER. *Rough endoplasmic reticulum (RER)* is characterized by the presence of numerous ribosomes located on the outer surface of the membranes. Proteins formed by ribosomes on the RER usually are exported from the cell, while those formed by ribosomes scattered throughout the cytoplasm usually are used within the cell. *Smooth endoplasmic reticulum (SER)* lacks ribosomes and serves as a site for the synthesis of lipids (figure 3.4).

Golgi Complex

This organelle appears as a stack of flattened membranous sacs that are usually located near the nucleus and in close association with the nucleus and ER. The **Golgi** (Gol′-jē) **complex** processes and sorts synthesized substances, such as proteins, into vesicles for transport to other parts of the cell or outside of the cell (figure 3.5).

Mitochondria

The **mitochondria** (mī″-to-kon′-drē-ah) are relatively large organelles that are characterized by having an

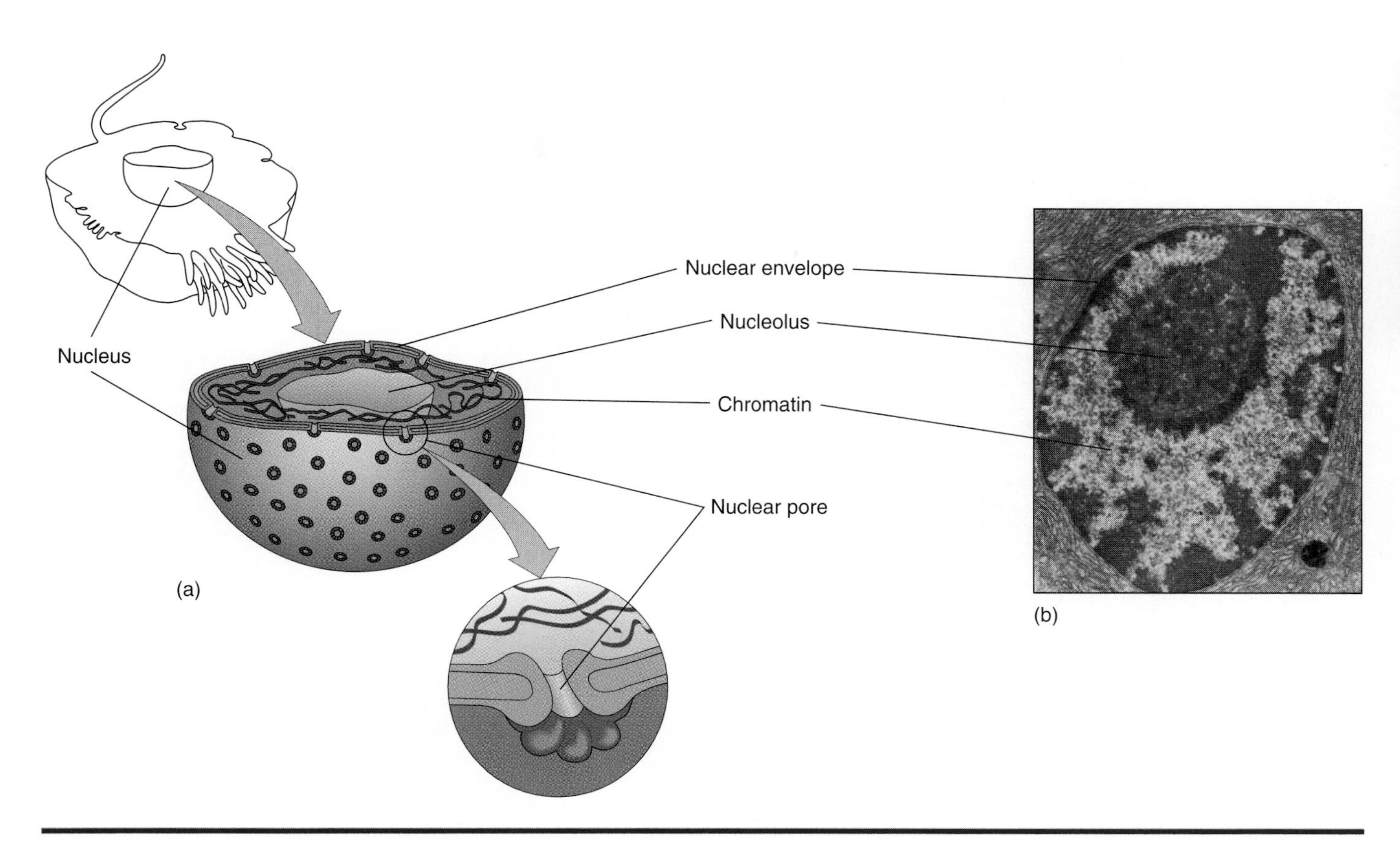

Figure 3.3 (*a*) The nuclear envelope is selectively permeable and allows certain substances to pass between the nucleus and the cytoplasm. (*b*) Transmission electron micrograph of a cell nucleus (8,000×). It contains a nucleolus and masses of chromatin.

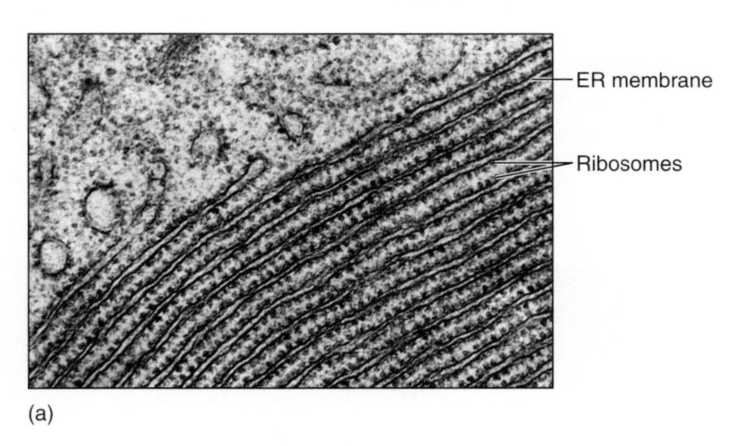

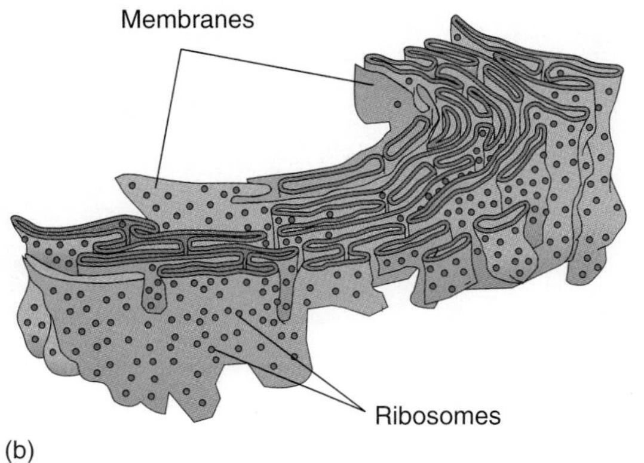

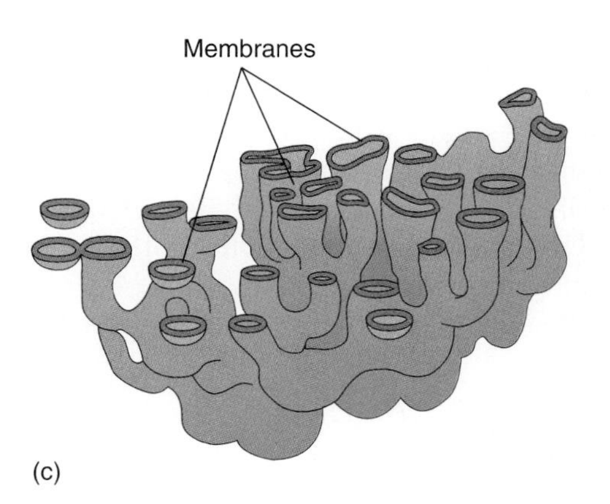

Figure 3.4 (*a*) Transmission electron micrograph of rough endoplasmic reticulum (ER) (100,000×). (*b*) Rough ER is dotted with ribosomes, while (*c*) smooth ER lacks ribosomes.

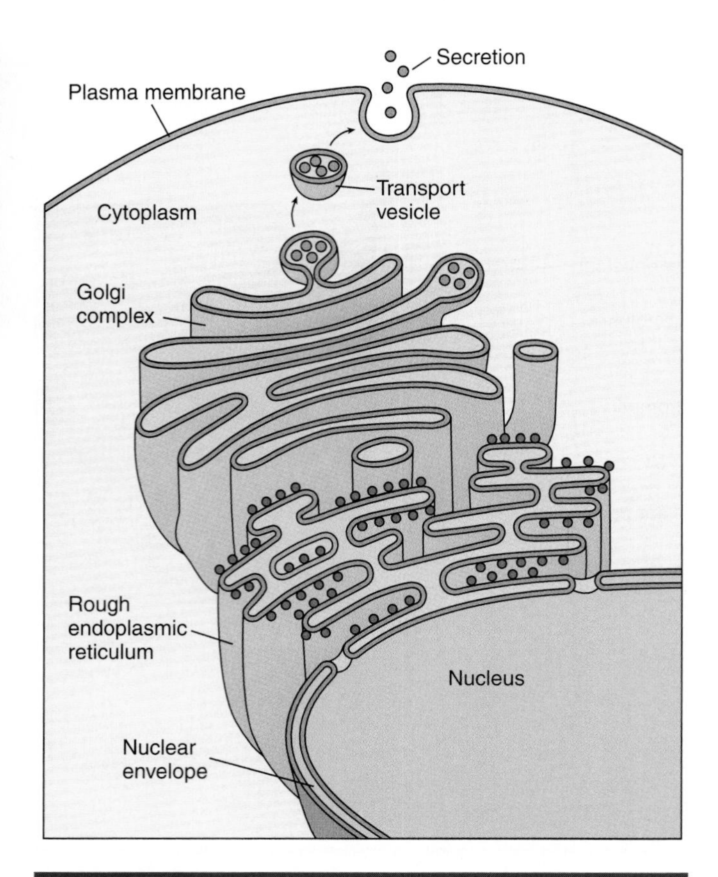

Figure 3.5 The Golgi complex packages substances in vesicles that may move to the plasma membrane to release the substance outside the cell.

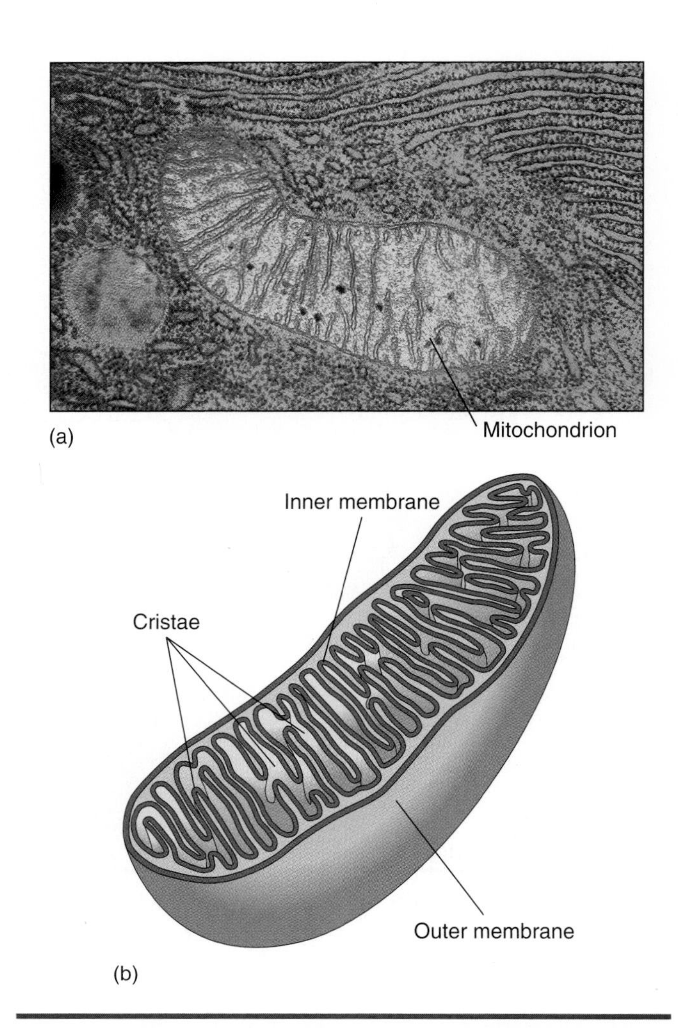

Figure 3.6 A mitochondrion. (*a*) A transmission electron photomicrograph (40,000×) and (*b*) a drawing of a mitochondrion shows the cristae formed by the inner membrane.

inner folded membrane surrounded by an outer smooth membrane. The inner membranous folds are called *cristae,* and on their surfaces are the enzymes involved in aerobic cellular respiration.

The release of energy from nutrients and the formation of ATP by aerobic cellular respiration occur within mitochondria. For this reason, mitochondria are sometimes called the "powerhouses" of the cell. Mitochondria can replicate themselves if the need for additional ATP production increases (figure 3.6).

Lysosomes

Lysosomes (lī′-sō-sōms) are formed by the Golgi complex. They are small vesicles that contain powerful digestive enzymes. These enzymes are used to digest (1) bacteria that may have entered the cell, (2) cell parts that need replacement, and (3) entire cells that have become damaged or worn out. Thus, they play an important role in cleaning up the cellular environment.

The Cytoskeleton

Microtubules and microfilaments compose the cytoskeleton. **Microtubules** are long, thin tubules that

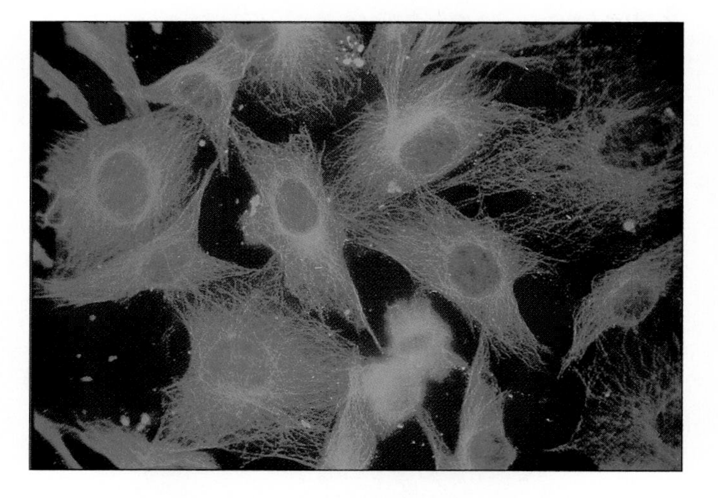

Figure 3.7 A false-color electron photomicrograph (750×) shows the cytoskeleton composed of microtubules and microfilaments.

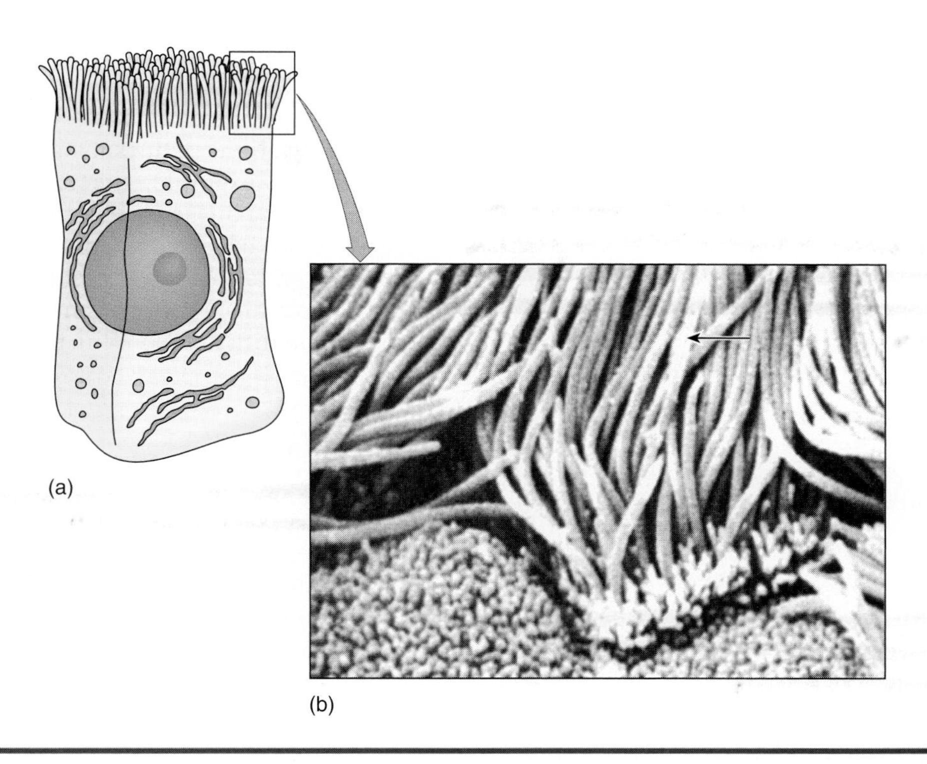

Figure 3.8 (*a*) Cilia are located on the free surface of certain cells, such as those lining the respiratory tract. Since the cells are stationary, beating cilia move substances across the surfaces of the cells. (*b*) An electron photomicrograph of cilia (10,000×).

provide support for the cell, and they are involved in movement of organelles. The thinner **microfilaments** are tiny rods of contractile protein that not only support the cell but enable cell movement (figure 3.7).

Centrioles

The **centrioles** (sen′-trē-olz) are two short cylinders that are located near the nucleus and are oriented at right angles to each other. Nine triplets of microtubules are arranged in a circular pattern to form the wall of each cylinder. Centrioles form and organize the spindle fibers during cell division, and they are involved in the formation of microtubules found in cilia and flagella.

Cilia and Flagella

Cilia and flagella are small hairlike projections from cells that are capable of wavelike movement. **Cilia** (sil′-ē-ah) are numerous, short hairlike projections from cells that, in humans, are used to move substances along the cell surface (figure 3.8). **Flagella** (flah-jel′-ah) are long whiplike projections from cells. In humans, only sperm cells possess flagella, and each sperm has a single flagellum that enables movement. Both cilia and flagella contain microtubules that originate from centrioles positioned at the base of these flexible structures (figure 3.9).

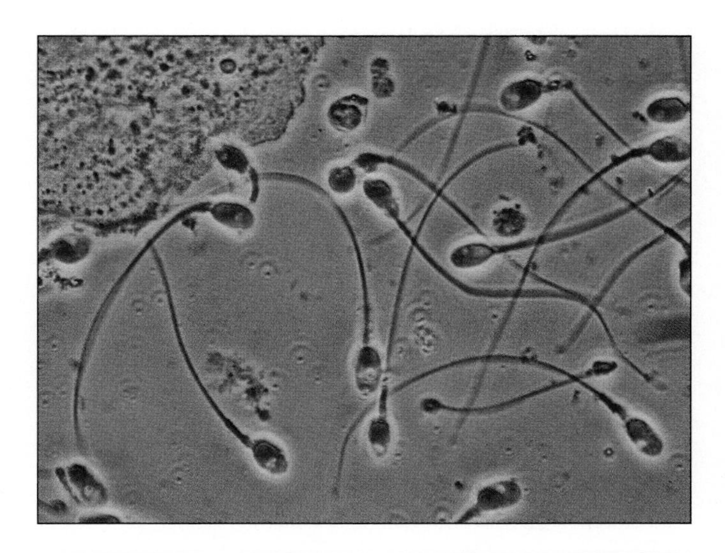

Figure 3.9 A light photomicrograph of human sperm cells (1,000×). Sperm cells are the only human cells that have a flagellum. Each sperm is propelled by the beating of its whiplike flagellum.

What are the distinguishing features and functions of a mitochondrion, a nucleus, the Golgi complex, and rough endoplasmic reticulum? What organelles enable cell movement or move substances across a cell surface?

Transport Through Plasma Membranes

Substances move through a plasma membrane by either passive or active processes. In **passive transport**, substances move across the membrane without assistance by the cell. **Active transport** requires the expenditure of energy (ATP) as the cell assists in the transport process.

Passive Transport

There are three major types of passive transport: diffusion, osmosis, and filtration.

Diffusion

Diffusion (di-fū′-zhun) is the net movement of molecules from an area of their higher concentration to an area of their lower concentration. Thus, the movement of molecules is along a *concentration gradient,* the difference between the concentration of the specific molecules in the two areas.

Diffusion occurs in both gases and liquids and results from the constant, random motion of molecules. Diffusion is not a living process; it occurs in both living and nonliving systems. For example, if a pellet of a water-soluble dye is placed in a beaker of water, the dye molecules will slowly diffuse from the pellet (the area of higher concentration) throughout the water (the area of lower concentration) until the dye molecules are equally distributed, that is, at equilibrium. Figure 3.10 shows diffusion in a liquid environment.

In a similar way, molecules of some substances are able to diffuse through a plasma membrane along a concentration gradient due to differences between the concentrations of the substance inside and outside the cell. For example, the exchange of respiratory gases occurs by diffusion. Air in the lungs has a greater concentration of oxygen and a lower concentration of carbon dioxide than in the blood. Therefore, oxygen diffuses from air in the lungs into the blood, while carbon dioxide diffuses from the blood into the air in the lungs.

Osmosis

The diffusion of water through a selectively permeable membrane is called **osmosis** (os-mo′-sis). Water molecules move through the membrane from an area of higher water concentration (lower solute concentration) into an area of lower water concentration (higher solute concentration).

Figure 3.11 illustrates the concept of osmosis. The beaker is divided into two compartments (A and B) by a selectively permeable membrane that allows water molecules, but not sugar molecules, to pass through it. Since the higher concentration of water is in compartment B, water diffuses from compartment B into compartment A. Since sugar molecules cannot pass through the membrane, water molecules from compartment B continue to diffuse into compartment A, causing the volume of its solution to increase as the volume of water in compartment B decreases.

When aqueous solutions of unequal solute concentrations are separated by a selectively permeable membrane, water *always* diffuses from a *hypotonic*

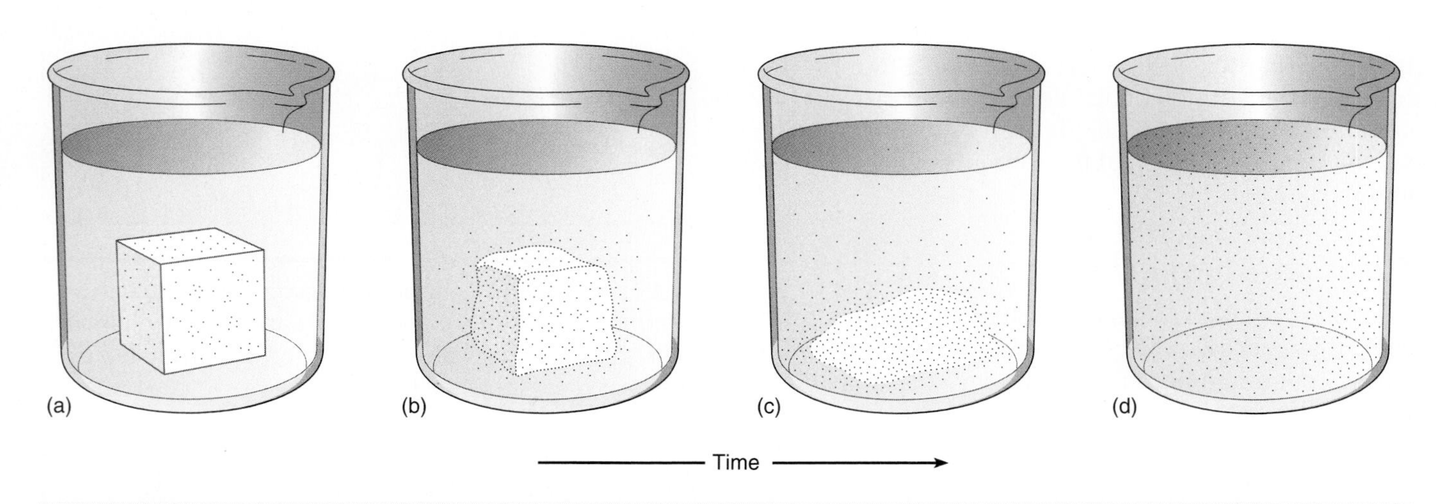

Figure 3.10 An example of diffusion. (*a*) A sugar cube placed in water gradually dissolves, (*b*) and (*c*), when sugar molecules diffuse from the region of their higher concentration to a region of their lower concentration. (*d*) In time, the sugar molecules are equally distributed throughout the water.

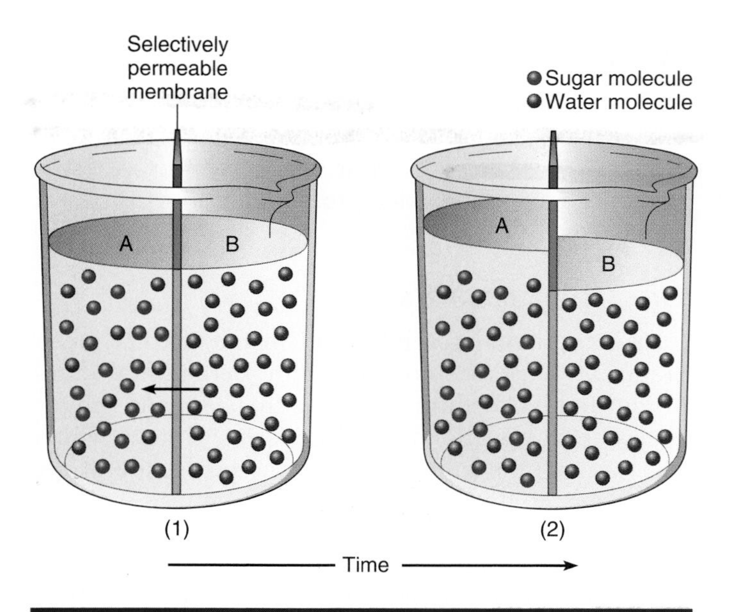

Figure 3.11 Osmosis (1) A selectively permeable membrane separates the container into two compartments. At first, compartment *A* contains water and sugar molecules, while compartment *B* contains only water. As a result of molecular motions, water diffuses by osmosis from compartment *B* into compartment *A*. Sugar molecules remain in compartment *A* because they are too large to pass through the pores of the membrane. (2) Because more water enters compartment *A* than leaves it, water accumulates in this compartment. The level of liquid in compartment *A* rises above that of compartment *B*.

solution, a solution with a lower concentration of solutes and therefore a higher concentration of water, into a *hypertonic solution,* a solution with a higher concentration of solutes and a lower concentration of water. The greater the difference in water (or solute) concentration between hypotonic and hypertonic solutions, the faster osmosis occurs. Solutions that have the same concentration of solutes are *isotonic solutions.* There is no net movement of water between isotonic solutions separated by a selectively permeable membrane. The ability of a solution to affect the tone or shape of living cells by altering the cells' water content is called *tonicity* (figure 3.12).

Filtration

Recall that a solution consists of a *solvent,* such as water, and one or more dissolved substances, collectively called *solutes,* whose molecules may vary in size. **Filtration** is the forcing of smaller molecules in a solution through a membrane due to greater *hydrostatic pressure* on one side of the membrane. In the body, filtration is powered by blood pressure. For example, smaller molecules are forced out of a capillary and into the tissue fluid surrounding cells because blood pressure in the capillary is greater than the hydrostatic pressure of the tissue fluid. Larger molecules are too big to pass through the capillary wall (figure 3.13).

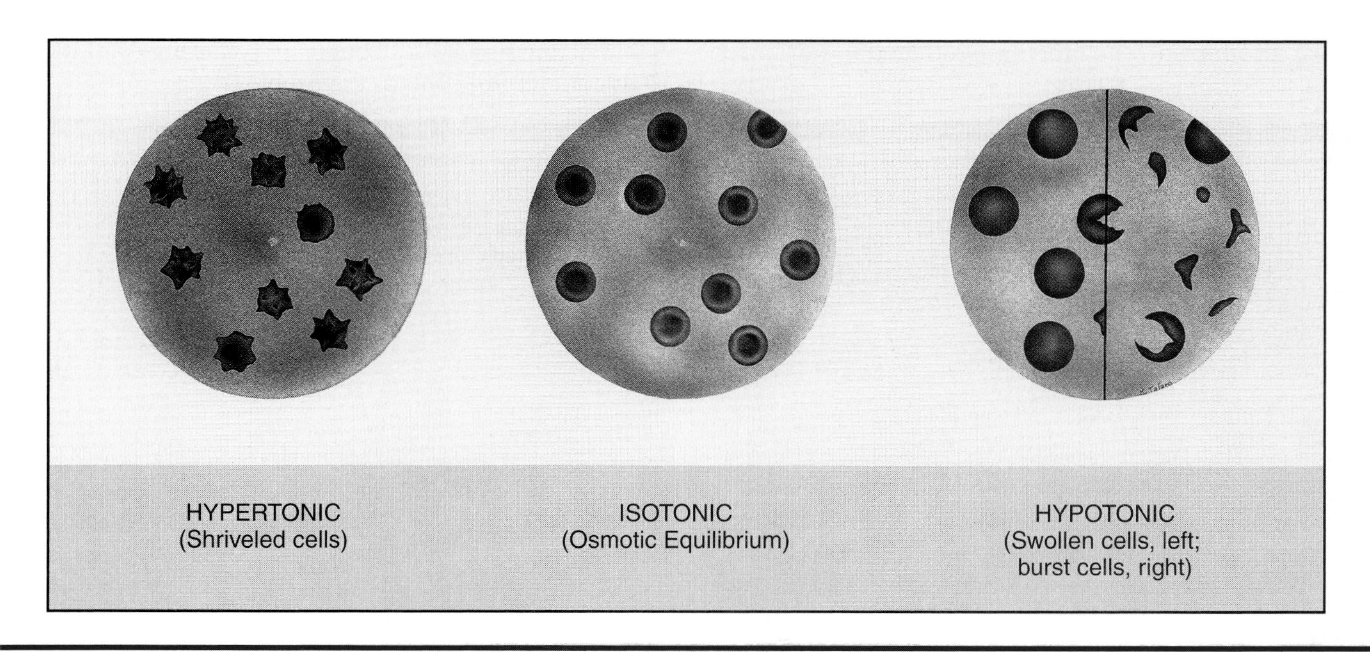

Figure 3.12 The effect of tonicity on human cells. When a cell is placed in an isotonic solution, there is no net movement of water into or out of the cell. When a cell is in a hypertonic solution, water diffuses out of the cell and the cell crumples. When a cell is in a hypotonic solution, water diffuses into the cell, causing it to swell and burst.

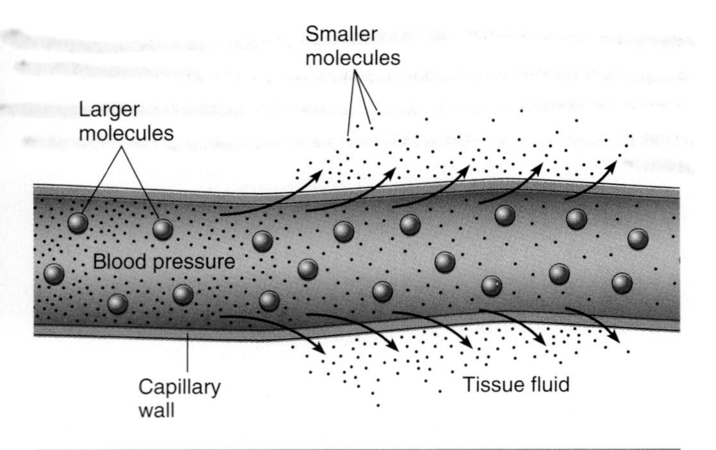

Figure 3.13 In this example of filtration, blood pressure forces smaller molecules through tiny openings in the capillary wall. Larger molecules remain inside.

Dialysis involves the application of diffusion to remove small solute molecules through a selectively permeable membrane from a solution containing both small and large molecules. Dialysis is the process that is used in artificial kidney machines. As blood is passed through a chamber with a selectively permeable membrane, small waste molecules diffuse from the blood across the membrane into an aqueous solution that has a low concentration of these waste molecules. In this way, waste products in the blood are reduced to normal levels.

Active Transport

Unlike passive transport, active transport requires the cell to expend energy (ATP) to move molecules across a plasma membrane. There are three basic active transport mechanisms: carrier proteins, endocytosis, and exocytosis.

Carrier Proteins

Carrier proteins are embedded in the plasma membrane and have receptors that allow certain molecules or ions to bind with them. When this happens, energy is used to change the shape of the carrier protein, which releases the molecule or ion on the other side of the membrane. In this way, substances are moved against a concentration gradient, that is, from an area of lower concentration to an area of higher concentration (figure 3.14). There is a different carrier protein for each type of molecule or ion moved by active transport.

Solutions that are administered to patients intravenously usually are isotonic with body fluids. Sometimes hypertonic solutions are given intravenously to patients with severe edema (swollen, waterlogged tissues) to move water from the tissue fluid into the blood, where it can be removed by the kidneys and excreted in urine. Severely dehydrated patients may be given a hypotonic solution intravenously to increase the water concentration of blood and tissue fluid, but usually simply drinking a hypotonic fluid solves the problem.

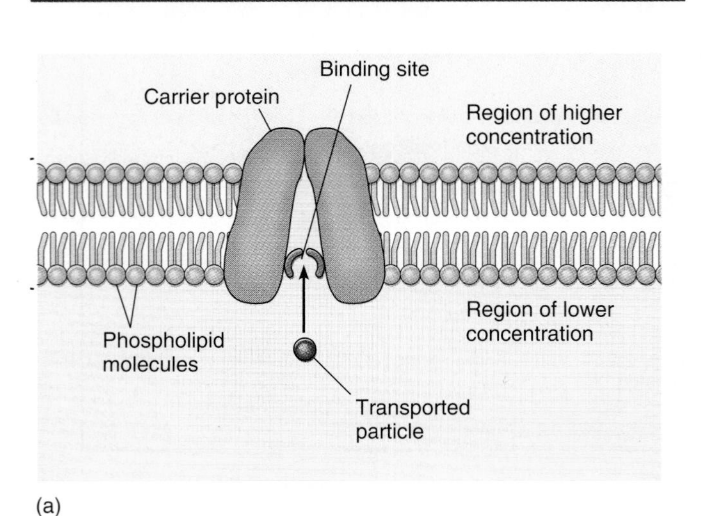

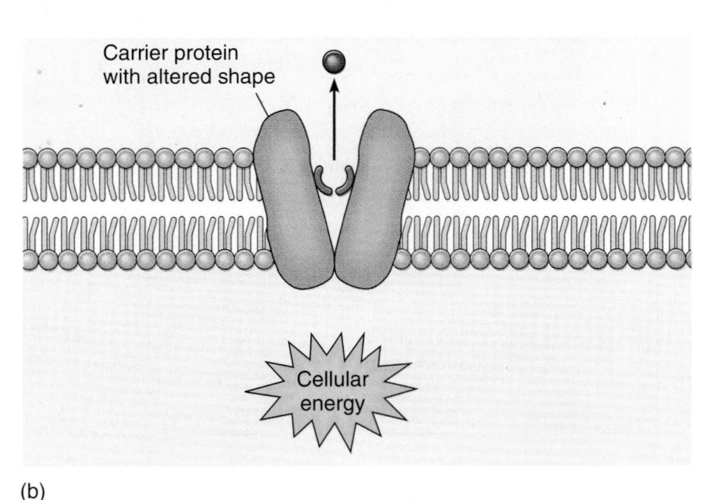

Figure 3.14 (*a*) During active transport, a molecule or an ion combines with a carrier protein, whose shape is altered as a result. (*b*) This process, which requires energy, transports the particle through the cell membrane.

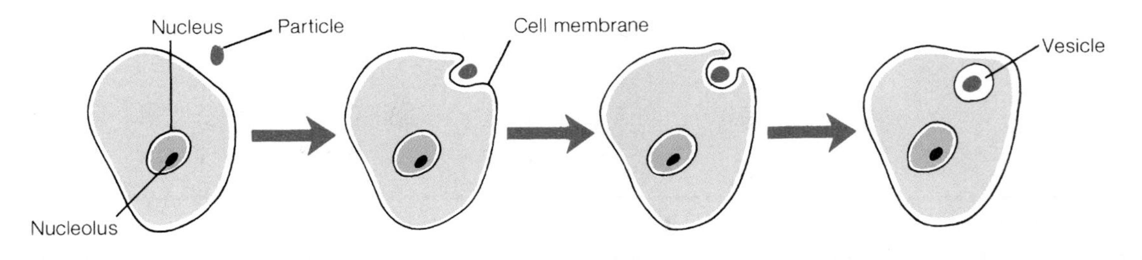

Figure 3.15 Endocytosis. A particle or liquid droplet is engulfed in a vesicle and is brought inside the cell.

Table 3.2 Summary of How Materials Enter and Leave Cells

Type	Process
Passive transport	Diffusion (solutes)
	Osmosis (water)
	Filtration (water and solutes)
Active transport	Transport by carrier proteins (solutes)
	Endocytosis
	Phagocytosis (particles)
	Pinocytosis (liquids)
	Exocytosis (solutes, liquids)

For example, the pumping of sodium ions (Na^+) out of a cell occurs via carrier proteins. Since Na^+ is pumped out faster than it can diffuse in, there is a greater concentration of Na^+ outside of cells than inside cells.

Endocytosis and Exocytosis

Materials that are too large to be actively transported by carrier proteins enter and exit a cell by totally different mechanisms. **Endocytosis** (en″-dō-sī-tō′-sis) is the engulfment of particles and droplets of liquid. **Exocytosis** is a reverse process in which substances are expelled (secreted) from the cell.

Figure 3.15 illustrates how substances are engulfed during endocytosis. The plasma membrane flows around the substance to be engulfed, forms an enveloping vesicle around the substance, and re-forms the plasma membrane exterior to the vesicle so that the vesicle and substance are brought inside the cell. There are two types of endocytosis: pinocytosis and phagocytosis. *Pinocytosis* (pī″-nō-sī-tō′-sis) is the engulfment of small droplets of fluid. *Phagocytosis* (fag″-ō-sī-tō′-sis) is the engulfment of small particles. Many types of cells use these processes, but phagocytosis is especially important for certain white blood cells that engulf and destroy bacteria as a defense against disease.

Exocytosis is the reverse of endocytosis, and it is used to remove substances from cells. A vesicle containing the substance forms within the cell. It then moves to the plasma membrane and empties its contents outside of the cell. The secretion of enzymes and hormones by cells involves exocytosis.

Table 3.2 summarizes transport through plasma membranes.

By what means do substances enter and exit living cells?

Cellular Respiration

Cells require a constant supply of energy to power the chemical reactions of life. This energy is directly supplied by ATP molecules, as noted in chapter 2. Since cells have a limited supply, ATP molecules must constantly be produced by cellular respiration in order to sustain life.

Cellular respiration is the process that breaks down nutrients to release energy held in their chemical bonds and transfers some of this energy into the high-energy phosphate bonds of ATP. About 40% of the energy in a nutrient molecule is "captured" in this way; the remainder is lost as heat energy. The building units of carbohydrates, proteins, and lipids may be used as nutrients in cellular respiration, but glucose is the most common nutrient.

The summary equation for the cellular respiration of glucose is shown in figure 3.16. Note that the reaction requires oxygen and yields carbon dioxide and water in addition to energy (ATP). The actual process is complex, and each step requires a special enzyme. But it may be simplified as shown in figure 3.17.

Cellular respiration involves two sequential processes: anaerobic respiration and aerobic respiration. **Anaerobic** (an-a-rō′-bik) **respiration** (1) does not require oxygen and (2) occurs in the cytoplasm. As shown in figure 3.17, it breaks down a six-carbon glucose molecule into two three-carbon pyruvic acid molecules to yield a net of two ATP molecules. As you

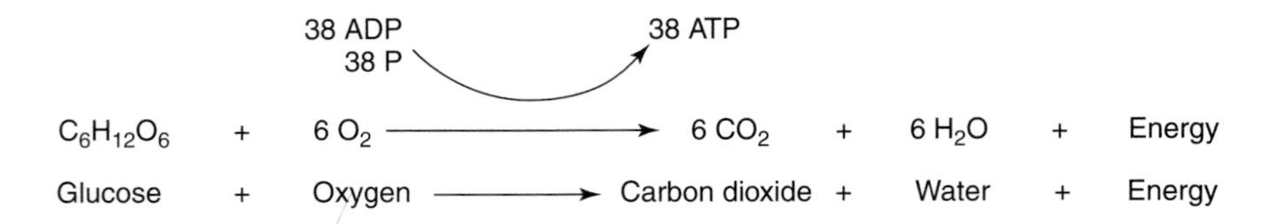

Figure 3.16 Cellular respiration of glucose. Glucose is broken down to release energy that is used to form ATP from ADP and —P. Oxygen is an essential reactant, and the reaction produces carbon dioxide and water.

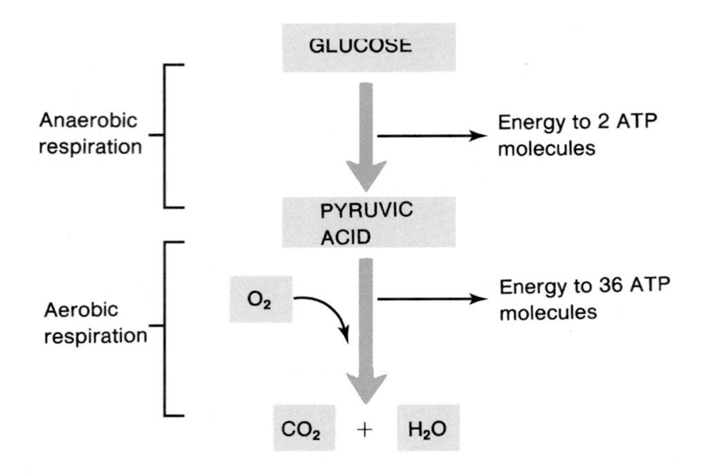

Figure 3.17 Cellular respiration of a glucose molecule occurs in two steps. Anaerobic respiration occurs in the cytoplasm, does not require oxygen, and yields a net of 2 ATP. Aerobic respiration occurs in mitochondria, requires oxygen, and yields a net of 36 ATP. About 40% of the energy in the chemical bonds of glucose is captured to form ATP molecules.

recall from chapter 2, energy, ADP, and a low-energy phosphate group (—P) are required to synthesize ATP. The low level of ATP production by anaerobic respiration is insufficient to keep a person alive.

Aerobic respiration, the second part of cellular respiration, (1) requires oxygen, (2) occurs only within mitochondria, and (3) is essential for human life. Aerobic respiration breaks down the two pyruvic acid molecules produced by anaerobic respiration into carbon dioxide and water and yields a net of 36 ATP molecules. Thus, the respiration of a molecule of glucose yields a net total of 38 ATP.

A person deprived of oxygen or of the ability to use oxygen in cellular respiration (as in cyanide poisoning) quickly dies because anaerobic respiration does not provide sufficient ATP to sustain life.

What is cellular respiration, and why is it important?

Protein Synthesis

Proteins play a vital role in the body. Structural proteins compose significant portions of all cells, and functional proteins, such as enzymes and hormones, directly regulate cellular activities. The synthesis of proteins is carefully controlled by DNA while RNA is directly involved in protein synthesis.

DNA and the Genetic Code

DNA consists of two coiled strands of nucleotides held together by hydrogen bonds that form between the organic bases. It resembles a coiled ladder with the rungs of the ladder formed by the pairs of organic bases. Each DNA nucleotide contains one of four kinds of organic bases: **adenine, thymine, cytosine,** and **guanine.** Adenine pairs only with thymine, and cytosine pairs only with guanine. This arrangement is known as *complementary base pairing* (figure 3.18).

The sequence of organic bases in DNA forms the **genetic code,** which contains encoded information that controls cell functions and determines inherited traits, such as the color of eyes and hair. A **gene,** the unit of inheritance, consists of a specific sequence of bases in DNA and is one small portion of the genetic code. Genes direct cell functions and determine hereditary traits by controlling protein synthesis.

The Role of RNA

Recall that RNA consists of a single strand of nucleotides. Each RNA nucleotide contains one of four organic bases: **adenine, guanine, cytosine,** and **uracil.** Note that uracil occurs in RNA instead of thymine, which is found in DNA. RNA is synthesized in a cell's nucleus by using a strand of DNA as a template. Complementary pairing of RNA organic bases with DNA organic bases produces a strand of RNA nucleotides whose bases are complementary to those in the DNA molecule.

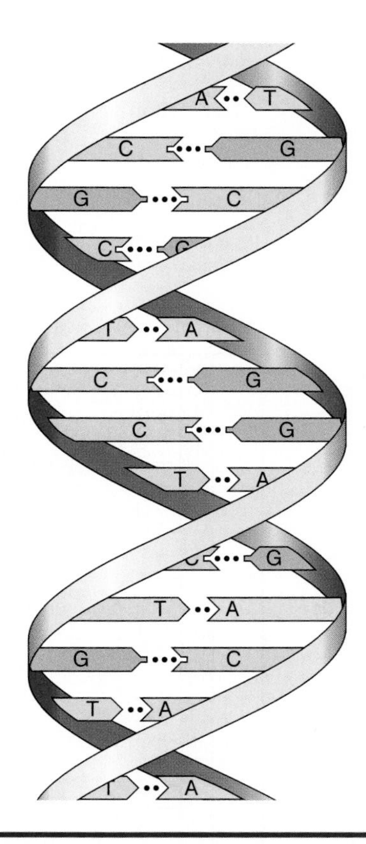

Figure 3.18 A portion of a DNA molecule. The helical shape and bonding between complementary organic bases of each strand are evident. Note that adenine bonds only with thymine and that cytosine bonds only with guanine.

There are three types of RNA, and each plays a vital role in protein synthesis.

Messenger RNA (mRNA) carries the genetic information from DNA into the cytoplasm to the ribosomes, the sites of protein synthesis. This information is carried by the sequence of bases in mRNA, which is complementary to the sequence of bases in the DNA template.

Ribosomal RNA (rRNA) and protein compose ribosomes, the sites of protein synthesis. Ribosomes contain the enzymes required for protein synthesis.

Transfer RNA (tRNA) carries amino acids to the ribosomes, where the amino acids are joined like a string of beads to form a protein. There is a different tRNA for transporting each of the 20 kinds of amino acids used to build proteins.

Transcription and Translation

The process of protein synthesis involves two successive events: transcription and translation.

In *transcription,* the sequence of bases in DNA determines the sequence of bases in mRNA due to complementary base pairing. Thus, transcription copies the encoded information of DNA into the sequence of bases in mRNA.

In *translation,* the encoded information in mRNA is used to produce a specific sequence of amino acids to form the protein. Three successive mRNA bases form a *codon,* which codes for a specific amino acid. So the sequence of codons in mRNA determines the sequence of amino acids in the synthesized protein, as shown in figure 3.19.

By transcription and translation, DNA controls the structure of synthesized proteins, which, in turn, determines the functions of the synthesized proteins. Transcription and translation may be summarized as follows:

$$\text{DNA} \xrightarrow[\text{Transcription}]{} \text{mRNA} \xrightarrow[\text{Translation}]{} \text{Protein}$$

Table 3.3 summarizes the characteristics of DNA and RNA.

Table 3.3 Distinguishing Characteristics of DNA and RNA

	DNA	RNA
Strands	Two strands joined by the complementary pairing of their organic bases	One strand
Sugar	Deoxyribose	Ribose
Bases	Adenine Thymine Cytosine Guanine	Adenine Uracil Cytosine Guanine
Shape	Helix	Straight

How does chromosomal DNA control protein synthesis?

Cell Division

Two types of cell division occur in the body: mitotic cell division and meiotic cell division. **Mitotic** (mī-tot′-ik) **cell division** is the replication of a parent cell so that the two new daughter cells have the same number of chromosomes (46) and composition of chromosomes as the parent cell. It enables growth and the repair of tissues. **Meiotic** (mī-ot′-ik) **cell division** occurs only in the production of ova and sperm. It produces four sex cells that contain only half the number of chromosomes (23) as found in the parent cell. In this chapter, we consider mitotic cell division only. Meiotic cell division is studied in chapter 17.

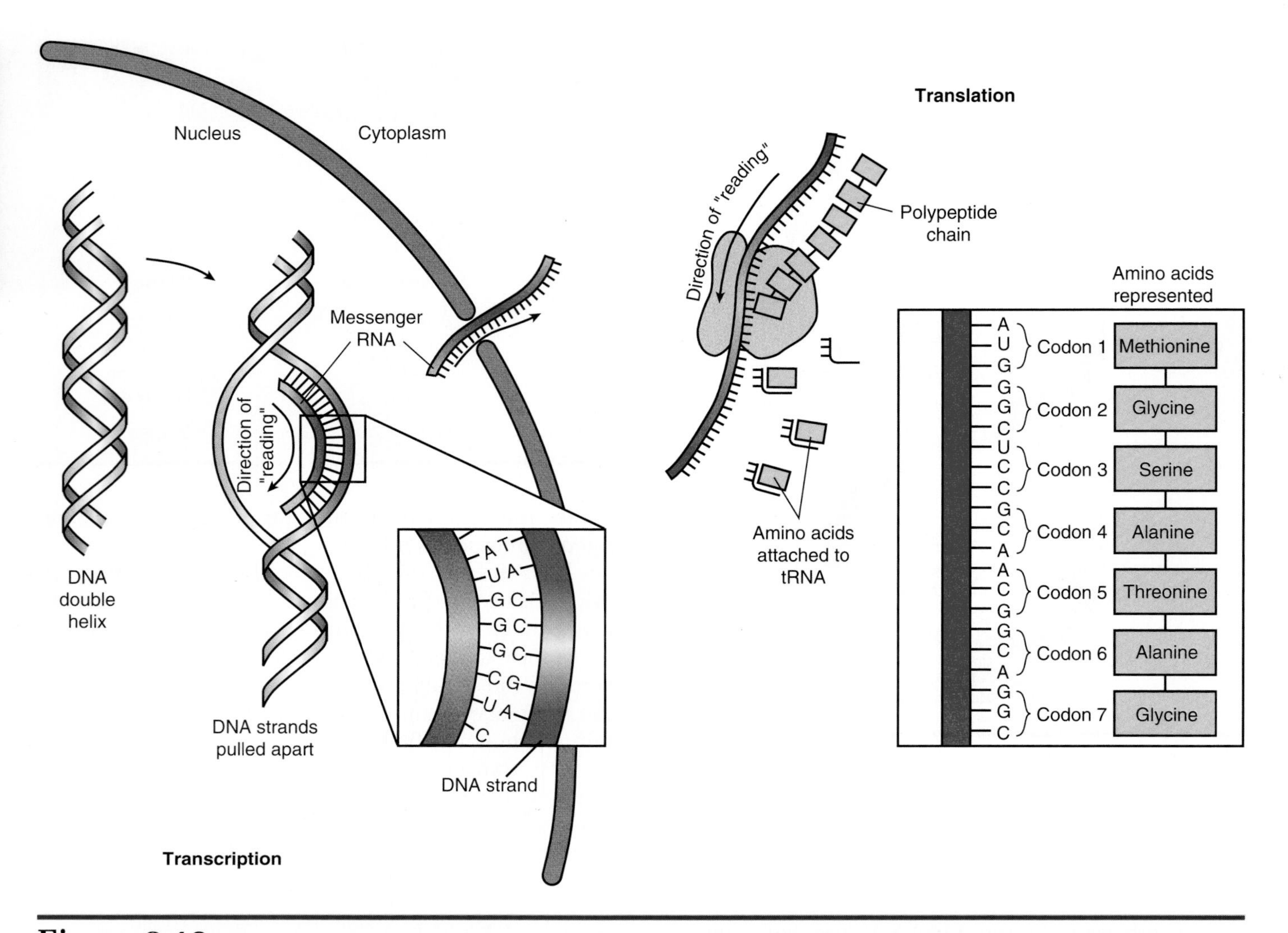

Figure 3.19 Protein synthesis. In transcription, the sequence of bases (coded information) in a DNA strand is copied into the sequence of bases (coded information) of mRNA. Three sequential bases in mRNA form a codon that codes for a specific amino acid. Translation uses the codons in mRNA to make a specific protein molecule from amino acids carried to ribosomes by tRNA.

Mitotic Cell Division

Starting with the first division of the fertilized egg, mitotic cell division is the process that produces new cells for growth of the new individual and the replacement of worn or damaged cells. Three processes are involved in mitotic cell division: (1) replication (production of exact copies) of chromosomes, (2) mitosis, and (3) division of the cytoplasm.

In dividing cells, the time period from the separation of daughter cells of one division to the separation of daughter cells of the next division is called the *cell cycle.* Mitosis constitutes only 5% to 10% of the cell cycle. Most of the time a cell is not dividing but is carrying out its normal functions (figure 3.20).

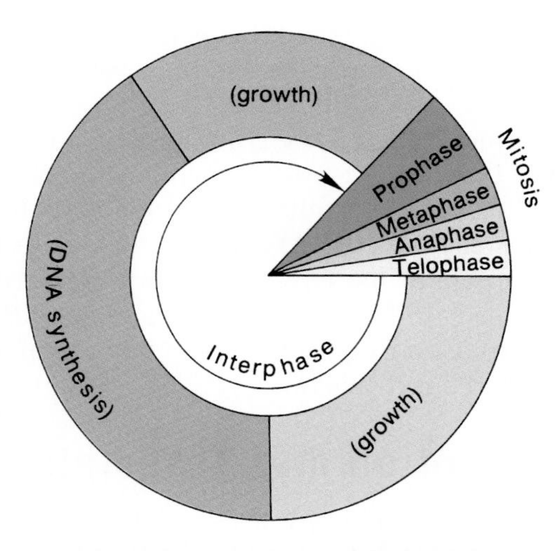

Figure 3.20 The cell cycle. Interphase occupies most of the cell cycle. Only 5% to 10% of the time is used in mitotic division. DNA and chromosome replication occur during interphase.

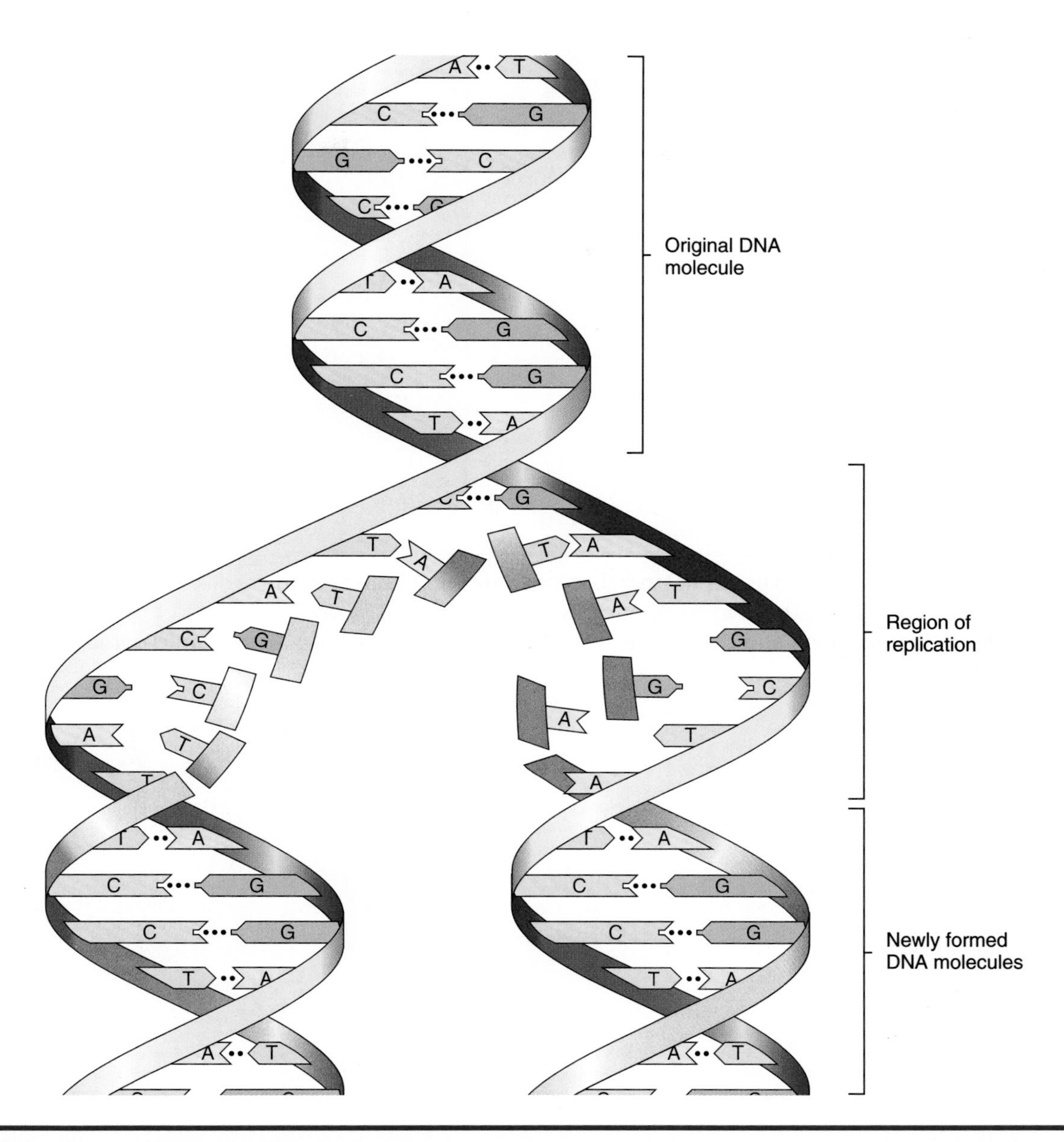

Figure 3.21 When a DNA molecule replicates, the two strands unzip. Then a new complementary strand of nucleotides forms along each "old" strand to produce two new DNA molecules.

Interphase is defined as the phase when the cell is not involved in mitosis. A cell in interphase has an intact nucleus and visible chromatin granules. In cells that are destined to divide, both the centrioles and chromosomes replicate during interphase. There is a growth period before and after replication of each of the 46 chromosomes.

A chromosome consists of a very long DNA molecule coated with proteins. During interphase, chromosomes are uncoiled and resemble very thin threads within the cell nucleus. Chromosomes replicate during interphase in order to provide one copy of each chromosome for each of the two daughter cells that will be formed by mitotic cell division. Chromosome replication is dependent upon the replication of the DNA molecule in each chromosome. Figure 3.21 illustrates the process of DNA replication.

The two DNA strands "unzip," and new nucleotides are joined in a complementary manner by their bases to the bases of each DNA strand. When the new nucleotides are in place and joined together, each new DNA molecule consists of one "new" strand of nucleotides joined to one "old" strand of nucleotides. In this way, a DNA molecule is precisely replicated so that both new DNA molecules are identical.

Mitotic Phases

Once it begins, mitosis is a continuous process that is arbitrarily divided into four sequential phases: prophase, metaphase, anaphase, and telophase. Each phase is characterized by specific events that occur.

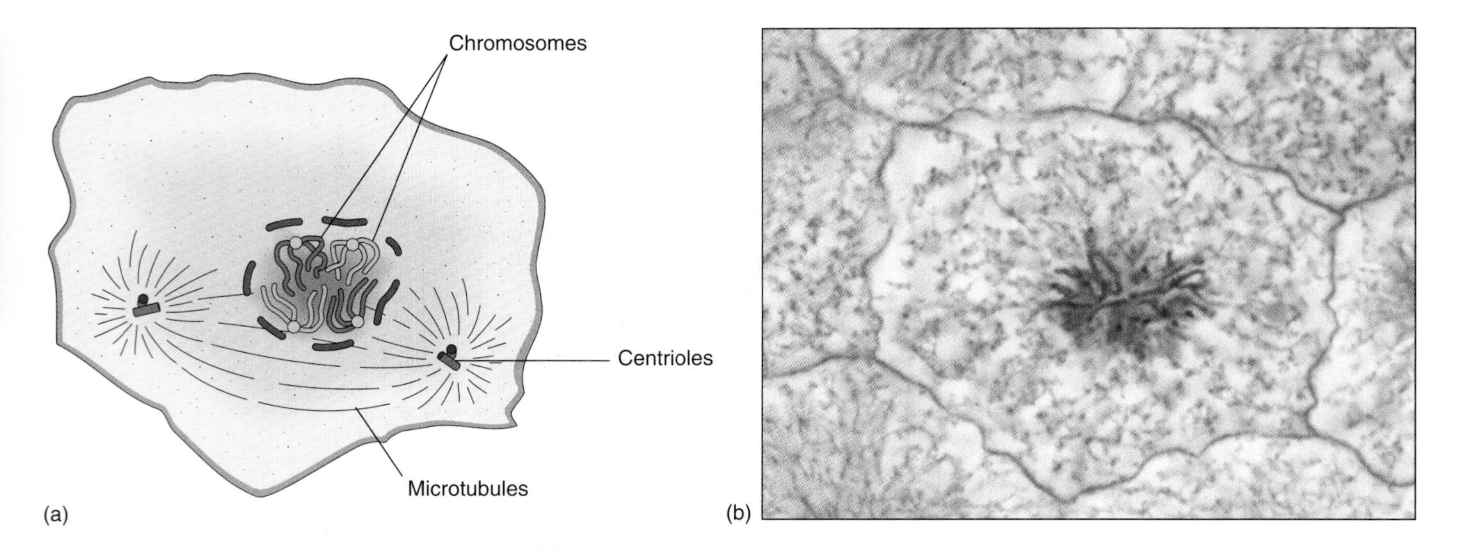

Figure 3.22 Prophase. (*a*) The replicated chromosomes first appear as threadlike strands and then shorten further to appear rod-shaped; the two pairs of centrioles move toward opposite sides of the cell, forming a spindle of microtubules between them; and the nuclear membrane disintegrates. (*b*) A photomicrograph of a cell in prophase (1,000×).

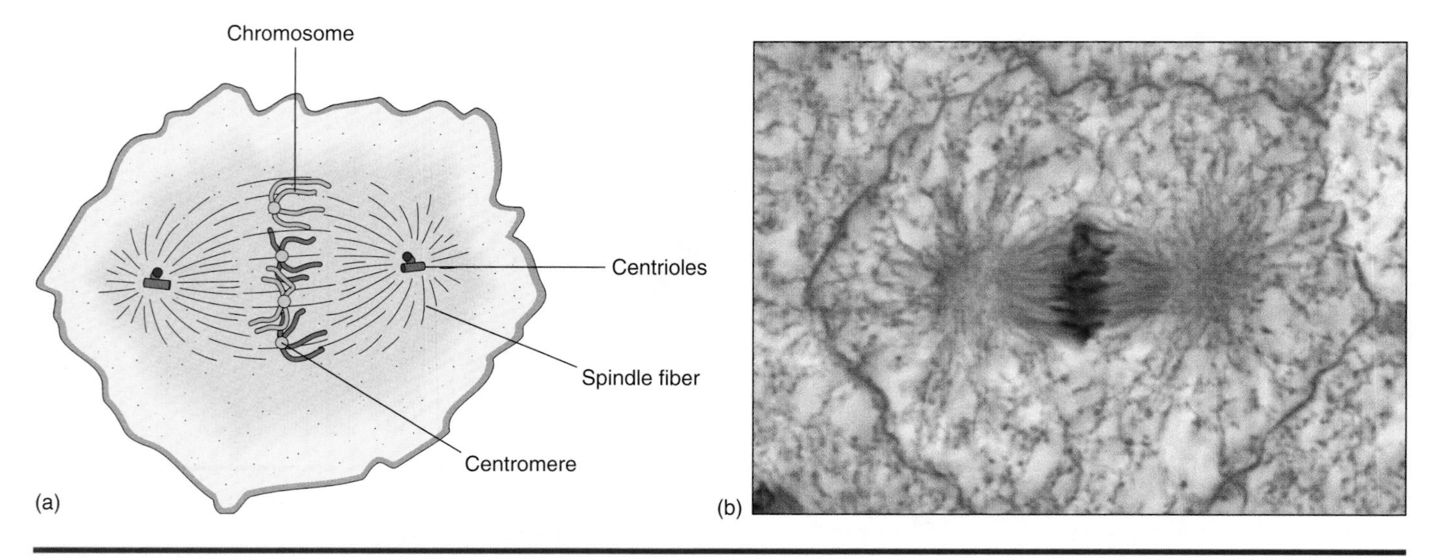

Figure 3.23 Metaphase. (*a*) The replicated chromosomes line up at the equator (midline) of the spindle. Note that each replicated chromosome consists of a pair of chromatids joined by a centromere. (*b*) A photomicrograph of a cell in metaphase (1,000×).

Prophase During **prophase,** the replicated chromosomes coil, appearing first as threadlike structures and finally shortening sufficiently to become rod-shaped. Each replicated chromosome consists of two *chromatids* joined at their *centromeres.* Simultaneously, the nuclear envelope gradually disappears, and each pair of centrioles migrate toward opposite ends of the cell. A *spindle* is formed between the migrating centrioles. The spindle consists of *spindle fibers* that are formed of microtubules (figure 3.22).

Metaphase During the brief **metaphase,** the replicated chromosomes line up at the equator of the spindle. Each replicated chromosome consists of a pair of chromatids joined at the centromere. The centromeres are attached to spindle fibers (figure 3.23).

Anaphase During **anaphase,** separation of the centromeres results in the separation of the paired chromatids. The members of each pair then migrate toward opposite ends of the spindle. They appear to be pulled along by the spindle fibers attached to their cen-

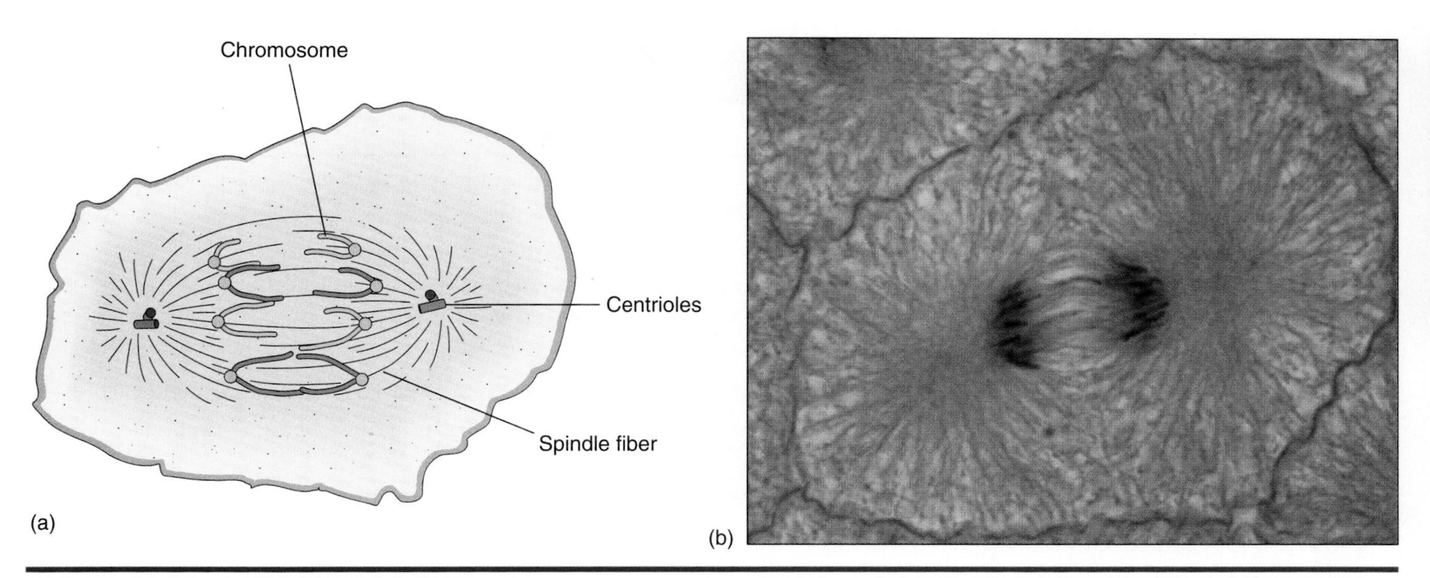

Figure 3.24 Anaphase. (*a*) The centromeres divide, and spindle fibers pull members of each pair of chromatids, now called chromosomes, to opposite ends of the spindle. The two sets of chromosomes are identical. (*b*) A photomicrograph of a cell in anaphase (1,000×).

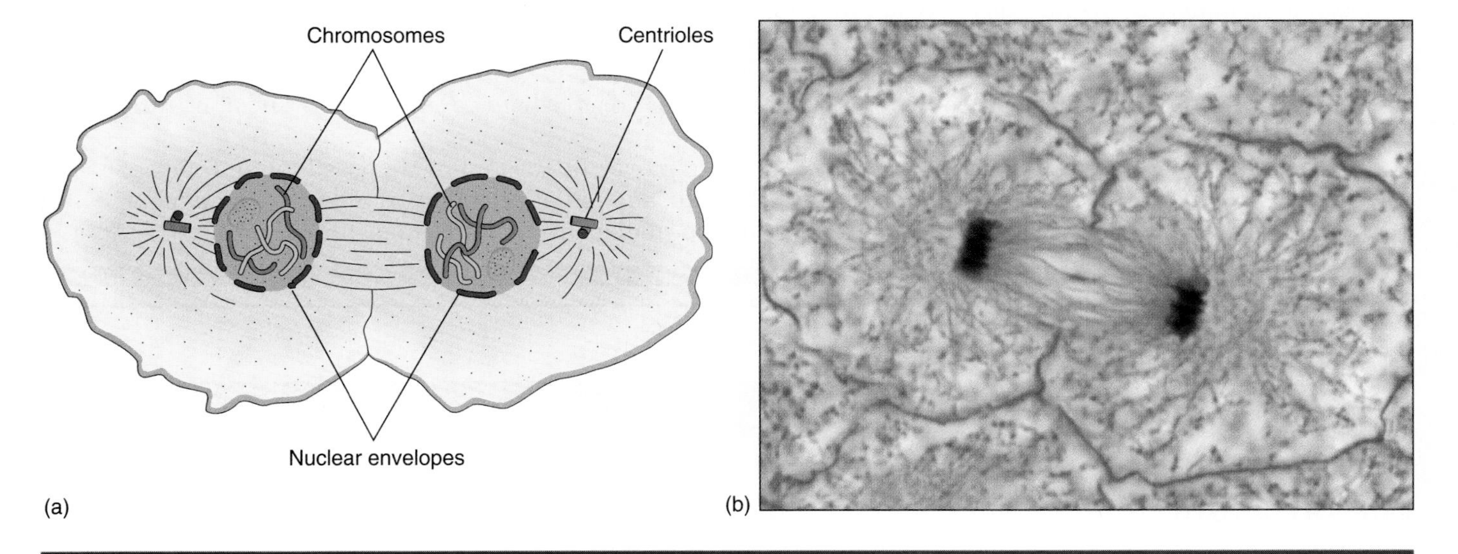

Figure 3.25 Telophase. (*a*) The chromosomes in each set uncoil, first appearing threadlike and ultimately as chromatin granules; a nuclear envelope forms around each set of chromosomes; and cytokinesis begins leading to the formation of two daughter cells. (*b*) A photomicrograph of a cell in telophase (1,000×).

tromeres. The separated chromatids are now called chromosomes, and each new set of chromosomes is identical (figure 3.24).

Telophase During **telophase,** a new nuclear envelope starts forming around each set of chromosomes as the new nuclei begin to take shape. The chromosomes start to uncoil, and they will ultimately become visible only as chromatin granules. The new daughter nuclei are completely formed by the end of telophase.

During this time, the most obvious change is the division of the cytoplasm, which is called **cytokinesis** (si″-to-ki-nē′-sis). It is characterized by a furrow that forms in the plasma membrane across the equator of the spindle and deepens until the parent cell is separated into two daughter cells. The formation of two daughter cells, each having identical chromosomes in the nuclei, marks the end of mitotic cell division (figure 3.25).

Table 3.4 summarizes the cell cycle.

What are the phases of mitosis, and how is each phase distinguished?

Table 3.4 Summary of the Cell Cycle

Phase	Events
Interphase	A nondividing cell carries on its normal functions. A cell that is destined to divide also undergoes replication of the centrioles and chromosomes.
Mitosis	
Prophase	Replicated chromosomes coil and shorten; nuclear envelope disappears; centriole pairs start migration toward opposite ends of the cell, forming the spindle between them.
Metaphase	Chromosomes line up at equator of spindle.
Anaphase	Chromatids separate; members of each chromatid pair migrate toward opposite ends of the spindle.
Telophase	Nuclear envelopes form around each set of chromosomes; spindle fibers disappear; chromosomes uncoil and extend; cytoplasm divides.

Mitotic cell division is normally a controlled process that ceases when it is not necessary to produce additional cells. Occasionally, control is lost and cells undergo continuous division, which leads to the formation of tumors. Tumors may be benign or malignant. *Benign tumors* do not spread to other parts of the body and may be surgically removed if they cause health or cosmetic problems. *Malignant tumors* or *cancers* may spread to other parts of the body by a process called metastasis (me-tas′-ta-sis). Cells break away from the primary tumor and are often carried by blood or lymph to other areas, where continued cell divisions form secondary tumors.

Treatment of malignant tumors involves surgical removal of the tumor, if possible, and subsequent chemotherapy and/or radiation therapy. Both chemotherapy and radiation therapy tend to kill malignant cells because dividing cells are more sensitive to treatment, and malignant cells are constantly dividing.

CHAPTER SUMMARY

Cell Structure

1. The plasma membrane is composed of a double layer of phospholipid molecules along with associated cholesterol and protein molecules. It is selectively permeable and controls the movement of the materials into and out of cells.
2. The cytoplasm lies external to the nucleus and is enveloped by the plasma membrane.
3. The nucleus is a large, spherical organelle surrounded by the nuclear envelope.
4. Chromosomes, composed of DNA and protein, are found in the cell nucleus. The uncoiled chromosomes appear as chromatin granules in nondividing cells.
5. The nucleolus is the site of rRNA synthesis.
6. Ribosomes are tiny organelles formed of rRNA and protein. They are sites of protein synthesis.
7. The endoplasmic reticulum (ER) consists of membranes that form channels for transport of materials within the cell. RER is studded with ribosomes that synthesize proteins for export from the cell. SER lacks ribosomes and is involved in lipid synthesis.
8. The Golgi complex packages materials into vesicles for secretion from the cell.
9. Mitochondria are large, double-membraned organelles within which aerobic cellular respiration occurs.
10. Lysosomes are small vesicles of digestive enzymes used to digest foreign particles, worn-out parts of a cell, or an entire damaged cell.
11. The cytoskeleton is formed by microtubules and microfilaments.
12. A pair of centrioles is present near the cell's nucleus. The wall of each centriole is composed of microtubules arranged in groups of three.
13. Cilia are short, hairlike projections on certain cells. The beating of cilia moves materials along the cell surface.
14. Each sperm cell swims by the beating of a flagellum, a long whiplike organelle.

Transport Through Plasma Membranes

1. Passive transport does not require the expenditure of energy by the cell.
2. Diffusion is the movement of molecules from an area of higher concentration to an area of lower concentration. It results from the constant motion of molecules in gases and liquids.
3. Osmosis is the diffusion of water through a selectively permeable membrane.
4. Hypotonic solutions have a greater water concentration than hypertonic solutions. Isotonic solutions have the same water concentration.
5. Water always diffuses from hypotonic solutions into hypertonic solutions when they are separated by selectively permeable membranes. There is no net movement of water between isotonic solutions.
6. Filtration is the forcing of small molecules through a membrane by hydrostatic pressure when the pressure is greater on one side of the membrane.
7. Active transport requires the cell to expend energy.
8. Molecules are transported through the plasma membrane by carrier proteins in one form of active transport.
9. The engulfment of particles and liquid droplets by endocytosis and the expulsion of cell products by exocytosis are specialized forms of active transport.

Cellular Respiration

1. Cellular respiration is the breakdown of nutrients to release energy and form ATP molecules, which power cellular processes.
2. Cellular respiration involves two phases: anaerobic and aerobic.
3. Anaerobic respiration occurs in the cytoplasm and yields 2 ATP; aerobic respiration occurs in mitochondria and yields 36 ATP.

Protein Synthesis

1. Protein synthesis involves the interaction of DNA, mRNA, rRNA, and tRNA.
2. The sequence of bases in DNA determines the sequence of codons in mRNA, which, in turn, determines the sequence of amino acids in a protein.

DNA ——→ mRNA ——→ Protein
Transcription *Translation*

Cell Division

1. Mitotic cell division produces two daughter cells that have the same number and composition of chromosomes. It enables growth and tissue repair.
2. Meiotic cell division occurs in production of ova and sperm. Four daughter cells are formed that have half the number of chromosomes as the parent cell.
3. Most of a cell cycle is spent in interphase, where cells carry out normal metabolic functions. In cells destined to divide, chromosomes and centrioles are replicated in interphase.
4. After chromosome replication, mitosis is the orderly process of separating and distributing chromosomes equally to the daughter cells.
5. Mitosis consists of four phases: prophase, metaphase, anaphase, and telophase.

BUILDING YOUR VOCABULARY

1. Selected New Terms

anaphase, p. 56
cilia, p. 47
cytokinesis, p. 57
flagella, p. 47
Golgi complex, p. 44
hypertonic, p. 49
hypotonic, p. 48
interphase, p. 55
isotonic, p. 49
lysosome, p. 46
microfilament, p. 47
microtubule, p. 46
messenger RNA, p. 53
metaphase, p. 56
phagocytosis, p. 51
pinocytosis, p. 51
prophase, p. 56
ribosomal RNA, p. 53
telophase, p. 57
transcription, p. 53
transfer RNA, p. 53
translation, p. 53

2. Related Clinical Terms

dialysis (dī-al′-i-sis) The process separating small molecules from larger molecules in a solution by diffusion through a selectively permeable membrane.

dysplasia (dis-plās′-ia) Abnormal change in cell shape or size.

hypertrophy (hī′-per-trōph-ē) Abnormal enlargement of a cell, tissue, or organ.

necrosis (ne-krō′-sis) Death of a group of cells.

STUDY ACTIVITIES

1. Complete the **Chapter 3 Study Guide,** which begins on page 407.
2. Complete the **Learning Objectives** listed on the first page of this chapter.

CHECK YOUR UNDERSTANDING

(Answers are located in appendix B.)

1. Movement of materials in and out of cells is controlled by the _____ .
2. Molecules of _____ located in chromosomes control the activities of cells.
3. Aerobic cellular respiration occurs within _____ .
4. The sites of protein synthesis are _____ .
5. The _____ assembles protein and RNA to form ribosomes.
6. The _____ consists of intracellular membranous channels for material transport.
7. Movement of molecules from an area of their higher concentration to an area of their lower concentration is known as _____ .
8. The forcing of small molecules through a membrane by blood pressure is _____ .
9. Movement of molecules through a membrane by carrier proteins is a form of _____ .
10. Breakdown of organic nutrients to release energy and form ATP is called _____ .
11. Instructions for synthesizing a protein are carried from DNA to ribosomes by _____ .
12. The equal distribution of chromosomes to daughter nuclei occurs by _____ .

C H A P T E R

Tissues and Membranes

FOUR

4

Chapter Preview & Learning Objectives

SELECTED KEY TERMS

Adipose tissue (adip = fat) A connective tissue that stores fat.
Chondrocyte (chondro = cartilage; cyt = cell) A cartilage cell.
Connective tissue (connect = to join) A tissue that binds other tissues together.
Epithelial tissue (epi = upon, over; thel = delicate) A thin tissue that covers body and organ surfaces and lines body cavities.
Fibroblast (fibro = fiber; blast = germ) A cell that produces fibers in connective tissue.
Matrix The intercellular substance in connective tissue.
Muscle tissue (mus = mouse) A tissue whose cells are specialized for contraction.
Nerve tissue A tissue that forms the brain, spinal cord, and nerves.
Neuron (neur = nerve) A nerve cell.
Osteocyte (oss = bone; cyt = cell) A bone cell.
Pseudostratified (pseudo = false; stratum = layer) Cells appear to be arranged in layers, but they are not.
Tissue (tissu = woven) A group of similar cells performing similar functions.

The different types of cells found in the body result from the specialization of cells during embryonic development. These specialized cells are organized into tissues to carry out their specific functions. A **tissue** is a group of similar cells that work together to perform a particular function. Since structure is a reflection of function, each type of tissue is distinguished by both structural and functional characteristics.

The different tissues of the body are classified into four basic groups: epithelial, connective, muscle, and nerve tissues.

1. **Epithelial** (ep″-i-thē′-lē-al) **tissue** covers the surfaces of the body, lines body cavities and cavities of organs, and forms the secretory portions of glands.
2. **Connective tissue** binds organs together and provides protection and support for organs and the entire body.
3. **Muscle tissue** contracts to enable movement of the body and body parts.
4. **Nerve tissue** coordinates body functions via the transmission of nerve impulses.

Epithelial Tissues

Epithelial tissues may be composed of one or more layers of cells. The number of cell layers and the shape of the cells provide the basis for classifying epithelial tissues. Epithelial tissues are distinguished by the following five characteristics:

1. Epithelial cells are packed closely together with very little intercellular material between them.
2. The sheetlike tissue is firmly attached to the underlying connective tissue by a thin, noncellular *basement membrane.*
3. The surface of the tissue opposite the basement membrane is always free and exposed, that is, not attached to other tissues.
4. Blood vessels are absent, so epithelial cells must receive nourishment from blood vessels in the underlying connective tissue.
5. Epithelial tissues regenerate rapidly by mitotic division of the cells. Large numbers of epithelial cells are destroyed and replaced each day.

The functions of epithelial tissues vary with the specific location and type of tissue, but they include *protection, absorption, filtration,* and *secretion.* Certain epithelial cells form *glandular epithelium,* the cells in glands that produce the gland's secretions. Two basic types of glands are contained in the body: exocrine and endocrine glands. **Exocrine glands** have ducts (small tubes) that carry their secretions to specific areas; sweat glands and salivary glands are examples. **Endocrine glands** lack ducts. Instead, their secretions are released into the blood, which carries them to target tissues. The thyroid gland and adrenal glands are examples of endocrine glands.

Simple Epithelium

Simple epithelial tissues consist of a single layer of cells that may be flat, cubelike, or columnlike in shape. These tissues occur where diffusion, secretion, or filtration occur in the body.

Simple squamous (skwā′-mus) **epithelium** consists of thin, flat cells that have an irregular outline and a centrally located nucleus. In a surface view, the cells somewhat resemble tiles arranged in a mosaic pattern. Simple squamous epithelium functions to aid diffusion, osmosis, and filtration. Its location in the body includes (1) the air sacs of the lungs, where O_2 and CO_2 diffuse into and out of the blood, respec-

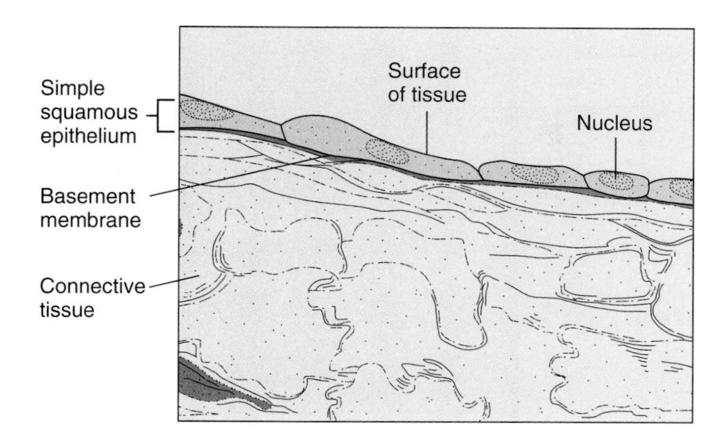

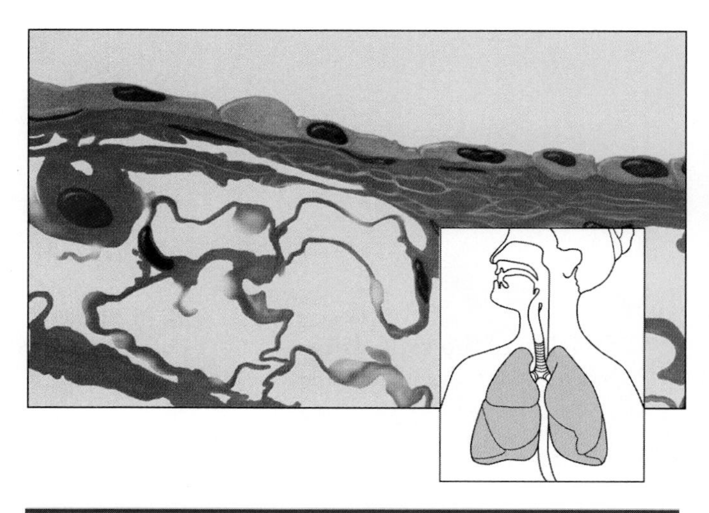

Figure 4.1 Simple squamous epithelium (250×).
Structure: A single layer of flattened cells.
Location: Inner lining of the heart, blood vessels, and the ventral body cavity; forms air sacs of the lungs and glomeruli of the kidneys.
Function: Absorption, secretion, and filtration.

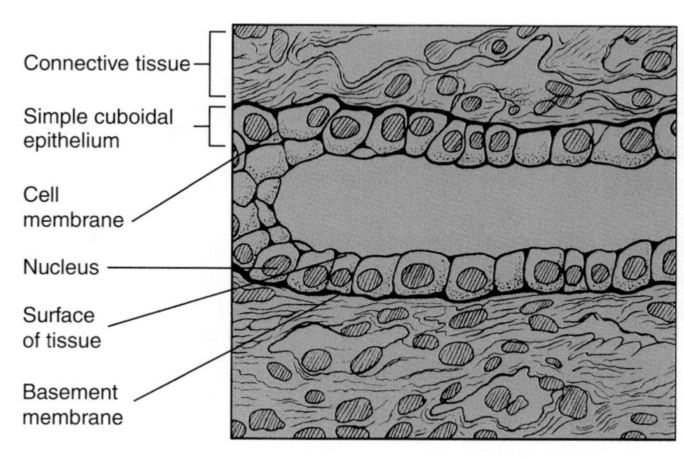

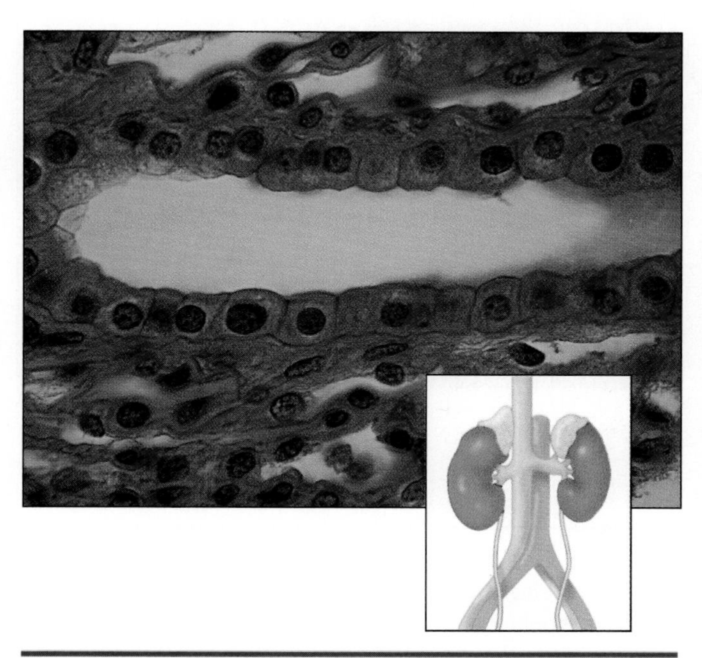

Figure 4.2 Simple cuboidal epithelium (250×).
Structure: A single layer of cube-shaped cells.
Location: Forms kidney tubules, ducts of some glands, and the surface layer of the ovaries.
Function: Absorption and secretion.

tively; (2) parts of the kidneys, where blood is filtered in urine production; and (3) the serous membranes that line the ventral cavity. The squamous epithelium that lines the interior of the heart, blood vessels, and lymphatic vessels is called **endothelium** because of its interior location (figure 4.1).

Simple cuboidal epithelium consists of a single layer of cube-shaped cells. The cells have a single, centrally located nucleus. Its basic functions are absorption and secretion. Simple cuboidal epithelium forms the (1) secretory portion of glands, such as the thyroid and salivary glands; (2) kidney tubules where reabsorption of materials occurs; and (3) the surface layer of the ovaries (figure 4.2).

Simple columnar epithelium consists of a single layer of elongated, column-shaped cells. Scattered among these cells are *goblet cells,* specialized mucus-secreting cells whose shape resembles a goblet or wine glass. This tissue lines the interior of the stomach and intestines, where it secretes digestive juices and absorbs nutrients (figure 4.3).

Pseudostratified ciliated columnar epithelium looks like it consists of more than one layer of cells, but it does not. All cells are joined to the basement membrane, but all of them do not reach the surface of the tissue. Mucus-secreting goblet cells are scattered throughout the tissue. This epithelium lines the interior of the upper respiratory passageways, where it collects and removes airborne particles. The particles

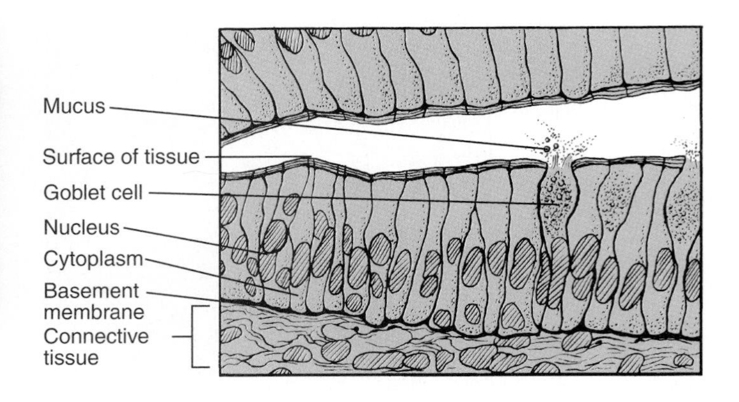

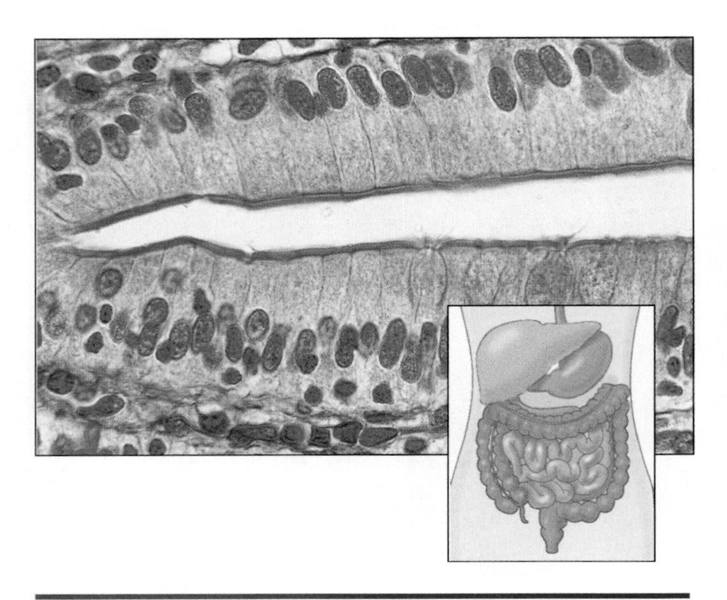

Figure 4.3 Simple columnar epithelium (400×).
Structure: A single layer of column-shaped cells; contains scattered goblet cells.
Location: Lines the interior of the stomach and intestines and the ducts of many glands.
Function: Absorption and secretion.

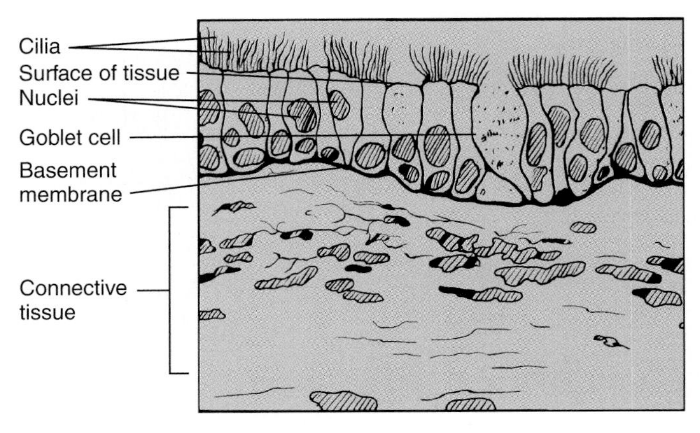

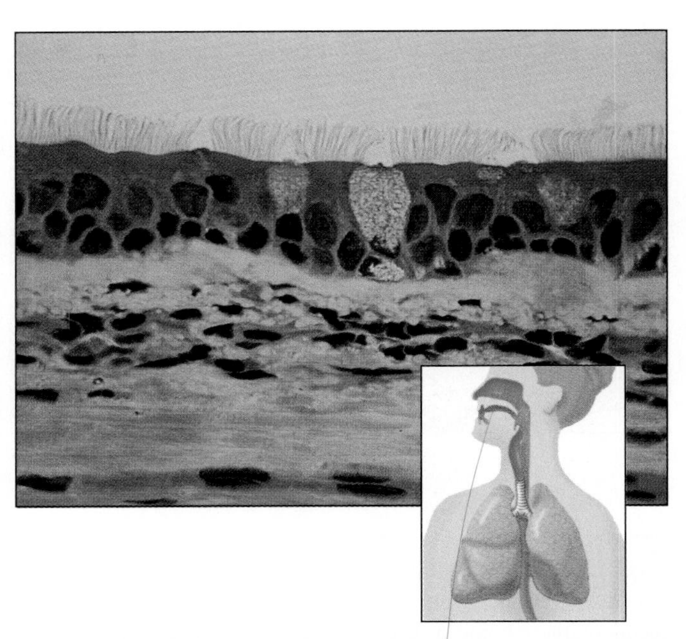

Figure 4.4 Pseudostratified ciliated columnar epithelium (500×).
Structure: A single layer of ciliated columnar cells that appears to be more than one layer of cells; contains scattered goblet cells.
Location: Lines most of the upper respiratory tract.
Function: Secretion of mucus; beating cilia remove secreted mucus and entrapped particles.

are trapped in the secreted mucus, which are moved by the beating cilia to the throat, where they are either swallowed or coughed up (figure 4.4).

Stratified Epithelium

Stratified epithelial tissues consist of more than one layer of cells, which makes them more durable to abrasion. Only the innermost layer of cells produces new cells by mitotic cell division. The cells are pushed toward the surface of the tissue as more new cells are formed below them. Cells in the surface layer are continuously lost as they die and are rubbed off by abrasion. Protection of underlying tissues is an important function of stratified epithelia. These tissues are named according to the shape of cells on their free surfaces.

Stratified squamous epithelium occurs in two distinct forms: keratinized and nonkeratinized. The keratinized type forms the outer layer (epidermis) of the skin. Its cells become impregnated with a waterproofing substance, **keratin** (ker′-ah-tin), as they migrate to the surface of the tissue. The nonkeratinized version lines the mouth, esophagus, and vagina. Both types provide resistance to abrasion (figure 4.5).

Transitional epithelium lines the interior of the urinary bladder and stretches as the bladder fills with urine. It consists of multiple layers of cells, and the surface cells of the unstretched tissue are large and

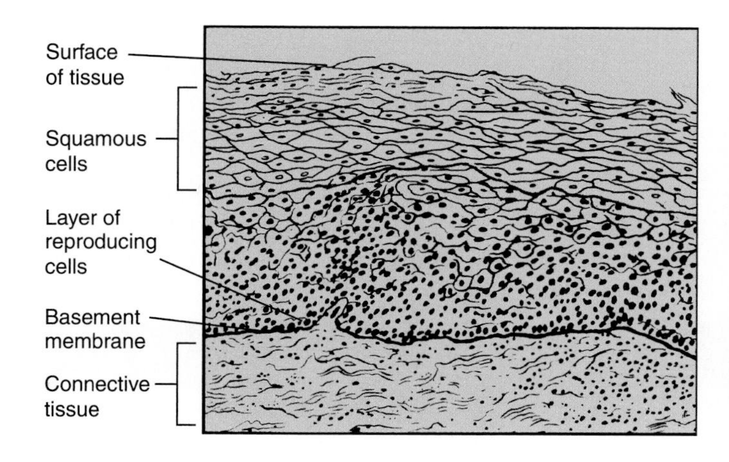

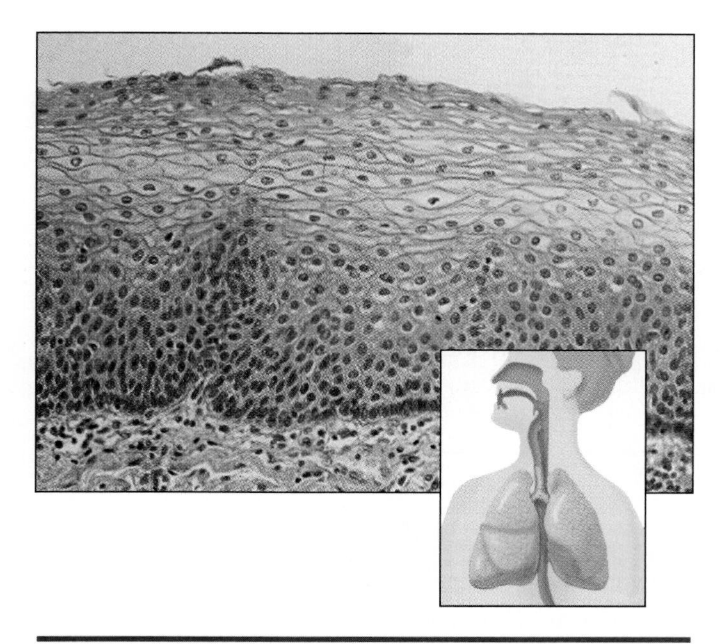

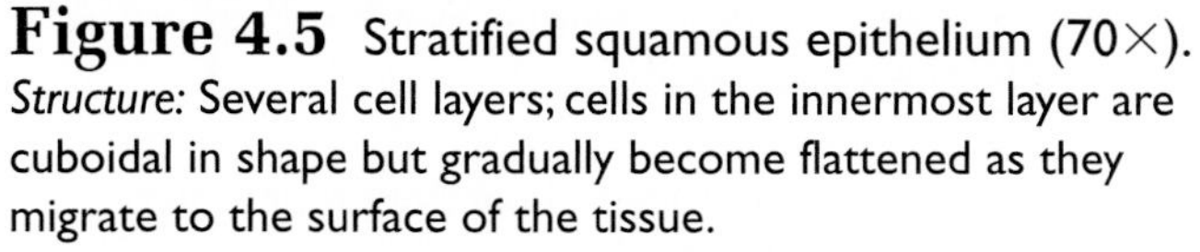

Figure 4.5 Stratified squamous epithelium (70×).
Structure: Several cell layers; cells in the innermost layer are cuboidal in shape but gradually become flattened as they migrate to the surface of the tissue.
Location: The keratinized type forms the epidermis of the skin; nonkeratinized type lines the mouth, esophagus, and vagina.
Function: Protection.

rounded. When stretched as the bladder fills with urine, the surface cells become thin, flat cells resembling squamous epithelial cells (figure 4.6).

Table 4.1 summarizes the types of epithelial tissues.

> ***What are the characteristics and general functions of epithelial tissues?***
> ***How do the types of epithelial tissues differ as to structure, location, and function?***

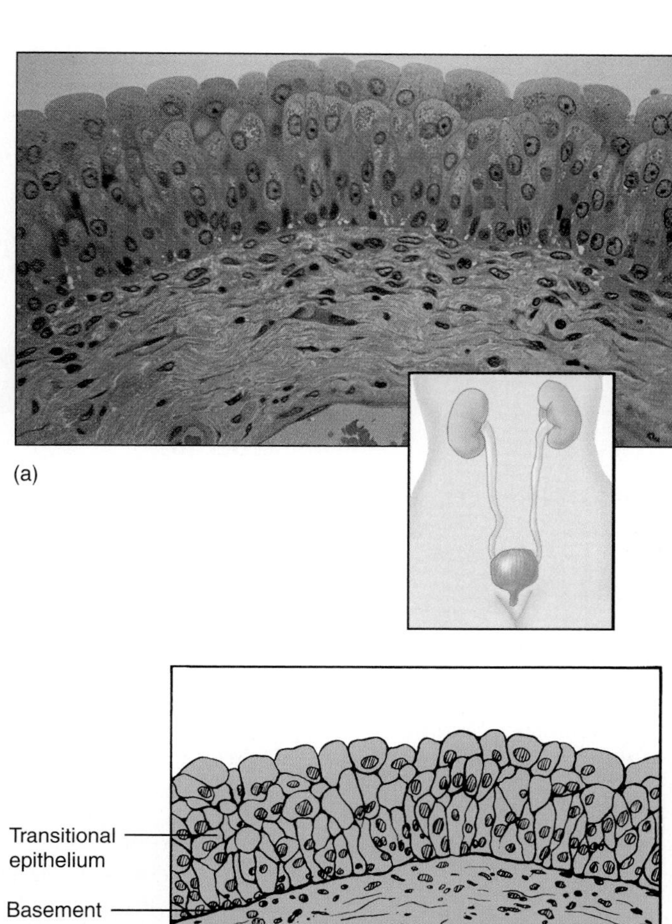

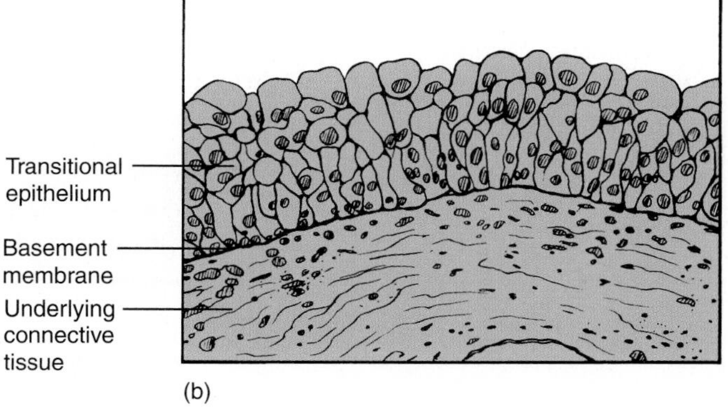

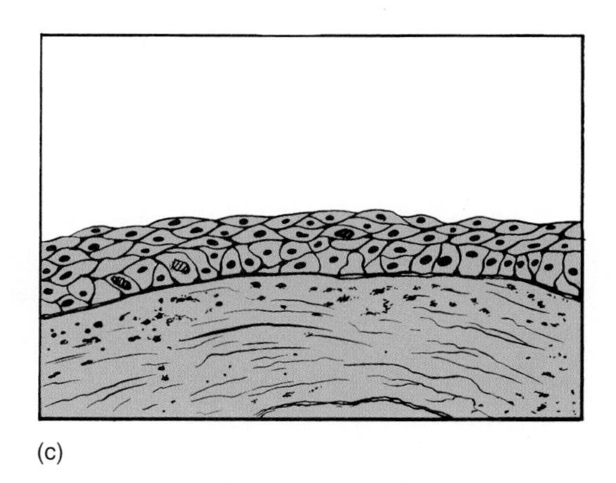

Figure 4.6 Transitional epithelium.
(*a*) A photomicrograph (250×) and (*b*) drawing showing several layers of rounded cells when the urinary bladder wall is contracted. (*c*) When the bladder wall is stretched, the tissue and cells become flattened.
Structure: Several layers of large, rounded cells that become flattened when stretched.
Location: Lines the interior of the urinary bladder.
Function: Protection; permits stretching of the bladder wall.

Connective Tissues

These widely distributed tissues are the most abundant tissues in the body. As the name implies, connective tissues support and bind together other tissues so they are never found on exposed surfaces. Like epithelial cells, connective tissue cells have retained the ability to reproduce by mitotic cell division.

Connective tissues consist of a diverse group of tissues subdivided into two broad categories: (1) connective tissue proper consists of the common connective tissues that bind other tissues and organs together and (2) connective tissues with more specialized functions—cartilage, bone, and blood (table 4.2).

All connective tissues consist of relatively few, loosely arranged cells and a large amount of intercel-

Table 4.1 Epithelial Tissues

Tissue	Structure	Location and Function
Simple Epithelia		
Squamous	Single layer of flat cells	Walls of capillaries, air sacs in lungs; diffusion, osmosis, filtration
Cuboidal	Single layer of cubelike cells	Glands, surface of ovaries, kidney tubules; secretion, absorption
Columnar	Single layer of columnlike cells	Lining of digestive tract; secretion, absorption, protection
Pseudostratified ciliated columnar	Single layer of ciliated columnlike cells that appears as multiple layers	Lining of upper air passages; sweeps away mucus and foreign particles, secretion, protection
Stratified Epithelia		
Squamous	Multiple layers of flat cells	Outer portion of the skin; protection
Transitional	Multiple layers of oval cells that flatten when the tissue is stretched	Lining of urinary bladder; protection

Table 4.2 Connective Tissues

Tissue	Structure	Location and Function
Loose connective tissue	Matrix with loose arrangement of collagenous and elastic fibers in semifluid ground substance; fibroblasts scattered	Attaches skin to underlying muscles; forms supporting framework for organs
Adipose tissue	Closely packed fat cells	Forms protective cushion around organs; insulates body; energy storage
Fibrous connective tissue	Matrix contains densely packed collagenous fibers; fibroblasts arranged in rows	Covers skeletal muscles; forms tendons binding muscles to bones and ligaments binding bones to bones
Elastic connective tissue	Matrix with many elastic fibers; scattered fibroblasts	Walls of blood vessels, air passages, and lungs; allows expansion and contraction; imparts flexibility
Cartilage		
Hyaline	Matrix glassy and smooth; few collagenous fibers	Covers ends of bones at joints; forms walls of air passages; supports part of nose; protection and support
Elastic	Matrix with many elastic fibers	Supports external ear; forms part of larynx (voice box); support, flexibility
Fibrocartilage	Matrix with densely packed collagenous fibers	Forms intervertebral disks and pads in the knee joint; absorbs shocks
Bone	Matrix is hard and solid; osteocytes arranged in concentric circles around osteonic canals	Forms bones of the skeleton; provides support and protection
Blood	Matrix is liquid with suspended blood cells	Transports materials

lular substance called **matrix** (mā′-triks). Matrix is produced by the cells and may be liquid, gel-like, or solid, and it may contain fibers formed of protein. The type of matrix is used to classify connective tissues.

Connective Tissue Proper

These connective tissues are the "ordinary" connective tissues in the body that bind together other tissues and form the basic supporting framework of organs. They are characterized by a matrix that consists of a semifluid or jellylike *ground substance* in which fibers and cells are embedded. Most of the cells are **fibroblasts** that produce the ground substance and the protein fibers in the matrix.

There are two basic types of fibers found in the matrix of connective tissues. The white **collagenous** (kol-laj′-e-nus) **fibers** are formed of a protein called collagen. They are relatively thick, very strong, nonelastic, and flexible. The yellow **elastic fibers** are not as strong as collagenous fibers, but they provide elasticity to tissues that contain them.

Scar tissue results from an excess production of collagenous fibers by fibroblasts. *Adhesions* are formed of scar tissue that joins tissues together abnormally. They are not uncommon following abdominal surgery. Sometimes, additional surgery is necessary to remove the adhesions.

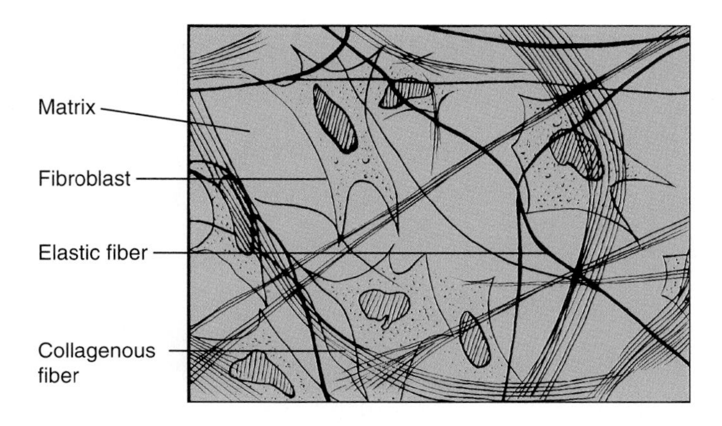

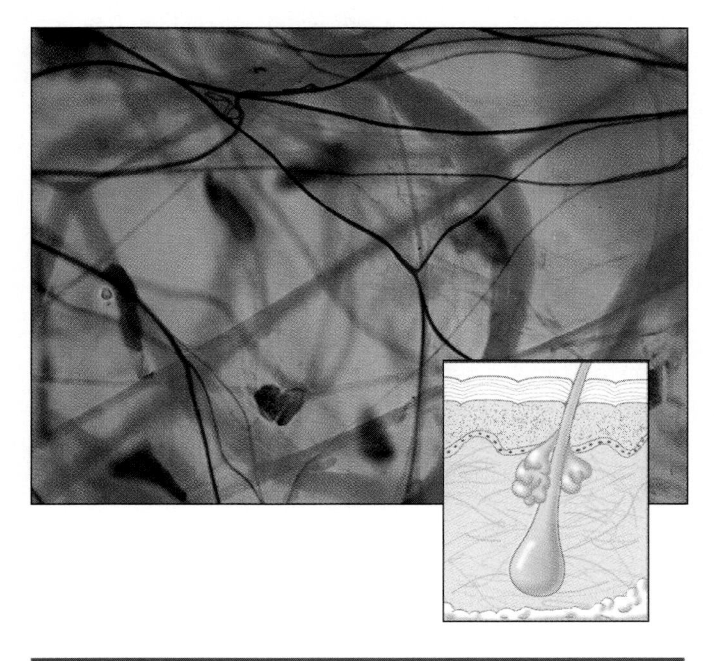

Figure 4.7 Loose connective tissue. (250×).
Structure: Formed of scattered fibroblasts and a loose network of collagenous and elastic fibers embedded in a gel-like ground substance.
Function: Attaches the skin to underlying muscles; supports internal organs, blood vessels, and nerves.

Loose Connective Tissue

Loose, or **areolar** (ah-rē′-ō-lar), **connective tissue** derives its name from the loose arrangement of the collagenous and elastic fibers within the matrix. It is the most abundant connective tissue in the body. Fibroblasts are the most numerous cells, but certain white blood cells migrate into the tissue. A semifluid ground substance fills the spaces between the cells and fibers. Loose connective tissue (1) attaches the skin to underlying muscles; (2) provides a supporting framework for internal organs, nerves, and blood vessels; and (3) is a site for many immune reactions (figure 4.7).

Since epithelial and connective tissue cells are active in cell division, they are prone to the formation of tumors when normal control of cell division is lost. The most common types of cancer arise from epithelial cells, possibly because these cells are often in contact with *carcinogens,* cancer-causing agents, in the environment. A cancer derived from epithelial cells is called a *carcinoma.* Malignant tumors that originate in connective tissue are also common types of cancer. A cancer of connective tissue is called a *sarcoma.*

Adipose Tissue

Large accumulations of fat cells form **adipose** (ad′-i-pōs) **tissue,** a special type of loose connective tissue. It occurs throughout the body but is more common beneath the skin and around internal organs. The vacuoles of fat cells are filled with fat droplets that push the nucleus and cytoplasm to the edge of the cells. In addition to fat storage, adipose tissue serves as a protective cushion for internal organs, especially around the kidneys and behind the eyeballs, and it helps to insulate the body from abrupt temperature changes (figure 4.8).

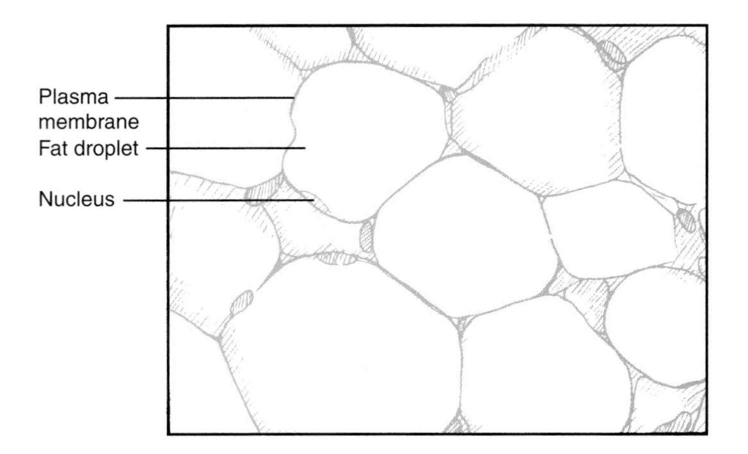

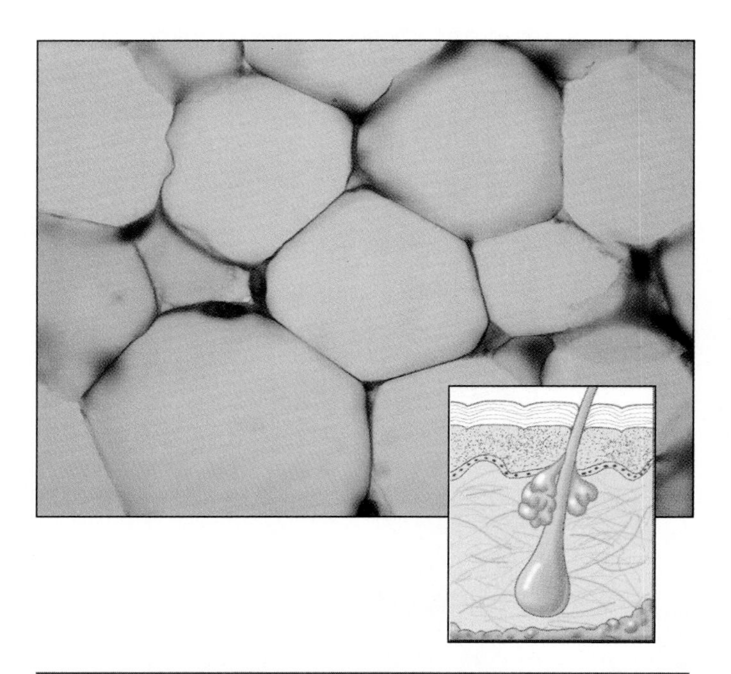

Figure 4.8 Adipose tissue (250×).
Structure: Formed of closely packed fat cells with little intercellular material. Large fat-containing vacuole pushes the cytoplasm and nucleus to the edge of the cell.
Function: Storage of excess nutrients as fat; occurs primarily under the skin and around internal organs.

Fibrous Connective Tissue

Fibrous connective tissue is characterized by an abundance of tightly packed collagenous fibers and relatively few cells. Nearly all of the cells are fibroblasts that are arranged in rows between the bundles of fibers. Fibrous connective tissue has great strength, but it is also flexible. It is the main component of ligaments, which attach bones to bones; tendons, which attach muscles to bones; and the inner layer (dermis) of the skin. This tissue has a poor blood supply, which causes injuries to ligaments and tendons to heal slowly (figure 4.9).

Elastic Connective Tissue

An abundance of elastic fibers in the matrix distinguishes **elastic connective tissue.** Collagenous fibers are also present, and fibroblasts are scattered between the fibers. Elastic connective tissue occurs where elasticity is advantageous, such as in the lungs, air passages, and arterial walls. For example, elastic connective tissue enables the expansion of the lungs as air is inhaled and the contraction of the lungs as air is exhaled.

Cartilage

Cartilage consists of a semisolid matrix in which cartilage cells, or **chondrocytes** (kon′-drō-sītz), are embedded. The spaces in the matrix that contain the chondrocytes are called **lacunae** (lah-kū′-nē). The major functions of cartilage are support and protection. Three types of cartilage are present in the body: hyaline cartilage, elastic cartilage, and fibrocartilage.

Hyaline Cartilage

The matrix of **hyaline** (hī′-a-lin) **cartilage** has a smooth, glassy, bluish white appearance. It contains collagenous fibers, but they are not easily visible. Numerous chondrocytes in lacunae are present. Hyaline cartilage is the most abundant cartilage in the body. It provides (1) a protective covering on bone surfaces forming joints, (2) forms part of the nose, and (3) supports the walls of air passages. During embryonic development, most bones of the body are initially formed of hyaline cartilage. Subsequently, the cartilage is gradually replaced by bone tissue (figure 4.10).

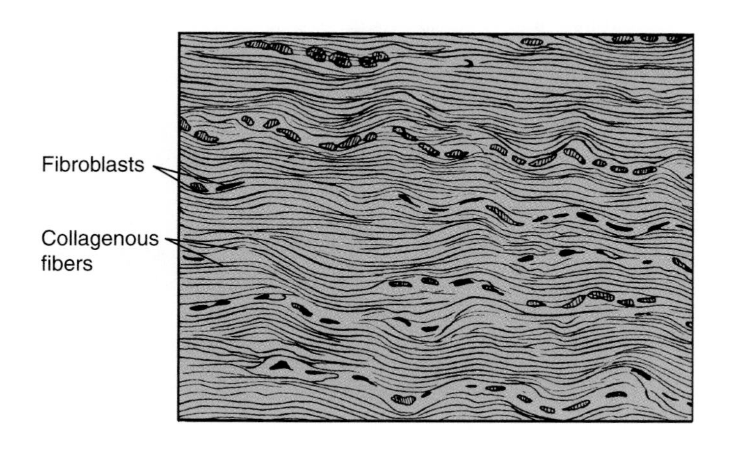

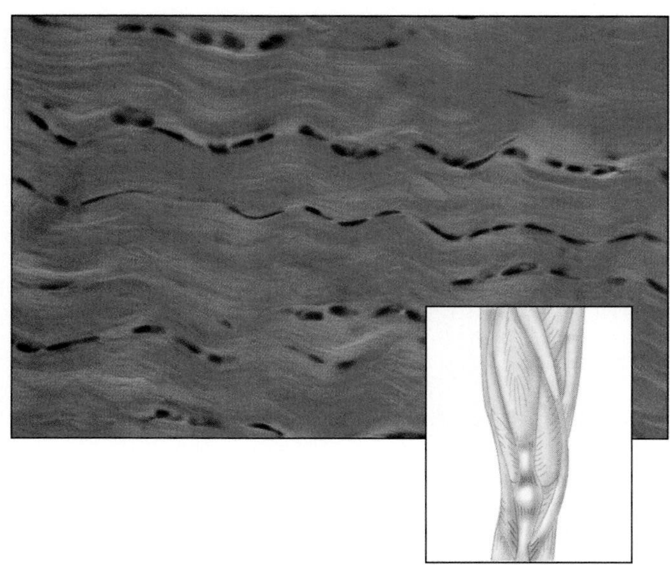

Figure 4.9 Fibrous connective tissue.
Structure: Consists of tightly packed collagenous fibers that are separated by scattered rows of fibroblasts.
Function: Strong attachment; forms ligaments attaching bones to bones at joints, tendons attaching muscles to bones, fibrous capsules around organs, and the dermis of the skin.

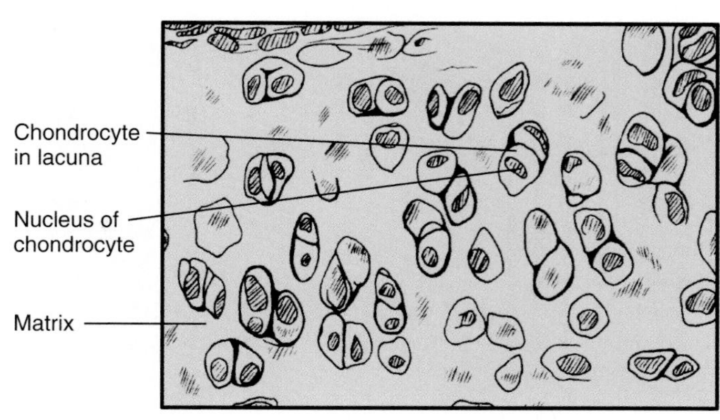

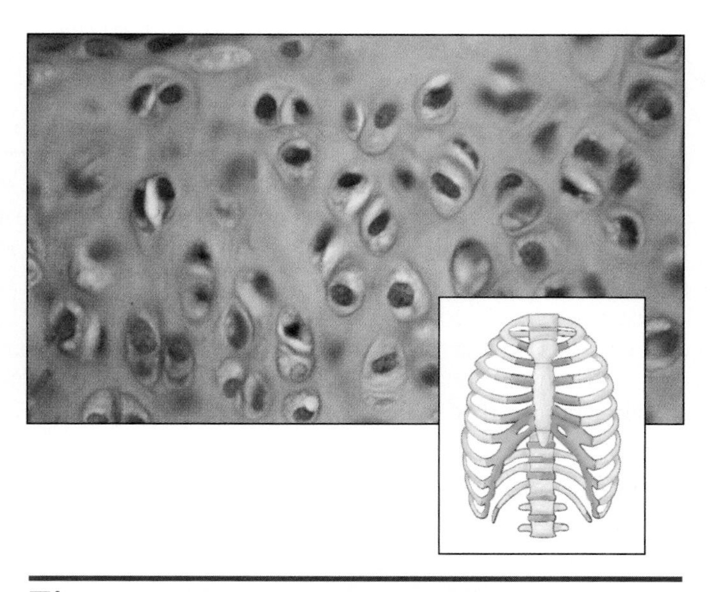

Figure 4.10 Hyaline cartilage (250×).
Structure: Smooth glassy matrix with many chondrocytes in lacunae.
Function: Forms protective covering of bones at joints; attaches ribs to breastbone, and supports walls of air passages.

Elastic Cartilage

This tissue is similar to hyaline cartilage, but **elastic cartilage** contains an abundance of elastic fibers that impart resiliency to the tissue. The distribution in the body is rather limited. Elastic cartilage forms part of the larynx (voice box), and it forms the supportive framework for the external ear (figure 4.11).

Fibrocartilage

The matrix of **fibrocartilage** contains many tightly packed collagenous fibers that lie between short rows or clumps of chondrocytes. This cartilage forms (1) the intervertebral disks that are located between vertebrae, (2) the cartilaginous pads in the knee joints, and (3) the protective cushion of the symphysis pubis (anterior union of the pelvic bones). Fibrocartilage is especially tough, and it is adapted to absorb shocks by resisting compression and tension (figure 4.12).

Bone

Of all the supportive connective tissues, **bone,** or **osseous tissue,** is the hardest and most rigid. This results from the minerals, mostly calcium salts, that compose the matrix along with some collagenous fibers. Bone

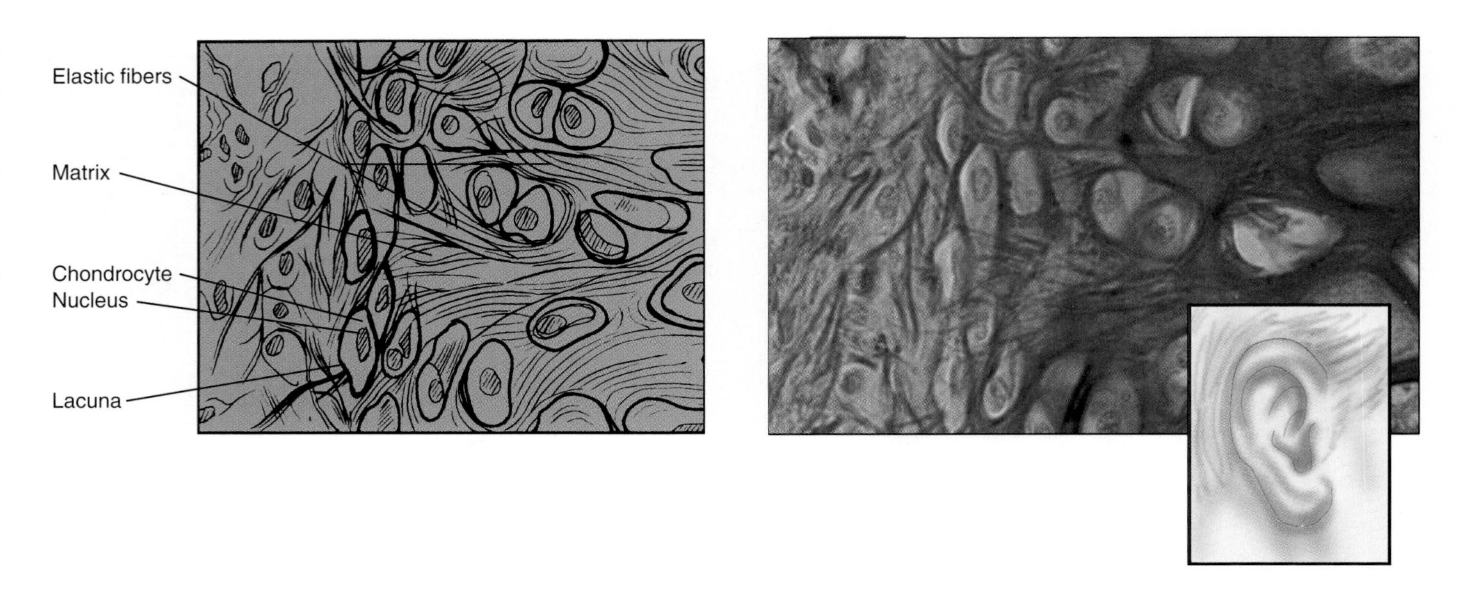

Figure 4.11 Elastic cartilage (100×).
Structure: Consists of numerous cartilage cells occupying lacunae in a gel-like matrix containing numerous elastic fibers.
Function: Provides the supporting framework for the external ears.

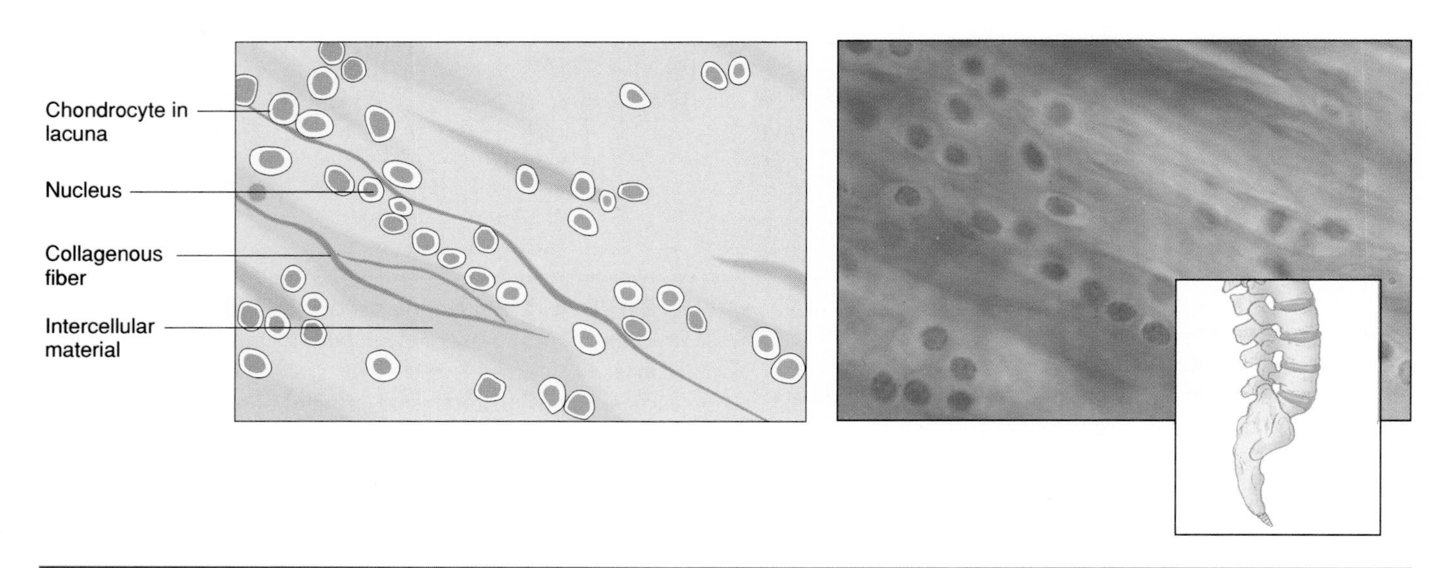

Figure 4.12 Fibrocartilage (250×).
Structure: Consists of rows or clusters of cartilage cells occupying lacunae in a matrix containing tightly packed collagenous fibers.
Function: Composes the intervertebral disks between vertebrae and cartilaginous pads in the knee joint where it serves as a protective shock absorber.

tissue provides the rigidity and strength necessary for the skeletal system to support the body.

The bone matrix is deposited in concentric rings, or **lamellae** (lah-mel′-ē), around microscopic tubes, the **osteonic** or **haversian** (ha-ver′-sē-an), **canals.** These canals contain blood vessels and nerves. The bone cells, or **osteocytes** (os′-tē-ō-sītz), are located in lacunae that are arranged in concentric rings between the lamellae. The tiny canals extending from the lacunae are called **canaliculi** (kan″-ah-lik′-ū-li), and they contain cell processes that extend from the osteocytes. Canaliculi serve as passageways for the movement of materials between the blood vessels in the haversian canals and the osteocytes (figure 4.13).

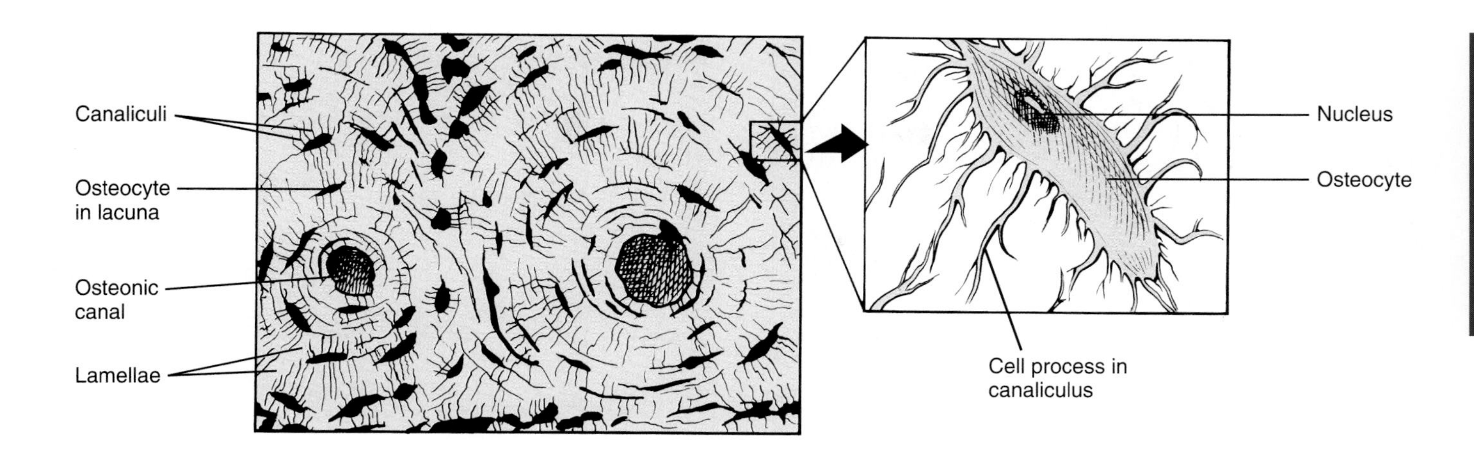

Figure 4.13 Compact bone.
Structure: Consists of rings of bone cells located in lacunae within a hard matrix of calcium salts. The matrix is arranged in concentric layers around osteonic (haversian) canals. Canaliculi, minute channels between lacunae, enable movement of materials between bone cells.
Function: Forms bones of the skeleton that provide support for the body and protection for vital organs.

Blood

Blood is a specialized type of connective tissue. It consists of numerous blood cells that are suspended in the plasma, the liquid matrix of the blood. There are three basic types of blood cells: red blood cells, white blood cells, and platelets. Blood plays a vital role in carrying materials throughout the body (figure 4.14).

Table 4.2 (page 66) summarizes the characteristics of connective tissue.

What are the general characteristics and functions of connective tissue?
What are the characteristics, locations, and functions of the different connective tissues?

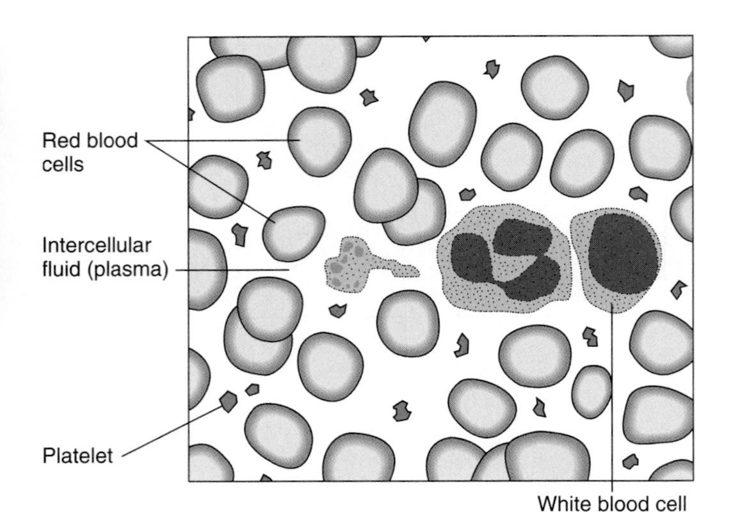

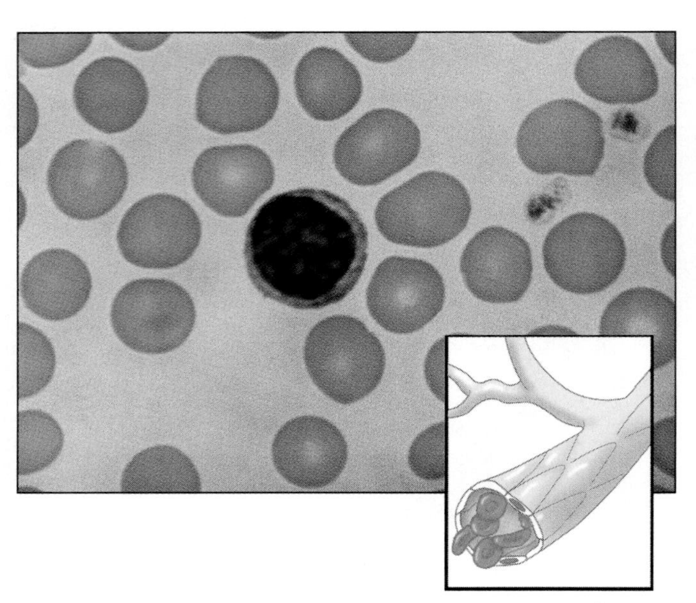

Figure 4.14 Blood (1,000×).
Structure: Consists of red blood cells, white blood cells, and platelets that are carried in a liquid matrix, the plasma.
Function: Transports materials throughout the body and provides a defense against disease.

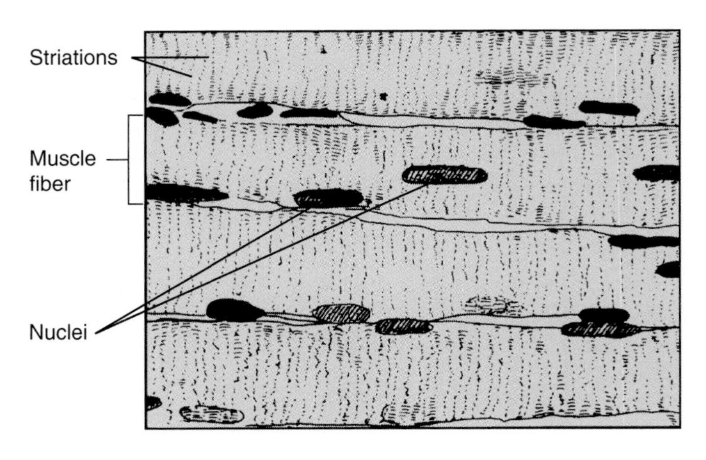

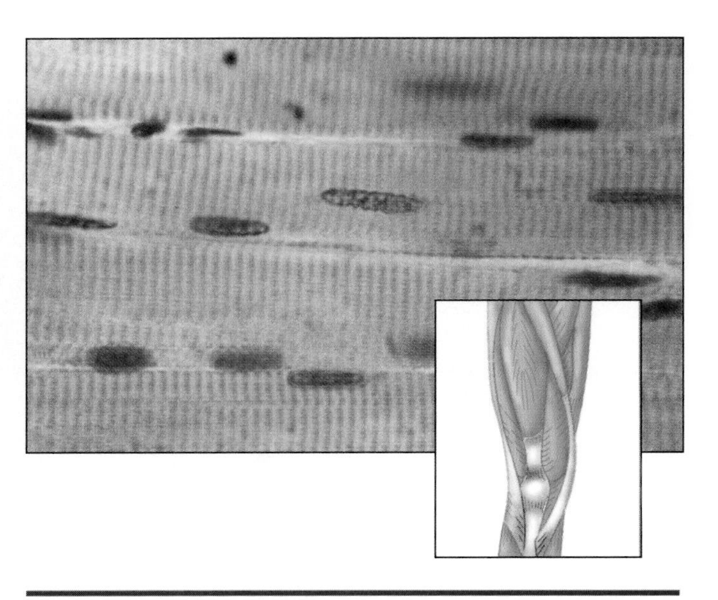

Figure 4.15 Skeletal muscle tissue (250×).
Structure: Consists of cylindrical muscle cells (fibers) that have striations and multiple, peripherally located nuclei.
Function: Voluntary, rapid contractions; composes skeletal muscles.

Muscle Tissues

Muscle tissue consists of muscle cells. Muscle cells have lost the ability to divide, so destroyed muscle cells cannot be replaced. Since muscle cells are elongated, they are often called muscle fibers, and they are adapted for *contraction* (shortening). Contraction is enabled by the interaction of specialized protein fibers. The contraction of muscle cells enables the movement of the body and body parts and also produces heat energy.

Three types of muscle tissue—skeletal, cardiac, and smooth muscle tissue—are classified according to their (1) location in the body, (2) structural features, and (3) functional characteristics.

Skeletal Muscle Tissue

Named for its location, **skeletal muscle tissue** is attached to bones. Its contractions enable movement of the head, trunk, and extremities. The cells are elongate and cylindrical, and they possess many nuclei that are located at the periphery of the cells. *Striations,* alternating light and dark bands, extend across the width of the fibers. Functionally, skeletal muscle tissue is considered to be *voluntary* muscle because its rapid contractions are under conscious control (figure 4.15).

Cardiac Muscle Tissue

The muscle tissue that occurs in the walls of the heart is **cardiac** (kar′-dē-ak) **muscle tissue.** It consists of branching cells or muscle fibers that interconnect in a netlike arrangement. *Intercalated* (in-ter-kah′-lā-ted) *disks* are present where the cells join together, and a single, centrally located nucleus is present in each cell. Like skeletal muscle, cardiac muscle cells are striated. The rhythmic contractions of cardiac muscle are *involuntary* since they are not consciously controlled (figure 4.16).

Smooth Muscle Tissue

Smooth muscle tissue derives its name from the absence of striations in its cells. It occurs in the walls of hollow internal organs, such as the digestive tract and blood vessels. The cells are long and spindle-shaped with a single, centrally located nucleus. The slow contractions of smooth muscle tissue are *involuntary* (figure 4.17).

Table 4.3 summarizes the characteristics of muscle tissue.

Nerve Tissue

The brain, spinal cord, and nerves are composed of **nerve tissue,** which consists of **neurons** (nū′-ronz), or nerve cells, and numerous supporting cells that are collectively called *neuroglia* (nū-rog′-lē-ah). Neurons are the functional units of nerve tissue.

A neuron consists of a *cell body,* the portion of the cell containing the nucleus, and one or more *neuron processes,* or nerve fibers that extend from the cell body (figure 4.18). Neurons are specialized to respond to environmental changes by forming and transmitting neural impulses. The complex interconnecting network of neurons enables the nervous system to coordinate body functions.

> ***What are the distinguishing characteristics of the three types of muscle tissue?***
> ***What types of cells form nerve tissue, and what are their functions?***

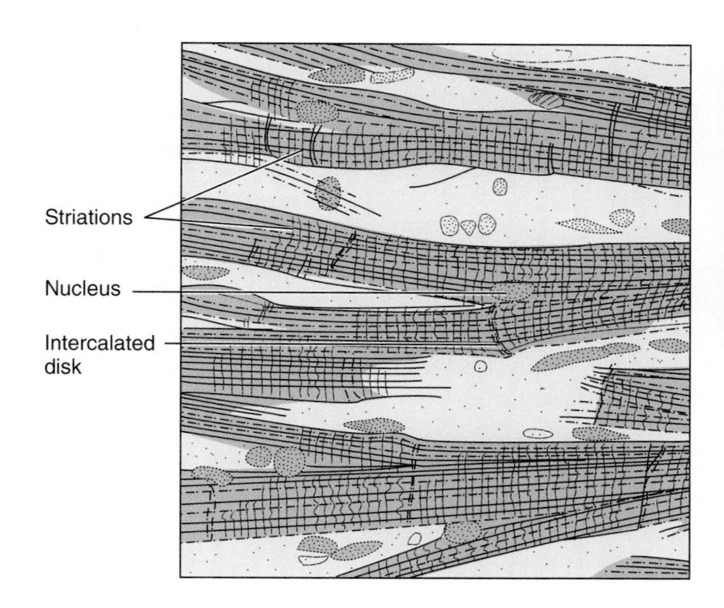

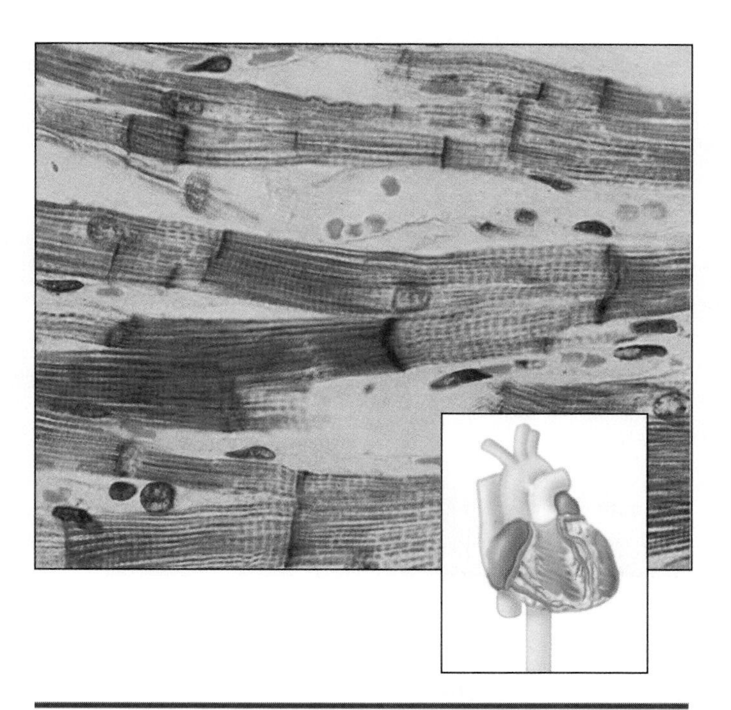

Figure 4.16 Cardiac muscle tissue.
Structure: Consists of striated cells that are arranged in an interwoven network. Intercalated disks are present at the junctions between cells. A single, centrally located nucleus is present in each cell.
Function: Involuntary, rhythmic contractions; forms the muscular walls of the heart.

Body Membranes

Membranes of the body are thin sheets of tissue that line cavities, cover surfaces, or separate tissues or organs. Some are composed of both epithelial and connective tissues; others consist of connective tissue only.

Epithelial Membranes

Sheets of epithelial tissue overlying a thin supporting framework of loose connective tissue form the epithelial membranes in the body. Blood vessels in the connective tissue serve both connective and epithelial

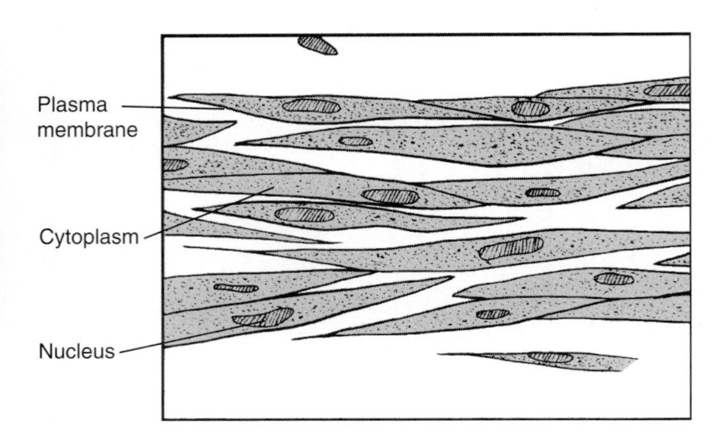

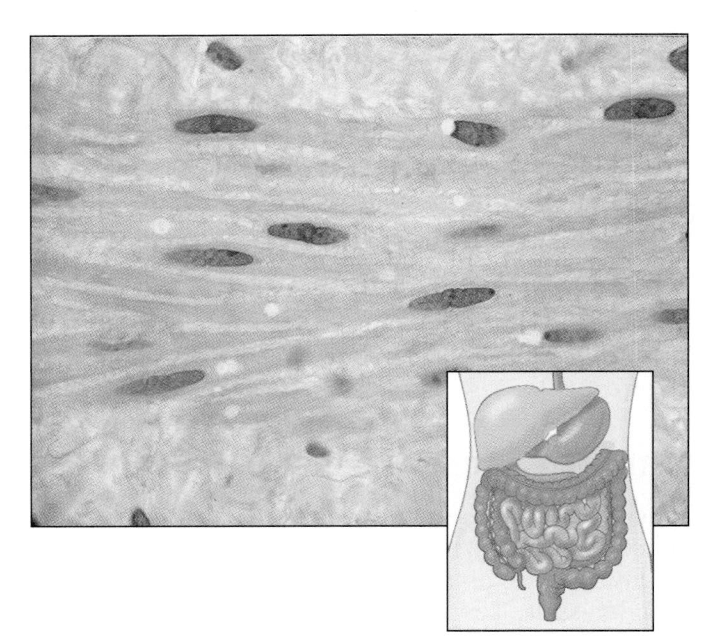

Figure 4.17 Smooth muscle tissue.
Structure: Consists of elongate, tapered cells that lack striations and have a single, centrally located nucleus.
Function: Involuntary, slow contractions; forms muscle layers in the walls of hollow internal organs.

Table 4.3 Muscle Tissues

Tissue	Structure	Location and Function
Skeletal muscle	Striated; long fibers with many peripherally located nuclei	Attached to bones; moves body and appendages; rapid voluntary contractions
Cardiac muscle	Striated; interwoven fibers joined by intercalated disks; single nucleus in each cell	Muscle of heart wall; involuntary rhythmic contractions
Smooth muscle	Nonstriated; single nucleus in each cell; slow contractions	Muscle in walls of hollow internal organs; involuntary

tissues. There are three types of epithelial membranes: serous, mucous, and cutaneous membranes.

Serous membranes line the ventral body cavity and cover most of the internal organs. They secrete **serous fluid,** a watery fluid, which reduces friction between the membranes. The pleurae, pericardium, and peritoneum are serous membranes.

Mucous membranes line tubes or cavities of organ systems, which have openings to the exterior. Their goblet cells secrete **mucus,** which coats the surface of the membranes, keeping the cells moist and lubricating their surfaces. The digestive, respiratory, reproductive, and urinary tracts are lined with mucous membranes.

The **cutaneous membrane** is the skin that covers the body. Unlike other membranes, its free surface is dry, composed of nonliving cells, and exposed to the environment. The skin is discussed in detail in chapter 5.

Connective Tissue Membranes

Some specialized membranes are formed only of connective tissue, usually fibrous tissue. These are considered with their respective organ systems, but here are four examples:

1. The *meninges* are three fibrous membranes that envelop the brain and spinal cord.
2. The *perichondrium* is a fibrous membrane covering the surfaces of cartilage. It contains blood vessels, which are the only blood supply for cartilage.

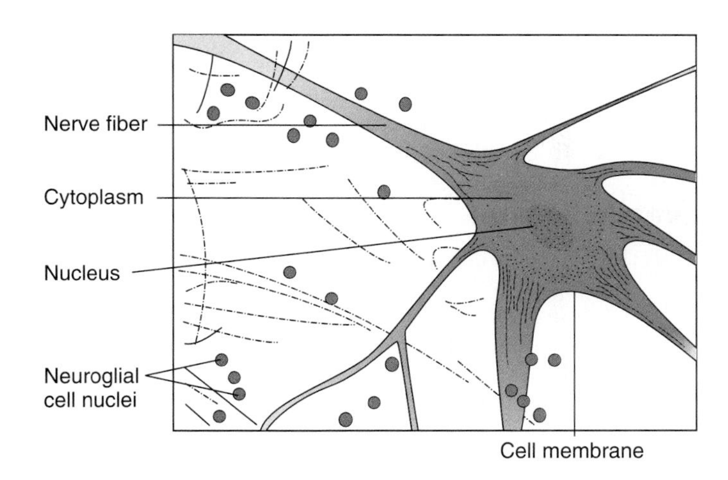

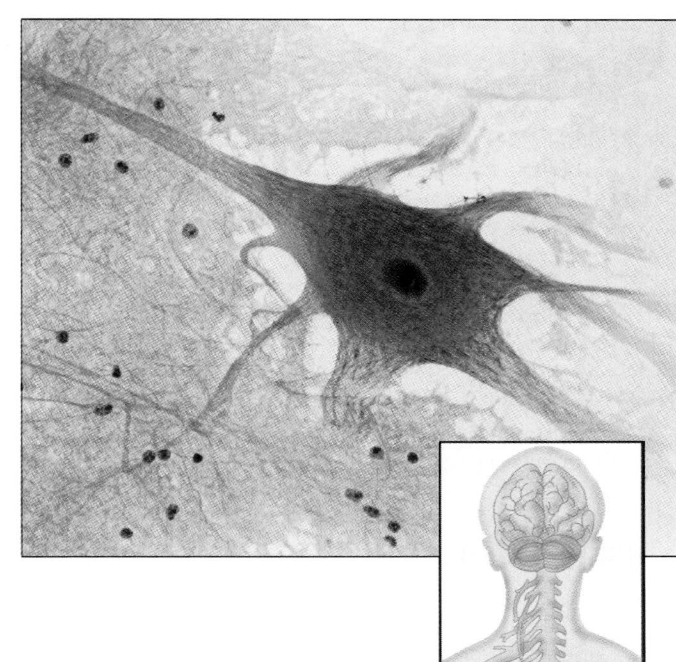

Figure 4.18 Nerve tissue (50×).
Structure: Consists of neurons and supportive neuroglial cells. Each neuron consists of a cell body, which houses the nucleus, and one or more nerve fibers.
Function: Impulse formation and transmission; forms the brain, spinal cord, and nerves.

3. The *periosteum* is a fibrous membrane that covers the surfaces of bones. It contains blood vessels that also serve the bone.
4. *Synovial membranes* line the cavities of freely movable joints, such as the knee. They secrete watery synovial fluid, which reduces friction in the joint.

What are the two kinds of body membranes? How do the types of epithelial membranes differ?

CHAPTER SUMMARY

1. As cells specialize during embryonic development, they form groups of similar cells called tissues.
2. The body is formed of four basic types of tissues: epithelial, connective, muscle, and nerve tissues.

Epithelial Tissues

1. Epithelial tissue covers surfaces of organs and of the body and lines the body cavities.
2. Epithelial tissue is composed of closely packed cells with little intercellular material.
3. The cells are attached to underlying connective tissue by a noncellular basement membrane.
4. Epithelial tissue lacks blood vessels.
5. Epithelial tissues function in absorption, secretion, filtration, and protection.
6. Epithelial tissues are classified according to the number of cell layers and the shape of the surface cells. The epithelial tissues are:

Simple
- squamous
- cuboidal
- columnar
- pseudostratified ciliated columnar

Stratified
- stratified squamous
- transitional

Connective Tissues

1. Connective tissue is composed of relatively few cells located within a large amount of matrix.

2. All but cartilage are supplied with blood vessels.
3. Connective tissue binds other tissues together and provides support and protection for organs and the body.
4. Connective tissue is classified according to the nature of the matrix. The connective tissues are:
   Connective tissue proper
   - loose connective tissue
   - adipose tissue
   - fibrous connective tissue
   - elastic connective tissue

   Cartilage
   - hyaline
   - elastic
   - fibrocartilage

   Bone, or osseous, tissue
   Blood

Muscle Tissues

1. Muscle tissue is composed of muscle cells that are specialized for contraction.
2. Contraction of muscle cells enables movement of the body and body parts.
3. Muscle tissue is classified according to its location in the body, the characteristics of the muscle cells, and the type of contractions (voluntary or involuntary).
4. Three types of muscle tissue are skeletal, cardiac, and smooth muscle tissue.

Nerve Tissue

1. Nerve tissue consists of neurons and supporting neuroglial cells.
2. Neurons consist of a cell body and long, thin cell processes, and they are adapted to form and conduct impulses.
3. Nerve tissue forms the brain, spinal cord, and nerves.

Body Membranes

1. Membranes in the body are either epithelial membranes or connective tissue membranes.
2. There are three types of epithelial membranes: serous, mucous, and cutaneous.
3. Examples of connective tissue membranes are meninges, perichondrium, periosteum, and synovial membranes.

BUILDING YOUR VOCABULARY

1. Selected New Terms.

canaliculi, p. 70
cartilage, p. 68
cutaneous membrane, p. 74
endocrine gland, p. 62
endothelium, p. 63
exocrine gland, p. 62
keratin, p. 64
lamellae, p. 70
mucous membrane, p. 74
neuroglia, p. 73
osteonic canals, p. 70
serous membrane, p. 74
striations, p. 72

2. Related Clinical Terms

adhesions (ad-hē′-shuns) Excessive scar tissue binding organs together abnormally.
carcinoma (kar-si-nō′-mah) Cancer arising from epithelial cells.
keloid (kē′loid) Mass of scar tissue at skin surface due to excessive formation of connective tissue during healing of a wound.
sarcoma (sar-kō′-mah) Cancer arising from connective tissue cells.

STUDY ACTIVITIES

1. Complete the **Chapter 4 Study Guide,** which begins on page 411.
2. Complete the **Learning Objectives** listed on the first page of this chapter.

CHECK YOUR UNDERSTANDING

(Answers are located in appendix B.)

1. Simple epithelial tissues consist of _____ layer(s) of cells.
2. Epithelial tissue contains _____ (little/much) intercellular material.
3. The upper air passages of the respiratory system are lined with _____ epithelium.
4. The digestive tract is lined with _____ epithelium.
5. Filtration, secretion, and absorption are functions of _____ tissue.
6. The intercellular substance in connective tissues is called _____ .
7. The supporting framework of internal organs is provided by _____ connective tissue.
8. Cushioning pads in the knee joint are composed of _____.
9. Muscle tissue in the wall of the heart is _____ muscle tissue.
10. Muscle tissue lacking striations and found in walls of the digestive tract is _____ muscle tissue.
11. Nerve tissue consists of neurons and supporting _____ cells.
12. Membranes lining digestive, respiratory, and urinary tracts are classified as _____ membranes.

FIVE

5

CHAPTER

The Integumentary System

Chapter Preview & Learning Objectives

Functions of the Skin
- Describe the functions of the skin.

Structure of the Skin
- Describe the structure and role of the two tissue layers forming the skin.
- Describe the structure and role of the subcutaneous layer.

Skin Color
- Describe how skin color is determined.
- Explain how the skin provides protection from ultraviolet radiation.

Accessory Structures
- Identify the accessory structures formed by the epidermis and describe their structures.
- Describe the function of each accessory structure.

Temperature Regulation
- Describe how the skin aids the regulation of body temperature.

Aging of the Skin
- Describe how aging affects the skin.

Disorders of the Skin
- Describe the common infectious and noninfectious disorders of the skin.

Chapter Summary
Building Your Vocabulary
Study Activities
Check Your Understanding

SELECTED KEY TERMS

Apocrine sweat gland (apo = detached; crin = separate off) A sweat gland opening into a hair follicle.
Cutaneous (cutane = skin) Pertaining to the skin.
Dermis (derm = skin) The inner layer of the skin.
Eccrine sweat gland (ec= out from) A sweat gland opening on the skin surface.
Epidermis (epi = upon) The outer layer of the skin.
Hair follicle (folli = bag) A saclike epidermal ingrowth in which a hair develops.
Integument (integere = to cover) The skin.
Keratin (kerat = horny, hard) Waterproofing, abrasion-resistant protein in epidermal cells.
Melanin (melan = black) The brown-black pigment formed by melanocytes.
Sebaceous gland (seb = grease, oil) A sebum-producing gland associated with a hair follicle.
Subcutaneous (sub = under, beneath) Pertaining to the layer of loose connective tissue beneath the dermis.
Sudoriferous gland (sudori = sweat) A sweat gland.

The skin and the structures that develop from it—hair, nails, and glands—form the **integumentary** (in-teg-ū-men′-tar—ē) **system.** The skin, or **integument,** is also known as the **cutaneous membrane,** one of the three types of membranes as noted in chapter 4. The skin is pliable and tough, and it provides a waterproof, self-repairing barrier separating underlying tissues and organs from the external environment. Although it often gets little respect, the skin is vital to our existence.

Physicians often find the condition of the skin helpful in making diagnoses, even of diseases that do not directly affect the skin. For example, dry, coarse skin may indicate dietary or glandular disorders. Of course, certain childhood diseases (e.g., measles and chicken pox), allergies, and local infections produce characteristic skin **lesions** (lē′-zhuns), that is, localized tissue damage.

Functions of the Skin

The skin performs five important functions:

1. **Protection.** The skin provides a physical barrier between underlying tissues and the external environment. It provides protection from abrasion, dehydration, ultraviolet radiation, and bacterial invasion.
2. **Excretion.** The production of perspiration by sweat glands removes small amounts of organic wastes, salts, and water.
3. **Temperature regulation.** During periods of excessive heat production by the body, the secretion of perspiration and its evaporation from the body surface helps to lower body temperature. During periods of excessive heat loss, blood vessels near the body surface constrict to reduce heat loss.
4. **Sensory perception.** The skin contains nerve endings and receptors that detect stimuli associated with touch, pressure, temperature, and pain.
5. **Synthesis of vitamin D.** Exposure to ultraviolet radiation converts precursor molecules in the skin into vitamin D.

Structure of the Skin

The skin is thickest in areas subjected to wear and abrasion, such as the soles of the feet, where it may be 6 mm in thickness. It is thinnest on the eyelids, eardrums, and external genitalia, where it averages about 0.5 mm in depth.

The skin consists of two major layers: the epidermis and the dermis. The *epidermis* is the thinner outer layer, which is composed of stratified squamous epithelium. The *dermis* is the thicker inner layer, and it is composed of fibrous connective tissue. Beneath the dermis is the *subcutaneous layer,* which is not part of the skin but is considered here because of its close association with the skin. Figure 5.1 shows the arrangement of epidermis, dermis, and the subcutaneous layer as well as accessory organs of the skin. Table 5.1 summarizes these skin layers.

Epidermis

The **epidermis** is formed of stratified squamous epithelium. Since they lack blood vessels, epidermal

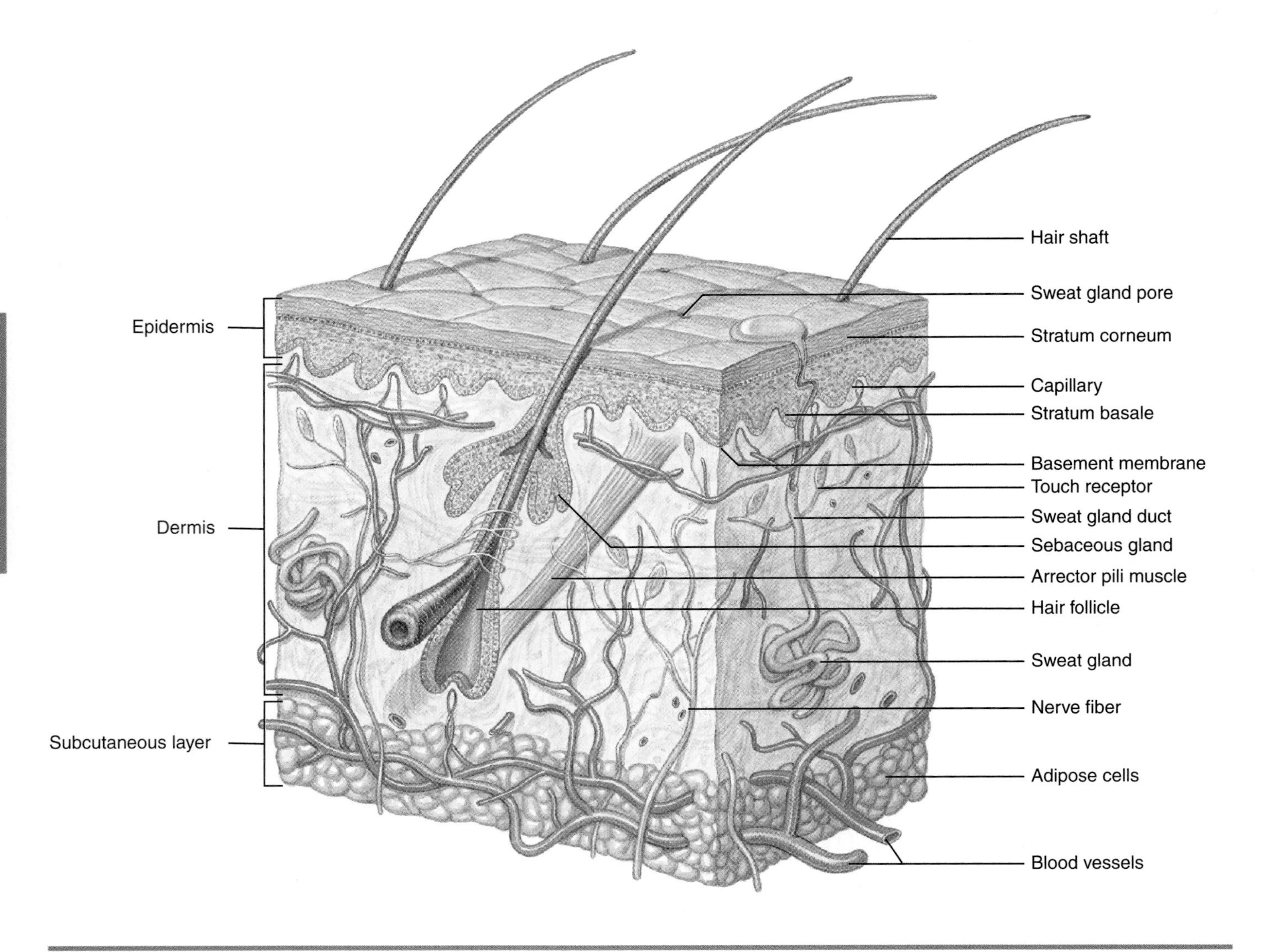

Figure 5.1 A section of skin.

Table 5.1 Layers of the Skin

Layer	Structure	Function
Epidermis	Stratified squamous epithelium; forms hair and hair follicles, sebaceous glands, and sweat glands that penetrate into the dermis or subcutaneous layers	Protects against abrasion, evaporative water loss, and bacterial or viral invasion
Dermis	Fibrous connective tissue containing blood vessels, nerves, and sensory receptors	Provides strength and elasticity; sensory receptors enable detection of touch, pressure, pain, cold, and heat; blood vessels supply nutrients to the epidermis
Subcutaneous	Loose connective tissue and adipose tissue; contains abundant blood vessels and nerves	Provides insulation and fat storage; attaches skin to underlying organs

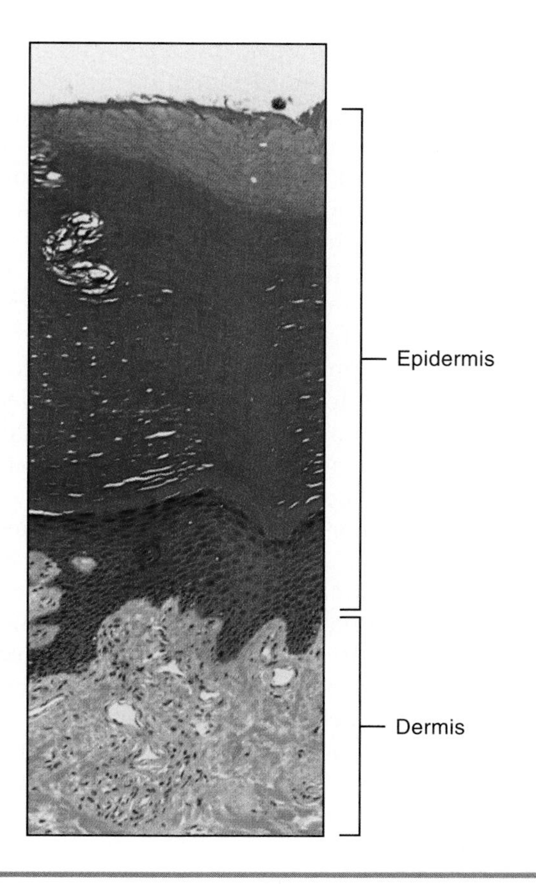

Figure 5.2 A photomicrograph of skin from the palm of the hand (50×).
Note the thick stratum corneum at the surface and the stratum basale next to the dermis.

cells must obtain nutrients from blood vessels in the underlying dermis. The epidermis is the boundary between the body and the environment. It protects the body against (1) the entrance of pathogenic organisms, (2) ultraviolet radiation, and (3) excessive water loss.

The innermost layer of cells, the **stratum basale** (ba-sah′-le), or basal layer, continuously produces new cells by mitotic cell division. The older cells are gradually pushed toward the surface, as new cells are formed beneath them. They are gradually deprived of essential nutrients and die.

As cells migrate to the surface, they change in composition so that several distinctive layers of cells are recognizable when sections of epidermis are viewed microscopically (figure 5.2). During this transition, the cells accumulate deposits of **keratin** (ker′-ah-tin), a waterproofing protein that is resistant to abrasion. The outermost layer, the **stratum corneum** (kor′nē-um), consists of numerous layers of dead, flat, keratinized cells. The superficial cells of the stratum corneum are constantly sloughed off and replaced by underlying cells moving to the surface. It takes about two weeks for a newly formed cell to migrate to the surface and be sloughed off.

Whenever possible, surgeons try to make incisions parallel to the dominant direction of collagen fibers in the dermis because the healing of such incisions forms little scar tissue. Incisions across numerous fibers tend to leave larger scars.

Dermis

The **dermis** is composed of fibrous connective tissue, which provides strength, extensibility (ability to stretch), and elasticity (ability to return to its original shape) for the skin. Both collagen and elastic fibers are abundant in the dermis. Collagen provides the strength and toughness of the dermis, while elastic fibers provide the extensibility and elasticity.

Where the dermis and epidermis join are numerous **dermal papillae** (pah-pil′-ē), conelike projections of the dermis that fit into recesses in the epidermis. These dermal papillae contain blood vessels, and many contain touch receptors (see figure 5.1). Pressure, pain, heat, and cold receptors are located in the deeper portions of the dermis. The epidermal ridges and grooves that produce fingerprints and toe prints unique to each person result from the pattern of dermal papillae.

The dermis is well supplied with nerves and blood vessels. Blood vessels in the papillary layer supply nutrients to the adjacent epidermal cells. The dermal blood vessels play an important role in temperature regulation, which is considered later in this chapter.

Subcutaneous Layer

The **subcutaneous layer,** or superficial fascia (fash′ē-ah), attaches the skin to underlying tissues and organs. It consists primarily of loose connective tissue and adipose tissue. Subcutaneous adipose tissue serves as a heat insulator and a storage site for fat. It conserves body heat and retards the penetration of external heat into the body. Blood vessels and nerves within the subcutaneous layer give off branches that supply the dermis.

What are the general functions of the skin?
What changes occur in epidermal cells after they are formed?
What are the functions of the dermis and subcutaneous layers?

Skin Color

The color of the skin results from the interaction of three different pigments: hemoglobin, carotene, and melanin. **Hemoglobin** (he′mō-glō″-bin) is the red pigment in red blood cells that gives the red color to blood. **Carotene** (kair′-ō-tēn) is a yellowish pigment found in the stratum corneum and dermis. **Melanin** (mel′-ah-nin) is the brown-black pigment that is formed by **melanocytes** (mel-an′-ō-sītz), special cells found in the deeper layers of the epidermis (figure 5.3).

When melanin is produced by melanocytes, it is picked up by adjacent epidermal cells. The amount of melanin that a person can produce is inherited. The greater the production of melanin, the darker is the skin color. Melanin protects the body from harmful ultraviolet radiation that is present in sunlight. Generally, melanocytes are equally distributed, but when they occur in small clumps, melanin is concentrated in small patches called *freckles.*

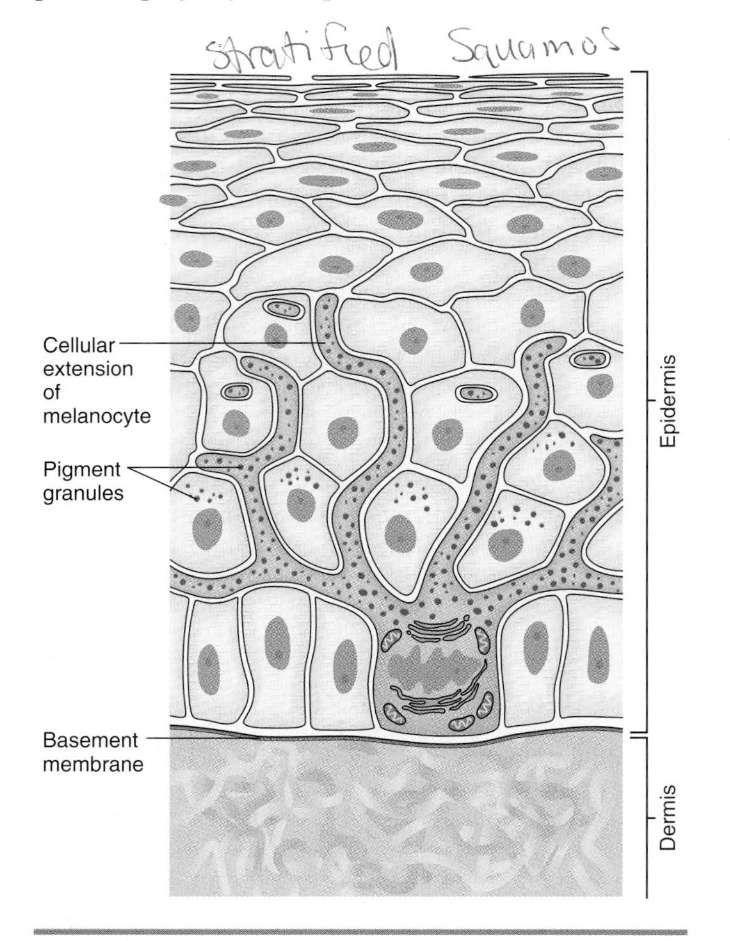

Figure 5.3 A melanocyte with cellular extensions that transfer melanin granules to adjacent epidermal cells.

Exposure to ultraviolet light increases melanin production and produces a tanned appearance in light-skinned individuals. The increased production of melanin is a protective, homeostatic mechanism. When exposure to ultraviolet radiation decreases, melanin production also decreases and the tan is lost in a few weeks when the tanned cells migrate to the surface and are sloughed off.

The skin color that is characteristic of the various human races results primarily from varying amounts of carotene and melanin in the skin. The effect of hemoglobin is relatively constant, but it is often masked by high concentrations of melanin. Dark-skinned races produce abundant melanin and have a greater protection from ultraviolet radiation. Asians tend to produce more carotene than other races, which gives their skin a yellowish tinge. Caucasians produce relatively little melanin and are susceptible to the harmful effects of ultraviolet radiation. The reduced amount of melanin in light-skinned Caucasians allows the hemoglobin of blood within dermal blood vessels to show through and give the skin a pinkish hue. These differences in skin color are inherited.

What relationship exists between skin color and protection against ultraviolet radiation?

Accessory Structures

The accessory structures of the skin—hair, nails, and glands—develop from the epidermis. These structures are located in either the dermis or subcutaneous layer because they originate from inward growths of the epidermis. Table 5.2 summarizes these accessory structures.

Hair

A **hair** is formed of keratinized cells and consists of two parts: a shaft and a root. The *shaft* is the portion that projects above the skin surface. The *root* lies below the skin surface in a **hair follicle,** an inward, tubular extension of the epidermis. The follicle penetrates into the dermis and usually into the subcuta-

Table 5.2 Accessory Structures

Structure	Origin	Function
Hair	Fused keratinized epidermal cells formed at base of hair follicle	Scalp hair protects scalp from excessive heat loss and mechanical injury Eyelashes and eyebrows protect eyes from sunlight and dust Nasal hairs keep dust and insects out of the nose
Nails	Fused keratinized epidermal cells formed within an ingrowth of the epidermis	Protect the dorsal tips of fingers and toes
Sebaceous glands	Formed as an epidermal outpocketing from a hair follicle	Sebum prevents excessive dryness and keeps hair and skin pliable and soft; reduces evaporative water loss
Sudoriferous glands	Apocrine glands formed as an epidermal outpocketing from hair follicles in axillary and genital regions	No value in humans
	Eccrine glands formed as epidermal ingrowth	Evaporation of perspiration cools body and eliminates excessive heat
Ceruminous glands	Modified sudoriferous glands	Waxy cerumen keeps eardrum moist and captures dust entering ear canal

neous layer (figures 5.4 and 5.5). The region of cell division, where the stratum basale forms new hair cells, is located in the *bulb* (base) of the follicle. The bulb is enlarged where it fits over a dermal papilla, which contains blood vessels that nourish the dividing epidermal cells. The hair cells become keratinized, die, and become part of the root as they move away from the source of nourishment. The continuing production of new cells causes growth of the hair.

Normal hair loss from the scalp is about 75 to 100 hairs per day. Hair loss is increased and regeneration is decreased by poor diet, major illnesses, emotional stress, high fever, certain drugs, chemical therapy, radiation therapy, and aging. Baldness, an inherited trait, is much more common in males than in females.

Each hair follicle has an associated **arrector pili muscle** that is attached at one end to the lower portion of a hair follicle and to the papillary layer of the dermis at the other end. Each muscle is formed of a small group of smooth muscle fibers. When a person is frightened or very cold, the arrector pili muscles contract and produce goose bumps. These muscles raise the hairs on end but have little value in humans. In many mammals, the erect hairs express rage to enemies or increase the thickness of the insulating layer of hair in cold weather.

Hair is present over most of the body, but it is absent on the soles, palms, nipples, lips, and portions of the external genitalia. Its primary function is protection, although the tiny hairs over much of the body have little function. Eyelashes and eyebrows shield the eye from sunlight and foreign particles. Hair in the nostrils and ear canals protects against the entrance of foreign particles and insects. Hair protects the scalp from sunlight and mechanical injury.

Nails

The hard, hornlike **nails** cover the dorsal surface of the terminal portions of the fingers and toes. They are formed of keratinized epidermal cells. A nail consists of a *body,* the portion that is visible, and a *root,* the proximal portion that is inserted into the dermis. Nails are colorless, but they normally appear pinkish due to the blood vessels in the underlying nail bed. The nails appear bluish in persons suffering from severe anemia or oxygen deficiency. Near the root is a whitish, crescent-shaped area that is called the *lunula.* The major function of nails is protection, but

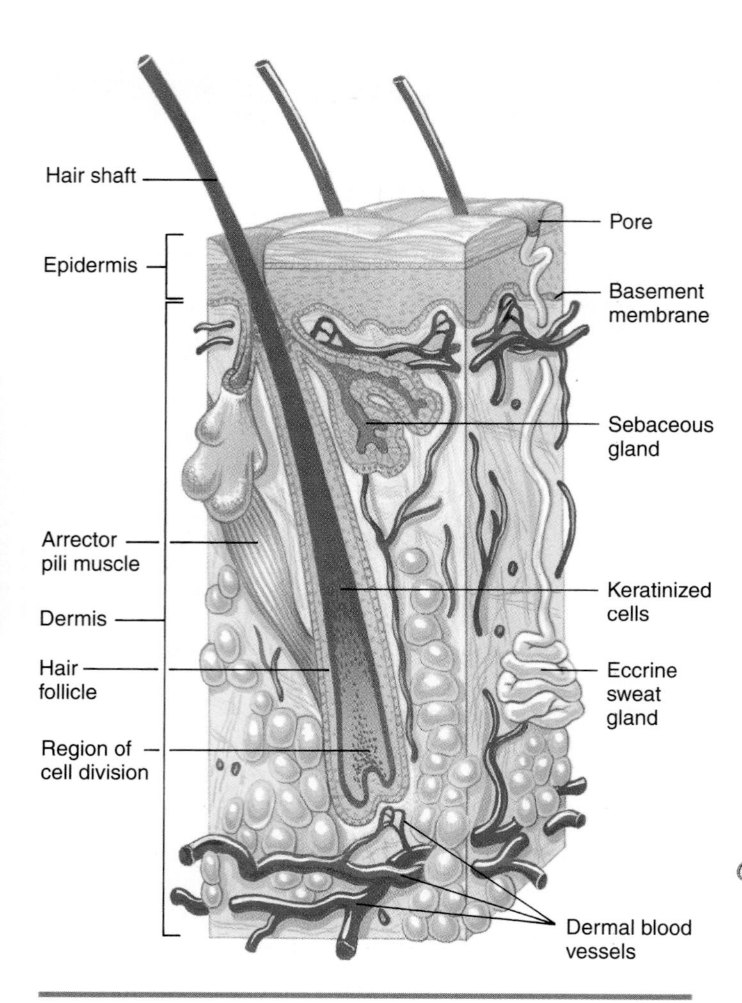

Figure 5.4 A section of skin showing the association of a hair follicle, arrector pili muscle, and sebaceous gland. A hair grows when cells at the base of the follicle divide, pushing older epidermal cells outward, where they become keratinized.

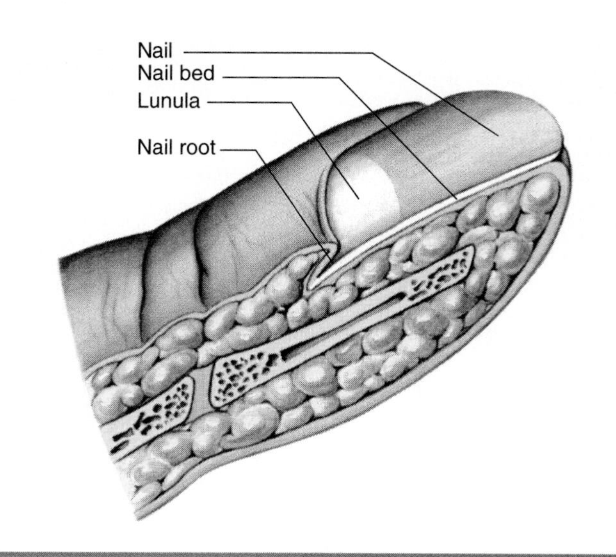

Figure 5.5 The structure of a nail.
A nail is composed of heavily keratinized epidermal cells that are formed and undergo keratinization in the nail root.

> Sebaceous glands increase their production of sebum at puberty. Accumulated sebum in enlarged hair follicles may form blackheads, whose color comes from oxidized sebum and melanin and not from dirt as is commonly believed. Invasion of certain bacteria may result in pimples or boils.

fingernails are also useful in manipulating small objects (figure 5.5).

Glands

Exocrine glands associated with the skin are of three types: sebaceous (oil), sudoriferous (sweat), and ceruminous (wax) glands. Each type of gland is formed by an inward growth of the epidermis during embryonic development.

Sebaceous Glands

Sebaceous (se-bā-shus) **glands** are oil-producing glands that empty their secretions into hair follicles. This oily secretion is called **sebum,** and it helps to keep the hair and skin pliable and soft. It also reduces evaporative water loss and inhibits the growth of some bacteria.

Sudoriferous Glands

Sudoriferous (sū-dō-rif′-er-us) **glands** are sweat glands. Sweat glands are richly supplied with blood vessels, and they play an important role in regulating the body temperature.

The two types of sweat glands are *apocrine* and *eccrine glands.* Both types are tubular in shape with a coiled secretory portion located deep in the dermis or subcutaneous layer and a narrow duct that carries the secretion toward the surface. Apocrine (ap′-ō-krin) glands have a duct that empties the secretion into a hair follicle. They occur primarily in the axillary and genital regions and produce a somewhat whitish perspiration that contains water, mineral salts, urea, and proteins. Bacteria metabolizing this secretion produce waste products that cause body odor. Eccrine (ek′-rin) glands occur over all of the body, and they produce a clear, watery perspiration lacking proteins. (See figure 5.1 for the structural similarities and differences of apocrine and eccrine sweat glands.)

Physicians have known for a long time that the ultraviolet (UV) radiation in sunlight produces damaging changes in the skin. The damage is cumulative, although a single day's exposure can produce noticeable changes. Short-term changes include sunburn and a tan as the body tries to protect itself from UV radiation by producing more melanin. Long-term UV damage ranges from increased wrinkling and loss of elasticity to liver spots and skin cancer.

Since the long-term effects of UV radiation do not appear for a number of years, the summer tans so important to many light-skinned teenagers and young adults will accelerate aging of the skin and may produce skin cancers in later life. The advent of tanning salons only increases potential problems for unwary users.

Skin cancer is by far the most common type of cancer. Fortunately, most skin cancers are *carcinomas* involving basal or squamous cells and are usually curable by surgical removal. However, *melanoma,* cancer of melanocytes, tends to spread rapidly to other organs and can be lethal if it is not detected and removed in an early stage. Melanoma is fatal in about 45% of the cases.

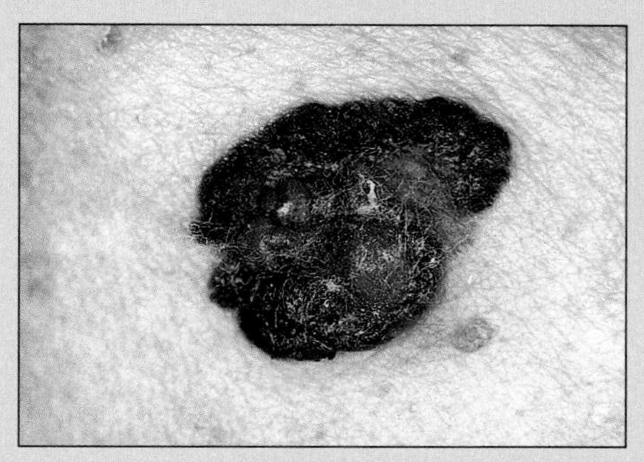

A Cutaneous Malignant Melanoma.

The undesirable effects of overexposure to UV radiation can easily be prevented by reducing the exposure of the skin to sunlight and by the liberal use of sunscreens. UV radiation is at its peak between 11:00 A.M. and 3:00 P.M., so avoiding exposure during these hours is especially helpful.

Sunscreens are available with different levels of protection, and they are labeled according to the sun protection factor (SPF) provided. This allows the selection of a sunscreen that is appropriate for a particular type of skin. For example, sunscreens with an SPF of 15 or higher are available for fair-skinned persons who burn easily. Sunscreens with an SPF of 8 to 15 may give adequate protection for olive-skinned persons who rarely burn.

Ceruminous Glands

Ceruminous (se-rū′mi-nus) **glands** are modified sweat glands that produce a waxy secretion called *cerumen.* They are simple coiled, tubular glands that occur in the external auditory canal. The sticky, waxy nature of cerumen helps to keep foreign particles and insects out of the auditory canal.

How is a hair formed?
What is the function of sebaceous glands?

Occasionally, excessive production of cerumen causes a buildup of wax that becomes impacted in the auditory canal. This condition may cause a slight hearing loss as well as pain. Once the impacted wax is removed by irrigation and/or mechanical means, hearing returns to normal.

Temperature Regulation

Humans are able to maintain a normal body temperature of 37°C (98.6°F), although the surrounding environmental temperature may vary widely. Variations in body temperature of 0.5° to 1.5°F during a 24-hour activity cycle are normal, but variations of more than a few degrees can be life threatening. The brain controls the regulation of body temperature, and the skin plays a key role in conserving or dissipating heat. Of course, the source of body heat is cellular respiration, especially in metabolically active organs like the liver and muscles. Figure 5.6 illustrates the major aspects of temperature regulation.

When the body temperature is higher than normal, the brain increases the flow of blood to the skin. This increases heat loss by radiation from the body surface and causes the sweat glands to produce perspiration. As perspiration spills out on the surface of the skin and evaporates, heat loss is accelerated, which lowers the body temperature to normal levels.

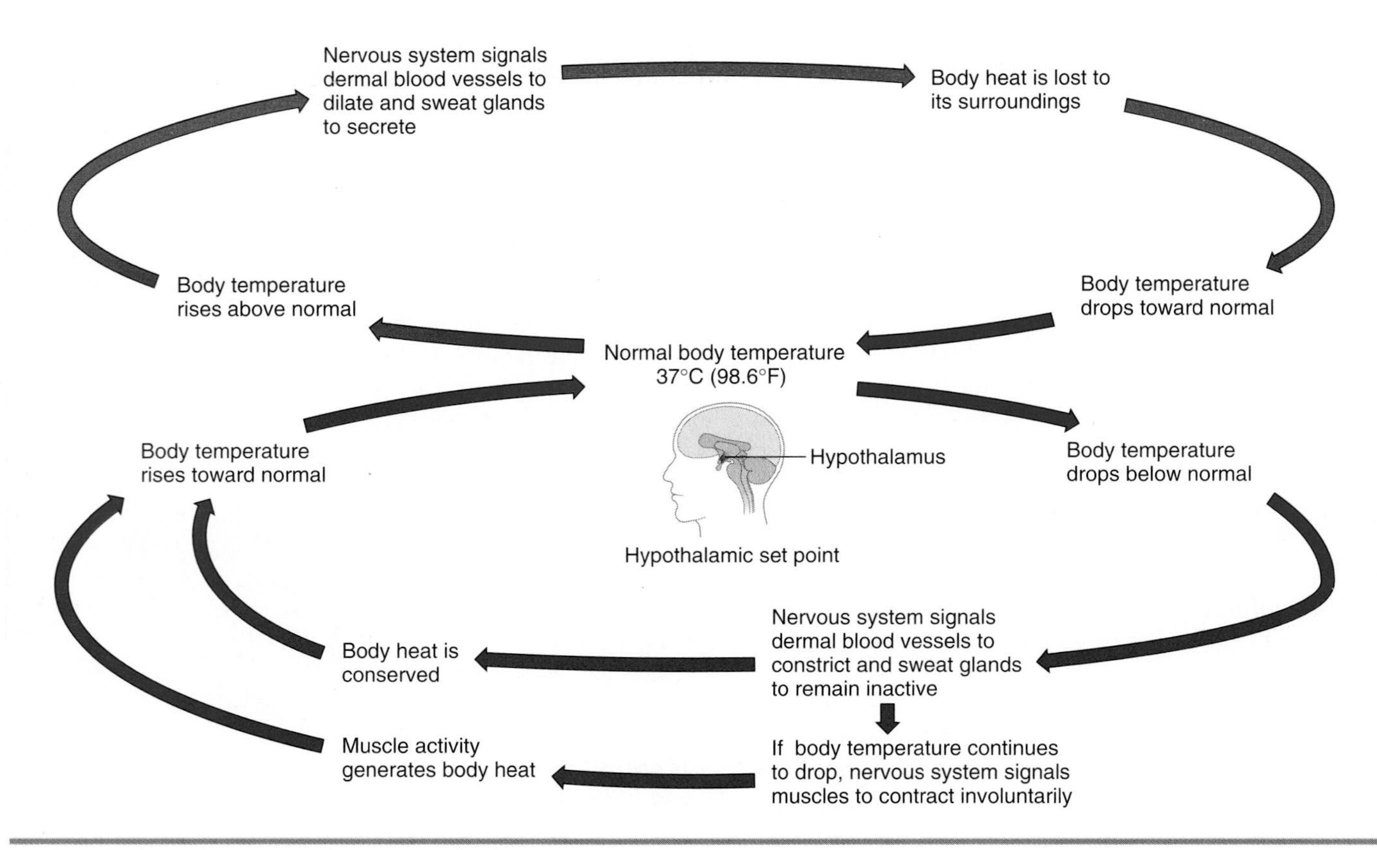

Figure 5.6 A flow diagram showing how the skin helps to regulate body temperature. Temperature regulation is controlled by the hypothalamus, which "sets the thermostat."

When body temperature drops below normal, the brain reduces the flow of blood to the skin. The decreased blood flow to the skin reduces heat loss by radiation from the body surface and prevents sweat glands from secreting perspiration that would cool the body further. If heat loss becomes excessive, the nervous system stimulates small groups of skeletal muscles to produce involuntary, rapid, small contractions (shivering). This increases cellular respiration and generates additional heat to raise the body temperature.

How is the skin involved in the regulation of body temperature?

Aging of the Skin

A newborn baby's skin is very thin, and there is not a lot of subcutaneous fat present. During infancy a baby's skin thickens, and more subcutaneous fat is deposited, producing the soft, smooth skin typical of infants.

As a child grows into adulthood, the skin continues to become thicker, and it is subjected to numerous harmful conditions: sunlight, wind, abrasions, chemical irritants, and invasions of bacteria. The continued exposure of the skin through the adult years produces damaging effects. However, noticeable changes usually are not apparent until a person approaches 50 years of age. Thereafter, continued aging of the skin is more noticeable.

Typical changes in aging skin are as follows: (1) a breakdown of collagen and elastic fibers (hastened by exposure to sunlight) causes wrinkles and sagging skin; (2) a decrease in subcutaneous fat makes a person more sensitive to temperature changes; (3) a decrease in sebum production by sebaceous glands may cause dry, itchy skin; (4) a decrease in melanin production produces gray hair and sometimes a splotchy pattern of pigmentation; and (5) a decrease in hair replacement results in thinning hair or baldness, especially in males.

Disorders of the Skin

Since the skin is in contact with the environment, it is especially susceptible to injuries, such as abrasions (scraping), contusions (bruises), and cuts. Other common disorders of the skin may be subdivided into infectious and noninfectious disorders. Some inflammatory disorders may fall into either group, depending upon the specific cause of the disorder. Common childhood diseases, such as chickenpox and measles, are not listed here but produce skin lesions that characterize the particular disease.

Infectious Disorders

Acne (ak′-nē) is a chronic skin disorder characterized by plugged hair follicles that often form pimples (pustules) due to infection by certain bacteria. It often appears at puberty, when reproductive hormones stimulate increased sebum secretion.

Athlete's foot is a slightly contagious infection that is caused by a fungus growing on the skin. It produces reddish, flaky, and itchy patches of skin, especially between and under the toes, where moisture persists.

Boils are acute, painful *Staphylococcus* infections of hair follicles and their sebaceous glands as well as the surrounding dermis and subcutaneous tissue. The union of several boils forms a *carbuncle.*

Fever blisters, or cold sores, are clusters of fluid-filled vesicles that occur on the lips or oral membranes. They are caused by a *Herpes simplex* virus (type 1) and are transmitted by oral or respiratory exposure. *Genital herpes* is a similar disorder in which painful blisters form on the genitals as a result of infection transmitted by sexual intercourse.

Impetigo (im-pe-tī′-gō) is a highly contagious skin infection caused by bacteria. It typically occurs in children and is characterized by fluid-filled pustules that rupture, forming a yellow crust over the infected area.

Warts are small skin tumors that have a normal skin color. They are caused by a virus infection that stimulates excessive growth of epithelial cells.

Noninfectious Disorders

Alopecia (al-ō-pāy′-shē-ah) is the loss of hair. It is most common in males who have inherited male pattern baldness, but it may result from noninherited causes, such as poor nutrition, sensitivity to drugs, and eczema.

Bedsores (decubitis ulcers) result from a chronic deficiency of blood circulation in the dermis and subcutaneous layers. Bedsores form over bones that are subjected to prolonged pressure against a bed or cast. They are most common in bedridden patients. Frequent turning of bed patients helps to prevent bedsores.

Burns are damage to the skin caused by heat, chemicals, or radiation. Burns are classified according to the degree of damage. A *first-degree burn* involves only the surface of the epidermis so that skin functions are not affected. Healing is usually rapid. A *second-degree burn* produces damage to the deeper layers of the epidermis and the outer portion of the dermis. Blisters form within the epidermis, and there is usually no infection. Healing requires five to seven days, and scarring is usually not a problem. A *third-degree burn* destroys the epidermis, dermis, glands, hair follicles, and nerve endings. Normal skin functions are lost, so care must be given in order to control fluid loss and bacterial infection. Skin grafting is often necessary, and scarring usually results.

Calluses and **corns** are thickened areas of skin that result from chronic pressure. Calluses are larger and often occur on the palms of the hands at the metacarpal joints and on the balls of the feet. Corns are smaller and usually occur on the upper surface of the toes. Improperly fitting shoes are frequent causes of corns on the feet.

Cancers result from excessive cell division by cells that have lost control of the process. Cancers of the skin are usually *carcinomas* (basal cell carcinomas or squamous cell carcinomas) involving epidermal cells. Less frequently, skin cancers are *melanomas,* cancers of the melanocytes. Melanomas are especially dangerous because they tend to grow and metastasize (spread to other parts of the body) rapidly. Excessive exposure to sunlight is the primary cause of skin cancers.

Dandruff is the excessive shedding of dead epidermal cells from the scalp as a result of excessive cell production. It is usually caused by seborrheic eczema of the scalp, a noninfectious dermatitis.

Eczema (ek-zē′-mah) is an inflammation producing redness, itching, scaling, and sometimes cracking of the skin. It is noninfectious and noncontagious, and it may result from exposure to irritants or from allergic reactions. *Seborrheic eczema* is characterized by hyperactivity of sebaceous glands and patches of red, scaling, and itching skin. It may occur at the corners of the mouth, in hairy areas, or in skin exposed to irritants.

Hives are red, itchy bumps or wheals that usually result from an allergic reaction to certain foods, drugs, or pollens.

Moles are slow-growing skin tumors that are usually brown or black in color.

Psoriasis (sō-rī′-ah-sis) is a chronic, noncontagious dermatitis that is characterized by reddish, raised patches of skin that are covered with whitish scales. It results from excessive cell production that may be triggered by emotional stress or poor health. The affected area may be slightly sore or may itch. Psoriasis occurs most often on the scalp, elbows, knees, buttocks, and lumbar areas.

CHAPTER SUMMARY

1. The integumentary system is composed of the skin and its epidermal derivatives.
2. The skin is also called the cutaneous membrane.

Functions of the Skin

1. Protection
2. Excretion
3. Temperature regulation
4. Sensory perception
5. Synthesis of vitamin D

Structure of the Skin

1. The skin is composed of an outer epidermis that covers the inner dermis.
2. The epidermis consists of stratified squamous epithelium, which lacks blood vessels and nerves.
3. New epithelial cells are constantly formed by the basal layer. As the cells migrate toward the surface, they become keratinized, die, and finally are sloughed off.
4. The dermis consists of fibrous connective tissue that contains both collagen and elastic fibers.
5. The dermis contains blood vessels, nerves, and sensory receptors.
6. The subcutaneous layer attaches the skin to underlying tissues and organs.
7. The subcutaneous layer consists of loose connective tissue and adipose tissue. It contains blood vessels and nerves.

Skin Color

1. The color of the skin is inherited and results from the presence of three pigments: hemoglobin in dermal blood vessels, carotene in the epidermis and dermis, and melanin in the epidermis.
2. Melanin is a brown-black pigment produced by melanocytes in the deeper layers of the epidermis and is incorporated into the epidermal cells.
3. Melanin protects the body from ultraviolet radiation.

Accessory Structures

1. Accessory structures are formed from the epidermis.
2. Hair consists of keratinized epidermal cells that are formed at the base of a hair follicle.
3. An arrector pili muscle is attached to the side of each hair follicle at one end and to the papillary layer of the dermis at the other end. Its contraction pulls the hair into a more erect position.
4. Hair occurs over most of the body.
5. Nails protect the dorsal surface of the ends of fingers and toes.
6. Nails are formed of layers of heavily keratinized epidermal cells.
7. Glands associated with the skin are the sebaceous, sudoriferous, and ceruminous glands.
8. Sebaceous glands produce sebum, an oily secretion that is emptied into hair follicles.
9. Sudoriferous glands are sweat glands. Apocrine sweat glands secrete a relatively thick perspiration into hair follicles. Eccrine sweat glands secrete a watery perspiration that is carried directly to the surface of the skin.
10. Ceruminous glands are located in the external auditory canal and secrete a waxy substance called cerumen.

Temperature Regulation

1. Normal human body temperature is 37°C (98.6°F).
2. Body heat is produced as a by-product of cellular respiration.
3. When heat loss is excessive, blood vessels in the dermis are constricted to reduce heat loss and arrector pili muscles contract. Under extreme heat loss, spontaneous weak muscle contractions (shivering) produce additional heat.
4. When heat production is excessive, blood vessels in the dermis dilate to increase heat loss and perspiration is produced. The evaporation of perspiration increases heat loss.

Aging of the Skin

1. After 50 years of age, wrinkles and sagging skin become noticeable.
2. The effects of aging are caused by a breakdown of collagen and elastic fibers, a decrease in sebum production, a decrease in melanin production, and a decrease in subcutaneous fat.

Disorders of the Skin

1. Infectious disorders of the skin include acne, athlete's foot, boils, fever blisters, impetigo, and warts.
2. Noninfectious disorders of the skin include alopecia, bedsores, burns, calluses and corns, cancers, dandruff, eczema, hives, moles, and psoriasis.

BUILDING YOUR VOCABULARY

1. **Selected New Terms**
   arrector pili muscle, p. 83
   carotene, p. 82
   cerumen, p. 85
   hair follicle, p. 82
   hemoglobin, p. 82
   lunula, p. 83
   melanocyte, p. 82
   sebum, p. 84
2. **Related Clinical Terms**
   birthmark A tumor or localized discoloration of the skin that is visible at birth.
   cyst (sist) A fluid-filled sac in epithelial tissue.
   dermatology (der-mah-tol′-ō-jē) Medical specialty that treats diseases of the skin.
   melanoma (mel-a-nō′-mah) Cancer of melanocytes.
   pustule (pus′-tūl) A localized, raised, pus-filled area of skin.

STUDY ACTIVITIES

1. Complete the **Chapter 5 Study Guide,** which begins on page 415.
2. Complete the **Learning Objectives** listed on the first page of this chapter.

CHECK YOUR UNDERSTANDING

(Answers are located in appendix B.)

1. New epidermal cells are formed by the stratum _____ .
2. Resistance to abrasion and waterproofing of the epidermis are due to the presence of _____ .
3. The strength and elasticity of the skin are due to protein fibers within the _____ .
4. The skin is attached to underlying tissues and organs by the _____ layer.
5. Epidermal cells are nourished by blood vessels located in dermal _____ .
6. The _____ glands produce an oily secretion that keeps the hair and skin moist, soft, and pliable.
7. Watery perspiration is produced by _____ sweat glands.
8. Goose bumps are produced when _____ muscles contract.
9. Constriction of dermal blood vessels _____ heat loss.
10. As the skin ages, a breakdown of collagen and elastic fibers leads to the formation of _____ .
11. _____ is a common skin disorder caused by a fungus.
12. _____ and _____ are thickened areas of epidermis resulting from chronic pressure.

C H A P T E R

The Skeletal System

6

Chapter Preview & Learning Objectives

SELECTED KEY TERMS

Articulation (articul = joint) A joint formed between two bones.
Articular cartilage Cartilage covering the ends of bones forming a joint.
Compact bone Dense bone tissue formed of numerous osteons.
Diaphysis (dia = through, apart; physis = to grow) The shaft of a long bone.
Endochondral ossification (endo = inside; chondr = cartilage; oss = bone) The formation of bone tissue by the replacement of cartilage.
Epiphysis (epi = upon) The enlarged ends of a long bone.
Intramembranous ossification (intra = inside) The formation of bone tissue within a fibrous membrane.
Ligament A band or cord of fibrous connective tissue that joins bones together at movable joints.
Medullary cavity (medulla = marrow) The cavity within the shaft of a long bone that is filled with yellow marrow.
Osteoblast (osteo = bone; blast = bud) A bone cell that deposits matrix.
Osteoclast (clast = break) A bone cell that breaks down bone matrix.
Osteocyte (cyt = cell) A bone cell occupying a lacuna.
Paranasal sinus (para = beside) An air-filled cavity in a bone located near the nasal cavity.
Periosteum (peri = around; os = bone) The fibrous membrane that covers bones.
Spongy bone Bone tissue that contains numerous spaces containing red marrow.

The skeletal system serves as the supporting framework of the body, and it performs several other important functions as well. The body shape, mechanisms of movement, and erect posture observed in humans would be impossible without the skeletal system. Two very strong tissues, bones and cartilages, compose the skeletal system.

Functions of the Skeletal System

The skeletal system performs five major functions:

1. **Support.** The skeleton provides support for the entire body.
2. **Protection.** The arrangement of bones in the skeleton provides protection for many internal organs. The thoracic cage provides protection for the internal thoracic organs including the heart and lungs; the cranial bones form a protective case around the brain, ears, and all but the anterior portion of the eyes; vertebrae protect the spinal cord; and the pelvic girdle protects reproductive, urinary, and digestive organs.
3. **Attachment sites for skeletal muscles.** Skeletal muscles are attached to bones and span across joints between bones. Bones function as levers, enabling movement at joints when skeletal muscles contract.
4. **Blood cell production.** The red marrow in spongy bone forms red blood cells, white blood cells, and platelets.
5. **Mineral storage.** The matrix of bones serves as a storage area for large amounts of calcium phosphate, which may be removed for use in other parts of the body when needed.

Bone Structure

Each bone is an organ composed of a number of tissues. Bone tissue forms the bulk of each bone and consists of both living cells and a nonliving matrix formed primarily of calcium salts. Other tissues include cartilage, blood, nerve, and fibrous connective tissue.

Structure of a Long Bone

The femur, the bone of the thigh, will be used as an example in considering the structure of a long bone. Refer to figure 6.1 as you study the following section.

At each end of the bone, there is an enlarged portion called an **epiphysis** (ē-pif′-e-sis). The epiphyses (plural) articulate with adjacent bones to form joints. The articular surface of each epiphysis is covered by an **articular** (hyaline) **cartilage** that protects and cushions the end of the bone and provides a smooth surface for movement. The long shaft of bone that extends between the two epiphyses is the **diaphysis** (dī-af′-e-sis). Each epiphysis is joined to the diaphysis by an **epiphyseal disk,** in immature bones, or by an *epiphyseal line,* a line of fusion, in mature bones.

Except for the region covered by articular cartilages, the entire bone is covered by the **periosteum** (per-ē-os′-tē-um), a fibrous connective tissue membrane that is firmly attached to the underlying bone.

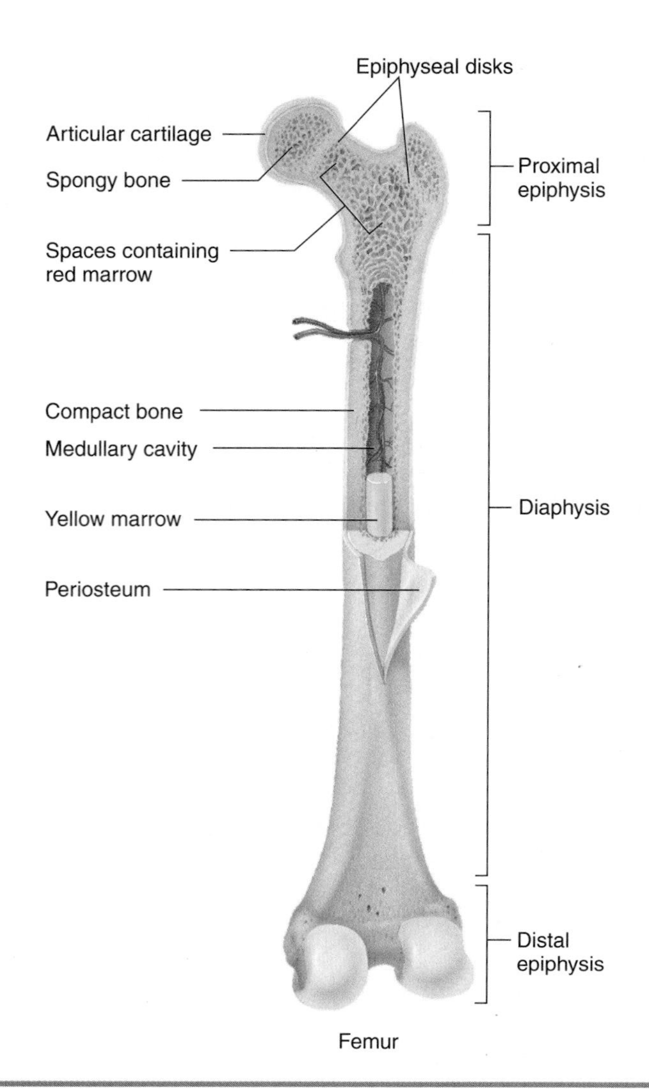

Figure 6.1 Major parts of a long bone.

The periosteum provides protection and also is involved in the formation and repair of bone tissue. Tiny blood vessels from the periosteum help to nourish the bone tissue.

The interior structure of a long bone is revealed by a longitudinal section. **Spongy bone** forms the interior of small bones, skull bones, and the epiphyses in long bones. Spongy bone reduces the weight of a bone without reducing its supportive strength. The numerous spaces in spongy bone are filled with *red marrow.*

Compact bone forms the wall of the diaphysis and a thin surface layer over the epiphyses. As the name implies, compact bone is formed of tightly packed bone tissue that lacks the spaces found in spongy bone. Compact bone is very strong, and it provides the supportive strength of long bones. The cavity that extends the length of the diaphysis is the **medullary cavity.** It is lined by a thin membrane, the **endosteum** (en-dos′-tē-um), and it is filled with fatty *yellow marrow.*

Microscopic Structure

As discussed in chapter 4, the bone matrix is deposited by bone cells that come to occupy the **lacunae** (lah-kū′-nē). In compact bone, the lacunae and bone cells are arranged in concentric rings around **osteonic** (os-tē-on′-ik) **canals,** which causes the bone matrix to be formed in concentric layers, or **lamellae** (lah-mel′-lē). An osteonic canal plus its concentric rings of lamellae and osteocytes form an **osteon** (os′-tē-on), the structural unit of compact bone.

Blood vessels and nerves enter a bone through a **foramen** (fō-rā′-men), a channel entering or passing through a bone. The blood vessels form branches that pass through communicating canals and enter the osteonic canals to supply nutrients to the bone cells. Materials are exchanged between bone cells and the blood vessels via numerous **canaliculi** (kan-ah-lik′-ū-lī), tiny channels that extend from the lacunae (figure 6.2).

The bony plates of spongy bone lack osteons, so bone cells receive nutrients by diffusion of materials through canaliculi from blood vessels in the red marrow surrounding the bony plates.

> ***What are the general functions of the skeletal system?***
> ***What is the basic structure of a long bone?***

Bone Formation

The process of bone formation is called **ossification** (os-i-fi-kā′-shun). It begins during the sixth or seventh week of embryonic life. Bones are formed by the replacement of existing connective tissues with bone tissue (figure 6.3). There are two types of bone formation: intramembranous ossification and endochondral ossification. Table 6.1 summarizes these processes.

In both types of ossification, some primitive connective tissue cells are changed to become bone-forming cells called **osteoblasts** (os′-tē-ō-blasts). Osteoblasts deposit bone matrix around themselves and soon become imprisoned in lacunae. Once this occurs, they are called **osteocytes,** or bone cells.

Intramembranous Ossification

Most skull bones are formed by **intramembranous ossification.** Connective tissue membranes form early in embryonic development at sites of future intramembranous bones. Later, some connective tissue cells become osteoblasts and deposit spongy bone within the membranes starting in the center of the bone.

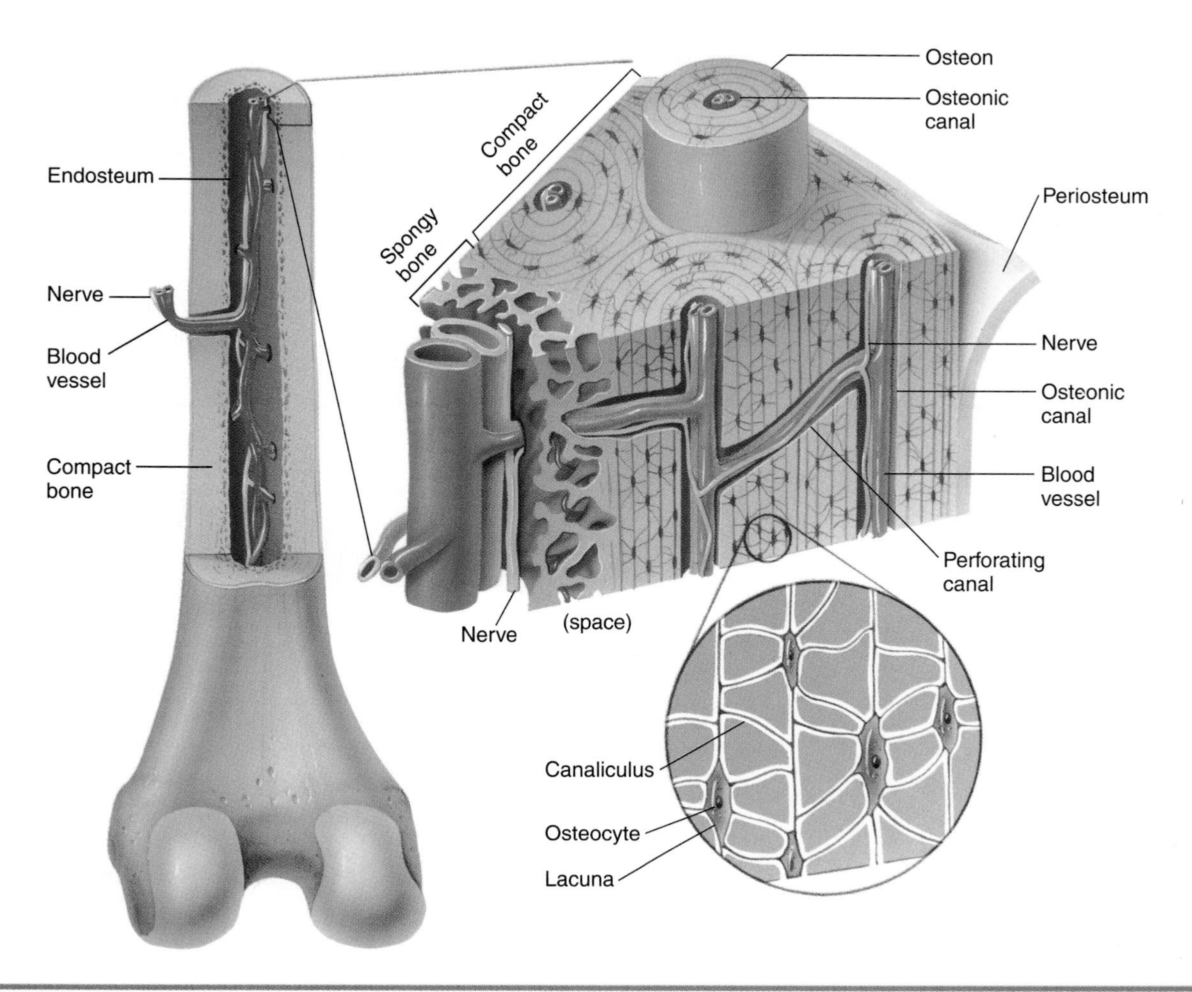

Figure 6.2 Compact bone is composed of osteons cemented together.

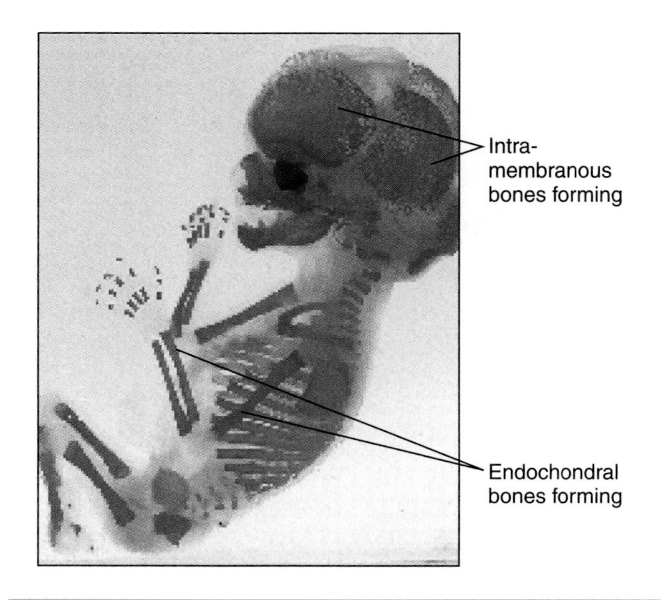

Figure 6.3 Note the stained, developing bones of this fourteen-week fetus.

Osteoblasts from the covering membrane (periosteum) deposit a layer of compact bone over the spongy bone.

Some bone tissue must be removed and re-formed in order to produce the correct shape of the bone as it develops and grows. Cells that remove bone matrix are called **osteoclasts.** The opposing actions of osteoblasts and osteoclasts ultimately produce the shape of the mature bone.

Endochondral Ossification

Most bones of the body are formed by **endochondral** (en-dō-kon′-drul) **ossification.** Future endochondral bones are preformed in hyaline cartilage early in embryonic development. Figure 6.4 illustrates the ossification of a long bone.

In long bones, a *primary ossification center* forms in the middle of the bone, and osteoblasts from the periosteum form a collar of compact bone around the ossification center. Cartilage in the primary ossification

Table 6.1 Comparison of Intramembranous and Endochondral Ossification

Intramembranous	Endochondral
1. Membranes of embryonic connective tissue form at sites of future bones.	1. Bone is preformed in hyaline cartilage.
2. Some connective tissue cells become osteoblasts, which deposit spongy bone within the membrane.	2. Cartilage is calcified, and osteoblasts derived from the periosteum form spongy bone, which replaces cartilage in ossification centers.
3. Osteoblasts from the enclosing membrane, now called the periosteum, deposit a layer of compact bone over the spongy bone.	3. Osteoblasts of periosteum form collar of compact bone that thickens and grows toward each end of the bone.

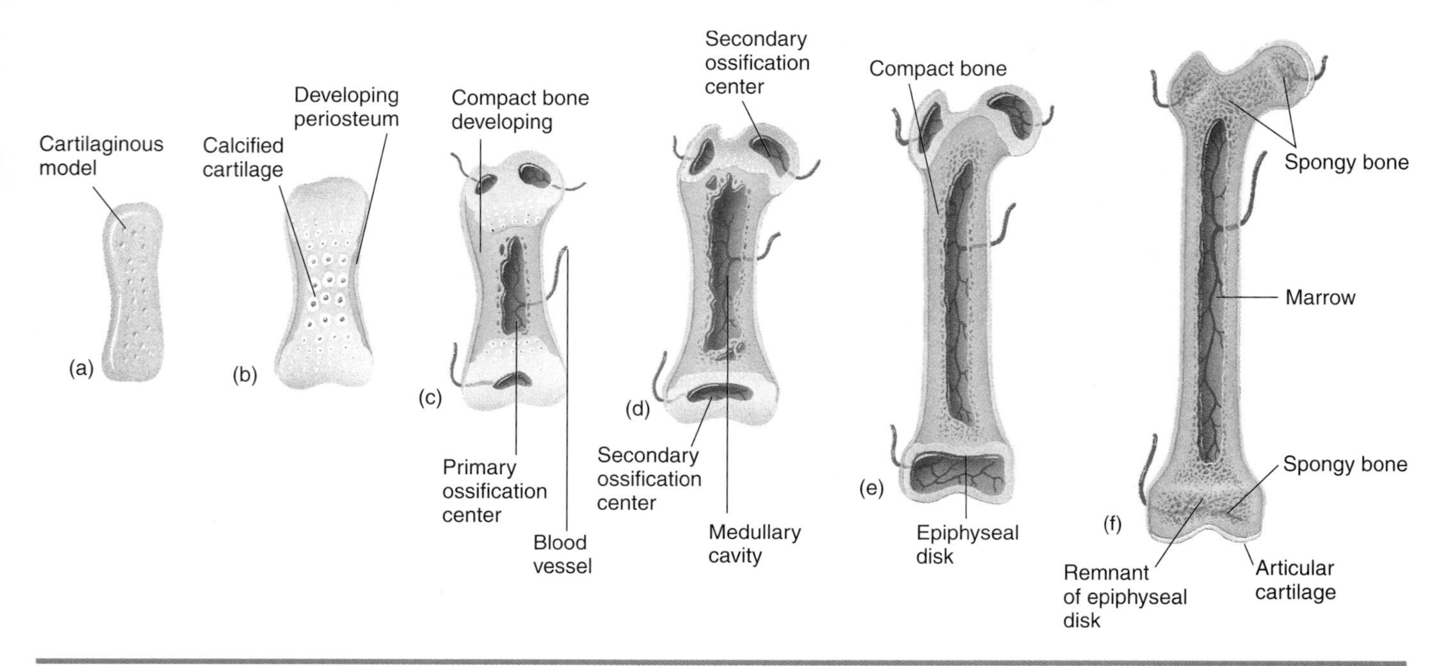

Figure 6.4 Major stages (*a–f*) in the development of an endochondral bone. (Bones are not shown to scale.)

center calcifies, and the chondrocytes die. Blood vessels and nerves penetrate into the primary ossification center carrying along osteoblasts from the periosteum, which form spongy bone. A *secondary ossification center* forms in each epiphysis, and osteoclasts begin to remove spongy bone to form the medullary cavity. The bone continues to grow as ossification progresses. In time, each epiphysis becomes separated from the diaphysis by a disk of cartilage called the *epiphyseal disk.*

Subsequent growth in diameter results from continued formation of compact bone by osteoblasts from the periosteum. Growth in length occurs as bone replaces cartilage on the diaphysis side of each epiphyseal disk while new cartilage is formed on the epiphysis side. The opposing actions of osteoblasts and osteoclasts continually reshape the bone as it grows.

Growth usually continues until about age 25, when the epiphyseal disks are completely replaced by bone tissue. After this, growth in the length of a bone is not possible. The visible lines of fusion between the epiphyses and the diaphysis are called **epiphyseal lines.**

Homeostasis of Bone

Bones are dynamic, living organs, and they are continually restructured throughout life. This occurs by the removal of calcium salts by osteoclasts and by the deposition of new bone matrix by osteoblasts. Physical activity causes the density and volume of bones to be maintained or increased, while inactivity results in a reduction in bone density and volume.

Calcium salts may be removed from bones to meet body needs when dietary calcium is inadequate. When dietary calcium salts return to a sufficient level, they are used to form new bone matrix.

Children have a relatively large amount of protein fibers in their bone matrix, which makes their bones somewhat flexible. But as people age, the

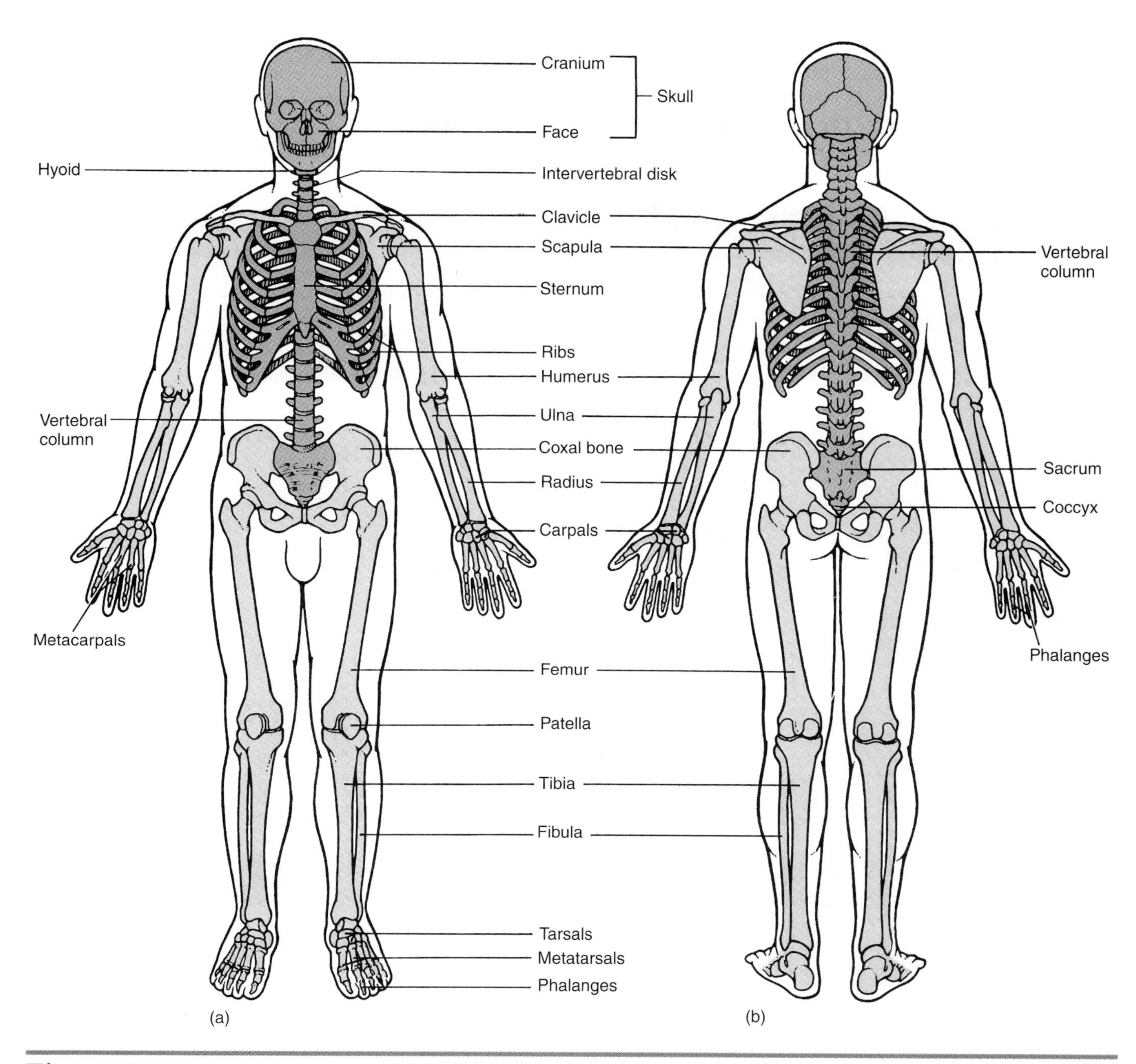

Figure 6.5 The major bones of the body: (*a*) anterior view; (*b*) posterior view. The axial and appendicular portions of the skeleton are distinguished by color.

amount of protein gradually decreases. This trend causes older people to have brittle bones that are prone to fractures. Older persons may also experience a gradual loss of calcium salts (osteoporosis), which reduces the strength of the bones.

How do intramembranous ossification and endochondral ossification differ?
How does physical activity affect the homeostasis of bones?

Divisions of the Skeleton

The human adult skeleton is composed of two distinct divisions: the axial skeleton and the appendicular skeleton. The **axial** (ak′-sē-al) **skeleton** consists of the bones along the longitudinal axis of the body that support the head, neck, and trunk. The **appendicular** (ap-en-dik′-ū-lar) **skeleton** consists of the bones of the upper extremities and pectoral girdle and of the lower extremities and pelvic girdle. Figure 6.5 illustrates the axial and appendicular divisions of the skeleton as well as the major bones of each.

A study of the skeleton includes the various surface features of bones, such as projections, depressions, ridges, grooves, and holes. Specific names are given to each type of surface feature. The names of the major surface features are listed in table 6.2 for easy reference as you study the bones of the skeleton.

The Axial Skeleton

The major components of the axial skeleton are the skull, vertebral column, and thoracic cage. Bones of the axial skeleton are listed in table 6.3.

Table 6.2 Surface Features of Bones

Feature	Description
Processes Forming Joints	
Condyle	A rounded or knucklelike process
Head	An enlarged, rounded end of a bone supported by a constricted neck
Facet	A smooth, nearly flat articulating surface
Processes for Attachment of Ligaments and Tendons	
Crest	A prominent ridge or border
Epicondyle	A prominence above a condyle
Trochanter	A very large process found only on the femur
Tubercle	A small, rounded process
Tuberosity	A large, roughened process
Depressions and Openings	
Alveolus	A deep pit or socket
Canal	A tubelike passageway into a bone
Foramen	An opening or passageway into or through a bone through which blood vessels or nerves pass
Fossa	A small depression
Groove	A furrowlike depression
Sinus	An air-filled cavity within a bone

Table 6.3 Bones of the Axial Skeleton

Region and Bones	Number of Bones	Region and Bones	Number of Bones
Skull		***Hyoid***	1
Cranium			1
Ethmoid	1	**Vertebral column**	
Frontal	1	Cervical	7
Occipital	1	Thoracic	12
Parietal	2	Lumbar	5
Sphenoid	1	Sacrum	1
Temporal	2	Coccyx	1
Ear ossicles*	6		26
	14	**Thoracic cage**	
Face		Ribs	24
Inferior nasal concha	2	Sternum	1
Lacrimal	2		25
Mandible	1		Total = 80
Maxilla	2		
Nasal	2		
Palatine	2		
Vomer	1		
Zygomatic	2		
	14		

*These tiny bones of the middle ear are involved in hearing and are studied in chapter 9.

Skull

The **skull** is subdivided into the *cranium,* which is formed of eight fused bones encasing the brain, and the *facial bones,* which consist of 13 fused bones and the movable lower jaw. The fused bones are joined by immovable joints called *sutures* (sū′-churs) because they resemble stitches. Several bones in the skull contain air-filled spaces called **paranasal sinuses** that reduce the weight of the skull (table 6.4). The bones of the skull are shown in figures 6.6 to 6.10. Locate the bones on these figures as you study this section.

Table 6.4 Skull Bones Containing Paranasal Sinuses

Bone	Number
Ethmoid	2 groups
Frontal	2
Maxillae	1 each
Sphenoid	2

Cranium

The cranium is formed of the following bones: one frontal bone, two parietal bones, one sphenoid bone, two temporal bones, one occipital bone, and one ethmoid bone (figure 6.6).

The **frontal bone** forms the anterior part of the cranium, including the superior portion of the eye orbits (sockets), the forehead, and the roof of the nasal cavity. There are two large sinuses in the frontal bone, one located over each eye (figure 6.7).

The two **parietal** (pah-rī′-e-tal) **bones** form the sides and roof of the cranium. They are joined at the midline by the *sagittal suture* and to the frontal bone by the *coronal sutures* (figure 6.8).

The **occipital** (ok-sip′-i-tal) **bone** forms the posterior portion and floor of the cranium. It contains a large opening, the *foramen magnum,* through which the brain stem extends to join with the spinal cord. On each side of the foramen magnum are the *occipital condyles* (kon′-dīls), large knucklelike surfaces that articulate with the first vertebra of the vertebral column. The occipital bone is joined to the parietal bones by the *lambdoidal* (lam-doy′-dal) *sutures* (figures 6.9 and 6.10).

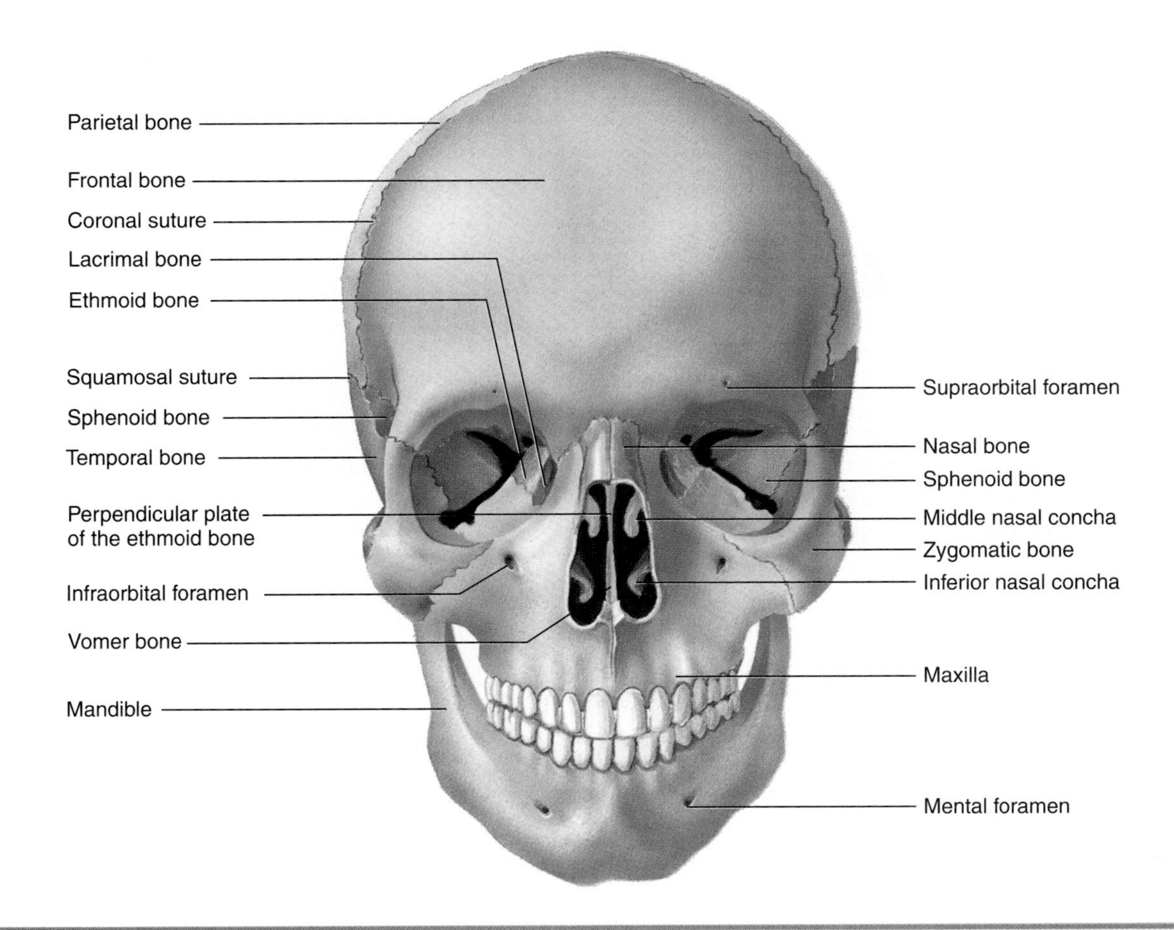

Figure 6.6 An anterior view of the skull.

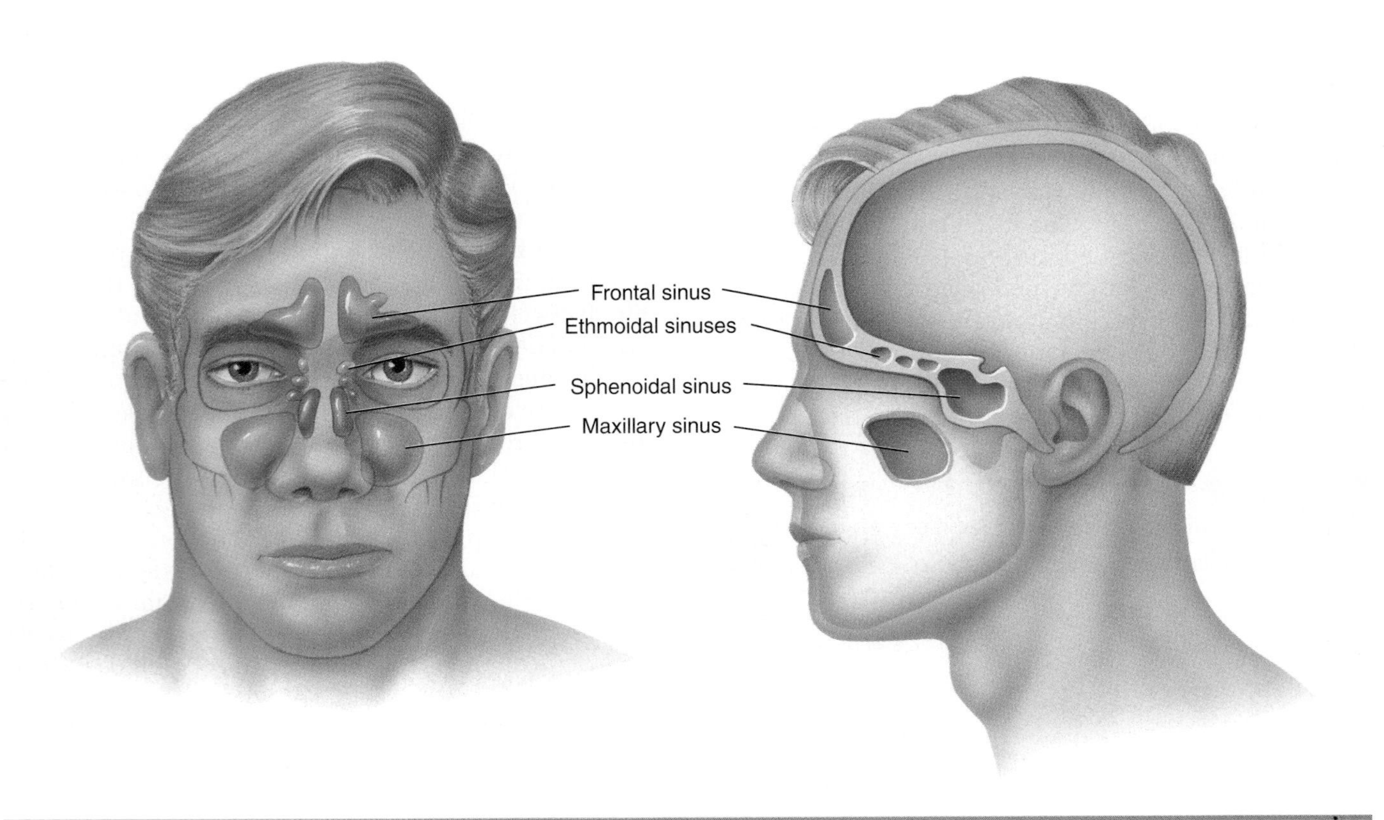

Figure 6.7 The locations of the paranasal sinuses.

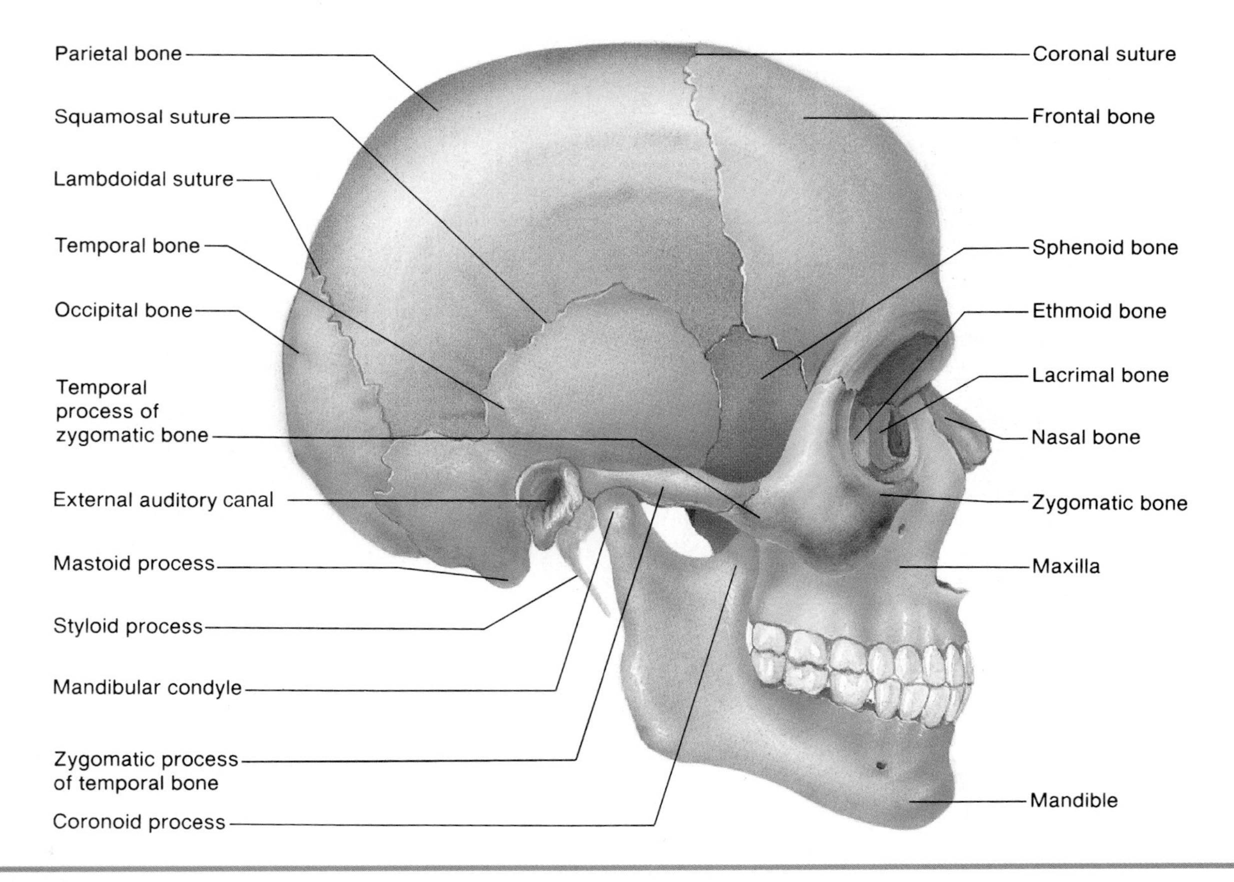

Figure 6.8 The lateral view of the skull.

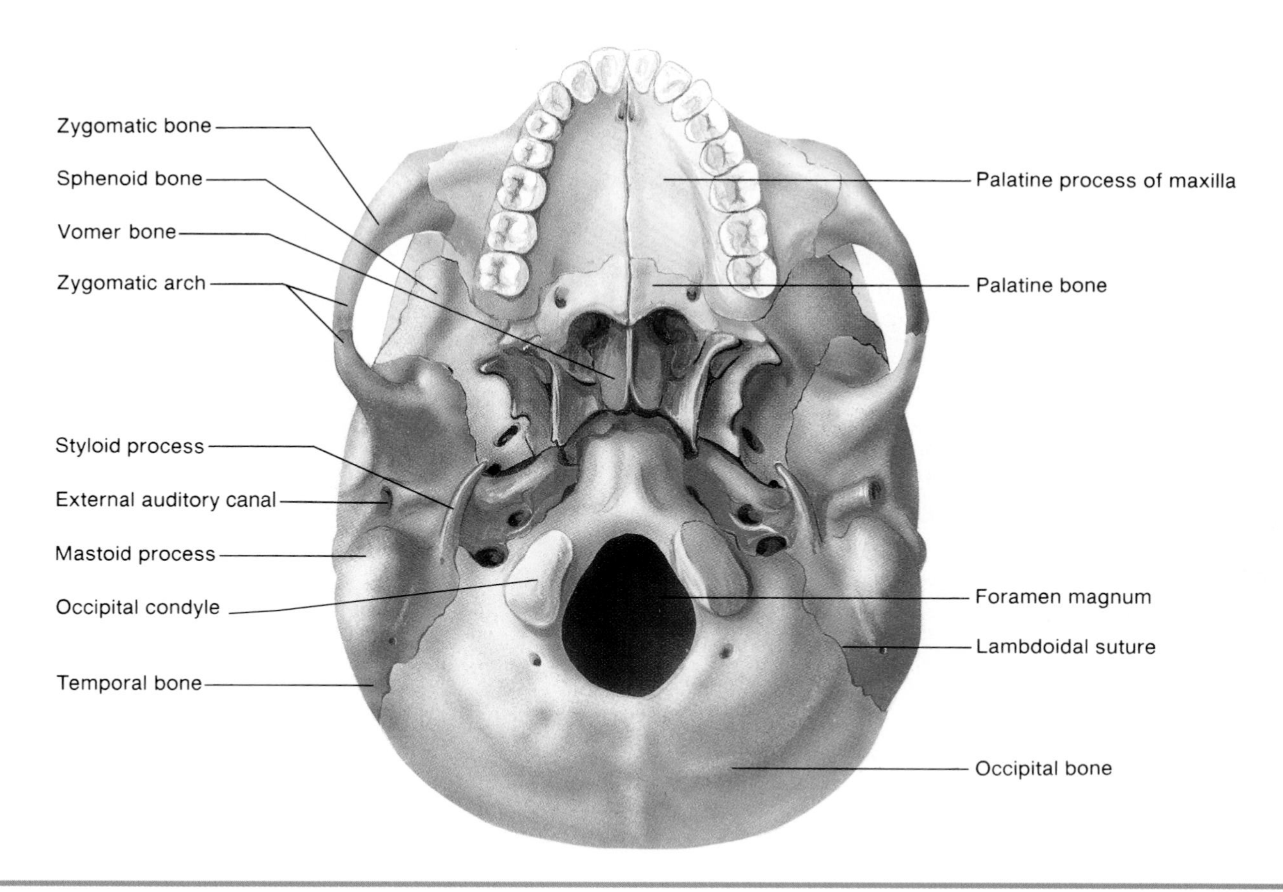

Figure 6.9 An inferior view of the skull.

The **temporal bones** are located inferior to the parietal bones on each side of the cranium. They are joined to the parietal bones by *squamosal* (skwa-mō′-sal) *sutures* and to the occipital bone by the lambdoidal suture. In each temporal bone, an *external auditory canal* leads inward to the eardrum. Just anterior to the auditory canal is the *mandibular fossa,* a depression that receives the mandibular condyle to form the **temporomandibular joint.**

Three processes are located on each temporal bone. The *zygomatic* (zī-gō-mat′-ic) *process* projects anteriorly to join with the zygomatic bone. The *mastoid* (mas′-toyd) *process* is a large, rounded projection that is located inferior to the auditory canal. It contains small spaces called air cells and serves as an attachment site for some neck muscles. The *styloid process* lies just medial to the mastoid process. It is a long, spikelike process to which muscles and ligaments of the tongue and neck are attached.

The **sphenoid** (sfē′-noid) **bone** forms part of the floor of the cranium, the lateral posterior portions of the eye orbits, and the lateral portions of the cranium just anterior to the temporal bones. It articulates with all other cranial bones, so it is the "keystone" of the cranium. On its superior surface at the midline is a saddle-shaped structure called the *sella turcica* (ter′-si-ka), or Turk's saddle. It has a depression that contains the pituitary gland. Two sphenoidal sinuses are located just below the sella turcica.

The **ethmoid** (eth′-moid) **bone** forms the anterior portion of the cranium, including part of the medial surface of each eye orbit and part of the roof of the nasal cavity. The lateral portions contain several air-filled sinuses. The *perpendicular plate* extends downward to form most of the nasal septum, which separates the right and left portions of the nasal cavity. It joins the sphenoid and vomer bones posteriorly and the nasal and frontal bones anteriorly.

The superior and middle **nasal conchae** (kong′-kē) extend from the lateral portions of the ethmoid toward the perpendicular plate. These delicate, scroll-like bones support the mucous membrane and increase the surface area of the nasal wall. The roof of the nasal cavity is formed by the *cribriform plates* of the ethmoid. On the superior surface where these plates join at the midline is a prominent projection called the *crista galli,* or cock's comb. The meninges that envelop the brain are attached to the crista galli.

Facial Bones

The paired bones of the face are the maxillae, palatine bones, zygomatic bones, lacrimal bones, nasal bones, and inferior nasal conchae. The single bones are the vomer and mandible.

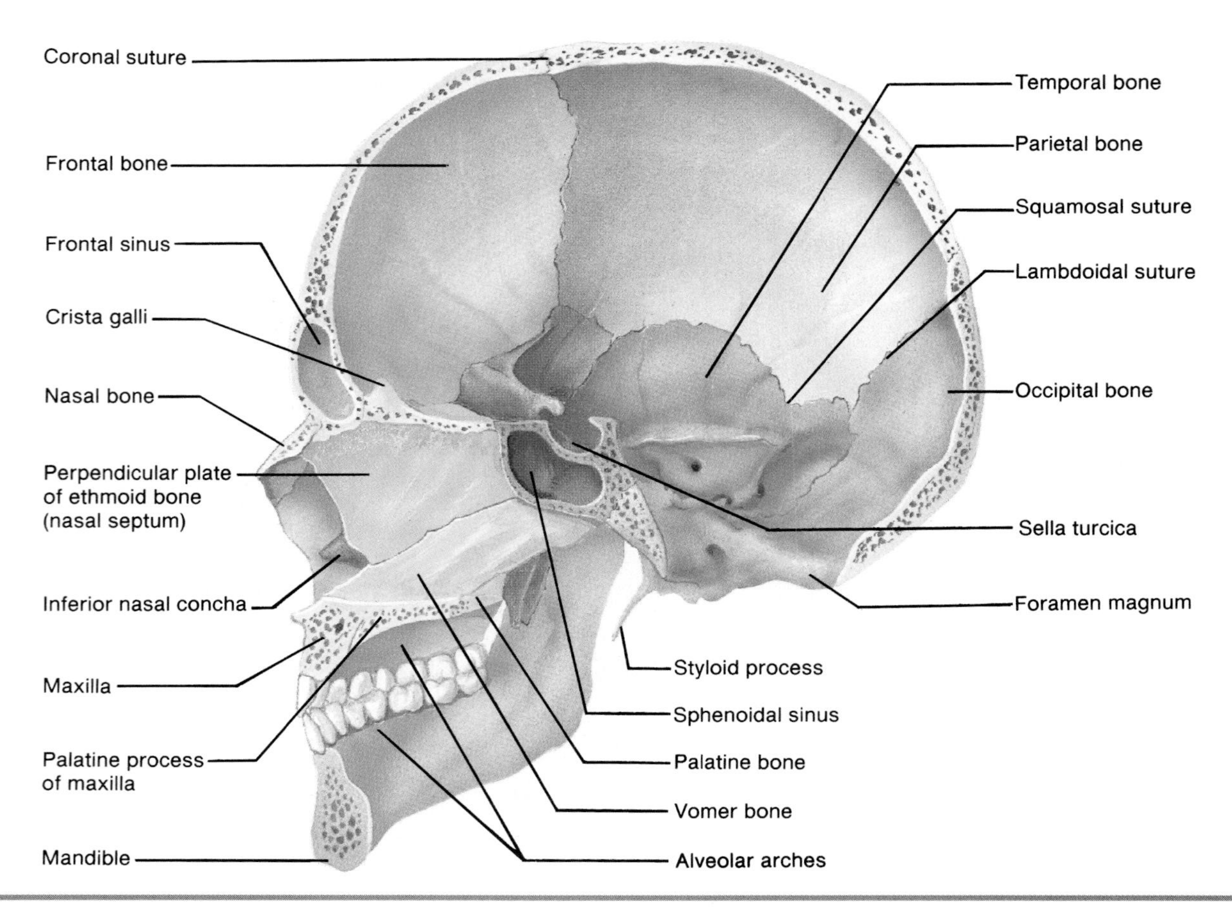

Figure 6.10 A midsagittal section of the skull.

The **maxillae** (mak-sil′-ē) form the upper jaw. Each maxilla is formed separately, but they are joined at the midline during embryonic development. The maxillae articulate with all of the other facial bones except the mandible. The palatine processes of the maxillae form the anterior portion of the hard palate (roof of the mouth and floor of the nasal cavity), part of the lateral walls of the nasal cavity, and the floors of the eye orbits.

A *cleft palate* results when the palatine processes of the maxillae and usually the palatine bones do not join before birth to form the hard palate. A *cleft lip,* a split upper lip, is often associated with a cleft palate. These conditions may be corrected surgically after birth.

Each maxilla possesses a downward projecting, curved ridge of bone that contains the teeth. This ridge is the *alveolar process,* and the sockets containing the teeth are called *alveoli.* The alveolar processes unite at the midline to form the U-shaped maxillary *alveolar,* or *dental, arch.* A large **maxillary sinus** is present in each maxilla just below the orbits.

The **palatine** (pal′-ah-tīn) **bones** are fused at the midline to form the posterior portion of the hard palate. Each bone has a lateral portion that projects upward (superiorly) to form part of a lateral wall of the nasal cavity.

The **zygomatic bones** are the cheek bones. They form the prominences of the cheeks and the floors and lateral walls of the eye orbits. Each zygomatic bone has a posteriorly projecting process, the *temporal process,* that extends to unite with the zygomatic process of the adjacent temporal bone. Together, they form the *zygomatic arch.*

The **lacrimal** (lak′-ri-mal) **bones** are small, thin bones that form part of the medial surfaces of the eye orbits. Each lacrimal bone is located between the ethmoid and maxilla.

The **nasal** (nā′-zal) **bones** are thin bones fused at the midline to form the bridge of the nose.

The **vomer** is a thin, flat bone located on the midline of the nasal cavity. It joins posteriorly with the perpendicular plate of the ethmoid, and these two bones form the nasal septum.

The **inferior nasal conchae** are scroll-like bones attached to the lateral walls of the nasal cavity inferior to the medial nasal conchae of the ethmoid bone. They project medially into the nasal cavity and support the mucous membrane.

Table 6.5 Fontanels in an Infant Skull

Fontanel	Number	Location
Anterior	1	On the midline at the junction of the frontal and parietal bones
Mastoid	2	Superior to the mastoid process at the junction of the occipital, parietal, and temporal bones
Posterior	1	On the midline at the junction of the occipital and parietal bones
Sphenoid	2	Superior to the temporomandibular joint at the junction of the frontal, parietal, sphenoid, and temporal bones

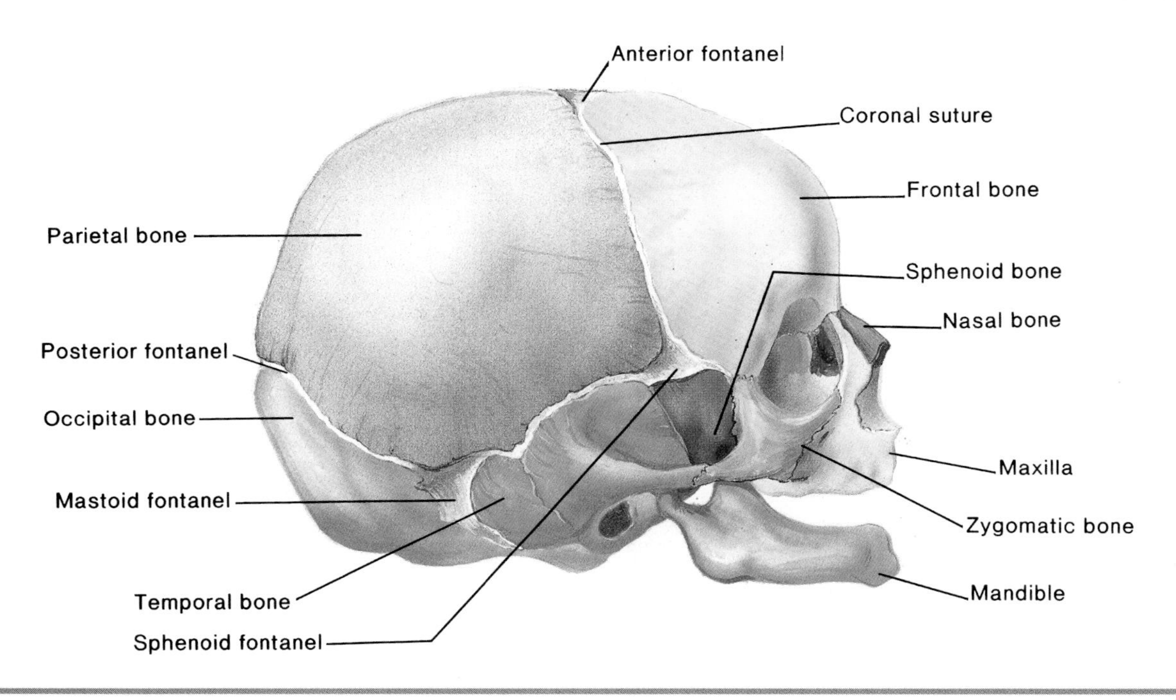

Figure 6.11 A lateral view of a newborn infant's skull. Note the fontanels and the membranes between the cranial bones.

The **mandible** is the lower jawbone, and it is the only movable bone of the skull. It consists of a U-shaped *body* with a superiorly (upward) projecting portion, a *ramus,* extending from each end of the body. The superior portion of the body forms the mandibular *alveolar arch,* which contains the teeth of the lower jaw. The superior part of each ramus is Y-shaped and forms two projections: an anterior *coronoid process* and a posterior *mandibular condyle.* The coronoid process is a site of attachment for muscles used in chewing. The mandibular condyle articulates with the mandibular fossa of the temporal bone to form a *temporomandibular joint.* These joints are sometimes involved in a variety of dental problems associated with an improper bite.

Hyoid Bone

The **hyoid** (hī′-oyd) is a small, U-shaped bone located in the anterior portion of the neck, inferior to the mandible. It does not articulate with any bone. Instead, it is suspended from the styloid processes of the temporal bones by ligaments. Muscles of the tongue are attached to the hyoid.

The Infant Skull

The skull of a newborn infant is incompletely developed. The face is relatively small with large eye orbits, and the bones are thin and incompletely ossified. The bones of the cranium are separated by fibrous membranes, and there are six rather large, nonossified areas called **fontanels** (fon′-tah-nels), or soft spots (table 6.5). The frontal bone is formed of two separate parts that fuse later in development. Incomplete ossification of the skull bones and the abundance of fibrous membranes make the skull somewhat flexible and allow partial compression of the skull during birth; they enable growth thereafter. Compare the infant skull in figure 6.11 with the adult skull in figure 6.8.

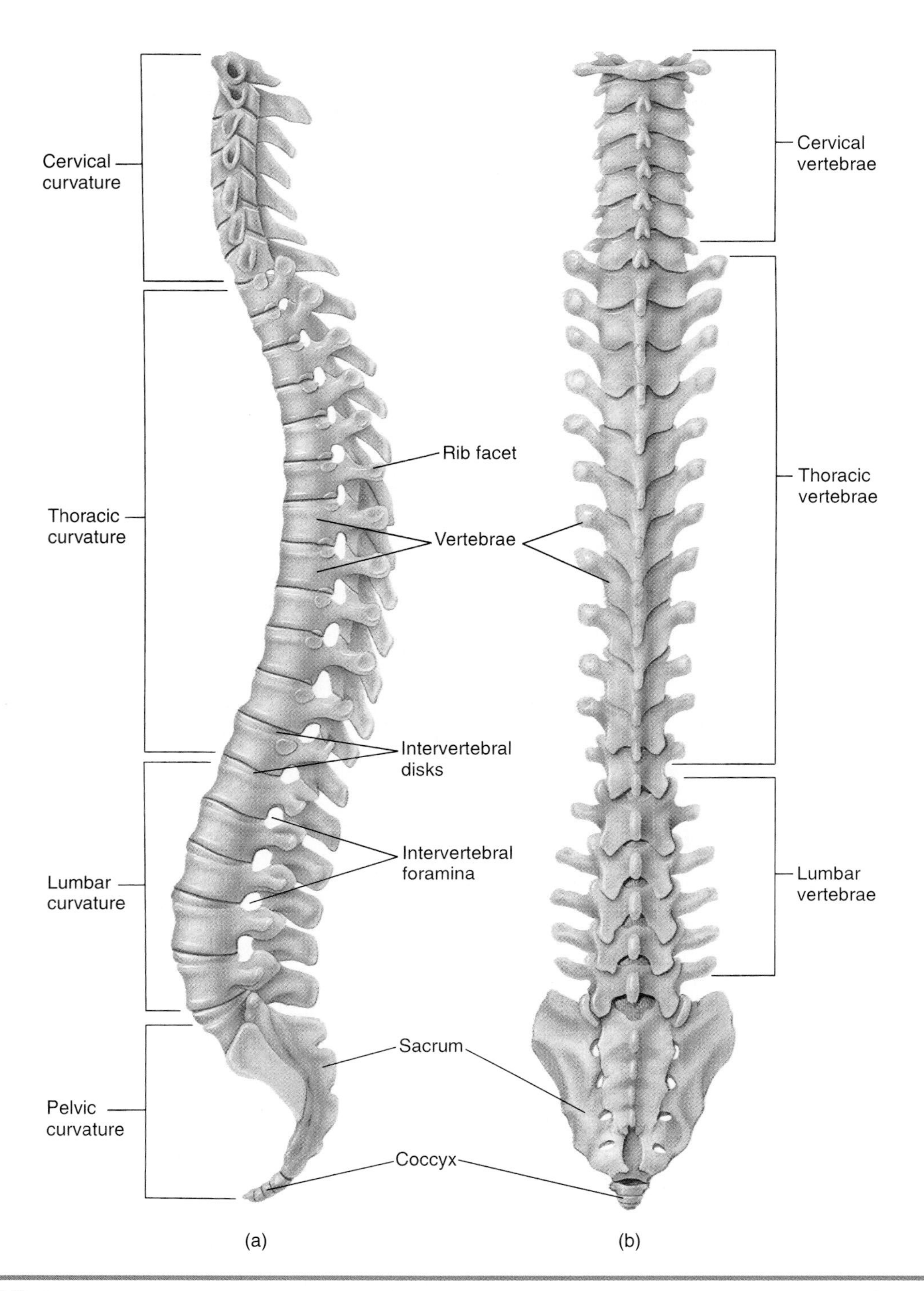

Figure 6.12 The vertebral column consists of 24 movable vertebrae, separated by intervertebral disks, sacrum, and coccyx. (*a*) Lateral view. (*b*) Posterior view.

What bones form the cranium?
What bones form the face?

Vertebral Column

The vertebral column extends from the skull to the pelvis and forms a somewhat flexible, but sturdy, longitudinal support for the trunk. It is formed of 24 slightly movable vertebrae, the sacrum, and the coccyx. The vertebrae are separated from each other by **intervertebral disks** that serve as shock absorbers and allow bending of the spinal column. The upright posture of humans produces four distinct curvatures in the vertebral column. From top to bottom they are the *cervical, thoracic, lumbar,* and *pelvic curvatures.* Locate these curvatures and the basic components of the vertebral column in figure 6.12.

Structure of a Vertebra

Vertebrae are divided into three groups: cervical, thoracic, and lumbar vertebrae. Although each type has a distinctive anatomy, they have many features in common (figure 6.13).

The anterior, drum-shaped mass is the *body,* which serves as the major load-bearing portion of a vertebra. A bony *neural arch* surrounds the large *vertebral foramen* through which the spinal cord passes. A *spinous process* projects posteriorly and *transverse processes* project laterally from each vertebra.

A pair of *superior articulating facets* (fa′-sets) of one vertebra articulates with the *inferior articulating facets* of the adjacent vertebra above it. When joined by ligaments, the vertebrae form a flexible bony cylinder that protects the spinal cord.

Small *intervertebral foramina* occur between adjacent vertebrae. They serve as lateral passageways for spinal nerves that exit the spinal cord (see figure 6.12).

Cervical Vertebrae

The first seven vertebrae are the **cervical** (ser′-vi-kul) **vertebrae** that support the neck. They are unique in having a *transverse foramen* in each transverse process. It serves as a passageway for blood vessels and nerves.

The first two cervical vertebrae are distinctly different than the rest. The first vertebra is the **atlas,** whose superior facets articulate with the condyles of the occipital bone and support the head. The second vertebra is the **axis,** which has a prominent *odontoid* (ō-don′-toyd) *process* (*dens*) that projects upwards from the vertebral body, providing a pivot point for the atlas. When the head is turned, the atlas rotates on the axis (figure 6.14).

Thoracic Vertebrae

The 12 *thoracic vertebrae* are larger than the cervical vertebrae, and their spinous processes are longer and slope downward. The ribs articulate with *facets* on the transverse processes and bodies of thoracic vertebrae.

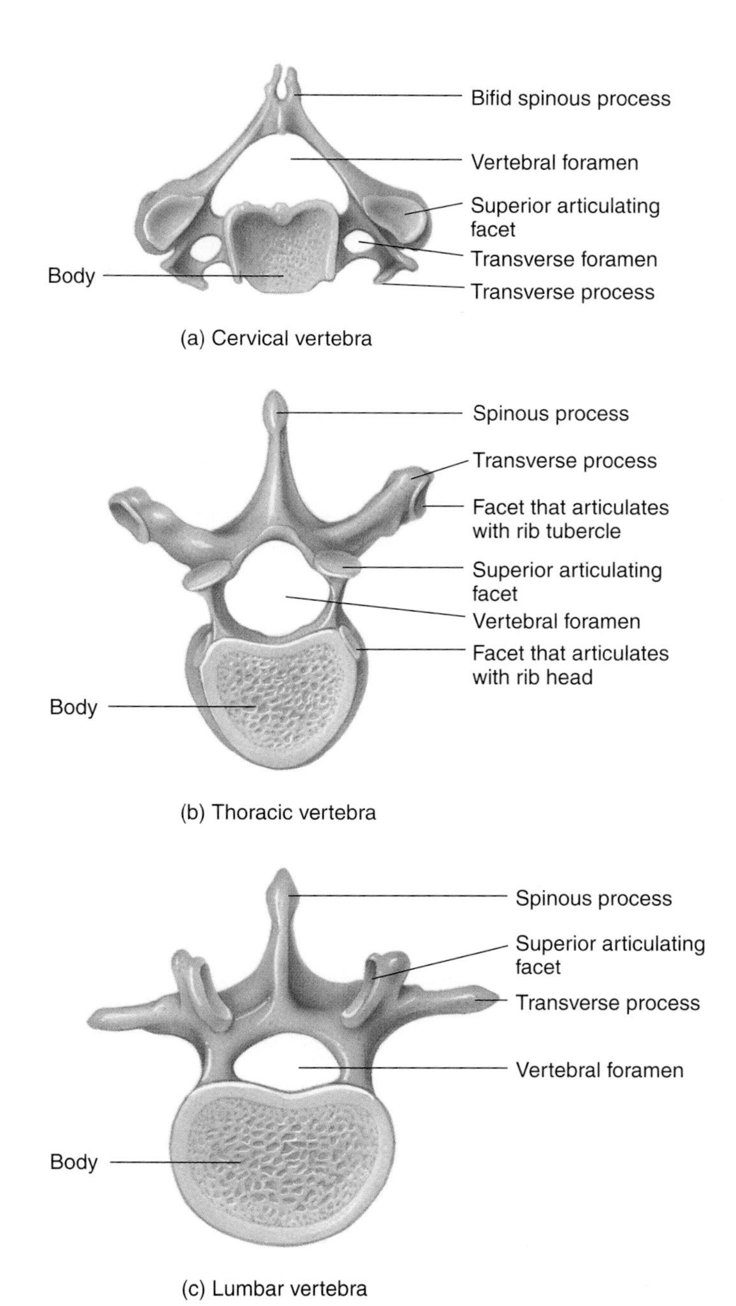

Figure 6.13 Superior view of (*a*) a cervical vertebra, (*b*) a thoracic vertebra, and (*c*) a lumbar vertebra.

Lumbar Vertebrae

The five **lumbar vertebrae** have heavy, thick bodies to support the greater stress and weight that is placed on this region of the vertebral column. The spinous processes are blunt and they provide a large surface area for the attachment of heavy back muscles.

Sacrum

The **sacrum** (sā-k′rum) is composed of five fused vertebrae (figure 6.15). It articulates with the fifth lumbar vertebra and forms the posterior wall of the pelvic girdle. The spinous processes of the fused vertebrae are reduced to a series of *tubercles* on the posterior midline. On either side of the tubercles are the *posterior sacral foramina,* passageways for blood vessels and nerves. Foramina on the anterior surface serve a similar function. The *sacral canal* is formed by the fused neural arches, and it continues to an inferior opening, the *sacral hiatus,* proximal to the coccyx.

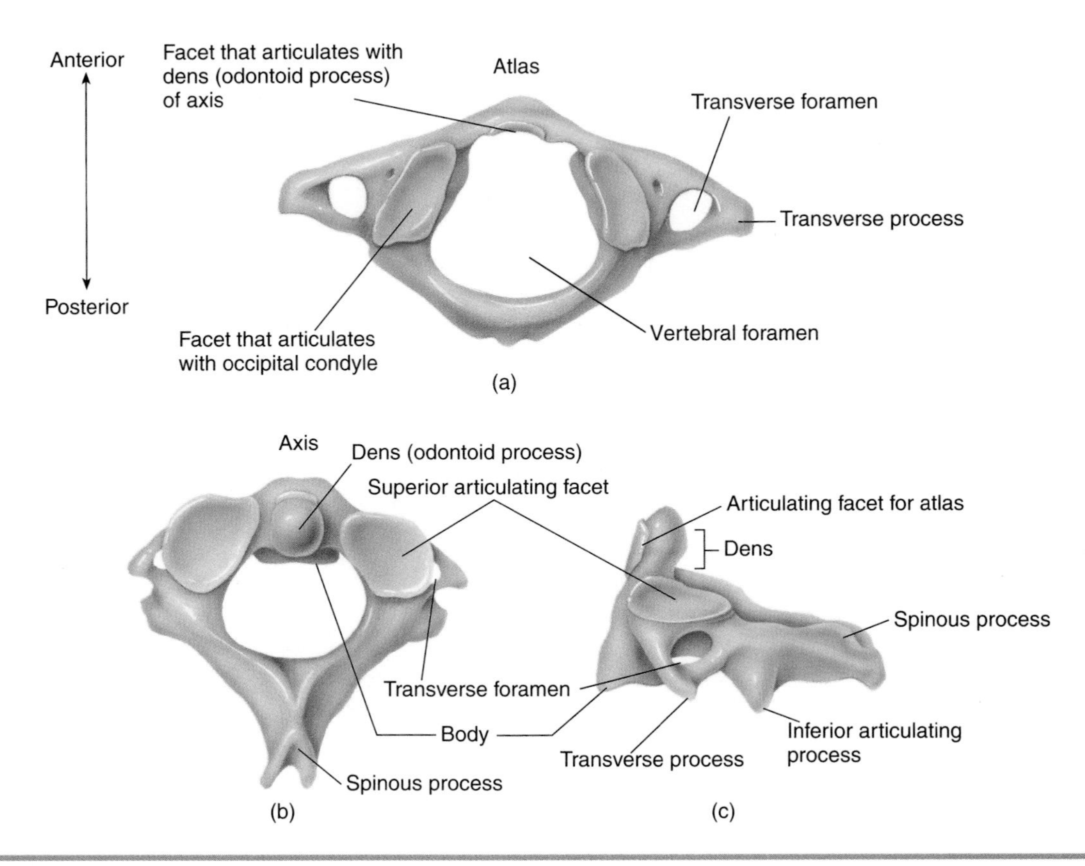

Figure 6.14 The structures of the (*a*) atlas and (*b*) axis function together to allow movement of the head. (*c*) Lateral view of the axis.

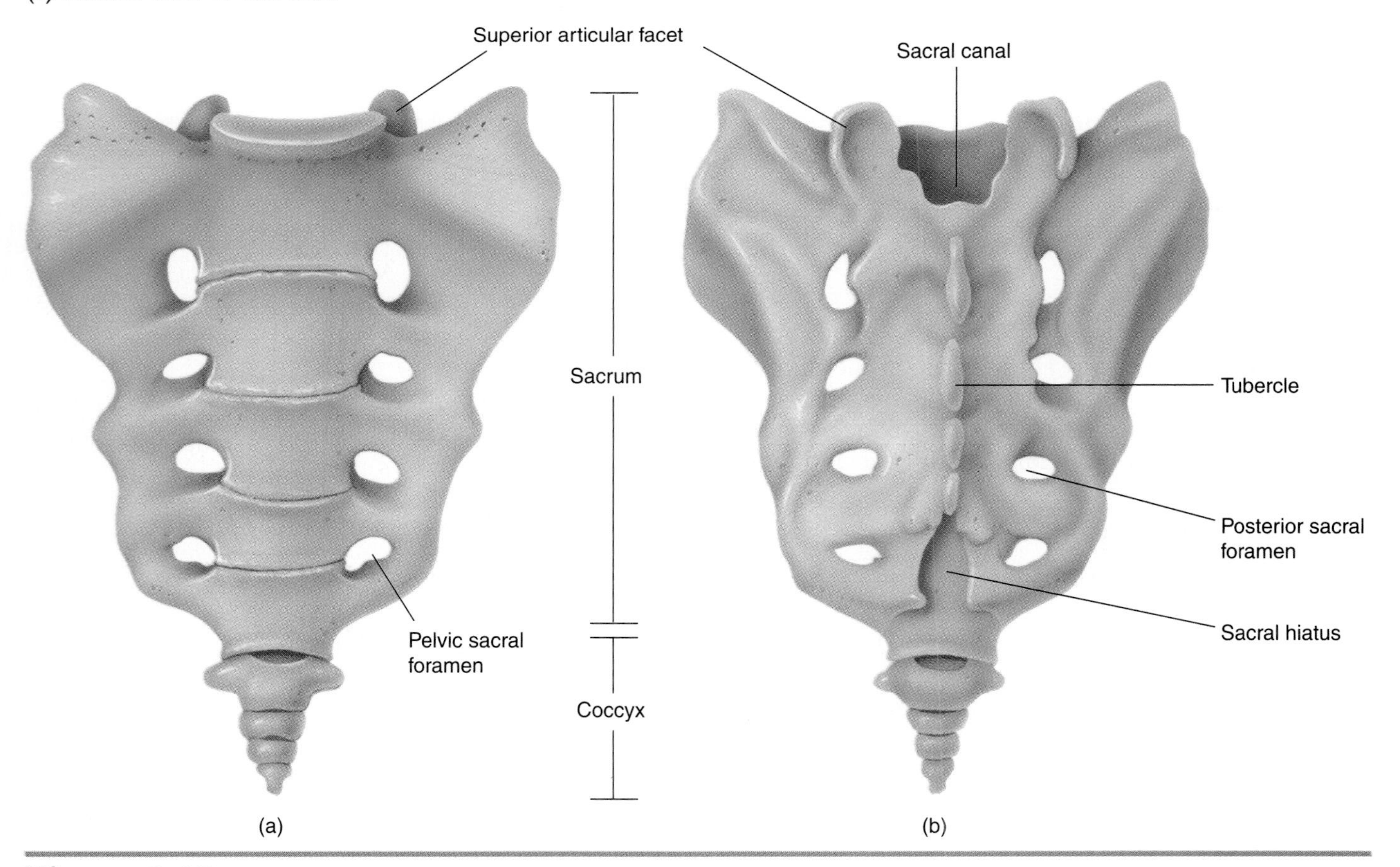

Figure 6.15 (*a*) Anterior view and (*b*) posterior view of the sacrum and coccyx.

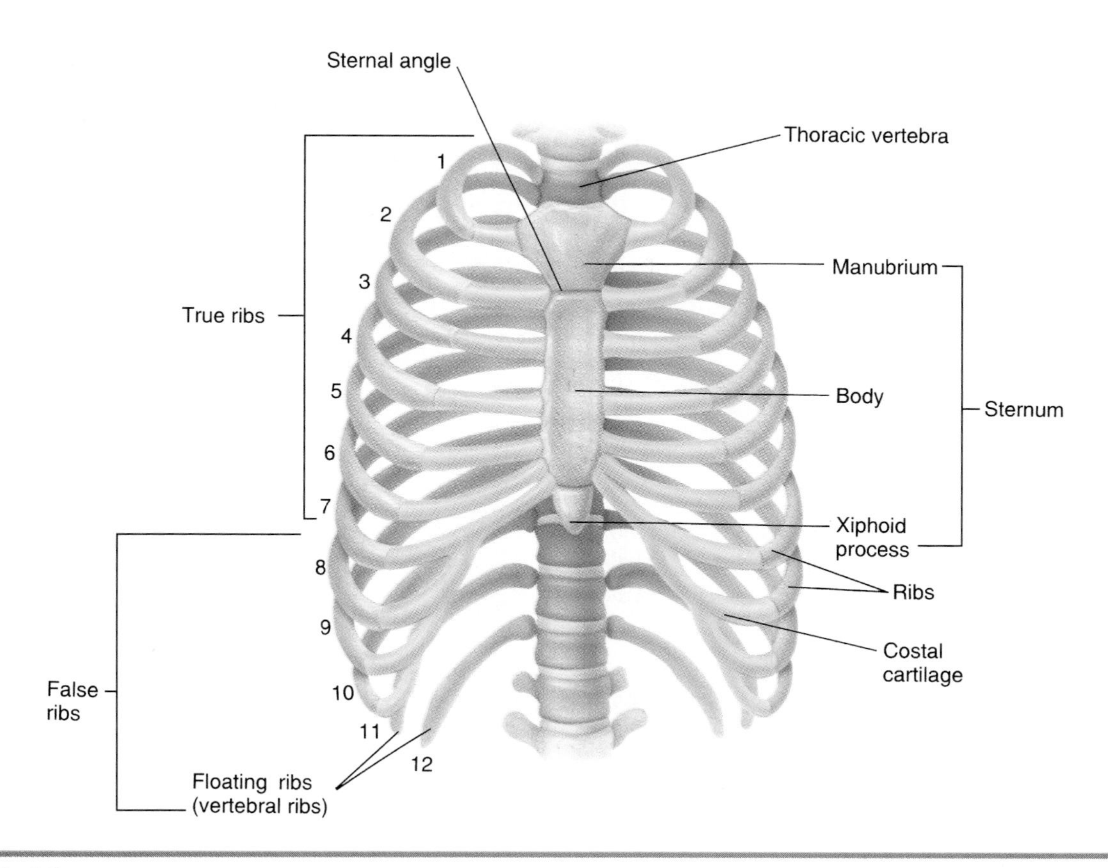

Figure 6.16 The thoracic cage is formed by thoracic vertebrae, ribs, costal cartilages, and the sternum. Note the difference between true and false ribs.

Coccyx

The most inferior part of the vertebral column is the **coccyx** (kok′-six), or tailbone, which is formed of three to five fused rudimentary vertebrae.

Thoracic Cage

The thoracic vertebrae, ribs, costal cartilages, and sternum form the **thoracic, or rib, cage.** It provides protection for the internal organs of the thorax and supports the upper trunk, shoulder girdle, and upper extremities (figure 6.16).

Ribs

Twelve pairs of **ribs** are attached to the thoracic vertebrae. The *head* of each rib articulates with the body of its own vertebra, and a *tubercle* near the head articulates with the transverse process. The head also articulates with the body of the next higher vertebra. The *shaft* of each rib curves around the thorax and slopes slightly downward.

The upper seven pairs of ribs are attached directly to the sternum by the **costal** (kos′-tal) **cartilages,** which extend medially from the ends of the ribs. These ribs are the *true ribs.* The remaining five pairs are the *false ribs.* The first three pairs of false ribs are attached by cartilages to the costal cartilages of the ribs just superior to them. The last two pairs of false ribs are called *floating ribs* because they lack cartilages and are not attached anteriorly. The costal cartilages give some flexibility to the thoracic cage.

How do cervical, thoracic, and lumbar vertebrae differ in structure and location?
How does the axial skeleton protect vital organs?

Sternum

The **sternum,** or breastbone, is a flat, elongated bone located at the midline in the anterior portion of the thoracic cage. It consists of three bones that are fused together. The *manubrium* (mah-nū′-brē-um) is the T-shaped upper portion that articulates with the first two pairs of ribs; the *body* is the larger middle segment; and the *xiphoid* (zīf′-oyd) *process* is the small inferior portion.

A biopsy of red bone marrow may be made by a *sternal puncture* since the sternum is covered only by skin and connective tissue. Under local anesthetic, a large bore hypodermic needle is inserted into the sternum, and red marrow is drawn into a syringe.

The Appendicular Skeleton

The appendicular skeleton consists of (1) the pectoral girdle and the bones of the upper extremities and (2) the pelvic girdle and the bones of the lower extremities. Bones of the appendicular skeleton are listed in table 6.6.

Table 6.6 Bones of the Appendicular Skeleton

Region and Bones	Number of Bones
Pectoral Girdle	
Clavicle	2
Scapula	2
	4
Upper Extremities	
Humerus	2
Ulna	2
Radius	2
Carpals	16
Metacarpals	10
Phalanges	28
	60
Pelvic Girdle	
Coxal bones	2
	2
Lower Extremities	
Femur	2
Patella	2
Tibia	2
Fibula	2
Tarsals	14
Metatarsals	10
Phalanges	28
	60
	Total = 126

Pectoral Girdle

The **pectoral** (pek′-to-ral) **girdle,** or shoulder girdle, consists of two clavicles (collarbones) and two scapulae (shoulder blades) (figure 6.17). Each S-shaped **clavicle** (klav′-i-cul) articulates with the acromion process of a scapula laterally and with the sternum medially. The **scapulae** (skap′-ū-le) are flat, triangular-shaped bones located on each side of the vertebral column, but they do not articulate with the axial skeleton (figure 6.18). Instead, they are held in place by muscles, an arrangement that enables freedom of movement for the shoulders.

The anterior surface of each scapula is flat and smooth where it moves over the ribs. The scapular *spine* runs diagonally across the posterior surface from the *acromion* (ah-krōm′-ē-on) *process* to the medial margin. On its lateral margin is the shallow *glenoid cavity,* which articulates with the head of the humerus. The *coracoid* (kor′-ah-koyd) *process* projects anteriorly from the superior margin and extends under the clavicle.

Upper Extremity

The skeleton of each **upper extremity** is composed of a humerus, an ulna, a radius, carpals, metacarpals, and phalanges. See figures 6.17 to 6.21 as you study this section.

Humerus

The **humerus** (hū′-mer-us) articulates with the scapula at the shoulder, and the ulna and radius at the elbow (figure 6.19). The rounded *head* of the humerus fits into the glenoid cavity of the scapula. Just inferior to the head are two large tubercles where muscles attach. The *greater tubercle* (tū′-ber-cul) is on the lateral surface, and the *lesser tubercle* is on the anterior surface. An *intertubercular groove* lies between them. Just below these tubercles is the *surgical neck,* which gets its name from the frequent fractures that occur in this area. Near the midpoint on the lateral surface is the

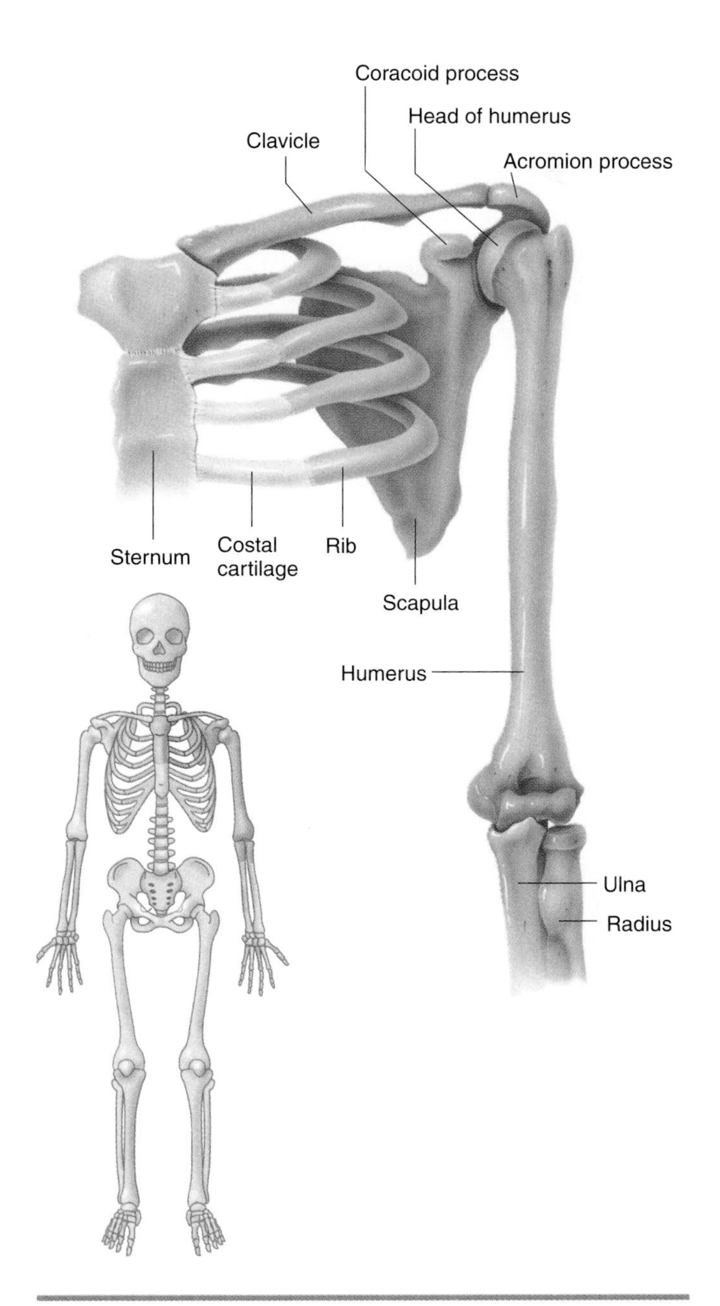

Figure 6.17 The pectoral girdle consists of a scapula and clavicle on each side of the body. Note how the head of the humerus fits into the glenoid fossa of the scapula.

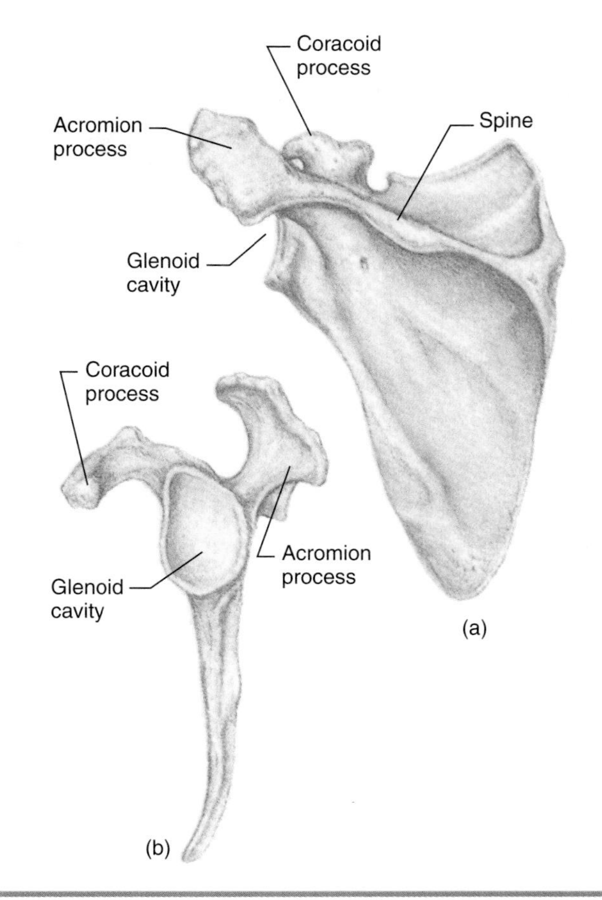

Figure 6.18 The left scapula: (*a*) posterior view; (*b*) lateral view.

deltoid tuberosity (tū-be-ros′-i-tē), a rough, raised area where the deltoid muscle attaches.

The distal end of the humerus has two condyles. The *trochlea* (trok′-lē-ah) is the medial condyle, which articulates with the ulna. The *capitulum* (kah-pit′-ū-lum) is the lateral condyle, which articulates with the radius. Just above these condyles are two enlargements that project laterally and medially: the *lateral epicondyle* (ep-i-kon′-dīl) and the *medial epicondyle.* On the anterior surface between the epicondyles is a depression, the *coronoid* (kor′-o-noyd) *fossa,* that receives the coronoid process of the ulna whenever the forearm is flexed at the elbow. The *olecranon* (o-lek′-rah-non) *fossa* is in a similar location on the posterior surface of the humerus, and it receives the olecranon process of the ulna when the arm is straightened at the elbow.

Ulna

The **ulna** (ul′-na) is the medial bone (little finger side) of the forearm (figure 6.20). The proximal end of the ulna forms the *olecranon process,* the bony point of the elbow. The large, half-circle depression just distal to the olecranon process is the *trochlear notch,* which articulates with the trochlea of the humerus. This articulation is secured by the *coronoid process* on the distal lip of the notch.

At the distal end, the knoblike *head* of the ulna articulates with the medial surface of the radius and with a fibrocartilaginous disk that separates it from the

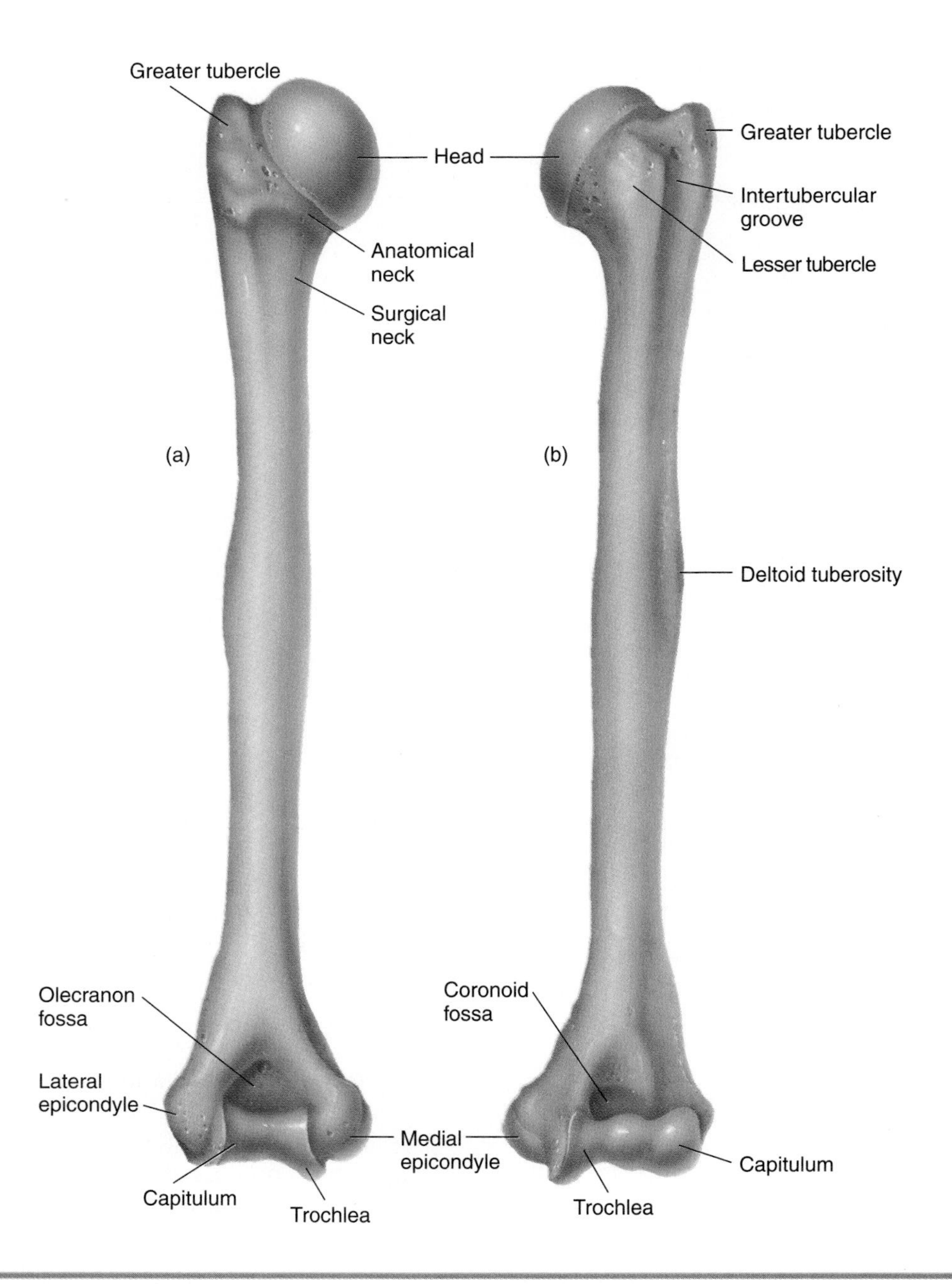

Figure 6.19 The left humerus: (*a*) posterior view; (*b*) anterior view.

wrist bones. The *styloid process* is a small medial projection to which ligaments of the wrist are attached.

Radius

The **radius** (rā′-dē-us) is the lateral bone (thumb side) of the forearm. The disklike *head* of the radius articulates with the capitulum of the humerus in a way that enables the head to rotate freely when the hand is rotated. A short distance distally from the head is the *radial tuberosity,* a raised, roughened area where the biceps muscle attaches. At its distal end, the radius articulates with the carpal bones. A small lateral *styloid process* serves as an attachment site for ligaments of the wrist.

Carpals, Metacarpals, and Phalanges

The skeleton of the hand consists of the carpals, metacarpals, and phalanges (figure 6.21). The **carpals** (kar′-pulz), or wrist bones, consist of eight small carpal bones that are arranged in two transverse rows of four

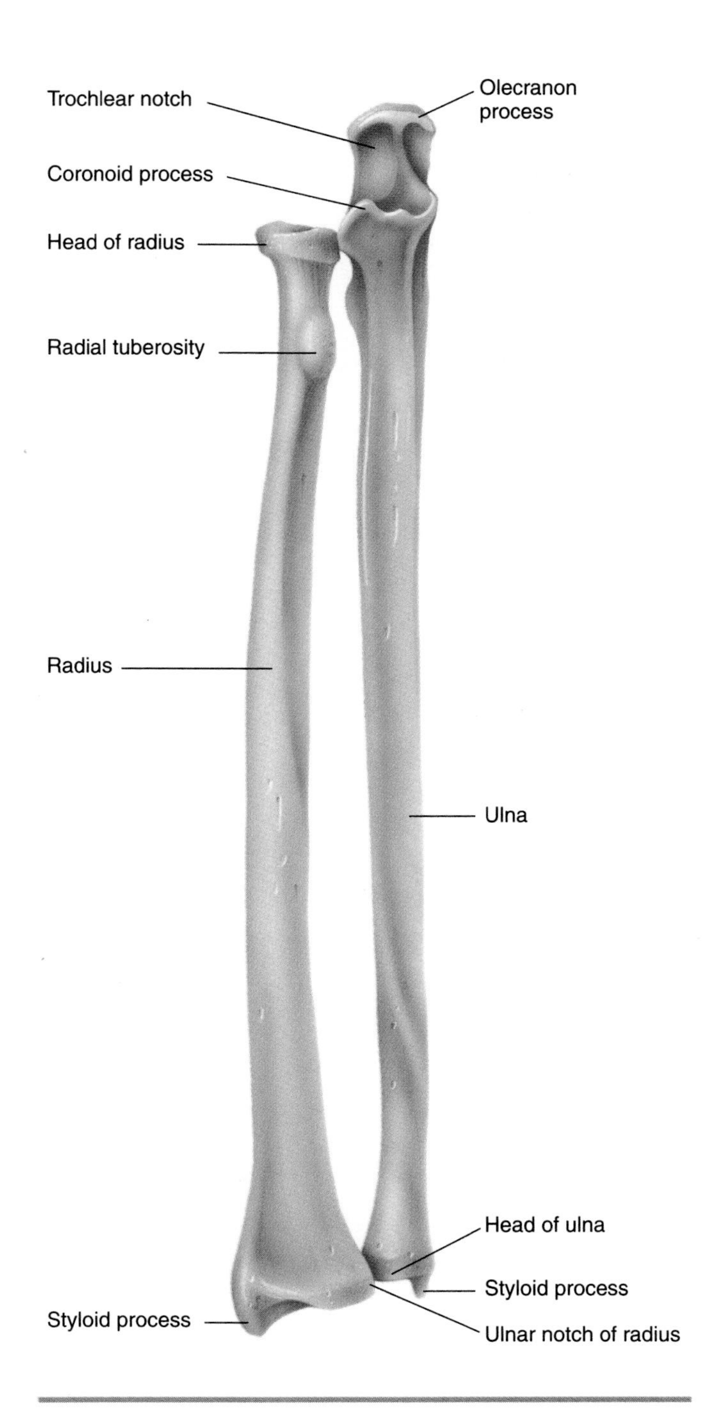

Figure 6.20 An anterior view of the right radius and ulna.

bones each. They are joined by ligaments that allow limited gliding movement.

The **metacarpals,** bones of the palm of the hand, consist of five metacarpal bones that are numbered 1 to 5 starting with the thumb. The bones of the fingers are the **phalanges** (fah-lan′-jēz). Each finger consists of three phalanges—proximal, middle, and distal—except the thumb, which has proximal and distal phalanges only.

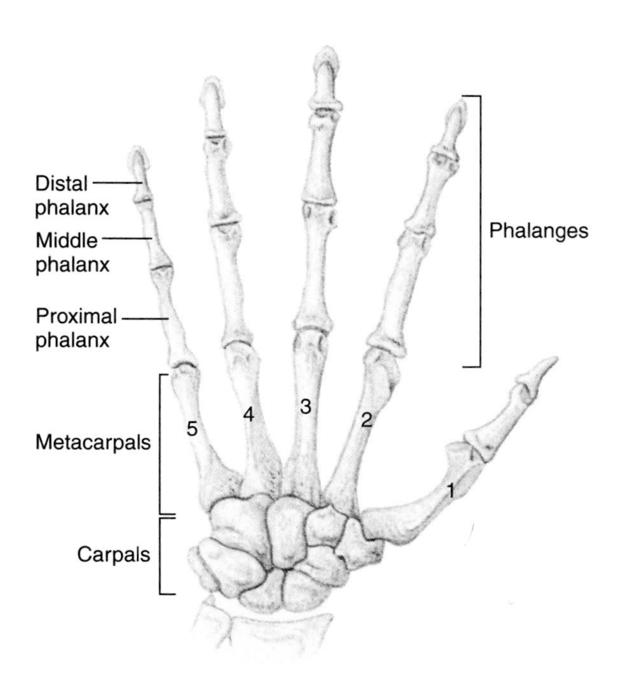

Figure 6.21 A posterior view of the bones of the left hand.

Pelvic Girdle

The **pelvic** (pel′-vik) **girdle** consists of two **coxal** (kok′sal) **bones,** or hipbones, that support the attachment of the lower extremities. The coxal bones articulate with the sacrum posteriorly and with each other anteriorly to form an almost rigid, bony **pelvis,** as shown in figure 6.22. Recall that the coccyx is attached to the inferior end of the sacrum.

Coxal Bones

Each coxal bone is formed by three fused bones—ilium, ischium, and pubis—that join at the *acetabulum* (as-e-tab′-ū-lum), the cup-shaped socket on the lateral surface (figure 6.23). The **ilium** is the broad upper portion whose superior margin forms the *iliac crest,* the prominence of the hip. Interior to the posterior superior iliac spine is the *greater sciatic* (sī-at′-ik) *notch,* which allows the passage of blood vessels and nerves from the pelvis to the thigh. Each ilium joins with the sacrum to form a *sacroiliac joint.*

When giving intramuscular injections in the hip, the region near the greater sciatic notch must be avoided to prevent possible injury to the large blood vessels and nerves in this area.

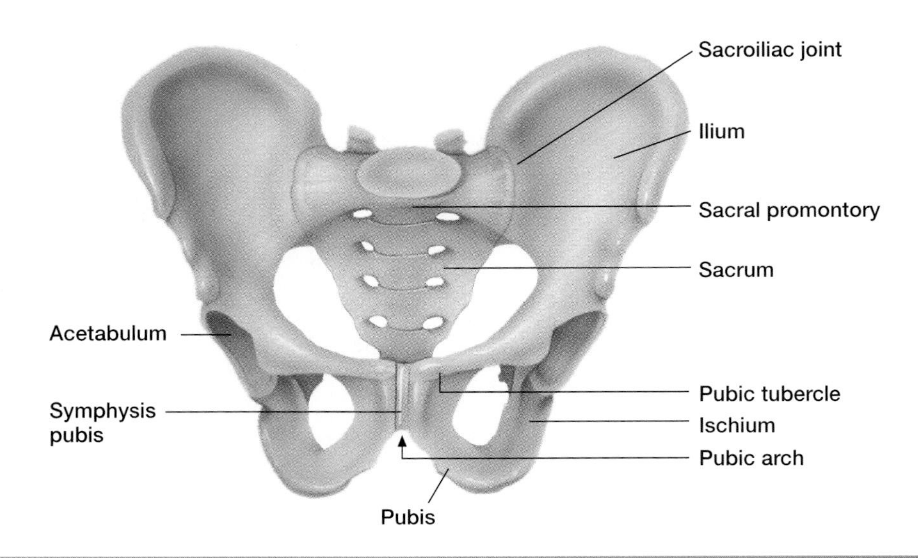

Figure 6.22 An anterior view of the bony pelvis.

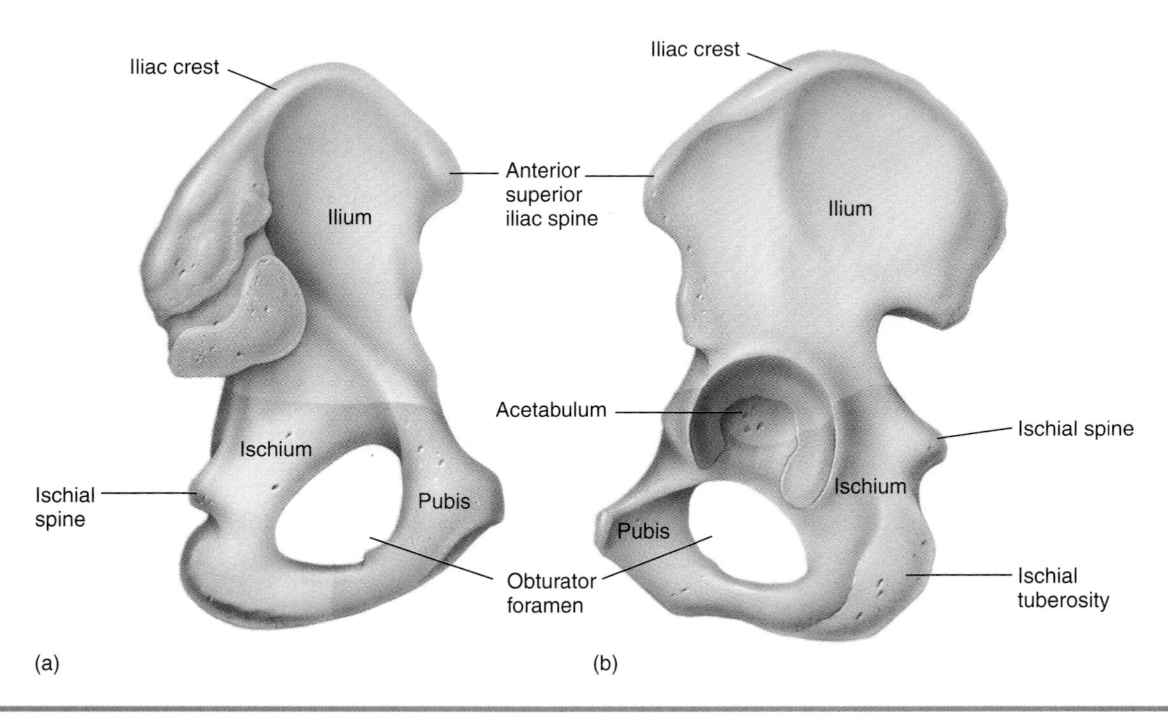

Figure 6.23 The left coxal bone. (*a*) Medial view. (*b*) Lateral view.

The **ischium** forms the inferior, posterior portion of a coxal bone and supports the body when sitting. The roughened projection at the posterior, inferior angle of the ischium is the *ischial tuberosity.* Just above this tuberosity is the *ischial spine,* which projects medially. The size of this spine in females is important during childbirth because it determines the diameter of the pelvic opening.

The **pubis** is the lower, anterior portion of a coxal bone. A portion of the pubis extends posteriorly to fuse with the anterior extension of the ischium. The large opening above this junction is the *obturator* (ob-

Table 6.7 Sexual Differences of the Pelvic Girdle

Characteristic	Male	Female
General structure	Heavier; processes prominent	Lighter; processes not so prominent
Pelvic opening	Narrower and heart-shaped	Wider and oval
Pubic arch angle	Less than 90°	More than 90°
Relative width	Narrower	Wider
Acetabulum	Faces laterally	Faces laterally but more anteriorly

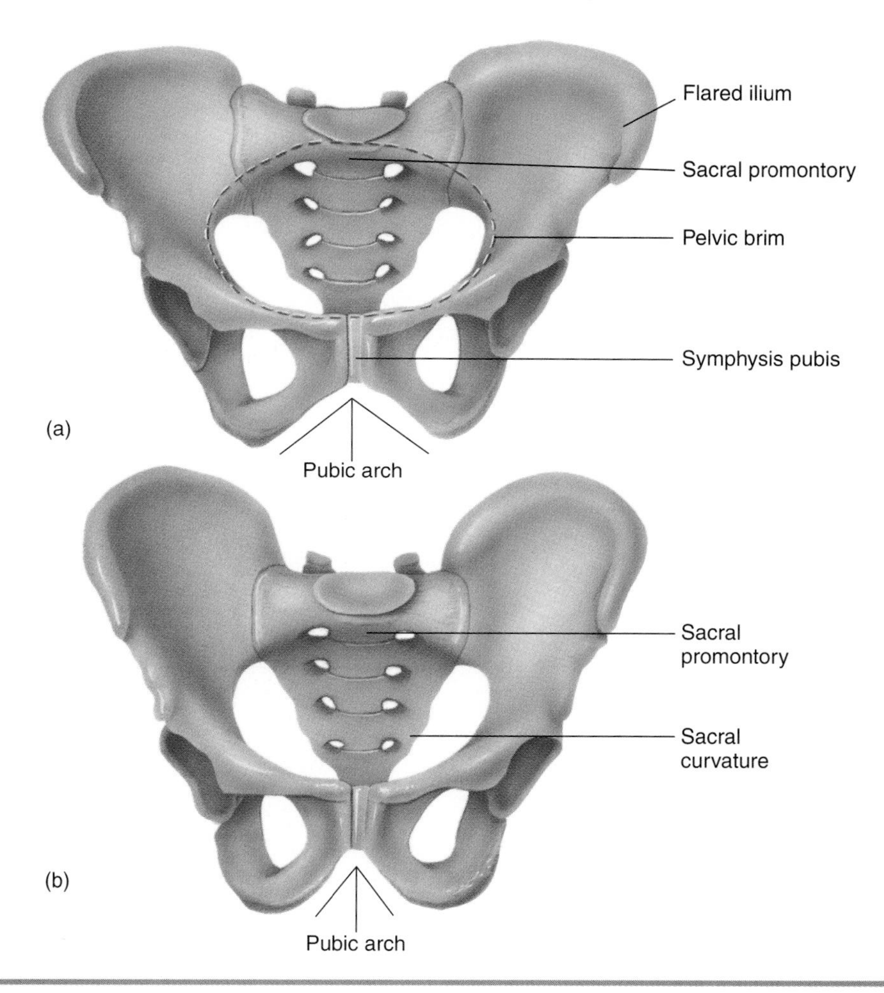

Figure 6.24 (*a*) Female and (*b*) male pelvic girdles. Note the obtuse angle of the pubic arch and greater pelvic width in the female pelvic girdle.

tū-rā′-ter) *foramen,* through which blood vessels and nerves pass into the thigh. The pubic bones unite anteriorly to form the **symphysis pubis** where the bones are joined by a pad of fibrocartilage.

Table 6.7 lists the major differences between the male and the female pelvic girdles. Compare them with the male and female pelvic girdles in figure 6.24 and note the adaptations of the female pelvic girdle for childbirth.

> The fetus must pass through the pelvic opening during birth. Physicians carefully measure this opening before delivery to be sure that it is of adequate size. If not, the baby is delivered via a *Cesarean section.* In a Cesarean section, a transverse incision is made through the pelvic and uterine walls to remove the baby.

Lower Extremity

The bones of each **lower extremity** consist of a femur, a patella, a tibia, a fibula, tarsals, metatarsals, and phalanges. See figures 6.25 to 6.28 as you study this section.

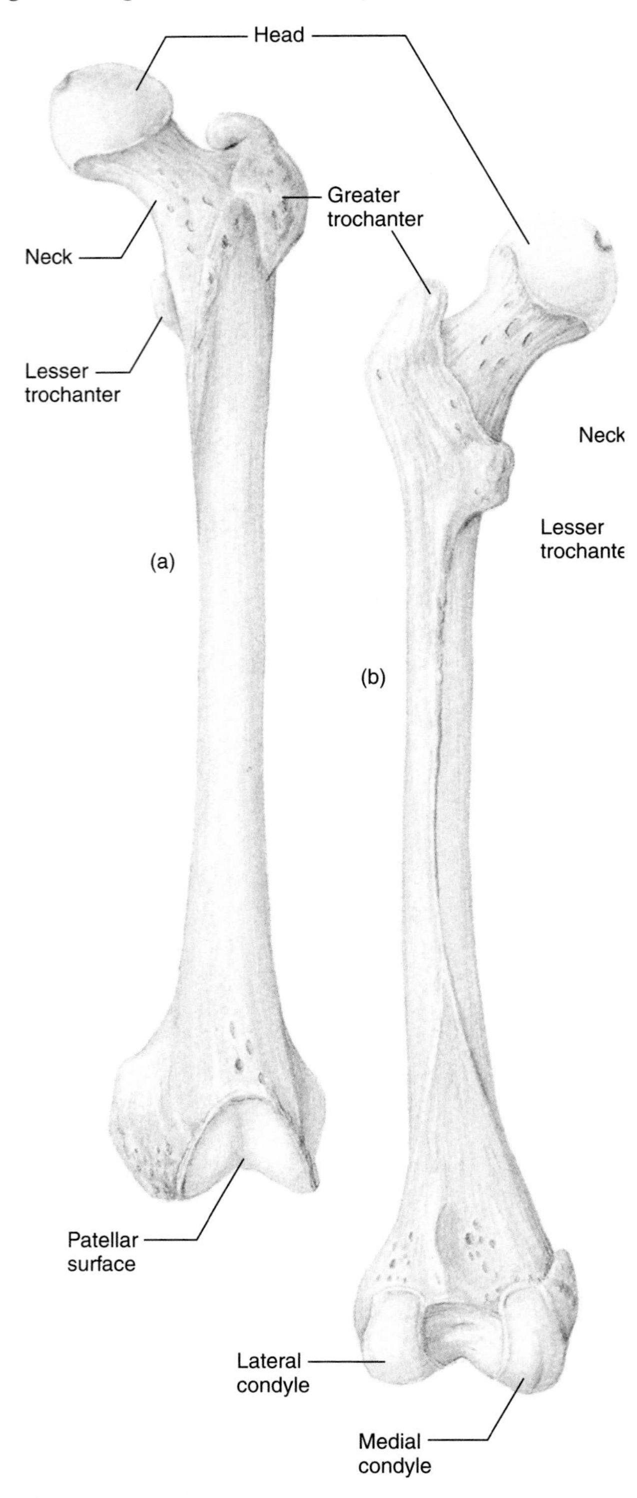

Figure 6.25 The left femur: (*a*) anterior view; (*b*) posterior view.

Femur

The **femur** is the thighbone (figure 6.25). It is the largest and strongest bone of the body. Structures at the proximal end include the rounded *head,* a short *neck,* and two large processes that are sites of muscle

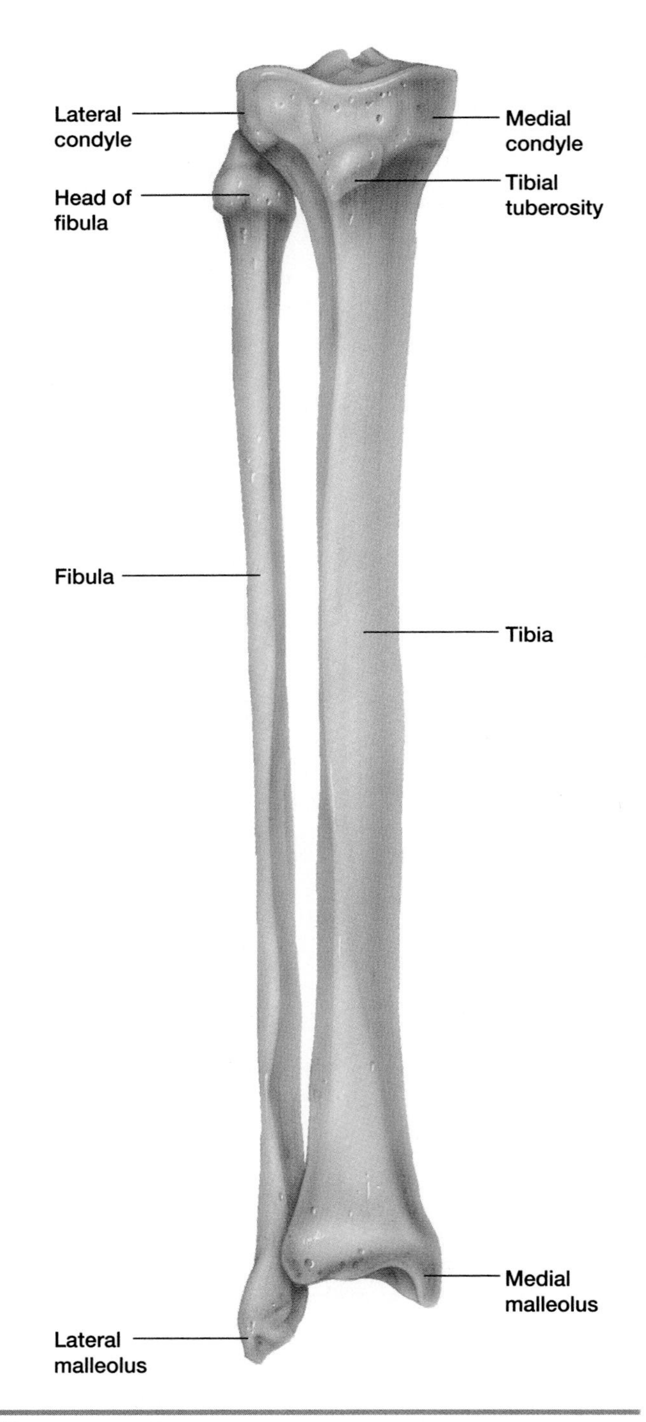

Figure 6.26 An anterior view of the right tibia and fibula.

attachment: a superior, lateral *greater trochanter* (trō-kan′-ter) and an inferior, medial *lesser trochanter.* The head of the femur fits into the acetabulum of the coxal bone. The neck is a common site of fractures in older people. At the enlarged distal end are the *lateral* and *medial condyles,* surfaces that articulate with the tibia.

Patella

The **patella,** or kneecap, is located anterior to the knee joint. It is embedded in the tendon of the thigh muscle (quadriceps femoris), which extends over the anterior of the knee to insert on the tibia. Bones embedded in tendons are called *sesamoid bones.*

Tibia

The **tibia,** or shinbone, is the larger of the two bones of the lower leg (figure 6.26). It bears the weight of the body. Its enlarged proximal portion consists of the *lateral* and *medial condyles,* which articulate with the femur to form the knee joint. The *tibial tuberosity,* a roughened area on the anterior surface just below the condyles, is the attachment site for the patellar ligament. The distal end of the tibia articulates with the talus, a tarsus bone, and laterally with the fibula. The *medial malleolus* (mah-lē-ō′-lus) forms the medial prominence of the ankle.

Fibula

The **fibula** is the slender, lateral bone in the lower leg. Both ends of the bone are enlarged. The proximal *head* articulates with the lateral surface of the tibia but is not involved in forming the knee joint. The distal end articulates with the tibia and tarsus. The *lateral malleolus* forms the lateral prominence of the ankle.

Tarsals, Metatarsals, and Phalanges

The skeleton of the foot consists of the tarsals (ankle), metatarsals (instep), and phalanges (toes) (figures 6.27 and 6.28). Seven bones compose the **tarsals.** The most prominent tarsal bones are the *talus,* which articulates with the tibia and fibula, and the *calcaneus* (kal-kā′n-ē-us), or heel bone. Five **metatarsals** support the instep. They are numbered 1 to 5, starting with the great toe. The tarsals and metatarsals are bound together by ligaments to form strong, resilient arches of the foot. Each toe consists of three **phalanges**—proximal, middle, and distal—except for the great toe, which has only two.

> ***What bones form the pectoral girdle and upper extremities?***
> ***What bones form the pelvic girdle and lower extremities?***

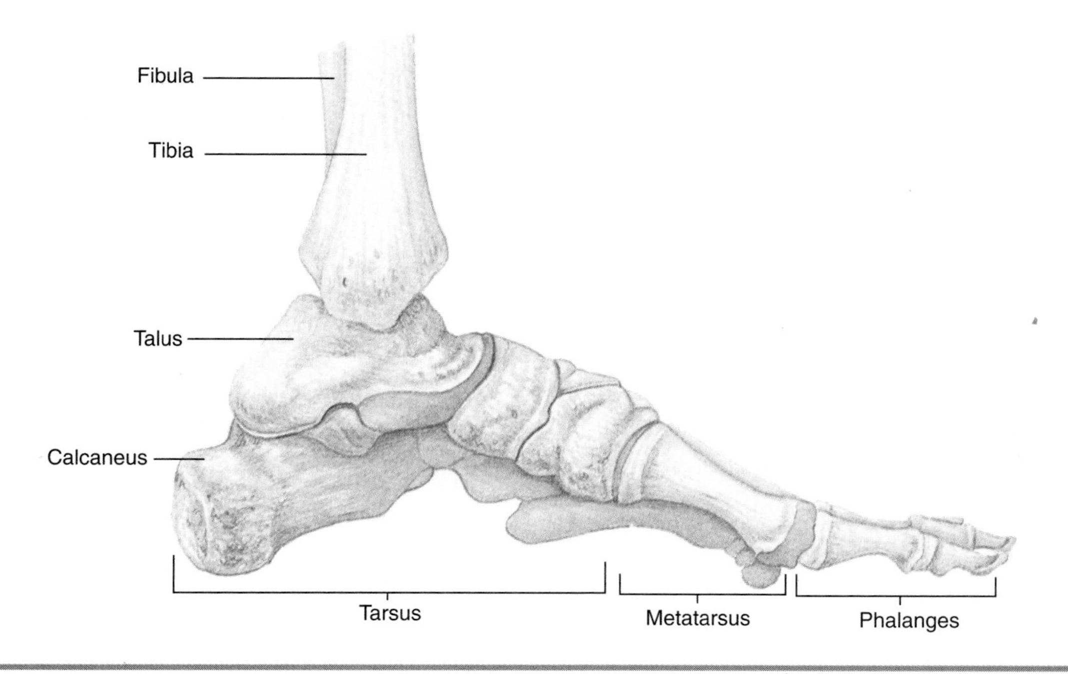

Figure 6.27 A medial view of the bones of the left foot showing articulation of the talus with the tibia.

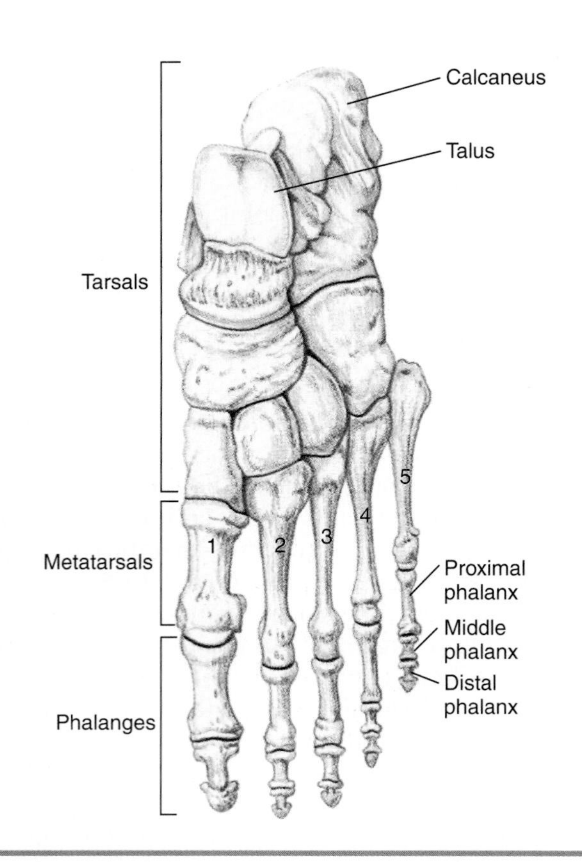

Figure 6.28 A superior view of the bones of the left foot.

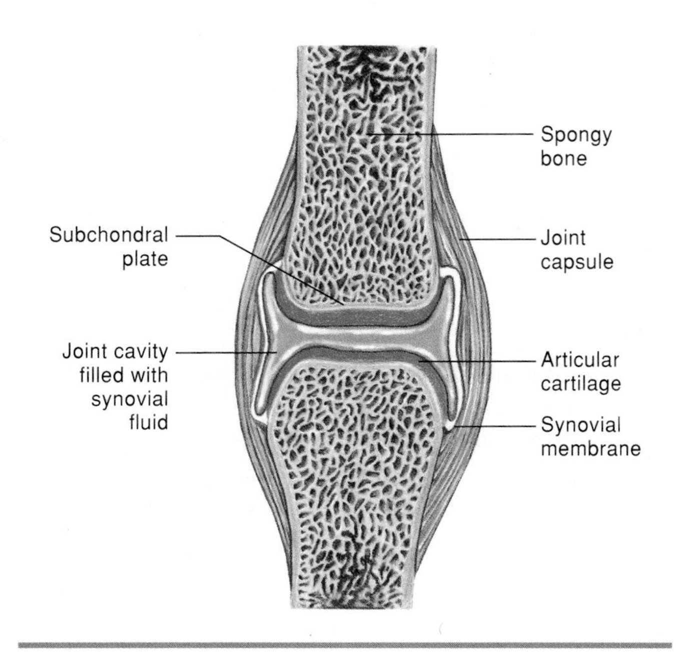

Figure 6.29 The basic structure of a freely movable joint.

Articulations

The junction between two bones forms an **articulation,** or **joint.** Joints allow varying degrees of movement. Joints are categorized as immovable, slightly movable, or freely movable.

Immovable Joints

Bones forming **immovable,** or **synarthrosis** (sin-ar-thrō′-sis), **joints** are tightly joined and are separated by a thin layer of fibrous connective tissue. For example, skull bones, except the mandible, are joined by immovable joints called *sutures* because they resemble stitches (see figure 6.8).

Slightly Movable Joints

Bones forming **slightly movable,** or **amphiarthrosis** (am-fē-ar-th-rō′-sis), **joints** are separated by a layer of cartilage or fibrous connective tissue. For example, the joints formed by adjacent vertebrae contain intervertebral disks formed of fibrocartilage. The limited flexibility of the disks allows slight movement between adjacent vertebrae. Other examples include the symphysis pubis and sacroiliac joints.

Freely Movable Joints

Most articulations are freely movable. The structure of **freely movable,** or **diarthrosis** (di-ar-thro′-sis), **joints** is more complex. The ends of the bones forming the joint are bound together by an *articular capsule* formed of ligaments. A *synovial* (si-nō′-vē-al) *membrane* lines the interior of the capsule and secretes *synovial fluid* that lubricates the joint. The ends of the bones are covered with *articular cartilage,* which protects bones and reduces friction (figure 6.29). Freely movable joints are categorized into several types based on their structure and types of movements.

Gliding Joints

Gliding joints occur between small bones that slide over one another. They occur between carpal bones, between tarsal bones, and between the clavicle and scapula.

Condyloid Joints

Condyloid joints allow movements in two planes: side to side or back and forth. The joints between the

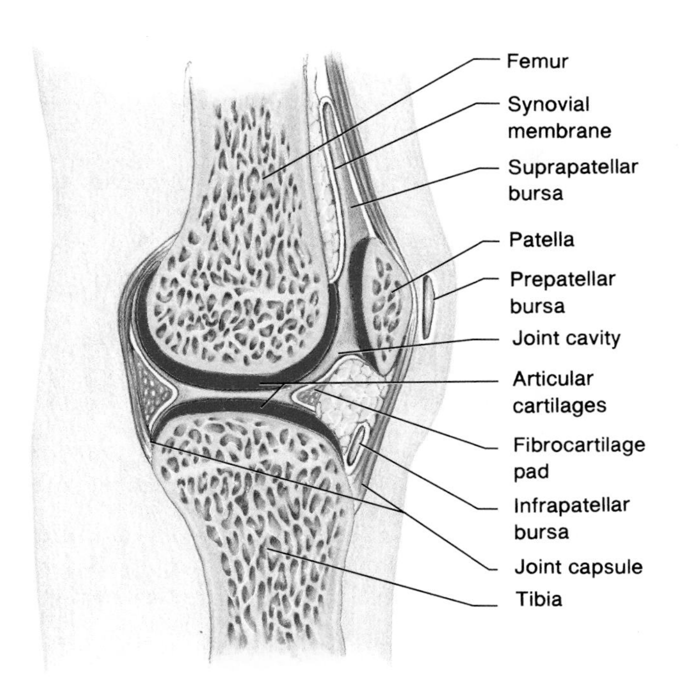

Figure 6.30 A sagittal section of a knee joint, a hinge joint. Note the articular cartilages, synovial membrane, fibrocartilage pads, bursae, and joint capsule.

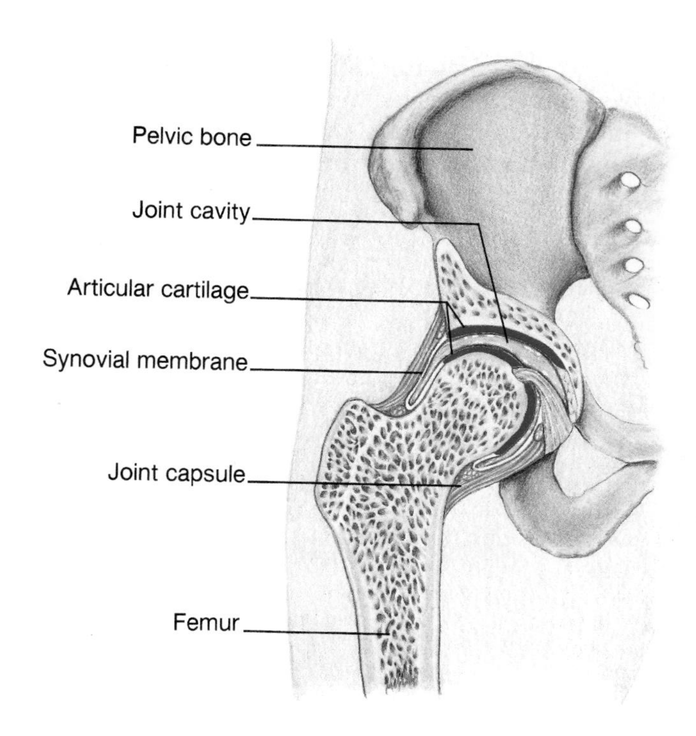

Figure 6.31 A coronal section of a hip joint, a ball-and-socket joint. Note the joint capsule formed of ligaments.

carpals and bones of the forearm (radius and ulna) and the metacarpals and first phalanges are examples.

Hinge Joints

Hinge joints allow movement in one plane only, like a door hinge. The elbow, knee, and joints between phalanges are all hinge joints. Examine the sagittal section of a knee joint in figure 6.30. Note the articular capsule (formed of ligaments), which binds the bones together; the bursae (sacs filled with synovial fluid); the articular cartilages, which reduce friction; and the cartilaginous pads, which cushion the ends of the bones.

Saddle Joints

Saddle joints occur where the ends of each bone are saddle-shaped: convex in one direction and concave in the other. Movement is side to side and back and forth. A joint of this type occurs between the trapezium (a carpal bone) and the metacarpal bone of the thumb.

Pivot Joints

Pivot joints allow rotational movement in a single plane. The rotation of the atlas on the axis is an example of a pivot joint.

A "torn cartilage" refers to a damaged cartilaginous pad (meniscus) in a knee joint, a common athletic injury. Treatment usually requires removal of the damaged cartilage by *arthroscopic surgery*. An arthroscope is a pencil-sized instrument containing lenses and a light source. It may be inserted into a joint through a small incision to enable a physician to observe the interior of the joint. While viewing the interior of the joint, the surgeon may insert small instruments to remove damaged tissue, make repairs, and flush out the joint.

Ball-and-Socket Joints

Ball-and-socket joints occur where a rounded head of one bone fits into a concavity of the other bone. These joints allow the greatest freedom of movement. Movement may be rotational or in any plane. The shoulder and hip joints are ball-and-socket joints. Examine the structure of a hip joint in figure 6.31.

Where are immovable, slightly movable, and freely movable joints found in the skeleton?
What types of freely movable joints occur in the body, and where are they located?

Older persons are prone to "breaking a hip," which means that a weakened femur breaks at the neck. This usually is a consequence of osteoporosis, the excessive loss of calcium from bones. Osteoporosis is caused by a combination of factors: insufficient calcium in the diet, lack of minimal exercise, and a decline in sex hormones, especially in postmenopausal women. Older persons are not only more prone to fractures, but healing of fractures takes much longer than in younger persons.

Movements at Freely Movable Joints

Movement at a joint results from the contraction of skeletal muscles that span across the joint. The type of movement that occurs is determined by the type of joint and the location of the muscle or muscles involved. The more common types of movements are listed in table 6.8 and illustrated in figure 6.32.

Disorders of the Skeletal System

Common disorders of the skeletal system may be categorized as disorders of bones or disorders of joints. **Orthopedics** (or-thō-pē-diks) is the branch of medicine that specializes in treating diseases and abnormalities of the skeletal system.

Disorders of Bones

Fractures are broken bones. Fractures are the most common type of bone injury. A few types of fractures are:

- **Complete.** The broken bone is separated into two parts.
- **Incomplete.** The broken bone is not separated into two parts.
- **Compound.** Part of the broken bone pierces and protrudes through the skin.
- **Simple.** Part of the broken bone does not pierce the skin.
- **Greenstick.** An incomplete break on one side of the bone with the other side bowed.
- **Comminuted.** The bone is broken into several fragments.
- **Impacted.** One bone is forced into another bone.

Osteoporosis (os-tē-ō-pō-rō′-sis) is a weakening of bones due to the removal of calcium salts. This is

Table 6.8 Movements at Freely Movable Joints

Movement	Description
Flexion	Decrease in the angle of bones forming joint
Extension	Increase in the angle of bones forming joint
Dorsiflexion	Flexion of the foot at the ankle
Plantar flexion	Extension of the foot at the ankle
Abduction	Movement of a bone away from the midline
Adduction	Movement of a bone toward the midline
Rotation	Movement of a bone around its longitudinal axis
Circumduction	Movement of the distal end of a bone in a circle while the proximal end forms the pivot joint
Eversion	Movement of the sole of the foot laterally
Inversion	Movement of the sole of the foot medially
Pronation	Rotation of the forearm when the palm is turned downward or posteriorly
Supination	Rotation of the forearm when the palm is turned upwards or anteriorly
Protraction	Movement of the mandible anteriorly
Retraction	Movement of the mandible posteriorly
Elevation	Movement of a body part upward
Depression	Movement of a body part downward

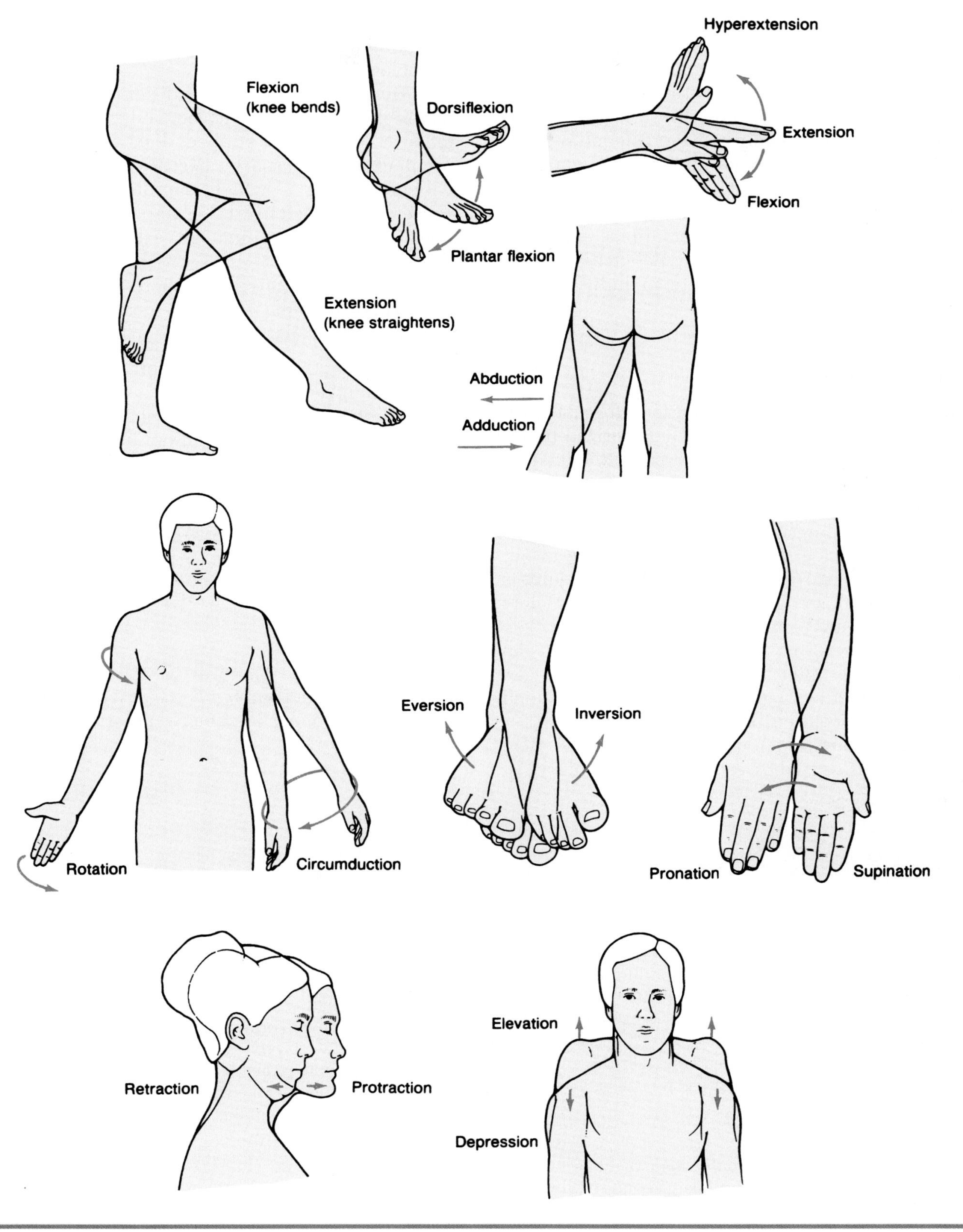

Figure 6.32 Common movements at freely movable joints.

a common problem in older persons due to inactivity and a decrease in hormone production. It is more common in postmenopausal women because of the lack of estrogen. Estrogen (hormone replacement therapy) and calcium are often prescribed for postmenopausal women to help prevent osteoporosis.

Rickets is a disease of children that is characterized by a deficiency of calcium salts in the bones. Affected children have a bowlegged appearance due to the bending of weakened femurs, tibiae, and fibulae. Rickets results from a dietary deficiency of calcium and/or vitamin D. It is rare in industrialized nations.

Disorders of Joints

Arthritis (ar-thrī′-tis) is the general term for many different diseases of joints that are characterized by inflammation, swelling (edema), and pain. Rheumatoid arthritis and osteoarthritis are the most common types.

Rheumatoid (rū′-mah-toid) *arthritis* is the most painful and crippling type. The synovial membrane thickens, synovial fluid accumulates causing swelling, and articular cartilages are destroyed. The joint is invaded by fibrous connective tissue that ultimately ossifies, making the joint immovable.

Osteoarthritis, the most common type, is a degenerative disease that results from aging and wear. The articular cartilages gradually disintegrate, which causes pain and restricts movement.

Bursitis (bur-sī′-tis) is the inflammation of a bursa. It may be caused by excessive use or injury.

Dislocation is the displacement of bones forming a joint. Pain, swelling, and reduced movement are associated with a dislocation.

Herniated disk is a condition in which an intervertebral disk protrudes beyond the edge of a vertebra. A ruptured, or slipped, disk refers to the same problem. It is caused by excessive pressure on the vertebral column, which compresses a disk and causes it to bulge outward. The protruding disk may place pressure on a spinal nerve and cause considerable pain.

Spinal curvatures are usually congenital disorders. There are three major types:

1. *Kyphosis* (kī-fō-sis) is an excessive thoracic curvature of the vertebral column, which produces a humpback condition.
2. *Lordosis* is an excessive lumbar curvature of the vertebral column, which produces a swayback condition.
3. *Scoliosis* is an abnormal lateral curvature of the vertebral column. For some reason, it is more common in adolescent girls.

Sprains result from tearing or excessive stretching of the ligaments and tendons at a joint without a dislocation.

What are some common types of fractures? How do osteoarthritis and rheumatoid arthritis differ?

CHAPTER SUMMARY

Functions of the Skeletal System

1. The skeletal system provides support for the body and protection for internal organs.
2. The bones of the skeleton serve as sites for the attachment of skeletal muscles.
3. Blood cells are formed by red bone marrow.
4. Bones serve as reservoirs for calcium salts.

Bone Structure

1. The diaphysis is a long shaft of bone that lies between the epiphyses, the enlarged ends of the bone.
2. Each epiphysis is joined to the diaphysis by an epiphyseal plate in immature bones, or by fusion at the epiphyseal line in mature bones.
3. Articular cartilages protect and cushion the articular surfaces of the epiphyses.
4. The periosteum covers the bone surface except for the articular cartilages.
5. Compact bone forms the diaphysis and the surface layer of the epiphyses.
6. Spongy bone forms the interior of the epiphyses.
7. Red marrow fills the spaces in spongy bone.
8. The diaphysis contains a medullary cavity filled with fatty yellow marrow.
9. Compact bone is formed of numerous osteons fused together.
10. Osteonic canals contain blood vessels and nerves.
11. Spongy bone lacks osteons; its cells are nourished by diffusion from blood vessels in the red marrow.

Bone Formation

1. Intramembranous bones are first formed by connective tissue membranes, which are replaced by bone tissue.
2. Connective tissue cells are transformed into osteoblasts, which deposit the spongy bone within the membrane.
3. Osteoblasts from the periosteum form a layer of compact bone over the spongy bone.
4. Endochondral bones are first formed of hyaline cartilage, which is later replaced by bone tissue.
5. In long bones, a primary ossification center forms in the center of the diaphysis and extends toward the epiphyses.
6. Secondary ossification centers form in the epiphyses.
7. An epiphyseal disk of cartilage remains between the epiphyses and the diaphysis in immature bones.
8. Growth in length occurs at the epiphyseal disk, which is gradually replaced by bone.
9. Compact bone is deposited by osteoblasts from the periosteum, and they are responsible for growth in the diameter of a bone.
10. Osteoclasts hollow out the medullary cavity and reshape the bone.
11. Bones are dynamic, living organs that are reshaped throughout life by the actions of osteoclasts and osteoblasts.
12. Calcium salts may be removed from bones for other body needs and redeposited in bones later on.
13. The concentration of protein fibers decreases with age. The bones of older persons tend to be brittle and weak due to the loss of fibers and calcium salts, respectively.

Divisions of the Skeleton

1. The skeleton is divided into the axial and appendicular divisions.
2. The axial skeleton includes the bones that support the head, neck, and trunk.
3. The appendicular skeleton includes the bones of the pectoral girdle and upper extremities and the bones of the pelvic girdle and the lower extremities.

The Axial Skeleton

1. The axial skeleton consists of the skull, vertebral column, and thoracic cage.
2. The skull consists of cranial and facial bones; all are joined by immovable joints except the mandible.
3. The cranial bones are the frontal bone (1), parietal bones (2), sphenoid bone (1), temporal bones (2), occipital bone (1), and ethmoid bone (1).
4. The frontal, sphenoid, and ethmoid bones contain sinuses.
5. The facial bones are the maxillae (2), palatine bones (2), zygomatic bones (2), lacrimal bones (2), nasal bones (2), inferior nasal conchae (2), vomer bone (1), and mandible (1).
6. Each maxilla contains a large sinus.
7. Cranial bones of an infant skull are separated by membranes and several fontanels, which allow some flexibility of the skull during birth.
8. The vertebral column consists of 24 vertebrae, the sacrum, and the coccyx.
9. Vertebrae are separated by intervertebral disks and are categorized as cervical (7), thoracic (12), and lumbar (5) vertebrae.
10. The first two cervical vertebrae are unique. The atlas rotates on the axis when the head is turned.
11. Thoracic vertebrae have facets on the body and transverse processes for articulation with the ribs.
12. The bodies of lumbar vertebrae are heavy and strong.
13. The sacrum is formed of five fused vertebrae and forms the posterior portion of the pelvic girdle.
14. The coccyx is formed of three to five rudimentary vertebrae and forms the inferior end of the vertebral column.
15. The thoracic cage consists of thoracic vertebrae, ribs, and sternum. It supports the upper trunk and protects internal thoracic organs.
16. There are seven pairs of true ribs and five pairs of false ribs. The inferior two pairs of false ribs are floating ribs.
17. The sternum is formed of three fused bones: manubrium, body, and xiphoid process.

The Appendicular Skeleton

1. The appendicular skeleton consists of the pectoral and pelvic girdles and of the bones of the extremities.
2. The pectoral girdle consists of clavicles (2) and scapulae (2), and it supports the upper extremities.
3. The bones of the upper extremity are the humerus, the ulna, the radius, carpals, metacarpals, and phalanges.
4. The humerus articulates with the glenoid cavity of the scapula to form the shoulder joint and with the ulna and radius to form the elbow joint.
5. The ulna is the medial bone of the forearm. It articulates with the humerus at the elbow and with the radius and a fibrocartilaginous disk at the wrist.
6. The radius is the lateral (thumb side) bone of the forearm. It articulates with the humerus at the elbow and with the ulna and carpus at the wrist.
7. The bones of the hand are the carpals (8), metacarpals (5), and phalanges (14).
8. The carpal bones are joined by ligaments to form the bones of the wrist; metacarpal bones support the palm of the hand; and the phalanges are the bones of the fingers.
9. The pelvic girdle consists of two coxal bones that are joined to the sacrum posteriorly and to each other anteriorly. It supports the lower extremities.

10. Each coxal bone is formed by the fusion of three bones: the ilium, ischium, and pubis.
11. The ilium forms the upper portion of a coxal bone and joins with the sacrum to form a sacroiliac joint.
12. The ischium forms the inferior, posterior portion of a coxal bone and supports the body when sitting.
13. The pubis forms the lower, anterior part of a coxal bone. The pubic bones unite anteriorly to form the symphysis pubis joint.
14. Each lower extremity consists of a femur, a patella, a tibia, a fibula, tarsals, metatarsals, and phalanges.
15. The head of the femur is inserted into the acetabulum of a coxal bone to form a hip joint. Distally, it articulates with the tibia at the knee joint.
16. The patella is a sesamoid bone in the anterior portion of the knee joint.
17. The tibia (shinbone) articulates with the femur at the knee joint and with the talus to form the ankle joint.
18. The fibula lies lateral to the tibia. It articulates proximally with the tibia and distally with the talus.
19. The skeleton of the foot consists of tarsals (7), metatarsals (5), and phalanges (14).
20. Tarsal bones form the ankle, metatarsal bones support the instep, and phalanges are the bones of the toes.

Articulations

1. There are three types of joints: immovable, slightly movable, and freely movable.
2. Bones forming immovable joints are closely joined and are separated by a thin layer of fibrous connective tissue. Articulations of skull bones, except the mandible, are examples.
3. Bones forming slightly movable joints are separated by fibrocartilage pads or fibrous connective tissue. Joints between vertebrae are examples.
4. Bones forming freely movable joints are bound together by a ligamentous articular capsule that is lined by a synovial membrane. The articular surfaces of the bones are covered by articular cartilages. Synovial fluid lubricates the joint, and bursae may be present to reduce friction.
5. There are several types of freely movable joints: gliding, condyloid, hinge, saddle, pivot, and ball-and-socket.
6. Movements at freely movable joints include flexion, extension, dorsiflexion, plantar flexion, hyperextension, abduction, adduction, rotation, circumduction, inversion, eversion, protraction, retraction, elevation, and depression.

Disorders of the Skeletal System

1. Disorders of bones include fractures, osteoporosis, and rickets.
2. Disorders of joints include arthritis, bursitis, dislocation, herniated disk, spinal curvatures, and sprains.

BUILDING YOUR VOCABULARY

1. **Selected New Terms**
   abduction, p. 116
   acetabulum, p. 109
   adduction, p. 116
   amphiarthrosis, p. 114
   appendicular, p. 106
   axial, p. 95
   circumduction, p. 116
   condyle, p. 96
   diaphysis, p. 91
   diarthrosis, p. 114
   epicondyle, p. 96
   epiphysis, p. 91
   extension, p. 116
   facet, p. 96
   flexion, p. 116
   foramen, p. 92
   head, p. 96
   pronation, p. 116
   red marrow, p. 92
   supination, p. 116
   suture, p. 97
   synarthrosis, p. 114
   trochanter, p. 96
   tubercle, p. 96
   tuberosity, p. 96

2. **Related Clinical Terms**
   ankylosis (ang-ki-lō′-sis) Abnormal stiffness of a joint.
   arthralgia (ar-thral′-jē-ah) Pain in a joint.
   osteomalacia (os-tē-ō-mah-lā′-she-ah) Softening of bone tissue due to abnormal calcium and phosphorus metabolism.
   osteomyelitis (os-tē-ō-mī-e-lī′-tis) Inflammation of a bone caused by a fungus or bacteria.

STUDY ACTIVITIES

1. Complete the **Chapter 6 Study Guide,** which begins on page 419.
2. Complete the **Learning Objectives** listed on the first page of this chapter.

CHECK YOUR UNDERSTANDING

(Answers are located in appendix B.)

1. The skeletal system provides _____ for the body and _____ for many internal organs.
2. The enlarged ends of a long bone are the _____ , which are composed of _____ bone that is coated with a thin layer of compact bone.
3. Blood vessels and nerves enter a bone through a _____ .
4. Cranial bones are formed by _____ ossification.
5. Growth in diameter of a long bone occurs by deposition of bone tissue by osteoblasts from the _____ .
6. The skull, vertebral column, and sacrum are part of the _____ skeleton.
7. The bone forming the lower jaw is the _____ , and it articulates with the _____ bone.
8. The first vertebra, the _____ , articulates with the _____ bone of the skull.
9. True ribs are attached directly to the sternum by the _____ .
10. The clavicles and scapulae form the _____ .
11. The upper arm bone, the _____ , articulates with two forearm bones, the _____ and the _____ .
12. Each coxal bone is formed of three fused bones: the _____ , _____ , and _____ .
13. The thigh bone is the _____ , and it articulates distally with the _____ .
14. Among freely movable joints, the elbow is an example of a _____ joint, while the shoulder is an example of a _____ joint.
15. _____ is a weakening of bones due to removal of calcium salts.

SEVEN

7

CHAPTER

The Muscle System

Chapter Preview & Learning Objectives

SELECTED KEY TERMS

Antagonist (anti = against) A muscle whose contraction opposes an action of another muscle.

Aponeurosis (apo = from; neur = cord) A broad sheet of fibrous connective tissue that attaches a muscle to another muscle or connective tissue.

Creatine phosphate A molecule that stores a small amount of energy in a muscle fiber.

Insertion The end of a muscle that is attached to a movable body part.

Motor unit A motor neuron and the muscle fibers that it controls.

Muscle fiber A single muscle cell.

Muscle tone The state of slight contraction in a skeletal muscle.

Myofibril (myo = muscle; fibril = little fiber) One of many contractile fibers within a muscle cell.

Myoglobin (myo = muscle) A molecule that stores a small amount of oxygen in a muscle fiber.

Neurotransmitter (neuro = nerve; transmit = to send across) A chemical released by axon tips of neurons that activates a muscle fiber, gland, or another neuron.

Origin The end of a muscle that is attached to an immovable body part.

Sarcomere (sarc = flesh) The smallest contractile unit of a myofibril.

Tendon A narrow band of fibrous connective tissue that attaches a muscle to a bone.

Tetanic contraction (tetan = rigid, stiff) A sustained muscle contraction.

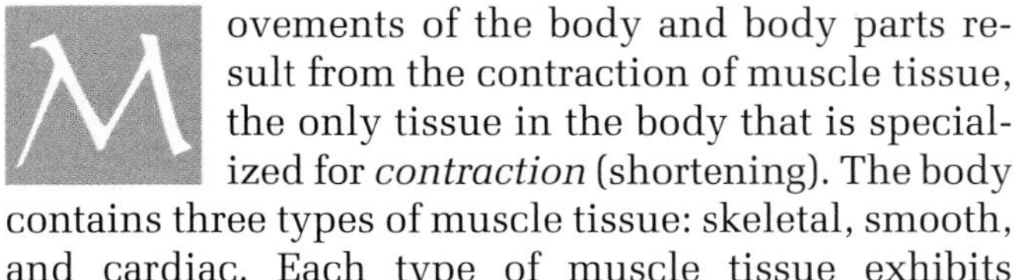

Movements of the body and body parts result from the contraction of muscle tissue, the only tissue in the body that is specialized for *contraction* (shortening). The body contains three types of muscle tissue: skeletal, smooth, and cardiac. Each type of muscle tissue exhibits unique structural and functional characteristics. Refresh your understanding of these tissues by referring to the discussion of muscle tissue in chapter 4. Table 7.1 summarizes the characteristics of muscle tissues.

Structure of Skeletal Muscle

Skeletal muscles are the organs of the muscular system. They are composed primarily of skeletal muscle tissue that is bound together by connective tissue. Connective tissue forms an integral part of skeletal muscles.

Each skeletal muscle contains many individual **muscle fibers,** or cells, that are arranged in small bundles called *fasciculi* (fah-sik′-ū-lē). Fibrous connective tissue envelopes and separates each fiber in a fasciculus and each fasiculus from other fasciculi. The entire muscle is enveloped by tough fibrous connective tissue, the **deep fascia** (fash′-ē-ah) (figure 7.1). Connective tissues continue to the ends of the muscle, where they usually form tough, cordlike **tendons,** which attach the muscle to bones. The fibers of both the tendon and the periosteum merge to form a secure attachment. A few muscles attach to other muscles or connective tissues rather than to bones. In these muscles, the connective tissue forms a sheetlike attachment called an **aponeurosis** (ap″-ō-nū-rō′-sis).

Table 7.1 Types of Muscle Tissue

Characteristic	Skeletal	Smooth	Cardiac
Striations	Present	Absent	Present
Nucleus	Many peripherally located nuclei	Single centrally located nucleus	Single centrally located nucleus
Fibers	Long and parallel	Short; tapered ends; parallel	Short and branching; intercalated disks join fibers end-to-end to form network
Neural control	Voluntary	Involuntary	Involuntary
Contractions	Rapid; easily fatigued	Slow; resistant to fatigue	Rhythmic; resistant to fatigue
Location	Attached to bones	Walls of hollow visceral organs and blood vessels	Wall of the heart

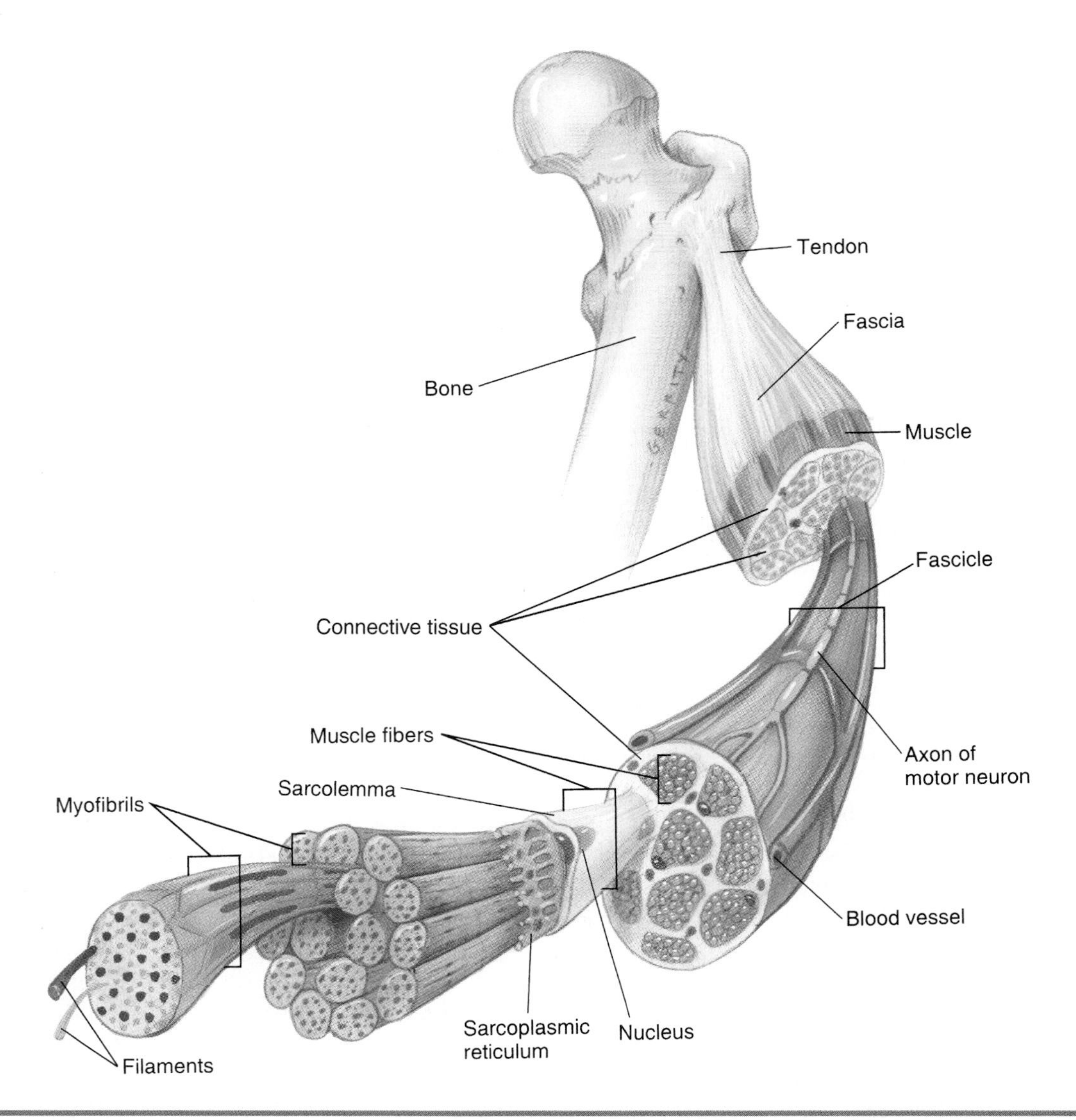

Figure 7.1 A muscle is composed of a variety of tissues, but it consists mostly of muscle tissue supported by connective tissue. Fascia covers the muscle surface, and interior connective tissues bind muscle cells into groups called fascicles and separate individual muscle fibers within each fascicle. Muscle fibers are composed of myofibrils.

Skeletal Muscle Fibers

Each muscle fiber is a multinucleated skeletal muscle cell. It is a long, thin cylinder with rounded ends that usually extends the full length of the muscle. The plasma membrane of a muscle fiber is called the **sarcolemma** (sar-kō-lem′-ah), and its cytoplasm is the **sarcoplasm.**

Each muscle fiber contains many threadlike **myofibrils,** which extend the length of the fiber, as shown in figure 7.1. Myofibrils are the contractile elements of a muscle fiber. They are composed of two basic types of protein filaments: thin **actin filaments** and thick **myosin filaments.** The arrangement of actin and myosin filaments produces the light and dark striations that are characteristic of skeletal muscle fibers (figure 7.2).

The contractile unit of a myofibril is a **sarcomere.** Many sarcomeres are arranged end-to-end in a myofibril. As shown in figure 7.2, a sarcomere is a portion of a myofibril between two successive Z lines, which are composed of proteins arranged transverse to the axis of the filament. Actin filaments are attached to each side of the Z lines and extend toward the middle of the sarcomeres. Note that the actin filaments do not meet. Myosin filaments occur only in the A band of the sarcomere.

How are muscle tissue and connective tissue arranged in a skeletal muscle?
What composes a muscle fiber?

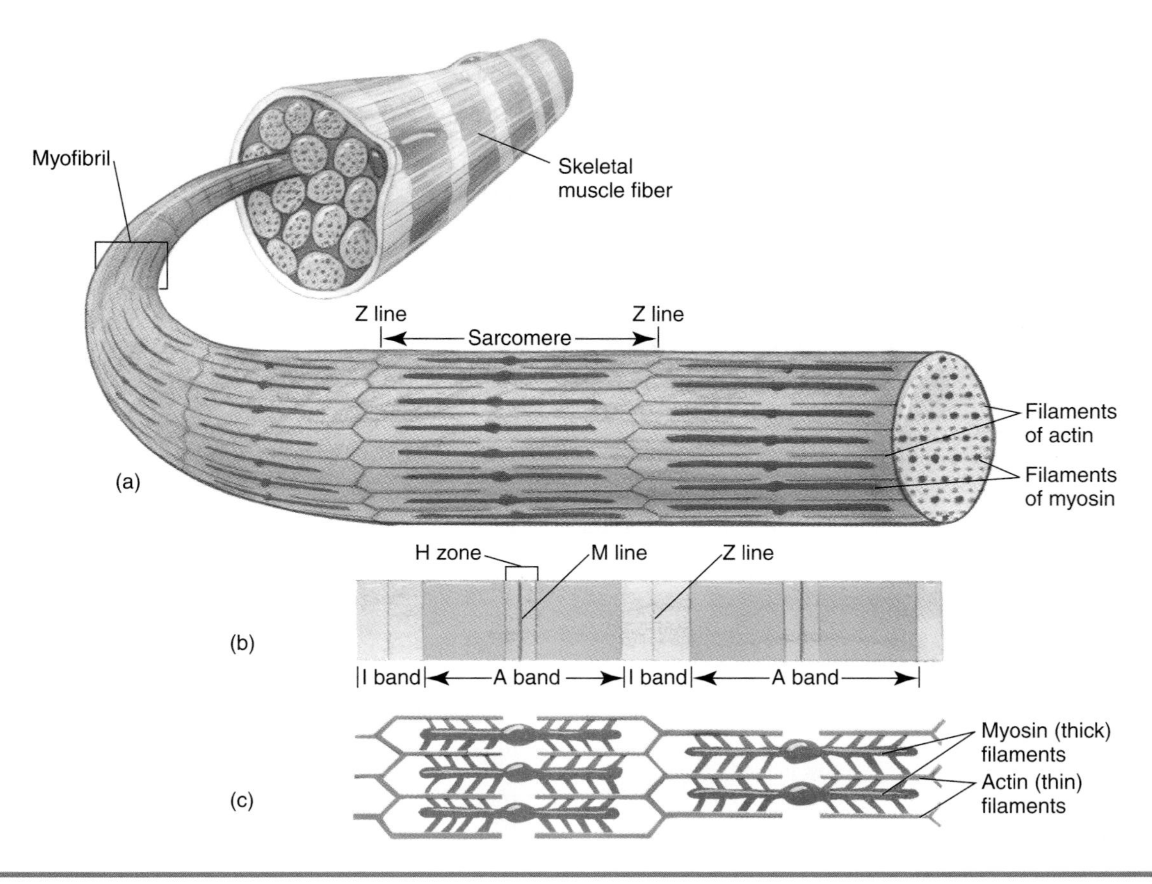

Figure 7.2 (*a*) A skeletal muscle fiber contains numerous myofibrils, each consisting of (*b*) units called sarcomeres. (*c*) The characteristic striations of a sarcomere are due to the spatial arrangement of actin and myosin filaments.

The number of skeletal muscle fibers cannot be increased after birth. However, heavy exercise, such as weight training, increases the number of myofibrils, which increases the diameter and strength of the muscle fibers and of the whole muscle itself.

Anabolic steroids, substances similar to the male sex hormone testosterone, have been used by some athletes to promote muscle development and strength. However, physicians have warned that such use can produce a number of harmful side effects, including damage to kidneys, increased risk of heart disease and liver cancer, and increased irritability. Other side effects include decreased testosterone and sperm production in males and increased facial hair and deepening of the voice in females.

Neuromuscular Interaction

Contraction of a muscle fiber involves the interaction of a motor neuron and a muscle fiber. A motor neuron is attached to each muscle fiber, and the fiber contracts only when stimulated by the motor neuron. The motor neuron originates in either the brain or the spinal cord.

Motor Units

A **motor unit** consists of a motor (action-causing) neuron and all of the muscle fibers to which it is attached (figure 7.3). While a muscle fiber is attached to only one motor neuron, a single motor neuron may innervate from 3 to 2,000 muscle fibers. Where precise muscle control is needed, such as in the fingers, a motor unit contains very few muscle fibers. Large numbers of such motor units are involved in the manipulative movements of the fingers. In contrast, where precise control is not needed, such as in the postural muscles, a motor unit controls hundreds of muscle fibers.

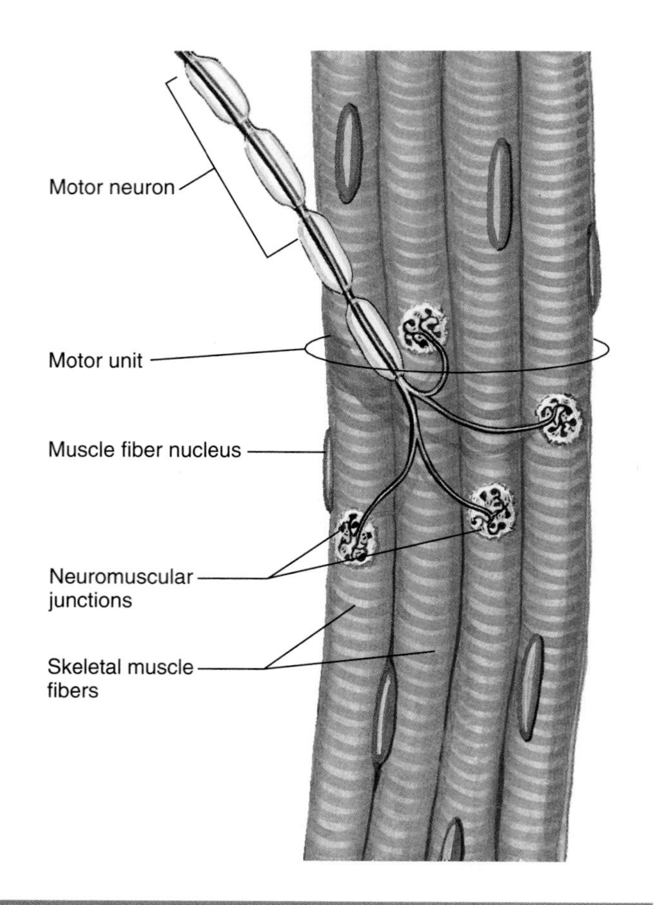

Figure 7.3 A motor unit consists of one motor neuron and all the muscle fibers that it innervates. Note the attachment of the axon tips to the muscle fibers.

Whenever a motor neuron is activated, it stimulates contraction of all of the muscle fibers that it controls.

Neuromuscular Junction

The part of a motor neuron that leads to a muscle fiber is called an *axon.* The connection between the terminal branches of an axon and the sarcolemma of a muscle fiber is known as a **neuromuscular junction.** As shown in figure 7.4, the axon tips fit into depressions in the sarcolemma known as *synaptic clefts.* Numerous vesicles in the axon tip contain the **neurotransmitter** (nū-rō-trans′-mit-er) *acetylcholine* (as″-ē-til-kō′-lēn). When a motor neuron is activated, acetylcholine is released into the synaptic cleft, which, in turn, stimulates the muscle fiber to contract.

Physiology of Muscle Contraction

Contraction of a skeletal muscle fiber is a complex process that involves a number of rapid structural and chemical changes within the fiber. The molecular mechanism of contraction is explained by the **sliding filament model.**

Mechanism of Contraction

When activated, a motor neuron releases acetylcholine into the synaptic cleft of the neuromuscular junction.

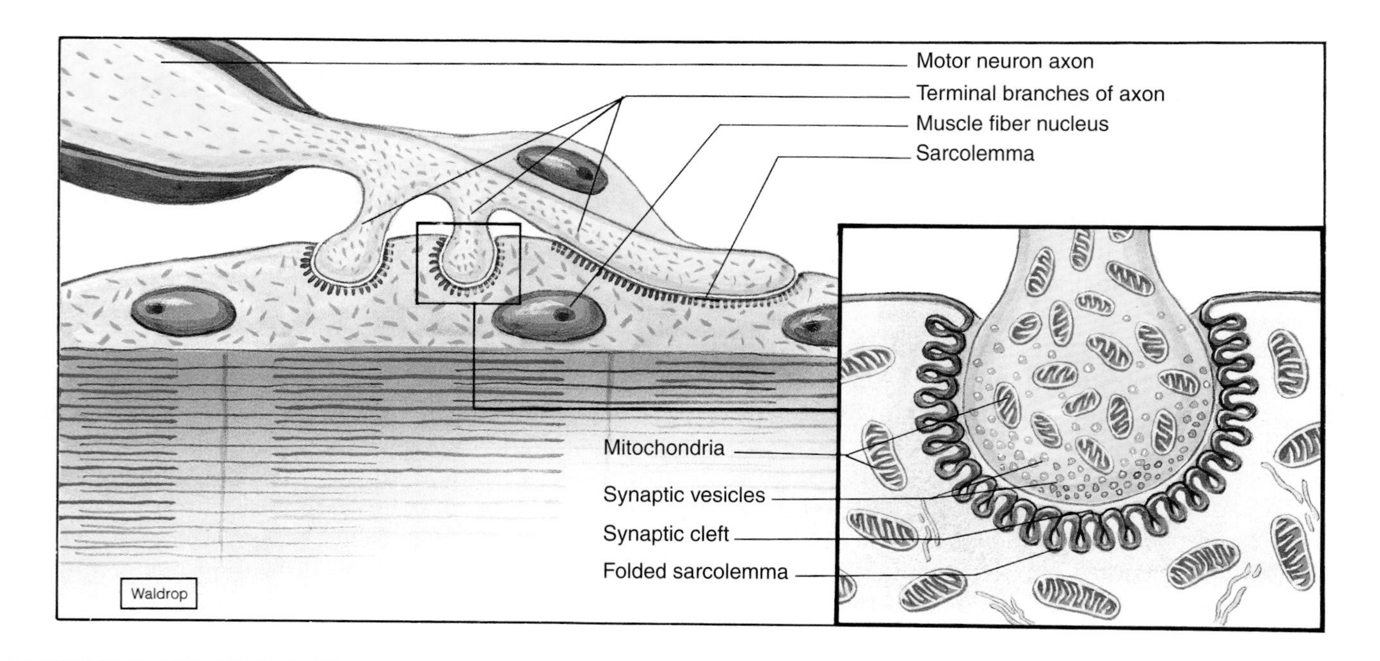

Figure 7.4 A neuromuscular junction is formed by the tip of a motor axon and the sarcolemma of a muscle fiber. The detailed insert shows the synaptic vesicles, the synaptic cleft, and the folded surface of the sarcolemma.

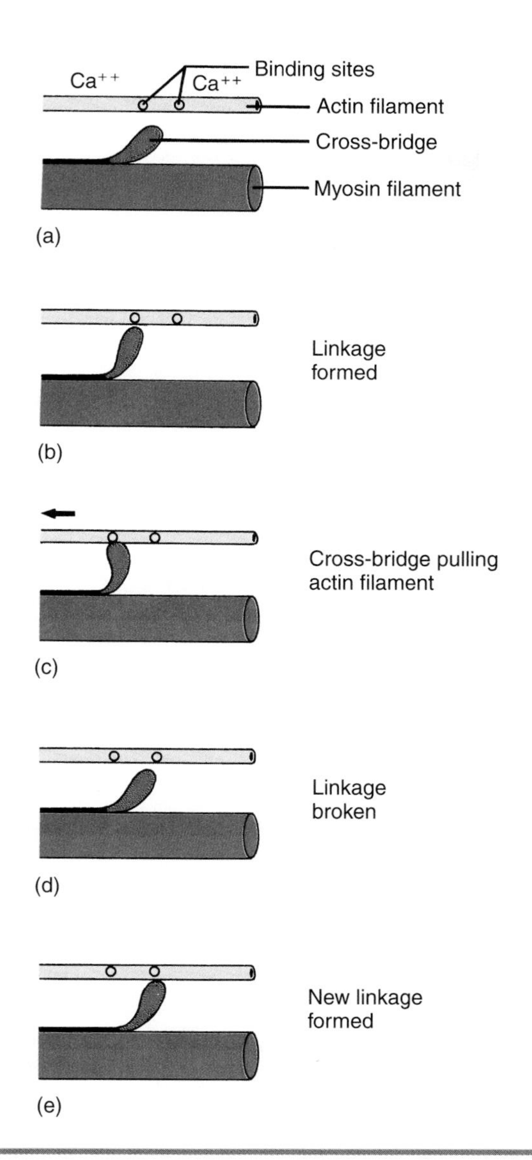

Figure 7.5 According to the sliding filament theory, (*a*) when calcium ions are present, binding sites on an actin filament are exposed. (*b*) Cross-bridges on a myosin filament form linkages at the binding sites. (*c*) A myosin cross-bridge bends slightly, pulling an actin filament, using energy from ATP. (*d*) The linkage breaks, and (*e*) the myosin cross-bridge forms a linkage with the next binding site.

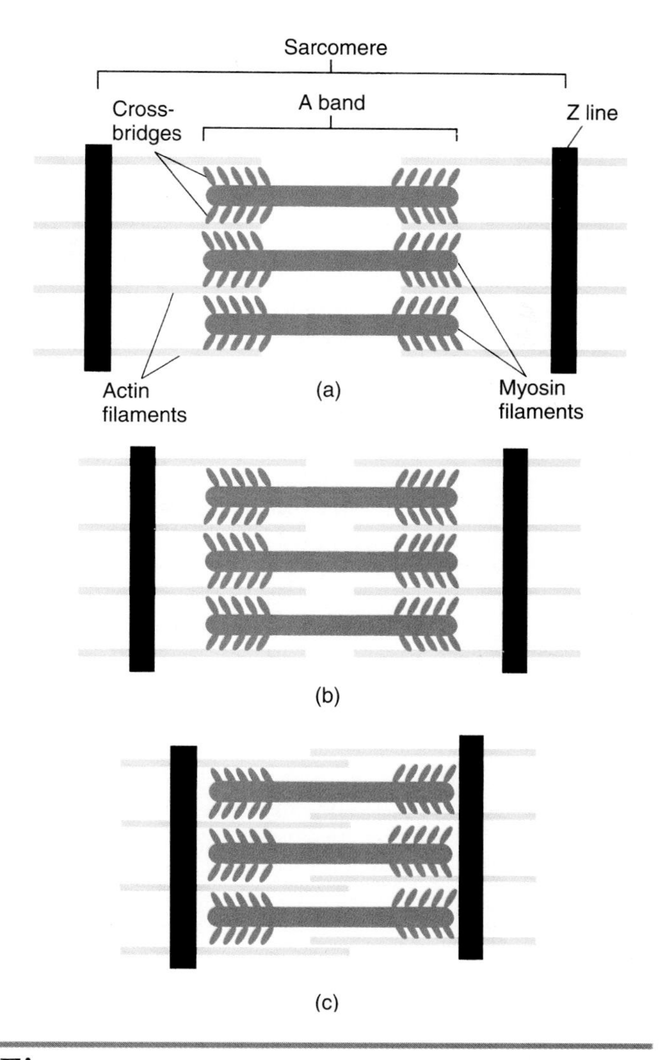

Figure 7.6 Orientation of actin and myosin filaments in a sarcomere during (*a*) relaxation, (*b*) partial contraction, and (*c*) full contraction.

Acetylcholine binds to receptors on the sarcolemma, which triggers a series of events leading to the release of Ca^{++} from storage areas within the muscle fiber. The presence of Ca^{++} exposes the active sites on actin filaments, enabling cross-bridges from myosin filaments to attach to the exposed active sites. Energy from ATP causes the cross-bridges to bend, pulling the actin filaments toward the center of the sarcomere. Then the cross-bridges detach and reattach to the next active sites and, powered by ATP, pull the actin filament farther toward the center of the sarcomere. The process repeats itself, as long as sufficient ATP is present, to attain maximum contraction (figures 7.5 and 7.6).

Once contraction occurs, the sarcolemma secretes the enzyme *cholinesterase* (kō-lin-es′-ter-ās), which decomposes acetylcholine in the synaptic cleft. This action of cholinesterase is necessary to prevent continued stimulation of the muscle fiber and to prepare the fiber for the next stimulus. For each contraction, acetylcholine is released into the synaptic cleft, and then it is decomposed by cholinesterase.

What is the structure and function of a neuromuscular junction?
How do actin and myosin filaments interact during muscle contraction?

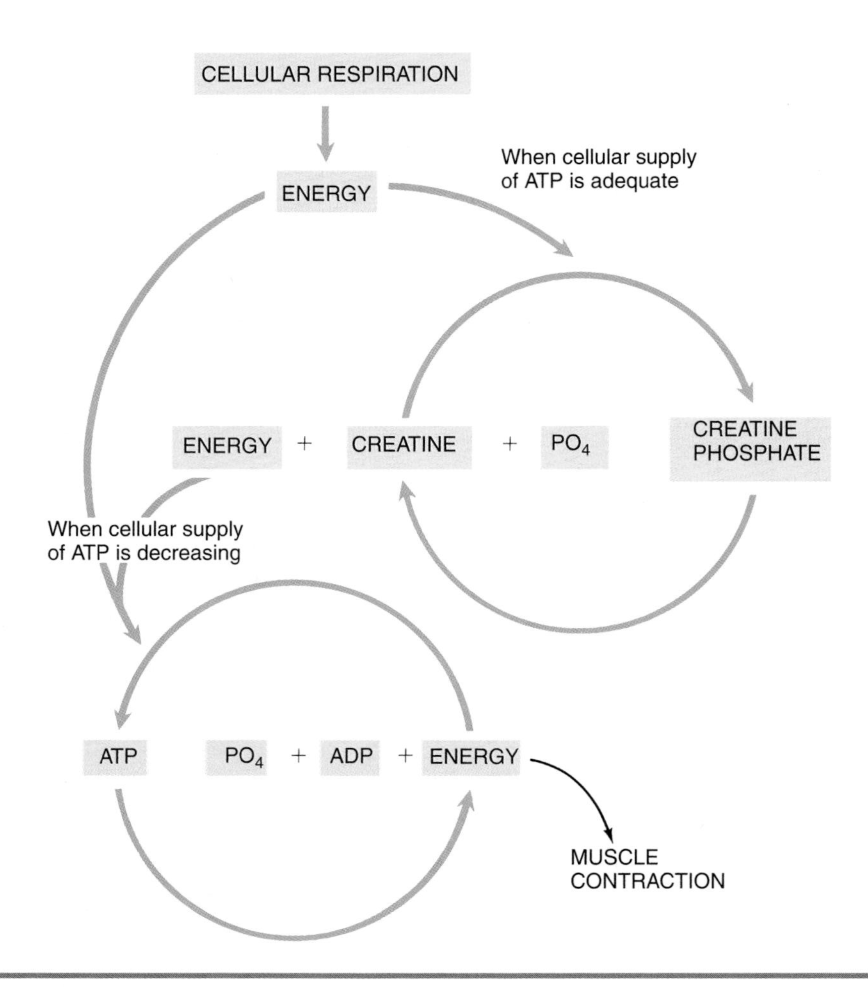

Figure 7.7 Muscle contraction is powered by energy from ATP molecules that are formed by cellular respiration. When ATP supplies are filled, energy from surplus ATP is transferred to creatine to form creatine phosphate, which temporarily stores the energy. As ATP is depleted, creatine phosphate releases the stored energy to form additional ATP.

Energy for Contraction

The energy for muscle contraction comes from ATP molecules in the muscle cell. Recall that ATP is a product of cellular respiration. However, there is only a small amount of ATP in each muscle fiber. Once it is used up, more ATP must be formed in order for additional contractions to occur.

Figure 7.7 shows two possible pathways for energy held in high-energy phosphate bonds. During cellular respiration, the energy released from nutrients is transferred to the high-energy phosphate bonds of ATP. Once there are sufficient amounts of ATP available in the cell, the high-energy phosphate is transferred to creatine to form *creatine phosphate,* which serves as a limited storage form of readily available energy.

When ATP supplies are decreased, a phosphate group containing this energy is transferred to ADP to form ATP, which can then be used to power additional contractions. There is four to six times more creatine phosphate than ATP in a muscle cell, so it is an important source for rapid ATP formation without waiting for the slower process of cellular respiration. However, it can also be depleted rather quickly in a muscle that is contracting repeatedly.

Oxygen and Cellular Respiration

ATP is formed by cellular respiration. Recall from chapter 3 that cellular respiration takes place in two steps: (1) an anaerobic phase in the cytoplasm (sarcoplasm in muscle cells) and (2) an aerobic phase in mitochondria. The small amount of ATP produced by anaerobic respiration cannot keep a person alive, except for very brief periods. Oxygen is life-sustaining because it enables the aerobic phase of cellular respiration to operate and produce adequate amounts of ATP.

Oxygen is transported to body cells in loose combination with **hemoglobin,** the red pigment in red blood cells. Muscle cells have another pigment,

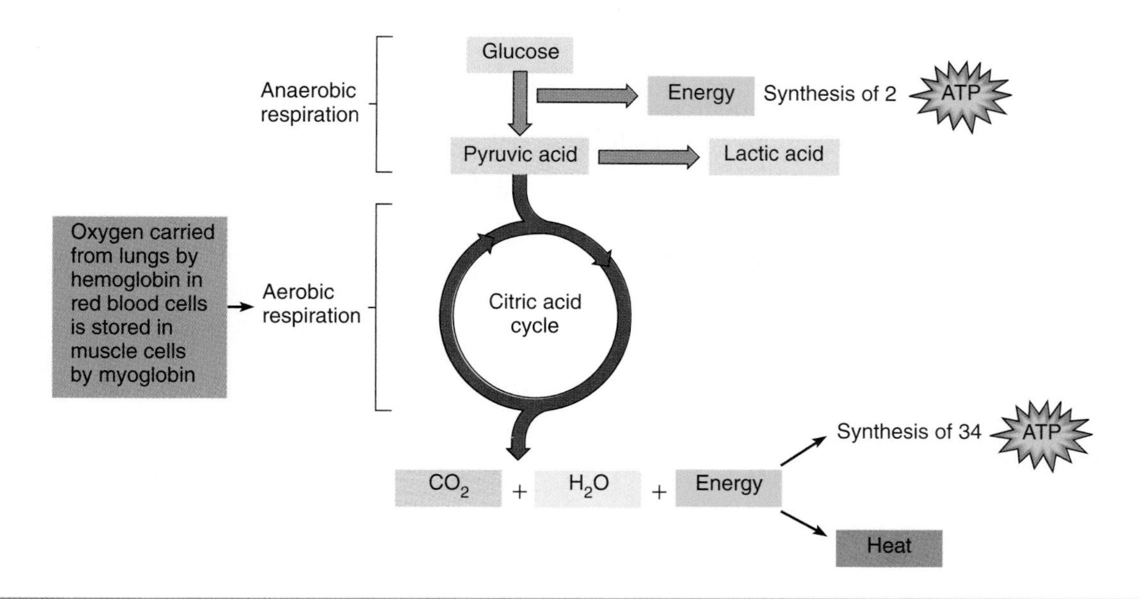

Figure 7.8 A summary of anaerobic and aerobic phases of cellular respiration. Oxygen bound to myoglobin provides a reserve supply within muscle cells. When the oxygen supply is inadequate, lactic acid accumulates in muscle cells, producing an oxygen debt. ATP production varies with cell type, and 36 ATP molecules per glucose molecule is maximum for skeletal muscle cells.

myoglobin, that can also combine loosely with oxygen. In times of muscle inactivity, some of the oxygen carried to muscle cells can be stored in combination with myoglobin, and the oxygen can be released later for aerobic cellular respiration during periods of muscle activity. This function of myoglobin reduces, for a brief period, the muscle cell's dependence on oxygen carried to it by the blood.

During inactivity or moderate physical activity, muscle cells receive sufficient oxygen via the blood to carry on the aerobic phase of cellular respiration. As shown in figure 7.8, this process involves the breakdown of *pyruvic acid* into carbon dioxide and water. However, during periods of strenuous exercise, the respiratory and circulatory systems cannot provide oxygen to muscle cells fast enough to maintain aerobic respiration. This causes some of the pyruvic acid to be converted to *lactic acid* as anaerobic respiration continues to break down glucose to form pyruvic acid and small amounts of ATP.

The accumulation of lactic acid causes discomfort and rapid, deep breathing in an attempt to provide adequate oxygen for aerobic respiration. About 80% of the lactic acid is carried by blood from the muscle cells to the liver, where it is later converted back into glucose and is reused. Lactic acid must be either broken down by aerobic respiration or converted back into glucose, and since oxygen is required for both processes, an **oxygen debt** develops as lactic acid accumulates. The oxygen debt is the amount of oxygen required to metabolize the accumulated lactic acid and to resupply the normal amounts of ATP and creatine phosphate in the muscle cells. When the strenuous muscle activity ceases, rapid, deep breathing continues until the oxygen debt is fully paid. Endurance training enhances the efficiency of aerobic cellular respiration in muscle cells by increasing (1) the number of mitochondria, (2) the efficiency of obtaining oxygen from the blood, and (3) the concentration of myoglobin.

Fatigue

If a muscle is stimulated to contract for a long period, its contractions will gradually decrease until it no longer responds to stimulation. This condition is called **fatigue.** Fatigue seems to result primarily from an accumulation of lactic acid and carbon dioxide, which cause chemical changes in the muscle fiber that prevent it from responding to stimulation.

Heat Production

Heat production by muscular activity is an important mechanism in maintaining a normal body temperature because muscles are active organs and form a large proportion of the body weight. Heat produced by muscles results from cellular respiration and other chemical reactions within the cells. Recall that 60% of the energy released by cellular respiration is heat energy, with only 40% of the released energy captured in ATP. The importance of muscle activity in maintaining body temperature is evident by the fact that a major response to a marked decrease in body temperature is shivering, involuntary muscle contractions.

> ***What are the roles of ATP and creatine phosphate in muscle contraction?***
> ***What is the relationship between cellular respiration, lactic acid, and oxygen debt?***

Contraction Characteristics

When studying muscle contraction, physiologists consider both single fiber contraction and whole muscle contraction.

Contraction of a Single Fiber

It is possible to remove a single muscle fiber in order to study its contraction in the laboratory. By using electrical stimuli to initiate contraction and by gradually increasing the strength (voltage) of each stimulus, it has been shown that the fiber will not contract until the stimulus reaches a certain minimal strength. This minimal stimulus is called the **threshold stimulus.**

Whenever a muscle fiber is stimulated by a threshold stimulus, or by a stimulus of greater strength, it always contracts *completely.* Thus, a muscle fiber either contracts completely or not at all—contraction is *not* proportional to the strength of the stimulus. This characteristic of individual muscle fibers is known as the **all-or-none response.**

Contraction of Whole Muscles

Much information has been gained by studying the contraction of a whole muscle of an experimental animal. In such studies, electrical stimulation is used to cause contraction, and the contraction is recorded to produce a tracing called a *myogram.*

If a single threshold stimulus is applied, some of the muscle fibers will contract to produce a single, weak contraction (a muscle twitch) and then relax, all within a fraction of a second. The myogram will look like the one shown in figure 7.9. After the stimulus is applied, there is a brief interval before the muscle starts to contract. This interval is known as the *latent period.* Then, the muscle contracts (shortens) rapidly during the *period of contraction* and relaxes (returns to its former length) during the longer *period of relaxation.* If a muscle is stimulated again after it has re-

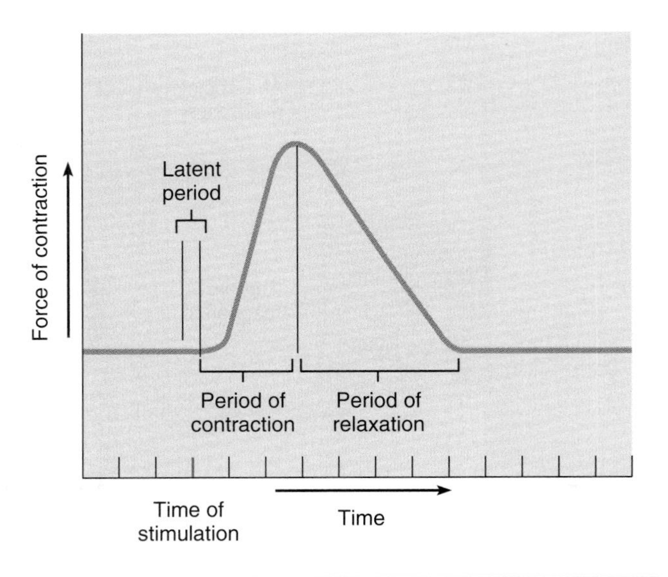

Figure 7.9 A myogram of a single muscle twitch. Note the brief latent period, rapid contraction period, and longer relaxation period.

laxed completely, it will contract and produce a similar myogram. A series of single stimuli applied in this manner will yield a myogram like the one in figure 7.10*a*.

If the interval between stimuli is shortened so that the muscle fibers cannot completely relax, the force of individual twitches combines by **summation** which increases the force of contraction (figure 7.10*b*). If stimuli are so frequent that relaxation is not possible, a **tetanic contraction** results (figure 7.10*c*).

Graded Responses Unlike individual muscle fibers that exhibit all-or-none responses, whole muscles exhibit *graded responses,* that is, varying degrees of contraction. Graded responses enable the degree of muscle contraction to fit the task being performed. Obviously, more muscle fibers are required to lift a 14 kg (30 lb) weight than to lift a feather. Yet, both activities can be performed by the same muscles.

Graded responses are possible because a muscle is composed of many different *motor units,* each responding to different thresholds of stimulation. In the laboratory, a weak stimulus that activates only low-threshold motor units produces a minimal contraction. As the strength of the stimulus is increased, the contractions get stronger as more motor units are activated until a **maximal stimulus** (one that activates all motor units) is applied, which produces a maximal contraction. Further increases in the

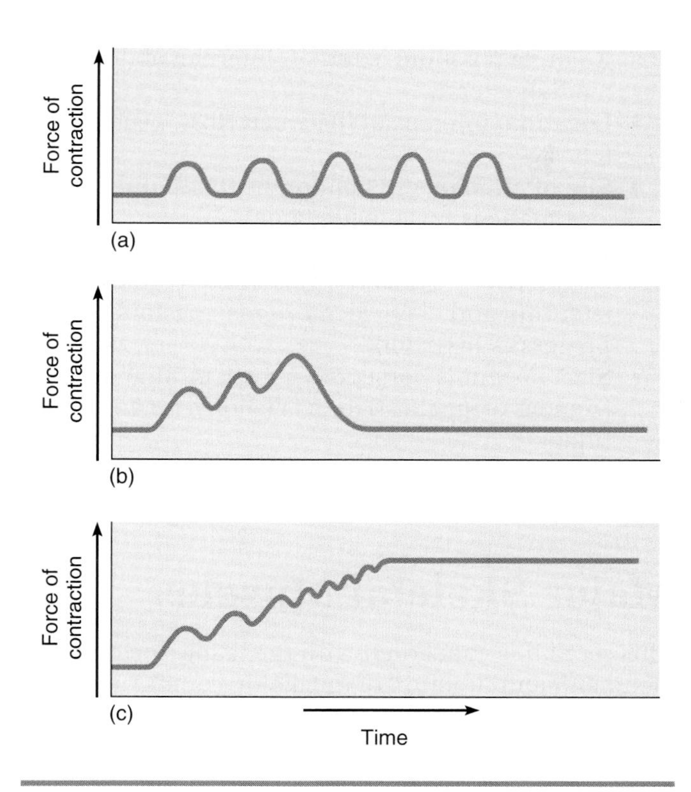

Figure 7.10 Myograms of (*a*) a series of simple twitches, (*b*) summation caused by incomplete relaxation between stimuli, and (*c*) tetanic contraction when there is no relaxation between stimuli.

strength of the stimulus cannot produce a greater contraction. The same results occur in a normally functioning body, but it is the nervous system that provides the stimulation and controls the number of motor units activated in each muscle contraction. The activation of more and more motor units is known as *recruitment.*

Sustained Contractions A sustained (tetanic) contraction lacks even partial relaxation and produces a myogram as shown in figure 7.10*c*. Sustained contractions for short time periods is the usual way in which muscles contract to produce body movements. Sustained contractions result from a rapid series of impulses carried by motor neurons to the muscle fibers.

Muscle Tone Even when a muscle is relaxed, some of its fibers are contracting. At any given time, some of the fibers in a muscle are involved in a sustained contraction that produces a constant partial, but slight, contraction of the muscle. This state of constant partial contraction is called **muscle tone** and keeps a muscle ready to respond. Muscle tone results from the alternating activation of different motor units by the nervous system so that some muscle fibers are always in sustained contraction. Muscle tone of postural muscles plays an important role in maintaining erect posture.

If a nerve to a muscle is severed, the muscle is not only paralyzed but becomes soft and flabby due to the absence of muscle tone. Such a muscle will decrease in size due to lack of use, a process known as *atrophy.*

What is meant by the all-or-none response? How are muscles able to make graded responses?

Actions of Skeletal Muscles

Skeletal muscles are usually arranged so that the ends of a muscle are attached to bones on each side of a joint. Thus, a muscle usually extends across a joint. The type of movement produced depends upon the type of joint and the locations of the muscle attachments. Common movements at joints were discussed in chapter 6.

Origin and Insertion

During contraction, a bone to which one end of the muscle is attached moves, but the bone to which the other end is attached does not. The movable attachment of a muscle is called the **insertion,** and the immovable attachment is called the **origin.** When a muscle contracts, the insertion is pulled toward the origin.

Consider the *biceps brachii* muscle in figure 7.11. It has two origins, and both are attached to the scapula. The insertion is on the radius, and the muscle lies along the anterior surface of the humerus. When the biceps contracts, the insertion is pulled toward the origin, which results in the flexion of the forearm at the elbow.

Most muscle contractions are *isotonic contractions,* which cause movement at a joint. Walking and lifting weights are examples. However, some contractions may not produce movement but only increase tension within a muscle. Pushing against an immovable object is an example. Such contractions are *isometric contractions,* and they are used by body builders to strengthen and enlarge muscles.

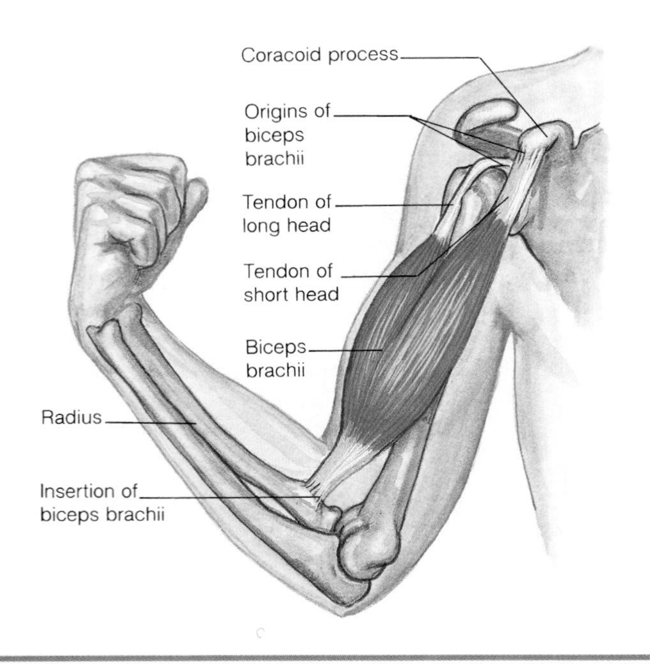

Figure 7.11 The biceps brachii muscle showing its two origins and single insertion. Note how the muscle extends across the elbow joint so that contraction flexes the forearm.

Muscle Interactions

Muscles function in groups rather than singly, and the groups are arranged to provide opposing movements. For example, if one group of muscles produces flexion, the opposing group produces extension. A group of muscles producing an action are called *agonists* and the opposing group of muscles are called *antagonists*. When agonists contract, antagonists must relax, and vice versa, for movement to occur. If both groups contract simultaneously, the movable body part remains rigid.

Naming of Muscles

Learning the complex names and functions of muscles can sometimes be confusing. However, the names of muscles are informative if their meaning is known. A few of the criteria used in naming muscles and examples of terms found in the names of muscles are listed following:

- **Function:** extensor, flexor, adductor, and pronator.
- **Shape:** trapezius (trapezoid), rhomboideus (rhomboid), deltoid (delta-shaped or triangular), biceps (two heads).
- **Relative position:** external, internal, abdominal, medial, lateral.
- **Location:** intercostal (between ribs), pectoralis (chest).
- **Site of attachment:** temporalis (temporal bone), zygomaticus (zygomatic bone).
- **Origin and insertion:** sternohyoid (sternum = origin; hyoid = insertion), sternocleidomastoid (sternum and clavicle = origins; mastoid process = insertion).
- **Size:** maximus (larger or largest), minimus (smaller or smallest), brevis (short), longus (long).
- **Orientation of fibers:** oblique (diagonal), rectus (straight), transverse (across).

Major Skeletal Muscles

This section is concerned with the name, location, attachment, and function of the major skeletal muscles. There are over 600 muscles in the body, but only a few of the major muscles are considered here. Nearly all of them are superficial muscles. Most of this information is presented in tables and figures to aid your learning. The tables are organized according to the primary actions of the muscles. The pronunciation of each muscle is included, since being able to pronounce the names correctly will help you memorize the names of the muscles.

As you study this section, locate each muscle listed in the tables on the related figures 7.12 to 7.26. This will help you visualize the location and action of each muscle. Also, if you visualize the locations of the origin and insertion of a muscle, its action can be determined since contraction pulls the insertion toward the origin. It may help to refresh your understanding of the skeleton by referring to appropriate figures in chapter 6.

Muscles of Facial Expression and Mastication

Muscles of the face and scalp produce the facial expressions that help communicate feelings, such as anger, sadness, happiness, fear, disgust, pain, and surprise. Most have origins on skull bones and insertions on connective tissue of the skin (table 7.2 and figure 7.14).

The *epicranius* is an unusual muscle. It has a large aponeurosis that covers the top of the skull and two contractile portions: the *frontalis* over the frontal bone and the *occipitalis* over the occipital bone.

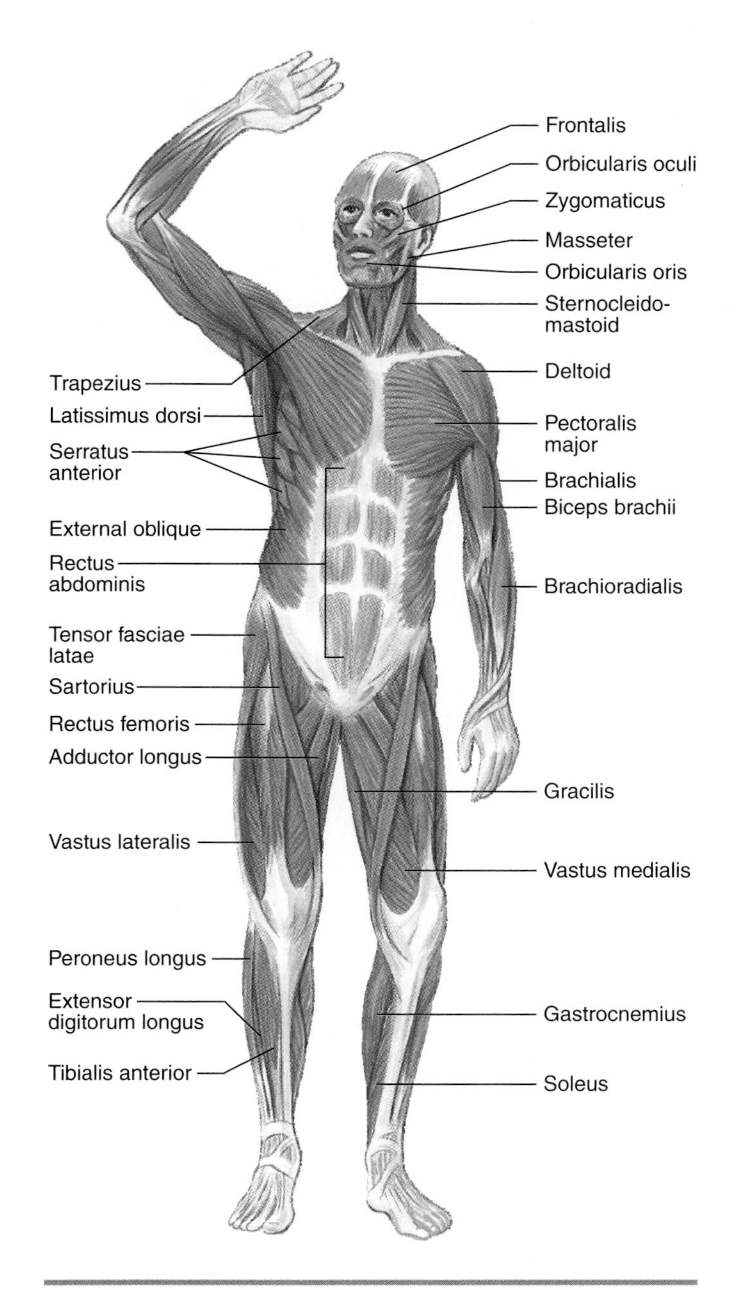

Figure 7.12 Anterior view of superficial skeletal muscles.

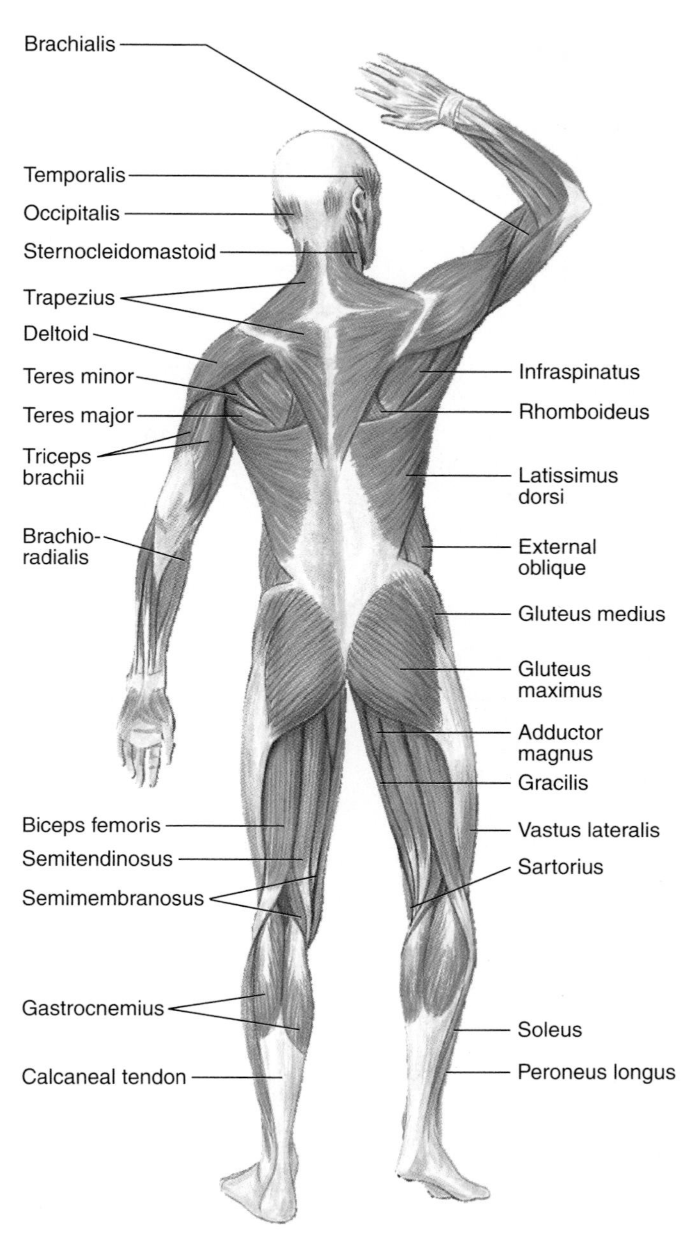

Figure 7.13 Posterior view of superficial skeletal muscles.

Two major pairs of muscles raise the mandible in the process of mastication: the *masseter* and the *temporalis* (table 7.3 and figure 7.14).

Muscles That Move the Head

Several pairs of neck muscles are responsible for flexing, extending, and turning the head. Table 7.4 lists two of the major muscles that perform this function: the *sternocleidomastoid* and the *splenius capitis.* As noted in table 7.7, the *trapezius* can also extend the head, although this is not its major function (figures 7.14, 7.15, and 7.16).

Muscles of the Abdominopelvic Wall

These paired muscles provide support for the anterior and lateral portions of the abdominal and pelvic

Table 7.2 Muscles of Facial Expression

Muscle	Origin	Insertion	Action
Buccinator (buk′-si-nā-tor)	Lateral surfaces of maxilla and mandible	Orbicularis oris	Compresses cheeks inward
Epicranius (ep-i-krā′-nē-us)	This muscle consists of two parts: the frontalis and the occipitalis. They are joined by the epicranial aponeurosis, which covers the top of the skull.		
Frontalis (fron-ta′-lis)	Epicranial aponeurosis	Skin and muscles above the eyes	Raises eyebrows and wrinkles forehead horizontally
Occipitalis (ok-sip-i-tal′-is)	Base of occipital bone	Epicranial aponeurosis	Pulls scalp posteriorly
Orbicularis oculi (or-bik′-ū-lar-is ok′-ū-li)	Frontal bone and maxillae	Skin around eye	Closes eye
Orbicularis oris (or-bik′-ū-lar-is- o′-ris)	Muscles around mouth	Skin around lips	Closes and puckers lips; shapes lips during speech
Platysma (plah-tiz′-mah)	Fascia of upper chest	Mandible and muscles around mouth	Draws angle of mouth downward; opens mouth
Zygomaticus (zī-gō-mat′-ik-us)	Zygomatic bone	Orbicularis oris at corners of the mouth	Pulls angle of mouth upward

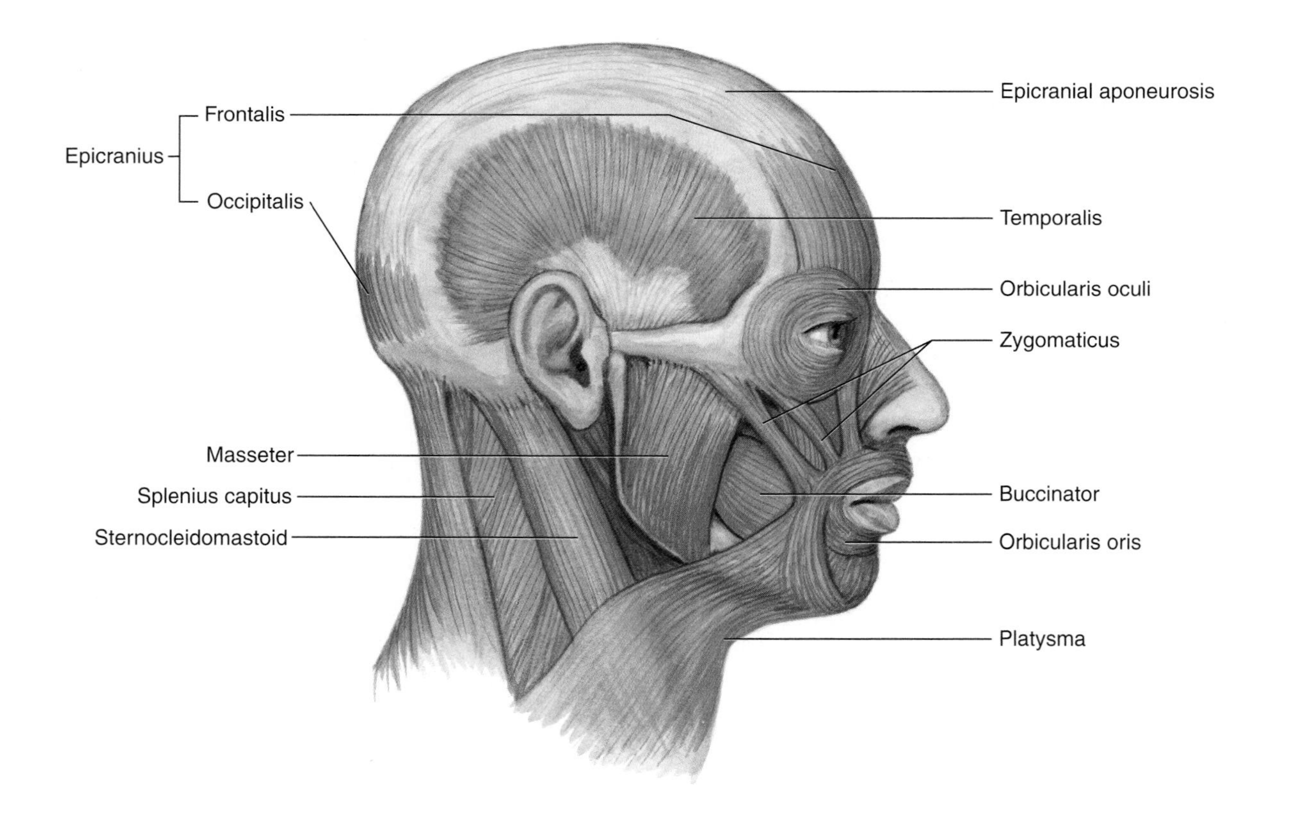

Figure 7.14 Muscles of facial expression and mastication.

Table 7.3 Muscles of Mastication

Muscle	Origin	Insertion	Action
Masseter (mas-se′-ter)	Zygomatic arch	Lateral surface of mandible	Raises mandible
Temporalis (tem-po-ra′-lis)	Temporal bone	Coronoid process of mandible	Raises mandible

Table 7.4 Muscles That Move the Head

Muscle	Origin	Insertion	Action
Sternocleidomastoid (ster-nō-klī-dō-mas′-toid)	Clavicle and sternum	Mastoid process of temporal bone	Contraction of both muscles flexes head toward chest; contraction of one muscle turns head away from contracting muscle
Splenius capitus (splē′-nē-us kap′-i-tis)	Lower cervical and upper thoracic vertebrae	Mastoid process of temporal bone	Contraction of both muscles extends head; contraction of one muscle turns head toward same side as contracting muscle

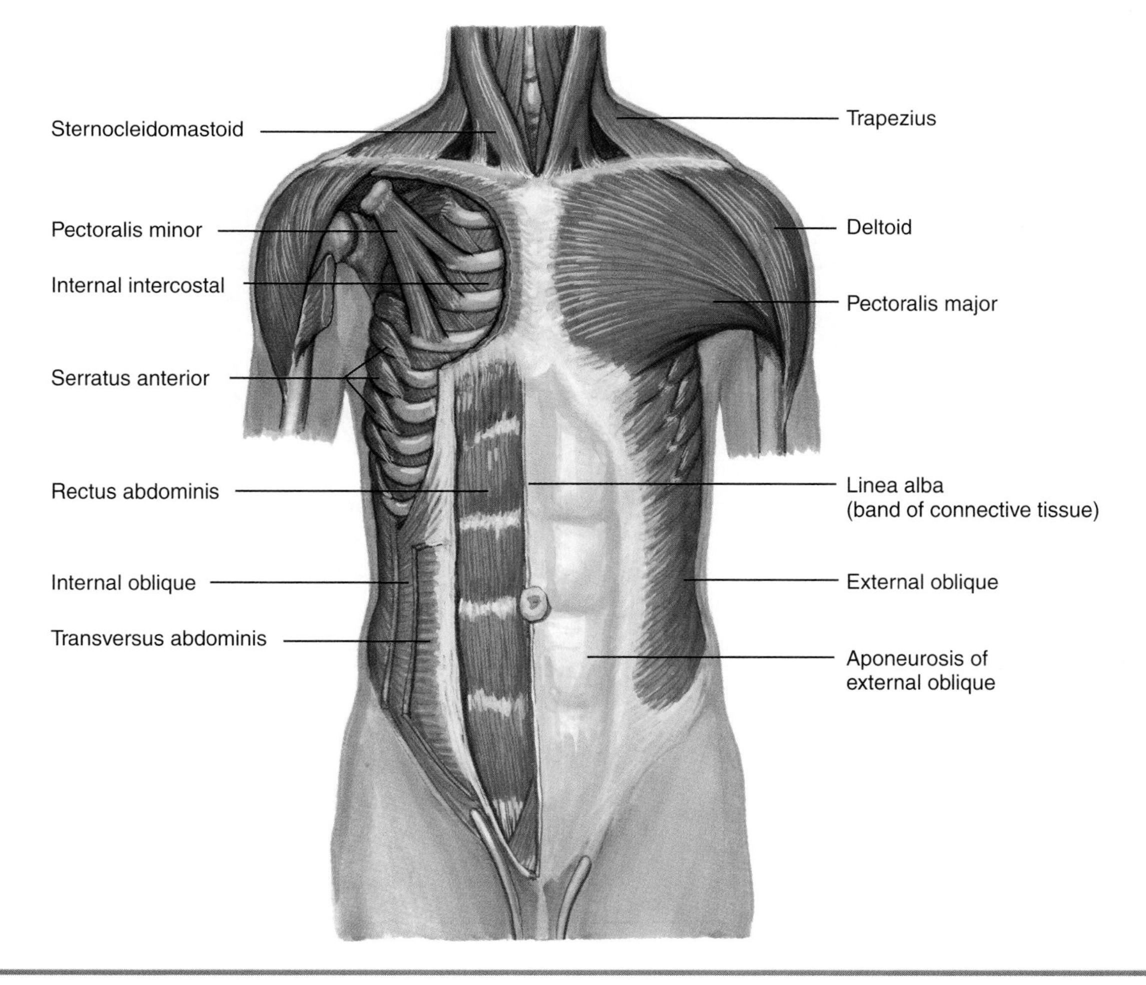

Figure 7.15 Muscles of the anterior chest and abdominal wall. The right pectoralis major is removed to show the pectoralis minor.

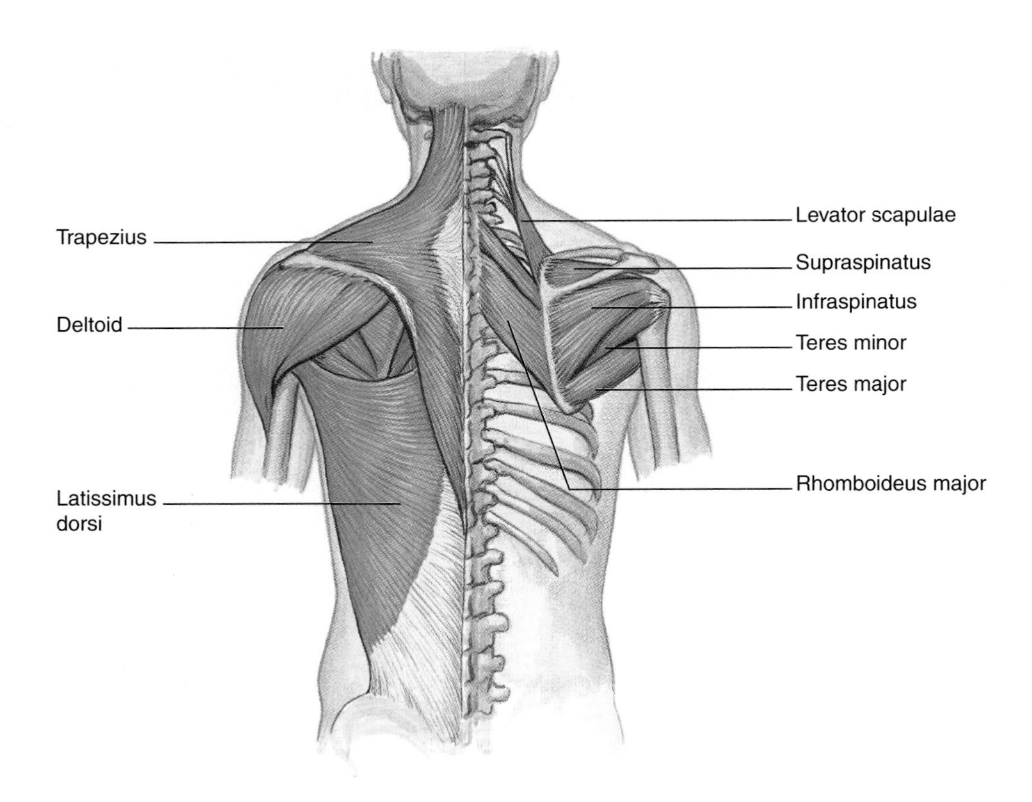

Figure 7.16 Muscles of the posterior shoulder. The right trapezius is removed to show underlying muscles.

Table 7.5 Muscles of the Abdominopelvic Wall

Muscle	Origin	Insertion	Action
Rectus abdominis (rek′-tus ab-dom′-i-nis)	Pubic symphysis and pubic crest	Xiphoid process of sternum and costal cartilages of ribs 5 to 7	Tightens abdominopelvic wall; compresses internal organs; flexes the vertebral column
External oblique (eks-ter′-nal o-blēk′)	Anterior surface of lower eight ribs	Iliac crest and linea alba	Tightens abdominopelvic wall and compresses internal organs
Internal oblique (in-ter′-nal o-blēk′)	Iliac crest and inguinal ligament	Cartilage of lower four ribs, pubic crest, and linea alba	Same as above
Transversus abdominis (trans-ver′-sus ab-dom′-i-nis)	Iliac crest, cartilages of lower six ribs, processes of lumbar vertebrae	Pubic crest and linea alba	Same as above

regions, including support for the internal organs. The muscles are named for the direction of their muscle fibers: *rectus abdominis, external oblique, internal oblique,* and *transversus abdominis.* They are arranged in overlapping layers and are attached by larger aponeuroses that merge at the anterior midline to form the *linea alba,* or white line (table 7.5 and figure 7.15).

Muscles of Breathing

Movement of the ribs occurs during breathing and is brought about by the contraction of two sets of muscles that are located between the ribs. The *external intercostals* lift the ribs upward and outward during inspiration, and the *internal intercostals* draw the ribs down and inward during expiration (table 7.6 and

Table 7.6 Muscles of Breathing

Muscle	Origin	Insertion	Action
Diaphragm (dī-a-fram)	Lumbar vertebrae, costal cartilages of lower ribs, xiphoid process	Central tendon located at midpoint of muscle	Forms floor of thoracic cavity; lowers during contraction, causing inspiration
External intercostals (eks-ter′-nal in-ter-kos′-tals)	Inferior border of rib above	Superior border of rib below	Raises ribs upward and outward during inspiration
Internal intercostals (in-ter′-nal in-ter-kos′-tals)	Superior border of rib below	Inferior border of rib above	Draws ribs downward and inward during expiration

Table 7.7 Muscles That Move the Pectoral Girdle

Muscle	Origin	Insertion	Action
Trapezius (trah-pē-zē′-us)	Occipital bone; cervical and thoracic vertebrae	Clavicle; spine and acromion process of scapula	Elevates clavicle; adducts and elevates scapula; extends head
Rhomboideus major (rom-boid′-ē-us)	Upper thoracic vertebrae	Medial border of scapula	Adducts and elevates scapula
Levator scapulae (le-va′-tor skap′-ū-lē)	Cervical vertebrae	Superior medial margin of scapula	Elevates scapula
Serratus anterior (ser-ra′-tus)	Upper eight to nine ribs	Medial border of scapula	Pulls scapula downward and anteriorly
Pectoralis minor (pek-to-rah′-lis)	Anterior surface of upper ribs	Coracoid process of scapula	Pulls scapula anteriorly and downward

figure 7.15). The primary breathing muscle is the *diaphragm,* a thin sheet of muscle that separates the thoracic and abdominal cavity. It is not shown in the figures.

Muscles That Move the Pectoral Girdle

These muscles originate on bones of the axial skeleton and insert on the scapula or clavicle. Since the scapula is supported mainly by muscles, it may be moved more freely than the clavicle. The *trapezius* is a superficial trapezoid-shaped muscle that covers much of the upper back. The *rhomboideus major* and the *levator scapulae* lie under the trapezius. The *serratus anterior* muscles are located on the lateral surfaces of the upper ribs near the axillary regions. The *pectoralis minor* lies under the pectoralis major. It pulls the scapula forward and downward (table 7.7 and figures 7.15 to 7.18).

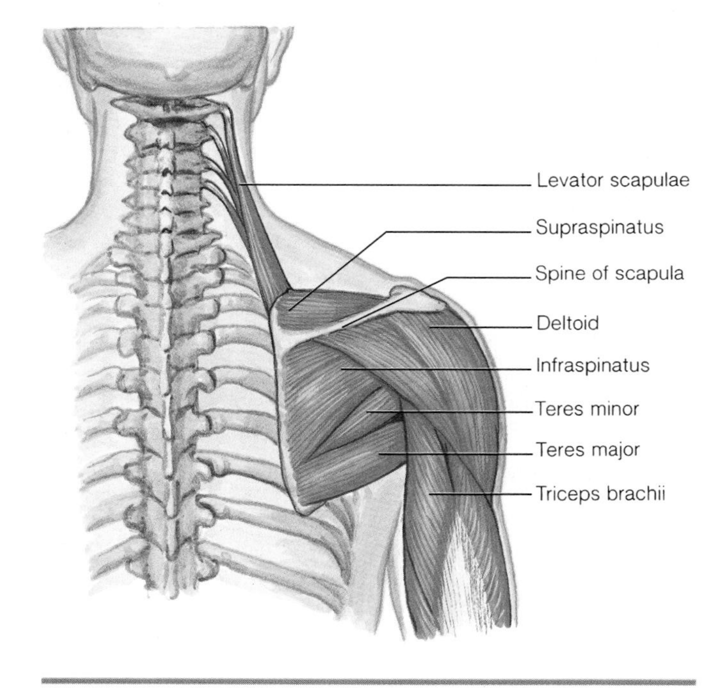

Figure 7.17 Muscles of the posterior surface of the scapula and arm.

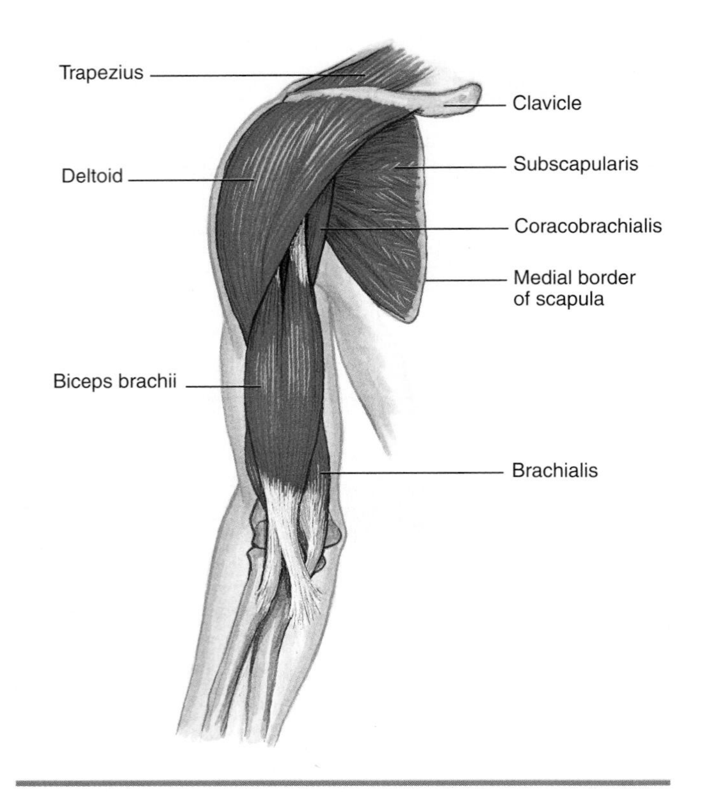

Figure 7.18 Muscles of the anterior shoulder and arm, with the rib cage removed.

Muscles That Move the Arm and Forearm

Movement of the humerus is enabled by the muscles that originate on the pectoral girdle, ribs, or vertebrae and insert on the humerus. The arrangement of these muscles and the ball-and-socket joint between the humerus and scapula enable great freedom of movement for the arm. The *pectoralis major* is the large superficial muscle of the chest. The *deltoid* is the thick muscle that caps the shoulder joint. The *coracobrachialis* is a relatively small muscle that lies under the pectoralis major and biceps brachii. It flexes and adducts the arm. The *supraspinatus, infraspinatus, teres major,* and *teres minor* cover the posterior surfaces of the scapulae (plural of *scapula*). The *latissimus dorsi* is a broad, sheetlike muscle that covers the lower back (table 7.8 and figures 7.15 to 7.20).

Muscles moving the forearm originate on either the humerus or scapula and insert on either the radius or ulna. Three flexors occur on the anterior surface of the upper arm: the *biceps brachii, brachialis,* and *brachioradialis.* One extensor, the *triceps brachii,* is located on the posterior surface of the upper arm (table 7.9 and figures 7.18, 7.19, and 7.20).

Muscles That Move the Wrist and Fingers

Many muscles that produce the various movements of the wrist and fingers are located in the forearm. Only

Table 7.8 Muscles That Move the Arm

Muscle	Origin	Insertion	Action
Pectoralis major (pek-tō-rah′-lis)	Clavicle; sternum, and cartilages of upper ribs	Greater tubercle of humerus	Adducts humerus; draws it forward across chest
Coracobrachialis (kōr″-a-kō-brā-kē-ah′-lis)	Coracoid process of scapula	Shaft of humerus	Adducts and flexes humerus
Deltoid (del′-toid)	Clavicle and spine, and acromion of scapula	Deltoid tuberosity of humerus	Abducts, flexes, and extends humerus
Supraspinatus (su-prah-spī′-na-tus)	Posterior surface of scapula above spine	Greater tubercle of humerus	Assists deltoid in abducting humerus
Infraspinatus (in-frah-spī′-na-tus)	Posterior surface of scapula below spine	Greater tubercle of humerus	Rotates humerus laterally
Latissimus dorsi (lah-tis′-i-mus dor′sī)	Lower thoracic, lumbar, and sacral vertebrae; lower ribs; iliac crest	Intertubercular groove of humerus	Adducts and extends humerus; rotates humerus medially
Teres major (te′r-ez)	Inferior angle of scapula	Distal to lesser tubercle of humerus	Extends, adducts, and rotates humerus medially
Teres minor	Lateral border of scapula	Greater tubercle of humerus	Rotates humerus laterally

a few of the larger superficial muscles are considered here. They originate from the distal end of the humerus and insert on carpals, metacarpals, or phalanges. Flexors on the anterior surface include the *flexor carpi radialis, flexor carpi ulnaris,* and *palmaris longus.* Extensors on the posterior surface include the *extensor carpi radialis longus, extensor carpi ulnaris,* and *extensor digitorum* (table 7.10 and figure 7.20). Note that the tendons of these muscles are held in position by a circular ligament at the wrist.

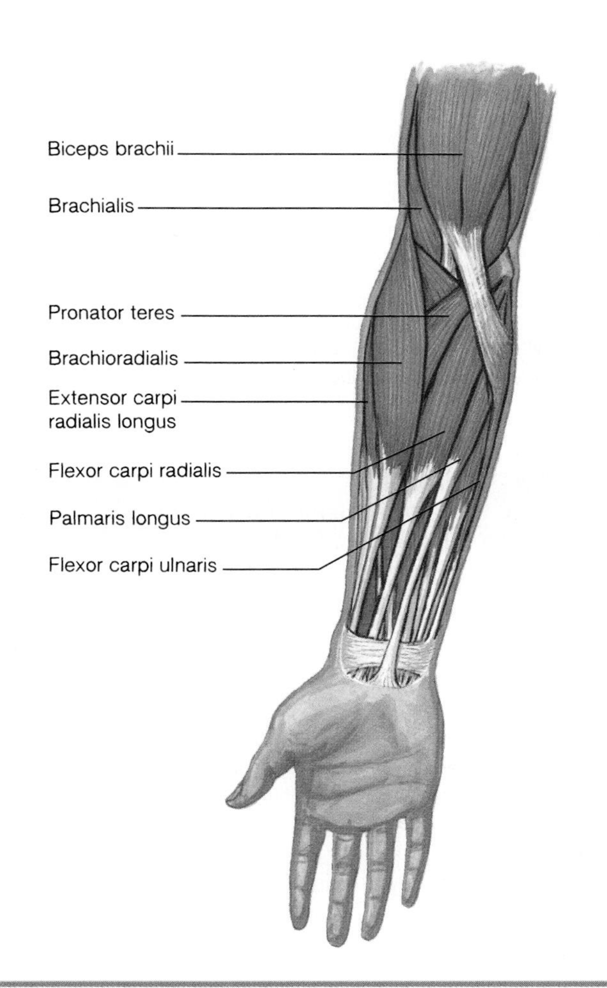

Figure 7.19 Muscles of the anterior forearm.

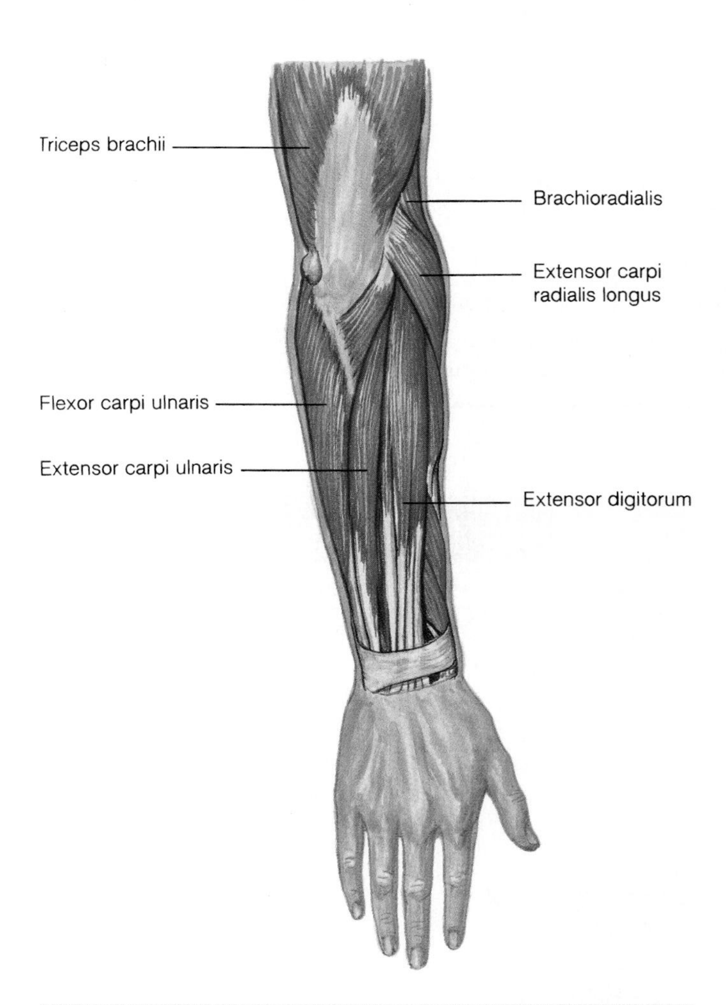

Figure 7.20 Muscles of the posterior forearm.

Table 7.9 Muscles That Move the Forearm

Muscle	Origin	Insertion	Action
Biceps brachii (bī′-seps brā′-kē-ī)	Coracoid process and tubercle above glenoid cavity of scapula	Radial tuberosity of radius	Flexes and rotates forearm laterally (supination)
Brachialis (brā′-kē-al-is)	Distal, anterior surface of humerus	Coronoid process of ulna	Flexes forearm
Brachioradialis (brā-kē-ō-rā-dē-a′-lis)	Lateral surface of distal end of humerus	Lateral surface of radius just above styloid process	Flexes forearm
Triceps brachii (trī′-seps brā′-kē-ī)	Lateral and medial surfaces of humerus and tubercle below glenoid cavity of scapula	Olecranon process of ulna	Extends forearm

Table 7.10 Muscles That Move the Wrist and Fingers

Muscle	Origin	Insertion	Action
Flexor carpi radialis (flek′-sor kar′-pī rā-dē-a′-lis)	Medial epicondyle of humerus	Second and third metacarpals	Flexes and abducts wrist
Flexor carpi ulnaris (flek′-sor kar′-pī ul-na′-ris)	Medial epicondyle of humerus and olecranon process of ulna	Carpal bones and fifth metacarpal	Flexes and adducts wrist
Palmaris longus (pal-ma′-ris long′-gus)	Medial epicondyle of humerus	Fascia of palm	Flexes wrist
Extensor carpi radialis longus (eks-ten′-sor kar′-pī rā-dē -a′-lis long′-gus)	Lateral epicondyle of humerus	Second metacarpal	Extends and abducts wrist
Extensor carpi ulnaris (eks-ten′-sor kar′-pī ul-na′-ris)	Lateral epicondyle of humerus	Fifth metacarpal	Extends and adducts wrist
Extensor digitorum (eks-ten′-sor dij-i-to′-rum)	Lateral epicondyle of humerus	Posterior surfaces of phalanges 2–5	Extends fingers

Table 7.11 Muscles That Move the Thigh

Muscle	Origin	Insertion	Action
Iliacus (il′-ē-ak-us)	Fossa of ilium	Lesser trochanter of femur	Flexes thigh
Psoas major (so′-as)	Lumbar vertebrae	Lesser trochanter of femur	Flexes thigh
Gluteus maximus (glū′-tē-us mak′-si-mus)	Posterior surfaces of ilium; sacrum; and coccyx	Posterior surface of femur and lateral fascia of thigh	Extends and rotates thigh laterally
Gluteus medius (glū′-tē-us mē′-dē-us)	Lateral surface of ilium	Greater trochanter of femur	Abducts and rotates thigh medially
Tensor fasciae latae (ten′-sor fash′-ē-ē lah-tē′)	Anterior iliac crest	Lateral fascia of thigh	Flexes and abducts thigh
Adductor longus (ad-duk′-tor long′-gus)	Pubic bone near symphysis pubis	Posterior surface of femur	Adducts, flexes, and rotates thigh laterally
Adductor magnus (ad-duk′-tor mag′-nus)	Inferior portion of ischium and pubis	Same as above	Same as above

Muscles That Move the Thigh and Leg

Muscles moving the thigh span the hip joint. They insert on the femur and most originate on the pelvic girdle. The *iliacus* and *psoas major* are located anteriorly, the *gluteus maximus* is located posteriorly and forms the buttocks, the *gluteus medius* is located under the gluteus maximus posteriorly and extends laterally, and the *tensor fasciae latae* is located laterally. Both the *adductor longus* and *adductor magnus* are located medially (table 7.11 and figures 7.21, 7.22, and 7.23).

The leg is moved by muscles located in the thigh. They span the knee joint and originate on the pelvic girdle or thigh and insert on the tibia or fibula. The *quadriceps femoris* is composed of four muscles that have a common tendon that inserts on the patella. However, this tendon continues as the patellar ligament, which attaches to the tibial tuberosity—the functional insertion for these muscles. The *biceps femoris, semitendinosus,* and *semimembranosus* on the posterior surface of the thigh are often collectively called the "hamstrings." The medially located *gracilis* has two insertions that give it dual actions. The long, straplike *sartorius* extends diagonally

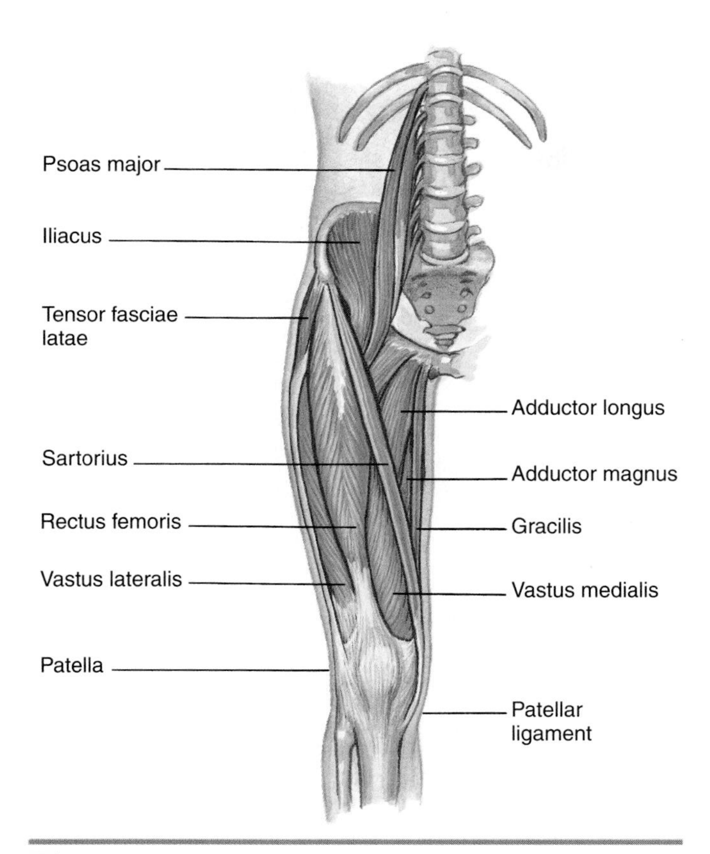

Figure 7.21 Muscles of the anterior right thigh. (Note that vastus intermedius is a deep muscle not visible in this view.)

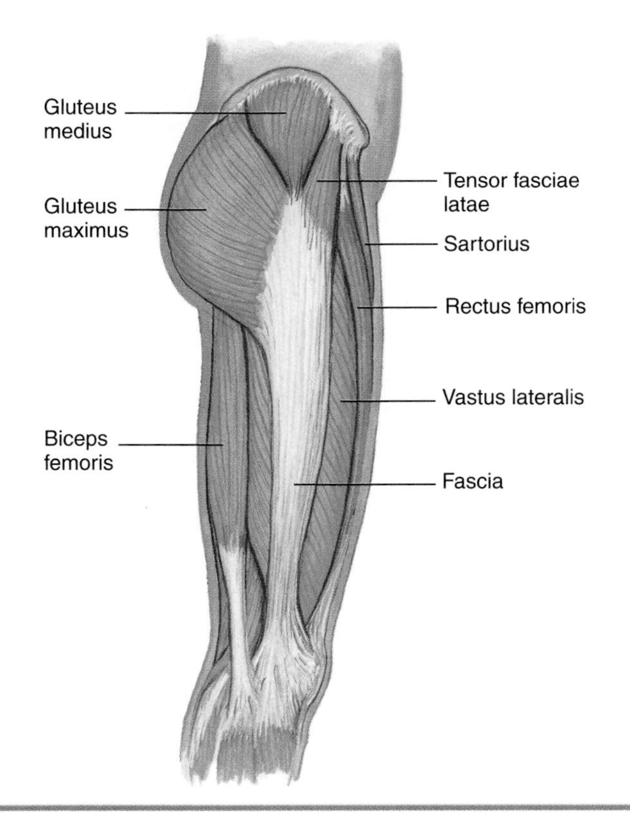

Figure 7.22 Muscles of the lateral right thigh.

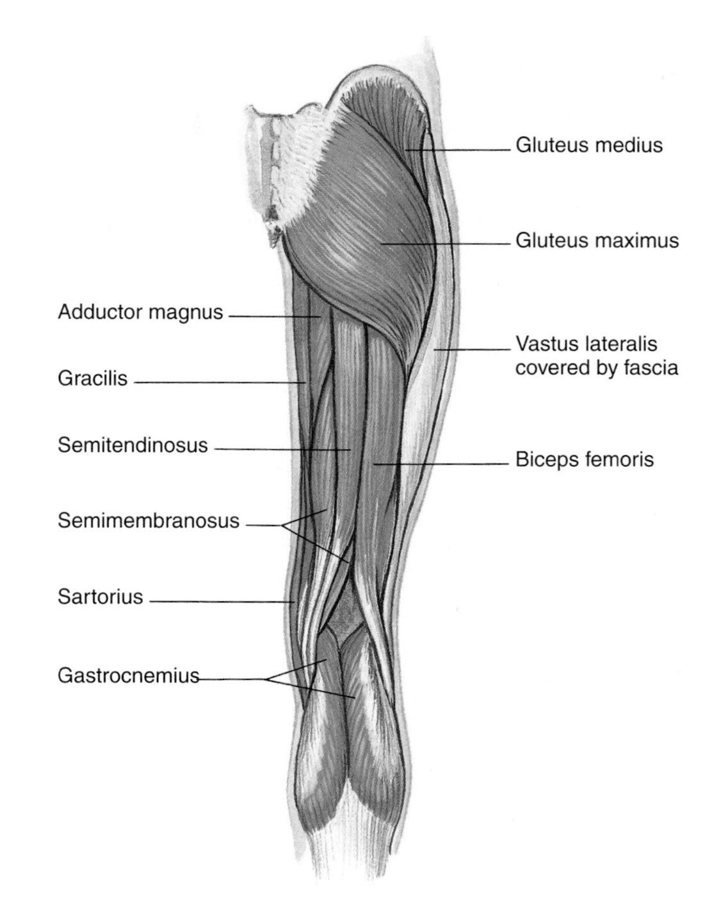

Figure 7.23 Muscles of the posterior right thigh.

across the anterior surface of the thigh and spans both the hip and knee joints. Its contraction enables the legs to cross (table 7.12 and figures 7.21, 7.22, and 7.23).

Intramuscular injections are commonly used when quick absorption is desired. Such injections are given in three sites: (1) the lateral surface of the deltoid; (2) the gluteus medius in the upper, outer portion of the buttock; and (3) the vastus lateralis near the midpoint of the lateral surface of the thigh. These injection sites are chosen because there are no major nerves or blood vessels present that could be damaged, and the muscles have a good blood supply to aid absorption. The site chosen may vary with the age and condition of the patient.

Table 7.12 Muscles That Move the Leg

Muscle	Origin	Insertion	Action
Quadriceps femoris (quad′-ri-seps fem′-or-is)	A composite thigh muscle formed of four parts that are usually described as separate muscles. Each muscle has a common tendon that attaches to the patella and continues as the patellar ligament to the tibial tuberosity.		
Rectus femoris (rek′-tus fem′-or-is)	Anterior inferior iliac spine and superior margin of acetabulum	Patella; tendon continues as patellar ligament, which attaches to tibial tuberosity	Extends leg and flexes thigh
Vastus lateralis (vas′-tus lat-er-a′lis)	Greater trochanter and posterior surface of femur	Same as above	Extends leg
Vastus medialis (vas′-tus me-de-a′lis)	Medial and posterior surfaces of femur	Same as above	Extends leg
Vastus intermedius (vas′-tus in-ter-mē′dē-us)	Anterior and lateral surfaces of femur	Same as above	Extends leg
Hamstrings	Three distinct muscles of the posterior thigh.		
Biceps femoris (bi′-seps fem′-or-is)	Ischial tuberosity and posterior surface of femur	Head of fibula and lateral condyle of tibia	Flexes and rotates leg laterally; extends thigh
Semitendinosus (sem-ē-ten-di-nō′-sus)	Ischial tuberosity	Medial surface of tibia	Flexes and rotates leg medially; extends thigh
Semimembranosus (sem-ē-mem-brah-nō′-sus)	Ischial tuberosity	Medial condyle of tibia	Flexes and rotates leg medially; extends thigh
Gracilis (gras′-il-is)	Pubis near symphysis	Medial surface of tibia	Flexes and rotates leg medially; adducts thigh
Sartorius (sar-to′r-ē-us)	Anterior superior iliac spine	Medial surface of tibia	Flexes thigh; rotates leg medially and thigh laterally as when crossing legs

Table 7.13 Muscles That Move the Foot and Toes

Muscle	Origin	Insertion	Action
Gastrocnemius (gas-trōk-nē′m-ē-us)	Medial and lateral condyles of femur	Calcaneus by the calcaneal tendon	Plantar flexes foot and flexes leg
Soleus (sō′l-ē-us)	Posterior surface of tibia and fibula	Calcaneus by the calcaneal tendon	Plantar flexes foot
Peroneus longus (per-ō-nē′-us long′-gus)	Lateral condyle of tibia and head and body of fibula	First metatarsal and tarsal bones	Plantar flexes and everts foot; supports arch
Tibialis anterior (tib-ē-a′l-is an-te′rē-or)	Lateral condyle and surface of tibia	First metatarsal and tarsal bones	Dorsiflexes and inverts foot
Extensor digitorum longus (eks-ten′-sor dig-i-tor′-um long′-gus)	Lateral condyle of tibia and anterior surface of fibula	Phalanges of toes 2–5	Dorsiflexes and everts foot; extends toes

Muscles That Move the Foot and Toes

Many muscles are involved in the movement of the foot and toes. They are located in the leg and originate on the femur, tibia, or fibula and insert on the tarsals, metatarsals, or phalanges. The posterior leg muscles include the *gastrocnemius* and *soleus,* which insert on a common tendon, the calcaneal tendon, which attaches to the calcaneus (heel bone). The *tibialis anterior* is anteriorly located, and the *extensor digitorum longus* lies lateral to it. Note that although the extensor digitorum

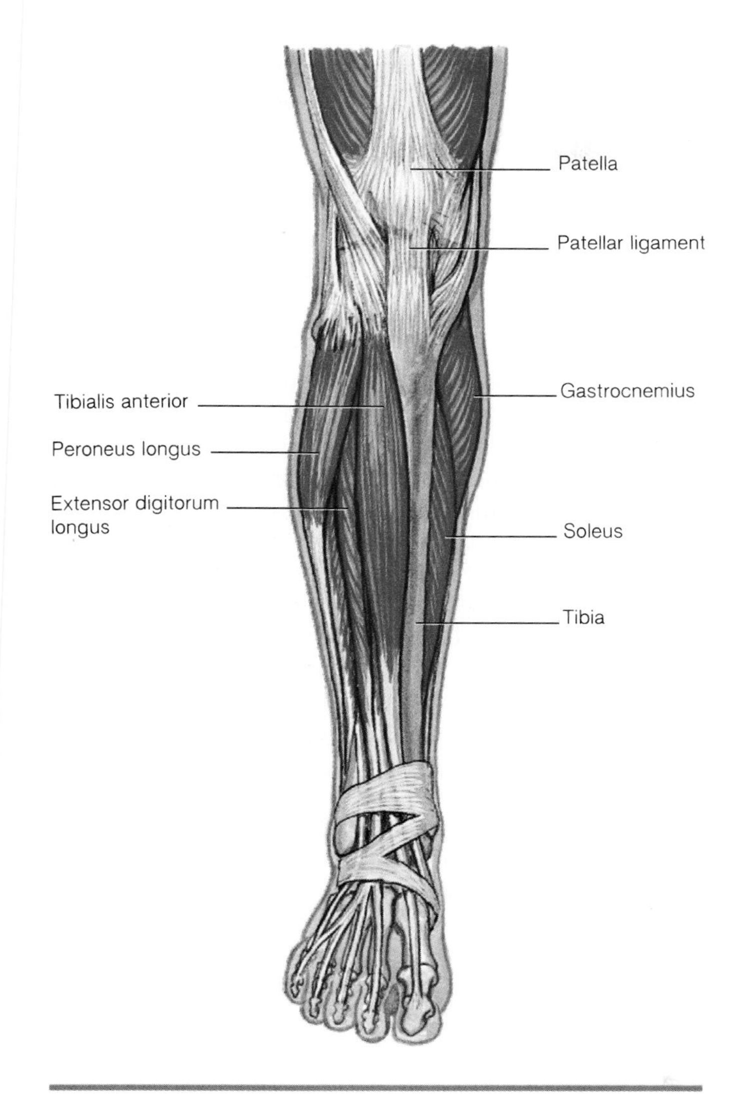

Figure 7.24 Muscles of the anterior right leg.

extends the toes, as its name implies, it also dorsiflexes the foot. The *peroneus longus* is located on the lateral surface of the leg (table 7.13 and figures 7.24 to 7.26).

Foot movements include moving the foot upward (dorsiflexion), moving the foot downward (plantar flexion), turning the sole inward (inversion), and turning the sole outward (eversion). Note how the tendons are held in position by the bands of ligaments at the ankle.

Repeated stress from athletic activities may cause inflammation of a tendon, a condition known as **tendonitis**. Tendons associated with the shoulder, elbow, hip, and knee joints are most commonly affected.

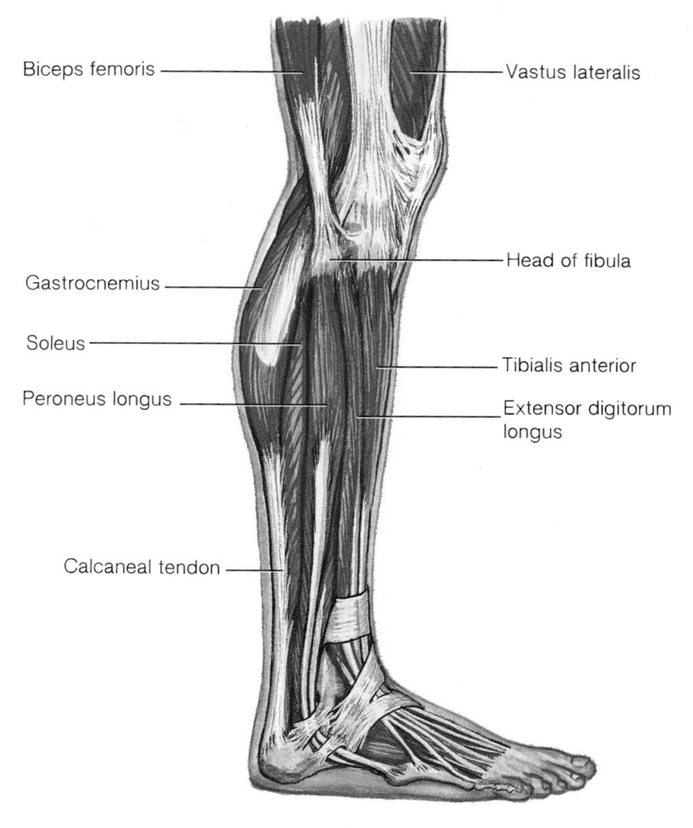

Figure 7.25 Lateral view of muscles of the right lower leg.

Disorders of the Muscle System

Some disorders of the muscle system may result from factors associated only with muscles, while others are caused by disorders of the nervous system. Certain neurological disorders are included here because of their obvious effect on muscle action.

Muscle Disorders

Cramps are involuntary, painful, sustained tetanic contractions of a muscle. The precise cause is unknown, but a cramp may result from chemical changes in the muscle, such as lactic acid accumulation or calcium deficiencies. Sometimes a severe blow to a muscle can produce a cramp.

Fibrosis (fī-brō′-sis) is an abnormal increase of fibrous connective tissue in a muscle. Usually, it results from connective tissue replacing dead muscle fibers following an injury.

Fibrositis (fī-brō-sī′-tis) is the inflammation of the connective tissue, especially muscle sheaths and fascia

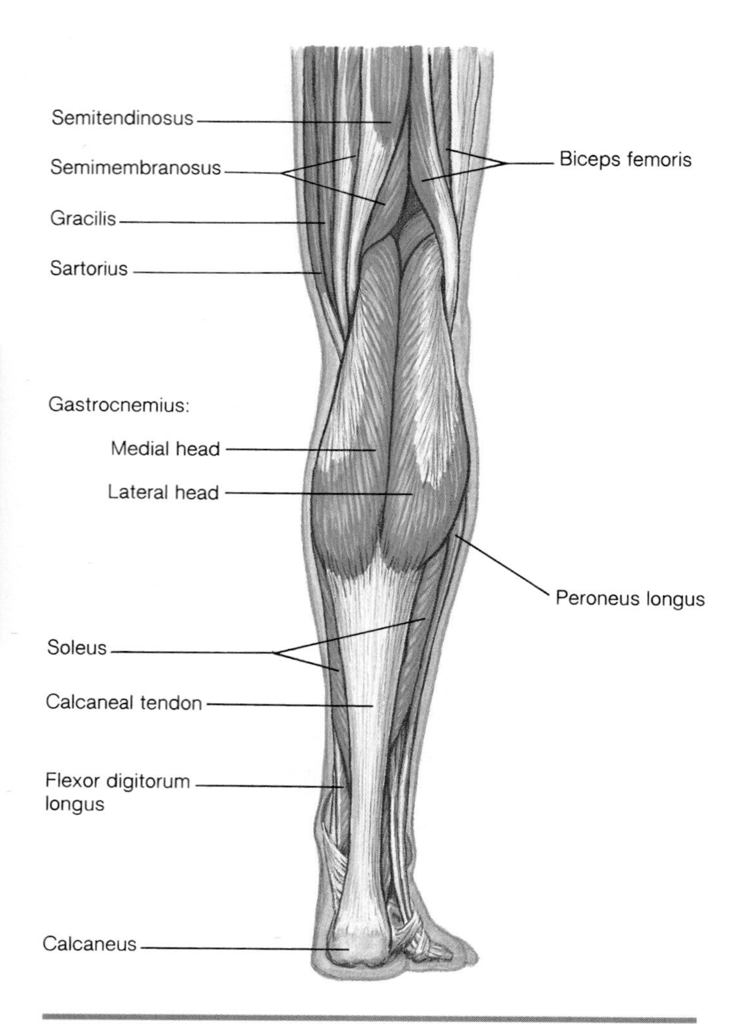

Figure 7.26 Muscles of the posterior right leg.

associated with muscles. It produces soreness and stiffness that is commonly called *muscular rheumatism.*

Muscular dystrophy (dis′-trō-fē) is a general term for a number of inherited muscular disorders that are characterized by the progressive degeneration of muscles. The affected muscles gradually weaken and atrophy (shrink), producing a progressive crippling of the patient. There is no specific drug cure, but patients are encouraged to keep active and are given muscle-strengthening exercises.

Myositis (mī-ō-sī′-tis) is the inflammation of muscle tissue. It produces soreness and stiffness similar to fibrositis of the muscles.

Strains, or "pulled muscles," result when a muscle is stretched excessively. This usually occurs when an antagonist has not relaxed quickly enough as an agonist contracts. The hamstrings are a common site of muscle strains. In mild strains, only a few muscle fibers are damaged. In severe strains, both connective and muscle tissues are torn, and muscle function may be severely impaired.

Neurological Disorders Affecting Muscles

Botulism (boch′-ū-lizm) poisoning is caused by a neurotoxin produced by the bacterium *Clostridium botulinum.* The toxin prevents release of acetylcholine from the tips of motor axons. Without prompt treatment with an antitoxin, death may result from paralysis of the muscles used in breathing. Poisoning results from eating improperly canned vegetables or meats that contain *C. botulinum* and the accumulated toxins.

Myasthenia gravis (mī-as-thē′-nē-ah grav′-is) is characterized by extreme muscular weakness caused by improper functioning of the neuromuscular junctions. It is an autoimmune disease in which antibodies are produced that attach to the acetylcholine receptors on the sarcolemma and reduce or block the stimulatory effect of acetylcholine. Myasthenia gravis occurs most frequently in women between 20 and 40 years of age. Usually, it first affects ocular muscles and other muscles of the face and neck, which may lead to difficulty in chewing, swallowing, and talking. Other muscles of the body may be involved later. Treatment typically involves the use of anticholinesterase drugs and immunosuppressive drugs, such as the steroid prednisone.

Poliomyelitis (pō-lē-ō-mī-e-lī′-tis) is a viral disease of motor neurons in the spinal cord. Destruction of the motor neurons leads to paralysis of skeletal muscles. It is now rare in industrialized countries due to the availability of a polio vaccine. Virtually all children in the United States receive this vaccine, which protects them from polio.

Spasms are sudden, involuntary contractions of a muscle or a group of muscles. They may vary from simple twitches to severe convulsions and may be accompanied by pain. Spasms may be caused by irritation of the motor neurons supplying the muscle, emotional stress, or neurological disorders. Spasms of smooth muscle in the walls of the digestive tract, respiratory passages, or certain blood vessels can be hazardous. Hiccupping is one type of spasm.

Tetanus (tet′-ah-nus) is a disease caused by the anaerobic bacterium *Clostridium tetani,* which is common in soil. Infection usually results from puncture wounds. *C. tetani* produces a neurotoxin that affects motor neurons in the spinal cord, resulting in continuous stimulation and tetanic contractions of certain muscles. Since the first muscles affected are those that move the mandible, this disease is often called "lockjaw." Without prompt treatment, mortality is high. Young children usually receive injections of tetanus toxoid to stimulate production of antibodies against the neurotoxin. Booster injections are given at regular intervals to keep the concentration of antibodies at a high level in order to prevent the disease.

CHAPTER SUMMARY

1. The three types of muscle tissue in the body are skeletal, smooth, and cardiac.
2. Each type of muscle tissue has unique structural and functional characteristics.

Structure of Skeletal Muscle

1. Each skeletal muscle is formed of many muscle fibers (cells) that are arranged in fasciculi.
2. Connective tissue envelops each fiber, each fasiculus, and the entire muscle.
3. Muscles are attached to bones or connective tissue by either tendons or aponeuroses.
4. The sarcolemma is the plasma membrane of a muscle fiber, and the sarcoplasm (cytoplasm) surrounds the myofibrils, the contractile elements.
5. Myofibrils consist of two protein filaments: actin and myosin. The arrangement of actin and myosin filaments produces the striations that are characteristic of skeletal muscle fibers.
6. Each myofibril consists of many sarcomeres joined end-to-end. A sarcomere is bounded by a Z line at each end.
7. The axon tip of a motor neuron is attached to each muscle fiber at the neuromuscular junction. The axon tip fits into a synaptic cleft in the sarcolemma. The neurotransmitter acetylcholine is contained in tiny vesicles in the axon tip.
8. A motor unit consists of a motor neuron and all muscle fibers to which it is attached.

Physiology of Muscle Contraction

1. An activated motor axon releases acetylcholine into the synaptic cleft. Acetylcholine attaches to receptors of the sarcolemma and causes the release of Ca^{++} within the muscle. This, in turn, causes the cross-bridges of myosin filaments to attach to actin filaments. A series of ratchetlike movements pulls the actin filaments toward the center of the sarcomere, producing contraction.
2. Cholinesterase quickly breaks down acetylcholine to prevent continued stimulation and to prepare the muscle fiber for the next stimulus.
3. Energy for contraction comes from high-energy phosphate bonds in ATP.
4. After cellular respiration has formed a muscle cell's normal supply of ATP, excess energy is transferred to creatine to form creatine phosphate, which serves as a small reserve supply of energy.
5. Small amounts of oxygen are stored in combination with myoglobin, which gives muscle cells a small reserve of oxygen for aerobic cellular respiration.
6. Vigorous muscular activity quickly exhausts available oxygen, leading to the accumulation of lactic acid and producing an oxygen debt. Heavy breathing after exercise provides the oxygen required to metabolize lactic acid and pay the oxygen debt.
7. Fatigue results primarily from the accumulation of lactic acid in a muscle fiber.
8. Large amounts of heat are produced by the chemical and physical processes of muscle contraction.
9. When stimulated by a threshold stimulus, individual muscle fibers exhibit an all-or-none contraction response.
10. A simple contraction consists of a (a) latent period, (b) period of contraction, and (c) period of relaxation.
11. Whole muscles provide graded contraction responses, which are enabled by the number of motor units that are recruited.
12. A sustained contraction of all motor units is a tetanic contraction.
13. Muscle tone is a state of partial contraction that results from alternating contractions of a few motor units.

Actions of Skeletal Muscles

1. The origin is the immovable attachment, and the insertion is the movable attachment.
2. Muscles are arranged in groups with opposing actions: agonists and antagonists.

Naming of Muscles

1. Several criteria are used in naming muscles.
2. These criteria include function, shape, relative position, location, site of attachment, origin and insertion, size, and orientation of fibers.

Major Skeletal Muscles

1. Muscles of facial expression originate on skull bones and insert on connective tissue of the skin. They include the epicranius, orbicularis oculi, orbicularis oris, buccinator, zygomaticus, and platysma.
2. Muscles of mastication originate on fixed skull bones and insert on the mandible. They include the masseter and the temporalis.
3. Muscles that move the head occur in the neck and upper back. They include the sternocleidomastoid and splenius capitis.
4. Muscles of the abdominopelvic wall connect the pelvic girdle, rib cage, and vertebral column. They include the rectus abdominis, external oblique, internal oblique, and transversus abdominis.
5. Muscles that move the ribs occur between the ribs. They are the external intercostals and internal intercostals.
6. Muscles that move the pectoral girdle originate on the rib cage or vertebrae and insert on the pectoral girdle. They include the trapezius, rhomboideus major, levator scapulae, pectoralis minor, and serratus anterior.
7. Muscles that move the arm originate on the rib cage, vertebrae, or pectoral girdle and insert on the humerus. They include the pectoralis major, coracobrachialis deltoid, supraspinatus, infraspinatus, latissimus dorsi, teres major, and teres minor.

8. Muscles that move the forearm originate on the scapula or humerus and insert on the radius or ulna. They include the biceps brachii, brachialis, brachioradialis, and triceps brachii.
9. Muscles that move the wrist and fingers are the muscles of the forearm. They include the flexor carpi radialis, flexor carpi ulnaris, palmaris longus, extensor carpi radialis longus, extensor carpi ulnaris, and extensor digitorum.
10. Muscles that move the thigh originate on the pelvic girdle and insert on the femur. They include the iliacus, psoas major, gluteus maximus, gluteus medius, tensor fasciae latae, adductor longus, and adductor magnus.
11. Muscles that move the leg originate on the pelvic girdle or femur and insert on the tibia or fibula. They include the quadriceps femoris, biceps femoris, semitendinosus, semimembranosus, gracilis, and sartorius.
12. Muscles that move the foot and toes are the muscles of the lower leg. They include the gastrocnemius, soleus, peroneus longus, tibialis anterior, and extensor digitorum longus.

Disorders of the Muscle System

1. Disorders of muscles include cramps, fibrosis, fibrositis, muscular dystrophy, myositis, and strains.
2. Neurological disorders that directly affect muscle action include botulism, myasthenia gravis, poliomyelitis, spasms, and tetanus.

BUILDING YOUR VOCABULARY

1. **Selected New Terms**
   acetylcholine, p. 126
   actin filament, p. 124
   cholinesterase, p. 127
   deep fascia, p. 123
   graded response, p. 130
   latent period, p. 130
   myogram, p. 130
   myosin filament, p. 124
   neuromuscular junction, p. 126
   recruitment, p. 131
   sarcolemma, p. 124
   sarcoplasm, p. 124
2. **Related Clinical Terms**
   fibromyositis (fī-brō-mī-ō-sī′-tis) A general term for inflammation of a muscle and its associated connective tissue.
   myalgia (mī-al′-jē-ah) Muscle pain from any cause.
   myopathy (mī-op′-ah-thē) Any disease of muscles.
   orthopedics (or-thō-pē′-diks) Branch of surgery dealing with repair of the skeletal and muscular systems.
   spasm An involuntary muscle contraction.

STUDY ACTIVITIES

1. Complete the **Chapter 7 Study Guide,** which begins on page 427.
2. Complete the **Learning Objectives** listed on the first page of this chapter.

CHECK YOUR UNDERSTANDING

(Answers are located in appendix B.)

1. A skeletal muscle consists of many _____ , which are arranged in fasciculi.
2. Muscles are attached to bones by _____ .
3. A contractile unit of a myofibril is a _____ .
4. A muscle contraction is triggered by _____ binding to receptors on the sarcolemma.
5. Contraction occurs when myosin filaments pull _____ filaments toward the center of a sarcomere.
6. Accumulation of _____ during vigorous exercise results in an oxygen debt.
7. The movable end of a muscle is its _____ .
8. The lower jaw is closed by the contraction of the temporalis and the _____ .
9. The abdominal muscle extending from the sternum to the pubis is the _____ .
10. The broad muscle of the lower back is the _____ .
11. The shoulder muscle that raises the arm is the _____ .
12. The arm muscle that extends the forearm is the _____ .
13. The large muscle that extends and rotates the thigh laterally is the _____ .
14. The four-part thigh muscle that extends the lower leg is the _____ .
15. The large calf muscle that plantar flexes the foot is the _____.

C H A P T E R

The Nervous System

EIGHT

8

Chapter Preview & Learning Objectives

Divisions of the Nervous System
- Name the anatomical and functional divisions of the nervous system and indicate their components.

Nerve Tissue
- Describe the structure of a neuron.
- Compare the three types of neurons.
- Compare the four types of neuroglial cells.

Neuron Physiology
- Describe the formation and conduction of a nerve impulse.
- Describe how impulses are transmitted across a synapse.

Protection for the Central Nervous System
- Describe how the brain and spinal cord are protected from injury.

The Brain
- Compare the major parts of the brain as to structure, location, and function.
- Describe the formation, circulation, absorption, and function of the cerebrospinal fluid.

The Spinal Cord
- Describe the structure and function of the spinal cord.

The Peripheral Nervous System (PNS)
- List the 12 cranial nerves and indicate their functions.
- Describe a spinal reflex and indicate the functions of the components involved.

The Autonomic Nervous System (ANS)
- Compare the structure and function of the sympathetic and parasympathetic divisions.

Disorders of the Nervous System
- Describe the common disorders of the nervous system.

Chapter Summary
Building Your Vocabulary
Study Activities
Check Your Understanding

SELECTED KEY TERMS

Autonomic nervous system (auto = self; nom = distribute) The involuntary portion of the nervous system that is involved in maintaining homeostasis.
Axon (ax = axis, central) A neuron process that carries impulses away from the cell body.
Central nervous system The portion of the nervous system composed of the brain and spinal cord.
Dendrite (dendr = tree) A neuron process that carries impulses toward the cell body.
Effector A muscle or gland.
Ganglion (gangli = a swelling) A group of neuron cell bodies.
Impulse A wave of depolarization passing over a neuron or muscle fiber.
Myelin sheath (myel = marrow) The fatty insulating substance around many neuron processes.
Nerve A bundle of neuron processes.
Neurilemma (neur = nerve; lemma = rind) The outer membrane covering the myelin sheath.
Neuroglial cells Supportive cells within the nervous system.
Neuron A nerve cell.
Neurotransmitter A chemical secreted by an axon that triggers the formation of an impulse in the postsynaptic neuron.
Peripheral nervous system (peri = around) Portion of the nervous system composed of cranial and spinal nerves.
Receptor Portion of the peripheral nervous system that responds to certain stimuli by forming impulses.
Reflex An involuntary response to a stimulus.
Somatic nervous system The voluntary portion of the nervous system that is involved in conscious activities.
Synapse (syn = together) The junction between two neurons.

The nervous system is the primary coordinating and controlling system of the body. It monitors internal and external changes, analyzes the information, makes conscious or unconscious decisions, and then takes appropriate action. All of these activities occur almost instantaneously because communication within the nervous system is by electrochemical **impulses** that flow rapidly over neurons and from neuron to neuron. Most of the activities of the nervous system occur below the conscious level and serve to maintain homeostasis.

The general functions of the nervous system may be summarized as

1. detection of internal and external changes;
2. analysis of the information received;
3. organization of the information for immediate and future use;
4. initiation of the appropriate actions.

Divisions of the Nervous System

Although the nervous system functions as a coordinated whole, it is divided into anatomical and functional divisions as an aid in understanding this complex organ system.

Anatomical Divisions

The nervous system has two anatomical divisions. The **central nervous system (CNS)** consists of the brain and spinal cord. The CNS is the body's neural control center. It receives incoming information (impulses), analyzes and organizes it, and initiates appropriate action. The **peripheral nervous system (PNS)** is located outside of the CNS and consists of nerves and sensory receptors. The PNS carries impulses formed by **receptors,** such as pain and sound receptors, to the CNS, and it carries impulses from the CNS to **effectors,** glands and muscles, that carry out actions directed by the CNS.

Functional Divisions

Similarly, the nervous system is divided into two functional divisions. The **somatic** (sō-mat′-ik) **nervous system (SNS)** is involved in *voluntary* or conscious actions, such as those involved in the contraction of skeletal muscles. The **autonomic** (aw-tō-nom′-ik) **nervous system (ANS)** is involved in *involuntary* or unconscious actions, such as controlling the heart rate, that maintain homeostasis.

Nerve Tissue

The nervous system consists of organs composed primarily of nerve tissue supported and protected by connective tissues. There are two types of cells that compose nerve tissue: neurons and neuroglial cells.

Neurons

Neurons (nū′-rahns) are delicate cells that are specialized to transmit neural impulses. They are the structural and functional units of the nervous system. Neurons may vary in size and shape, but they have many common features.

As shown in figures 8.1 and 8.2, the **cell body** is the portion of a neuron that contains the large,

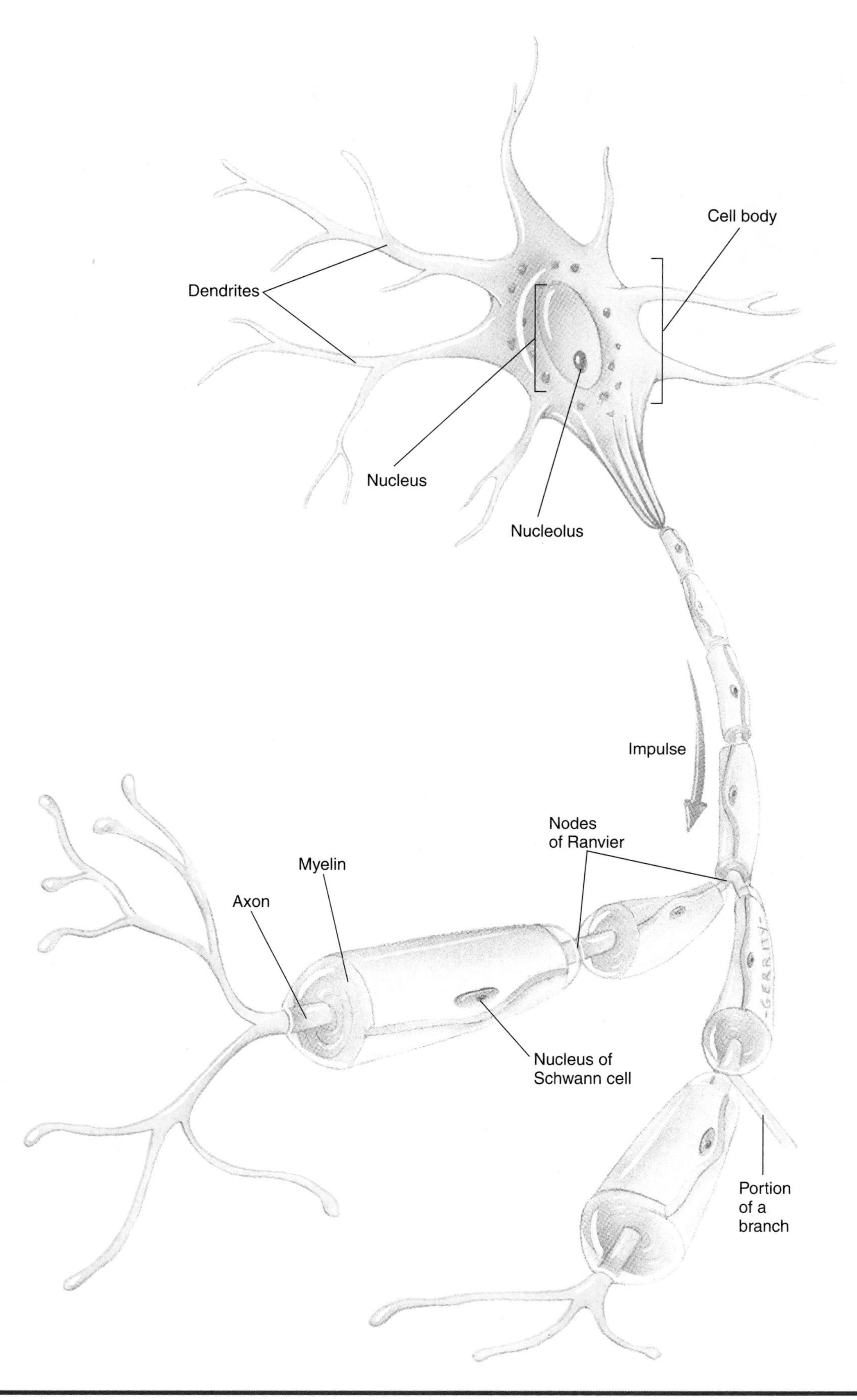

Figure 8.1 A common neuron.

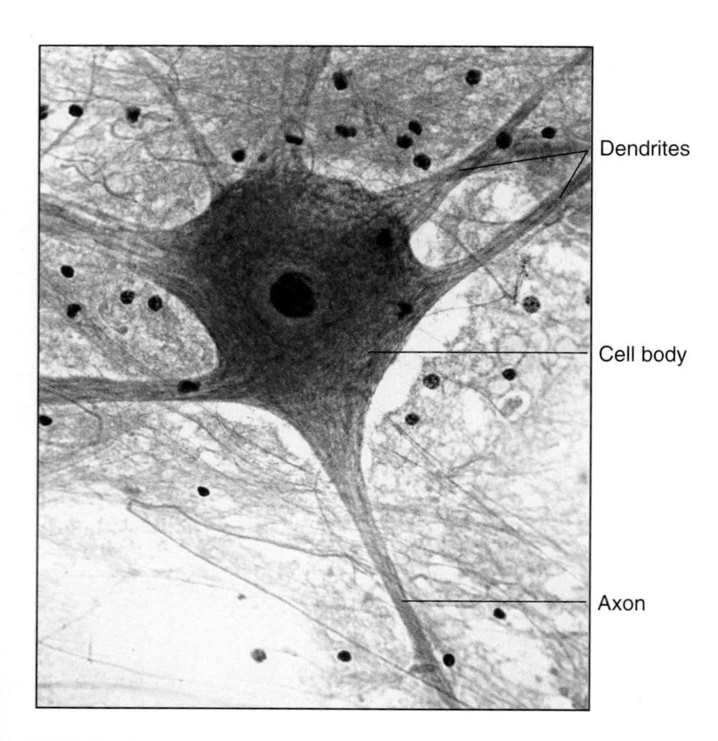

Figure 8.2 Neurons are the structural and functional units of the nervous system (50×). The dark spots in the area surrounding the neuron are nuclei of neuroglial cells. Note the location of nerve fibers (dendrites and a single axon).

spherical nucleus. The cell body also contains the usual cytoplasmic organelles. Two types of neuron processes extend from the cell body: dendrites and axons. A neuron may have many dendrites, but it has only one axon.

Dendrites (den′-drīts) may be short or long, and they are highly branched, which allows contact with many neighboring neurons. Dendrites are the primary sites for receiving impulses from adjacent neurons, and they carry impulses toward the cell body.

The **axon** (ak′-sahn) carries impulses away from the cell body. It may be a single fiber, or it may give off one or more side branches. It typically forms a number of small branches at its terminal end. Each small branch is attached to another neuron or to an effector, such as a muscle cell or gland.

Large dendrites and axons of neurons are usually enclosed in a sheath formed by neuroglial cells. In peripheral neurons, this sheath is formed by **Schwann cells,** a peripheral type of neuroglial cell. Schwann cells wrap tightly around the neuron process (also called a nerve fiber) to form a sheath that consists of multiple layers of plasma membranes with little cytoplasm between them. The outer layer of a Schwann cell, which contains the nucleus and most of the cytoplasm, forms the **neurilemmal sheath.** All layers inward from the neurilemma form the **myelin sheath.** The high fat content of the myelin sheath insulates the neuron process. The tiny spaces between adjacent Schwann cells are the **nodes of Ranvier** (figure 8.3).

After birth, neurons are *not* capable of mitotic cell division, so neurons whose cell bodies have been destroyed cannot be replaced. However, some damaged neuron processes are capable of regeneration if a neurilemma is present.

Classification of Neurons

There are three basic types of neurons: sensory neurons, motor neurons, and interneurons.

Sensory neurons carry impulses from receptors toward the CNS. These neurons are found in cranial and spinal nerves. Sensory neurons usually have a long dendrite and a short axon. Their cell bodies are located outside of the CNS in clusters called *ganglia* (singular, *ganglion*).

Motor neurons carry impulses from the CNS toward effectors (glands and muscles). Axons of motor neurons also are found in cranial and spinal nerves, but their cell bodies and dendrites are located within the CNS. Motor neurons have a long axon and short dendrites.

Interneurons are located entirely within the CNS. They carry impulses from place to place within the CNS and play an important role in transmitting impulses from sensory neurons to motor neurons. Interneurons usually have short dendrites and a long axon.

Neuroglia

The **neuroglia** (nū-rog′-lē-ah), or **neuroglial cells,** provide support and protection for neurons. Only one type of neuroglial cell—Schwann cells—occurs in the PNS. Four types of neuroglial cells occur in the CNS, where they are even more numerous than neurons (figure 8.4).

Oligodendrocytes (ōl-i-gō-den′-drō-sītz) form the myelin sheath of myelinated neurons within the CNS, but they do not form a neurilemma. Destruction of neuron processes in the brain and spinal cord is permanent, because neuron processes cannot regenerate without a neurilemma.

Astrocytes (as′-trō-sītz) bind neurons to blood vessels and play a role in the exchange of materials between the blood and neurons.

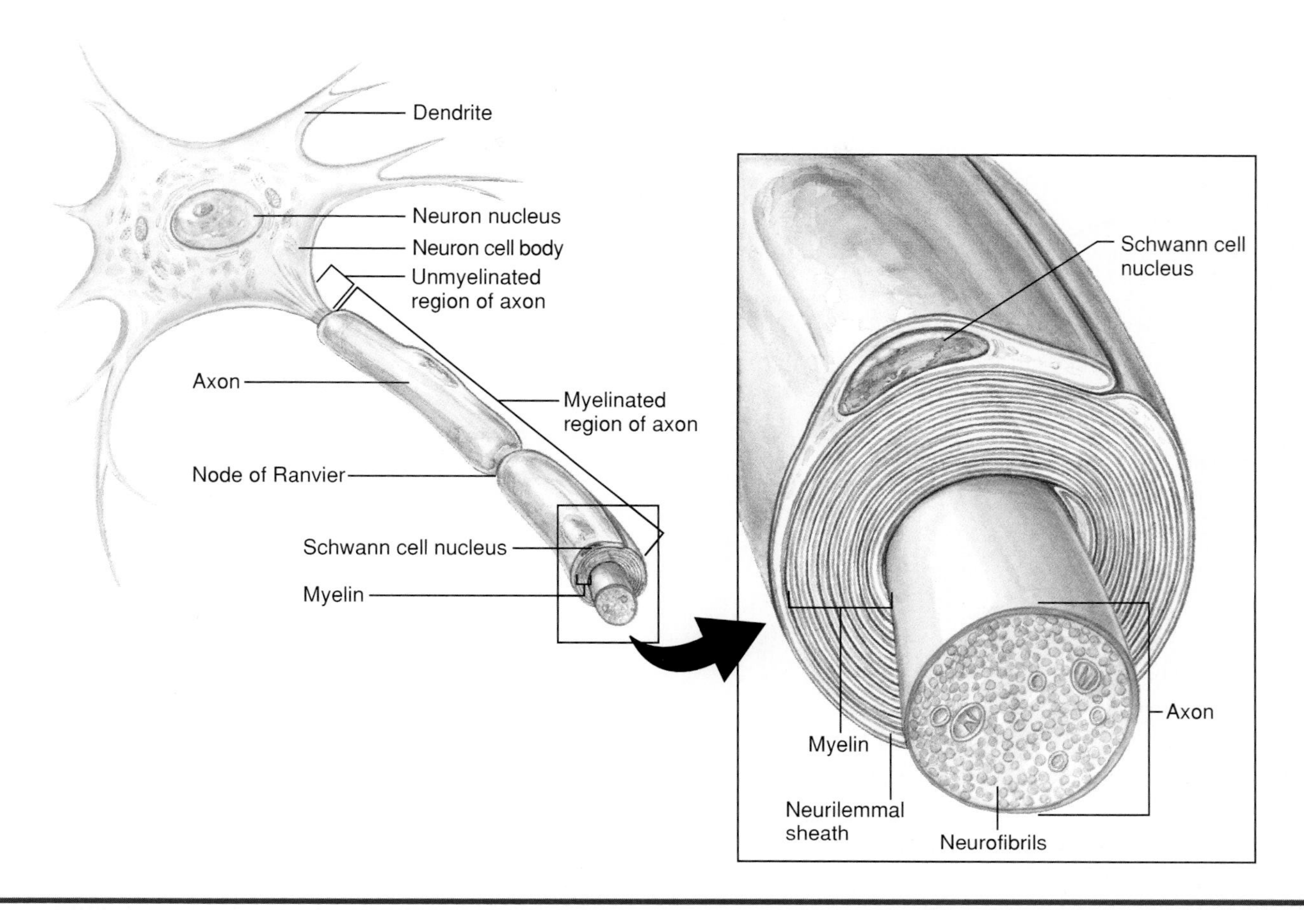

Figure 8.3 The portion of a Schwann cell that winds tightly around an axon forms a myelin sheath, and the cytoplasm and nucleus of the Schwann cell remaining outside form a neurilemmal sheath.

Microglial cells are scattered throughout the CNS, where they keep the tissues clean by engulfing and digesting cellular debris and bacteria.

Ependymal (e-pen-dī′-mal) **cells** form an epithelial-like lining of cavities in the brain and spinal cord.

> ***What are the general functions of the nervous system?***
> ***What is the structure and function of neurons?***
> ***What are the roles of the four types of neuroglial cells?***

Neuron Physiology

Neurons have two unique functional characteristics: irritability and conductivity. **Irritability** is the ability to respond to a stimulus by forming an impulse. **Conductivity** is the ability to transmit an impulse along a neuron to other neurons, muscles, or glands. These characteristics enable the functioning of the nervous system.

Resting Potential

An inactive or resting neuron actively pumps sodium ions (Na^+) out of the cell and potassium ions (K^+) into the cell. The result is that there are more Na^+ outside the neuron and more K^+ and *an excess of negative ions* inside the cell. So, the neuron membrane has a positive charge on the outside and a negative charge on the inside. This unequal distribution of electrical charges on each side of the plasma membrane makes the membrane polarized, and this condition is known as the *resting potential* (figure 8.5*a*). A "potential" is the difference in electrical charge between two sites. The polarization of the neuron membrane does not change as long as the neuron is inactive.

Impulse Formation

When stimulated, neurons exhibit an all-or-none response. They either form an impulse, or they do not respond. There are no strong or weak impulses. All impulses are alike.

When a neuron is activated by a stimulus, its plasma membrane instantly becomes permeable to

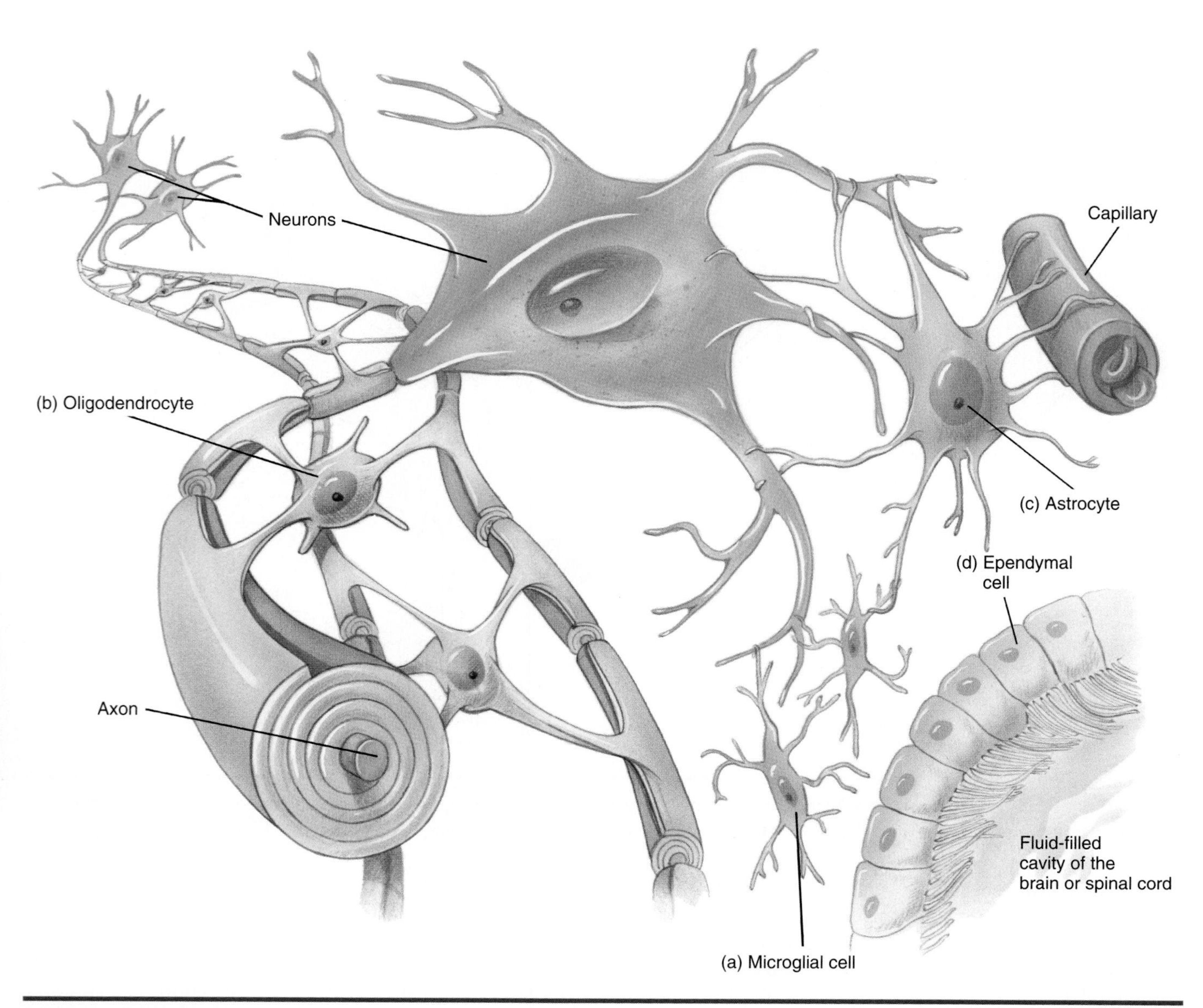

Figure 8.4 Types of neuroglial cells in the central nervous system include (*a*) microglial cell, (*b*) oligodendrocyte, (*c*) astrocyte, and (*d*) ependymal cell.

Na^+, so these ions quickly diffuse into the neuron. This causes the plasma membrane to be *depolarized,* that is, there is no net electrical charge on either side of the membrane. This sudden depolarization *is* the **nerve impulse,** or *action potential* (figure 8.5*b*). The wave of depolarization, that is, the nerve impulse, then flows along the neuron. You will see how this happens momentarily.

Repolarization

Immediately after depolarization, that is, the formation of an impulse, K^+ diffuses outward to reestablish the resting potential of the membrane, that is, an excess of positive charges outside and an excess of negative charges inside. In this way, the neuron membrane is repolarized. Then, Na^+ is pumped out and K^+ is pumped into the neuron to reestablish the resting-state distribution of ions. When this is accomplished, the neuron is ready to respond to another stimulus. Depolarization and repolarization are accomplished in about 1/1,000 of a second (figure 8.5*c*).

Impulse Conduction

When an impulse is formed at one point in a neuron, it triggers the depolarization of adjacent portions of the plasma membrane, which, in turn, depolarizes

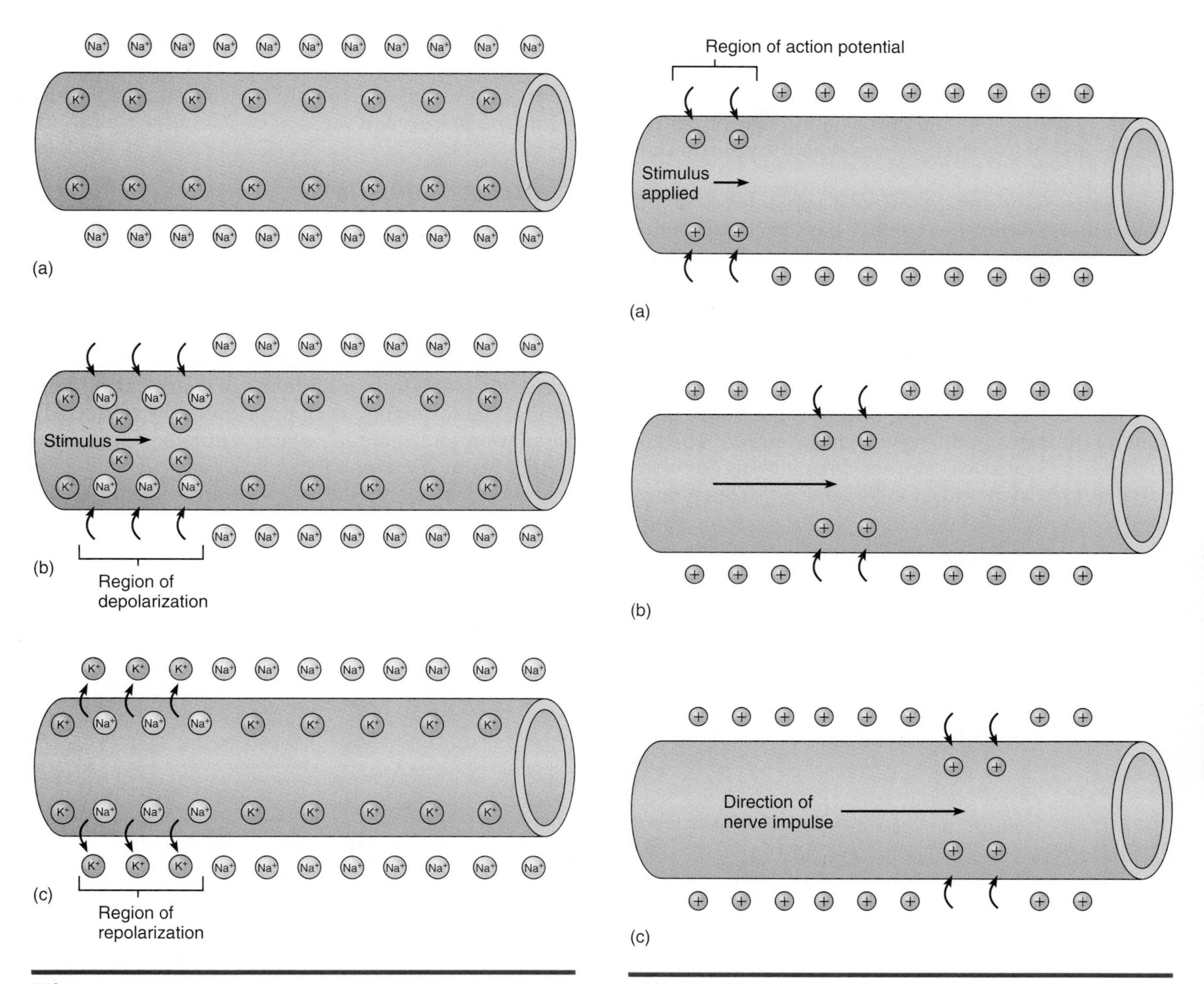

Figure 8.5 Distribution of Na^+ and K^+ (*a*) in a resting (polarized) neuron, (*b*) during depolarization, and (*c*) during repolarization. For simplicity, negative ions are not shown.

Figure 8.6 Distribution of excess positive charges during (*a*) impulse formation (depolarization), and (*b*) and(*c*) impulse conduction. Note that repolarization immediately follows depolarization.

still other regions of the membrane. The result is a wave of depolarization, the conduction of a nerve impulse, that sweeps along the neuron (figure 8.6). Repolarization immediately follows an impulse.

Conduction of nerve impulses is more rapid in myelinated neuron processes than in unmyelinated processes. Recall that a myelinated neuron process is exposed only at nodes of Ranvier. Because of this, an impulse jumps from node to node and does not have to depolarize the intervening segments of the neuron process.

Synaptic Transmission

Impulses are transmitted from neuron to neuron within the nervous system. The junction between neurons is called a **synapse** (sin′-aps). At a synapse, the *synaptic knob* (axon tip) of the presynaptic neuron fits into a small depression on the postsynaptic neuron's dendrite or cell body. There is a tiny space, the *synaptic cleft,* between the neurons, so they are not in physical contact (figure 8.7).

When an impulse reaches the synaptic knob of the presynaptic neuron, it causes the knob to secrete a neurotransmitter into the synaptic cleft. Then, the neurotransmitter binds to receptors on the postsynaptic neuron's plasma membrane, which triggers the formation of an impulse in the postsynaptic neuron. The impulse passes along the postsynaptic neuron until it reaches another synapse, where synaptic transmission takes place again.

Since only synaptic knobs can release neurotransmitters, impulses can only pass in one direction across a synapse—from the presynaptic neuron to the postsynaptic neuron. Thus, impulses always pass in the "correct" direction, which maintains order in the nervous system.

As soon as synaptic transmission occurs, the postsynaptic neuron secretes an enzyme that breaks down the neurotransmitter, preventing a flood of unwanted impulses in the postsynaptic neuron. The end products of neurotransmitter breakdown are reabsorbed by the synaptic knob for later use. The removal of the neurotransmitter from the synaptic cleft prepares the synapse for another synaptic transmission. From start to finish, synaptic transmission takes only a fraction of a second.

Neurotransmitters

There are many different neurotransmitters, but they may be categorized as two basic types: excitatory and inhibitory. **Excitatory neurotransmitters** cause the formation of an impulse in the postsynaptic neuron and include *acetylcholine* (as-e-til-kō′-lēn), *norepinephrine* (nor-ep-i-nef′-rin), *dopamine* (dō′pah-mēn), and *serotonin* (ser-ō-tō-′nin). **Inhibitory neurotransmitters** inhibit the formation of an impulse in the postsynaptic neuron and include several amino acids, such as *GABA* (gamma-aminobutyric acid) and *glycine*. Each neuron releases only one or two neurotransmitters.

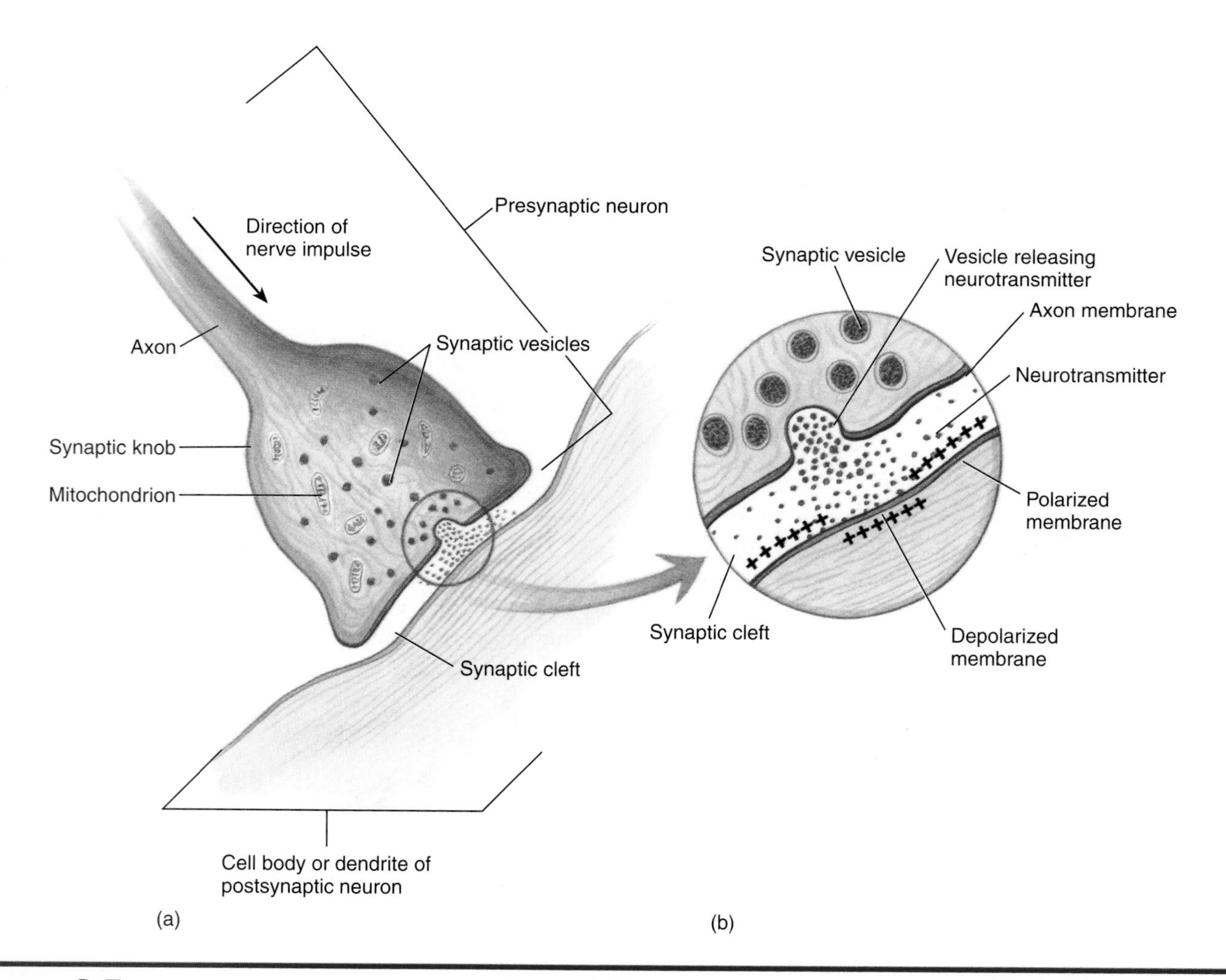

Figure 8.7 Synaptic transmission. (*a*) A synaptic knob of the presynaptic neuron fits into a depression on a dendrite or the cell body of the postsynaptic neuron. When an impulse reaches the synaptic knob, a neurotransmitter is released into the synaptic cleft. (*b*) The neurotransmitter binds with receptors on the membrane of the postsynaptic neuron, causing depolarization of the membrane.

Each postsynaptic neuron has synaptic knobs of hundreds of presynaptic neurons attached to its dendrites and cell body. Some presynaptic neurons secrete excitatory neurotransmitters, and others secrete inhibitory neurotransmitters. Whether or not an impulse is formed in the postsynaptic neuron depends on which type of neurotransmitter is predominant at any instant.

How are nerve impulses formed and conducted?
What is the mechanism of synaptic transmission?

Protection for the Central Nervous System

Both the brain and spinal cord are soft, delicate organs that would be easily damaged without adequate protection. Surrounding bones and fibrous membranes provide both protection and support. The brain occupies the cranial cavity formed by the cranial bones, and the spinal cord lies within the vertebral canal formed by the vertebrae. Three membranes are located between the central nervous system and the surrounding bones. These membranes are collectively called the meninges.

Meninges

The **meninges** (me-nin′-jēs) consist of three membranes arranged in layers. From innermost to outermost, they are the pia mater, arachnoid mater, and dura mater (figures 8.8 and 8.9).

The *pia mater* is the very thin, innermost membrane. It tightly envelops both the brain and spinal cord and penetrates into each groove and depression. It contains many blood vessels that nourish the underlying brain and spinal cord.

The *arachnoid* (ah-rak′-noyd) *mater* is the middle membrane. It is a thin, weblike membrane without blood vessels, and it does not penetrate into the small depressions as does the pia mater. Between the pia mater and arachnoid mater is the *subarachnoid space,* which contains *cerebrospinal fluid.* This clear, watery liquid serves as a shock absorber around the brain and spinal cord.

The *dura* (du′-rah) *mater* is the tough, fibrous outermost layer. In the cranial cavity, it is attached to the inner surfaces of the cranial bones and penetrates into fissures between some parts of the brain. In the vertebral canal, the dura mater forms a protective tube that extends to the sacrum. It does not attach to the bony surfaces of the vertebral canal but is separated from the bone by an *epidural space.* Fatty connective tissues fill the epidural space and serve as an additional protective cushion.

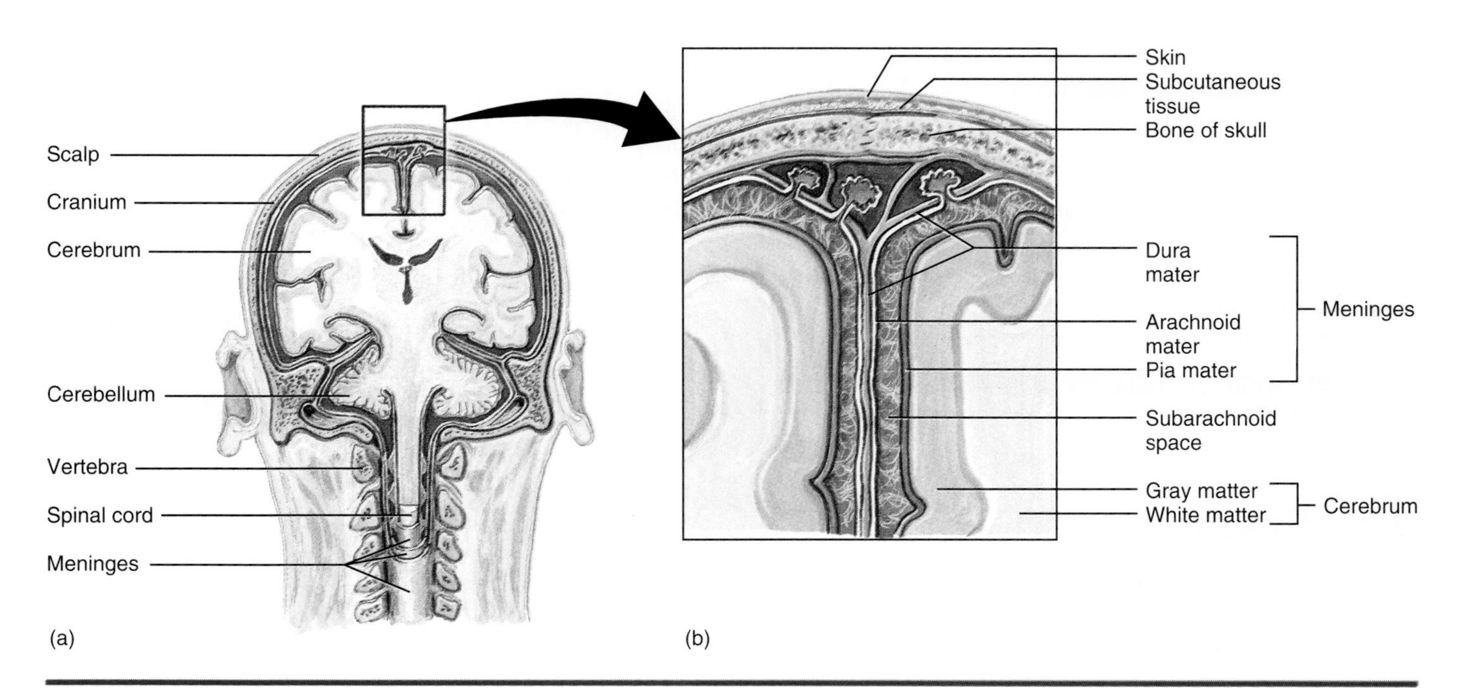

Figure 8.8 (*a*) Membranes called meninges enclose the brain and spinal cord. (*b*) The meninges include three layers: dura mater, arachnoid mater, and pia mater.

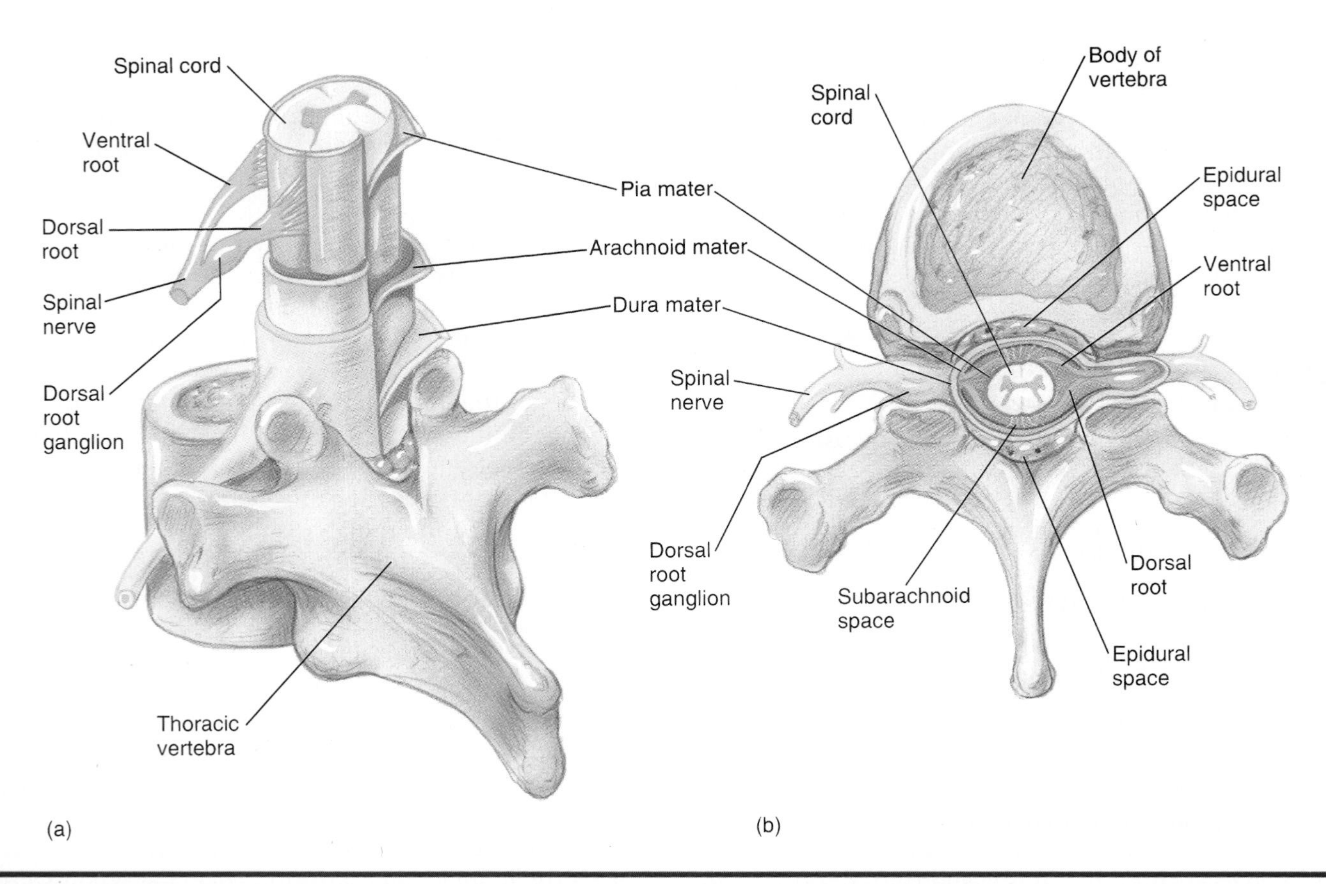

Figure 8.9 Meninges and spinal cord. (*a*) The meninges support and protect the spinal cord. (*b*) Fatty connective tissue fills the epidural space providing a protective cushion.

Inhibitory and stimulatory drugs act by affecting synaptic transmission. *Tranquilizers* and *anaesthetics* inhibit synaptic transmission by increasing the threshold of postsynaptic neurons. *Nicotine, caffeine,* and *benzedrine* promote synaptic transmission by decreasing the threshold of postsynaptic neurons.

The Brain

The brain is a large, exceedingly complex organ. It contains about 100 billion neurons and innumerable neuron processes and synapses. The brain consists of five basic parts: cerebrum, thalamus, hypothalamus, brain stem, and cerebellum. As you study this section on the brain, locate the structures on figures 8.10 to 8.12.

Cerebrum

The **cerebrum** is the largest portion of the brain. It performs the higher brain functions involved with sensations, voluntary actions, reasoning, planning, and problem solving.

Structure

The cerebrum consists of the left and right **cerebral hemispheres,** which are joined by a mass of myelinated neuron processes called the **corpus callosum.** The cerebral hemispheres are separated by the **longitudinal fissure,** which lies along the superior midline and extends down to the corpus callosum. The left cerebral hemisphere controls the right side of the body, and the right cerebral hemisphere controls the left side of the body. This results because neuron processes passing between the brain and spinal cord cross to the opposite side in the brain stem.

The surface of the cerebrum has numerous folds or ridges, called *gyri* (jī′-rē). The shallow grooves between the gyri are called *sulci* (sul′-sē). The surface

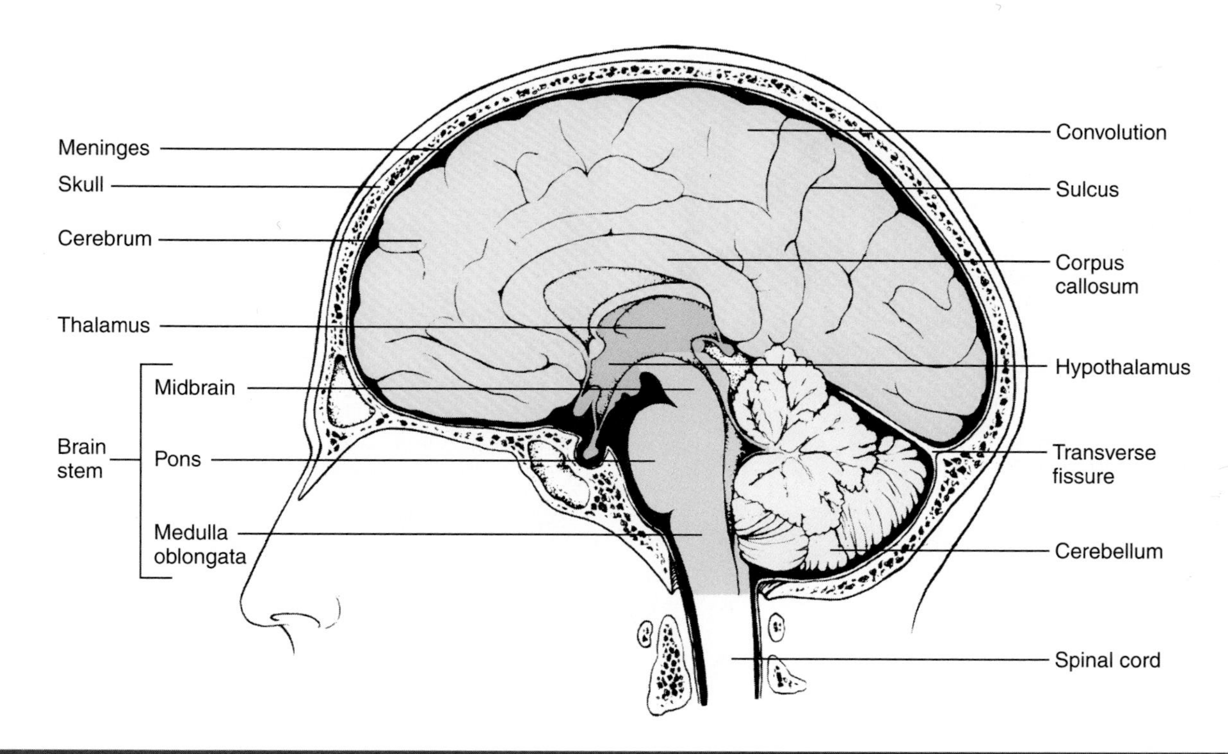

Figure 8.10 A midsagittal section of the brain shows its major components: cerebrum, thalamus, hypothalamus, cerebellum, and brain stem.

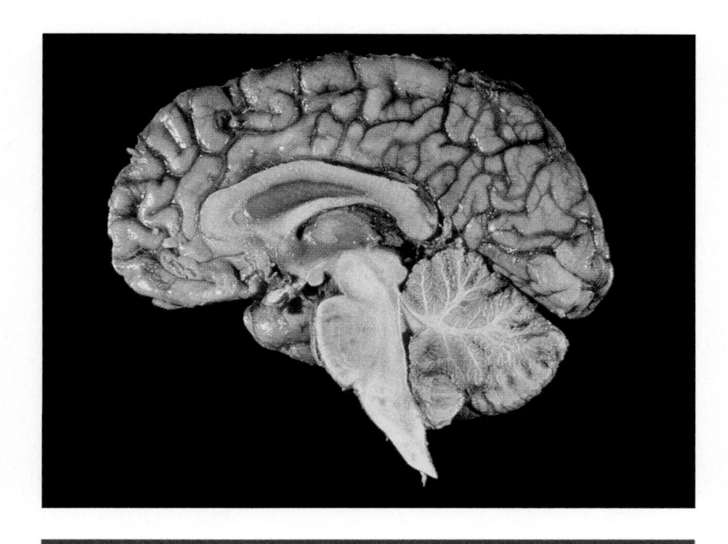

Figure 8.11 A midsagittal section of the brain. Compare it to figure 8.10.

layer of the cerebrum is composed of gray matter (neuron cell bodies and unmyelinated fibers) and is called the **cerebral cortex.** White matter (myelinated fibers) lies beneath the cortex and composes most of the cerebrum. Myelinated fibers transmit impulses between the cerebral hemispheres via the corpus callosum and between the cerebral cortex and lower brain centers. Several masses of gray matter are embedded deep within the white matter of each cerebral hemisphere.

Each cerebral hemisphere is divided into four lobes. Each lobe is named for the cranial bone under which it lies. Locate the cerebral lobes in figures 8.10 and 8.12.

1. The **frontal lobe** lies anterior to the *central sulcus* and superior to the *lateral sulcus.*
2. The **parietal lobe** lies posterior to the central sulcus, superior to the temporal lobe, and anterior to the occipital lobe.
3. The **temporal lobe** lies inferior to the frontal and parietal lobes and anterior to the occipital lobe.
4. The **occipital lobe** lies posterior to the parietal and temporal lobes. The boundaries between the parietal, temporal, and occipital lobes are not distinct.

Functions

The cerebrum is involved in the interpretation of sensory impulses as sensations and in controlling voluntary motor responses, intellectual processes, the will, and many personality traits. The cerebrum has three major types of functional areas: sensory, motor, and association areas (figure 8.12).

Sensory areas receive impulses formed by sensory receptors and interpret them as sensations. These areas occur in several cerebral lobes. For example, the sensory areas for vision are in the occipital lobes and those for hearing are found in the temporal lobes.

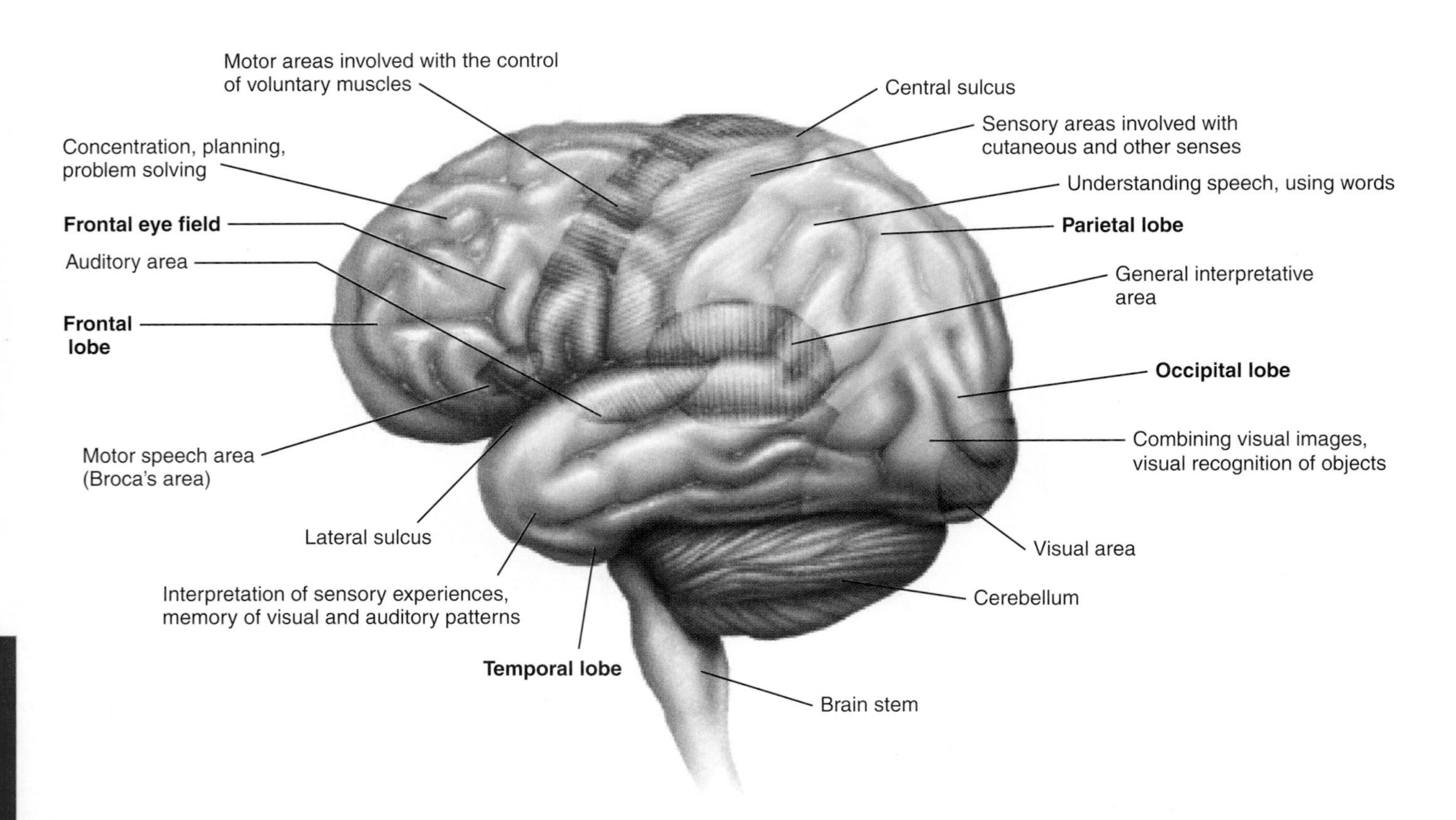

Figure 8.12 A lateral view of the brain showing the cerebral lobes and their functional areas.

Areas interpreting sensations from skin stimulation lie along the *postcentral gyri* (gyri just posterior to the central sulci) of the parietal lobes, and sensory areas for taste are located at the lower end of the postcentral gyri.

Motor areas are located in the frontal lobe. The primary motor areas that control skeletal muscles lie along the *precentral gyri* of the frontal lobes. The region anterior to the primary motor area is the *premotor area.* It is involved in complex learned activities, such as writing, problem solving, and planning. *Broca's area,* which controls the ability to speak, is located near the inferior end of the primary motor area. It is found in only one hemisphere; the left hemisphere in about 90% of people. Also in the premotor region is the *frontal eye field,* which controls eye movements.

Association areas occur in each cerebral lobe, where they interrelate sensory inputs and motor outputs. They play critical roles in the interrelationships of sensations, memory, will, and the coordination of motor responses. The *general interpretative area* is a major association area that is located at the junction of the temporal, parietal, and occipital lobes. It is involved with the interpretation of complex sensory experiences and thought processes.

The transmission of impulses by neurons in the brain produces electrical potentials that can be detected and recorded as brain waves. A recording of brain waves is called an *electroencephalogram (EEG).* The patterns of brain waves are used in the diagnosis of certain brain disorders. The cessation of brain wave production is one criterion of brain death.

Hemisphere Specialization

The two cerebral hemispheres perform different functions in most people, although each performs basic functions of receiving sensory input and initiating voluntary motor output. In about 90% of the population, the left cerebral hemisphere controls analytical and verbal skills, such as mathematics, reading, writing, and speech. In these persons, the right hemisphere controls musical, artistic, and spatial awareness, imagination, and insight. In some persons, this pattern is reversed, and in a few, there seems to be no specialization.

Thalamus

The **thalamus** (thal′-ah-mus) consists of two lateral masses of neural tissue that are joined by a narrow mass of neural tissue called the *intermediate mass.* Sensory impulses (except those for smell) coming from lower regions of the brain and the spinal cord are first received by the thalamus before being relayed to the cerebral cortex. The thalamus provides a general, but nonspecific, awareness of sensations such as pain, pressure, touch, and temperature. But, it is the cerebral cortex that interprets the precise sensation. The thalamus also serves as a relay station for motor impulses descending from the cerebral cortex to lower brain regions.

Hypothalamus

The **hypothalamus** (hī-pō-thal′-ah-mus) is located inferior to the thalamus, and it is the site of attachment for the hypophysis, or pituitary gland. The hypothalamus is a major control center of the autonomic nervous system, and it is a major link between the nervous and endocrine systems because of its relationship with the pituitary gland. It receives and sends impulses to the thalamus and cerebrum and to other parts of the brain.

The primary function of the hypothalamus is the maintenance of homeostasis, and this is accomplished through its regulation of

- body temperature;
- mineral and water balance;
- appetite and digestive processes;
- heart rate and blood pressure;
- sleep and wakefulness;
- emotions of fear and rage;
- secretion of hormones by the pituitary gland.

Limbic System

The thalamus and hypothalamus are associated with parts of the cerebral cortex and deep nuclei (concentrations of gray matter) of the cerebrum to form a complex known as the **limbic system.** The limbic system is involved in memory and in emotions such as sadness, happiness, anger, and fear. It seems to regulate emotional behavior, especially behavior that enhances survival. Mood disorders, such as depression, are usually a result of malfunctions of the limbic system.

How is the CNS protected from mechanical injuries?
What are the functional areas of the cerebrum?
What are the functions of the thalamus and hypothalamus?

Brain Stem

The **brain stem** is the stalklike portion of the brain that joins higher brain centers to the spinal cord. It contains several localized masses of *gray matter,* called **nuclei,** that are surrounded by *white matter.* Ascending and descending fibers between higher brain centers and the spinal cord pass through the brain stem. The components of the brain stem include the midbrain, pons, and medulla oblongata.

Midbrain

The **midbrain** is a small part of the brain stem located between the hypothalamus above and the pons below. It contains reflex centers for head, eye, and body movements in response to visual and auditory stimuli. For example, reflexively turning the head to enable better vision or better hearing is activated by the midbrain.

Pons

The **pons** lies between the midbrain and the medulla oblongata and is recognizable by its bulblike anterior portion. It consists primarily of nerve fibers. Longitudinal fibers connect lower and higher brain centers, and transverse fibers connect with the cerebellum. The pons works with the medulla oblongata to control the rate and depth of breathing.

Medulla Oblongata

The **medulla oblongata** (me-dūl′-ah ob-lon-ga′-ta) is the lowest portion of the brain, and it is the connecting link with the spinal cord. Ascending and descending fibers extending between the brain and spinal cord cross over to the opposite side of the brain within the medulla. The medulla contains three control centers that are vital for homeostasis:

1. The **respiratory control center** works with the pons to regulate the rate and depth of breathing. It is also involved in associated reflexes such as coughing and sneezing.
2. The **cardiac control center** regulates the rate of heart contractions.

Table 8.1 Summary of Brain Functions

Part	Function
Cerebrum	Sensory areas interpret impulses as sensations. Motor areas control voluntary muscle actions. Association areas interrelate various sensory and motor areas and are involved in intellectual processes, will, memory, emotions, and personality traits. The limbic system is involved with emotions as they relate to survival behavior.
Cerebellum	Controls posture, balance, and the coordination of skeletal muscle contractions.
Thalamus	Receives and relays sensory impulses (except smell) to the cerebrum and motor impulses to lower brain centers. Provides a general awareness of pain, touch, pressure, and temperature and determines sensations as pleasant or unpleasant.
Hypothalamus	Serves as a major control center for the autonomic nervous system. Controls water and mineral balance, heart rate and blood pressure, appetite and digestive activity, body temperature, and sexual response. Is involved in sleep and wakefulness and in emotions of anger and fear. Regulates functions of the pituitary gland.
Brain Stem	
Midbrain	Relays sensory impulses from the spinal cord to the thalamus and motor impulses from the cerebrum to the spinal cord. Contains reflex centers that move eyeballs in response to visual stimuli and head and neck in response to auditory stimuli.
Pons	Relays impulses between the midbrain and the medulla and between the cerebellar hemispheres. Helps medulla control breathing.
Medulla	Relays impulses between the brain and spinal cord. Reflex centers control heart rate, blood vessel diameter, breathing, swallowing, vomiting, coughing, sneezing, and hiccupping. Ascending and descending nerve tracts cross over to the opposite side. Reticular formation (also in midbrain and pons) controls wakefulness.

3. The **vasomotor center** regulates blood pressure and blood flow by controlling the diameter of blood vessels.

Reticular Formation

The **reticular** (re-tik′-ū-lar) **formation** is a network of nerve fibers and small nuclei of gray matter scattered in the brain stem and spinal cord. Its fibers connect centers in the hypothalamus, cerebrum, and cerebellum. This network generates and transmits impulses that arouse the cerebrum to wakefulness. A decrease in activity results in sleep. Damage to the reticular formation may cause unconsciousness or a coma.

Cerebellum

The **cerebellum** (ser-e-bel′-um) is the second largest portion of the brain. It is located below the occipital and temporal lobes of the cerebrum posterior to the pons and medulla oblongata. It is divided into two lateral hemispheres by a medial constriction, the **vermis** (ver′-mis). Gray matter forms a thin outer layer covering the underlying white matter, which forms most of the cerebellum.

The cerebellum is a reflex center that controls and coordinates the interaction of skeletal muscles. It controls posture, balance, and muscle coordination. Damage to the cerebellum may result in a loss of equilibrium, muscle coordination, and muscle tone.

Table 8.1 summarizes the major brain functions.

What are the functions of the medulla oblongata? How is the cerebellum involved in skeletal muscle contractions?

Ventricles and Cerebrospinal Fluid

There are four interconnecting **ventricles,** or cavities, within the brain. Each ventricle is lined by ependymal cells and is filled with **cerebrospinal fluid.** The largest ventricles are the two *lateral ventricles* (first and second ventricles), which are located within the cerebral hemispheres. The *third ventricle* is a narrow space that lies on the midline between the lateral masses of the thalamus and hypothalamus. The *fourth ventricle* is located on the midline in the posterior portion of the brain stem just anterior to the cerebellum. It is continuous with the central canal of the spinal cord. Observe the relative positions of the ventricles in figure 8.13.

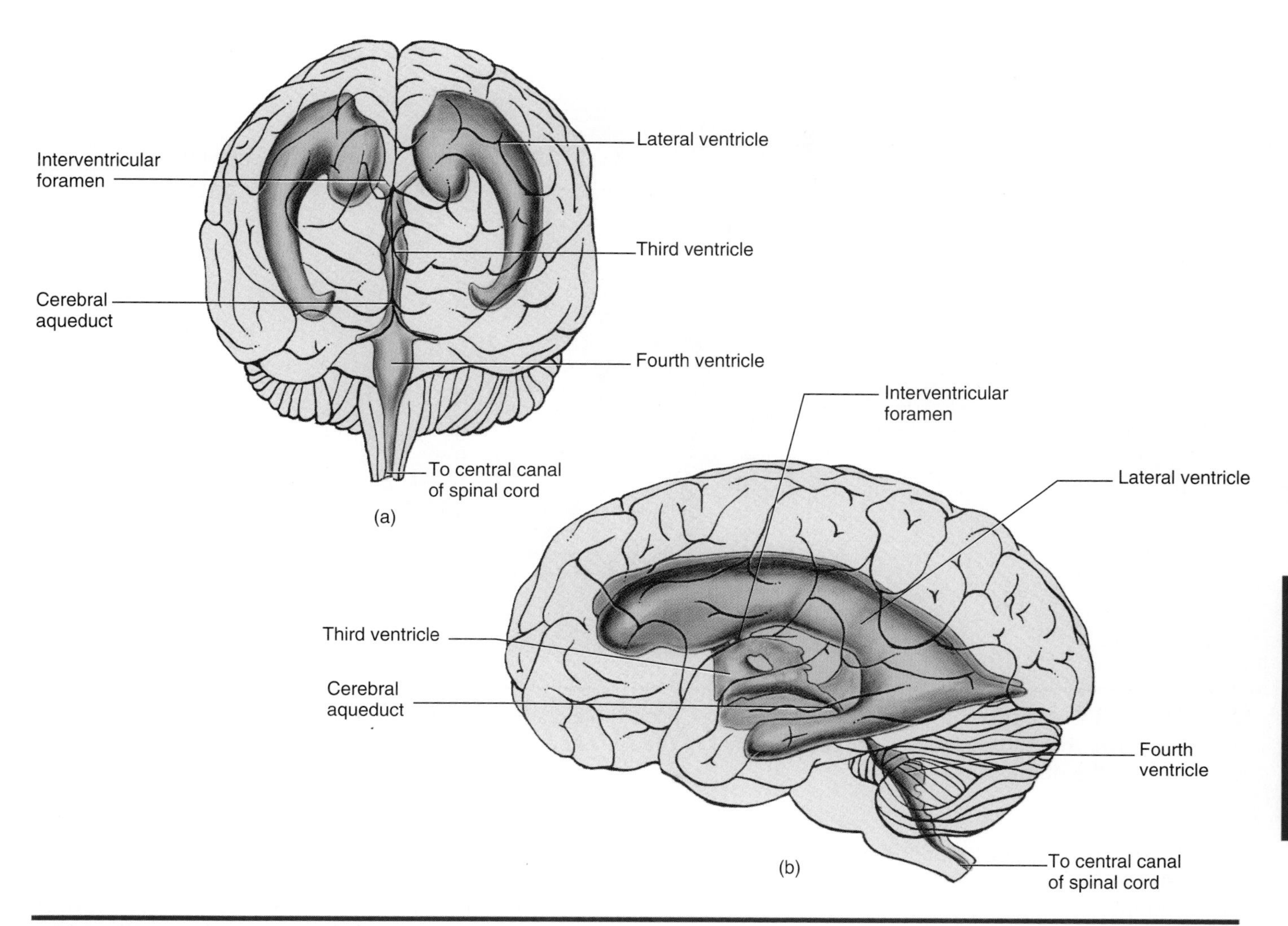

Figure 8.13 Anterior (*a*) and lateral (*b*) views of the ventricles of the brain. Note how they are interconnected.

Each ventricle contains a **choroid** (kō′-royd) **plexus,** a mass of special capillaries that secrete the cerebrospinal fluid, but most of the cerebrospinal fluid is produced in the lateral ventricles. From the lateral ventricles, the cerebrospinal fluid flows into the third ventricle and on into the fourth ventricle. From the fourth ventricle, some of the fluid flows down the central canal of the spinal cord, but most of it passes into the subarachnoid space of the meninges. Within the subarachnoid space, the cerebrospinal fluid flows in two directions. Some flows upward around the brain. The remainder flows down the posterior side of the spinal cord, returns upward along its anterior surface, and continues upward around the brain in the subarachnoid space. Cerebrospinal fluid is reabsorbed into the blood-filled dural sinus that is located along the superior midline within the dura mater (figures 8.8 and 8.14). The secretion and absorption of cerebrospinal fluid normally occurs at equal rates, which results in a rather constant hydrostatic pressure within the ventricles and subarachnoid space.

The Spinal Cord

The **spinal cord** is continuous with the brain. It descends from the medulla oblongata through the foramen magnum into the vertebral canal and extends to the second lumbar vertebra. Beyond this point, only the proximal portions of the lower spinal nerves occupy the vertebral canal.

Structure

The spinal cord is cylindrical in shape. It has two small grooves that extend throughout its length: the wider *anterior median fissure* and the narrower *posterior*

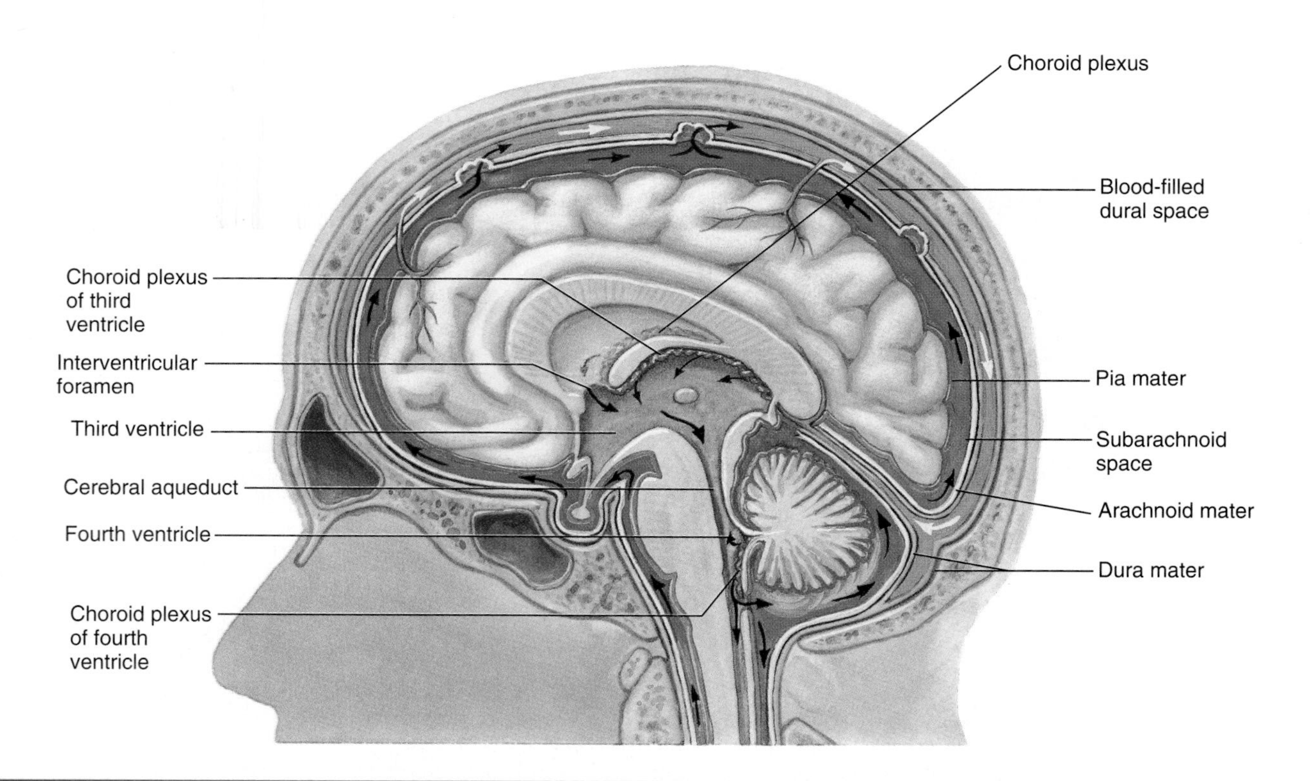

Figure 8.14 Circulation of cerebrospinal fluid. Choroid plexuses in ventricle walls secrete cerebrospinal fluid. The fluid flows through the ventricles, central canal of the spinal cord, and subarachnoid space. It is reabsorbed into the blood of the dural sinus.

median sulcus. These grooves divide the spinal cord into left and right portions. Thirty-one pairs of spinal nerves branch from the spinal cord.

The cross-sectional structure of the spinal cord is shown in figures 8.9 and 8.15. Gray matter, shaped like the outstretched wings of a butterfly, is centrally located and is surrounded by white matter. The *central canal* extends the length of the spinal cord and contains cerebrospinal fluid.

In *hydrocephalus,* a congenital defect restricts the movement of cerebrospinal fluid from the ventricles into the subarachnoid space. In severe cases, the buildup of hydrostatic pressure within an infant's brain causes a marked enlargement of the ventricles and brain and widens the fontanels of the cranium. Without treatment, death usually results within two to three years. Treatment involves surgical insertion of a small tube to drain the excess fluid from a ventricle into a blood vessel in the neck.

The pointed projections of the gray matter, as seen in cross section, are called horns. The *anterior horns* contain the cell bodies of motor neurons whose axons enter spinal nerves and carry impulses to muscles and glands. The *posterior horns* contain interneurons that receive impulses from sensory neurons in the spinal nerves and carry them to sites within the central nervous system. Interneurons form most of the gray matter in the central nervous system.

The horns of the gray matter divide the white matter into three regions: the *anterior, posterior,* and *lateral columns.* These columns are bundles of myelinated axons of interneurons that extend up and down the spinal cord. These bundles are called **nerve tracts.**

Functions

The spinal cord has two basic functions. It transmits impulses to and from the brain, and it serves as a reflex center for spinal reflexes. Impulses are transmitted to and from the brain by myelinated fibers composing the nerve tracts. *Ascending tracts* carry impulses to the brain; *descending tracts* carry impulses from the brain.

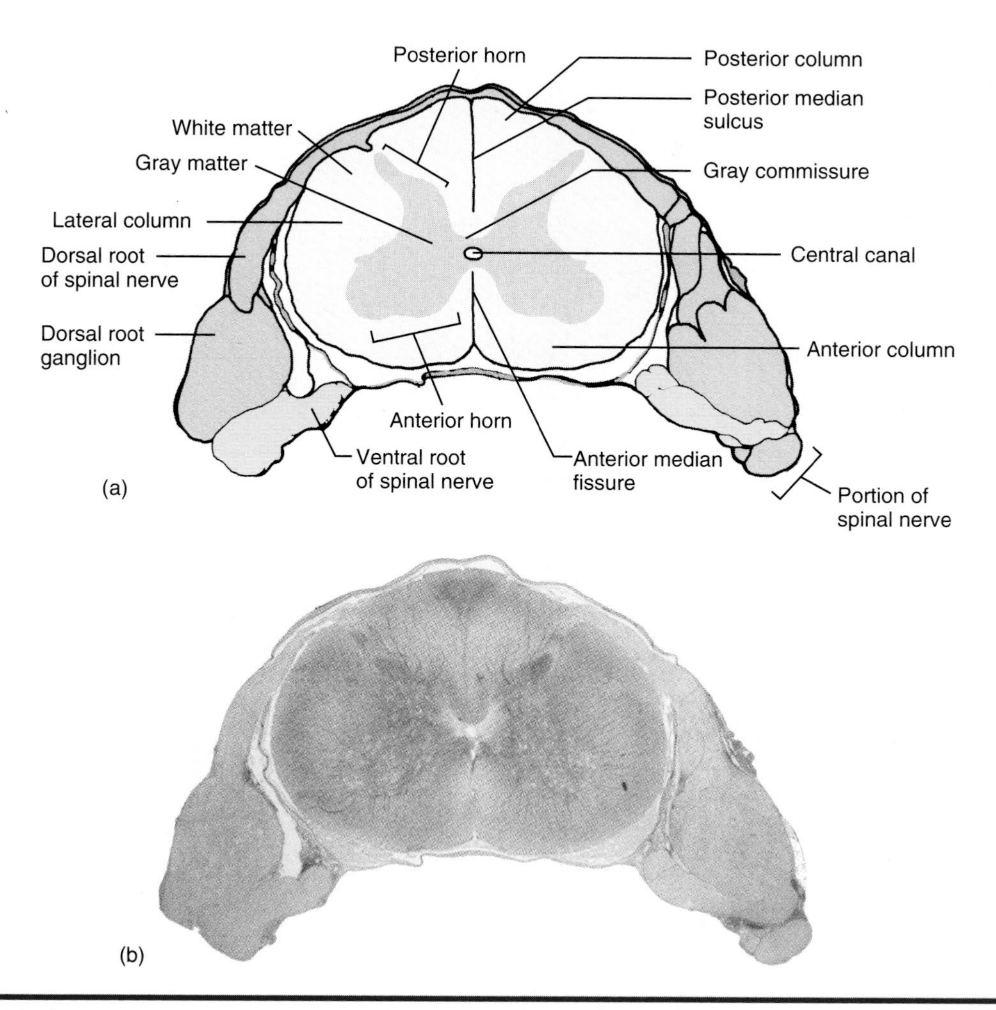

Figure 8.15 A drawing (*a*) and a photomicrograph (*b*) of the spinal cord in cross section show its basic structure.

> ***What is the relationship between the ventricles, meninges, and the cerebrospinal fluid?***
> ***What are the functions of the spinal cord?***

The Peripheral Nervous System (PNS)

The *peripheral nervous system (PNS)* consists of cranial and spinal nerves that connect the central nervous system to other portions of the body. A **nerve** consists of neuron processes—axons, dendrites, or both—that are bound together by connective tissue. **Motor nerves** contain only axons of motor neurons; **sensory nerves** contain only dendrons of sensory neurons; and **mixed nerves** contain both motor nerve axons and sensory nerve dendrons. Most nerves are mixed. Nerves may contain fibers of both the somatic nervous system, which is involved with voluntary responses, and the autonomic nervous system, which controls involuntary (automatic) responses.

Cranial Nerves

Twelve pairs of **cranial nerves** arise from the brain and connect the brain with organs and tissues that are primarily located in the head and neck. Most cranial nerves arise from the brain stem. Cranial nerves are identified by both Roman numerals and names. The numerals indicate the order in which the nerves arise from the inferior surface of the brain: I is most anterior, XII is most posterior (figure 8.16).

Most cranial nerves are mixed nerves. A few are sensory, and some are mixed but primarily motor nerves. Table 8.2 lists the numerals, names, types, and functions of the 12 pairs of cranial nerves.

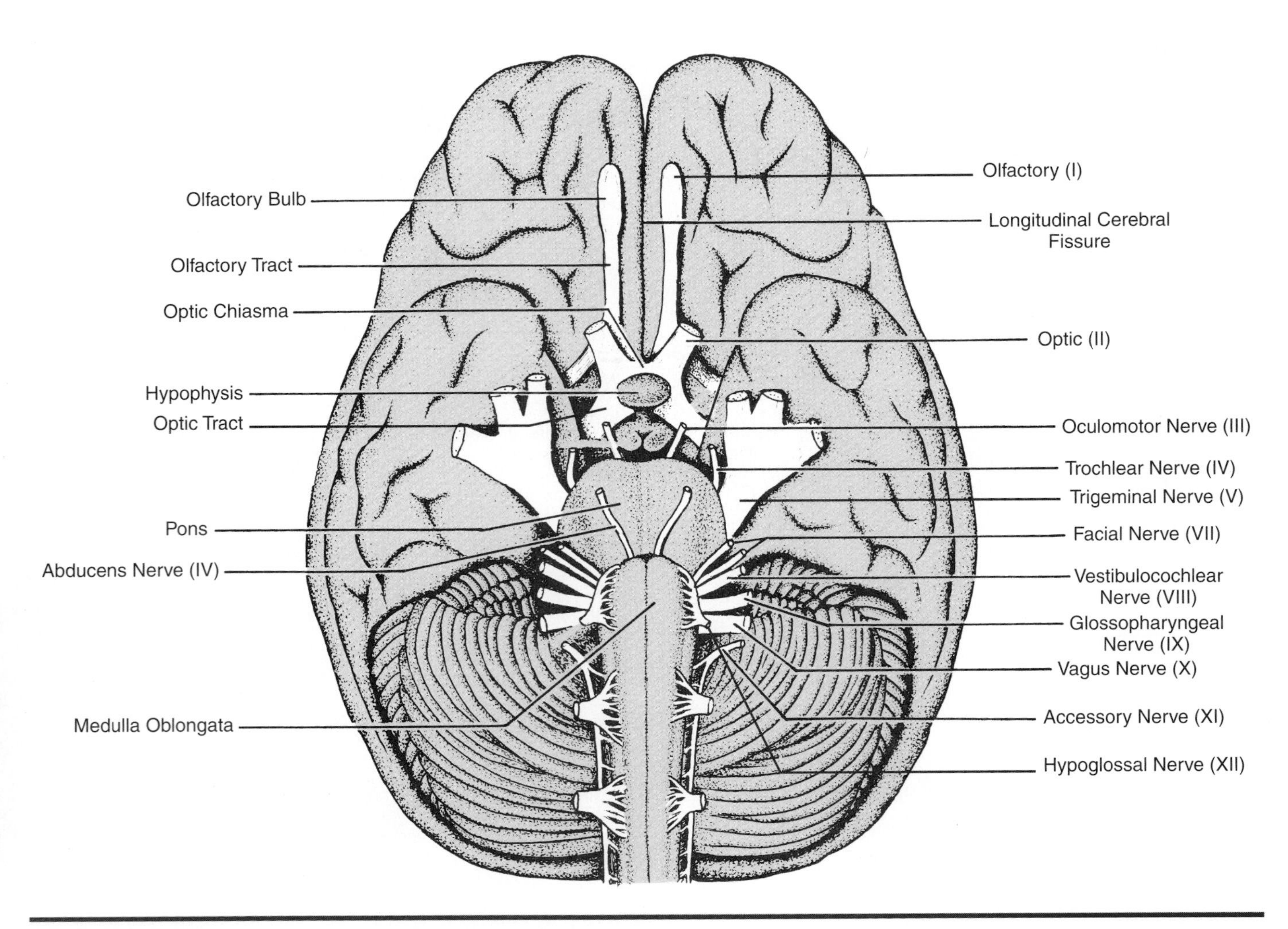

Figure 8.16 A ventral view of the brain showing the roots of the 12 pairs of cranial nerves. Cranial nerves are identified by both Roman numerals and names. Most cranial nerves arise from the brain stem.

Spinal Nerves

Thirty-one pairs of **spinal nerves** arise from the spinal cord. The first pair branches from the spinal cord between the atlas and the occipital bone. The remaining spinal nerves emerge from the spinal cord and pass between adjacent vertebrae through the *intervertebral foramina.* Spinal nerves are mixed nerves.

Spinal nerves are identified by the spinal region from which they branch and are numbered in sequence within each region. There are eight pairs of *cervical nerves* (C1–C8), twelve pairs of *thoracic nerves* (T1–T12), five pairs of *lumbar nerves* (L1–L5), five pairs of *sacral nerves* (S1–S5), and one pair of *coccygeal nerves* (figure 8.17).

As shown in figure 8.17, the spinal cord ends at the second lumbar vertebra. Proximal portions of lumbar, sacral, and coccygeal spinal nerves continue down the vertebral canal to exit between the appropriate vertebrae. The bases of these nerves form the *cauda equina,* or horse's tail, in the inferior portion of the vertebral canal.

Spinal nerves branch from the spinal cord by two short roots that merge a short distance from the spinal cord to form a spinal nerve. The **anterior root** contains axons of motor neurons whose cell bodies are located within the spinal cord. These neurons carry motor impulses from the spinal cord to effectors. The **posterior root** contains sensory neurons. The swollen region in a posterior root is a **spinal ganglion,** which contains cell bodies of sensory neurons. The long dendrites of these neurons carry sensory impulses to the cell bodies, and their short axons carry the impulses on into the spinal cord. Observe these structures and their relationships in figures 8.9, 8.15, and 8.18.

Table 8.2 Summary of the Cranial Nerves

Nerve	Type	Function
I Olfactory	Sensory	Transmits sensory impulses from olfactory receptors in nasal mucosa to the brain.
II Optic	Sensory	Transmits sensory impulses for vision from the retina of the eye to the brain.
III Oculomotor	Primarily motor	Transmits motor impulses to muscles that move the eyes upward, downward, and medially; the eyelids; adjust pupil size; and control the shape of the lens. Transmits sensory impulses associated with the degree of contraction of these muscles to the brain.
IV Trochlear	Primarily motor	Transmits motor impulses to muscles that rotate the eyes. Transmits sensory impulses associated with the degree of contraction of these muscles to the brain.
V Trigeminal	Mixed	Transmits sensory impulses from scalp, forehead, face, teeth, and gums to the brain. Transmits motor impulses to chewing muscles and muscles in floor of mouth.
VI Abducens	Primarily motor	Transmits motor impulses to muscles that move the eyes laterally. Transmits sensory impulses associated with the degree of contraction of these muscles to the brain.
VII Facial	Mixed	Transmits sensory impulses from the anterior part of the tongue to the brain. Transmits motor impulses to facial muscles, salivary glands, and tear glands.
VIII Vestibulocochlear	Sensory	Transmits sensory impulses from the inner ear associated with hearing and equilibrium.
IX Glossopharyngeal	Mixed	Transmits sensory impulses from posterior portion of the tongue, tonsils, pharynx, and carotid arteries to the brain. Transmits motor impulses to salivary glands and pharyngeal muscles used in swallowing.
X Vagus	Mixed	Transmits sensory impulses from thoracic and abdominal organs, esophagus, larynx, and pharynx to the brain. Transmits motor impulses to these organs and to muscles of speech and swallowing.
XI Accessory	Primarily motor	Transmits motor impulses to muscles of the palate, pharynx, and larynx, and to the trapezius and sternocleidomastoid muscles. Transmits sensory impulses associated with the degree of contraction of these muscles to the brain.
XII Hypoglossal	Primarily motor	Transmits motor impulses to the muscles of the tongue. Transmits sensory impulses associated with the degree of contraction of these muscles to the brain.

Spinal Plexuses

After a spinal nerve passes through an intervertebral foramen, it divides into two or three major parts: an anterior branch, a posterior branch, and, in thoracic and lumbar spinal nerves only, a visceral branch. The posterior branch continues directly to the innervated structures. The visceral branch passes to the sympathetic chain ganglia and is part of the autonomic system. The anterior branches of many spinal nerves merge to form a **plexus** (plek′-sus), a network of nerves, before continuing to the innervated structures.

In a plexus, the nerve fibers in the anterior branches are sorted and recombined so that fibers going to a specific body part are carried in the same spinal nerve, although they may originate in several different spinal nerves. There are four pair of plexuses: cervical, brachial, lumbar, and sacral plexuses. Since many fibers from the lumbar plexus contribute to the sacral plexus, these two plexuses are sometimes called the *lumbosacral plexuses.* The anterior branches of most thoracic spinal nerves (T2–T11) take a direct route to the skin and muscles of the trunk that they supply (figure 8.17).

Cervical Plexus The first four cervical nerves (C1–C4) merge to form a **cervical plexus** on each side of the neck. The nerves from these plexuses supply the muscles and skin of the neck and portions of the head and shoulders. The paired *phrenic* (fren′-ik) *nerves,* which stimulate the diaphragm to contract, producing inspiration, arise from the cervical plexus.

Brachial Plexus The last four cervical nerves (C5–C8) and the first thoracic nerve (T1) join to form a **brachial plexus** on each side of the vertebral column in the shoulder region. Nerves that serve skin and

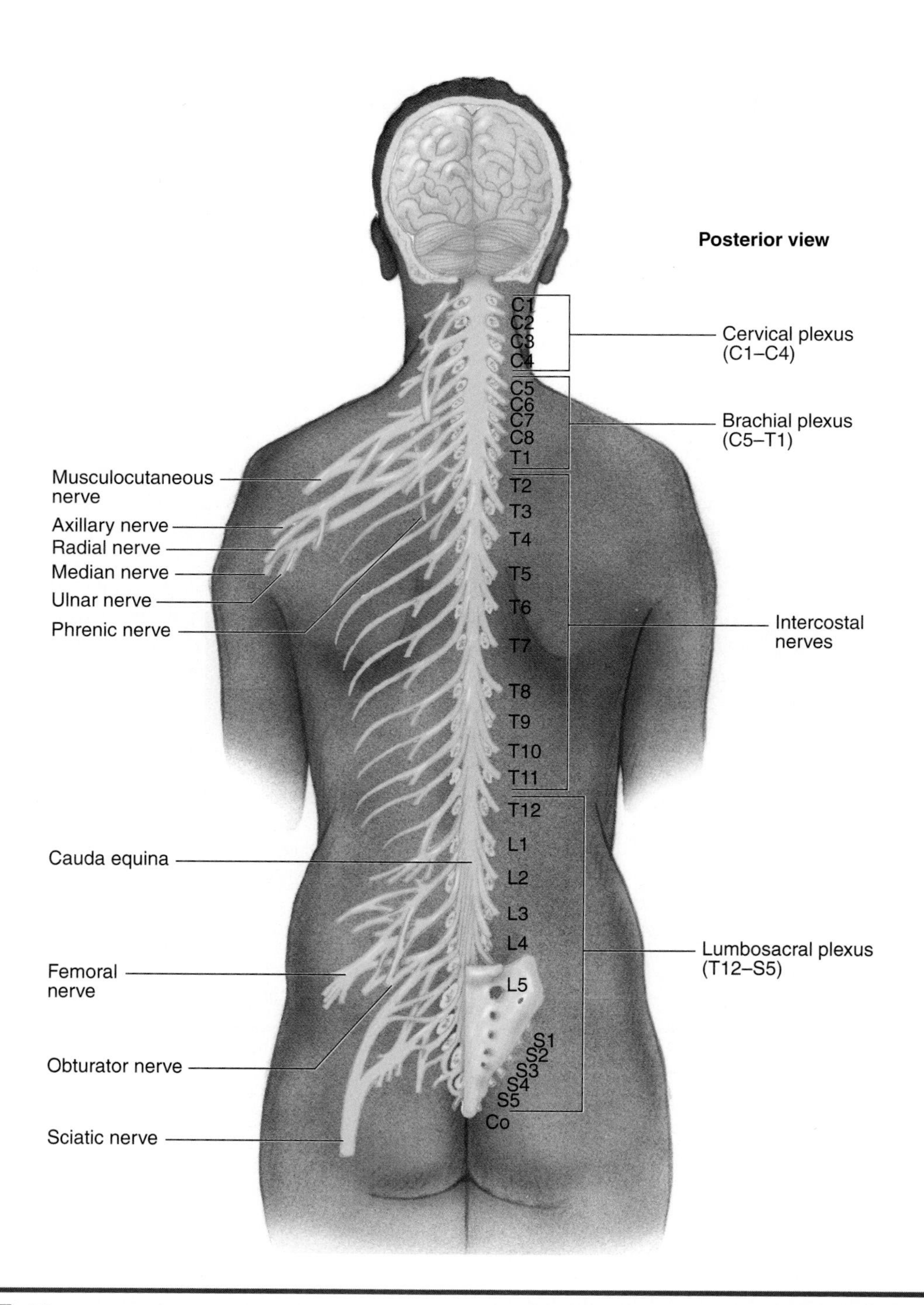

Figure 8.17 Thirty-one pairs of spinal nerves arise from the spinal cord. Anterior branches of spinal nerves in the thorax form the intercostal nerves. Those in other regions form nerve networks called plexuses before continuing on to their target tissues.

muscles of the arms and shoulders emerge from the brachial plexuses. The *musculocutaneous, axillary, radial, median,* and *ulnar nerves* arise here.

Lumbar Plexus The last thoracic nerve (T12) and the first four lumbar nerves (L1–L4) unite to form a **lumbar plexus** on each side of the vertebral column just above the hip bones. Nerves from the lumbar plexus supply the skin and muscles of the lower trunk, genitalia, and part of the thighs. The *femoral* and *obturator nerves* arise here.

Sacral Plexus The last two lumbar nerves (L4–L5) and the first four sacral nerves (S1–S4) merge to form a **sacral plexus** on each side of the sacrum within the pelvic girdle. Nerves from the sacral

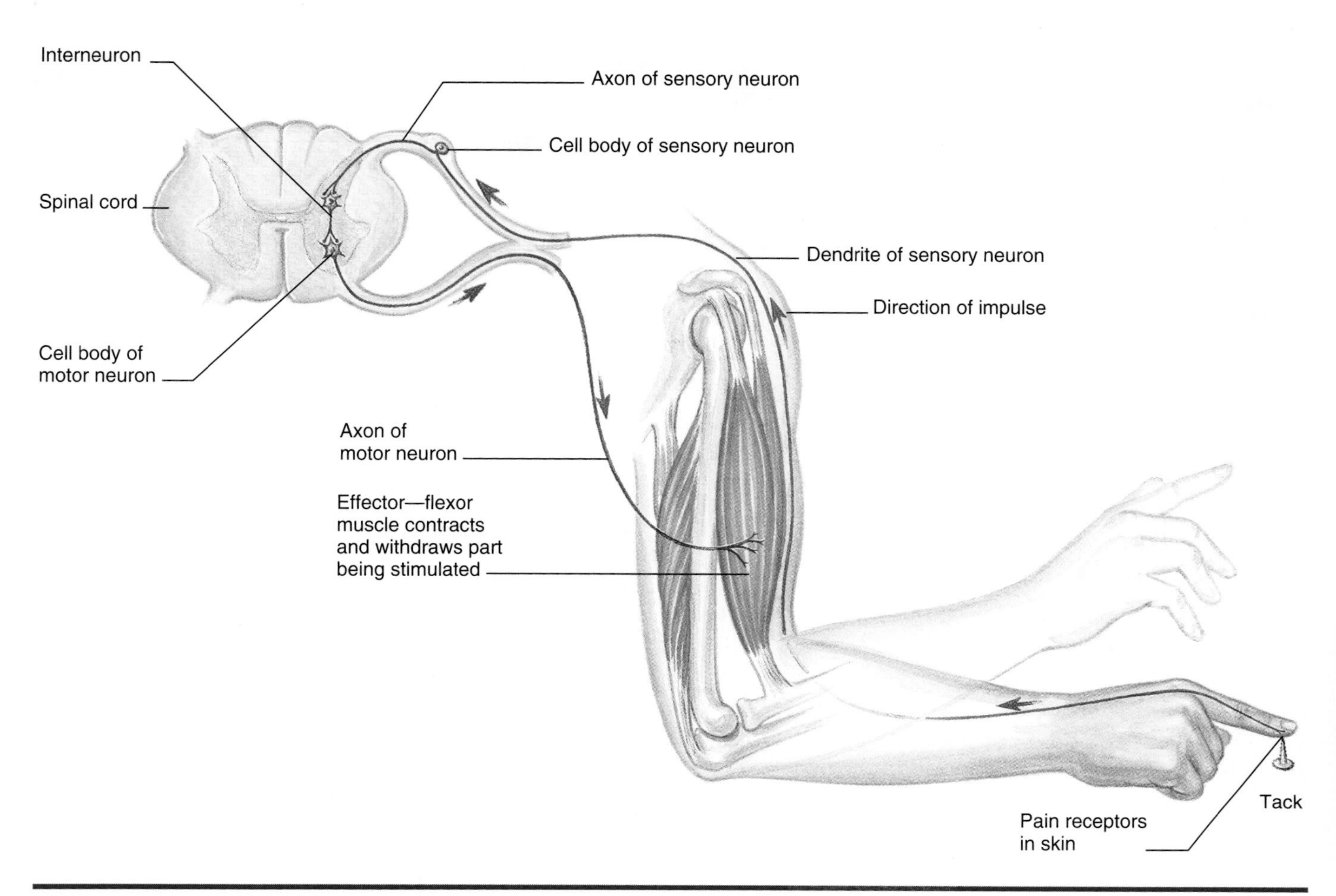

Figure 8.18 A withdrawal reflex involves a sensory neuron, an interneuron, and a motor neuron.

plexuses supply the skin and muscles of the buttocks and legs. The *sciatic nerves,* which emerge from the sacral plexuses, are the largest nerves in the body.

> ***What composes the peripheral nervous system? What is a spinal plexus?***

Reflexes

Reflexes are rapid, involuntary, and predictable responses to internal and external stimuli. Reflexes maintain homeostasis and enhance chances of survival. A reflex involves either the brain or spinal cord, a receptor, sensory and motor neurons, and an effector.

Most pathways of impulse transmission within the nervous system are complex and involve many neurons. In contrast, reflexes require few neurons in their pathways and therefore produce very rapid responses to stimuli. Reflex pathways are called **reflex arcs.**

Reflexes are divided into two types: autonomic and somatic reflexes. *Autonomic reflexes* act on smooth

> Since the spinal cord ends at the level of the second lumbar vertebra, spinal taps and epidural anesthetics are administered inferior to this point. For these procedures, a patient is placed in a fetal position in order to open the spaces between the posterior margins of the vertebrae. A hypodermic needle is inserted through the intervertebral disk either between the third and fourth lumbar vertebrae or between the fourth and fifth lumbar vertebrae. In a *spinal tap* (lumbar puncture), a hypodermic needle is inserted into the subarachnoid space to remove spinal fluid for diagnostic purposes. An *epidural anesthetic* is given by injecting an anesthetic into the epidural space with a hypodermic syringe. The anesthetic prevents sensory impulses from reaching the spinal cord via spinal nerves inferior to the injection. Epidurals are sometimes used to ease pain during childbirth.

Table 8.3 Comparison of Somatic and Autonomic Nervous Systems

	Somatic	Autonomic
Control	Voluntary	Involuntary
Neural Pathway	One motor neuron extends from the CNS to an effector	A preganglionic neuron extends from the CNS to an autonomic ganglion and synapses with a postganglionic neuron that extends to an effector
Neurotransmitters	Acetylcholine	Acetylcholine or norepinephrine
Effectors	Skeletal muscles	Smooth muscle, cardiac muscle, and glands
Action	Excitatory	Excitatory or inhibitory

muscle, cardiac muscle, and glands. They are involved in controlling homeostatic processes such as heart rate, blood pressure, and digestion. Autonomic reflexes maintain homeostasis and normal body functions at the unconscious level, which frees the mind to deal with those actions that require conscious decisions. *Somatic reflexes* act on skeletal muscles. They enable quick movements such as moving your hand away from a painful stimulus. A person is usually unaware of autonomic reflexes but is aware of somatic reflexes.

Figure 8.18 illustrates a somatic spinal reflex, which withdraws the hand after sticking a finger with a tack. Three neurons are involved in this reflex. Pain receptors are stimulated by the sharp pin and form impulses that are carried by a sensory neuron to an interneuron in the spinal cord. Impulses pass along the interneuron to a motor neuron, which carries the impulses to a muscle that contracts to move the hand. Although the brain is not involved in this reflex, it does receive sensory impulses that make a person aware of a painful stimulus resulting in the quick withdrawal of the hand.

What is a reflex?
What are the components of a spinal reflex?

Since the responses of reflexes are predictable, physicians usually test a patient's reflexes in order to determine the health of the nervous system. Exaggerated, diminished, or distorted reflexes may indicate a neurological disorder.

The Autonomic Nervous System (ANS)

The *autonomic* (aw-to-nom′-ik) *nervous system (ANS)* consists of portions of the central and peripheral nervous systems, and it functions involuntarily without conscious control. Its role is to maintain homeostasis in response to changing internal conditions. The effectors under autonomic control are cardiac muscle, smooth muscle, and glands. The ANS functions mostly by involuntary reflexes. Sensory impulses carried to the autonomic reflex centers in the hypothalamus, brain stem, or spinal cord cause motor impulses to be carried to effectors via cranial or spinal nerves. Higher brain centers, such as the limbic system and cerebral cortex, influence the ANS during times of emotional stress.

Table 8.3 compares the somatic and autonomic nervous systems.

Organization

Autonomic nerve fibers are motor fibers. Unlike the somatic nervous system, in which a single motor neuron extends from the central nervous system to an effector, the ANS uses two motor neurons in sequence to carry motor impulses to an effector. The first neuron, or *preganglionic neuron,* extends from the CNS to an **autonomic ganglion,** a mass of neuron cell bodies. The cell body of the preganglionic neuron is located in the brain or spinal cord. The second neuron, or *postganglionic neuron,* extends from the autonomic ganglion to the visceral effector, and its cell body is located within the autonomic ganglion (figure 8.19).

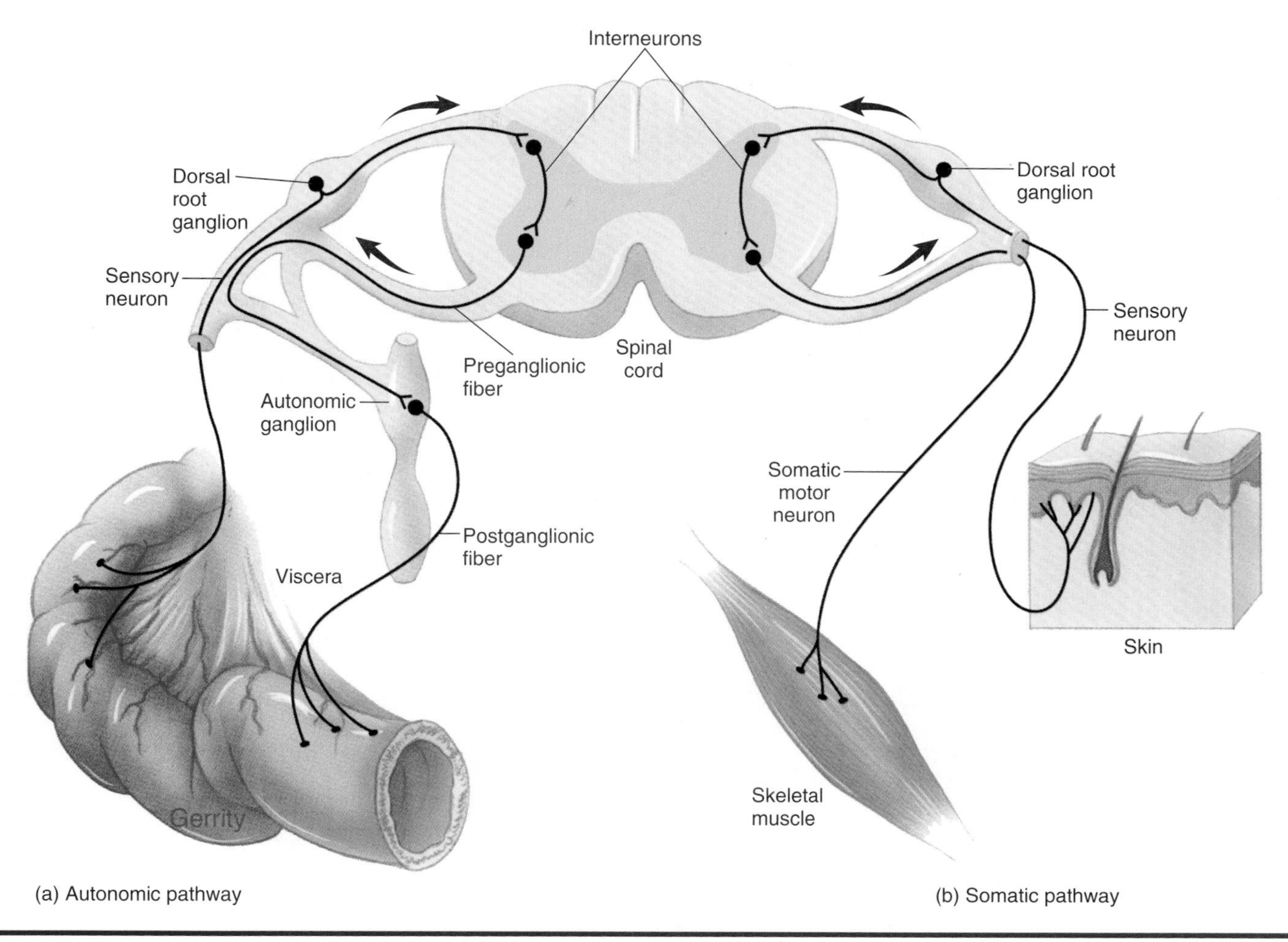

Figure 8.19 Comparison of autonomic and somatic motor pathways in spinal nerves. (*a*) An autonomic pathway involves a presynaptic neuron and a postsynaptic neuron that synapse at a ganglion outside the spinal cord. (*b*) A somatic pathway involves a single neuron.

The autonomic nervous system is subdivided into the **sympathetic division** and the **parasympathetic division.** The origin of their motor neurons and the organs innervated are shown in figures 8.20 and 8.21.

Preganglionic fibers of the sympathetic division arise from the thoracic and lumbar regions of the spinal cord—spinal nerves T1–L2. Some sympathetic fibers branch from the spinal nerves to synapse with postganglionic neurons in autonomic ganglia that are arranged in two chains, one on each side of the vertebral column. These ganglia are called *paravertebral* or *sympathetic chain ganglia.* Other sympathetic preganglionic fibers pass through a paravertebral chain ganglion without synapsing and extend to another type of ganglion, a *collateral ganglion,* before synapsing with a postganglionic neuron. Both pathways are shown in figure 8.20.

Preganglionic fibers of the **parasympathetic division** arise from the brain stem and sacral region (S2–S4) of the spinal cord. They extend through cranial or sacral nerves to synapse with postganglionic neurons within ganglia that are located very near or within visceral organs (figure 8.21).

Most visceral organs receive postganglionic fibers of both the sympathetic and parasympathetic divisions, but a few, such as sweat glands and most blood vessels, receive only sympathetic fibers. Figure 8.22 summarizes the organizational differences between the sympathetic and parasympathetic divisions.

Autonomic Neurotransmitters

Preganglionic fibers of both the sympathetic and parasympathetic divisions secrete *acetylcholine* to

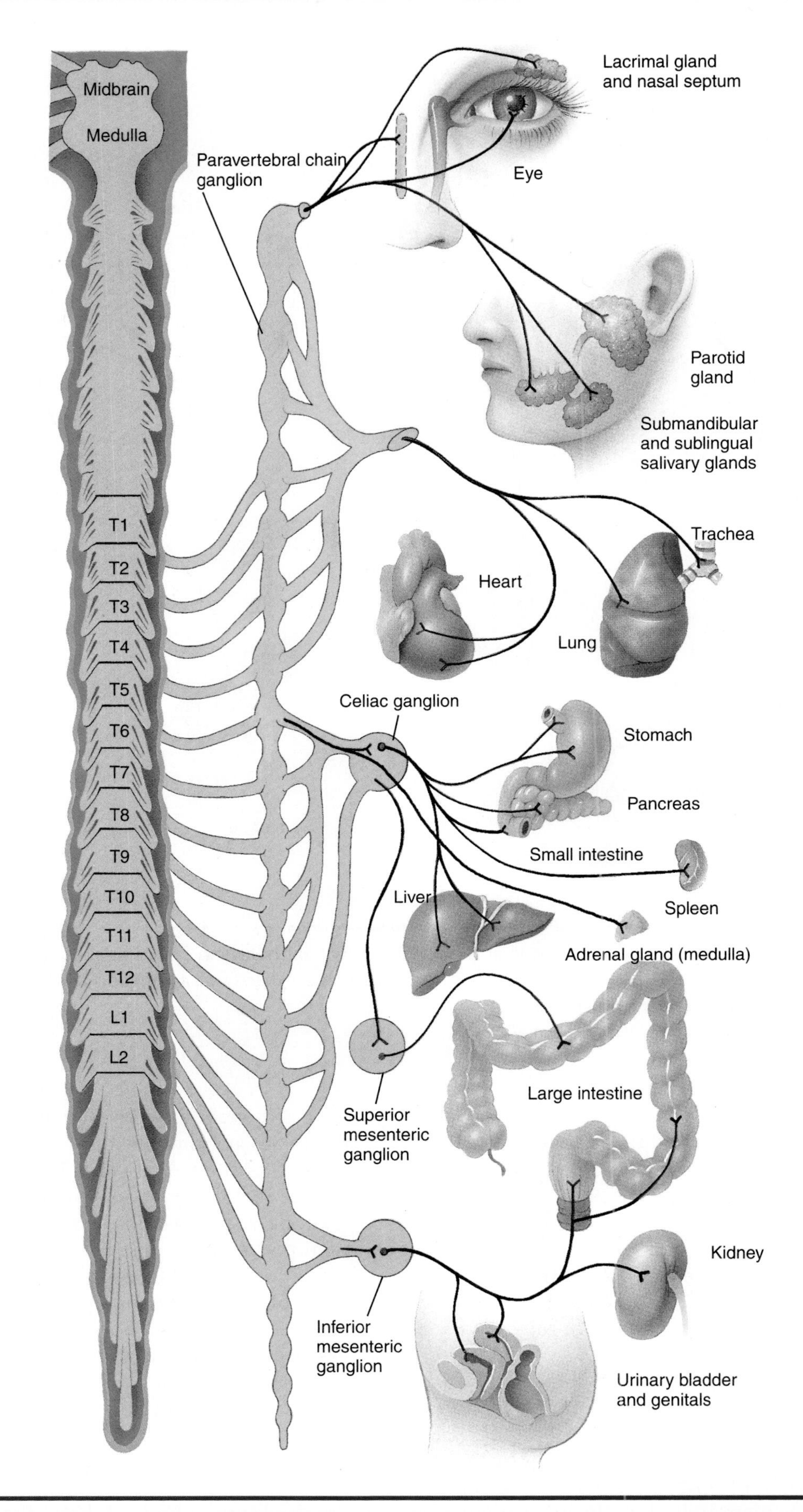

Figure 8.20 The preganglionic fibers of the sympathetic division of the autonomic nervous system arise from the thoracic and lumbar regions of the spinal cord. Note that the adrenal medulla is innervated directly by a preganglionic fiber.

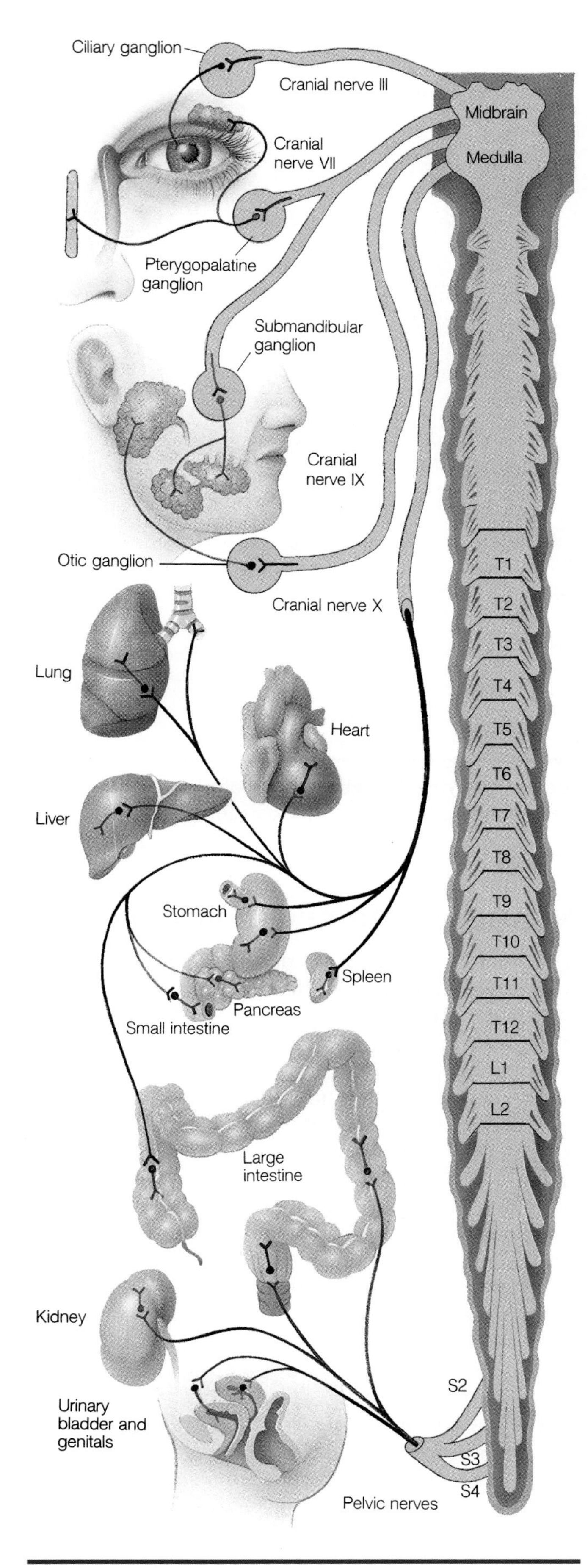

Figure 8.21 The preganglionic fibers of the parasympathetic division of the autonomic nervous system arise from the brain and sacral region of the spinal cord.

initiate impulses in postganglionic neurons, but their postganglionic fibers secrete different neurotransmitters. Sympathetic postganglionic fibers secrete *norepinephrine,* a substance similar to adrenalin, and they are called *adrenergic fibers.* Parasympathetic postganglionic fibers secrete acetylcholine, and thus are called *cholinergic fibers* (figure 8.22).

Functions

Both sympathetic and parasympathetic divisions stimulate some visceral organs and inhibit others, but their effects on a given organ are opposite. For example, the sympathetic division increases heart rate while the parasympathetic division slows heart rate. The contrasting effects are due to the different neurotransmitters secreted by postganglionic sympathetic and parasympathetic fibers and the receptors of the receiving organs.

The sympathetic division prepares the body for physical action to meet emergencies. Its action has been summarized as preparing the body for *fight or flight.* The parasympathetic division is dominant under the normal, nonstressful conditions of everyday life, and its actions are usually opposite those of the sympathetic division. Table 8.4 compares some of the effects of the sympathetic and parasympathetic divisions on visceral organs.

How do the somatic and autonomic nervous systems differ?
How does the autonomic nervous system maintain homeostasis?

Disorders of the Nervous System

The disorders of the nervous system may be categorized as either inflammatory or noninflammatory disorders.

Inflammatory Disorders

Meningitis (men-in-jī′-tis) results from a bacterial, fungal, or viral infection of the meninges. Bacterial meningitis cases are the most serious, with about 20% being fatal. If the brain is also involved, the disease is called *encephalitis.* Some viruses causing encephalitis are transmitted by bites of certain mosquitoes.

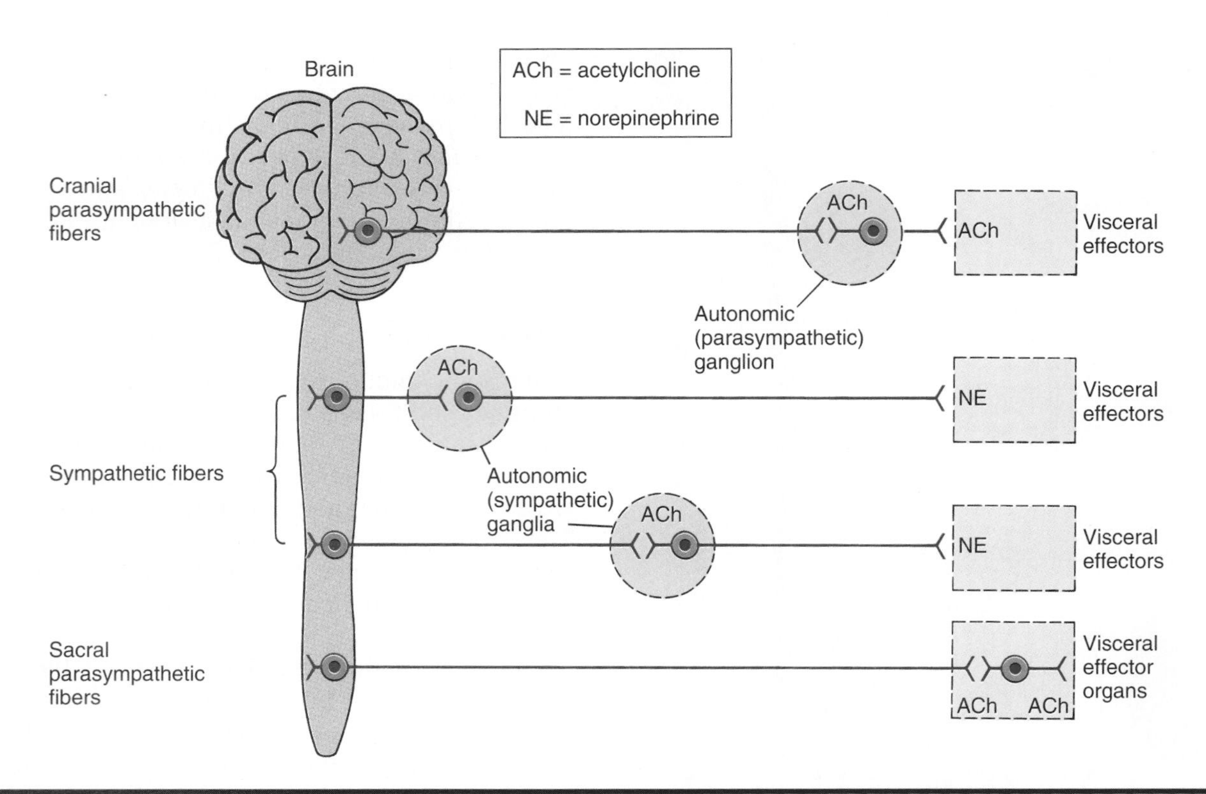

Figure 8.22 Parasympathetic fibers are cholinergic and secrete acetylcholine at the ends of postganglionic fibers. Most sympathetic fibers are adrenergic and secrete norepinephrine at the ends of postganglionic fibers.

Cocaine exerts a major effect on the autonomic nervous system. It not only stimulates the sympathetic division but also inhibits the parasympathetic division. In an overdose, this double-barreled action produces an erratic, uncontrollable heartbeat that may result in sudden death.

Neuritis is the inflammation of a nerve or nerves. It may be caused by several factors such as infection, compression, or trauma. Associated pain may be moderate or severe.

Sciatica (sī-at′-i-kah) is neuritis involving the sciatic nerve. Neuralgia may be severe, with the pain radiating down the thigh and leg.

Shingles is an infection of one or more nerves. It is caused by the reactivation of the chickenpox virus, which, until that time, was dormant in the nerve root. The virus causes painful blisters on the skin at the sensory nerve endings, followed by prolonged neuralgia.

Noninflammatory Disorders

Alzheimer's (alts′-hī-merz) **disease (AD)** is a progressively disabling disease affecting older persons. It is associated with a loss of certain cholinergic neurons in the brain and a reduced ability of neurons to secrete acetylcholine. AD is characterized by a progressive loss of memory, disorientation, and mood swings.

Cerebral palsy (ser-ē′-bral pawl-zē) is characterized by partial paralysis and sometimes a degree of mental retardation. It may result from damage to the brain during prenatal development, often from viral infections caused by German measles, or from trauma during delivery.

Cerebrovascular accidents (CVAs) are disorders of blood vessels serving the brain. They result from blood clots, aneurysms (an′-ū-rizm), or hemorrhage. Often called strokes, CVAs cause major damage to the brain. They are a major cause of disability and the third highest cause of death in the United States.

Comas are states of unconsciousness in which the patient cannot be aroused even with vigorous stimulation. Illness or trauma to the brain may alter the functioning of the reticular formation, resulting in a coma.

Table 8.4 Representative Actions of the Autonomic Nervous System

Effector	Sympathetic Stimulation	Parasympathetic Stimulation
Eye	Dilation of pupil; changes lens shape for far vision	Constriction of pupil; changes lens shape for near vision
Heart	Increases rate and strength of contraction	Decreases rate and strength of contraction
Arterioles	Constriction increases blood pressure	No known action
Blood distribution	Increases supply to skeletal muscles; decreases supply to digestive organs	Decreases supply to skeletal muscles; increases supply to digestive organs
Lungs	Dilates bronchioles	Constricts bronchioles
Digestive tract	Inhibits motility and secretion by glands	Promotes motility and secretion by glands
Liver	Decreases bile production; increases blood glucose	Increases bile production; decreases blood glucose
Gallbladder	Relaxation	Contraction
Kidneys	Decreases urine production	No known action
Pancreas	Decreases secretion of insulin and digestive enzymes	Increases secretion of insulin and digestive enzymes
Spleen	Constriction injects stored blood into circulation	No known action
Urinary bladder	Contraction of external sphincter; relaxation of bladder wall	Relaxation of external sphincter; contraction of bladder wall
Reproductive organs	Vasoconstriction; ejaculation in males; reverse uterine contractions in females; stimulates uterine contractions in childbirth	Vasodilation; erection in males; vaginal secretion in females

Concussion results from a severe jarring of the brain caused by a blow to the head. Unconsciousness, confusion, and amnesia may result in severe cases.

Dyslexia (dis-lek′-sē-ah) causes the afflicted person to reverse letters or syllables in words and words within sentences. It results from malfunctioning of the language center of the cerebrum.

Epilepsy (ep′-i-lep″-sē) may have a hereditary basis, or it may be triggered by injuries, infections, or tumors. There are two types of epilepsy. *Grand mal epilepsy* is the more serious form and is characterized by convulsive seizures. *Petit mal epilepsy* is the less serious form and is characterized by momentary loss of contact with reality without unconsciousness or convulsions.

Fainting is a brief loss of consciousness due to a sudden reduction in blood supply to the brain. It may result from either physical or psychological causes.

Headaches may result from several physical or psychological causes, but they often result from a dilation of blood vessels within the meninges of the brain. Migraine headaches have visual or digestive side effects and may result from stress, allergies, or fatigue. Sinus headaches may result from inflammation that causes increased pressure within the sinuses. Some headaches result from tension in muscles of the head and neck.

Mental illnesses may be broadly categorized as either neuroses or psychoses. *Neuroses* are mild maladjustments to life situations that may produce anxiety and interfere with normal behavior. *Psychoses* are serious mental disorders that sometimes cause delusions, hallucinations, or withdrawal from reality.

Multiple sclerosis (MS) is a progressive degeneration of the myelin sheath around neuron processes in the CNS, accompanied by the formation of plaques of scar tissue called *scleroses.* This destruction results in a short-circuiting of neural pathways and an impairment of motor functions.

Neuralgia (nū-ral′-jē-ah) is pain arising from a nerve regardless of the cause of the pain.

Paralysis is the permanent loss of motor control of body parts. It most commonly results from accidental injury to the CNS.

Parkinson's disease is caused by an insufficient production of *dopamine* by neurons in certain basal nuclei within the cerebrum. It produces tremors and impairs normal skeletal muscle contractions. Parkinson's disease is more common among older persons.

CHAPTER SUMMARY

Divisions of the Nervous System

1. Anatomical divisions are the central nervous system (CNS), composed of the brain and spinal cord, and the peripheral nervous system (PNS), composed of cranial and spinal nerves.
2. Functional divisions are the somatic nervous system (SNS), which is involved in voluntary responses, and the autonomic nervous system (ANS), which is involved in involuntary responses.

Nerve Tissue

1. Nerve tissue consists of neurons and neuroglial cells.
2. A neuron is composed of a cell body, which contains the nucleus, one or more dendrites that conduct impulses toward the cell body, and one axon that conducts impulses away from the cell body.
3. Myelinated neuron processes are covered by a myelin sheath. Schwann cells form the myelin sheath and neurilemma of peripheral myelinated nerve fibers. Oligodendrocytes form the myelin sheath of myelinated nerve fibers in the CNS; these fibers lack a neurilemma.
4. There are three functional types of neurons. Sensory neurons carry impulses toward the CNS. Interneurons carry impulses within the CNS. Motor neurons carry impulses from the CNS.
5. Four types of neuroglial cells occur in the CNS: oligodendrocytes, astrocytes, microglial cells, and ependymal cells.

Neuron Physiology

1. Neurons are specialized to form and conduct impulses.
2. The membrane of a resting neuron is polarized with an excess of positive ions outside the cell and negative ions inside the cell.
3. When a threshold stimulus is applied, the neuron membrane becomes permeable to sodium ions (Na^+), which quickly move into the neuron and cause depolarization of the membrane. This depolarization is the formation of a nerve impulse.
4. The depolarized portion of the membrane causes the depolarization of adjacent portions so that a depolarization wave (impulse) flows along the neuron.
5. Depolarization makes the neuron membrane permeable to potassium ions (K^+), and they quickly diffuse out of the neuron, which repolarizes the membrane.
6. In synaptic transmission, the synaptic knob of the presynaptic neuron secretes a neurotransmitter into the synaptic cleft. The neurotransmitter reacts with receptors on the postsynaptic neuron, causing the formation of an impulse. Then, the neurotransmitter is quickly removed by an enzymatic reaction.
7. The most common neurotransmitters are acetylcholine and norepinephrine. Some neurotransmitters are excitatory, and others are inhibitory.

Protection for the Central Nervous System

1. The brain is encased by the cranial bones, and the spinal cord is surrounded by vertebrae.
2. Both the brain and spinal cord are covered by the meninges: the pia mater, arachnoid mater, and dura mater.
3. Cerebrospinal fluid in the subarachnoid space serves as a fluid shock absorber.

The Brain

1. The brain consists of the cerebrum, thalamus, hypothalamus, brain stem, and cerebellum.
2. The cerebrum consists of two cerebral hemispheres joined by the corpus callosum. Gyri and sulci increase the surface area of the cerebral cortex. Each hemisphere is subdivided into four lobes: frontal, parietal, temporal, and occipital lobes.
3. The cerebrum interprets sensations, initiates voluntary motor responses, and is involved in will, personality traits, and intellectual processes. The left cerebral hemisphere is dominant in most people.
4. Sensory interpretation areas occur in the parietal, temporal, and occipital lobes. Motor areas occur in the frontal lobe. Association areas occur in all lobes of the cerebrum.
5. The thalamus is formed of two lateral masses connected by the intermediate mass. It is a relay station for sensory and motor impulses going to and from the cerebrum and provides an uncritical awareness of sensations.
6. The hypothalamus is located below the thalamus and forms the floor of the third ventricle. It is a major control center for the autonomic nervous system, and it regulates several homeostatic processes such as body temperature, mineral and water balance, appetite, digestive processes, and secretion of pituitary hormones.
7. The limbic system is associated with emotional behavior and memory.
8. The brain stem consists of the midbrain, pons, and medulla oblongata. Ascending and descending fibers between higher brain centers and the spinal cord pass through the brain stem.
9. The midbrain is a small, superior portion of the brain stem. It contains reflex centers for movements associated with visual and auditory stimuli.

10. The pons is the intermediate portion of the brain stem. It works with the medulla oblongata to control breathing.
11. The medulla oblongata is the lowest portion of the brain stem and is continuous with the spinal cord. It contains reflexive control centers that control breathing, heart rate, and blood pressure.
12. The reticular formation consists of nuclei in the brain stem and fibers that extend to the cerebrum. It is involved with wakefulness and sleep.
13. The cerebellum lies posterior to the fourth ventricle. It is composed of two hemispheres separated by the vermis, and it coordinates skeletal muscle contractions.
14. The ventricles of the brain, the central canal of the spinal cord, and the subarachnoid space around the brain and spinal cord are filled with cerebrospinal fluid. Cerebrospinal fluid is secreted by a choroid plexus in each ventricle.
15. Cerebrospinal fluid is absorbed into blood of the sagittal sinus in the dura mater.

The Spinal Cord

1. The spinal cord extends from the medulla oblongata down the vertebral canal to the second lumbar vertebra.
2. Gray matter is located interiorly and is surrounded by white matter. Anterior horns of gray matter contain motor neuron cell bodies; posterior horns contain interneuron cell bodies. White matter contains ascending and descending tracts of myelinated neuron processes.
3. The spinal cord serves as a reflex center and conducting pathway for impulses between the brain and spinal nerves.

The Peripheral Nervous System (PNS)

1. The PNS consists of cranial and spinal nerves. Most nerves are mixed nerves; a few cranial nerves are sensory only. A nerve contains bundles of nerve fibers (neuron processes) supported by connective tissue.
2. The 12 pairs of cranial nerves are identified by Roman numeral and name. The 31 pairs of spinal nerves are divided into 8 cervical, 12 thoracic, 5 lumbar, 5 sacral, and 1 coccygeal nerve.
3. Anterior branches of many spinal nerves form plexuses where nerve fibers are sorted and recombined so that all fibers to a particular organ are carried in the same nerve. The four pairs of plexuses are cervical, brachial, lumbar, and sacral plexuses.

The Autonomic Nervous System (ANS)

1. The ANS involves portions of the central and peripheral nervous systems that are involved in involuntary maintenance of homeostasis.
2. Two ANS neurons are used to activate an effector. The presynaptic neuron arises from the CNS and ends in an autonomic ganglion, where it synapses with a postsynaptic neuron that continues on to the effector.
3. The ANS is divided into two subdivisions that generally have antagonistic effects. Nerves of the sympathetic division arise from the thorax and lumbar regions of the spinal cord and prepare the body to meet emergencies. Nerves of the parasympathetic division arise from the brain and sacral region of the spinal cord and function mainly in nonstressful situations.

Disorders of the Nervous System

1. Disorders may result from infectious diseases, degeneration from unknown causes, malfunctions, and physical injury.
2. Inflammatory neurological disorders include meningitis, neuritis, sciatica, and shingles.
3. Noninflammatory neurological disorders include Alzheimer's disease, cerebral palsy, CVAs, comas, concussion, dyslexia, epilepsy, fainting, headaches, mental illness, multiple sclerosis, neuralgia, paralysis, and Parkinson's disease.

BUILDING YOUR VOCABULARY

1. **Selected New Terms**
acetylcholine, p. 154
action potential, p. 152
arachnoid mater, p. 155
cerebellum, p. 160
cerebrum, p. 156
dura mater, p. 155
gray matter, p. 159
interneuron, p. 150
limbic system, p. 159
medulla oblongata, p. 159
motor neuron, p. 150
nerve tracts, p. 162
norepinephrine, p. 154
pia mater, p. 155
resting potential, p. 151
reticular formation, p. 160
sensory neuron, p. 150
spinal plexus, p. 165
synaptic cleft, p. 153
synaptic knob, p. 153
ventricle, p. 160
white matter, p. 159

2. **Related Clinical Terms**

analgesic (an″-al-jē′-sik) A pain-relieving drug.
encephalopathy (en-se-fa-lop′-a-thē) Any disorder or disease of the brain.
hemiplegia (hem-i-plē′-jē-ah) Paralysis of one side of the body.
neurology (nū-rol′-ō-jē) Branch of medicine dealing with disorders of the nervous system.
paraplegia (par″-ah-plē′-jē-ah) Paralysis of both legs.
psychiatry (si-kī′-ah-trē) Branch of medicine that treats mental illness.

STUDY ACTIVITIES

1. Complete the **Chapter 8 Study Guide,** which begins on page 435.
2. Complete the **Learning Objectives** listed on the first page of this chapter.

CHECK YOUR UNDERSTANDING

(Answers are located in appendix B.)

1. Impulses are carried away from a neuron cell body by the _____ .
2. Neurons that conduct impulses from place to place within the CNS are _____ .
3. An impulse is formed by the sudden flow of _____ ions through the plasma membrane into a neuron.
4. Synaptic transmission is dependent upon the secretion of a _____ by an axon's synaptic knob.
5. Impulses pass from one cerebral hemisphere to the other over neuron processes composing the _____ .
6. Voluntary muscle contractions are controlled by the _____ lobe of the cerebrum.
7. Intelligence, will, and memory are functions of the _____ .
8. The _____ regulates appetite, water balance, and body temperature.
9. The _____ regulates heart and breathing rates.
10. Coordination of body movements is a function of the _____ .
11. Cerebrospinal fluid fills the ventricles of the brain and the _____ space of the meninges.
12. The _____ is a conducting pathway between the brain and spinal nerves.
13. The _____ roots of spinal nerves consist of axons of motor neurons.
14. The _____ nervous system is involved in involuntary responses that maintain homeostasis.
15. The _____ division prepares the body for physical responses to emergencies.

C H A P T E R

The Senses

NINE

9

Chapter Preview & Learning Objectives

SELECTED KEY TERMS

Accommodation The focusing of light rays on the retina by the lens.
Adaptation The decrease in the formation of impulses by a receptor when repeatedly stimulated by the same stimulus.
Chemoreceptor A sensory receptor that is stimulated by certain chemicals.
Choroid (choroid = membranelike) The middle layer of the eyeball.
Cochlea (cochlea = snail) The coiled portion of the inner ear containing the receptors for hearing.
Cones Photoreceptors for color vision.
Cornea The transparent window through which light enters the eye.
Dynamic equilibrium Maintenance of balance when the head and body are in motion.
Mechanoreceptor A sensory receptor that is stimulated by certain mechanical forces such as pressure or touch.
Organ of Corti The sense organ in the inner ear containing the receptors for hearing.
Photoreceptor (photo = light) A sensory receptor that is stimulated by light.
Projection The process by which the brain makes a sensation seem to come from the body part being stimulated.
Referred pain The projection of pain to a site different than the source of the pain.
Retina (retin = net) The inner layer of the eye, which contains the photoreceptors.
Rods Photoreceptors for black and white vision.
Sclera (scler = hard) The outer layer of the eye.
Semicircular canals The portion of the inner ear containing the receptors for dynamic equilibrium.
Static equilibrium The maintenance of balance when the head and body are not in motion.

Our senses constantly inform us of what is going on in our environment so that our body can take appropriate voluntary or involuntary action and maintain homeostasis. Several different types of sensory receptors are involved, and they send impulses to the central nervous system, which initiates the appropriate response.

The senses may be subdivided into two broad categories: general senses and special senses. *General senses* include pain, touch, pressure, cold, and heat. *Special senses* are taste, smell, vision, hearing, and equilibrium.

Sensations

Each type of sensory receptor is sensitive to a particular type of stimulus that causes the receptor to form impulses. These impulses are carried by cranial or spinal nerves to the central nervous system. **Sensations** result from an interpretation of impulses reaching sensory areas of the cerebral cortex. Since all impulses are alike, sensations are determined by the area of the cerebral cortex receiving the impulses rather than by the type of receptor forming the impulses. For example, all impulses reaching the visual center of the occipital lobe are interpreted as visual sensations. A strong blow to an eye or to the back of the head may produce a visual sensation (flashes of light), although the stimulus is mechanical.

The intensity of a sensation is dependent upon the frequency of impulses reaching the cerebral cortex. The greater the frequency of impulses, the greater is the intensity of the sensation. Of course, the frequency of impulses sent to the brain is, in turn, dependent upon the action of receptors. The greater the intensity of a stimulus, the greater the frequency of impulse formation by receptors.

Projection

Whenever a sensation occurs, the cerebral cortex projects the sensation back to the body region where the impulses originate so that the sensation seems to come from that region. This phenomenon is called **projection.** For example, if your thumb is injured, the pain is projected back to your thumb, and you are aware that your thumb hurts. Similarly, projection of visual and auditory sensations gives the feeling that eyes see and ears hear. The projection of sensations has obvious survival value in pinpointing the source of a sensation, because it allows corrective action to remove harmful stimuli.

Adaptation

If a receptor is repeatedly stimulated by the same stimulus, the rate of impulse formation may decline until impulses may not be formed at all. This phenomenon is called **sensory adaptation.** For example, when the odor of perfume is first encountered, it is very noticeable. But as the olfactory receptors become adapted to the stimulus, the strength of the sensation rapidly declines until the odor is hardly noticeable. Adaptation prevents overloading the nervous system with unimportant stimuli, such as clothes touching the body. Once a receptor is adapted, a stronger stimulus is needed to form impulses.

General Senses

Receptors for the general senses are widely distributed in the skin, muscles, tendons, and visceral organs.

Temperature

Two types of temperature receptors are located in the skin. **Heat receptors** are located deep in the dermis, and they are most sensitive to temperatures above 25°C (77°F). Temperatures above 45°C (113°F) stimulate the formation of impulses that are interpreted as painful, burning sensations. **Cold receptors** are located closer to the surface of the dermis. They are most sensitive to temperatures below 20°C (68°F). Temperatures below 10°C (50°F) stimulate impulses that are interpreted as very cold and painful sensations.

Pressure and Touch

Pressure and touch receptors are **mechanoreceptors** (mek-ah-nō-rē-cep′tors), which are sensitive to mechanical stimuli displacing the tissue in which they are located. *Pressure receptors* are located deep in the dermis and in ligaments and tendons of joints, where they respond to stimulation by pressure.

There are two kinds of touch receptors: Meissner's corpuscles and free nerve endings (figure 9.1). *Meissner's* (mīz′-nerz) *corpuscles* are located in the superficial layer of the dermis, often in dermal papillae. They are most abundant in hairless portions of the skin, such as fingertips, palms, and lips. Meissner's corpuscles are especially sensitive to very light stimuli, such as the motion of objects that barely touch the skin. *Free nerve endings* are abundant in epithelial and connective tissues. In the skin, they extend from the dermis upwards between the cells of the epidermis. Free nerve endings are most important as pain receptors, but they also function as touch receptors. Some free nerve endings are wrapped around the base of hair follicles, where they are stimulated by movement of the hair shaft.

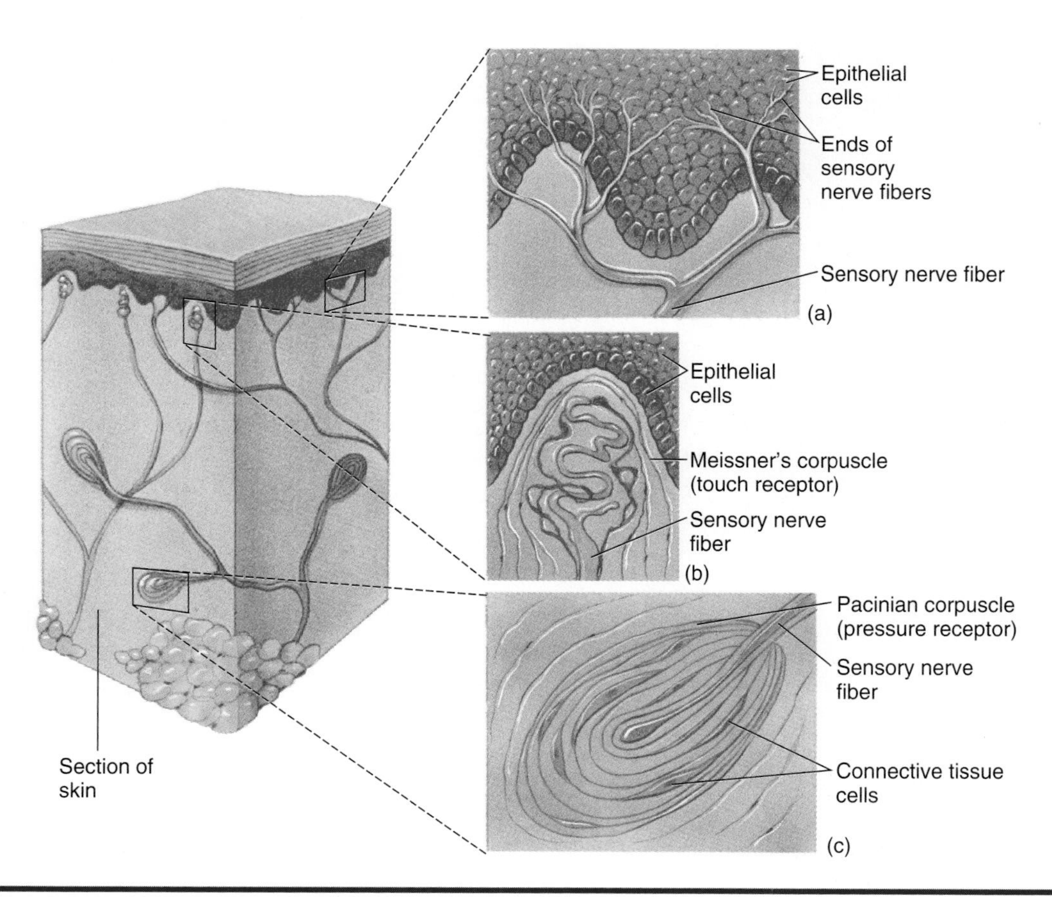

Figure 9.1 Touch receptors, (*a*) free nerve endings and (*b*) Meissner's corpuscles, are located near the surface of the skin. Pressure receptors, (*c*) Pacinian corpuscles, are located deep in the dermis.

Pain

Receptors for the sense of pain are **free nerve endings,** which are widespread in body tissues, except in the neural tissue of the brain. They are especially abundant in the skin, the organ that is in direct contact with the external environment. Pain receptors are stimulated whenever tissues are damaged, and the pain sensation initiates actions by the central nervous system to remove the source of the stimulation. Further, pain receptors do not easily adapt like many other receptors. They adapt slowly or not at all.

Referred Pain

Projection by the cerebral cortex is not always accurate when the impulses originate from pain receptors in visceral organs. When damage to visceral organs occurs, pain may seem to be located in some part of the body wall. Such pain is called **referred pain.**

Referred pain is consistent from person to person, and it is important in the diagnosis of many disorders. For example, pain caused by a heart attack seems to come from the left anterior chest wall, left shoulder, and left arm. Referred pain seems to be projected along common nerves used by neurons carrying impulses from both the body wall and visceral organs. For example, neurons carrying impulses from the heart use the same nerves as those from the left shoulder and arm (figure 9.2).

What parts of the nervous system are involved in the development of a sensation?
What sensory receptors are located within the skin?

Pain is unpleasant, but it has obvious survival value. It informs us of tissue damage and enables us to protect ourselves from further damage by either avoiding painful stimuli or seeking medical help.

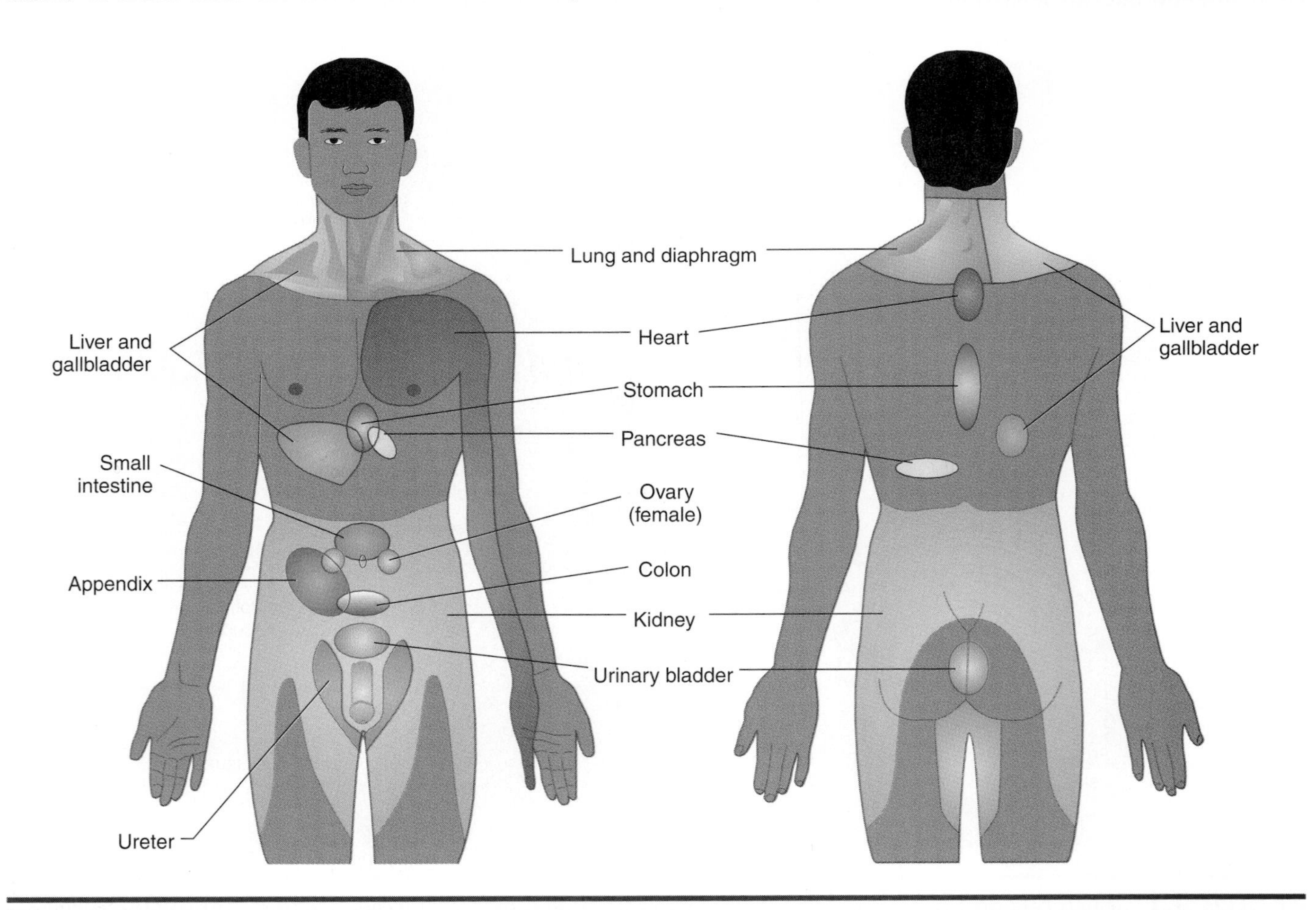

Figure 9.2 Surface regions to which visceral pain originating from various internal organs may be referred.

Special Senses

The receptors for special senses are localized rather than widely distributed, and they are specialized to respond to only certain types of stimuli. There are three different kinds of receptors for the special senses. Taste and smell receptors are **chemoreceptors,** which are sensitive to chemical substances. Receptors for hearing are mechanoreceptors, which are sensitive to vibrations formed by sound waves. Receptors for vision are **photoreceptors,** which are sensitive to light rays.

Taste

The chemoreceptors for taste are located in specialized microscopic organs called **taste buds.** Taste buds are located on the tongue in small, raised structures called *papillae* (figure 9.3).

A taste bud consists of several taste receptors called **taste cells** that are grouped in a bulblike arrangement within the epithelium of a papilla. *Taste hairs,* hairlike projections from the taste cells, protrude through the opening of the taste bud and are exposed to chemicals on the tongue. Nerve fibers leading to the brain are connected to the opposite ends of the taste cells. In order for a substance to activate the taste cells, it must be dissolved in a liquid—usually saliva.

There are four types of taste receptors, and each is associated with a distinct taste sensation: sweet, sour, salty, and bitter. These receptors have specific distributions on the tongue, as shown in figure 9.4. The many flavor sensations of foods result from the stimulation of one or more of these types of receptors and, even more important, the activation of olfactory receptors discussed in the next section. Taste receptors adapt rapidly to repeated stimulation.

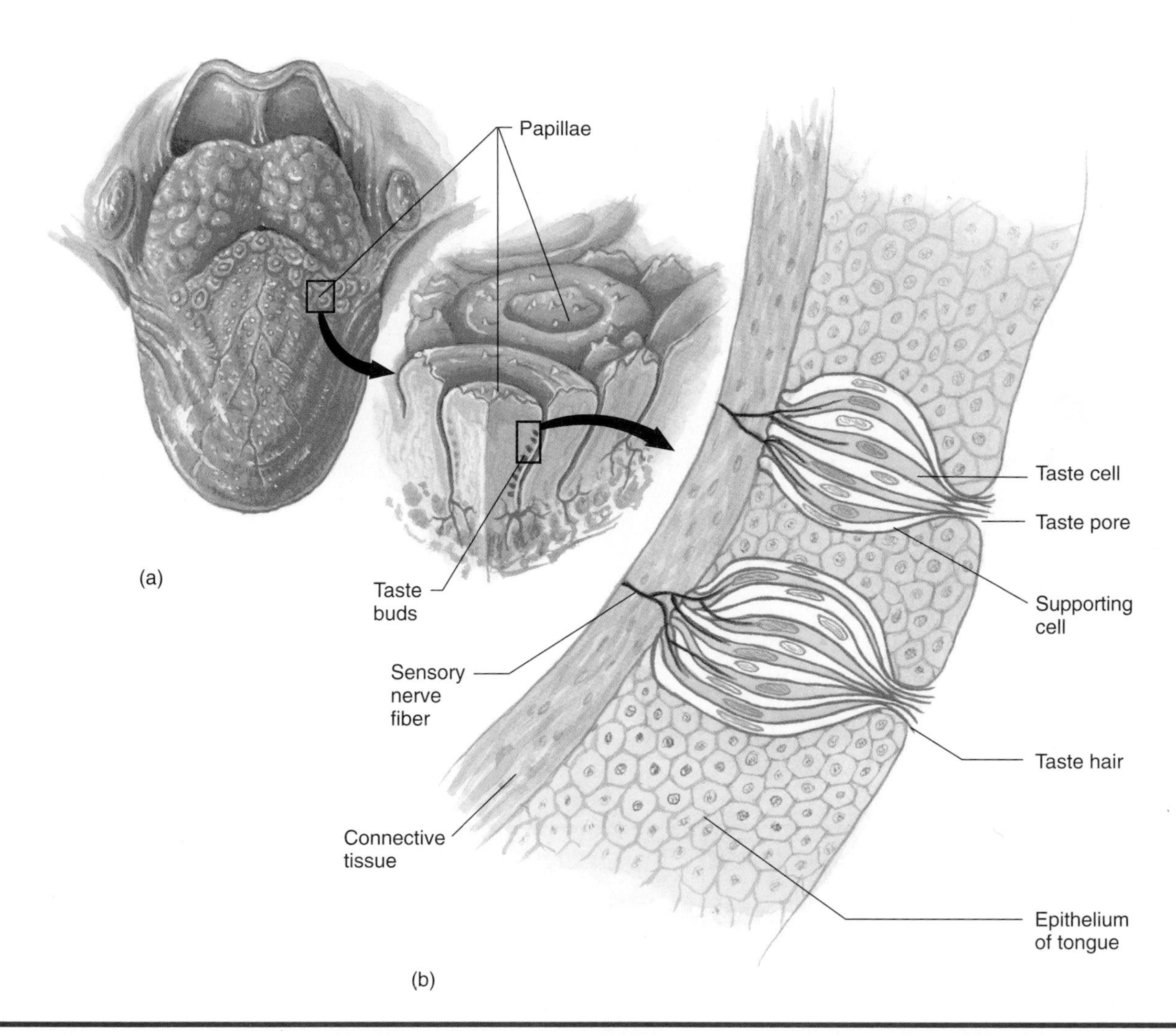

Figure 9.3 (*a*) Taste buds are located on papillae of the tongue. (*b*) A taste bud consists of taste cells whose hairlike tips protrude through the taste pore.

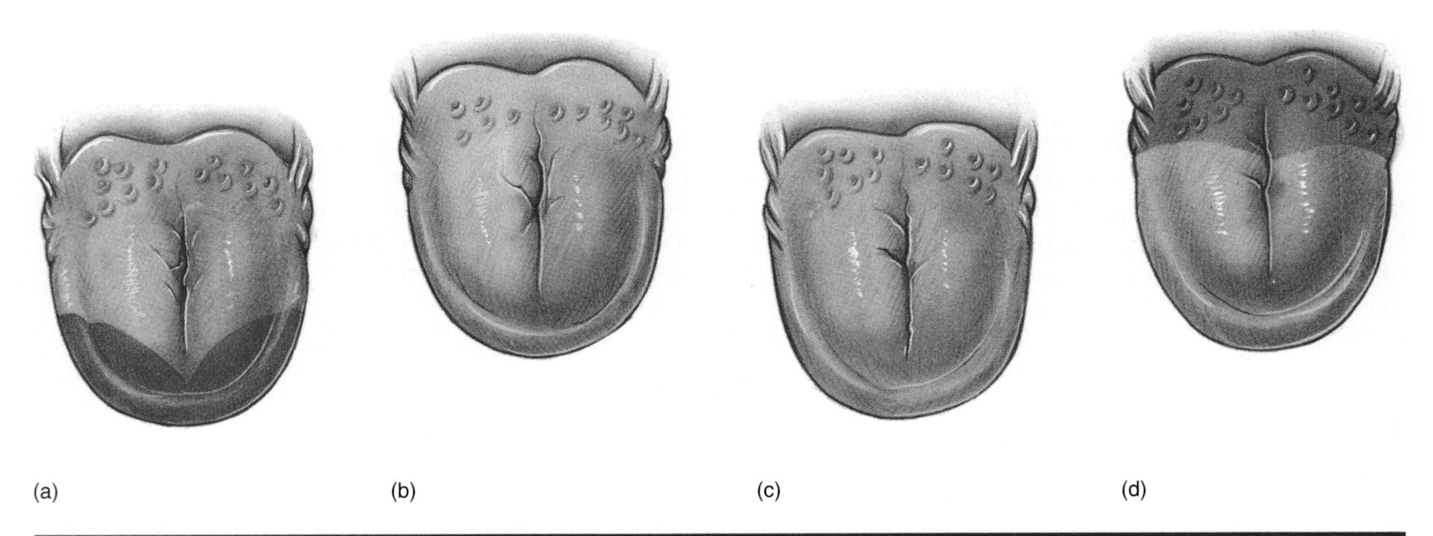

Figure 9.4 Distribution of the four primary types of taste receptors: (*a*) sweet, (*b*) sour, (*c*) salty, and (*d*) bitter.

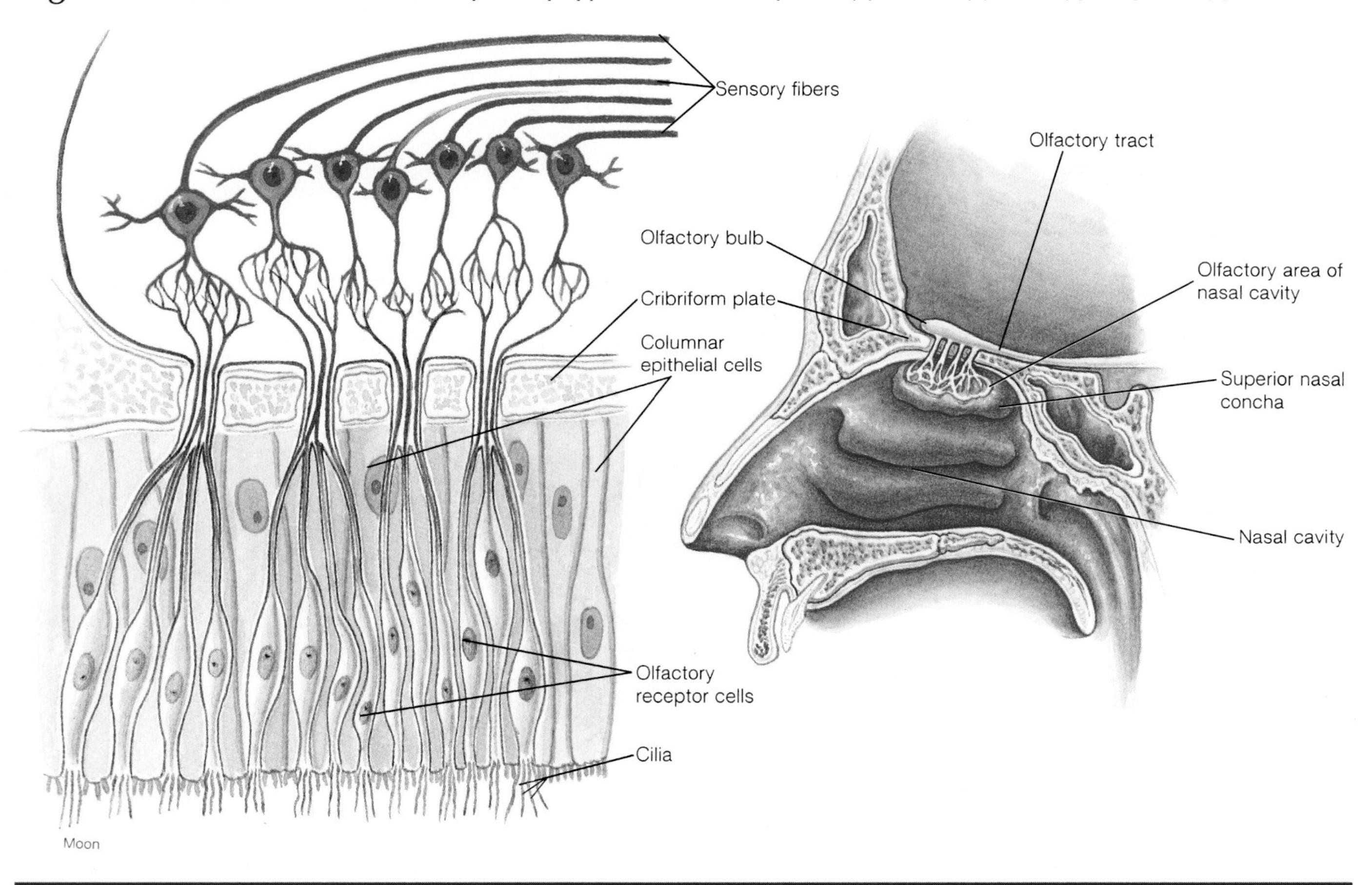

Figure 9.5 Olfactory receptors are located between columnar epithelial cells in the upper portion of the nasal cavity.

Smell

The **olfactory** (ōl-fak′-tō-rē) **organs** are located in the upper portion of the nasal cavity, including the superior conchae and nasal septum. They consist of olfactory **chemoreceptor cells,** which are supported by columnar epithelial cells of the nasal epithelium. The distal ends of the receptor cells are covered with cilia that project into the nasal cavity, where they can contact airborne molecules. Chemicals in inhaled air are in a gaseous state, but they must dissolve in the film of fluid covering the receptors in order to stimulate impulse formation (figure 9.5). When impulses are formed, they are carried by axons of receptor cells to

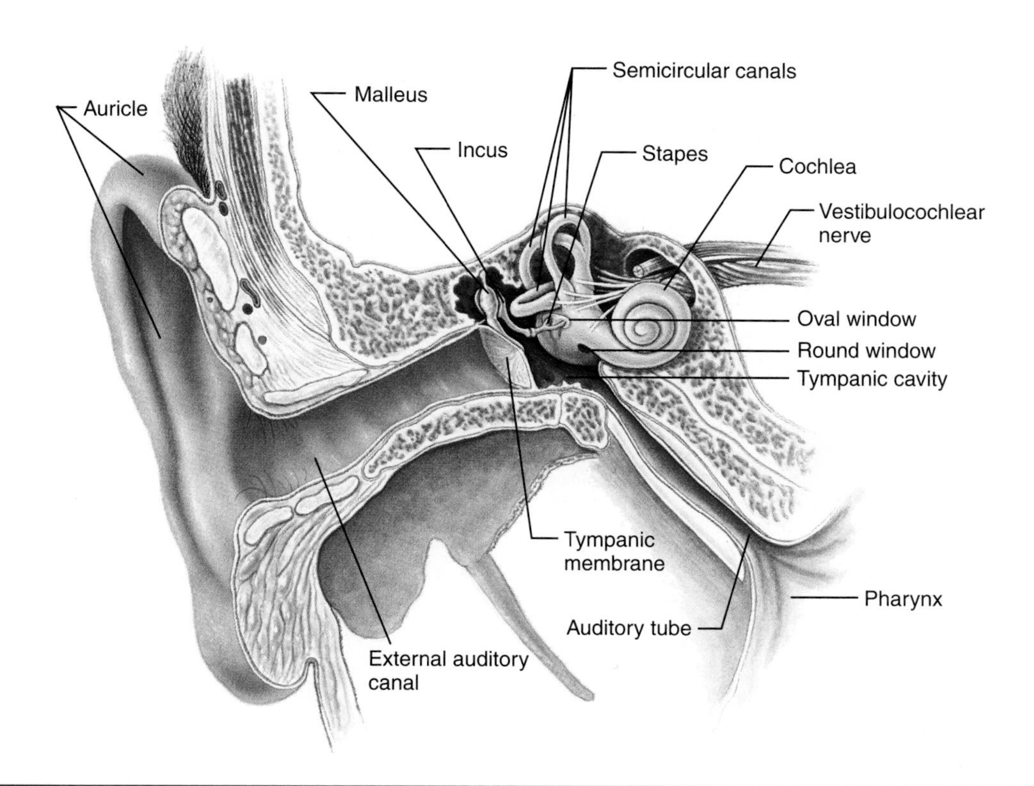

Figure 9.6 Anatomy of the ear.

the olfactory bulbs, where they synapse with neurons that form the olfactory tract and carry the impulses to the olfactory centers in the brain.

It is common for a person to sniff the air when trying to detect faint odors. This is because the olfactory receptors are located above the usual path of inhaled air, and additional force is needed to send larger amounts of air over the receptor cells. Like taste receptors, olfactory receptors rapidly adapt to a particular stimulus.

Our sense of smell enables us to distinguish about 10,000 different odors. But how our olfactory receptors enable so many olfactory sensations is poorly understood, since there don't seem to be 10,000 kinds of receptors. We'll have to wait for further research.

Where are taste and olfactory receptors located? How are the four basic taste sensations produced?

Hearing

The **ear** is the organ of hearing. It is also the organ of equilibrium. The ear is subdivided into three major parts: the external ear, middle ear, and inner ear (figure 9.6).

External Ear

The external ear consists of two parts: the auricle and the external auditory canal. The **auricle** (pinna) is the funnel-like structure composed primarily of cartilage and skin that is attached to the side of the head. The **external auditory canal** is a short tube that extends from the auricle through the temporal bone to the eardrum. Sound waves striking the auricle are channeled into the auditory canal. Earwax (cerumen) and hairs in the auditory canal help to prevent foreign particles from reaching the eardrum.

The ability to distinguish various foods relies predominantly on the sense of smell. This explains why foods seem to have little taste when one is suffering from a head cold. The taste and smell of appetizing foods prepare the digestive tract for digestion by stimulating the flow of saliva in the mouth and gastric juice in the stomach.

Middle Ear

The tympanic membrane, tympanic cavity, auditory tube, and ear ossicles are parts of the middle ear. The **tympanic** (tim-pan′-ik) **membrane,** or eardrum, is a membrane that closes the interior end of the external auditory canal. Its exterior surface is covered with skin, while the interior surface is coated with a mucous

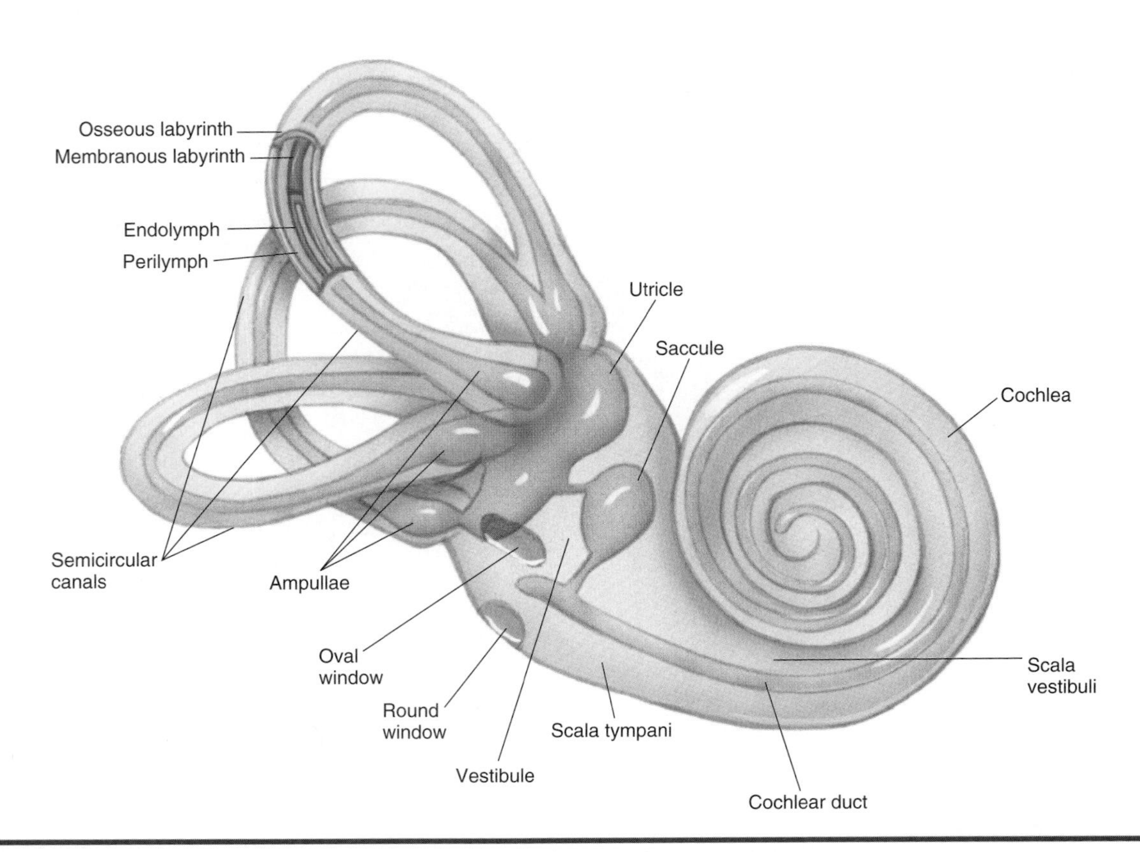

Figure 9.7 The bony and membranous labyrinths of the inner ear. Perilymph fills the space between the membranous labyrinth and the bony labyrinth. Endolymph fills the membranous labyrinth. Note that the ampullae of the semicircular canals, utricle, saccule, and cochlear duct are portions of the membranous labyrinth.

membrane. Sound waves, or air pressure waves, entering the auditory canal cause the tympanic membrane to vibrate in and out at the same frequency as the sound waves. The **tympanic cavity** is an air-filled space in the temporal bone that is separated from the external auditory canal by the tympanic membrane.

The **auditory** (eustachian) **tube** connects the tympanic cavity with the pharynx. Its function is to allow air to enter or exit the tympanic cavity in order to keep the air pressure within the tympanic cavity the same as the external air pressure. Equal air pressure on each side of the eardrum is essential for the eardrum to function properly. A valve at the pharyngeal end of the tube is usually closed, but it opens when a person swallows or yawns and allows air pressure to equalize. If you have experienced a rapid change in air pressure, you probably have noticed your ears "popping," as the air pressure is equalized and the tympanic membrane snaps back into place.

The **ear ossicles** (os′-si-kulz) are three tiny bones that articulate to form a lever system from the tympanic membrane to the inner ear. Each ossicle is named for its shape. The tip of the "handle" of the club-shaped *malleus* (mal′-ē-us), or hammer, is attached to the tympanic membrane while its head articulates with the *incus* (ing′-kus), or anvil. The base of the incus articulates with the *stapes* (stā′-pēz), or stirrup, whose foot plate is inserted into the oval window of the inner ear.

Sometimes the secretion of earwax is excessive and results in accumulations that block the auditory canal and decrease hearing. Removal of impacted earwax should be done by a physician. Irrigation with water is used to loosen the wax so that it can be safely removed.

The vibrations of the tympanic membrane cause corresponding movements of the ossicles, which result in the stapes vibrating in the oval window. In this way, vibrations of the eardrum are transmitted to the fluid-filled inner ear.

Inner Ear

The inner ear is embedded in the temporal bone. It consists of two series of connecting tubes and chambers, one within the other: an outer **bony labyrinth** (lab′-i-rinth) and an inner **membranous labyrinth.** Both labyrinths have a similar shape (figure 9.7). The

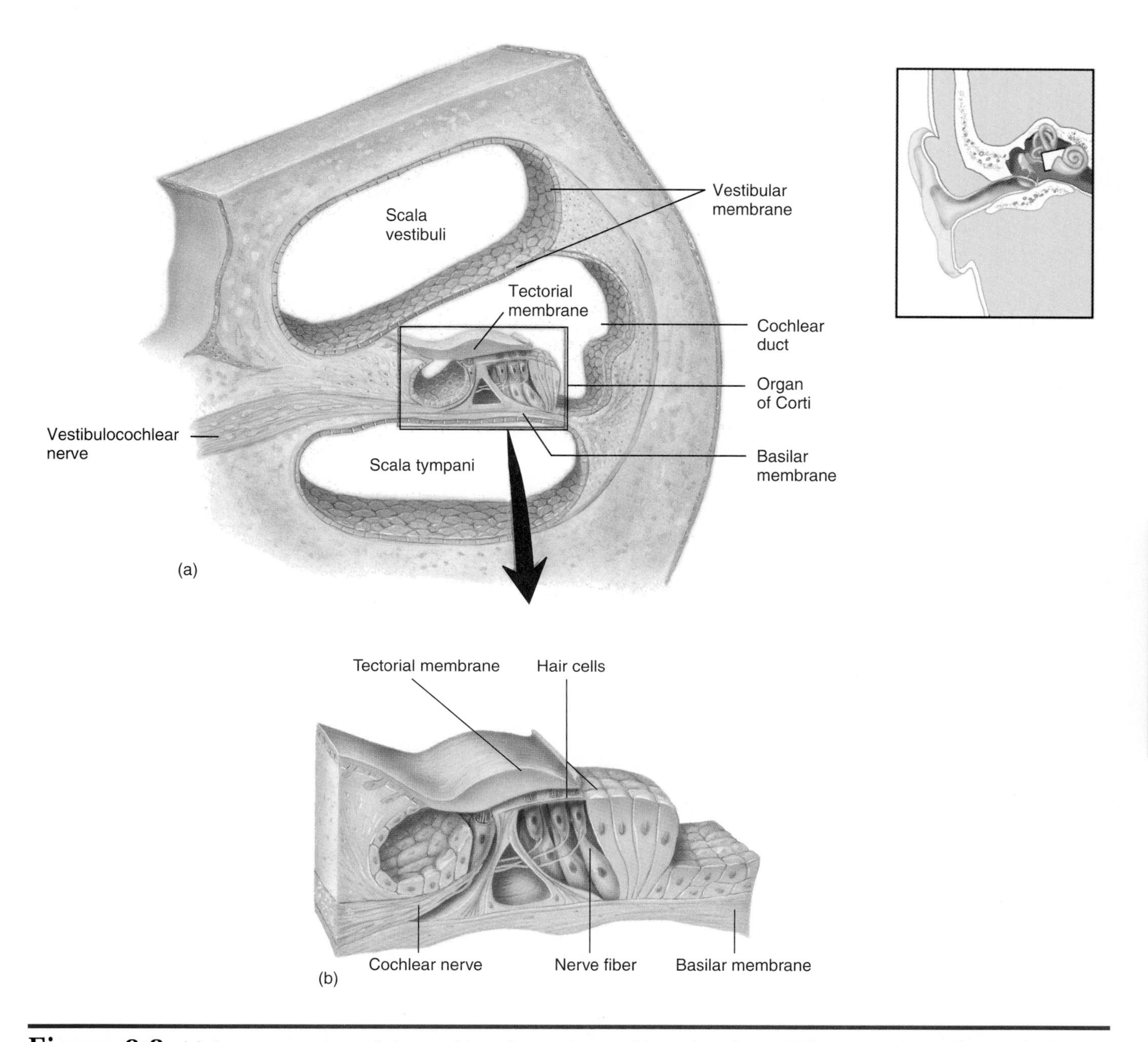

Figure 9.8 (*a*) A cross section of the cochlea shows the cochlear duct located between the scala vestibuli and the scala tympani and the organ of Corti resting on the basilar membrane. (*b*) Detail of the organ of Corti shows the tectorial membrane overlying the hair cells.

space between the bony and membranous labyrinths is filled with **perilymph,** while the membranous labyrinth contains **endolymph.** These fluids play important roles in the functions of the inner ear. The inner ear consists of three major parts: the cochlea, vestibule, and semicircular canals.

The **cochlea** (kok′-lē-ah) is the coiled portion of the inner ear. When viewed in cross section, as in figure 9.8, it can be seen that the cochlea is composed of three chambers that are separated from each other by membranes. The *scala vestibuli* (skā-la ves-tib′-ū-li) and the *scala tympani* extend the length of the cochlea and are continuous with each other at the apex of the cochlea. The scala vestibuli continues into the vestibule, which houses the *oval window.* The scala tympani extends toward the vestibule to the membrane-covered *round window.*

The *cochlear duct* extends nearly to the apex of the cochlea (see figure 9.7). As shown in figure 9.8, it is separated from the scala vestibuli by the vestibular membrane and from the scala tympani by the **basilar membrane.** The basilar membrane contains about 20,000 cross fibers that gradually increase in length from the base to the apex of the cochlea. The fibers are attached at one end to the bony center of the cochlea. This attachment allows them to vibrate like reeds of a harmonica when activated by vibrations generated by sound.

The **organ of Corti** (kor′-tī), which contains the receptors for sound stimuli, is located on the upper surface of the basilar membrane within the cochlear duct. The receptors are called **hair cells,** and they have hairlike cilia extending from their free surfaces toward the overlying *tectorial* (tek-to′-rē-al) *membrane.* Nerve fibers of the cochlear branch of the vestibulocochlear (VIII) nerve exit the hair cells and lead to the brain.

Physiology of Hearing

The human ear is able to detect sound waves with frequencies ranging from near 30 to 20,000 vibrations per second, but hearing is most acute between 2,000 and 3,000 vibrations per second. For hearing to occur, vibrations formed by sound waves must be transmitted to the hair cells of the organ of Corti. Then, the hair cells form impulses that are transmitted to the hearing centers of the brain, where they are interpreted as sound sensations.

Figure 9.9 shows the structure of the inner ear with the cochlea uncoiled to show more clearly the relationships of its parts. Refer to this figure as you study the following outline of hearing physiology.

1. Sound waves enter the external auditory canal and strike the tympanic membrane, causing it to vibrate in and out at the same frequency and comparable intensity as the sound waves. Loud sounds cause a greater displacement of the eardrum than do soft sounds.
2. Vibration of the tympanic membrane causes movement of the ear ossicles, resulting in the in-and-out vibration of the stapes in the oval window.
3. The vibration of the stapes causes a corresponding oscillatory (back-and-forth) movement of the perilymph in the scala vestibuli and scala tympani and a corresponding movement of the membrane over the round window. This movement of the perilymph causes vibrations in the vestibular and basilar membranes.
4. The vibration of the basilar membrane causes the hairs of the hair cells to contact the tectorial membrane, which stimulates the formation of impulses by the hair cells.
5. Impulses formed by the hair cells are carried by the cochlear branch of the vestibulocochlear (VIII) nerve to the hearing centers of the temporal lobes of the cerebrum, where the sensation is interpreted. Some of the nerve fibers cross over to the opposite side of the brain so that the hearing center in each temporal lobe interprets impulses originating in each ear.

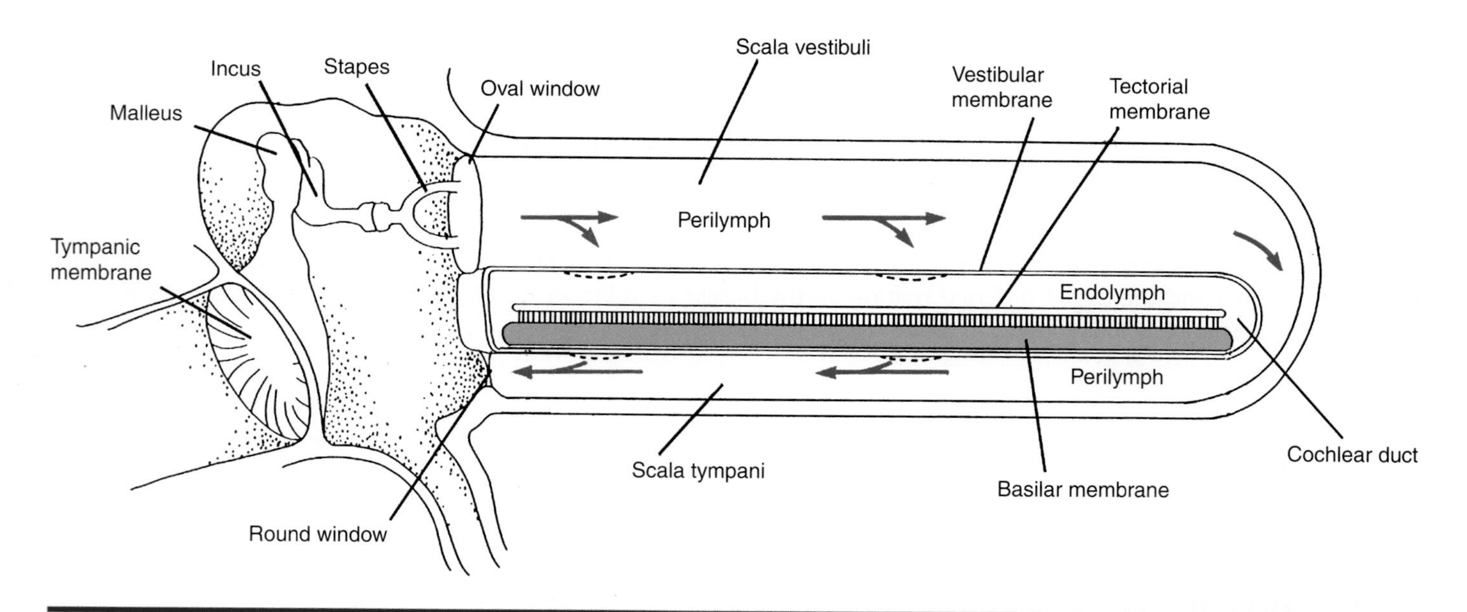

Figure 9.9 Transmission of vibrations produced by sound waves. Vibrations of the tympanic membrane are carried by the ear ossicles to the perilymph. Oscillating movements of the perilymph cause the vibration of the basilar membrane and organ of Corti, which, in turn, results in the formation of impulses by the hair cells.

Pitch and Loudness

Because of the gradually increasing length of the fibers in the basilar membrane, different portions of the basilar membrane vibrate in accordance with the different frequencies (pitch) of sound waves. Low-pitched sounds cause the longer fibers of the membrane near the apex of the cochlea to vibrate, while high-pitched sounds activate the shorter fibers of the membrane near the base of the cochlea. The pitch of a sound sensation is determined by the portion of the basilar membrane and the organ of Corti that are activated by the specific sound frequency and by the parts of the hearing centers that receive the impulses. Impulses from different regions of the organ of Corti go to slightly different portions of the hearing centers in the brain, which causes them to be interpreted as different pitches.

The loudness of the sound is dependent upon the intensity of the vibration of the basilar membrane and organ of Corti, which, in turn, determines the frequency of impulse formation. The greater the frequency of impulses sent to the brain, the louder is the sound sensation.

How do sound waves stimulate the formation of impulses?
How do the impulses produce a sound sensation?

The hair cells of the organ of Corti are easily damaged by high-intensity sounds, such as loud music and the noise of jet airplanes. Such damage produces a form of *nerve deafness* that may be partial or total, and it is permanent. Disorders of sound transmission by the tympanic membrane or ear ossicles cause *conduction deafness,* which may be repairable by surgical means or overcome by the use of hearing aids.

Equilibrium

Several types of sensory receptors provide information to the brain for the maintenance of equilibrium. Receptors in joints and muscles inform the brain about the movement of body parts, and the eyes are especially important in informing the brain about the position of the body parts and equilibrium. However, the crucial receptors are located in the inner ear. There are two types of equilibrium senses that involve different receptors in the inner ear. *Static equilibrium* occurs when the head is motionless. *Dynamic equilibrium* occurs when the head is moving.

Static Equilibrium

An organ of static equilibrium, the **macula** (mak′-u-lah), is located within the **utricle** (u′-tri-kul) and within the **saccule** (sak′-ul), enlarged portions of the membranous labyrinth (see figure 9.7). Each macula contains thousands of **hair cells,** the sensory receptors, whose hairlike cilia are embedded in a gelatinous material. *Otoliths* (ō′-tō-liths), crystals of calcium carbonate, are also embedded in the gelatinous mass. The otoliths increase the weight of the gelatinous mass and make it more responsive to the pull of gravity (figure 9.10).

The mechanism of static equilibrium may be summarized as follows:

1. The pull of gravity on the gelatinous mass stimulates the hair cells to form impulses that are carried by the vestibular portion of the vestibulocochlear nerve (VIII) to the brain. No matter the position of the head, impulses are formed that inform the brain of the head's position.
2. The cerebellum uses this information to maintain static equilibrium subconsciously.
3. Our awareness of static equilibrium results when the impulses are interpreted by the cerebrum.

Dynamic Equilibrium

The membranous labyrinths of the **semicircular canals** contain the receptors that detect motion of the head. Examine figure 9.7 and note that the semicircular canals are arranged at right angles to each other and that each occupies a different plane in space.

Near the attachment of each membranous canal to the utricle is an enlarged region called the *ampulla* (am-pūl′-lah). Each ampulla contains a sensory organ for dynamic equilibrium called the **crista ampullaris** (kris′-ta am-pūl-lar′-is). Each crista ampullaris contains a number of hair cells whose hairlike projections extend into a dome-shaped gelatinous mass, the cupula. Nerve fibers of the vestibular branch of the vestibulocochlear (VIII) nerve lead from the hair cells to the brain (figure 9.11).

The mechanism of dynamic equilibrium may be described as follows:

1. When the head is turned, the endolymph pushes against the cupula, bending the hairs of hair cells, which stimulates the formation of impulses. The impulses are carried to the brain

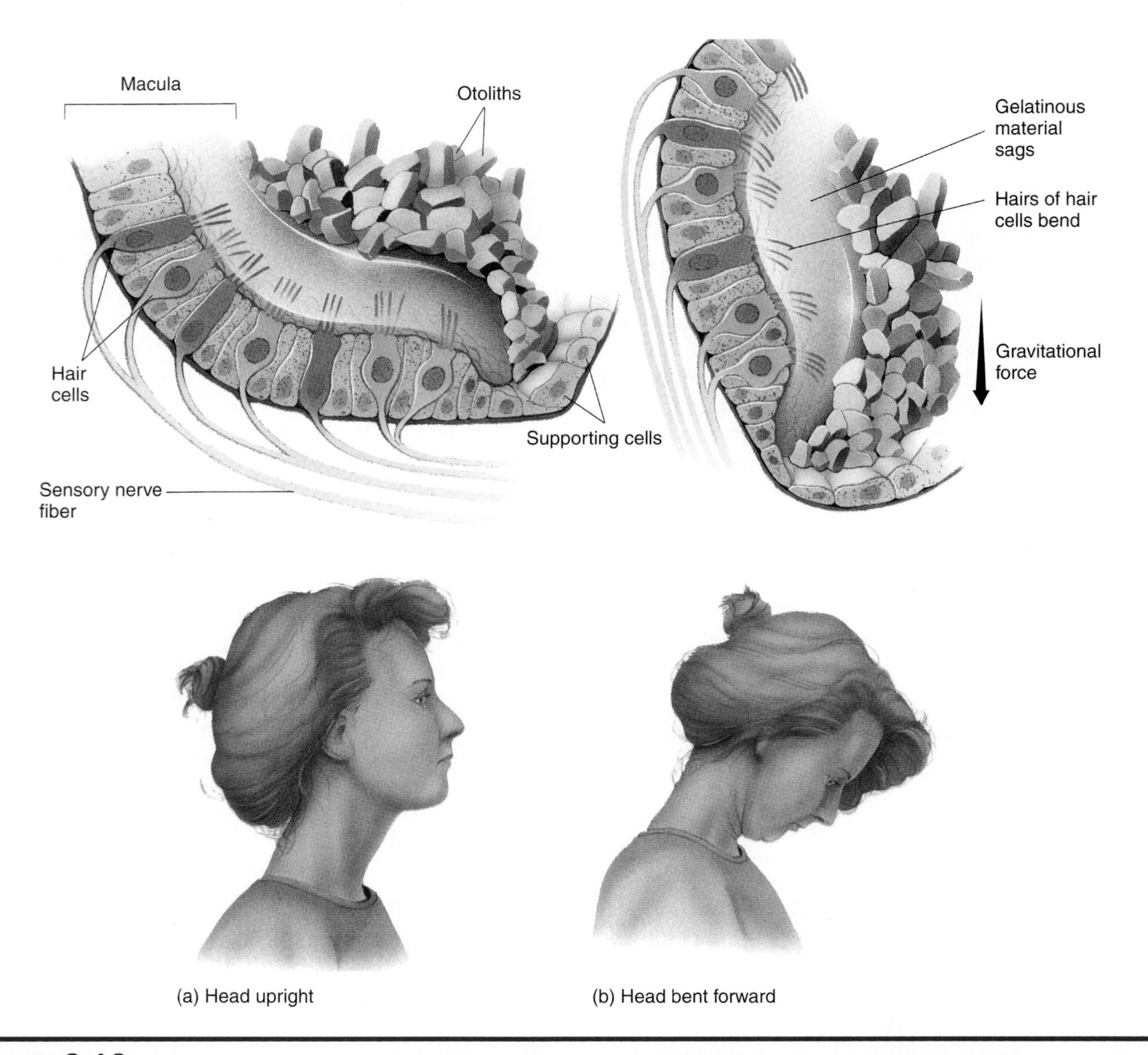

Figure 9.10 A macula is a receptor for static equilibrium. Note how bending the head forward causes bending of the hairs of the hair cells.

via the vestibular branch of the vestibulocochlear (VIII) nerve (see figure 9.6).

2. Since each semicircular canal is oriented in a different plane, the cristae and hair cells are not stimulated equally with a given head movement. Thus, the brain receives a different pattern of impulses for each type of head movement.
3. The cerebellum uses the impulses to make adjustments below the conscious level to maintain dynamic equilibrium.
4. Awareness of dynamic equilibrium, or lack of it, results from the cerebrum interpreting the pattern of impulses it receives (figure 9.12).

Table 9.1 summarizes the functions of the ear.

What structures are involved in static and dynamic equilibrium?

Vision

Vision is one of the most important senses supplying information to the brain. The receptors for light stimuli are located within the **eyes,** the organs of vision. The eyes are located within the **eye orbits,** where they are protected by the surrounding bones. Connective tissues provide support, and fatty tissues behind the eyes provide a protective cushion.

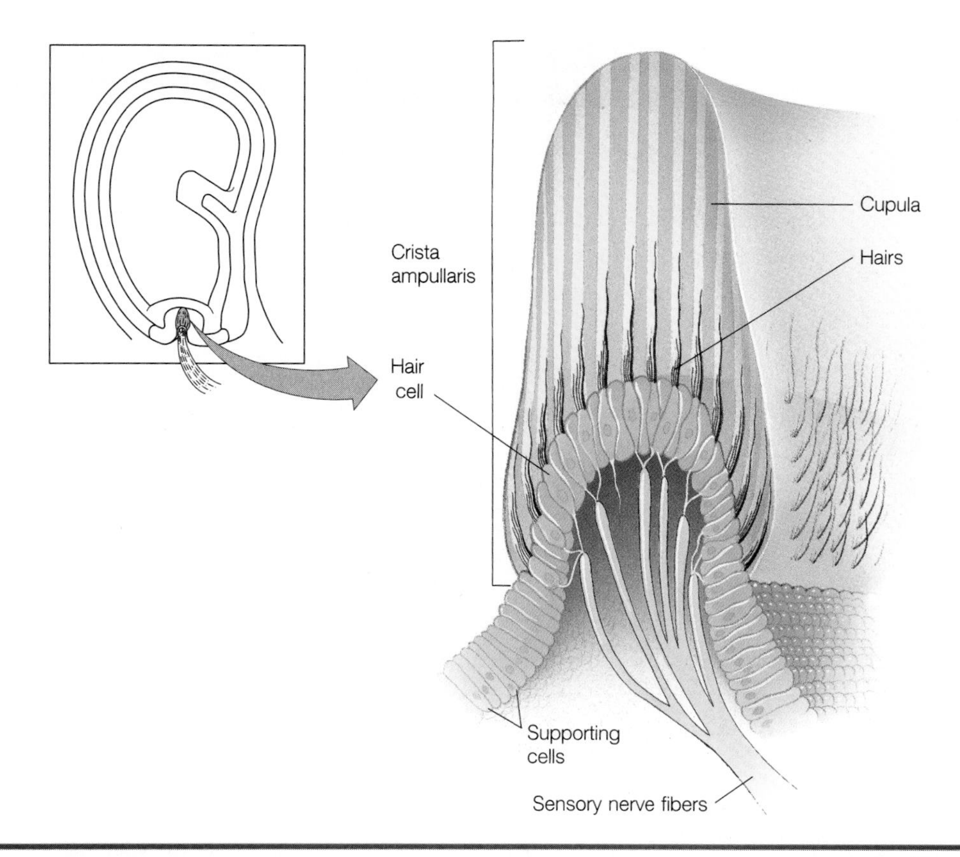

Figure 9.11 A crista ampullaris is located within the ampulla of each semicircular canal.

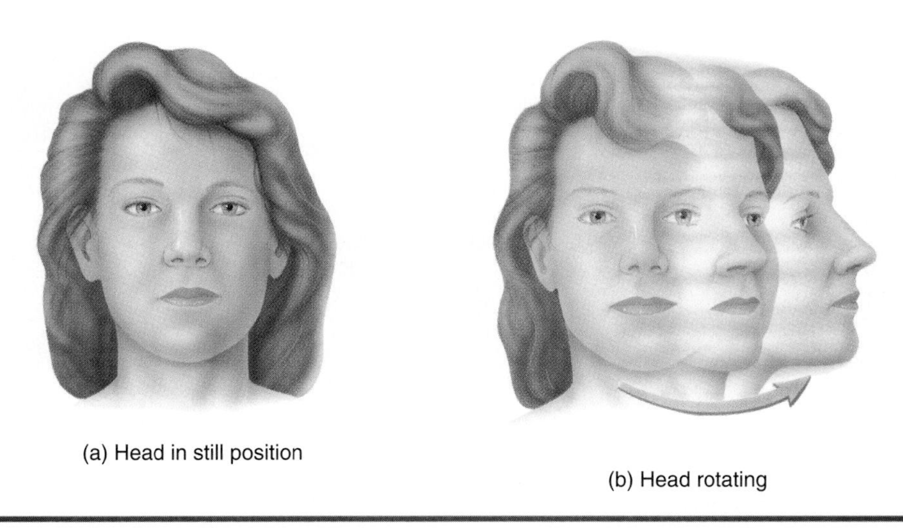

Figure 9.12 Stimulation of the crista ampullaris. (*a*) When the head is still, the cupula is upright and hair cells are not stimulated. (*b*) When the head is turned, the endolymph bends the cupula and the hairs, causing the hair cells to form impulses.

Eyelids, Eyelashes, and Eyebrows

The exposed anterior surface of the eye is protected by the **eyelids.** The eyelids are blinked frequently to spread tears and mucus over the anterior eye surface to keep it moist. The inner surface of the eyelids is lined with a mucous membrane, the **conjunctiva** (kon-junk-tī′-vah), that continues across the anterior surface of the eye. Only its transparent outer epithelium covers the cornea. Mucus from the conjunctiva helps to lubricate the eye and keep it moist. The conjunctiva contains many blood vessels and pain receptors.

Eyelashes help to keep airborne particles from reaching the eye surface and provide some protection

Table 9.1 Summary of Ear Function

Structure	Function
External Ear	
Auricle	Channels sound waves into external auditory canal
External auditory canal	Directs sound waves to tympanic membrane
Tympanic membrane	Vibrates when struck by sound waves
Middle Ear	
Tympanic cavity	Air-filled space that allows tympanic membrane to vibrate freely when struck by sound waves
Ear ossicles	Transmit vibrations produced by sound waves from the tympanic membrane to the perilymph within the cochlea
Auditory tube	Equalizes air pressure on each side of tympanic membrane
Inner Ear	
Cochlea	Fluids and membranes transmit vibrations initiated by sound waves to the organ of Corti, whose hair cells generate impulses associated with hearing
Saccule	Hair cells of the macula form impulses associated with static equilibrium
Utricle	Hair cells of the macula form impulses associated with static equilibrium
Semicircular canals	Hair cells of the cristae form impulses associated with dynamic equilibrium

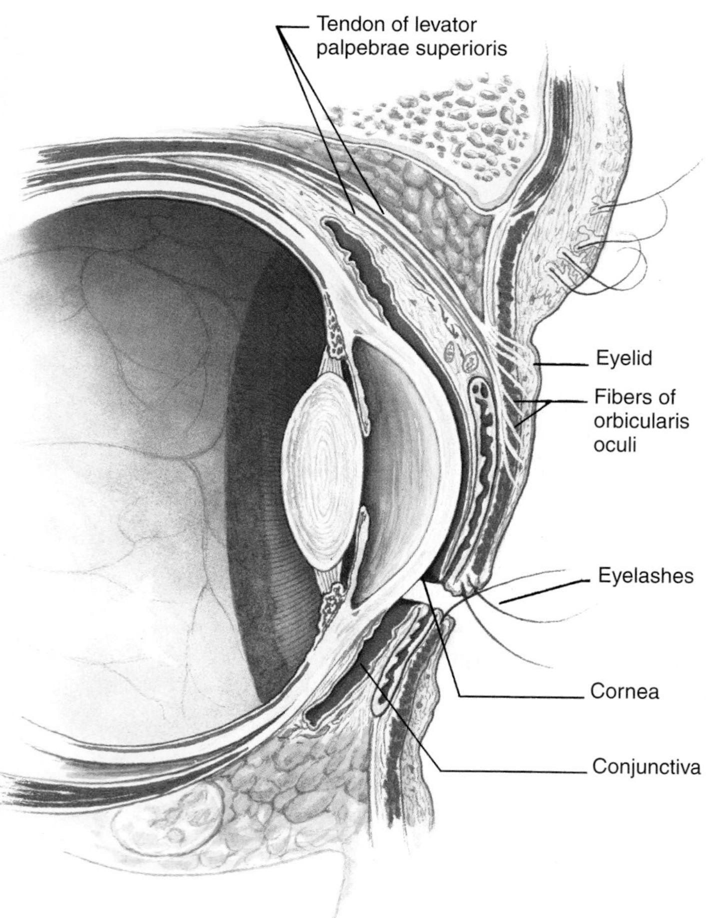

Figure 9.13 Accessory structures and the anterior portion of the eye as shown in a sagittal section.

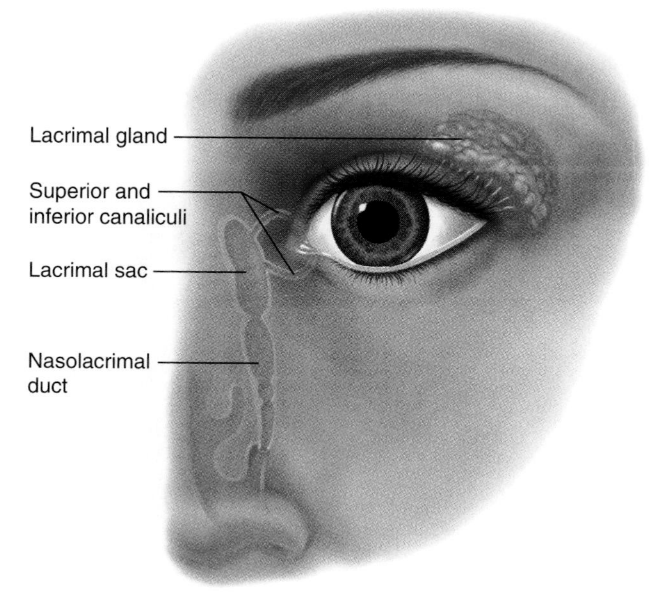

Figure 9.14 The lacrimal apparatus consists of a tear-secreting lacrimal gland and a series of ducts.

from excessive light. **Eyebrows,** located on the brow ridges, also shield the eyes from light from above. Observe the accessory structures in figure 9.13.

Lacrimal Apparatus

The **lacrimal** (lak′-ri-mal) **apparatus,** shown in figure 9.14, is involved in the production and removal of tears. Tears are secreted continuously by the **lacrimal**

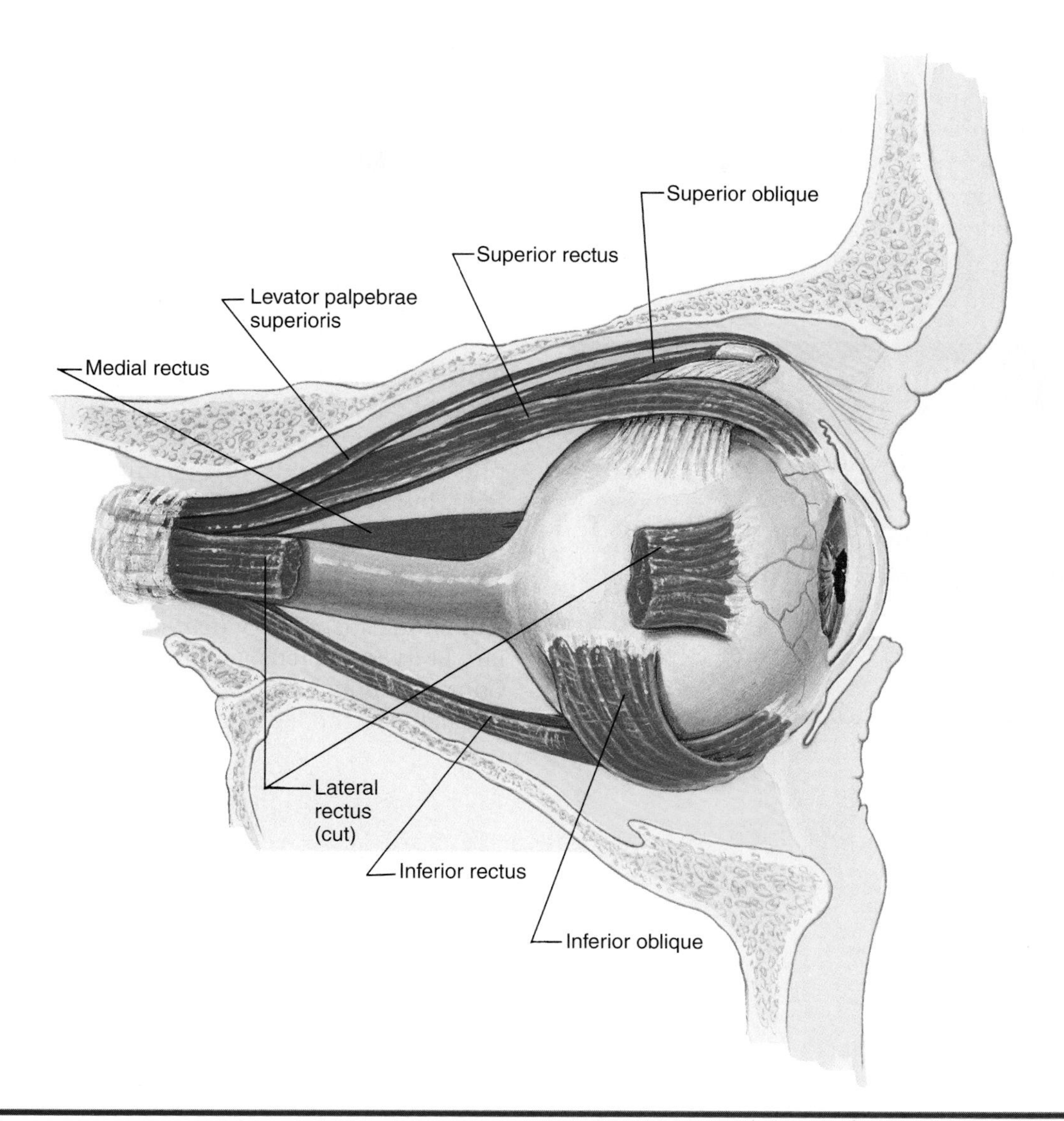

Figure 9.15 Six extrinsic muscles move the eyeball, and the levator palpebrae superioris raises the upper eyelid.

gland, which is located in the upper, lateral part of the eye orbit above the eye. Tears are carried to the surface of the eye by a series of tiny *lacrimal ducts.* The tears flow downward and medially across the eye surface as they are spread over the eye surface by the opening and closing of the eyelids. Tears are collected at the inner corner of the eye by the *superior* and *inferior canaliculi,* flow into the *lacrimal sac,* and on through the *nasolacrimal duct* into the nasal cavity.

Tears perform an important function in keeping the anterior surface of the eye moist and in washing away foreign particles. An antibacterial enzyme (lysozyme) in tears helps to reduce the chance of eye infections.

Extrinsic Muscles

Movement of the eyes must be precise and in unison to enable good vision. Each eye is moved by six **extrinsic** (external) **muscles** that originate from the back of the eye orbit and insert on the surface of the eye. Four muscles exert a direct pull on the eye, but two muscles pass through cartilaginous loops, enabling them to exert an oblique pull on the eyeball. Although each muscle has its own action, these muscles function as a coordinated group to enable eye movements. The locations of these muscles are shown in figure 9.15, and their functions are described in table 9.2.

What are the functions of the accessory structures of the eye?

Structure of the Eye

The eye is a hollow, spherical organ about 2.5 cm (1 in) in diameter. It has a wall composed of three layers, and its interior spaces are filled with fluids that support the walls and maintain the shape of the eye. The major parts of the eye are shown in figure 9.16.

Outer Layer The outer layer of the eye consists of two parts: the sclera and the cornea. The **sclera** (skle′-rah) is the opaque, white portion of the eye that forms most of the outer layer. The sclera is a tough, fibrous layer that provides protection for the delicate internal portions of the eye and for the optic nerve, which emerges from the posterior portion of the eye. The anterior portion of the sclera is covered by the conjunctiva. The **cornea** (kor′-nē-ah) is the anterior clear window of the eye. It has a greater convex curvature than the rest of the eyeball, and it bends the light rays as they pass through it. It lacks blood vessels and nerves.

Table 9.2 Action of Extrinsic Eye Muscles

Muscle	Function
Lateral rectus	Turns eye laterally
Medial rectus	Turns eye medially
Superior rectus	Turns eye upward and medially
Inferior rectus	Turns eye downward and medially
Superior oblique	Turns eye downward and laterally
Inferior oblique	Turns eye upward and laterally

Middle Layer The middle layer includes the choroid coat, ciliary body, and iris. The **choroid** (kō′-roid) **coat** contains blood vessels that nourish the eye, plus large amounts of melanin. The absorption of light by melanin prevents back-scattering of light, which would impair vision. The **ciliary** (sil′-ē-ar-ē) **body** contains the ciliary muscles and forms a ring around the **lens** just anterior to the choroid coat. Numerous *suspensory ligaments* extend from the ciliary body to the lens, and they hold the lens in place. Contraction and relaxation of ciliary muscles change the shape of the lens.

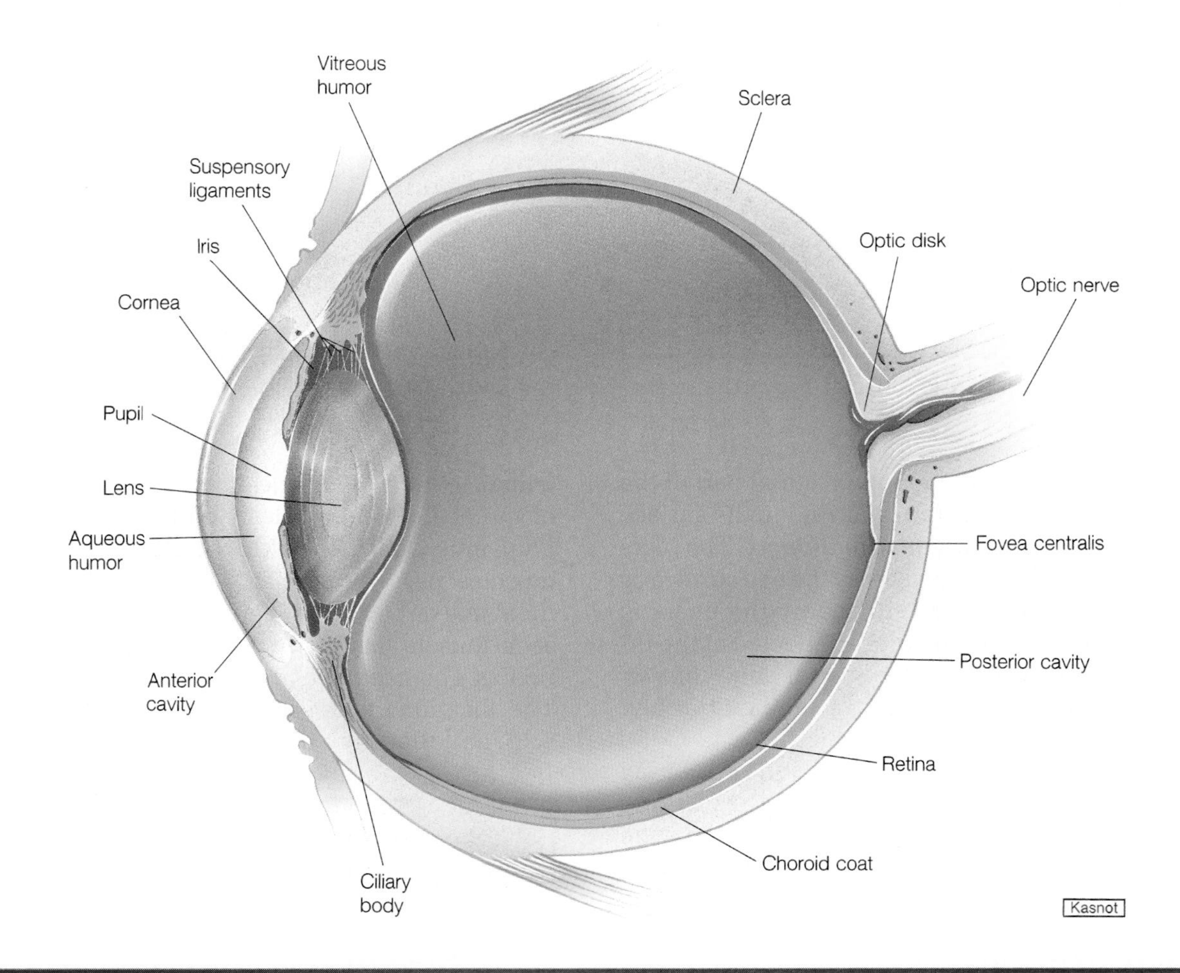

Figure 9.16 The structure of the left eye in transverse section.

Although entering light rays are bent by the cornea, it is the lens that focuses light rays precisely on the retina. The transparent, somewhat elastic lens is composed of protein fibers, and it lacks blood vessels and nerves. When tension is relaxed on the suspensory ligaments by the contraction of the ciliary muscles, the lens retracts to assume a thicker shape because of its elasticity. The relaxation of the ciliary muscles increases tension on the suspensory ligaments and causes the lens to take on a more flattened shape. In this way, the shape of the lens is adjusted for distance, intermediate, and near vision so that the image is focused precisely on the retina. This process is called **accommodation** (figure 9.17).

The colored portion of the eye is the **iris,** a thin disk of connective tissue and smooth muscle that extends from the ciliary body in front of the lens. The iris controls the amount of light entering the eye by controlling the size of the pupil. The **pupil** is the opening in the center of the iris through which light passes to the lens. The pupil is constricted in bright light and is dilated in dim light, and its size is constantly adjusted by the iris as lighting conditions change.

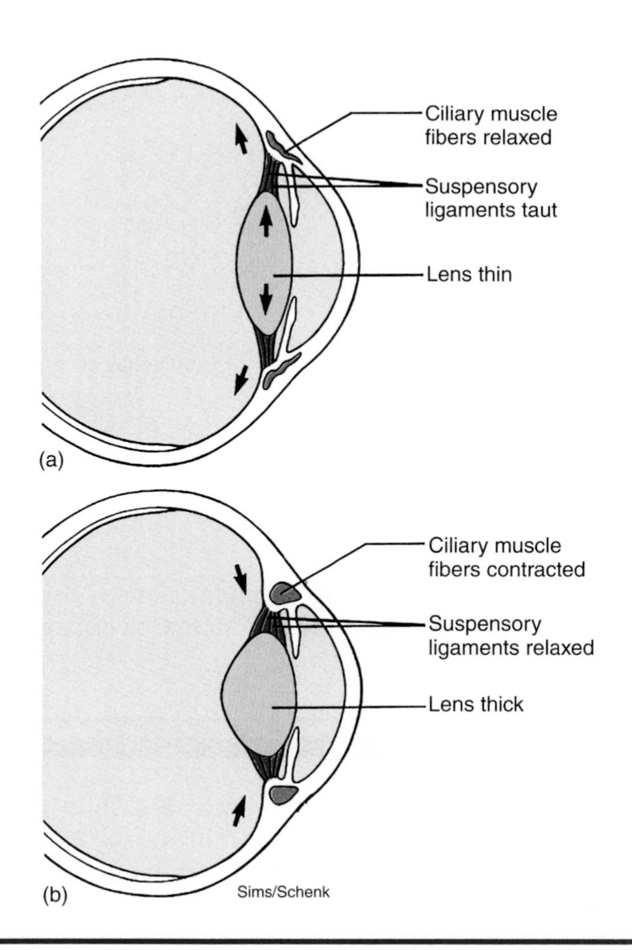

Figure 9.17 Light rays are focused on the retina by the lens. Note the shape of the lens during accommodation for(*a*) distance vision and (*b*) near vision.

Inner Layer The filmlike **retina** (ret′-i-nah) lines the interior of the eye posterior to the ciliary body. The retina contains two types of photoreceptor cells: rods and cones. The thin, elongate **rods** are photoreceptors for black-and-white vision since they are sensitive only to the presence of light. But it is rods that enable us to see in dim light. The shorter and thicker **cones** are photoreceptors for color vision, but they require bright light and do not function in dim light.

On the retina directly behind the lens is a yellowish disk called the **macula** (mak′-u-lah). In the center of the macula is a small depression called the **fovea centralis** (fō′-vē-ah sen-trah′-lis), which contains only densely packed cones. This makes the fovea the area for the sharpest, bright-light vision. The density of cones diminishes with distance from the fovea. In contrast, the density of rods, which are absent in the fovea, increases with distance from the fovea. Therefore, dimlight vision is best at the edge of the visual field (figures 9.16 and 9.18).

The retina contains neurons in addition to rods and cones (figure 9.19). Impulses formed by rods and cones are transmitted to *ganglion neurons,* whose fibers converge at the **optic disk** to form the optic nerve. The optic disk is located medial to the fovea. Since the optic disc lacks photoreceptors, it is also known as the "blind spot." However, we usually do not notice a blind spot in our field of vision, because the visual fields of our eyes overlap.

An artery enters the eye and a vein exits the eye via the optic disk. These blood vessels are continuous with capillaries that nourish the internal tissues of the eye. These blood vessels are the only ones in the body that can be viewed directly. A special instrument called an *ophthalmoscope* (of-thal′-mō-skōp) is used to look through the lens and observe these vessels. Figure 9.18 shows the appearance of blood vessels and the retina as viewed with an ophthalmoscope.

Internal Cavities The space between the cornea and the lens is known as the **anterior cavity,** and it is filled with a watery fluid called **aqueous** (ā′-kwē-us) **humor.** The aqueous humor is secreted by capillaries in the ciliary body, and it helps to maintain the shape of the cornea. The aqueous humor is absorbed into blood vessels located at the junction of the sclera and cornea. Aqueous humor is largely responsible for the internal pressure within the eye. Normally, it is secreted and absorbed at the same rate so that intraocular pressure is maintained at a constant level.

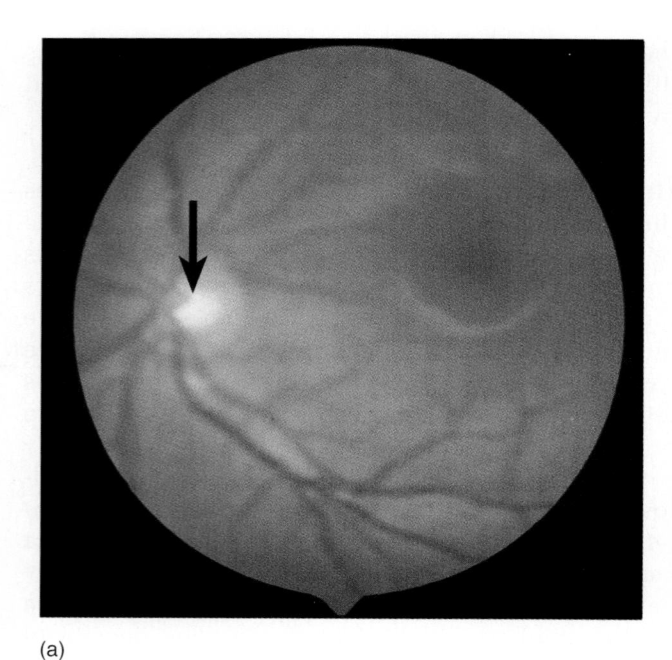

(a)

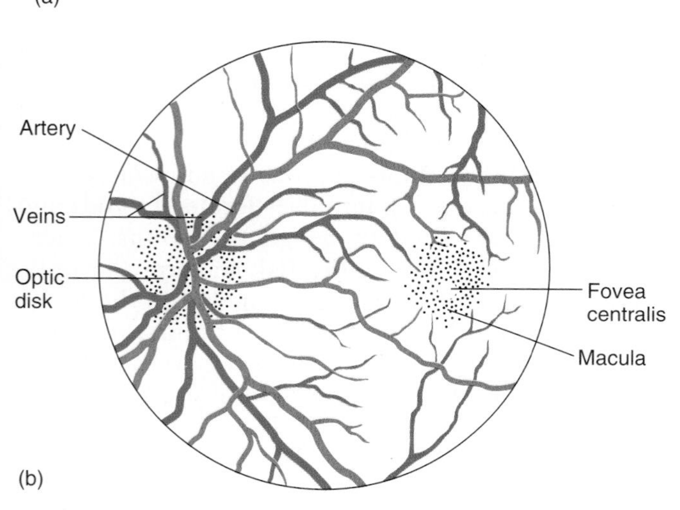

(b)

Figure 9.18 (*a*) Nerve fibers leave the eye in the area of the optic disk (arrow) to form the optic nerve (53X). (*b*) Major features of the retina.

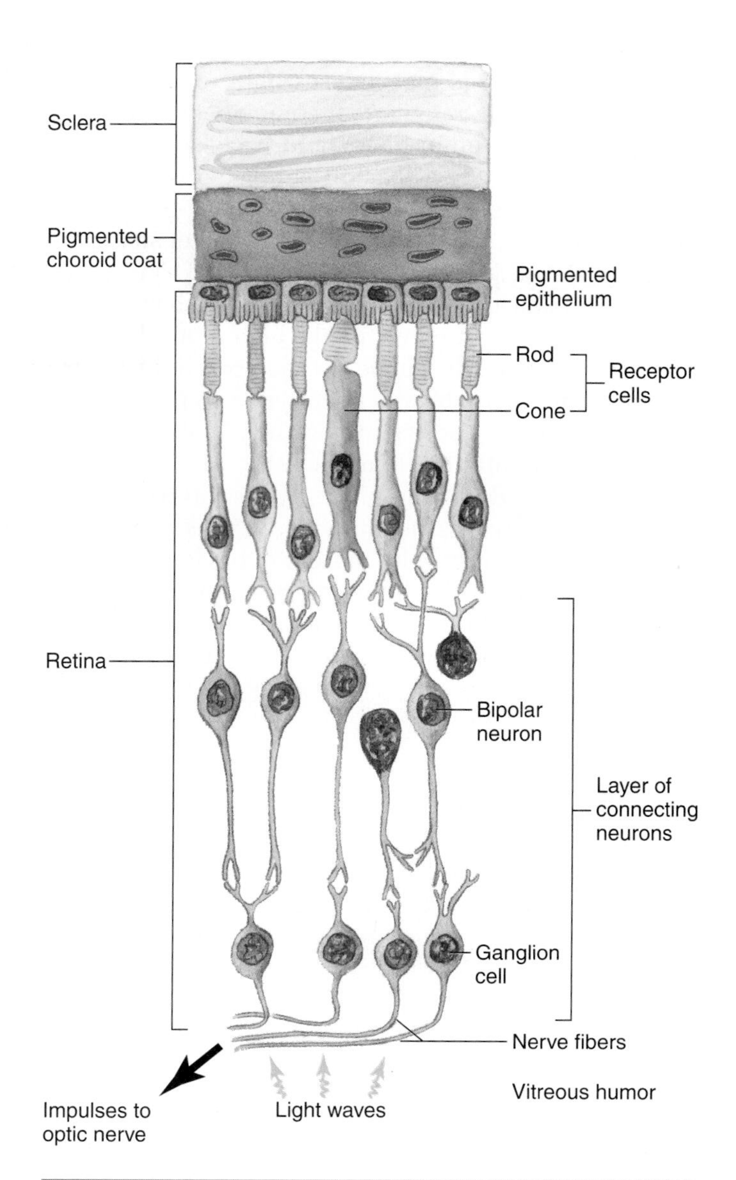

Figure 9.19 The retina consists of several cell layers.

The large **posterior cavity** is located behind the lens. It is filled with a clear, gel-like substance called **vitreous** (vit′-rē-us) **humor.** The vitreous presses the retina firmly against the wall of the eye and helps to maintain the shape of the eye.

Table 9.3 summarizes the functions of eye structures.

Physiology of Vision

Light rays coming to the eye must be precisely bent so that they are focused on the retina. This bending of the light rays is called *refraction* (rē-frak′-shun), and it is produced by the cornea and lens. The convex surface of the cornea produces the greatest refraction of light rays, while further bending (accommodation) by the lens provides a "fine adjustment" so that the image is focused precisely on the retina.

Glaucoma results when the rate of absorption of aqueous humor is less than its rate of secretion. This causes a buildup of intraocular pressure that, without treatment, can compress and close down the blood vessels nourishing the photoreceptors of the retina. If this occurs, the photoreceptors die and permanent blindness results.

Table 9.3 Summary of Eye Function

Structure	Function
Outer Layer	
Sclera	Provides protection and shape for eye
Cornea	Allows entrance of light and bends light rays
Middle Layer	
Choroid	Contains blood vessels that nourish interior structures and black pigment that absorbs excessive light
Ciliary body	Supports and changes shape of lens in accommodation; secretes aqueous humor
Iris	Regulates amount of light entering eye by controlling the size of the pupil
Inner Layer	
Retina	Contains photoreceptors that convert light rays into impulses; impulses transmitted to brain via optic (II) nerve
Lens	Bends light rays and focuses them on the retina
Anterior cavity	Aqueous humor controls intraocular pressure, which maintains shape of the eye and presses retina against choroid
Posterior cavity	Vitreous humor helps maintain shape of the eye and holds retina against choroid

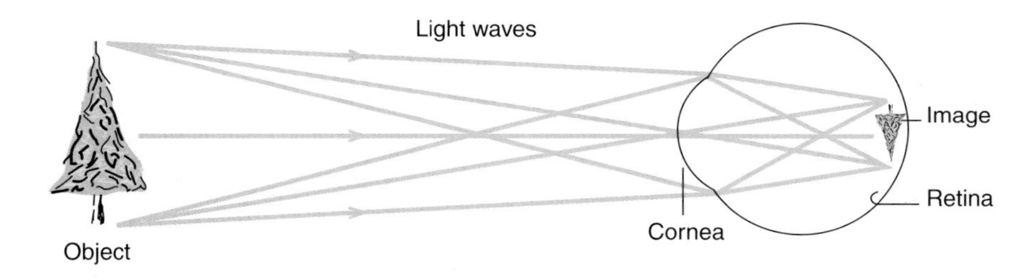

Figure 9.20 The optics of the eye cause the image to be upside down on the retina.

The optics of the eye cause the image to be inverted on the retina, as shown in figure 9.20. However, the visual centers of the cerebral cortex correct for this inversion so that objects are seen in their correct orientation. When images are incorrectly focused on the retina, poor vision results. Figure 9.21 shows common optical disorders and how they may be corrected with glasses or contact lenses.

When light rays strike the retina, the light stimuli must be converted into impulses that are sent to the brain. Both rods and cones contain light-sensitive pigments that break down into simpler substances when light is absorbed. The breakdown of these pigments results in the formation of impulses.

Rods contain a light-sensitive pigment called **rhodopsin** that breaks down into *opsin,* a protein, and *retinene,* which is derived from vitamin A. This breakdown triggers the formation of impulses that are carried via the optic nerve to the brain. Rhodopsin is resynthesized from opsin and retinene to prepare the rods for receiving subsequent stimuli. A deficiency of vitamin A may result in an insufficient amount of rhodopsin in the rods, which, in turn, may lead to *night blindness,* the inability to see in dim light.

Although the light-sensitive pigments are different in cones, they function in a similar way as rhodopsin. There seems to be three different types of cones, and each has a pigment that responds best to a different color (wavelength) of light. One type responds best to red light, another type responds best to green light, and the third type responds to blue light. The perceived color of objects results from the combination of the cones that are stimulated and the interpretation of the impulses that they form by the cerebral cortex.

Red-green color blindness, the most common type of color blindness, results whenever cones sensitive to red light or to green light are absent from the retina. Persons with this disorder cannot distinguish between reds and greens.

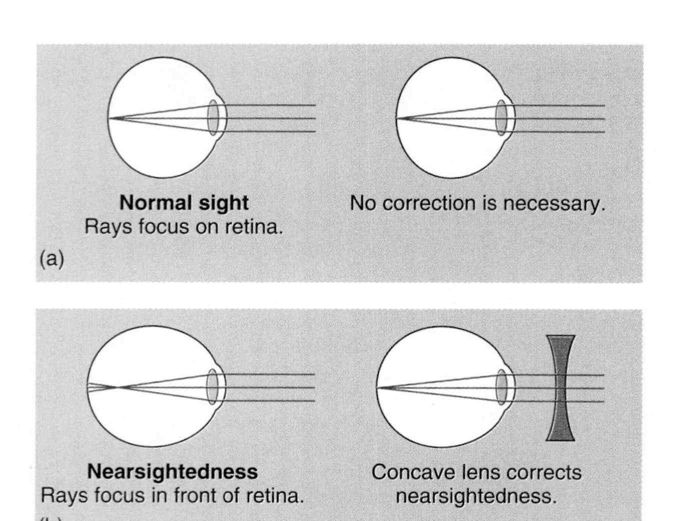

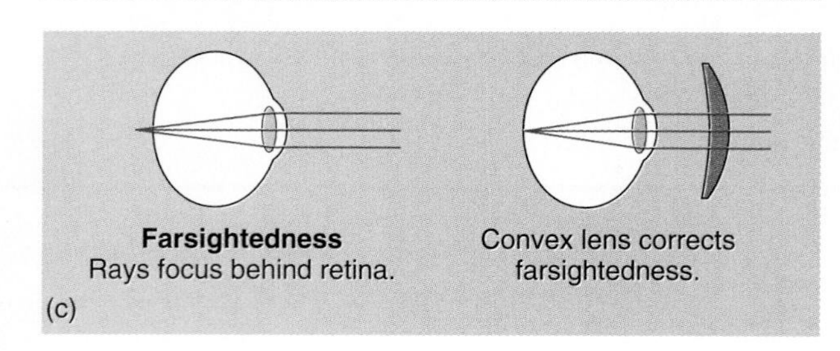

Figure 9.21 (*a*) Normal sight, (*b*) nearsightedness, and (*c*) farsightedness comparison.

Nerve Pathway Impulses formed by the photoreceptors are transmitted via nerve fibers of the optic nerve to the brain. The optic nerves (II) merge just anterior to the pituitary gland to form an X-shaped pattern called the **optic chiasma** (kī′-as-mah) (figure 9.22). Within the optic chiasma, the fibers from the medial half of the retina in each eye cross over to the opposite optic nerve. Thus, the medial fibers of the left eye and the lateral fibers of the right eye form the right optic tract leading from the optic chiasma. Similarly, the medial fibers of the right eye and the lateral fibers of the left eye form the left optic tract leading from the optic chiasma. The fibers of the optic tracts enter the thalamus, where they synapse with neurons that carry the impulses on to the visual centers of the occipital lobes.

The crossing of the medial fibers results in each visual center receiving images of the entire object but from slightly different views that enable stereoscopic (three-dimensional) vision. Depth perception is a result of stereoscopic vision.

The mechanism of vision may be summarized as follows:

1. Light rays are bent as they pass through the cornea.
2. The iris controls the amount of light passing through the pupil.
3. The ciliary body adjusts the shape of the lens to focus the light rays (image) on the retina.
4. Light absorbed by the rods and cones causes the formation of nerve impulses.
5. The impulses are transmitted to neurons whose fibers converge at the optic disk to form the optic nerve.
6. The medial fibers of the optic nerves cross over at the optic chiasma forming the optic tracts, which continue to the thalamus.
7. The impulses are then carried to the vision centers in the occipital lobes of the cerebrum, where they are interpreted as visual images.

How are light rays converted into visual sensations?

Disorders of Hearing and Vision

The disorders of hearing and vision are best considered separately, as noted next.

Disorders of the Ear

Deafness is a partial or total loss of hearing. There are two major types of hearing loss: neural deafness and conduction deafness. *Neural deafness* results from impairment of the cochlea or cochlear nerve. **Conduction deafness** results from impairment of the tympanic membrane or ear ossicles.

Labyrinthine disease is a term applied to disorders of the inner ear that produce symptoms of dizziness, nausea, ringing in the ears (tinnitis), and hearing loss. It may be caused by an excess of endolymph, infection, allergy, trauma, circulation disorders, and aging.

Motion sickness is a functional disorder that is characterized by nausea and is produced by repetitive stimulation of the equilibrium receptors in the inner ear by angular or vertical motion.

Otitis media (ō-tī′-tis mē′-dē-ah) is an acute infection of the tympanic cavity. It may cause severe pain and an outward bulging of the tympanic membrane due to accumulated fluids. Pathogens enter the middle ear from the pharynx via the auditory tube or through a perforated tympanic membrane. Young children are especially susceptible because their auditory tubes are short and horizontal, which aids the spread of bacteria from the pharynx to the tympanic cavity.

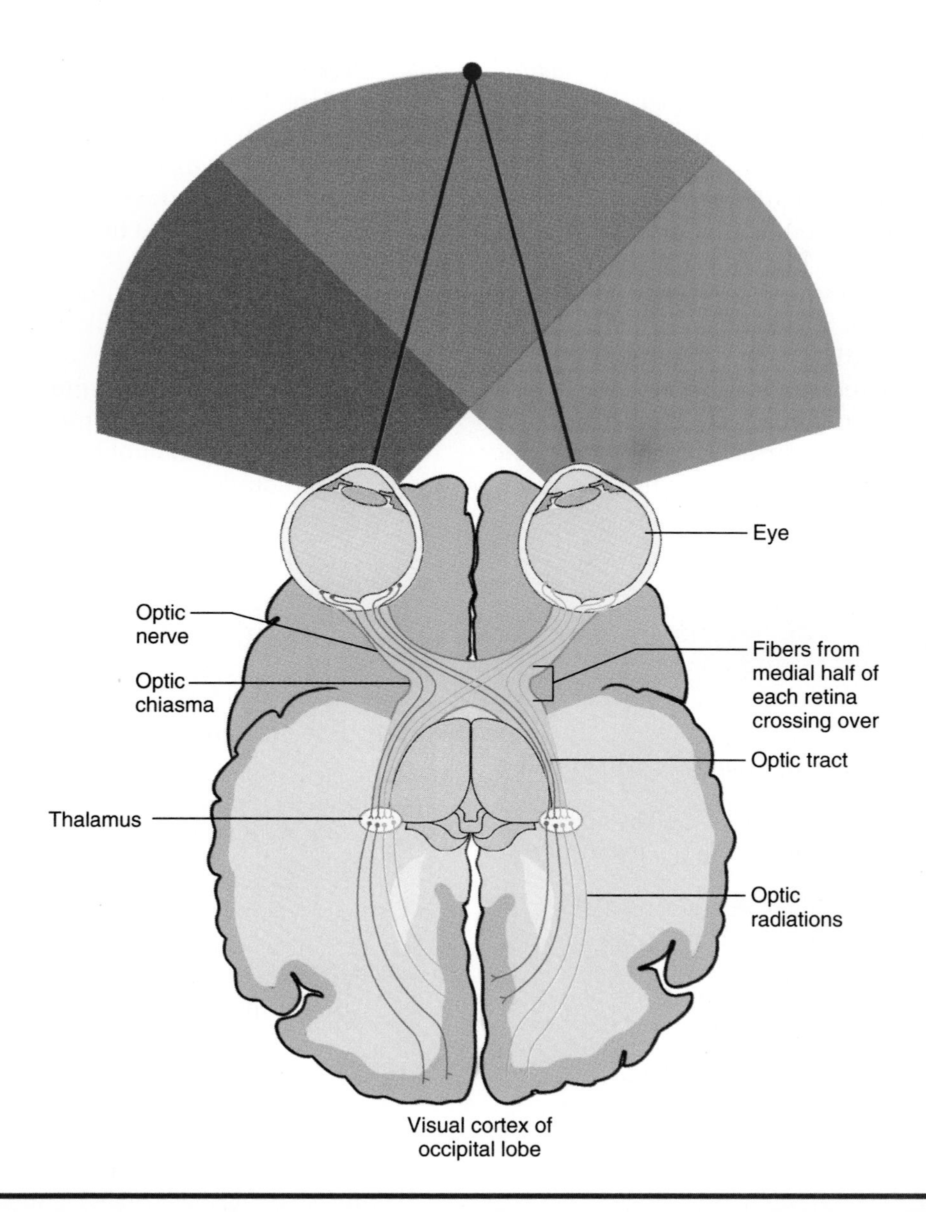

Figure 9.22 The optic nerve pathway. The medial fibers of the optic nerves cross over to the opposite side of the brain at the optic chiasma.

Disorders of the Eye

Astigmatism (a-stig′-mah-tizm) is the unequal focusing of light rays on the retina, which causes part of an image to appear blurred. It results from an unequal curvature of the cornea or lens.

Blindness is partial loss or lack of vision. It may be caused by a number of disorders such as cataract, glaucoma, and detachment or deterioration of the retina. It may also result from damage to the optic nerves or the visual centers in the occipital lobes of the cerebrum.

Cataract is cloudiness or opacity of the lens, which impairs or prevents vision. It is common in older people, and it is the leading cause of blindness. Surgical removal of the clouded lens and implantation of a plastic lens usually restore good vision.

Color blindness is the inability to perceive certain colors or, more rarely, all colors. Red-green color blindness, the most common type, is characterized by difficulty distinguishing reds and greens due to the absence of either red or green cones. Color blindness is inherited, and it occurs more often in males than females because it is a sex-linked hereditary trait.

Conjunctivitis (con-junk-ti-vī′-tis) is inflammation of the conjunctiva. It may be caused by allergic reactions, physical or chemical causes, or infections. Inflammation that results from a bacterial or viral infection is commonly called *pinkeye,* and it is highly contagious.

Farsightedness (hyperopia) is blurred vision caused by light rays being incorrectly focused behind the retina. It is commonly caused by the eye being shorter than normal.

Glaucoma (glaw-kō′-mah) is a condition in which aqueous humor is produced more rapidly than it is removed, resulting in a high intraocular pressure. The high intraocular pressure compresses blood vessels, which results in death of receptor cells. Glaucoma is the second leading cause of blindness. The tendency to develop glaucoma is inherited.

Nearsightedness (myopia) is blurred vision caused by light rays being incorrectly focused in front of the retina. It is commonly caused by the eye being longer than normal.

Presbyopia (prez-bē-ō′pē-ah) is the diminished ability of the lens to accommodate for near vision due to a decrease in its elasticity. It is a natural result of aging. At age 20, an object can be clearly observed about 10 cm (4 in) from the eye. At age 60, an object must be about 75 cm (30 in) from the eye to be clearly observed.

Retinoblastoma (ret-i-nō-blas-tō′-mah) is a cancer of immature retinal cells. It constitutes about 2% of the cancers in children.

Strabismus (strah-biz′-mus) is a disorder of the extrinsic eye muscles in which the eyes are not directed toward the same object simultaneously. Treatment may include eye exercises, corrective lenses, or corrective surgery.

CHAPTER SUMMARY

Sensations

1. Each type of sensory receptor is most sensitive to a particular type of stimulus.
2. Sensations result from impulses formed by receptors that are carried to the brain and interpreted by the cerebral cortex.
3. A particular sensory region of the cerebral cortex interprets all impulses that it receives as the same type of sensation.
4. The brain projects the sensation back to the body region from which the impulses seem to have originated.
5. Sensory receptors for touch, pressure, heat, cold, taste, and smell tend to adapt to repetitions of the same stimulus by decreasing the rate of impulse formation.

General Senses

1. Temperature receptors are located in the dermis of the skin. Heat receptors are located deeper than are cold receptors.
2. Pressure receptors are located in tendons and ligaments of joints and deep in the dermis of the skin.
3. There are two types of touch receptors. Meissner's corpuscles are located in the superficial portion of the dermis, and they are especially important in hairless skin. Free nerve endings function secondarily as touch receptors where they penetrate the epidermis or wrap around the base of hair follicles.
4. Free nerve endings serve as pain receptors, and they are widely distributed throughout the body. Pain receptors are especially abundant in the skin, and they are the only sensory receptors in visceral organs.
5. Referred pain is pain from visceral organs that is erroneously projected to the body wall. The pattern of referred pain is useful in diagnosing disorders of visceral organs.

Special Senses

1. Taste receptors are located in taste buds, which are located on papillae of the tongue.
2. There are four basic types of taste receptors: sour, sweet, salty, and bitter. Each type of taste receptor has a unique distribution on the superior surface of the tongue.
3. Chemicals must be in solution in order to stimulate the taste receptors.
4. Taste receptors rapidly adapt to a repeated stimulus.

Smell

1. Olfactory receptors are located within the epithelium of the superior portion of the nasal cavity.
2. Airborne molecules must be dissolved in the fluid film covering the olfactory epithelium in order to stimulate the olfactory receptors.
3. Impulses are carried from the receptors to the olfactory bulbs and via the olfactory tracts to the brain.
4. Olfactory receptors adapt quickly to a repeated stimulus.

Hearing

1. The ear is subdivided into the external, middle, and inner ear.
2. The external ear consists of (a) the auricle, which directs sound waves into the external auditory canal and (b) the external auditory canal, which channels sound waves to the tympanic membrane.
3. The middle ear consists of (a) the tympanic membrane, which vibrates when struck by sound waves, (b) the tympanic cavity, an air-filled space, (c) the auditory tube, which enables the equalization of air pressure on each side of the tympanic membrane, and (d) the ear ossicles, which transmit vibrations of the tympanic membrane to the oval window of the inner ear.

4. The inner ear is embedded in the temporal bone. It consists of (a) the membranous labyrinth, which is filled with endolymph and lies within (b) the bony labyrinth, which is filled with perilymph. The major portions of the inner ear are the cochlea, vestibule, and semicircular canals.
5. The cochlear duct, part of the membranous labyrinth, extends nearly to the apex of the cochlea. The organ of Corti rests on the basilar membrane within the cochlear duct, and it contains the hair cells, which are the receptors for vibrations produced by sound waves.
6. Sound waves vibrate the tympanic membrane. These vibrations are transmitted to the oval window of the inner ear by the ear ossicles, which then sets up oscillatory movements of the perilymph in the cochlea. The movements of the perilymph vibrate the basilar membrane and organ of Corti, resulting in the formation of impulses.
7. Impulses are carried to the brain by the cochlear branch of the vestibulocochlear (VIII) nerve.

Equilibrium

1. Receptors in joints, muscles, eyes, and the inner ear send impulses to the brain that are associated with equilibrium.
2. The maculae within the saccule and utricle contain hair cells that are receptors for static equilibrium.
3. The cristae within the ampullae of the semicircular canals contain hair cells that are the receptors for dynamic equilibrium.

Vision

1. The eyes contain the receptors for vision. Accessory organs include the extrinsic eye muscles, eyelids, eyelashes, eyebrows, and lacrimal apparatus.
2. The wall of the eye is composed of three layers. The outer layer consists of (a) the sclera, which supports and protects internal structures, and (b) the transparent cornea, which allows light to enter the eye. The middle layer consists of (a) the darkly pigmented choroid, which contains blood vessels nourishing internal structures and also absorbs excess light, (b) the ciliary body, which changes the shape of the lens to focus light rays on the retina, and (c) the iris, which controls the amount of light entering the posterior cavity via the pupil. The inner layer consists of the retina, which contains the photoreceptors.
3. Fluids fill the cavities of the eye and give it shape. Aqueous humor fills the anterior cavity and is primarily responsible for maintaining intraocular pressure. Vitreous humor fills the posterior cavity and helps to hold the retina against the choroid.
4. Light rays pass through the cornea, aqueous humor, pupil, lens, and vitreous humor to reach the retina. Light rays are primarily refracted by the cornea, but the lens focuses them on the retina.
5. The photoreceptors are rods and cones. Rods are adapted for dim-light, black-and-white vision. Cones are adapted for bright-light, color vision. There are three types of cones based on the color of light that they primarily absorb: red, green, and blue.
6. Impulses formed by the photoreceptors are carried to the brain by the optic (II) nerve. Fibers from the medial half of each eye cross over to the opposite side at the optic chiasma.
7. The slightly different retinal images of each eye enable stereoscopic vision when the two images are superimposed by the brain.

Disorders of Hearing and Vision

1. Disorders of the ears include deafness, labyrinthine disease, motion sickness, and otitis media.
2. Disorders of the eyes include astigmatism, blindness, cataract, color blindness, conjunctivitis, farsightedness, glaucoma, nearsightedness, presbyopia, retinoblastoma, and strabismus.

BUILDING YOUR VOCABULARY

1. **Selected New Terms**
   aqueous humor, p. 193
   auditory tube, p. 184
   basilar membrane, p. 185
   crista ampullaris, p. 187
   conjunctiva, p. 189
   ear ossicles, p. 184
   endolymph, p. 185
   external auditory canal, p. 183
   fovea centralis, p. 193
   labyrinth, p. 184
   lacrimal apparatus, p. 190
   macula, p. 187, 193
   Meissner's corpuscles, p. 179
   olfactory, p. 182
   perilymph, p. 185
   rhodopsin, p. 195
   saccule, p. 187
   semicircular canals, p. 187
   taste buds, p. 181
   tympanic membrane, p. 183
   utricle, p. 187
   vitreous humor, p. 194

2. **Related Clinical Terms**
   audiometry (aw′-dē-om′-e-trē) Measurement of hearing acuity at various sound frequencies.
   ophthalmology (off-thal-mol′-ah-jē) Medical specialty treating disorders of the eye.
   otolaryngology (ō-tō-lar-in-gol′-ah-jē) Medical specialty treating disorders of the ear, nose, and throat.
   otosclerosis (ō-tō-skle-rō′-sis) Formation of bone tissue in joints of the ear ossicles, which prevents their movement.
   tonometry (tō-nom′-e-trē) Measurement of the pressure within the eye,.
   vertigo (ver′-ti-gō) A sensation of dizziness.

STUDY ACTIVITIES

1. Complete the **Chapter 9 Study Guide,** which begins on page 445.
2. Complete the **Learning Objectives** listed on the first page of this chapter.

CHECK YOUR UNDERSTANDING

(Answers are located in appendix B.)

1. There are two types of touch receptors: _____ and _____ .
2. Pain originating from visceral organs but projected to parts of the body wall is called _____ .
3. Receptors for taste are localized in bulblike aggregations called _____ .
4. _____ receptors are located in the epithelium of the upper nasal cavity.
5. The decline in impulse formation when a receptor is repeatedly exposed to the same stimulus is called _____ .
6. The cochlea of the inner ear is involved in _____ , and the vestibule and semicircular canals are involved in _____ .
7. The membranous labyrinth is filled with a fluid called _____ .
8. The receptors for hearing are located in the organ of _____ , which rests on the _____ membrane within the cochlear duct.
9. Vibrations of the eardrum are transmitted to the inner ear by the _____ .
10. The hair cell receptors for static equilibrium are located in maculae in the _____ and _____ .
11. When light rays enter the eyes, they are refracted first by the _____ and focused on the retina by the _____ .
12. The fovea centralis contains only _____ , which are receptors for _____ vision.
13. Muscles within the _____ are responsible for changing the shape of the lens.
14. The retina is pressed against the choroid coat by the _____ in the posterior chamber.
15. The medial fibers of the optic nerve cross over at the _____.

C H A P T E R

The Endocrine System

TEN

Chapter Preview & Learning Objectives

The Nature of Hormones
- Distinguish between endocrine and exocrine glands.
- Distinguish between hormones and prostaglandins.
- Explain the negative feedback control of hormone secretion.

Pituitary Gland
- Describe the control and actions of anterior lobe hormones.
- Describe the control and actions of posterior lobe hormones.
- Describe the major disorders.

Thyroid Gland
- Describe the control and actions of thyroid hormones.
- Describe the major disorders.

Parathyroid Glands
- Describe the control and actions of parathyroid hormone.
- Describe the major disorders.

Adrenal Glands
- Describe the control and actions of adrenal hormones.

Pancreas
- Describe the control and actions of pancreatic hormones.
- Describe the major disorders.

Gonads
- Describe the control and actions of male and female gonadal hormones.

Other Endocrine Glands and Tissues
- Describe the action of melatonin.
- Describe the function of the thymus gland.

Chapter Summary
Building Your Vocabulary
Study Activities
Check Your Understanding

IO

SELECTED KEY TERMS

Endocrine gland (endo = within; crin = secrete) A ductless gland whose secretions diffuse into the blood for distribution.
Hormone (hormon = to set in motion) A chemical messenger secreted by an endocrine gland.
Hypersecretion (hyper = above) Production of an excessive amount of a secretion.
Hyposecretion (hypo = below) Production of an insufficient amount of secretion.
Negative feedback control A control mechanism in which an increase in the concentration of a substance inhibits the production of that substance.
Prostaglandin A chemical released by cells producing local actions by nearby cells.
Second messenger An intracellular substance, activated by a nonsteroid hormone, that produces the specific cellular effect associated with the hormone.
Target cell A cell whose functions are affected by a specific hormone.

Two interrelated regulatory systems coordinate body functions and maintain homeostasis: the nervous system and the **endocrine** (en′-do-krin) **system.** Unlike the almost instantaneous coordination by the nervous system, the endocrine system provides slower, but longer-lasting, coordination. The endocrine system consists of endocrine (ductless) glands and tissues that secrete **hormones** (chemical messengers) into the blood for transport to other tissues and organs, where they alter cellular functions. Figure 10.1 shows the locations of the major endocrine glands.

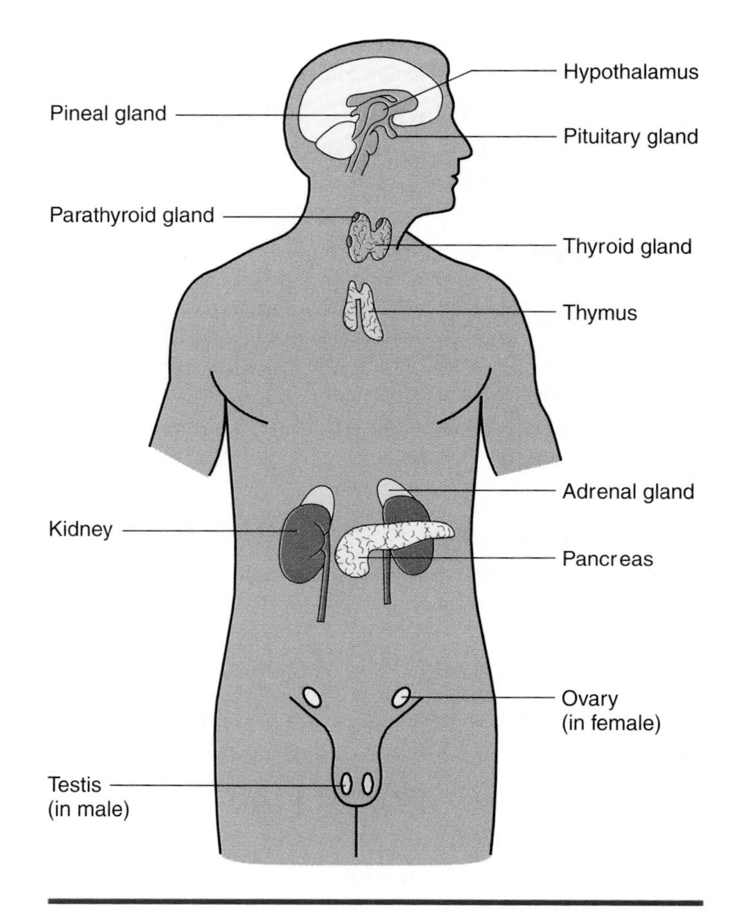

Figure 10.1 The major endocrine glands.

The Nature of Hormones

Since hormones are transported by the blood, virtually all body cells are exposed to a variety of hormones. But each hormone produces a specific effect only in its **target cells.** The target cells for a hormone may be numerous or few and may be widely distributed or localized. In addition, they may be nearby or a long distance from the endocrine gland that produces the hormone. The target cells of a particular hormone have one thing in common: specific receptors for that hormone. These receptors may be in the plasma membrane or inside the cell. Nontarget cells lack these receptors and, therefore, are unaffected by the hormone.

Hormones are secreted in very small amounts, so their concentrations in the blood are extremely low. However, since they act on cells that have specific receptors for particular hormones, large quantities are not necessary to produce effects. Chemically, hormones may be classified in two broad groups: **steroids,** which are derived from cholesterol, and **nonsteroids,** which are derived from amino acids.

Mechanisms of Hormone Action

All hormones affect target cells by altering their metabolic activities. For example, they may change the rate of cell processes in general, or they may promote or inhibit specific cellular processes. The end result is that homeostasis is maintained. But steroid and nonsteroid hormones produce their effects in different ways.

Steroid Hormones

Steroid hormones act on DNA in a cell's nucleus. Since they are fat-soluble (see chapter 2), steroid hormones can easily move through the lipid bilayers of

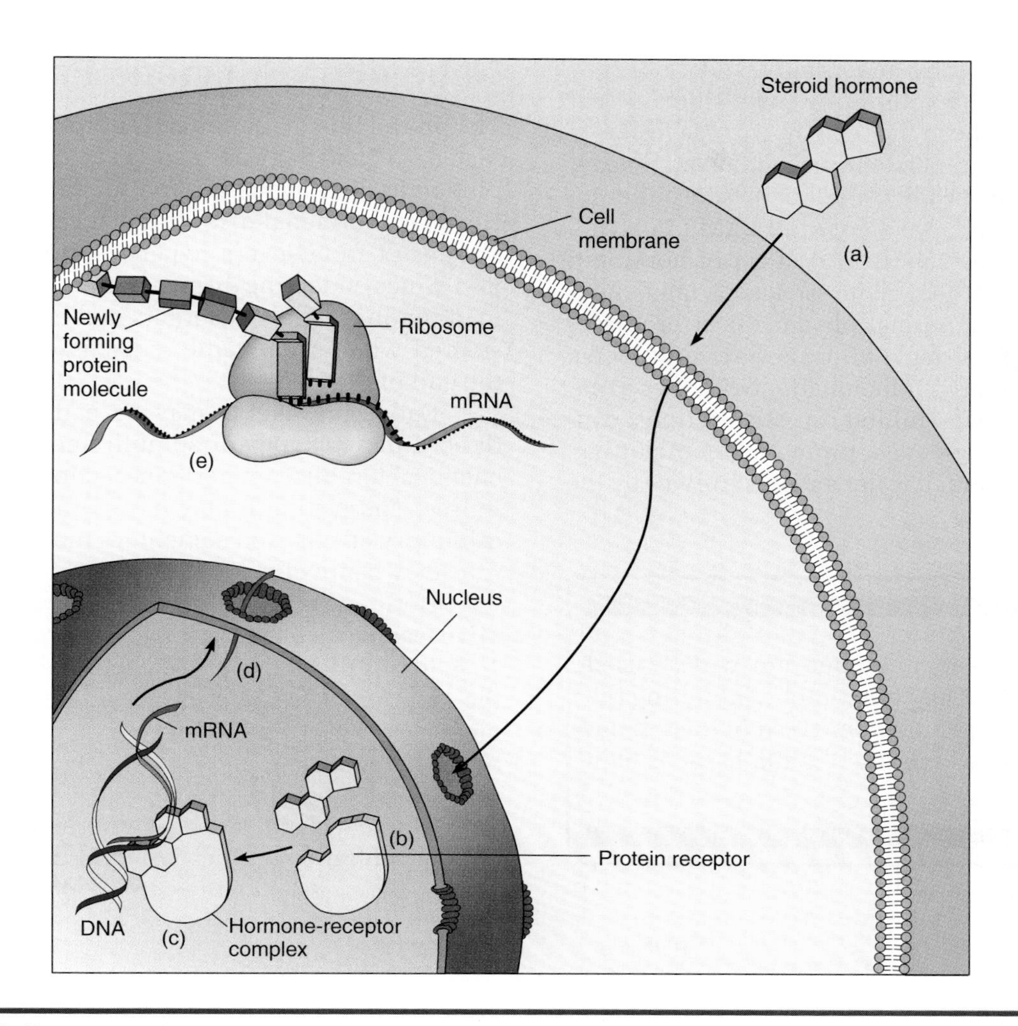

Figure 10.2 Action of a steroid hormone. (*a*) A steroid hormone crosses a cell membrane and (*b*) combines with a protein receptor, usually in the nucleus. (*c*) The hormone-receptor complex activates messenger RNA (mRNA) synthesis. (*d*) The mRNA leaves the nucleus and (*e*) guides protein synthesis.

plasma membranes to enter the nucleus. After a steroid hormone enters a cell, it combines with a receptor protein in the nucleus to form a hormone-receptor complex. The hormone-receptor complex interacts with DNA, activating specific genes that synthesize messenger RNA (mRNA). The mRNA exits the nucleus and interacts with ribosomes, which results in the synthesis of specific proteins, usually enzymes. Then the enzymes catalyze reactions that produce the specific effect that is characteristic of the particular hormone (figure 10.2).

Nonsteroid Hormones

Nonsteroid hormones are usually proteins or peptides, and they are not fat soluble. Two messengers are required for these hormones to produce their effect on a target cell. The first messenger is the nonsteroid hormone. The *second messenger* is often, but not always, **cyclic adenosine monophosphate (cAMP).** The second messenger is formed within the cell, and it activates enzymes that produce the characteristic effect for the hormone (figure 10.3). The sequence of events is as follows.

A nonsteroid hormone binds to a receptor on the target cell's plasma membrane. This releases a membrane enzyme that converts ATP into cAMP. The cAMP activates other enzymes that produce the specific changes in cell function.

Prostaglandins

Prostaglandins (pros-tah-glan′dins) are not true hormones, but they are considered here because they provide a chemical control for many cellular processes. Prostaglandins are often called *local,* or *tissue, hormones* because of the local nature of their actions.

They are not secreted by endocrine glands, but they seem to be formed by cells in nearly all organs. Although much has been learned about the roles of prostaglandins, our understanding is not complete.

Prostaglandins produce their effects by either increasing or decreasing the amount of cAMP in a cell. In this way, they can modify the effect of hormones that use cAMP as the second messenger. A large number of normal and abnormal physiological processes involve actions of prostaglandins. For example, some prostaglandins promote inflammation with its accompanying pain and fever, inhibit digestive actions, constrict or dilate blood vessels, promote blood clotting, and promote excretion of water and electrolytes by the kidneys.

Aspirin and acetaminophen are widely used pain relievers. They function by inhibiting the synthesis of prostaglandins involved in the inflammation response, which often is the basis of pain and fever.

Control of Hormone Production

The production of hormones is normally carefully regulated so that there is no *hypersecretion* (excessive production) or *hyposecretion* (deficient production). However, hormonal disorders do occur, and they usually result from severe hypersecretion or hyposecretion. Since endocrine disorders are specifically related to individual glands, disorders in this chapter are considered when each gland is discussed rather than at the end of the chapter.

Endocrine glands may be activated by one of three different types of stimuli: another hormone, a chemical product of hormone action, or neural impulses. These stimuli may have either stimulatory or inhibitory effects, and sometimes more than one stimulus may be involved.

Hormone production is usually regulated by **negative feedback control,** which is diagrammed in figure 10.4. In this process, the endocrine gland is sensitive to the blood concentration of a substance that the hormone controls. This substance may, in some cases, actually be another hormone. Whenever the concentration of that substance reaches a certain level, it causes the endocrine gland to decrease hormone secretion. As the concentration of the hormone declines, the con-

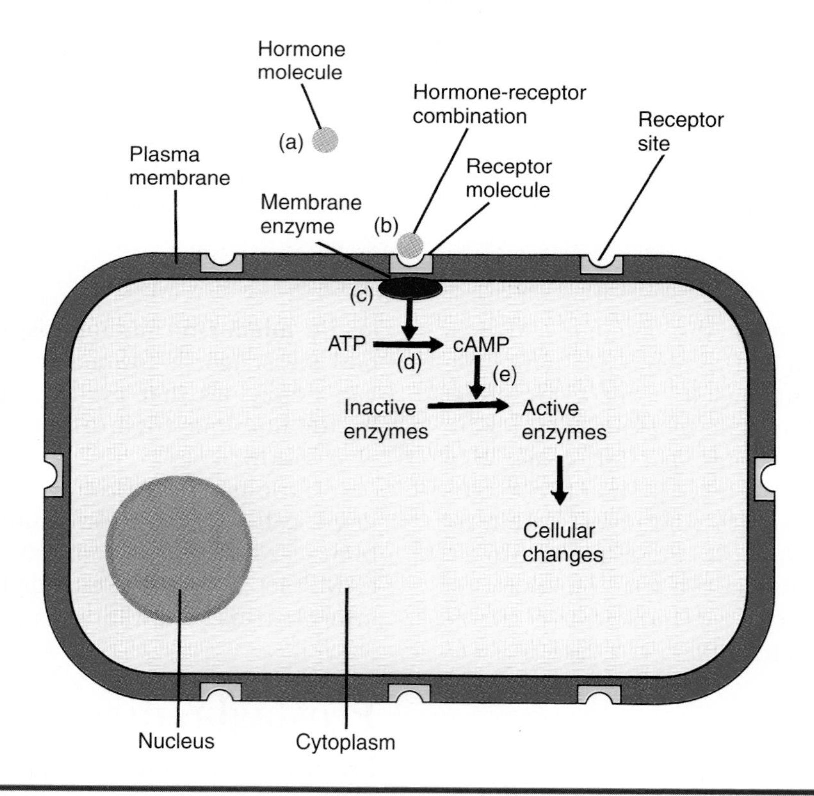

Figure 10.3 Action of a nonsteroid hormone. (*a*) Nonsteroid hormone in body fluids (*b*) binds with a receptor on the plasma membrane. (*c*) This releases a membrane enzyme that (*d*) converts ATP into cAMP. (*e*) Cyclic AMP is the second messenger that activates specific enzymes that, in turn, produce cellular changes.

centration of the product also declines. When the level of the hormone declines sufficiently, its concentration no longer inhibits hormone production, and hormone secretion increases. By this means, the concentrations of hormones in the blood are kept relatively constant, with only slight variations, as shown in figure 10.5.

> ***How do steroid and nonsteroid hormones produce their effects on target cells?***
> ***How are hormones and prostaglandins similar but different?***
> ***What is negative feedback control?***

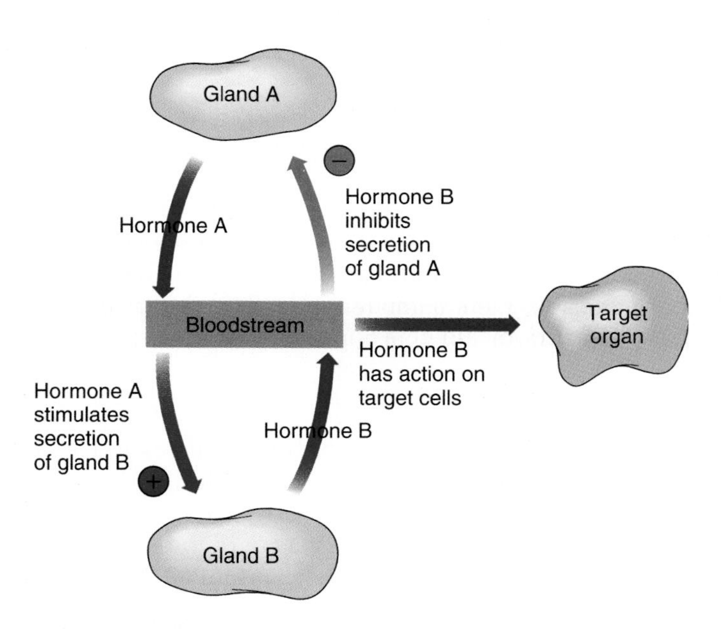

Figure 10.4 An example of a negative feedback system. Gland A secretes a hormone that stimulates gland B to increase secretion of another hormone. The hormone from gland B alters its target cells and inhibits activity of gland A. (Note: + = stimulation; − = inhibition.)

Pituitary Gland

The **pituitary** (pi-tū′-i-tar-ē) **gland,** or **hypophysis** (hī-pof′-i-sis), is attached to the hypothalamus by a short stalk. It rests in a depression of the sphenoid bone, the sella turcica, which provides protection. The pituitary consists of two major parts that have different functions: an **anterior lobe** and a **posterior lobe.** Although the pituitary gland is small, it regulates many body functions. The pituitary is controlled by neurons that originate in the hypothalamus, as shown in figure 10.6. Table 10.1 summarizes the hormones of the pituitary gland and their functions.

Control of the Anterior Lobe

Special neurons in the hypothalamus regulate the secretion of hormones from the anterior lobe by secreting **releasing hormones.** The releasing hormones enter the hypophyseal portal veins, which carry them directly into the anterior lobe without circulating throughout the body. In the anterior lobe, the releasing hormones exert their effects on specific groups of cells. There is a releasing hormone for each hormone produced by the anterior lobe. Most of the releasing hormones stimulate hormone production by the anterior lobe, but a few inhibit hormone production. The secretion of releasing hormones by the hypothalamus is regulated by a negative feedback control system.

Control of the Posterior Lobe

The posterior lobe is controlled by neural impulses. Neurons that originate in the hypothalamus have axons that extend into the posterior lobe of the pituitary. Impulses passed along these neurosecretory neurons cause the release of hormones from their axons within the posterior lobe, where they are absorbed into the blood.

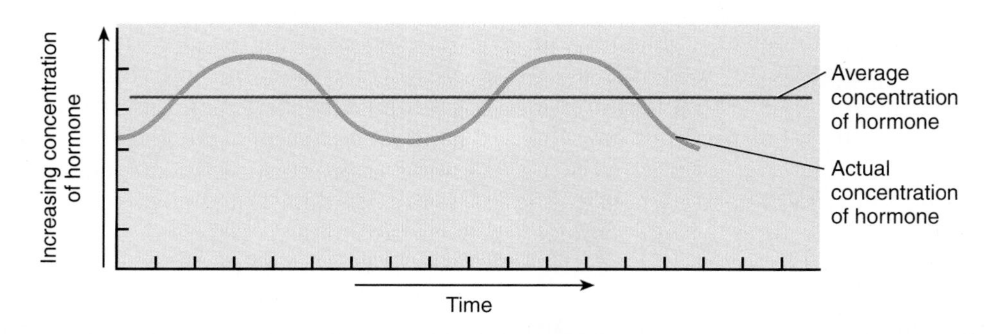

Figure 10.5 As a result of negative feedback, some hormone concentrations remain relatively stable, although they may fluctuate slightly above and below average concentrations.

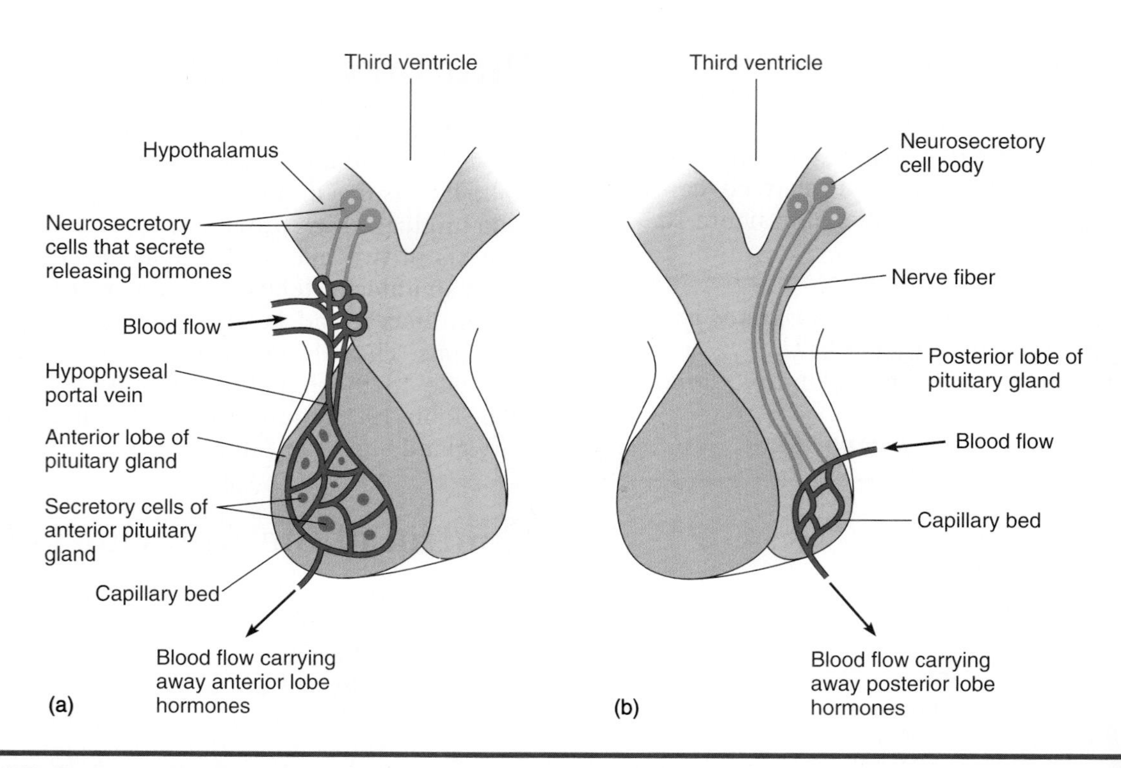

Figure 10.6 Control of pituitary secretions. (*a*) Hypothalamic releasing hormones are carried by the hypophyseal portal vein and stimulate the anterior lobe to secrete its hormones. (*b*) Neural impulses stimulate neurons, which originate in the hypothalamus, to release hormones within the posterior lobe.

Anterior Lobe Hormones

The anterior lobe of the pituitary gland is sometimes called the "master gland" because it affects so many body functions. It secretes six hormones: growth hormone (GH), thyroid-stimulating hormone (TSH), adrenocorticotropic hormone (ACTH), follicle-stimulating hormone (FSH), luteinizing hormone (LH), and prolactin (PRL).

Growth Hormone

As the name implies, **growth hormone (GH)** stimulates the division and growth of body cells. Increased growth results because GH promotes the synthesis of proteins and other complex biological molecules. GH also increases available energy for these synthesis reactions by promoting the release of fat from adipose tissue, the use of fat in cellular respiration, and the conversion of glycogen to glucose. Although GH is more abundant during the growth years, when it determines how tall a person will be, it is secreted throughout life.

Regulation of growth hormone secretion is by two hypothalamic releasing hormones with antagonistic functions. GH releasing hormone (GHRH) stimulates GH secretion, and GH inhibiting hormone (GHIH) inhibits GH secretion. Strenuous exercise, a low level of blood sugar (hypoglycemia), and an excess of amino acids in the blood trigger the secretion of GHRH. Conversely, high levels of blood sugar (hyperglycemia) stimulate the secretion of GHIH.

Disorders If there is an excess concentration of growth hormone during the growing years, the individual becomes very tall—sometimes nearly 2.5 m (8 ft) in height. This condition is known as **gigantism.** If the excessive concentration of growth hormone occurs in an adult after growth has been attained, it produces a condition known as **acromegaly** (ak-rō-meg′-ah-lē). Since the growth of long bones has been completed, only the bones of the face, hands, and feet continue to grow. Over time, the individual develops heavy, protruding brow ridges and a jutting lower jaw. Both gigantism and acromegaly may result from tumors of the anterior pituitary. Afflicted persons may have other health problems due to oversecretion of other anterior lobe hormones.

If the concentration of growth hormone is low during the growing years, body growth is limited. In extreme cases, this results in **hypopituitary dwarfism.** Afflicted persons have well-proportioned body parts but may be less than 1 m (3 ft) in height. They may suf-

Table 10.1 Hormones of the Pituitary Gland

Hormone	Control	Action	Disorders
Anterior Pituitary Hormones			
Growth hormone (GH)	Growth hormone releasing hormone (GHRH); growth hormone inhibiting hormone (GHIH)	Promotes growth of body cells and cell division; promotes protein synthesis	Hyposecretion in childhood causes hypopituitary dwarfism. Hypersecretion in childhood causes gigantism; in adults, it causes acromegaly.
Thyroid-stimulating hormone (TSH)	Thyrotropin releasing hormone (TRH)	Stimulates thyroid to produce thyroid hormone	
Adrenocorticotropic hormone (ACTH)	Corticotropin releasing hormone (CRH)	Stimulates adrenal cortex to secrete glucocorticoids	
Follicle-stimulating hormone (FSH)	Gonadotropin releasing hormone (GnRH)	In ovaries, stimulates development of follicles and secretion of estrogen; in testes, stimulates the production of sperm	
Luteinizing hormone (LH)	Gonadotropin releasing hormone (GnRH)	In females, promotes ovulation, secretion of progesterone by ovaries, preparation of uterus to receive embryo, and preparation of mammary glands for milk secretion; in males, stimulates testes to secrete testosterone	
Prolactin (PRL)	Prolactin releasing hormone (PRH); prolactin inhibiting hormone (PIH)	Helps prepare mammary glands for milk secretion and maintains milk production	
Posterior Pituitary Hormones			
Antidiuretic hormone (ADH)	Concentration of water in body fluids	Decreases excretion of water by kidneys	Hyposecretion causes diabetes insipidus
Oxytocin (OT)	Stretching of uterus; stimulation of nipples	Stimulates contractions of uterus in childbirth and contraction of milk glands when nursing infant	

fer from other maladies due to a deficient supply of other anterior lobe hormones.

Thyroid-Stimulating Hormone

Thyroid-stimulating hormone (TSH) stimulates the thyroid gland to produce thyroid hormone. Blood concentrations of thyroid hormone activate the negative feedback control mechanism for TSH production. Low thyroid hormone levels activate the hypothalamus to secrete *thyrotropin releasing hormone (TRH),* which stimulates release of TSH by the anterior pituitary. Conversely, high concentrations of thyroid hormone inhibit the secretion of TRH, which decreases production of TSH.

Adrenocorticotropic Hormone

Adrenocorticotropic (ad-re-nō-kor-ti-kō-trōp′-ik) **hormone (ACTH)** controls the secretion of hormones produced by the cortical (outer) portion of the adrenal gland. ACTH production is controlled by *corticotropin releasing hormone (CRH)* from the hypothalamus in a negative feedback control mechanism. Excessive stress may stimulate the production of excessive amounts of ACTH by overriding the negative feedback control.

Gonadotropins

The **follicle-stimulating hormone (FSH)** and **luteinizing** (lū-tē-in-īz-ing) **hormone (LH)** affect the sex glands. Their release is stimulated by *gonadotropin* *releasing hormone (GnRH)* from the hypothalamus. The onset of puberty in both sexes is caused by the start of FSH secretion. In females, FSH acts on the ovaries to promote the development of ovarian follicles, which contain ova and produce estrogen, the primary female sex hormone. In males, FSH acts on testes to promote sperm production. In females, LH stimulates ovulation and the secretion of progesterone, the secondary female sex hormone. In males, it stimulates the secretion of testosterone, the male sex hormone, by the testes.

Prolactin

Prolactin (prō-lak′-tin) **(PRL)** helps to initiate and maintain milk production by the mammary glands after the birth of a child. Prolactin stimulates milk secretion after the mammary glands have been prepared for milk production by other hormones, including female sex hormones. Prolactin secretion is regulated by the antagonistic actions of *prolactin releasing hormone (PRH)* and *prolactin inhibiting hormone (PIH)* produced by the hypothalamus.

Posterior Lobe Hormones

The posterior lobe produces two hormones: the antidiuretic hormone and oxytocin. Both of these hormones are secreted by neurons that originate in the hypothalamus and extend into the posterior pituitary. The hormones are released into the blood within the posterior lobe and are distributed throughout the body (see figure 10.6).

Antidiuretic Hormone

The **antidiuretic** (an-ti-dī-ū-ret′-ik) **hormone (ADH)** promotes water reabsorption by the kidneys to reduce the volume of water that is excreted in urine. ADH secretion is regulated by special neurons that detect changes in the water concentration of the blood. If water concentration decreases, secretion of ADH increases to promote water reabsorption by the kidneys. If water concentration increases, secretion of ADH decreases, causing more water to be excreted in urine. By controlling the water concentration of blood, ADH helps to control blood volume and blood pressure.

Disorders A severe hyposecretion of ADH results in the production of excessive quantities (20–30 liters per day) of dilute urine, a condition called **diabetes insipidus** (dī-ah-bē′-tēz in-sip′-i-dus). The afflicted person is always thirsty and must drink water almost constantly. This condition may be caused by head injuries that affect the hypothalamus or pituitary.

A *diuretic* is often prescribed for patients with high blood pressure or edema. A diuretic promotes the excretion of water by the kidneys, which decreases the water content of body fluids and tissues.

Oxytocin

Oxytocin (ok-sē-tō′-sin) **(OT)** is released in large amounts during childbirth. It stimulates and strengthens contraction of the smooth muscles of the uterus, which culminates in the birth of the baby. It also has an effect on the mammary glands. Stimulation of a nipple by a suckling infant causes the release of oxytocin, which, in turn, contracts the milk glands of the breast, forcing milk into the milk ducts, where it can be removed by the suckling infant.

How does the hypothalamus control the secretions of the pituitary gland?
What are the functions on anterior lobe and posterior lobe hormones?

Oxytocin may be used clinically to induce labor. It also may be used immediately after birth of a baby to increase the muscle tone of the uterus and to control uterine bleeding.

Thyroid Gland

The **thyroid gland** is located just below the larynx. It consists of two lobes, one on each side of the trachea, that are connected by an anterior isthmus of tissue (figure 10.7). Table 10.2 summarizes the control, action, and disorders of the thyroid gland.

Thyroxine and T_3

Iodine atoms are essential for the formation and functioning of two similar thyroid hormones. **Thyroxine** is the primary hormone. It is also known as $\mathbf{T_4}$ since each molecule contains four iodine atoms. The other hormone, $\mathbf{T_3}$, contains three iodine atoms in each molecule. Both T_4 and T_3 exert their effect on every body cell, and they have similar functions. They increase the metabolic rate, promote protein synthesis, and enhance neuron function. Moderately low levels of T_4 and T_3 cause chronic fatigue, and very low levels lead to impaired physical and mental growth and development. Secretion of these hormones is stimulated by TSH from the anterior lobe of the pituitary, and TSH, in turn, is regulated by a negative feedback control system.

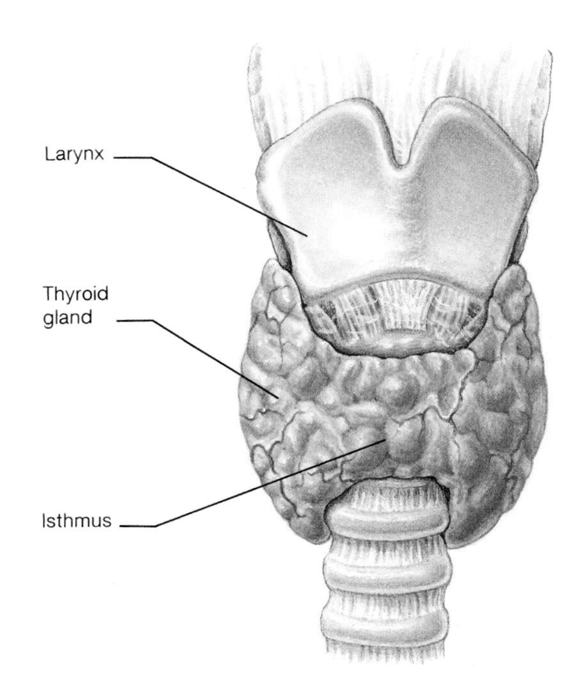

Figure 10.7 The thyroid gland consists of two lobes connected anteriorly by an isthmus.

Disorders Hypersecretion, hyposecretion, and iodine deficiencies are involved in the thyroid disorders: exophthalmic goiter, simple goiter, cretinism, and myxedema.

Exophthalmic goiter (ek-sof-thal′-mik goy′-ter), or **Grave's disease,** results from the hypersecretion of thyroxine (and T_3). It is thought to be an autoimmune disorder in which antibodies bind to TSH receptors, stimulating excessive hormone production. It is characterized by restlessness, increased metabolic rate with possible weight loss, and bulging eyes due to the swelling of tissues behind the eyes. Usually, the thyroid gland is also enlarged somewhat.

Simple goiter is the enlargement of the thyroid gland and results from a deficiency of iodine in the diet. Without adequate iodine, inadequate amounts of thyroxine and T_3 hormones are produced, and the thyroid enlarges in an attempt to produce more hormones. In some cases, the thyroid may become the size of an orange. Goiter can be prevented by including very small amounts of iodine in the diet. For this reason, salt manufacturers produce "iodized salt," which contains sufficient iodine to prevent simple goiter.

Cretinism (krē′-tin-izm) is caused by a severe deficiency of thyroxine and T_3 in infants. Without treatment, it produces severe mental and physical retardation. Cretinism is characterized by stunted growth, abnormal bone formation, mental retardation, and sluggishness.

Myxedema (mik-se-dē′-mah) is caused by severe thyroxine and T_3 deficiency in adults. It is characterized by sluggishness, weight gain, weakness, dry skin, and puffiness of the face.

Table 10.2 Hormones of the Thyroid Gland

Hormone	Control	Action	Disorders
Thyroxine (T_4) and T_3	TSH from anterior pituitary	Increases metabolic rate; accelerates growth; stimulates neural activity	Hyposecretion in growth years causes cretinism; in adult, it causes myxedema. Hypersecretion causes Grave's disease. Iodine deficiency causes simple goiter.
Calcitonin (CT)	Blood calcium level	Lowers blood calcium levels by promoting calcium deposition in bones; inhibits removal of calcium from bones	

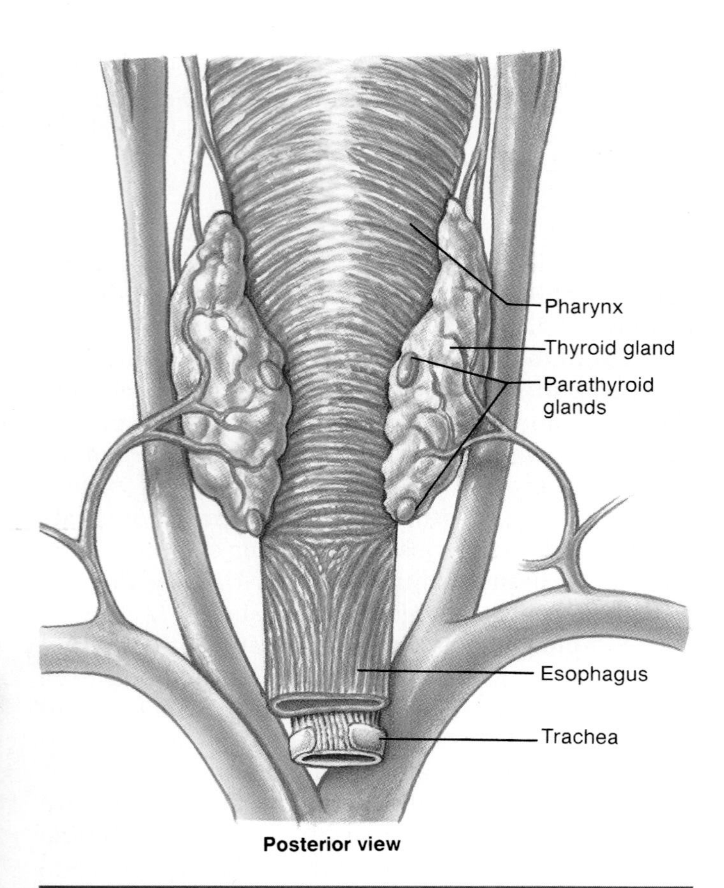

Figure 10.8 Two small parathyroid glands are located on the posterior surface of each lobe of the thyroid gland.

Calcitonin

The thyroid gland produces a third hormone with entirely different functions. **Calcitonin** (kal-si-tō′-nin) lowers blood calcium by stimulating calcium deposition by osteoblasts in bones. Its function is antagonistic to parathyroid hormone, which is discussed in the next section.

Parathyroid Glands

The **parathyroid glands** are small glands that are located on the posterior surface of the thyroid gland. There are usually four parathyroid glands, two glands on each lobe of the thyroid (figure 10.8).

Parathyroid Hormone

Parathyroid glands secrete the **parathyroid hormone (PTH),** the most important regulator of blood calcium levels. PTH increases the concentration of blood calcium by promoting the removal of calcium from bones by osteoclasts and by inhibiting calcium deposition by osteoblasts. It also inhibits excretion of calcium by the kidneys and promotes calcium absorption by the intestine (figure 10.9 and table 10.3). The antagonistic actions of PTH and calcitonin maintain a normal level of calcium in the blood.

Disorders **Hypoparathyroidism,** the secretion of insufficient amounts of PTH, can produce devastating effects. Without treatment, the concentration of blood calcium may drop to levels that impair neural and muscular activity. Tetanic contractions of skeletal muscles may occur, and death may result from a lack of oxygen due to the inability of breathing muscles to function normally.

Hyperparathyroidism, the secretion of excessive amounts of PTH, causes too much calcium to be removed from bones and raises blood calcium to abnormally high levels. Without treatment, calcium loss results in soft, weak bones that are prone to spontaneous fractures. The excess calcium in the blood may lead to the formation of kidney stones or may be deposited in abnormal locations.

> ***What is the function of thyroxine?***
> ***How is the level of blood calcium regulated?***

Adrenal Glands

There are two **adrenal glands;** one is located on top of each kidney (figure 10.10). Each adrenal gland consists of two portions that are distinct endocrine glands: an inner adrenal medulla and an outer adrenal cortex. Table 10.4 summarizes the control, action, and disorders of the adrenal gland.

Hormones of the Adrenal Medulla

The **adrenal medulla** secretes **epinephrine** and **norepinephrine,** two closely related hormones that have very similar actions on target cells. Epinephrine forms about 80% of the medullary secretions.

The sympathetic division of the autonomic nervous system regulates the secretion of adrenal medullary hormones. They are secreted whenever the body is under stress, and they duplicate the action of the sympathetic division. But the medullary hormones have a stronger and longer-lasting effect in preparing the body for "fight or flight." The effects of epinephrine and norepinephrine include (1) a decrease in blood flow to the viscera and skin; (2) an increase in blood flow to the skeletal muscles, lungs, and nervous system; (3) conversion of glycogen to glucose to raise the glucose level in the blood; and (4) an increase in the rate of cellular respiration.

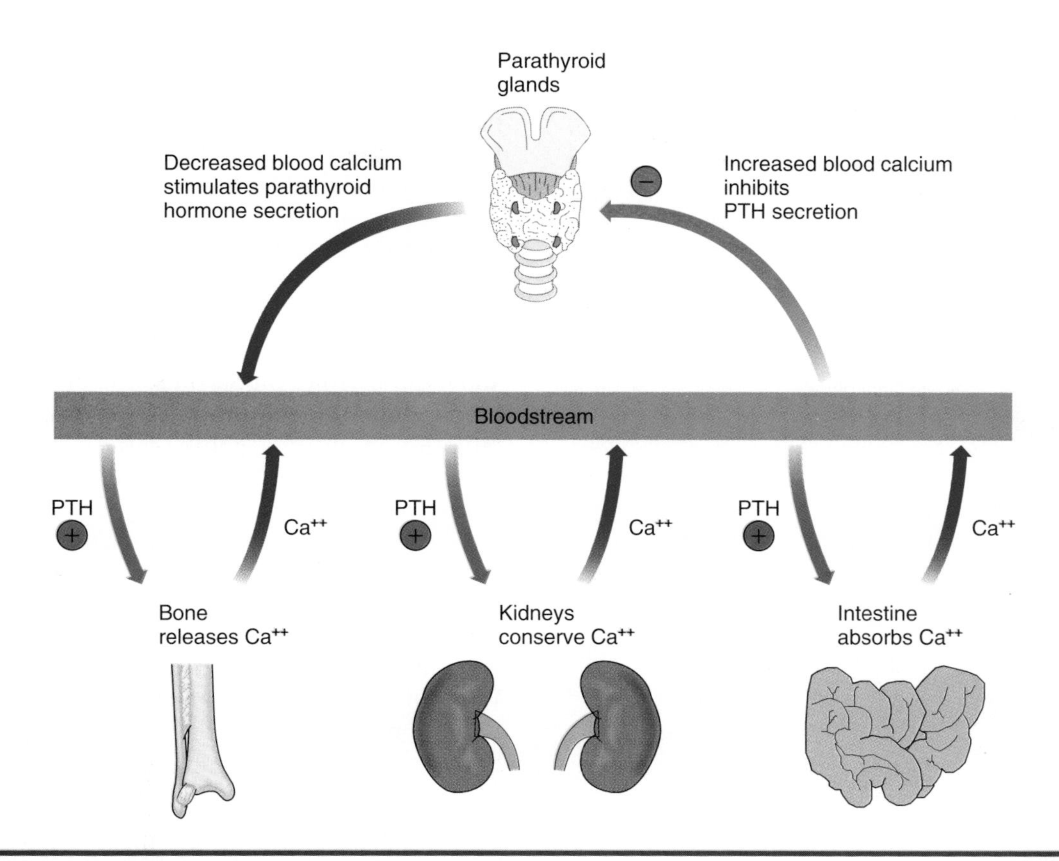

Figure 10.9 The parathyroid glands. The concentration of calcium in the blood controls parathyroid secretion. Parathyroid hormone increases the concentration of calcium in the blood.

Table 10.3 Parathyroid Hormone

Hormone	Control	Action	Disorders
Parathyroid hormone (PTH)	Blood calcium level	Increases blood calcium level by promoting calcium removal from bones and calcium reabsorption by kidneys	Hyposecretion causes tetany, which may result in death. Hypersecretion causes weak, soft, deformed bones that may fracture spontaneously.

Hormones of the Adrenal Cortex

Several different steroid hormones are produced by the **adrenal cortex,** but the most important ones are aldosterone, cortisol, and the sex hormones.

Aldosterone (al-dō-ster′-ōn) is the most important *mineralocorticoid* secreted by the adrenal cortex. Mineralocorticoids regulate the concentration of electrolytes (mineral ions) in body fluids. Aldosterone stimulates the kidneys to retain sodium ions (Na^+) and to excrete potassium ions (K^+). This action not only maintains the normal balance of sodium and potassium in body fluids, but it also maintains blood volume and blood pressure. The reabsorption of Na^+ into the blood causes negative ions, such as chloride (Cl^-) and bicarbonate (HCO_3^-), to be reabsorbed due to their opposing charges. And it causes water to be reabsorbed by osmosis, which maintains blood volume and blood pressure. Aldosterone secretion is stimulated by several factors, including (1) a decrease in Na^+, (2) an increase in K^+, or (3) a decrease in blood pressure.

Cortisol (kor′-ti-sol) is the most important of several *glucocorticoids* that are secreted by the adrenal cortex under the stimulation of ACTH. The blood levels of glucocorticoids are kept in balance because they exert a negative feedback control on the secretion of ACTH from the anterior pituitary.

Glucocorticoids are so named because they affect glucose metabolism. There are three major actions of glucocorticoids. (1) They help to keep blood glucose concentration within normal limits by promoting the conversion of noncarbohydrate nutrients into glucose. This is important since carbohydrate sources, such as glycogen, may be exhausted after several hours without food. (2) They help the body respond to stress by making more glucose available to body cells. (3) They reduce inflammation but depress immune reactions.

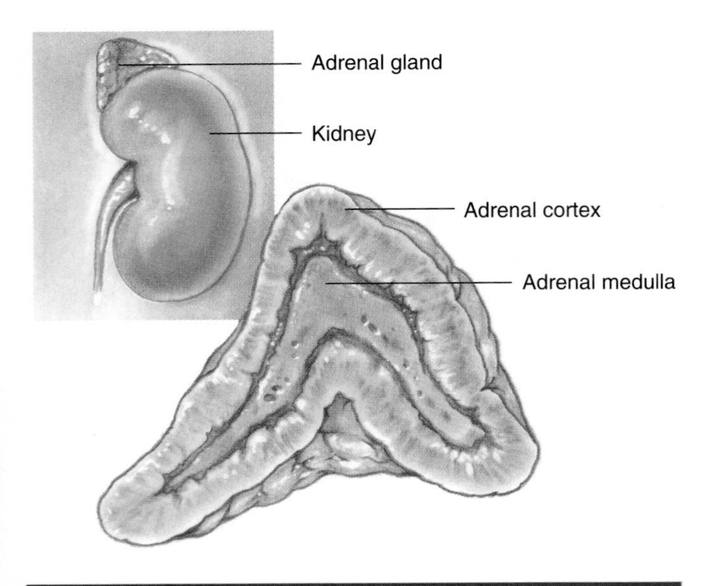

Figure 10.10 An adrenal gland consists of an outer cortex and an inner medulla.

The adrenal cortex also secretes small amounts of male sex hormones (androgens) and female sex hormones (estrogens). Their concentrations are usually so low that the effects of these hormones in a normal adult are insignificant.

Disorders **Cushing's syndrome** results from hypersecretion by the adrenal cortex. It may be caused by an adrenal tumor or by excessive production of ACTH by the anterior pituitary gland. This syndrome is characterized by high blood pressure, an abnormally high blood glucose level, protein loss, osteoporosis, fat accumulation on the trunk, fatigue, edema, and decreased immunity. A person with this condition tends to have a full, round face.

Addison's disease results from a severe hyposecretion by the adrenal cortex. It is characterized by low blood pressure, low blood glucose and sodium levels, an increase in the blood potassium level, dehydration, muscle weakness, and increased skin pigmentation. Without treatment to control blood electrolytes, death may occur in a few days.

> ***How do secretions of the adrenal medulla prepare the body to react in emergencies?***
> ***How does the adrenal cortex help to maintain blood pressure?***

Table 10.4 Hormones of the Adrenal Glands

Hormone	Control	Action	Disorders
Adrenal Medulla			
Epinephrine and norepinephrine	Sympathetic division of the autonomic nervous system	Prepares body to meet emergencies; increases heart rate, cardiac output, blood pressure, and metabolic rate; increases blood sugar by converting glycogen to glucose; dilates respiratory passages	Hypersecretion causes prolonged responses. Hyposecretion causes no major disorders.
Adrenal Cortex			
Aldosterone	Blood electrolyte levels	Increases blood levels of sodium and water; decreases blood levels of potassium	Hypersecretion inhibits neural and muscular activity, and also causes edema.
Cortisol	ACTH from anterior pituitary	Promotes normal metabolism; provides resistance to stress and inhibits inflammation; promotes formation of glucose from non-carbohydrate sources	Hyposecretion causes Addison's disease. Hypersecretion causes Cushing's syndrome.
Adrenal sex hormones		Effects are insignificant in normal adults	Hypersecretion as a result of tumors causes masculinization in females.

Pancreas

The **pancreas** (pan′-krē-as) is an elongate organ that is located posterior to the stomach (figure 10.11). It is both an exocrine gland and an endocrine gland. Its exocrine functions are performed by secretory cells that secrete digestive enzymes into tiny ducts within the gland. These ducts merge to form the pancreatic duct, which carries the secretions into the small intestine. Its endocrine functions are performed by secretory cells that are arranged in clusters or clumps called the **islets of Langerhans.** Their secretions diffuse into the blood. The islets contain two types of cells. One type produces the hormone glucagon; the other type forms the hormone insulin. Table 10.5 summarizes the control, action, and disorders of the pancreas.

Persons with inflamed joints often receive injections of **cortisone,** a glucocorticoid, to temporarily reduce inflammation and the associated pain. Such a procedure is fairly common in sports medicine.

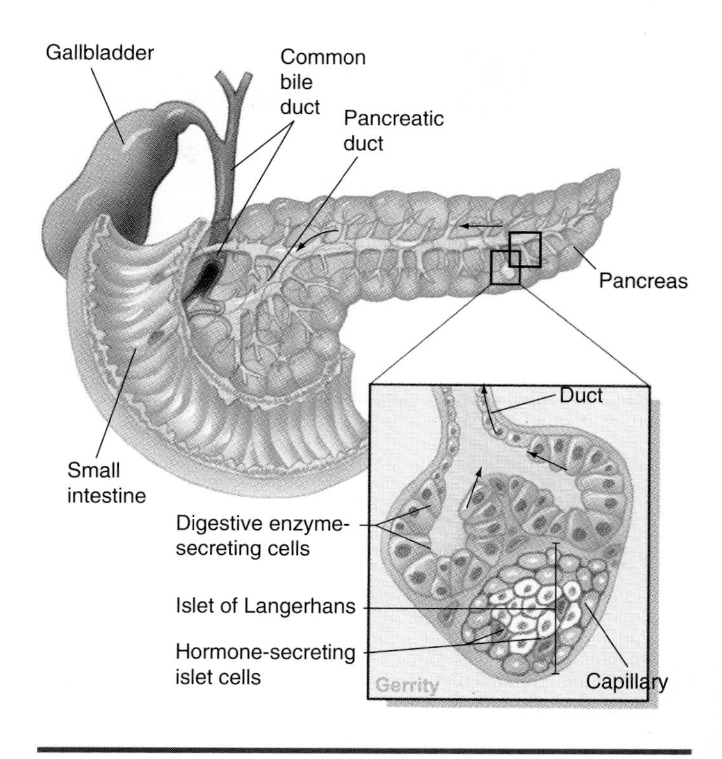

Figure 10.11 The pancreas is both an endocrine and an exocrine gland. The hormone-secreting cells are grouped in clusters, called islets. Other pancreatic cells secrete digestive enzymes.

Everyone experiences stressful situations. Stress may be caused by physical or psychological stimuli that are perceived as potentially threatening, if not life-threatening. While mild stress can stimulate creativity and productivity, severe and prolonged stress can have serious consequences.

The hypothalamus is the initiator of the stress response. When stress occurs, the hypothalamus activates the sympathetic division of the autonomic nervous system and the secretion of epinephrine and norepinephrine by the adrenal medulla. Thus, both neural and hormonal activity prepare the body to meet the stressful situation by increasing blood glucose, heart rate, breathing rate, blood pressure, and blood flow to the muscle and nervous systems.

Simultaneously, the hypothalamus stimulates the release of ACTH from the anterior pituitary. ACTH, in turn, causes the secretion of glucocorticoids by the adrenal cortex. Glucocorticoids increase the levels of amino acids and fatty acids in the blood and promote the formation of additional glucose from noncarbohydrate sources.

All of these responses prepare the body for an immediate response to remove the stressful stimulus. Unfortunately, the complexity of modern society often prevents removing stressful stimuli promptly and tends to create prolonged stress. For example, if your boss is the cause of psychological stress, a physical or verbal assault or walking away from your job may create more problems and stress than it solves.

Prolonged stress may cause several undesirable side effects from the constant secretion of large amounts of epinephrine and glucocorticoids, such as decreased immunity and high blood pressure—problems that are common in our society. Many corporations have established counseling services that allow a person to talk out stressful problems. This helps minimize the damaging effects of long-term psychological stress on both employees and the productivity of the corporation.

Table 10.5 Hormones of the Pancreas

Hormone	Control	Action	Disorders
Glucagon	Blood glucose level	Increases blood glucose by stimulating the liver to convert glycogen and other nutrients into glucose	
Insulin	Blood glucose level	Decreases blood glucose by aiding movement of glucose into cells and promoting the conversion of glucose into glycogen	Hyposecretion causes diabetes mellitus. Hypersecretion may cause hypoglycemia.

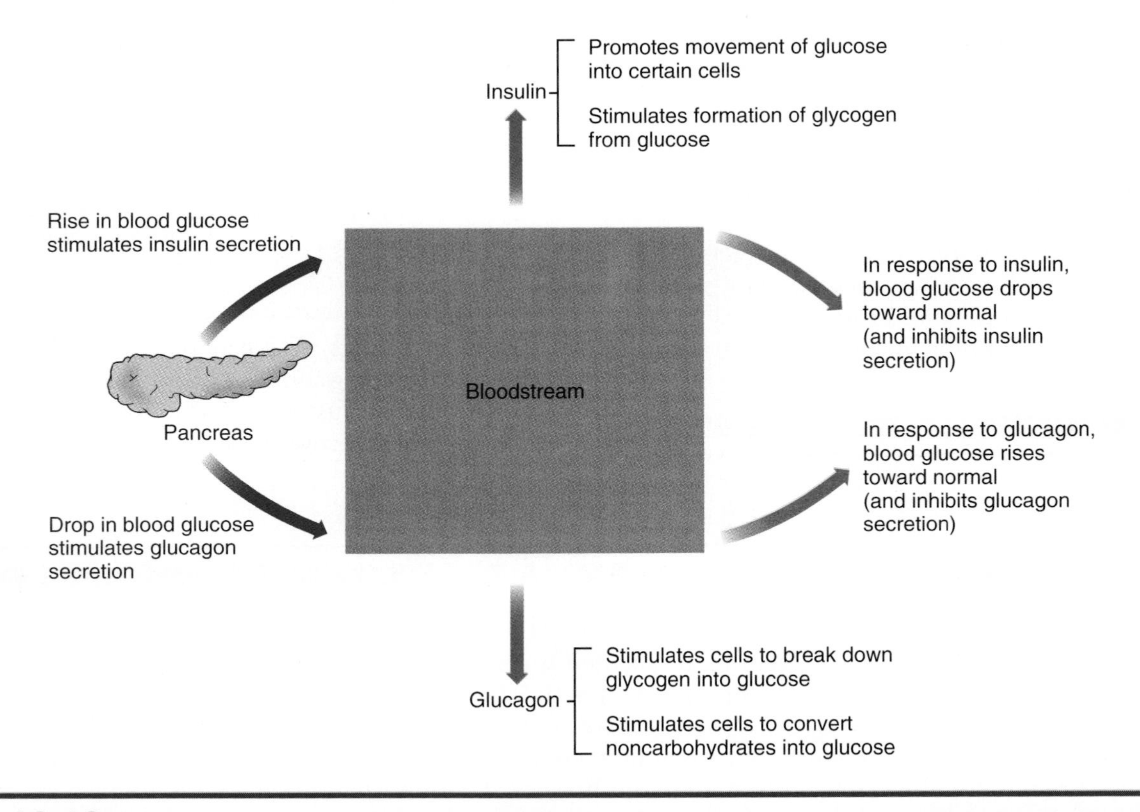

Figure 10.12 Insulin and glucagon function together to help maintain a relatively stable blood glucose concentration. Negative feedback responding to blood glucose concentration controls the levels of both hormones.

Glucagon

Glucagon (glū′-kah-gon) increases the concentration of glucose in the blood. It does this by activating the liver to convert glycogen and certain noncarbohydrates, such as amino acids, into glucose. Glucagon helps to maintain the blood level of glucose within normal limits even when carbohydrates are depleted due to long intervals between meals. Epinephrine stimulates a similar action, but glucagon is more effective. Glucagon secretion is controlled by the blood level of glucose via a negative feedback mechanism. A low level of blood glucose stimulates glucagon secretion, and a high level of blood glucose inhibits glucagon secretion.

Insulin

The effect of **insulin** on the level of blood glucose is opposite that of glucagon. Insulin decreases blood glucose by aiding the movement of glucose into body cells, where it can be used as a source of energy. Without insulin, glucose is not readily available to cells for cellular respiration. Insulin also stimulates the liver to convert glucose into glycogen for storage. Figure 10.12 shows how the antagonistic functions of glucagon and insulin maintain the concentration of glucose in the blood within normal limits. Like glucagon, the level of blood glucose regulates the secretion of insulin. High blood glucose levels stimu-

Table 10.6 Hormones of Ovaries and Testes

Hormone	Control	Action
Ovaries		
Estrogen	FSH	Development of female sex organs, secondary sex characteristics, and sex drive; prepares uterus to receive an embryo and helps maintain pregnancy
Progesterone	LH	Prepares uterus to receive an embryo and maintains pregnancy; prepares mammary glands for milk production
Testes		
Testosterone	LH	Development of male sex organs, secondary sex characteristics, and sex drive

late insulin secretion; low levels inhibit insulin secretion.

Disorders **Diabetes mellitus** (dī-ah-bē′-tēz mel-lī′-tus) is caused by the hyposecretion of insulin. It is characterized by excessively high levels of glucose in the blood. With insufficient insulin, glucose cannot get into cells easily, and cells must rely more heavily on fat molecules as an energy source for cellular respiration. The products of this reaction tend to decrease the pH of the blood (acidosis), which can inactivate vital enzymes and may lead to death.

There are two types of diabetes. *Type 1,* or *insulin-dependent, diabetes* appears in persons less than 20 years of age and persists throughout life. These persons must give themselves insulin injections at regular intervals. *Type 2,* or *noninsulin-dependent, diabetes* appears later in life, usually in persons who are over 40 years of age and overweight. This form of diabetes may be controlled by diet and medication.

An excessive production of insulin may lead to **hypoglycemia,** a condition characterized by excessively low blood glucose levels. Symptoms include acute fatigue, weakness, increased irritability, and restlessness. In extreme conditions, it may lead to an insulin-triggered coma.

How does the pancreas regulate the level of blood glucose?

Gonads

The gonads are the sex glands: the ovaries and testes. They not only produce ova and sperm, respectively, but also secrete the sex hormones. The gonads and the actions of sex hormones are covered in more detail in chapter 17. Table 10.6 summarizes the actions of the sex hormones.

Female Hormones

The **ovaries** are the female gonads. They are small, almond-shaped organs located in the pelvic cavity. The ovaries begin to function at the onset of puberty, when the gonadotropic hormones (FSH and LH) are released from the anterior pituitary. Subsequently, ovarian hormones, FSH, and LH interact in an approximately 28-day *ovarian cycle* in which their concentrations increase and decrease in a rhythmic pattern.

Estrogens (es′-trō-jens), the primary female sex hormones, are several related compounds that are secreted by developing ovarian follicles that also contain an ovum (egg cell). Estrogens stimulate the development and maturation of the female sex organs and the secondary sexual characteristics (e.g., female fat distribution, breasts, and broad hips). They also help to prepare the uterus to receive an embryo and to maintain pregnancy.

Progesterone (prō-jes′-te-rōn) is secreted by an empty follicle after the ovum has been released by ovulation. It helps prepare the uterus for receiving an embryo and maintains the pregnancy. It also helps to prepare the mammary glands for milk production.

Male Hormone

The **testes** are paired, ovoid organs located outside the pelvic cavity in the scrotum, a sac of skin located posterior to the penis. The seminiferous tubules of the testes produce sperm, the male sex cell, and the interstitial cells (cells between the tubules) secrete the male hormone **testosterone** (tes-tos′-te-rōn). Testosterone stimulates the development and maturation of the male sex organs, the secondary sexual characteristics (e.g., growth of facial and body hair, low voice, narrow hips, and heavy muscles and bones), and the male sex drive.

Other Endocrine Glands and Tissues

There are a few other glands and tissues of the body that secrete hormones and are part of the endocrine system. These include the pineal gland, the thymus gland, and certain small glands in the lining of the stomach and small intestine. In addition, the placenta is an important temporary endocrine organ during pregnancy. It is considered in chapter 18.

Pineal Gland

The **pineal** (pin′-ē-al) **gland** is a small, cone-shaped nodule of endocrine tissue that is located in the brain between the cerebral hemispheres near the roof of the third ventricle. It secretes the hormone **melatonin** (mel-ah-tō′-nin), which seems to inhibit the secretion of gonadotropic hormones and may help control the onset of puberty. Melatonin seems to regulate wake-sleep cycles and other biorhythms associated with the cycling of day and night. The secretion of melatonin is regulated by exposure to light and darkness. When exposed to light, impulses from the retinas of the eyes are sent to the pineal gland, causing a decrease in melatonin production. Secretion is greatest at night and lowest in the day.

Thymus Gland

The **thymus gland** is located in the mediastinum over the heart. It is large in infants and children but is greatly reduced in adults. It plays a crucial role in the development of immunity, which is discussed in chapter 13. The thymus produces the hormone **thymosin** (thī-mō′-sin), which is involved in the maturation of T lymphocytes.

Sunlight has a marked effect on the biological clock, which controls biorhythms, such as sleep and wakefulness cycles and seasonal moods. Since light affects the secretion of melatonin, it is believed that the pineal gland is involved in the controlling process.

Normally, melatonin levels are lowest during daylight hours and highest at night, keeping our sleep-wakefulness cycle in harmony with the day-night cycle. If a person flies to a time zone where the day-night cycle is reversed, that person experiences jet lag until the melatonin cycle and the day-night cycle gradually become synchronized.

Studies have shown that exposing a person with jet lag to bright light, with wavelengths similar to sunlight, will reset the biological clock and quickly resynchronize the melatonin cycle with the new day-night cycle. This technique may also be a benefit for workers whose shifts alternate between day and night every few weeks.

Some persons become depressed and irritable during the short days and long nights of winter, apparently because of the increased output of melatonin. This seasonal depression may be overcome by exposing the patient to bright light for an appropriate amount of time, which decreases the secretion of melatonin.

CHAPTER SUMMARY

1. The endocrine system is composed of the endocrine glands and tissues. The major endocrine glands are the adrenal glands, gonads, pancreas, parathyroid glands, pineal gland, pituitary gland, thymus gland, and thyroid gland. In addition, the hypothalamus functions like an endocrine gland in some ways.
2. The endocrine system provides a chemical control of cellular functions.

The Nature of Hormones

1. Hormones are chemical messengers that are carried by the blood throughout the body, where they modify cellular functions of target cells.
2. Target cells may be many or few, widely distributed or localized, and nearby or distant from the hormone source. All target cells have receptors for hormones that affect them.
3. Hormones may be classified chemically as either steroids or nonsteroids.
4. Steroid hormones combine with a receptor within the cell and move into the nucleus to affect production of mRNA. Nonsteroid hormones combine with a receptor in the plasma membrane, which activates a membrane enzyme that promotes synthesis of cAMP, a second messenger. Cyclic AMP, in turn, activates other enzymes that bring about cellular changes.

5. Prostaglandins are not true hormones. They are formed by most body cells and have a distinctly local effect. They modify the amount of cAMP available in a cell, so they affect the action of a hormone whose action involves cAMP.
6. Production of most hormones is controlled by a negative feedback control mechanism.
7. Hormone production may be stimulated by another hormone, a substance whose production is controlled by the hormone, or by nerve impulses.

Pituitary Gland

1. The pituitary gland is attached to the hypothalamus by a short stalk. It consists of an anterior lobe and a posterior lobe.
2. The hypothalamus secretes releasing hormones and inhibiting hormones that are carried to the anterior lobe by a portal vein. The releasing and inhibiting hormones regulate the secretion of anterior lobe hormones.
3. Anterior lobe hormones are
   a. growth hormone (GH), which stimulates growth and division of body cells;
   b. thyroid-stimulating hormone (TSH), which activates the thyroid to secrete thyroxine;
   c. adrenocorticotropic hormone (ACTH), which stimulates the secretion of hormones by the adrenal cortex;
   d. follicle-stimulating hormone (FSH) and luteinizing hormone (LH), which affect the gonads. In females, FSH stimulates production of estrogens by the ovaries; in males, it activates sperm production by the testes. In females, LH promotes ovulation and stimulates progesterone production by the ovaries; in males, it stimulates testosterone production.
   e. Prolactin (PRL), which initiates and maintains milk production by the mammary glands.
4. Hormones of the posterior lobe are formed by neurons in the hypothalamus and are released within the posterior lobe.
5. There are two posterior lobe hormones:
   a. antidiuretic hormone (ADH) reduces excretion of water by the kidneys;
   b. oxytocin stimulates contraction of the uterus during childbirth.

Thyroid Gland

1. The thyroid gland is located just below the larynx, with lobes on each side of the trachea.
2. TSH stimulates the secretion of thyroxine (T_4) and T_3, which increases cellular metabolism, protein synthesis, and neural activity.
3. Iodine is an essential component of the T_4 and T_3 molecules.
4. Calcitonin lowers the level of blood calcium by promoting calcium deposition in bones. Its secretion is controlled by the level of calcium in the blood.

Parathyroid Glands

1. The parathyroid glands are embedded in the posterior surface of the thyroid gland.
2. Parathyroid hormone increases the level of blood calcium by promoting calcium removal from bones, calcium absorption from the intestine, and calcium retention by the kidneys.
3. Parathyroid secretion is controlled by the level of blood calcium.

Adrenal Glands

1. An adrenal gland is located on top of each kidney. Each gland consists of two parts: an inner medulla and an outer cortex.
2. The adrenal medulla secretes epinephrine and norepinephrine, which prepare the body to deal with stressful situations. They increase the heart rate, circulation to nervous and muscle systems, and glucose level in the blood.
3. The adrenal cortex secretes a number of hormones that are classified as mineralocorticoids, glucocorticoids, and sex hormones.
4. Aldosterone is the most important mineralocorticoid. It helps to regulate the concentration of electrolytes in the blood, especially sodium and potassium ions.
5. Cortisol is the most important glucocorticoid. It promotes the formation of glucose from noncarbohydrate sources and inhibits inflammation. Its secretion is regulated by ACTH.
6. Small amounts of male and female sex hormones are secreted, but their effects are usually minimal in normal adults.

Pancreas

1. The pancreas is both an exocrine and an endocrine gland. Its hormones are formed by the islets of Langerhans, and their secretions are controlled by the level of blood glucose.
2. Glucagon increases the level of blood glucose by stimulating the liver to form glucose from glycogen and some noncarbohydrate sources.
3. Insulin decreases the level of blood glucose by aiding the movement of glucose into cells.
4. The antagonistic functions of glucagon and insulin keep the level of blood glucose within normal limits.

Gonads

1. Gonads are the sex glands: the ovaries in females and the testes in males. They secrete sex hormones, in addition to secreting sex cells. The secretion of these hormones is controlled by FSH and LH.
2. Estrogen is secreted by follicles in the ovaries, and they stimulate development of female sex organs and secondary sex characteristics. Estrogens also help to prepare the uterus for an embryo and help to maintain pregnancy.
3. Progesterone is secreted by the ovaries, mostly after ovulation. It prepares the uterus for the embryo, maintains pregnancy, and prepares the mammary glands for milk production.
4. The testes secrete testosterone, the male hormone that stimulates the development of the male sex organs and secondary sex characteristics.

Other Endocrine Glands and Tissues

1. The pineal gland is located near the roof of the third ventricle of the brain. It secretes melatonin, which seems to inhibit the secretion of FSH and LH by the anterior pituitary. The pineal gland also seems to be involved in biorhythms.
2. The thymus gland is located in the thorax over the heart. It secretes thymosin, which is involved in the maturation of white blood cells called T lymphocytes.
3. Endocrine disorders are associated with severe hyposecretion or hypersecretion of various hormones. Hyposecretion may result from injury. Hypersecretion is sometimes caused by a tumor.
4. Major endocrine disorders include acromegaly, Addison's disease, cretinism, Cushing's syndrome, diabetes insipidus, diabetes mellitus, exophthalmic goiter, gigantism, hypopituitary dwarfism, simple goiter, and myxedema.

BUILDING YOUR VOCABULARY

1. **Selected New Terms**
   adrenocorticotropic hormone, p. 207
   aldosterone, p. 211
   antidiuretic hormone, p. 208
   cAMP, p. 203
   cortisol, p. 211
   estrogens, p. 215
   glucagon, p. 214
   glucocorticoids, p. 211
   growth hormone, p. 206
   insulin, p. 214
   melatonin, p. 216
   mineralocorticoids, p. 211
   oxytocin, p. 208
   parathyroid hormone, p. 210
   progesterone, p. 215
   prolactin, p. 208
   releasing hormones, p. 205
   testosterone, p. 215
   thymosin, p. 216
   thyroid-stimulating hormone, p. 207
   thyroxine, p. 209
2. **Related Clinical Terms**
   goiter (goi'-ter) An enlargement of the thyroid gland.
   exophthalmos (ek-sof-thal'-mōs) An abnormal bulging of the eyes.
   hypercalcemia (hī-per-kal-sē'-mē-ah) An excess of calcium ions in the blood.
   hypoglycemia (hī-pō-glī-sē'-mē-ah) A deficiency of blood glucose.

STUDY ACTIVITIES

1. Complete the **Chapter 10 Study Guide,** which begins on page 451.
2. Complete the **Learning Objectives** listed on the first page of this chapter.

CHECK YOUR UNDERSTANDING

(Answers are located in appendix B.)

1. Chemical coordination of body functions is the function of the _____ system, whose glands secrete _____ that serve as chemical messengers.
2. A particular hormone affects only those cells that have _____ for that hormone.
3. _____ hormones use a second messenger to produce their characteristic effects on cells.
4. The secretion of most hormones is regulated by a _____ control system.
5. The secretion of pituitary hormones is regulated by a part of the brain called the _____.
6. The pituitary secretes three hormones that regulate secretion of other endocrine glands. _____ acts on the thyroid gland; ACTH acts on the _____; _____ and _____ act on the gonads.
7. Metabolic rate is regulated by _____ secreted by the _____.
8. The concentration of calcium in the blood is regulated by two hormones with antagonistic actions: _____ promotes calcium deposition in bones; _____ promotes calcium removal from bones.
9. Secretions of the adrenal _____ prepare the body to react in emergencies.
10. The primary hormone regulating the concentration of mineral ions in the blood is _____.
11. The pancreatic hormone that increases the concentration of blood sugar is _____.
12. The primary sex hormone in females is _____ ; the male sex hormone is _____ .

C H A P T E R

Blood

ELEVEN

11

Chapter Preview & Learning Objectives

General Characteristics of Blood
- Describe the general characteristics and functions of blood.

Red Blood Cells
- Describe the appearance and normal concentration of red blood cells in blood.
- Explain the role of hemoglobin in red blood cell function.
- Explain where production and destruction of red blood cells occur and the factors controlling these processes.

White Blood Cells
- Describe the types of white blood cells and explain the functions of each type.
- Indicate the normal concentration of white blood cells in blood.

Platelets
- Indicate the origin and normal concentration of platelets in blood.
- Describe the function of platelets.

Plasma
- Identify the normal components of plasma and explain their importance.

Hemostasis
- Describe the sequence of events in hemostasis.

Human Blood Types
- Explain the basis of blood typing and why it is important.
- Identify the blood typing antigens and antibodies in each ABO blood type.

Disorders of the Blood
- Describe the major blood disorders.

Chapter Summary
Building Your Vocabulary
Study Activities
Check Your Understanding

SELECTED KEY TERMS

Agglutination (agglutin = to stick together) The clumping of red blood cells in an antigen-antibody reaction.
Antigen A substance recognized as foreign by the immune system that reacts with specific antibodies or immune cells.
Antibody A protein produced by immune cells that reacts with a specific antigen.
Coagulation The formation of a blood clot.
Erythrocyte (erythr = red) A red blood cell.
Formed elements The cellular components of blood: erythrocytes, leukocytes, and thrombocytes.
Hemoglobin (hemo = blood) The red, oxygen-carrying pigment in red blood cells.
Hemostasis (hemo = blood; stasis = standing still) The stoppage of bleeding.
Leukocyte (leuk = white) A white blood cell.
Plasma The liquid portion of blood.
Thrombocyte (thromb = clot) A platelet; a formed element that initiates the clotting process.
Thrombus A blood clot.

Blood is normally confined within the heart and blood vessels as it transports materials from place to place within the body. Substances carried by blood include nutrients, waste products, hormones, mineral ions, and water. Blood does more than just transport substances. For example, blood helps to (1) regulate the pH of body tissues, (2) prevent excessive blood loss by hemorrhage, and (3) fight infections. In a variety of ways, blood helps to maintain a stable environment for body cells.

General Characteristics of Blood

Blood is classified as a connective tissue that is composed of **formed elements** (blood cells and *platelets*) suspended in **plasma,** the liquid portion (matrix) of the blood. It is the only fluid tissue in the body. Blood is heavier and about four times more viscous than water. It is slightly alkaline, with a pH between 7.35 and 7.45. The volume of blood varies with the size of the individual, but it averages 5 to 6 liters in males and 4 to 5 liters in females. Blood comprises about 8% of the body weight.

About 55% of the blood volume consists of plasma, while 45% is made up of blood cells and platelets. Red blood cells are so numerous that they form almost 45% of the blood volume, nearly all of the formed elements. White blood cells and platelets combined form less than 1% of the blood volume (figure 11.1).

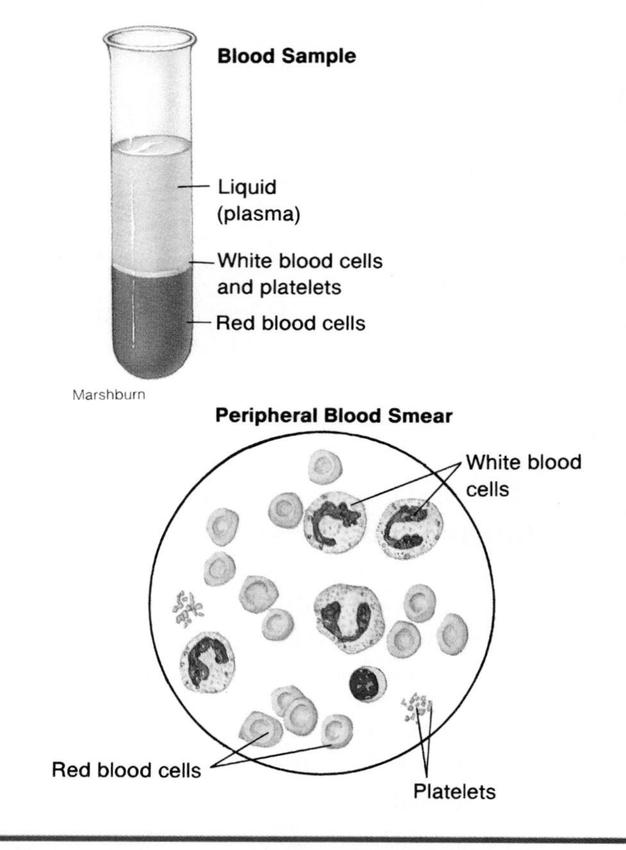

Figure 11.1 Blood consists of plasma and formed elements. If blood is centrifuged, the red blood cells sink to the bottom of the tube, the plasma forms the top layer, and the white blood cells and platelets form a thin layer between the two.

Red Blood Cells

Erythrocytes (eh-rith′-rō-sīts), or **red blood cells (RBCs),** are tiny, biconcave disks that transport oxygen to body cells and carbon dioxide from body cells. The biconcave shape exposes maximal surface area of the cell for the diffusion of these gases through the plasma membrane. Mature erythrocytes lack a nucleus, although a nucleus is present in immature RBCs.

Hemoglobin

About 33% of each erythrocyte, by volume, consists of **hemoglobin** (hē′-mō-glō-bin), the red pigment that gives the red color to blood. Hemoglobin consists of *heme,* a pigment containing an iron atom in its molecule, and *globin,* a protein. Hemoglobin combines reversibly with oxygen and plays a vital role in the transport of oxygen by red blood cells. It also plays a minor role in carbon dioxide transport.

When blood flows through the lungs, oxygen diffuses from air spaces in the lungs into the blood. Oxygen enters RBCs and combines with hemoglobin to form *oxyhemoglobin,* which gives a bright red color to blood. After the release of oxygen from oxyhemoglobin to body cells, the resultant *deoxyhemoglobin* carries a small amount of carbon dioxide from body cells back to the lungs for removal. Deoxyhemoglobin gives a dark red color to blood. The mechanism of transporting oxygen and carbon dioxide is covered in chapter 14.

Concentration of RBCs

Erythrocytes are by far the most abundant blood cells. Since they are vital in the transport of oxygen, changes in the concentration of RBCs can have clinical significance. A *red blood cell count* is a routine clinical test to determine the number of red blood cells in a cubic millimeter (mm^3) of blood. For adult males, normal values range from 4.5 million to 6.2 million RBCs per mm^3. For adult females, normal values range from 4.2 million to 5.4 million RBCs per mm^3. The higher count in males results because males have a higher metabolic rate than females, and a greater concentration of RBCs is required to provide needed oxygen for the rapidly metabolizing cells.

Normal values of RBCs per mm^3 of blood also vary with altitude. The concentration of RBCs is greater in persons living at higher altitudes because of the reduced atmospheric pressure and oxygen concentration. This reduces the rate at which oxygen can enter the blood, causing a decline in the concentration of oxygen in the blood, which, in turn, stimulates RBC production.

Production

Prior to birth, RBCs are produced largely by the liver and spleen, but after birth production occurs only in the red bone marrow. In infants, RBCs are formed in the red marrow of all bones, but in adults RBC formation primarily occurs in the red marrow of the skull bones, ribs, sternum, vertebrae, and pelvic bones.

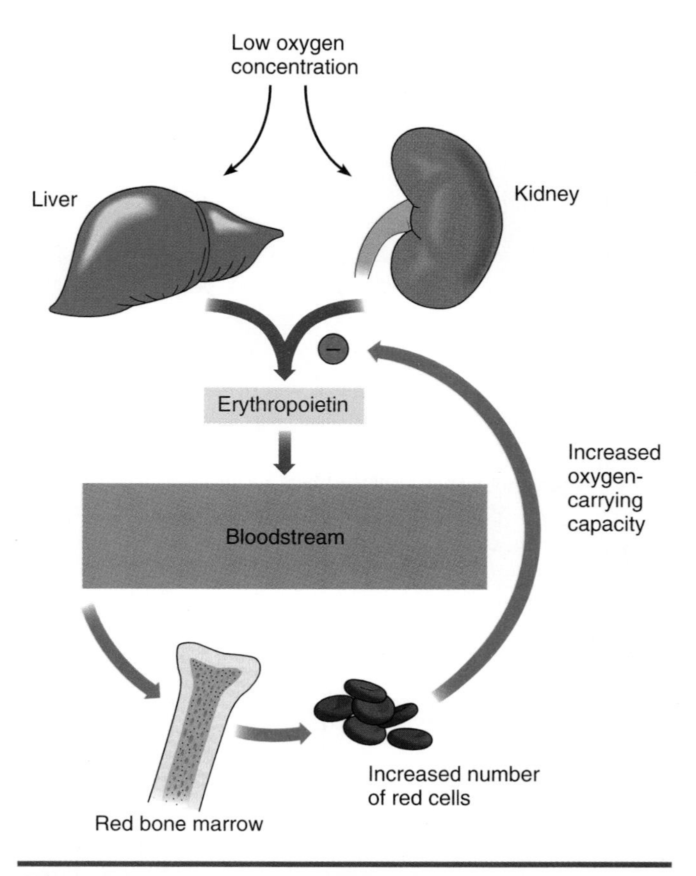

Figure 11.2 Control of erythrocyte production. A low blood level of oxygen causes the liver and kidneys to produce erythropoietin, which stimulates red bone marrow to produce more red blood cells. The added red blood cells increase the oxygen concentration of the blood, which decreases erythropoietin production.

Erythrocyte production varies with the oxygen concentration of the blood in a negative feedback control mechanism. If blood oxygen concentration is low (hypoxia), such as after a hemorrhage, the kidneys and liver release **erythropoietin** (e-rith-ro-poi′-e-tin), a hormone that stimulates red bone marrow to produce more RBCs. When the new RBCs are added to the blood and the oxygen concentration rises to normal levels, production of erythropoietin declines, causing a decrease in RBC production (figure 11.2). A low level

The oxygen-carrying capacity of the blood is commonly measured by two clinical tests: hemoglobin percentage and hematocrit. *Hemoglobin percentage* is the hemoglobin content expressed in grams per 100 ml of blood. Average normal values are 14.9 ± 1.5 g for adult males and 13.7 ± 1.5 g for adult females. The *hematocrit,* or *volume of packed red cells,* determines the percentage by volume of erythrocytes in the blood. Average normal values are 47% in adult males and 42% in adult females.

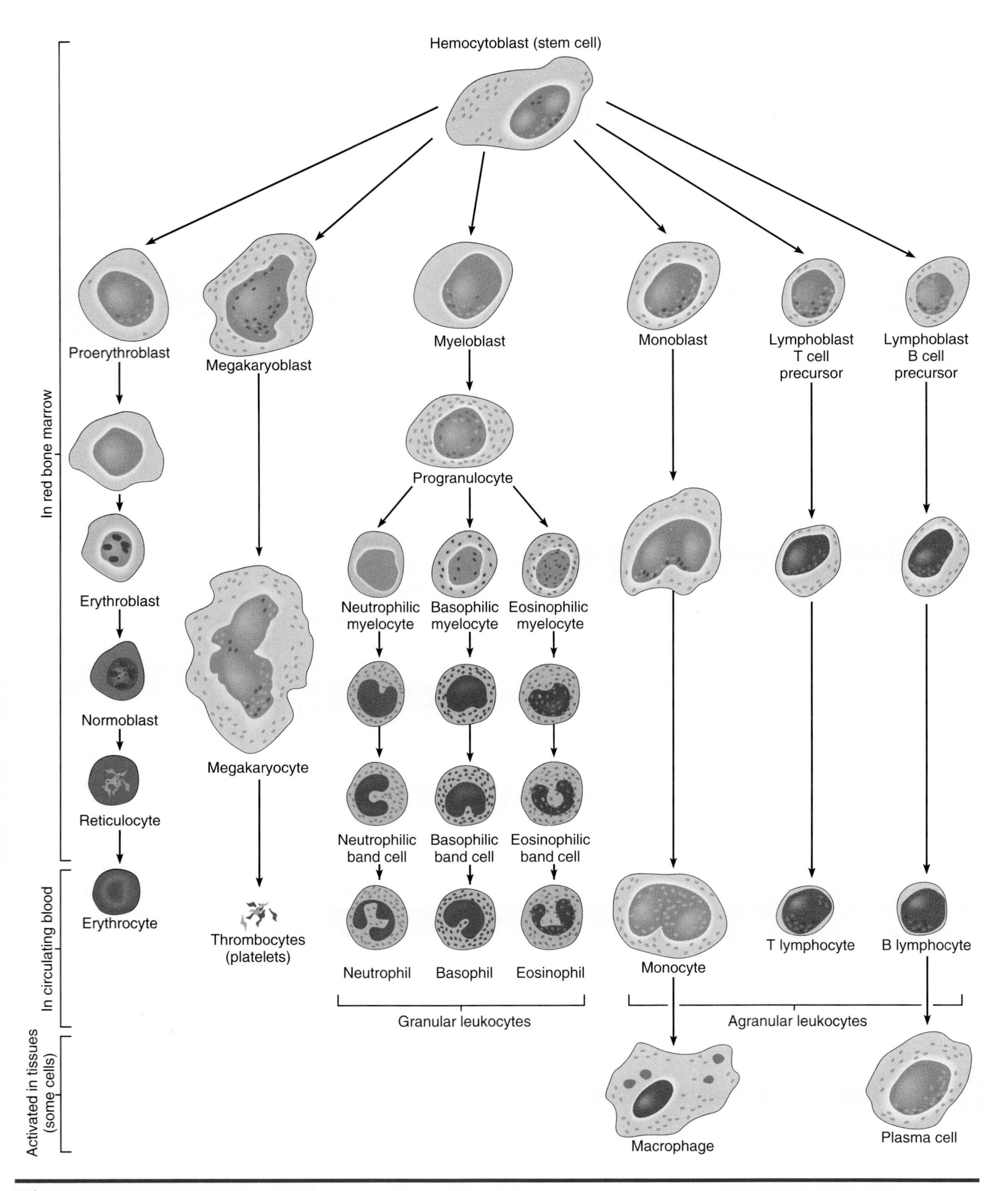

Figure 11.3 Origin and development of the formed elements from a hemocytoblast (stem cell) in red bone marrow.

of erythropoietin is always present to maintain RBC production at a basal rate.

Iron, folic acid, and vitamin B_{12} are required for erythrocyte production. Iron is required for hemoglobin synthesis since each hemoglobin molecule contains four iron atoms. Folic acid and vitamin B_{12} are required for DNA synthesis during early stages of erythrocyte formation in red bone marrow. Vitamin B_{12} is sometimes called the *extrinsic factor* since it is obtained external to the body. Effective absorption of vitamin B_{12} from the digestive tract into the blood depends upon the stomach lining secreting the *intrinsic factor.* The intrinsic factor is a chemical that combines with vitamin B_{12}, enabling its absorption into the blood.

All blood cells, including erythrocytes, are formed from special red bone marrow cells called **hemocytoblasts** and pass through several developmental stages before maturing into the cells found in circulating blood. Erythrocytes lose their nuclei in the final stages of their development (figure 11.3).

Life Span and Destruction

The life span of red blood cells is about 120 days, and billions of RBCs are destroyed and produced each day—about 2 million per second! Normally, destruction and production are kept in balance.

The plasma membranes of newly formed cells are flexible, which allows them to change shape as they pass through small blood vessels. However, with age the membranes lose their elasticity and become fragile and damaged. Worn-out RBCs are removed from circulation by **macrophages** (mak′-rō-fāj-es), phagocytic cells in the liver and spleen. Macrophages engulf and digest old and damaged erythrocytes.

The heme portion of hemoglobin is broken down into an iron-containing compound and a yellow pigment, *bilirubin* (bil-i-rū′-bin). The iron-containing compound may be temporarily stored in the liver before being recycled to the red bone marrow and used to form more hemoglobin. Bilirubin (a bile pigment) is secreted by the liver in bile, which is carried by the bile duct into the small intestine.

How does hemoglobin contribute to the function of erythrocytes?
How is erythrocyte production regulated?

White Blood Cells

Leukocytes (lū′-kō-sīts), or **white blood cells (WBCs)**, are derived from hemocytoblasts in the red bone marrow and undergo several developmental stages in the maturation process. All WBCs retain their nuclei. Note the developmental stages and mature white blood cells in figure 11.3. Leukocytes normally number 4,500 to 10,000 per mm^3 of blood.

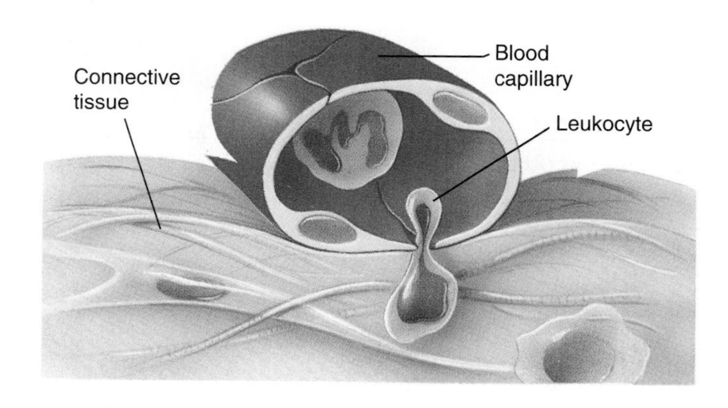

Figure 11.4 Leukocytes can squeeze between the cells of a capillary wall and enter the tissue space outside the bloodstream.

Production of a particular type of white blood cell is increased whenever the cells encounter disease-causing organisms or chemicals that they destroy. These encounters seem to cause the involved leukocytes to secrete a hormone that is carried to red bone marrow, where it stimulates increased production of that particular type of leukocyte.

Function

Leukocytes help provide a defense against disease organisms, and certain cells either promote or decrease inflammatory responses. Most of the functions of white blood cells are performed within the tissues, rather than in blood vessels. WBCs have the ability to move through capillary walls into tissues in response to chemicals released by damaged tissues (figure 11.4). They are able to follow a "chemical trail" through the tissue spaces to reach the damaged tissues, where they congregate to destroy dead cells and foreign substances. Leukocytes move by ameboid movement, a motion characterized by flowing extensions of cytoplasm that pull the cell along.

Types of Leukocytes

There are five types of white blood cells. In contrast to RBCs, WBCs are colorless. WBCs may be distinguished from RBCs by microscopic examination of fresh blood, but white blood cells must be stained in order to distinguish them from each other.

Leukocytes are broadly grouped into two categories: granulocytes and agranulocytes. **Granulocytes** (gran′-ū-lō-sīts) have small granules in their

Table 11.1 Formed Elements in Blood

Formed Element	Description	Number	Function
Erythrocytes (RBCs)	Biconcave disks; no nucleus	4.2–5.4 million/mm^3 in females; 4.5–6.2 million/mm^3 in males	Transport O_2 and CO_2
Leukocytes (WBCs)		4,500–10,000/mm^3	Help defend against invasion by microorganisms
Granulocytes	Cytoplasmic granules present		
Neutrophils	Nucleus with three to five lobes; tiny cytoplasmic granules stain lavender	60%–70% of total WBCs	Phagocytosis
Eosinophils	Nucleus bilobed; cytoplasmic granules stain red	2%–4% of total WBCs	Counteract histamine released in allergic reactions; destroy parasitic worms; phagocytosis
Basophils	Nucleus U-shaped or lobed; cytoplasmic granules stain blue	0.5%–1% of total WBCs	Intensify inflammation response in allergic reactions by releasing histamine and heparin
Agranulocytes	Cytoplasmic granules absent		
Lymphocytes	Very little cytoplasm around spherical nucleus; slightly larger than RBCs	20%–25% of total WBCs	Produce antibodies and provide immunity
Monocytes	Nucleus usually U- to kidney-shaped; two to three times larger than RBCs	3%–8% of total WBCs	Phagocytosis
Thrombocytes (platelets)	Tiny cytoplasmic fragments	150,000–400,000 per mm^3	Form platelet plugs and start clotting of the blood

cytoplasm that are distinctively colored after staining. Neutrophils, eosinophils, and basophils are granulocytes, and they are about twice as large as red blood cells. In contrast, **agranulocytes** lack cytoplasmic granules. Lymphocytes and monocytes are agranulocytes. The five types of white blood cells are distinguished by the shape of the nucleus, the presence or absence of cytoplasmic granules, and the size of the cells. See table 11.1 and figures 11.5 to 11.9.

Neutrophils

Neutrophils (nū′-trō-fils) are the most abundant leukocytes, and they form 60% to 70% of the total WBCs. They are distinguished by a nucleus with two to five lobes and inconspicuous lavender-staining granules. Neutrophils are attracted by chemicals to damaged tissues and are the first WBCs to respond to tissue damage. They engulf bacteria and cellular debris by a process called phagocytosis, and they release the enzyme lysozyme, which destroys some bacteria. The number of neutrophils increases dramatically in acute infections.

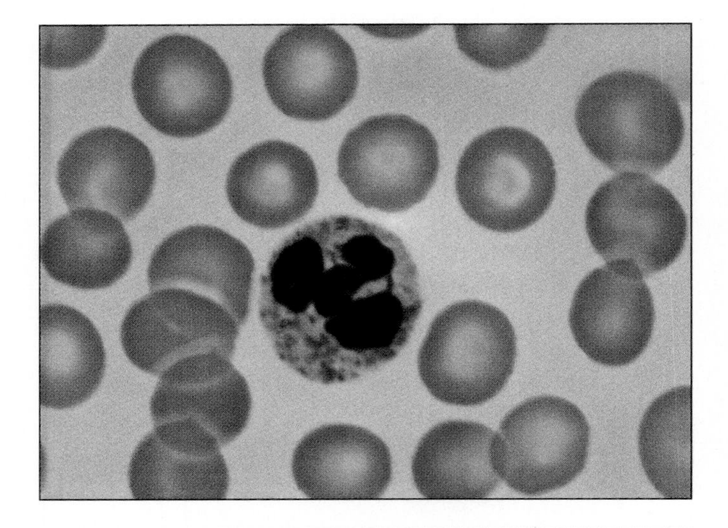

Figure 11.5 A neutrophil has a nucleus with two to five lobes and small lavender-staining cytoplasmic granules.

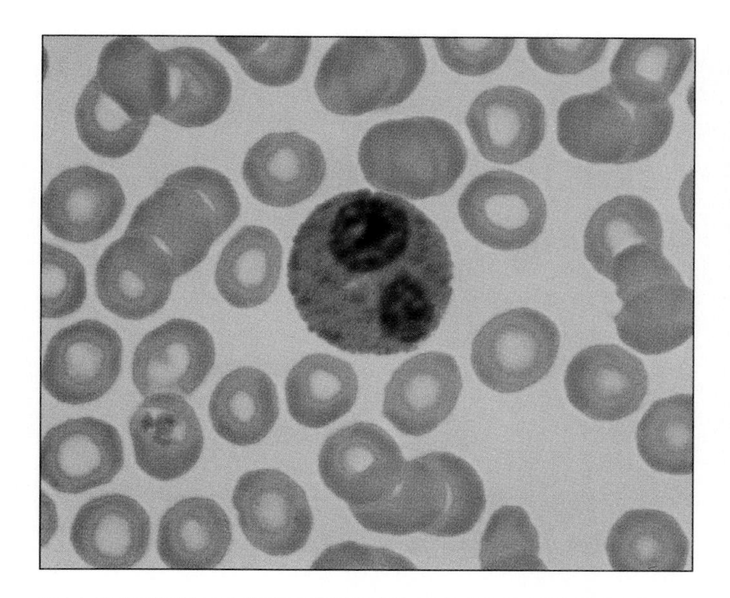

Figure 11.6 An eosinophil has red-staining cytoplasmic granules and a nucleus that usually has only two lobes. The narrow portion of the nucleus connecting the lobes is not shown here.

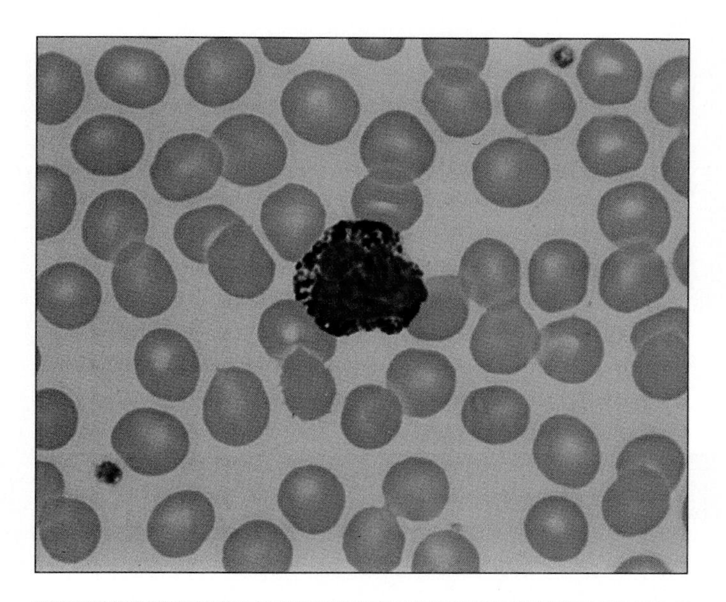

Figure 11.7 The basophil has cytoplasmic granules that stain deep blue.

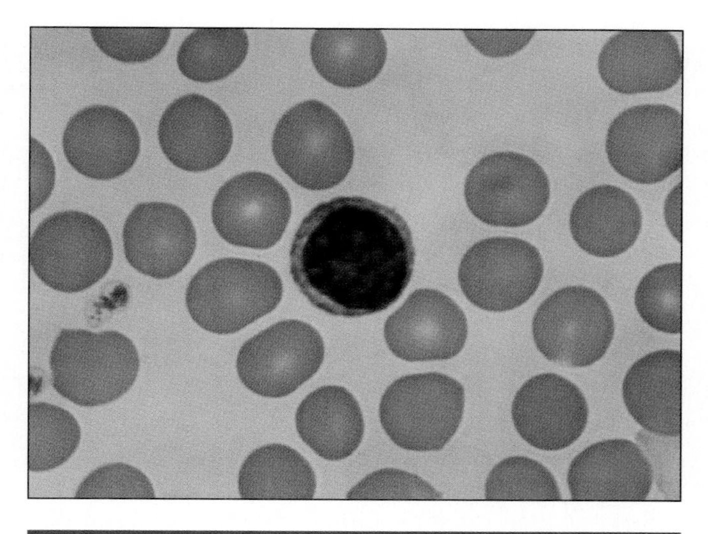

Figure 11.8 A lymphocyte is distinguished by its small size, a large spherical nucleus and very little cytoplasm.

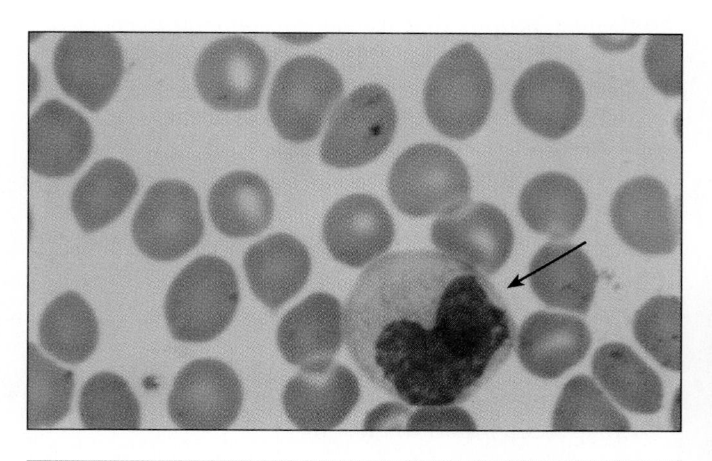

Figure 11.9 A monocyte is the largest white blood cell. It has a kidney-shaped nucleus and a large amount of cytoplasm that lacks granules.

Eosinophils

Eosinophils (ē-ō-sin′-ō-fils) constitute 2% to 4% of the WBCs. They are characterized by a bilobed nucleus and red-staining cytoplasmic granules. Eosinophils neutralize histamine, a chemical released by basophils during allergic reactions, and they also destroy parasitic worms. The number of eosinophils increases rapidly during allergic reactions and parasitic worm infections.

Basophils

Basophils (bā′-sō-fils) are the least numerous of the WBCs, forming only 0.5% to 1% of the WBCs. They are characterized by a nucleus that is U-shaped or lobed and by large, blue-staining cytoplasmic granules. Basophils that have moved into the tissues are called **mast cells.** Mast cells release histamine and heparin when tissues are damaged and in allergic reactions. Histamine dilates blood vessels to increase blood flow in the affected areas, while heparin inhibits clot formation.

Lymphocytes

Lymphocytes (lim´-fō-sīts) form 20% to 25% of the circulating WBCs. They are the smallest leukocytes and are distinguished by a spherical nucleus that is enveloped by very little cytoplasm. Lymphocytes play a vital role in immunity and are especially abundant in lymphoid tissues. There are two types of lymphocytes. *T lymphocytes* directly attack and destroy pathogens (bacteria and viruses), and *B lymphocytes* produce antibodies that attack bacteria and bacterial toxins (poisons).

Monocytes

Monocytes (mon′-ō-sīts) are the largest WBCs, and they comprise 3% to 8% of the WBCs. A U-shaped or kidney-shaped nucleus and abundant cytoplasm distinguishes monocytes. Monocytes are active in phagocytosis. They move into damaged tissues, where they engulf bacteria and cellular debris and complete the cleanup started by the neutrophils. After moving into the tissues, monocytes are called **macrophages.**

A **complete blood count (CBC)** is one of the most common and clinically useful blood tests. It consists of several different blood tests: RBC count, WBC count, platelet count, differential white cell count (the percentage of each type of leukocyte), hematocrit, and hemoglobin percentage. Abnormal values for these tests are associated with infectious and inflammatory processes and with specific blood disorders.

Platelets

Thrombocytes (throm′-bō-sīts), or **platelets,** are actually cytoplasmic fragments of *megakaryocytes,* large cells that develop from hemocytoblasts in red bone marrow. There are 150,000 to 400,000 thrombocytes per mm^3 of blood. Thrombocytes clump together to plug breaks in blood vessels, and they start the clotting process.

What are the functions of each type of leukocyte? What are the recognition characteristics of each type of leukocyte?

Plasma

Plasma is the straw-colored fluid portion of the blood, and it consists of over 90% water. Water is the liquid carrier of plasma solutes (dissolved substances) and formed elements, and it is the solvent of all living systems. Plasma contains a great variety of solutes, such as nutrients, enzymes, hormones, antibodies, waste products, electrolytes, and respiratory gases. Table 11.2 lists the major types of solutes in plasma. Plasma solutes are constantly being added and removed, so the solutes are normally in a state of dynamic balance that is maintained by a variety of homeostatic mechanisms.

Plasma Proteins

Plasma proteins are the most abundant solutes. They are not used as an energy source but remain in the plasma. The three major groups of plasma proteins are albumin, globulins, and fibrinogen. Except for gamma

Table 11.2 Major Solutes in Blood Plasma

Solute	Description
Proteins	
Albumins	60% of plasma proteins; help produce viscosity of blood; help maintain osmotic pressure and pH of blood
Globulins	Alpha and beta globulins transport lipids and fat-soluble vitamins; gamma globulins consist of antibodies
Fibrinogen	Soluble protein that is converted into insoluble fibrin during clot formation
Nonprotein nitrogen substances	Amino acids absorbed from intestine; urea and uric acid formed from protein and nucleoprotein breakdown
Nutrients	Products of digestion, such as amino acids, fatty acids, glycerol, and glucose, are carried from intestines to body cells
Enzymes and hormones	Help to regulate metabolic processes
Electrolytes	Mineral ions help to maintain osmotic pressure, pH, and ionic balance between body cells and blood
Respiratory gases	Both oxygen and carbon dioxide are dissolved in the plasma

globulins, plasma proteins are produced by the liver and are released into the blood.

Albumins form about 60% of the plasma proteins. They are buffers that help to keep the pH of the blood within narrow limits, and they play an important role in maintaining the osmotic pressure of the blood. Osmotic pressure determines the water balance between the blood and body cells. If osmotic pressure of the blood declines, water moves into the body tissues and causes the tissues to swell (edema). This also decreases blood volume and, in severe cases, may decrease blood pressure as well. If osmotic pressure of the blood increases, water moves into the blood, causing an increase in blood volume and in blood pressure while reducing the amount of water available to body cells.

Globulins form about 36% of plasma proteins. The three types of globulins are alpha, beta, and gamma globulins. Alpha and beta globulins carry lipids and fat-soluble vitamins in the blood. Gamma globulins are produced in lymphoid tissues and consist of antibodies that are involved in immunity.

Fibrinogen forms only 4% of the plasma proteins, but it plays a vital role in the blood-clotting process. Fibrinogen is a soluble protein that is converted to insoluble fibrin, which forms blood clots.

Nonprotein Nitrogen Substances

The plasma also contains **nonprotein nitrogen substances (NPN)** that contain nitrogen in their molecules. These substances include amino acids, urea, and uric acid. Amino acids are nutrients that are formed by the digestion of protein in the intestines and are absorbed into the blood. Urea and uric acid are waste products of protein and nucleoprotein breakdown. They are carried by the blood to the kidneys for removal by excretion in the urine.

Electrolytes

Most of the plasma **electrolytes** (ē-lek′-trō-līts) are ions of inorganic compounds that are either absorbed from the intestine or released from body cells. The common electrolytes include sodium (Na^+), potassium (K^+), calcium (Ca^{++}), chloride (Cl^-), bicarbonate (HCO_3^-) and phosphate (PO_4^{--}) ions. Electrolytes help to maintain the osmotic pressure and pH of the plasma, and most of them are essential to life processes.

What are the major components of blood plasma?

Hemostasis

Whenever blood vessels are broken, several homeostatic processes are implemented to prevent excessive blood loss. The stoppage of bleeding is called **hemostasis,** and it involves three separate but interrelated processes: blood vessel spasm, platelet plug formation, and blood clot formation.

Vascular Spasm

The first response of a damaged blood vessel is constriction, or spasm, of the vessel. A *vascular spasm* restricts blood loss from the damaged vessel and it lasts for several minutes, which allows time for formation of the platelet plug and clotting. As platelets accumulate at the site of the damage, they secrete *serotonin,* a chemical that continues the contraction of the smooth muscles in the damaged vessel.

Platelet Plug Formation

Platelets normally do not stick to each other or to the wall of the blood vessel because they are repelled by the positively charged vessel wall. However, when a vessel is damaged the underlying connective tissue is exposed. Platelets are attracted to the site and adhere to the negatively charged connective tissue and to each other so that a cluster of platelets accumulates to plug the break. Figure 11.10 shows the stages in this process. The formation of a *platelet plug* sets the stage for the next process, blood clot formation.

Coagulation

Coagulation (kō-ag-ū-lā′-shun), or blood clotting, is the third and most effective process of hemostasis. The formation of a blood clot is a complex process involving many substances. Blood contains both *procoagulants,* substances that promote clotting, and *anticoagulants,* substances that inhibit clotting. Normally, the anticoagulants predominate, and blood does not clot. However, when a vessel is broken the increase in procoagulant activity starts the clotting process.

Clot formation is a complex process but it is completed within three to six minutes after a blood vessel has been damaged. The clot is restricted to the site of damage because that is where procoagulants outnumber anticoagulants. The key steps in the clotting process are summarized here and shown in figure 11.11:

1. Damaged tissues release *thromboplastin* and aggregated platelets release *platelet factors,* which react with several clotting factors in the plasma to produce **prothrombin activator.**

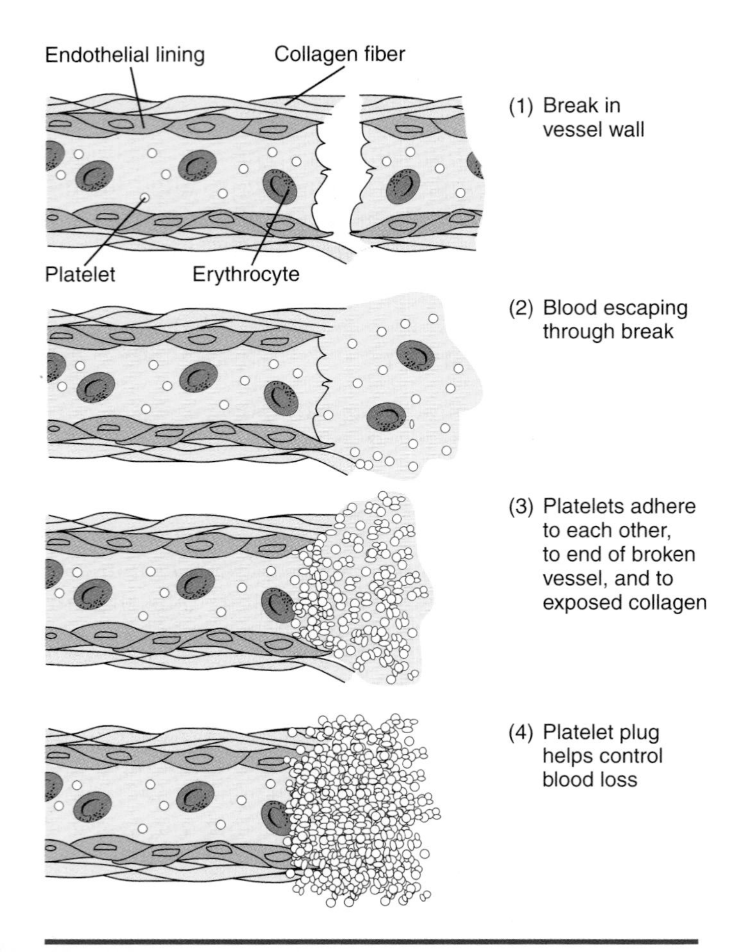

Figure 11.10 Steps in the formation of a platelet plug.

2. In the presence of calcium ions, prothrombin activator stimulates the conversion of **prothrombin,** an inactive enzyme, into the active enzyme **thrombin.**
3. In the presence of calcium ions, thrombin converts molecules of **fibrinogen,** a soluble plasma protein, into threadlike, interconnected strands of insoluble **fibrin.** Fibrin strands form a meshwork that entraps blood cells and sticks to the damaged tissue to form a **thrombus,** a blood clot.

After a clot has formed, the fibrin strands contract, which produces a more compact clot and pulls the damaged tissues closer to each other (figure 11.12). Subsequently, fibroblasts migrate into the clot and form fibrous connective tissue that repairs the damaged area. As healing occurs, the clot is gradually dissolved by enzymes present in the blood.

What are the three major steps in hemostasis? How are blood clots formed?

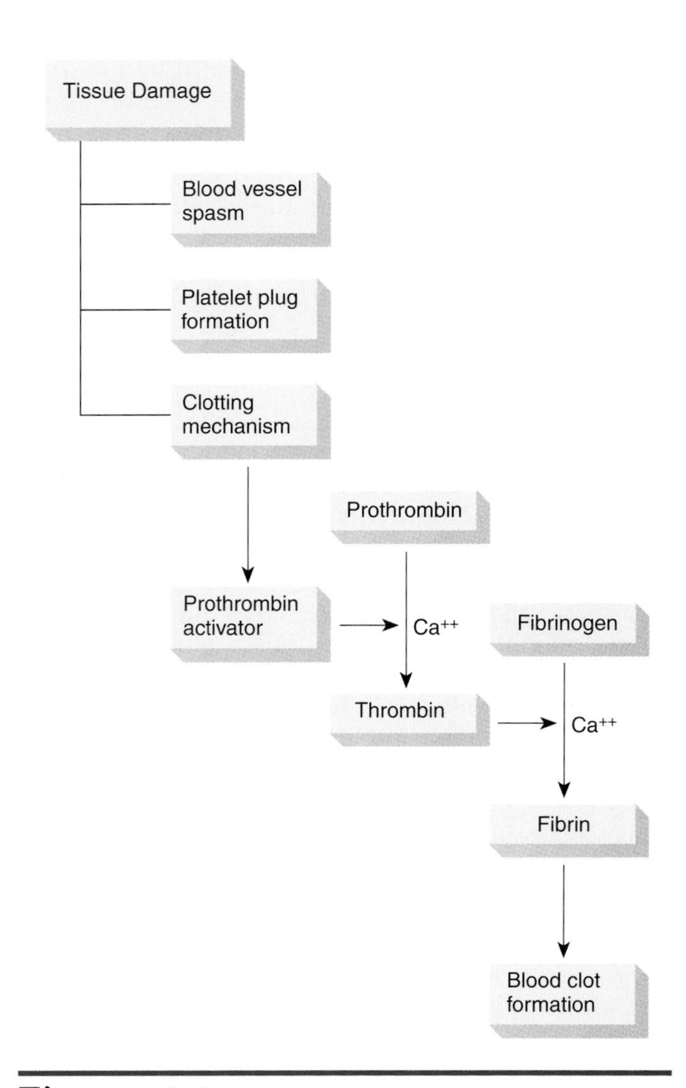

Figure 11.11 Hemostasis consists of three processes: blood vessel spasm, platelet plug formation, and blood clot formation.

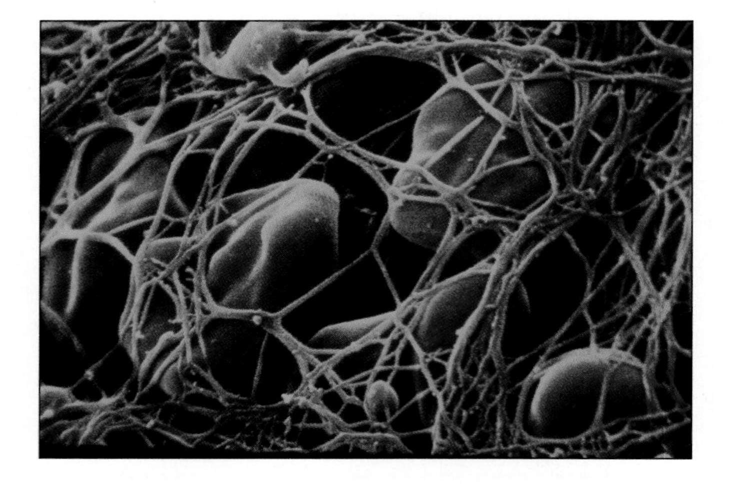

Figure 11.12 A false-colored scanning electron photomicrograph of a blood clot shows blood cells trapped in a meshwork of fibrin strands.

Sometimes unwanted blood clots form in unbroken blood vessels, where they may pose a serious health threat. Certain enzymes, such as streptokinase and urokinase, have been used for some time to help dissolve such clots. A newer, genetically engineered enzyme, *tissue plasminogen activator* (*t-PA*), has been developed. It is an engineered form of a clot-dissolving enzyme that naturally occurs in the body, so unwanted side effects are minimal. t-PA is less likely to trigger allergic reactions or antibody production.

Persons at risk for thrombus formation may be advised to take periodic low dosages of aspirin (e.g., one aspirin three times per week) as a preventive measure. Aspirin inhibits the release of prostaglandins by platelets, which are essential for clot formation. In this way, aspirin slows clot formation and helps to prevent thrombus formation. However, excessive use of aspirin may lead to a prolonged bleeding time.

Human Blood Types

Several different blood types occur in humans. The most familiar ones involve the ABO blood group (types A, B, AB, and O) and the Rh factor (Rh^+ and Rh^-).

Blood types are determined by specific **antigens,** special proteins located on the plasma membranes of erythrocytes. Only a few of the many antigens on the surface of erythrocytes are used in blood typing. They are sometimes called *blood typing antigens* since blood is typed by identifying these antigens on the red blood cells of a blood sample. These antigens are the only ones to cause adverse reactions in blood transfusions.

Whenever RBCs with one type of antigen are introduced into the blood of a person whose RBCs possess a different antigen, the introduced antigens (and RBCs) are recognized as foreign by the recipient's body, and they are attacked by **antibodies** present in the recipient's plasma. Antibodies are substances in plasma that react with foreign proteins in various ways to protect the body from foreign substances and pathogens. In this case, the antibodies bind to the foreign antigens, causing the foreign RBCs to clump together. The clumping of foreign RBCs is called **agglutination** (ah-glū-ti-nā′-shun).

Whenever blood loss is substantial, transfusions of whole blood are routinely given to replace the lost blood. The blood used in the transfusion is usually of the same blood type as that of the recipient. However, a compatible but different blood type may be used in an extreme emergency. Care must be taken to ensure that the transfused blood is compatible with the blood of the recipient. A transfusion of incompatible blood can be fatal because the transfused incompatible RBCs clump together and block small blood vessels, which deprives the tissues supplied by these vessels of nutrients and oxygen.

Table 11.3 Antigens and Antibodies in ABO Blood Types*

Blood Type	Antigen on RBCs	Antibodies in Plasma
A	A	b
B	B	a
AB	A, B	None
O	None	a,b

*Note that the antigen present determines the blood type.

ABO Blood Group

The ABO blood group is based on the presence or absence of two antigens on the surface of red blood cells: **antigen A** and **antigen B.** Like all antigens, they are inherited and remain unchanged from birth to death. The ABO blood group is divided into four possible types: A, B, AB, and O.

After birth, antibodies against the ABO antigen that is *not* present are produced and accumulate in the plasma. Table 11.3 summarizes the relationships of blood type, antigens, and antibodies, and figure 11.13 shows these relationships diagrammatically.

Since antibodies anti-A and anti-B cause clumping of red blood cells with antigens A and B, respectively, the ABO blood type is easily determined. This may be done by placing two separate drops of blood to be tested on a glass slide. A drop of serum containing anti-A antibodies is added to one drop and serum containing anti-B antibodies is added to the other. The pattern of agglutination that occurs in the separate drops of blood indicates the blood type, as

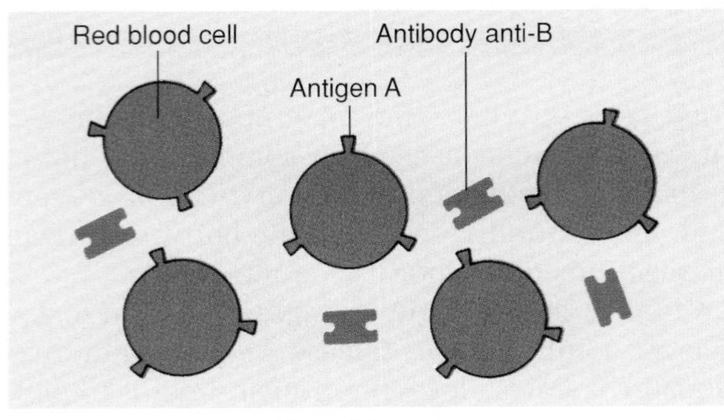

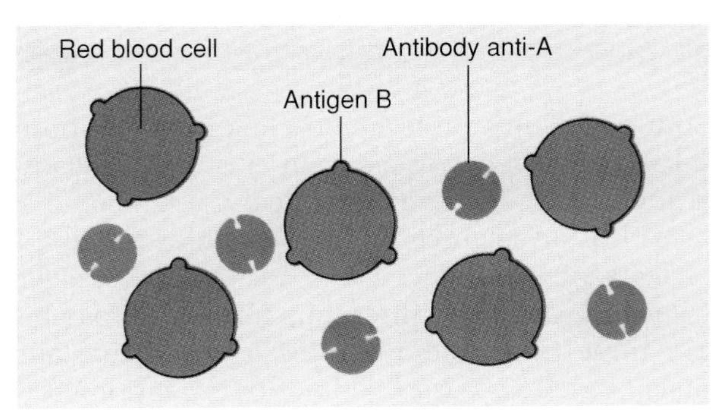

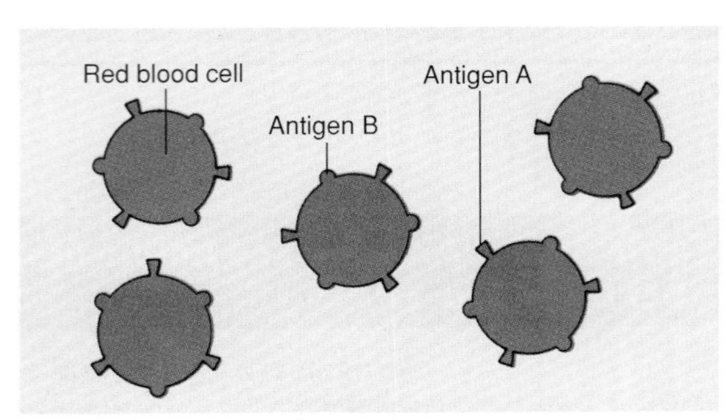

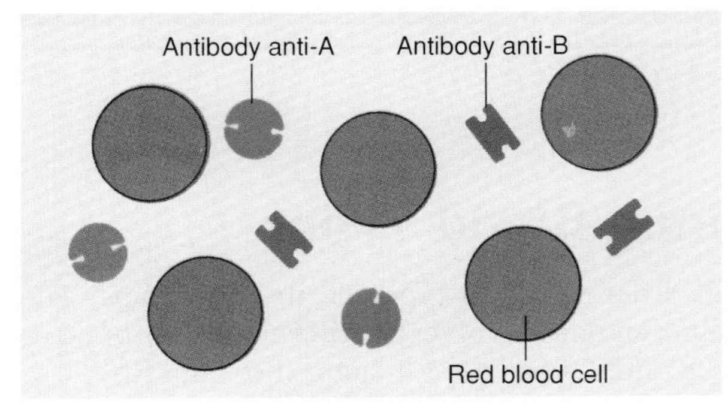

Figure 11.13 Each ABO blood type is distinguished by a different combination of antigens on red blood cells and antibodies in blood plasma.

Table 11.4 Agglutination Pattern in ABO Blood Typing

Blood Type	Anti-A Serum Plus Blood	Anti-B Serum Plus Blood
A	Clumping	No clumping
B	No clumping	Clumping
AB	Clumping	Clumping
O	No clumping	No clumping

shown in table 11.4. Figure 11.14 shows how agglutination occurs when anti-A antibodies are added to type A blood.

As noted earlier, blood of the patient's blood type is used for a transfusion except in extreme emergencies. When a different blood type must be used, it is crucial that the antigens of the transfused blood are compatible with the antibodies of the recipient's blood. For example, blood with antigen A or B cannot be given to a patient whose blood contains antibody anti-A or anti-B. Considering this and the pattern of antigens and antibodies in the ABO blood types, the compatibility of the blood types for transfusions may be determined.

Table 11.5 indicates the preferred ABO blood types that are used for transfusions and blood types that are "acceptable" in emergency situations when the preferred blood type is not available. Note that type AB blood may receive blood from all ABO blood types and that type O blood may be given to all ABO blood types. Therefore, type AB blood is sometimes called the *universal recipient,* and type O blood is known as the *universal donor.*

Rh Blood Type

Blood typing also routinely tests for the presence of the **Rh (D) antigen.** There are several Rh antigens, but it is the D antigen that is of prime significance. The Rh antigen is named after *Rhesus* monkeys, in which the blood group was first discovered.

If the Rh antigen is present on the red blood cells, the blood is typed as Rh positive (Rh^+). If the Rh antigen is absent, the blood is Rh negative (Rh^-). Like the A and B antigens, the presence or absence of the Rh antigen is inherited.

The Rh blood type is determined by adding serum containing anti-Rh antibodies to a drop of

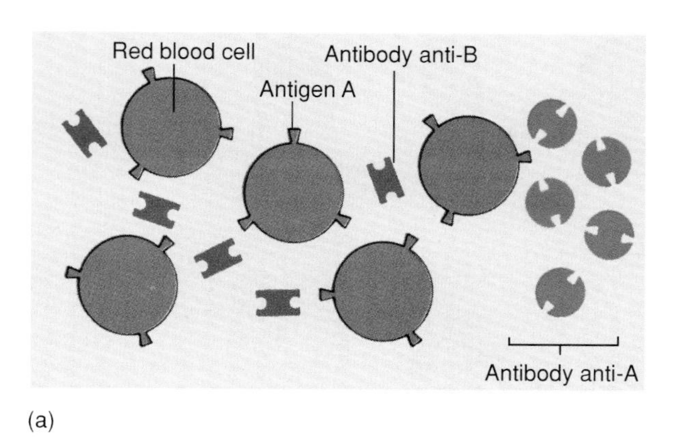

(a)

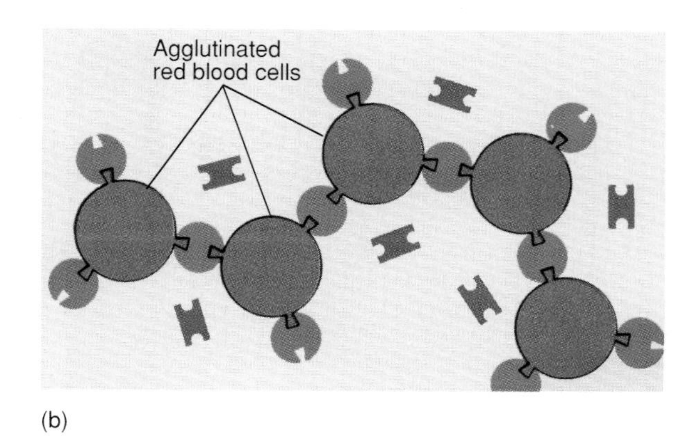

(b)

Figure 11.14 (*a*) If red blood cells with antigen A are added to blood containing antibody anti-A, (*b*) the antibodies react with the antigens causing clumping (agglutination).

Table 11.5 Preferred and Acceptable ABO Blood Types for Transfusions*

Blood Type of Recipient	Blood Types of Donors Preferred	Acceptable
A	A	O
B	B	O
AB	AB	A, B, O
O	O	None

*Acceptable blood types are used only in emergencies when the preferred blood type is not available.

blood on a glass slide. If agglutination occurs, the blood is Rh^+. If clumping does not occur, the blood is Rh^-.

Anti-Rh antibodies are not normally present in the plasma of Rh^- persons. Instead, they are formed only when Rh^+ red blood cells are introduced into a person with Rh^- blood. The first time this occurs, there is no clumping reaction, but the production of anti-Rh antibodies begins. The buildup of anti-Rh antibodies sensitizes the person to future introductions of Rh antigens. If a person with Rh^- blood, who has been sensitized to the Rh antigen, receives a subsequent transfusion of Rh^+ RBCs, the anti-Rh antibodies will cause agglutination of the transfused Rh^+ RBCs, usually with serious or fatal results.

Erythroblastosis Fetalis

A similar kind of problem occurs in **erythroblastosis fetalis,** a blood disorder of newborn infants that results from destruction of fetal erythrocytes by maternal antibodies.

When a woman with Rh^- blood is pregnant with her first Rh^+ fetus, some of the fetal Rh^+ RBCs may accidentally enter the maternal blood due to broken placental blood vessels. This occurs most often during childbirth. The introduction of fetal RBCs with Rh antigens triggers the buildup of anti-Rh antibodies in the woman's blood. The buildup is slow, but the mother has become sensitized to the Rh antigen.

Erythroblastosis fetalis may develop in a subsequent pregnancy with an Rh^+ fetus because the anti-Rh antibodies in maternal blood readily pass through the placenta into the fetal blood, where they agglutinate the fetal RBCs (figure 11.15). If a large number of RBCs are agglutinated and destroyed, the fetus has a decreased ability to transport oxygen.

In response to a decreased oxygen concentration, the fetal blood-forming tissues increase production of RBCs. In an attempt to rapidly produce RBCs, large numbers of nucleated, immature RBCs called *erythroblasts* are released into the blood. These immature cells are not as capable of carrying oxygen as are mature RBCs.

Also, the destruction of large numbers of RBCs produces other harmful effects. Hemoglobin freed from erythrocytes may interfere with normal kidney function and may cause kidney failure. The breakdown of large amounts of hemoglobin forms an excess of bilirubin, a yellow bile pigment that produces jaundice. Oxygen deficiency and excessive bilirubin concentrations in the fetal blood may cause brain damage in afflicted infants.

Upon birth of an infant suffering from erythroblastosis fetalis, the infant's total blood volume may be slowly replaced with Rh^- blood. The transfused blood provides functional RBCs that cannot be agglutinated by anti-Rh antibodies that may still be present, and it

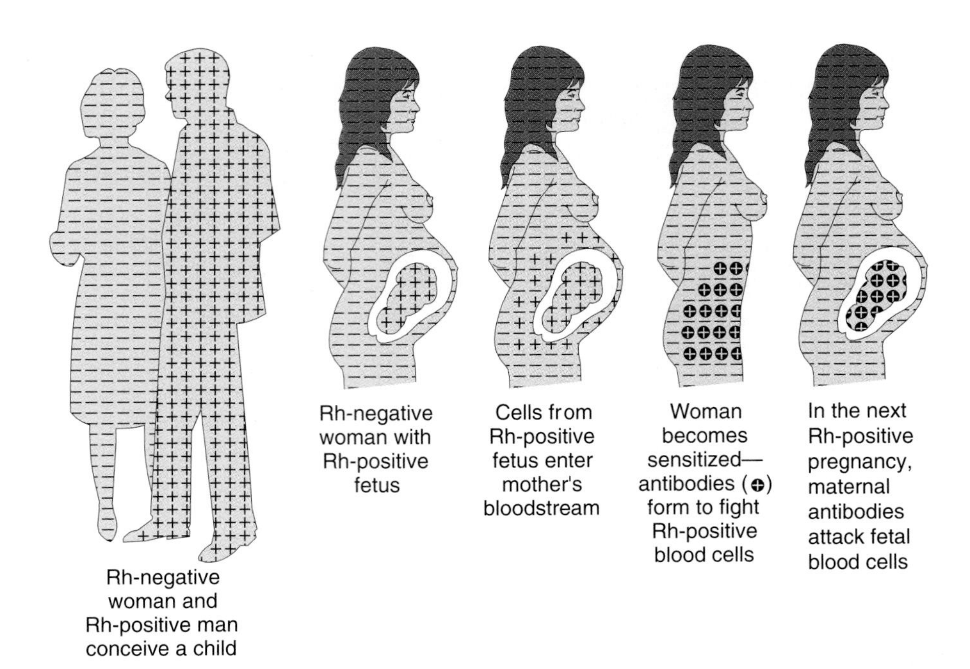

Figure 11.15 Development of erythroblastosis. A Rh^- woman pregnant with a Rh^+ fetus becomes sensitized to the Rh^+ antigen and forms anti-Rh^+ antibodies. In the next pregnancy with a Rh^+ fetus, the woman's antibodies attack the fetal Rh^+ red blood cells.

reduces the bilirubin concentration to eliminate the jaundice. Subsequently, the infant's own RBC production will again produce Rh^+ RBCs, but by then all anti-Rh antibodies have been removed from the blood.

What determines an individual's ABO blood type? Why is blood typing important in transfusions? What is the cause of erythroblastosis fetalis?

The cause of erythroblastosis fetalis is now preventable by injecting serum containing anti-Rh antibodies (trade name RhoGam) into the blood of Rh^- females within 72 hours after they have given birth to an Rh^+ child, or after miscarriage or abortion. The anti-Rh antibodies agglutinate and destroy any fetal Rh^+ RBCs that may have entered the mother's blood before they can stimulate the production of anti-Rh antibodies and sensitize the mother. Further, pregnant Rh^- mothers may be given an injection of RhoGam near the fifth month of subsequent pregnancies as a safety precaution.

Disorders of the Blood

Blood disorders may be grouped as erythrocyte disorders, leukocyte disorders, and clotting disorders. Normal values for common blood tests are located on inside back cover.

Erythrocyte Disorders

Anemia (ah-nē′-mē-ah) is the decreased oxygen-carrying capacity of the blood, and it is the most common blood disorder. A decreased number of RBCs or an insufficient amount of hemoglobin reduce the blood's capacity to carry oxygen. There are several different types of anemia:

- **Nutritional anemia** results from insufficient amounts of iron and vitamin B_{12} in the diet.
- **Hemorrhagic anemia** results from the excessive loss of RBCs through bleeding.
- **Pernicious anemia** results from a deficiency of intrinsic factor, which prevents absorption of sufficient vitamin B_{12} from the intestine to support adequate RBC production.
- **Hemolytic anemia** results from premature rupture of RBCs so that hemoglobin is released into the plasma.
- **Aplastic anemia** results from destruction of red bone marrow or its inability to produce a sufficient number of RBCs.

- **Sickle-cell anemia** results from abnormal hemoglobin that cannot carry sufficient oxygen. Sickled RBCs rupture easily, causing hemolytic anemia. Sickle-cell anemia is inherited and occurs primarily in descendents of African blacks who lived in malarial regions.

Erythroblastosis fetalis results from anti-Rh antibodies produced by a previously sensitized Rh^- woman carrying an Rh^+ fetus. The antibodies pass through the placenta into the fetal blood and agglutinate the fetal RBCs. It is characterized by anemia, jaundice, and the presence of erythroblasts in the blood.

Polycythemia (pol-ē-sī-thē-mē-ah) is a condition characterized by an excess of erythrocytes in the blood. The excess RBCs increase blood volume and viscosity, which impairs circulation. It may result from cancer of the RBC-forming cells.

Leukocyte Disorders

Infectious mononucleosis is a contagious disease of the lymphoid tissue caused by the Epstein-Barr virus (EBV). It occurs primarily in young adults, and kissing is a common mode of transmission. Three times more females contract the disease than males. It infects B lymphocytes, which enlarge and resemble monocytes. Symptoms include fever, headache, fatigue, sore throat, and swollen lymph glands. There is no cure, but infectious mononucleosis usually persists for only about four weeks. However, in some persons it may linger for months or years, and relapses may be frequent.

Leukemia (lū-kē′-mē-ah) is a group of cancers of the red bone marrow cells that form WBCs. It is characterized by an excess production of leukocytes and the crowding out of RBC- and platelet-forming cells. Acute forms affect primarily children or young adults; chronic forms occur more often in adults. The various types of leukemia are classified according to the predominant white cell involved. Treatment usually involves chemotherapy and sometimes a transplant of bone marrow from a compatible donor.

Clotting Disorders

Hemophilia (hē-mō-fil′-ē-ah) is a group of inherited diseases that occur almost always in males because they are sex-linked disorders. Hemophilia is characterized by spontaneous bleeding and a reduced ability to form blood clots. It may be caused by a deficiency of any one of several plasma clotting factors. There is no cure for hemophilia, but it is treated by transfusion of the missing clotting factors.

Thrombocytopenia (throm-bō-sī-tō-pē′-nē-ah) is a condition in which the number of platelets is so low ($< 50{,}000/mm^3$) that spontaneous bleeding cannot be prevented. Bleeding from many small vessels typically results in purplish blotches appearing on the skin.

Thrombosis is the condition resulting from the formation of a blood clot in an unbroken blood vessel. Such clots tend to form where the lining of a blood vessel is roughened or damaged. They can cause serious effects if they plug an artery and deprive vital tissues of blood. Blood clots form more frequently in veins than arteries, causing a condition known as *thrombophlebitis*.

Sometimes, a clot formed in a vein breaks free and is carried by the blood only to lodge in an artery, often a branch of a pulmonary artery. Such a moving clot is called an *embolus,* and when it blocks a blood vessel, the resulting condition is known as an **embolism.** An embolism can produce very serious and sometimes fatal results if it lodges in a vital organ and blocks the flow of blood.

CHAPTER SUMMARY

General Characteristics of Blood

1. Blood is composed of plasma (55%) and formed elements (45%). Erythrocytes constitute nearly all of the formed elements.
2. Blood is heavier and about four times more viscous than water, and it is slightly alkaline.
3. About 8% of the body weight consists of blood. Blood volume ranges between 4 and 6 liters.

Red Blood Cells

1. Erythrocytes, or red blood cells, are biconcave disks that lack nuclei and contain a large amount of hemoglobin. Their primary function is the transport of oxygen.
2. Hemoglobin is composed of heme, an iron-containing pigment, and globin, a protein. It plays a vital role in oxygen transport and participates in carbon dioxide transport.
3. Erythrocytes are very abundant in the blood. They number 4.5 million to 6.2 million per mm^3 in males and 4.2 million to 5.4 million per mm^3 in females.
4. Erythrocytes are formed from hemocytoblasts in the red bone marrow. The rate of production is controlled by the oxygen concentration of the blood via a negative feedback mechanism. A decreased

oxygen concentration stimulates kidney cells to release erythropoietin, which stimulates increased production of RBCs by red bone marrow.

5. Iron, amino acids, vitamin B_{12}, and folic acid are essential for RBC production.
6. RBCs live about 120 days before they are destroyed by macrophages in the spleen and liver. In hemoglobin breakdown, the iron-containing molecules are recycled for use in forming more hemoglobin. Bilirubin, a bile pigment, is a by-product of hemoglobin breakdown.

White Blood Cells

1. Leukocytes, or white blood cells, are also formed from hemocytoblasts in the red bone marrow. They retain their nuclei and number 4,500 to 10,000 per/mm^3 of blood.
2. White blood cells help to defend the body against invasion by disease organisms, and most of their activities occur within body tissues.
3. The five types of leukocytes are categorized into two groups. Granulocytes have cytoplasmic granules and include neutrophils, eosinophils, and basophils. Agranulocytes lack cytoplasmic granules and include lymphocytes and monocytes.
4. Neutrophils and monocytes are phagocytes that destroy bacteria and clean up cellular debris.
5. Eosinophils help to reduce inflammation and destroy parasitic worms.
6. Basophils promote inflammation.
7. Lymphocytes play vital roles in immunity.

Platelets

1. Thrombocytes, or platelets, are fragments of megakaryocytes in the red bone marrow. They number 150,000 to 400,000 per/mm^3 of blood.
2. Platelets play a crucial role in hemostasis by forming platelet plugs and starting the clotting process.

Plasma

1. Plasma, the liquid portion of the blood, consists of over 90% water along with a variety of solutes including nutrients, waste materials, proteins, electrolytes, and respiratory gases.
2. There are three types of plasma proteins. Albumins are most numerous, and they help to maintain the osmotic pressure and pH of the blood. Alpha and beta globulins transport lipids and fat-soluble vitamins. Gamma globulins consist of antibodies that are involved in immunity. Fibrinogen is a soluble protein that is converted into insoluble fibrin during the clotting process.
3. Nonprotein nitrogen substances in plasma include amino acids, urea, and uric acid.
4. Electrolytes include ions of sodium, potassium, magnesium, calcium, carbonate, bicarbonate, phosphate, and chlorine. Electrolytes help to maintain the pH and osmotic pressure of the blood.

Hemostasis

1. Hemostasis is the stoppage of bleeding. It consists of three processes: vascular spasm, platelet plug formation, and blood clotting.
2. Vascular spasm is the first process in hemostasis. It reduces blood loss until the other processes can occur.
3. Platelets stick to the damaged tissue and to each other to form a platelet plug.
4. Platelets and damaged tissues initiate clot formation by releasing thromboplastin, which causes the formation of prothrombin activator. Prothrombin activator converts prothrombin into thrombin, which, in turn, converts fibrinogen into fibrin. Fibrin strands form the clot.
5. After clot formation, fibroblasts invade the clot and gradually replace it with connective tissue as the clot is dissolved by enzymes.

Human Blood Types

1. Blood types are determined by the presence or absence of blood typing antigens on the plasma membranes of RBCs.
2. The four ABO blood types, A, B, AB, and O, are based on the presence or absence of antigen A and antigen B.
3. Antibodies anti-A and anti-B are spontaneously formed against the antigen(s) that is (are) not present in a person's blood.
4. Transfusions must be made using only compatible blood types. This means that the transfused RBCs must not have an antigen that reacts with the antibodies of the recipient's blood.
5. If incompatible blood is transfused, agglutination of the transfused RBCs occurs. The clumped RBCs plug small blood vessels, depriving tissues of nutrients and oxygen. The result may be fatal.
6. Blood with RBCs containing the Rh antigen is typed as Rh^+. Blood without the Rh antigen is typed as Rh^-.
7. Anti-Rh antibodies are produced only after Rh^+ RBCs are introduced into a person with Rh^- blood. Once a person is sensitized in this way, a subsequent transfusion of Rh^+ blood results in agglutination of the transfused RBCs.
8. Erythroblastosis fetalis occurs in newborn infants when a sensitized Rh^- woman is pregnant with an Rh^+ fetus. Her anti-Rh antibodies pass through the placenta into the fetus and agglutinate the fetal RBCs, producing anemia and jaundice.

Disorders of the Blood

1. Anemia is the most common disorder, and it may result from a variety of causes.
2. Other disorders include erythroblastosis fetalis, hemophilia, infectious mononucleosis, leukemia, polycythemia, thrombocytopenia, and thrombosis.

BUILDING YOUR VOCABULARY

1. **Selected New Terms**
   albumin, p. 227
   basophil, p. 225
   deoxyhemoglobin, p. 221
   eosinophil, p. 225
   erythropoietin, p. 221
   extrinsic factor, p. 223
   fibrin, p. 228
   fibrinogen, p. 228
   globulin, p. 227
   intrinsic factor, p. 223
   lymphocyte, p. 226
   macrophages, p. 223
   mast cell, p. 225
   monocyte, p. 226
   neutrophil, p. 224
   oxyhemoglobin, p. 221
   platelet, p. 226
   prothrombin, p. 228
   prothromnin activator, p. 227
   thromboplastin, p. 227
2. **Related Clinical Terms**
   hematoma (hē-mah-tō'mah) Black-and-blue marks (bruises) caused by coagulated blood in tissues.
   purpura (per'-pū-rah) Spontaneous bleeding into the tissues.
   septicemia (sep-ti-sē'mē-ah) Blood poisoning caused by bacteria or toxins in the blood.
   thalassemia (thal-ah-sē'mē-ah) A type of hemolytic anemia caused by abnormally fragile RBCs.

STUDY ACTIVITIES

1. Complete the **Chapter 11 Study Guide,** which begins on page 455.
2. Complete the **Learning Objectives** listed on the first page of this chapter.

CHECK YOUR UNDERSTANDING

(Answers are located in appendix B.)

1. About _____ % of blood consists of erythrocytes.
2. The red color of blood results from the presence of _____ in _____ .
3. All formed elements are derived from stem cells, the _____ , within red bone marrow.
4. A low blood concentration of _____ promotes the formation of the hormone _____ , which stimulates RBC production.
5. Erythrocytes are destroyed in the spleen and _____ .
6. Fighting disease invasion is the function of nucleated formed elements called _____ .
7. The two major phagocytic WBCs are _____ and _____ .
8. The release of histamine by _____ helps to promote inflammation.
9. WBCs that destroy parasitic worms and fight inflammation are the _____ .
10. Immunity is the prime function of _____ .
11. The fluid carrier of solutes and formed elements in blood is the _____ .
12. Damaged tissues and _____ start coagulation by releasing thromboplastin.
13. Blood clot formation involves converting _____ , a soluble plasma protein, into an insoluble protein called _____ .
14. ABO blood types are named for the _____ on the surface of erythrocytes.
15. Type _____ blood is the universal recipient because it lacks anti-A and anti-B antibodies.

TWELVE

12

CHAPTER Heart and Blood Vessels

Chapter Preview & Learning Objectives

Structure of the Heart
- Identify the protective coverings of the heart.
- Identify the parts of the heart and describe their functions.

Cardiac Cycle
- Describe the events of the cardiac cycle.
- Trace the flow of blood through the heart.

Heart Conduction System
- Identify parts of the heart conduction system and describe their functions.

Regulation of Heart Rate
- Explain how the heart rate is regulated.

Types of Blood Vessels
- Describe the structure and function of arteries, arterioles, capillaries, venules, and veins.
- Describe how materials are exchanged between capillary blood and tissue fluid.

Blood Flow
- Describe the mechanism of blood circulation.

Blood Pressure
- Compare systolic and diastolic blood pressure.
- Describe how blood pressure is regulated.

Circulation Pathways
- Compare the systemic and pulmonary circuits.

Systemic Arteries
- Identify the major systemic arteries and the organs or body regions that they supply.

Systemic Veins
- Identify the major systemic veins and the organs or body regions that they drain.

Disorders of the Heart and Blood Vessels
- Describe the common disorders of the heart and blood vessels.

Chapter Summary
Building Your Vocabulary
Study Activities
Check Your Understanding

SELECTED KEY TERMS

Arteries Blood vessels carrying blood from the heart to capillaries in body tissues.
Atrium (atrium = vestibule) A heart chamber that receives blood returned to the heart by veins.
Capillaries Tiny blood vessels in tissues where exchange of materials between the blood and tissue cells occurs.
Diastole The relaxation phase of the heart cycle.
Myocardium (myo = muscle; cardi = heart) The thick layer of cardiac muscle in the wall of the heart.
Pericardium (peri = around) The membranous sac enclosing the heart.
Pulmonary circuit (pulmo = lung) Transport of blood to and from the lungs.
Systemic circuit Transport of blood to and from all parts of the body except the lungs.
Systole The contraction phase of the heart cycle.
Vasoconstriction (vas = vessel) A decrease in the diameter of blood vessels.
Vasodilation An increase in the diameter of blood vessels.
Veins Blood vessels carrying blood from capillaries in body tissues toward the heart.
Ventricle (ventr = underside) A heart chamber that pumps blood into an artery.

The heart and blood vessels are part of the **cardiovascular** (kar-dē-ō-vas′-kū-lar) **system.** The heart pumps blood through a closed system of blood vessels. Arteries carry blood away from the heart to capillaries in body tissues. Veins carry blood from capillaries in body tissues back to the heart. Figure 12.1 shows the general scheme of circulation of blood in the body. Blood vessels colored blue carry deoxygenated (oxygen-poor) blood; those colored red carry oxygenated (oxygen-rich) blood.

Structure of the Heart

The heart is a four-chambered muscular pump that is located within the mediastinum in the thoracic cavity. It lies between the lungs and just superior to the diaphragm. The *apex* of the heart is the pointed end, which extends toward the left side of the thorax at the level of the fifth rib. The *base* of the heart is the superior portion, which is attached to several large blood vessels at the level of the second rib. The heart is about the size of a closed fist. Note the relationship of the heart with the surrounding organs in figure 12.2.

Protective Coverings

The heart and the bases of the attached blood vessels are enveloped by membranes that are collectively

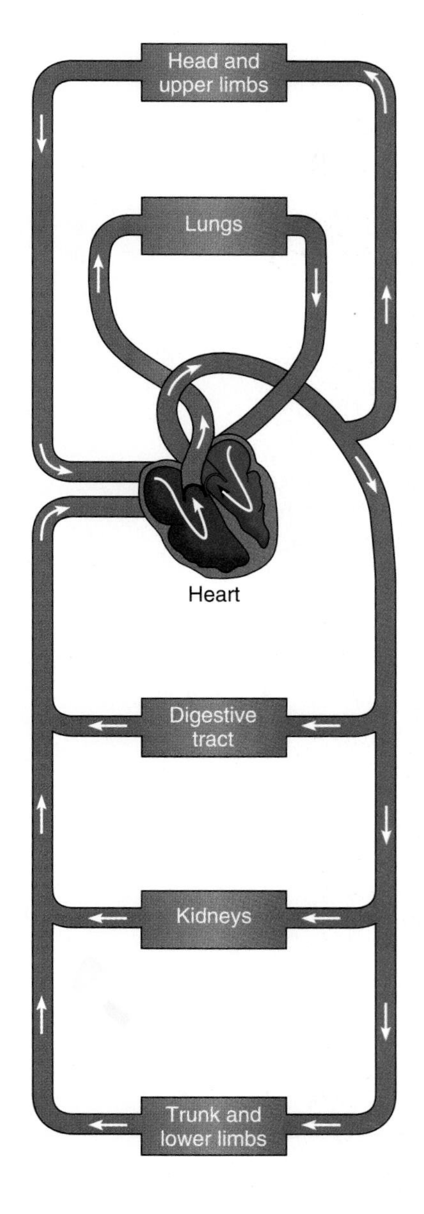

Figure 12.1 The general scheme of circulation of the blood. Vessels colored red carry oxygenated blood. Vessels colored blue carry deoxygenated blood.

called the **pericardium** (per-i-kar′-dē-um). An outer, loosely fitting sac is formed of two layers: an outer *fibrous pericardium* and an inner *parietal pericardium.* The fibrous pericardium is a tough, supporting membrane that is attached to the diaphragm, inner surfaces of the sternum and vertebral column, and adjacent connective tissues. The parietal pericardium lines the fibrous pericardium. At the bases of the great vessels, the parietal pericardium folds back on itself to form the *visceral pericardium,* which forms the thin outer layer of the heart wall (figure 12.3).

The parietal and visceral pericardia are serous membranes, and they secrete serous fluid into the *pericardial cavity,* the potential space between them. Serous fluid reduces friction as the membranes rub against each other during heart contractions.

The Heart Wall

The wall of the heart consists of a thick layer of cardiac muscle tissue, the **myocardium** (mī-ō-kar′-dē-um), sandwiched between two thin membranes. Contractions of the myocardium provide the force that pumps the blood through the blood vessels. The **epicardium,** or visceral pericardium, is the thin membrane that is firmly attached to the exterior surface of the myocardium. Blood vessels that nourish

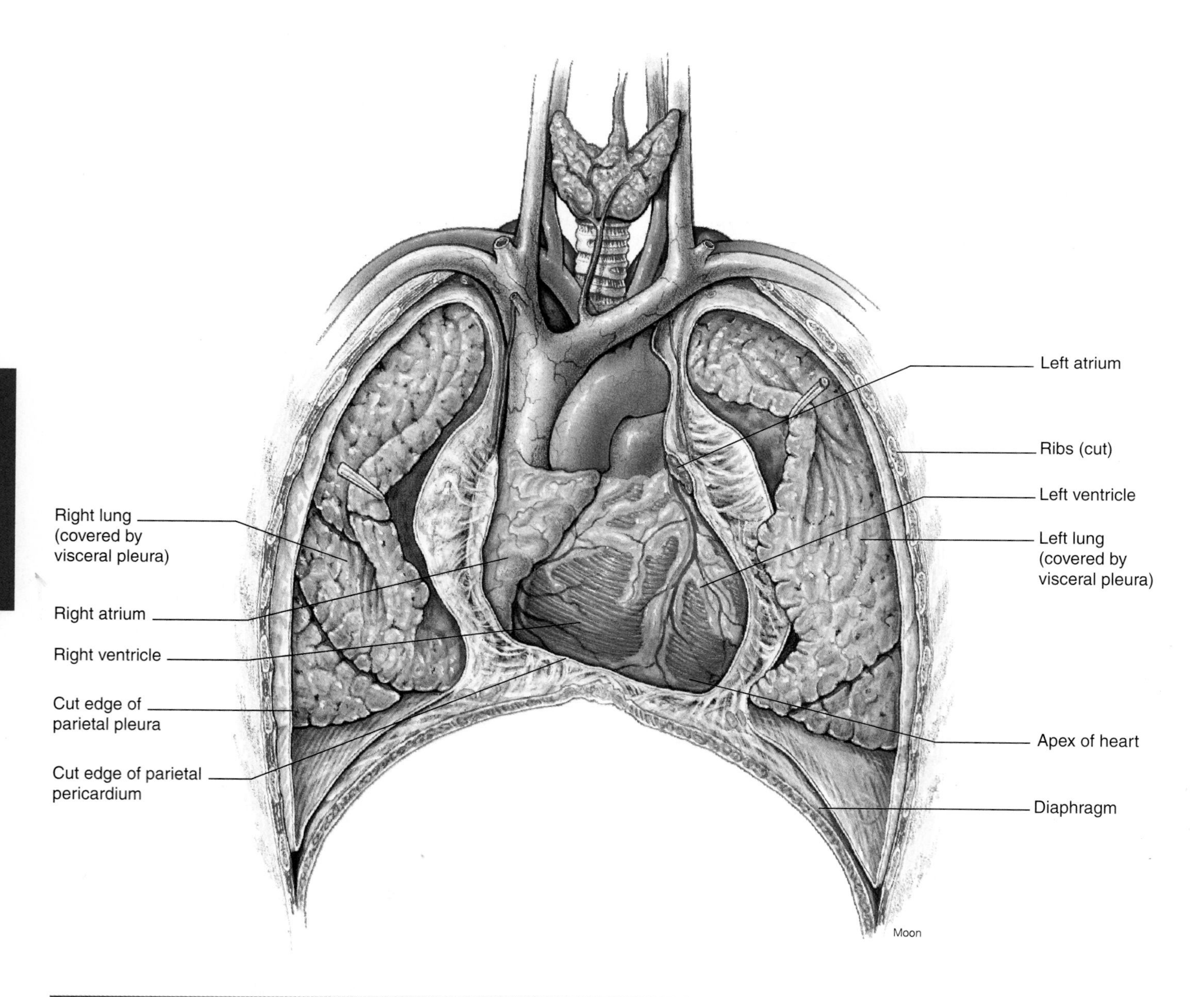

Figure 12.2 The heart is located within the mediastinum in the thoracic cavity.

the heart itself are located within the epicardium. The internal surface of the myocardium is covered with a single layer of squamous epithelium called the **endocardium.** The endocardium not only lines the chambers of the heart, but it also is continuous with the internal lining of the blood vessels attached to the heart.

Heart Chambers

The heart contains four chambers. The two upper chambers are the **atria** (ā′-trē-ah), which receive blood being returned to the heart by the veins. The two lower chambers are the **ventricles** (ven′-tri-kuls), which pump blood into the arteries carrying blood away from the heart. There is no opening between the two atria or between the two ventricles. The atria are separated from each other by a partition called the *atrial septum.* The ventricles are separated by the *interventricular septum,* a thick partition of cardiac muscle (figure 12.4). The heart is a double pump. The left atrium and left ventricle compose the left pump. The right atrium and right ventricle compose the right pump.

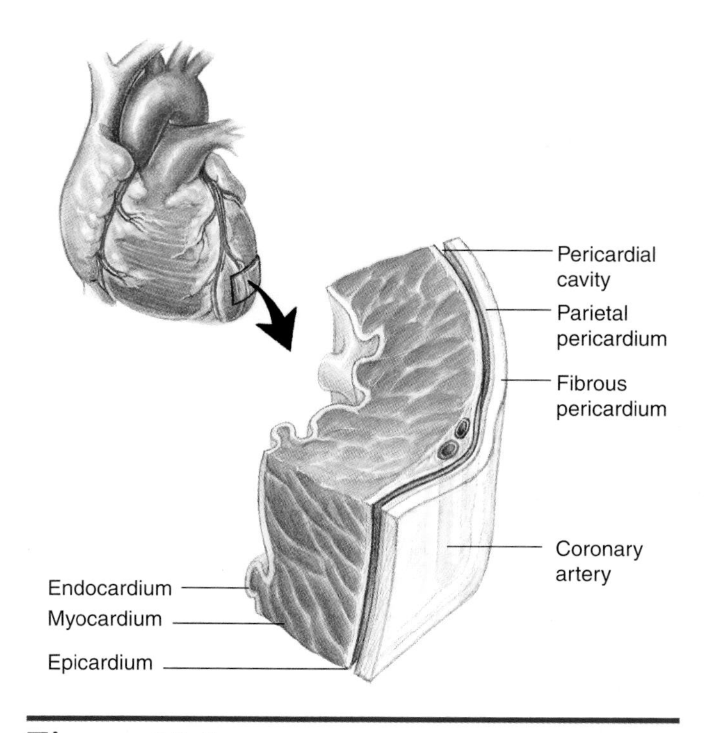

Figure 12.3 The heart wall has three layers: an endocardium, a myocardium, and an epicardium.

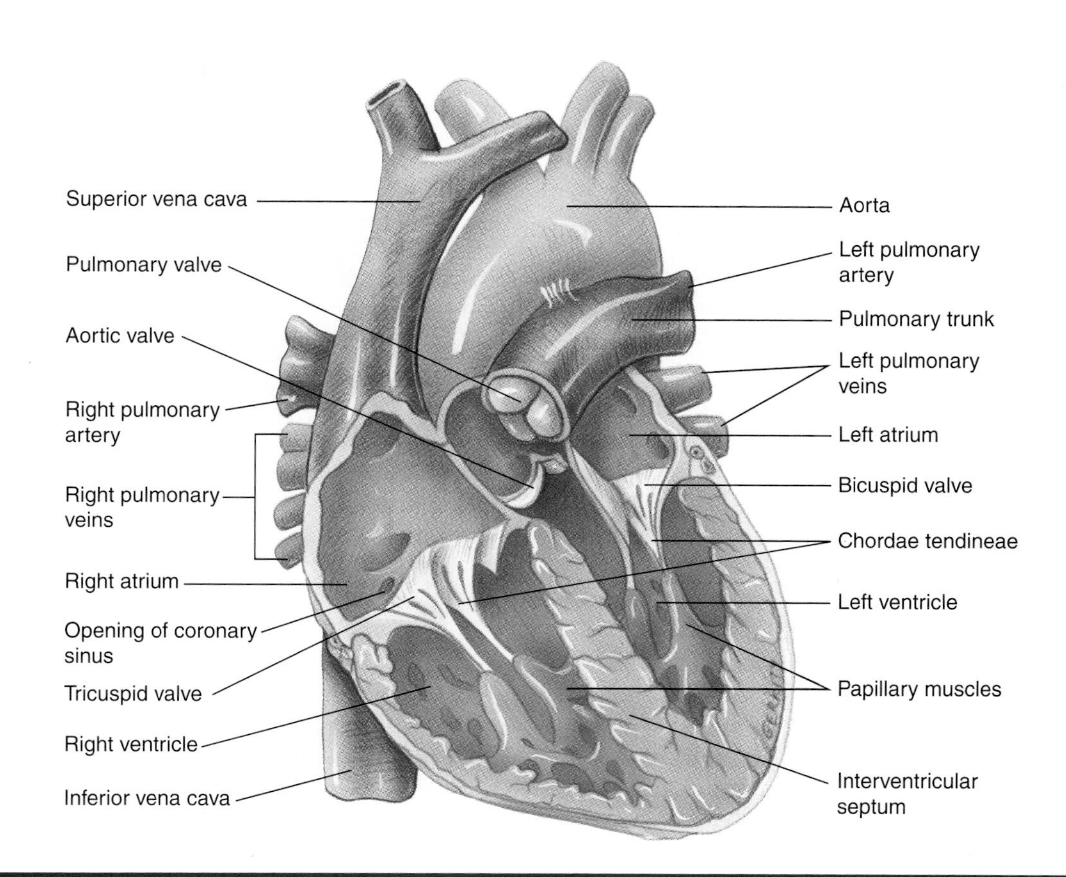

Figure 12.4 The internal structure of the heart is shown in coronal section.

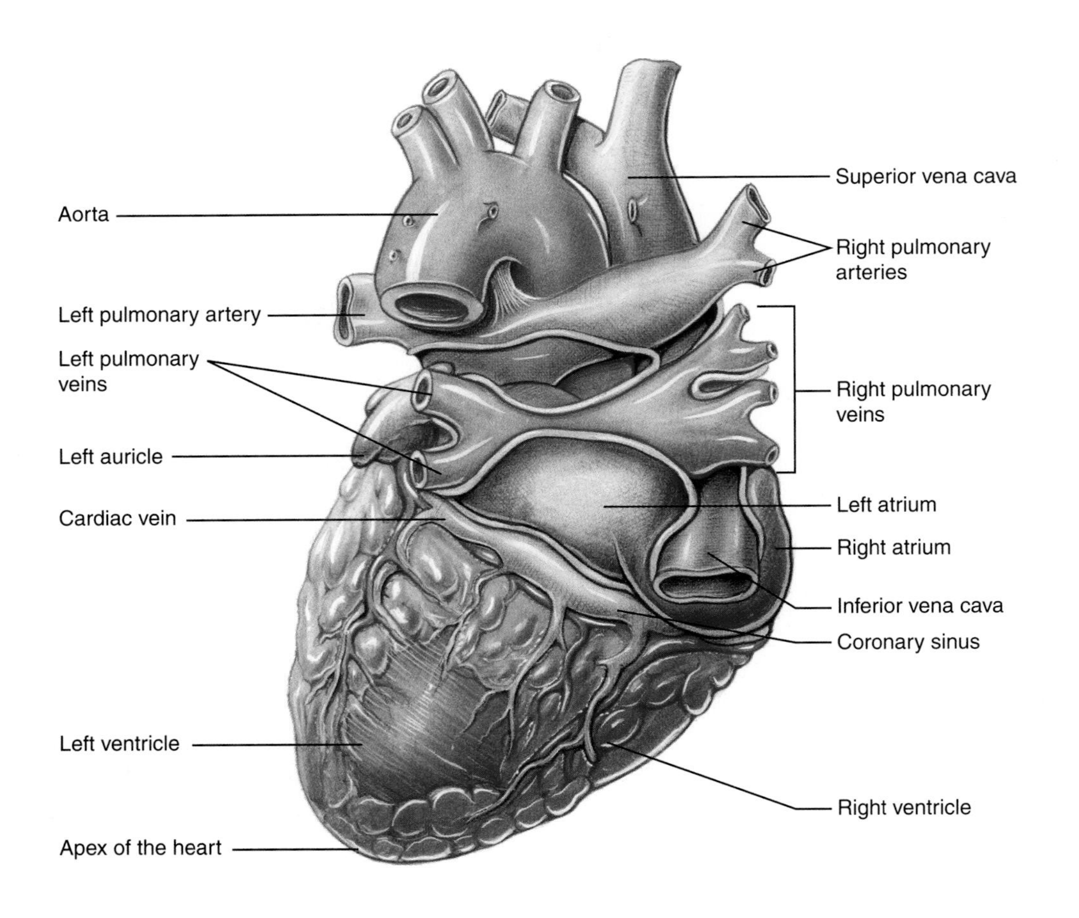

Figure 12.5 A posterior view of the heart and the associated blood vessels.

Table 12.1 Functions of the Heart Chambers

Chamber	Function
Right atrium	Receives deoxygenated blood from the superior and inferior venae cavae and passes this blood through the tricuspid A-V valve to the right ventricle
Right ventricle	Receives deoxygenated blood from the right atrium and pumps this blood through the pulmonary semilunar valve into the pulmonary trunk
Left atrium	Receives oxygenated blood from the pulmonary veins and passes this blood through the bicuspid A-V valve to the left ventricle
Left ventricle	Receives oxygenated blood from the left atrium and pumps this blood through the aortic semilunar valve into the aorta

The walls of the atria are much thinner than the walls of the ventricles. Differences in thickness are due to differences in the amount of cardiac muscle tissue that is present, which, in turn, reflects the work required of each chamber. The left ventricle has a thicker, more muscular wall than the right ventricle because it must pump blood to the entire body, except the lungs, while the right ventricle pumps blood only to the lungs. Locate the atria and ventricles in figure 12.5, which shows the external structure of the heart and the associated blood vessels. Table 12.1 summarizes the functions of the heart chambers.

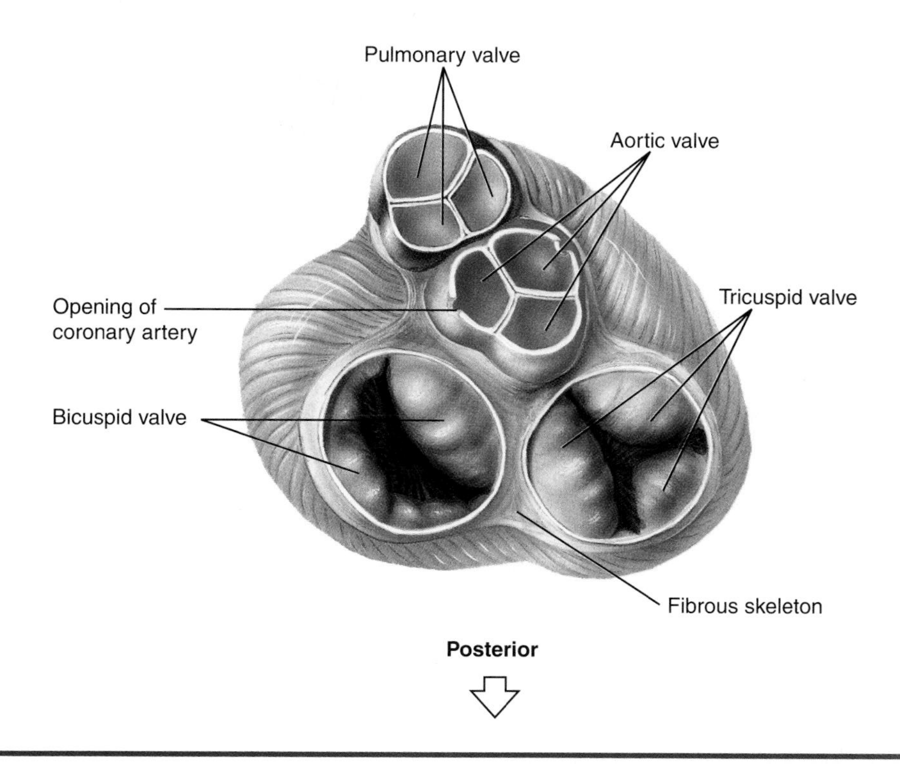

Figure 12.6 The valves of the heart as seen from above. Note the fibrous skeleton of the heart.

Heart Valves

Like all pumps, the heart contains valves that allow the blood to flow in only one direction through the heart. The two types of heart valves are atrioventricular (A-V) valves and semilunar valves. Observe the location and structure of the heart valves in figures 12.4 and 12.6.

Atrioventricular Valves

The opening between each atrium and its corresponding ventricle is guarded by an **atrioventricular** (ā-trē-ō-ven-trik′-ū-lar) **valve** that is formed of fibrous connective tissue. Each valve allows blood to flow from the atrium into the ventricle but prevents a backflow of blood from the ventricle into the atrium. The A-V valve between the right atrium and the right ventricle is the **tricuspid** (trī-kus′-pid) **valve.** Its name indicates that it is composed of three cusps, or flaps, of tissue. The **bicuspid valve,** or **mitral** (mī′-tral) **valve,** consists of two cusps and is located between the left atrium and the left ventricle.

The A-V valves originate from rings of thick, fibrous connective tissue that support the junction of the ventricles with the atria and the large arteries attached to the ventricles. This supporting fibrous tissue is sometimes called the *fibrous skeleton* of the heart (figure 12.6).

Thin strands of fibrous connective tissue, the *chordae tendineae* (kor′-dē ten′-di-nē), extend from the valve cusps to the *papillary muscles,* small mounds of cardiac muscle that project from the inner walls of the ventricles (see figure 12.4). The chordae tendineae prevent the valve cusps from being forced into the atria during ventricular contraction. In fact, they are normally just the right length to allow the cusps to press against each other and tightly close the opening during ventricular contraction.

Semilunar Valves

The **semilunar valves** are located in the bases of the large arteries that carry blood from the ventricles. The *pulmonary semilunar valve* is located at the base of the pulmonary trunk, which extends from the right ventricle. The *aortic semilunar valve* is located at the base of the aorta, which extends from the left ventricle.

Each semilunar valve is composed of three pocketlike cusps of fibrous connective tissue. They allow blood to be pumped from the ventricles into the arteries during ventricular contraction, but they prevent a backflow of blood from the arteries into the ventricles during ventricular relaxation.

What are the names and functions of the heart chambers?
What are the names and functions of the heart valves?

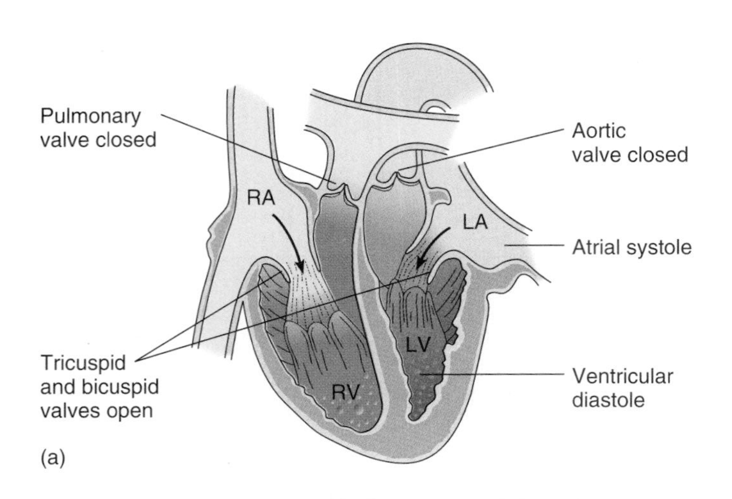

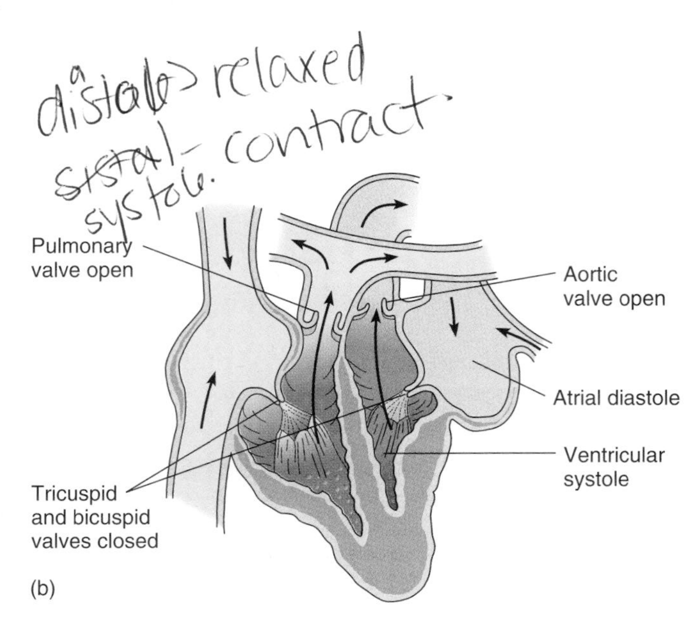

Figure 12.7 The cardiac cycle. (*a*) Blood flows from the atria into the ventricles during ventricular diastole. (*b*) Blood is pumped from the ventricles during ventricular systole.

Heart valves may become "leaky" due to the formation of scar tissue or lengthening of the chordae tendineae. Leaky valves make the heart work harder and may cause it to become weakened. If necessary, a malfunctioning valve may be surgically replaced with an artificial valve.

Cardiac Cycle

The cardiac cycle includes both contraction and relaxation phases. Contraction is known as **systole** (sis′-to-lē); relaxation is called **diastole** (dī′-as-to-lē). These phases are illustrated in figure 12.7. Note that the ventricles are relaxed when the atria contract, and the atria are relaxed when the ventricles contract.

When both the atria and ventricles are relaxed between beats, blood flows into the atria from the large veins leading to the heart and on into the ventricles. Then, the atria contract (atrial systole), forcing more blood into the ventricles so that they are filled. Immediately thereafter, the ventricles contract. Ventricular systole suddenly produces high blood pressure within the ventricles, and this blood pressure closes both atrioventricular valves and opens both semilunar valves as blood is pumped into the arteries leading from the heart. Ventricular diastole immediately follows, enabling the atrioventricular valves to open. Simultaneously, the semilunar valves close because of the greater blood pressure within the arteries. The cardiac cycle is then repeated. Study these relationships in figure 12.7.

Heart Sounds

The sounds of the heartbeat are usually described as *lub-dup* (pause) *lub-dup,* and so forth. These sounds are produced by the closing of the heart valves. The first sound results from the closing of the atrioventricular valves during ventricular systole. The second sound results from the closing of the semilunar valves during ventricular diastole. If any of the heart valves are defective and do not close properly, an additional sound, known as a heart murmur, may be heard.

Flow of Blood Through the Heart

Figure 12.8 diagrammatically shows the flow of blood through the heart and the great vessels attached to the heart. Chambers and vessels colored blue carry deoxygenated blood; those colored red carry oxygenated blood. Blood is oxygenated as it flows through the lungs, and it becomes deoxygenated as it releases oxygen to body tissues. Trace the flow of blood through the heart and major vessels in figure 12.8 as you read the following description.

The right atrium receives deoxygenated blood from all parts of the body except the lungs via two large veins: the superior and inferior venae cavae. The **superior vena cava** (vē′-nah kā′-vah) returns blood from the head, neck, shoulders, and arms. The **inferior vena cava** returns blood from body regions below the heart. Simultaneously, the left atrium receives oxygenated blood returning to the heart from the lungs via the **pulmonary veins.**

After blood has flowed from the atria into their respective ventricles, the ventricles contract. The right ventricle pumps deoxygenated blood into the

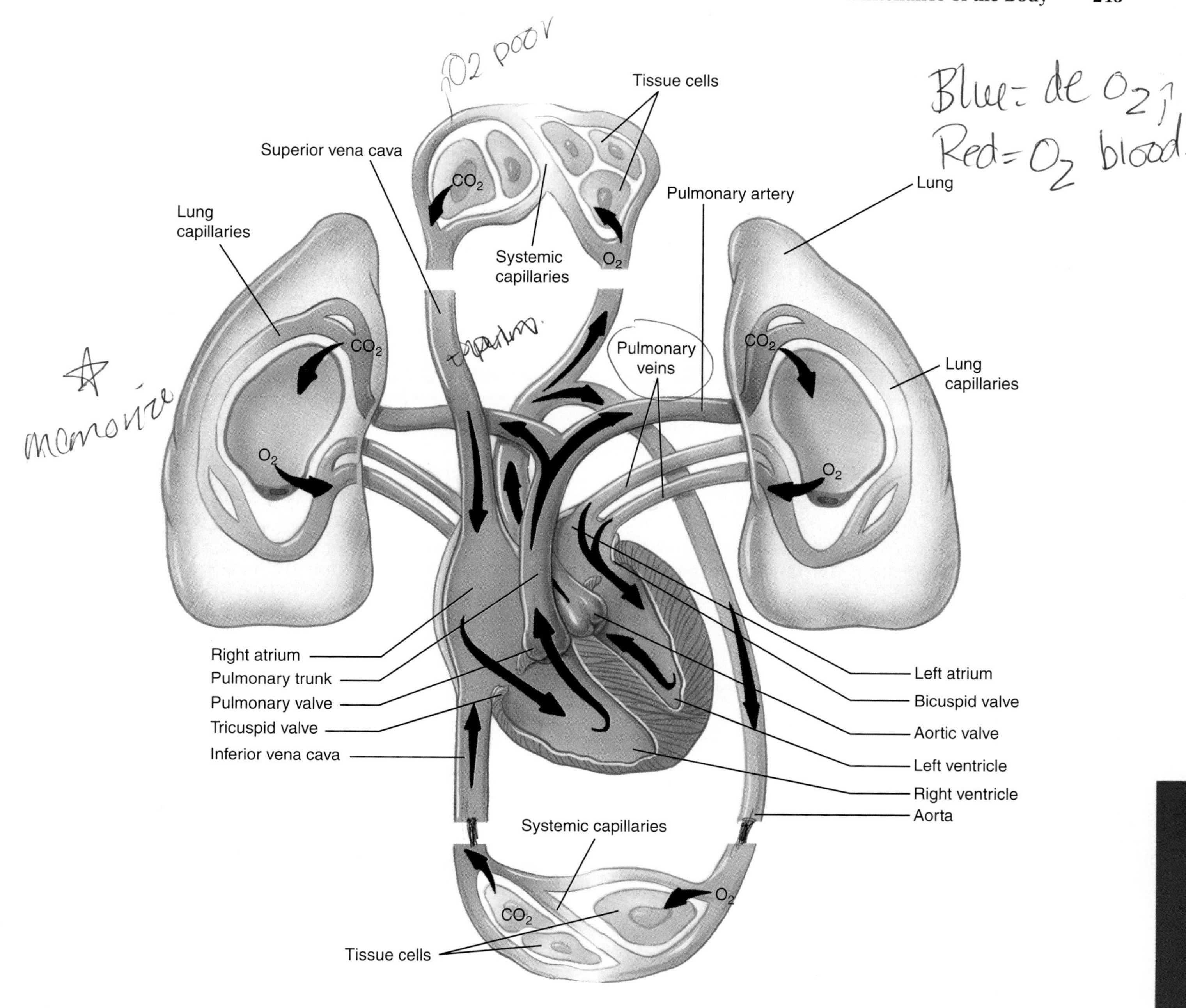

Figure 12.8 Circulation of blood through the heart and major blood vessels.

pulmonary trunk. The pulmonary trunk branches to form the **left** and **right pulmonary arteries,** which carry blood to the lungs. The left ventricle pumps oxygenated blood into the **aorta** (ā-or′-tah). The aorta branches to form smaller arteries that carry blood to all parts of the body except the lungs. Locate these major blood vessels associated with the heart in figure 12.5.

Since the heart is a double pump, there are two basic pathways, or circuits, of blood flow as shown in figure 12.8. The **pulmonary circuit** carries deoxygenated blood from the right ventricle to the lungs and returns oxygenated blood from the lungs to the left atrium. The **systemic circuit** carries oxygenated blood to all parts of the body except the lungs and returns deoxygenated blood to the right atrium.

Blood Supply to the Heart

The heart requires a constant supply of blood to nourish its own tissues. Blood is supplied by left and right **coronary** (kor′-ō-na-rē) **arteries,** which branch from the aorta just distal to the aortic semilunar valve. Blockage of a coronary artery is what causes a heart attack. After passing through capillaries in

If a coronary artery is partially obstructed by the fatty deposits of atherosclerosis, portions of the myocardium may be deprived of adequate blood. This produces chest pain known as *angina pectoris.* In severe cases, treatment may involve one of two approaches: coronary angioplasty or coronary bypass surgery.

In *coronary angioplasty,* a catheter that contains a balloon at its tip is inserted into an artery of an arm or leg and is threaded into the affected coronary artery. The balloon is positioned at the obstruction and is inflated for a few seconds to compress the fatty deposit and enlarge the lumen of the affected coronary artery.

In *coronary bypass surgery,* a portion of a vein from elsewhere in the body is removed and is surgically grafted, providing a bypass around the obstruction to supply blood to the distal portion of the affected coronary artery.

cardiac tissues, blood is returned via **cardiac** (kar′-dē-ak) **veins,** which lie next to the coronary arteries. These veins empty into the **coronary sinus,** which drains into the right atrium. Locate these blood vessels in figure 12.5 and note the fat deposits that lie alongside the vessels. Also, study the relationships of the atria, ventricles, and large blood vessels associated with the heart.

What are the events of a cardiac cycle?
What produces the heart sounds?
Trace a drop of blood as it flows through the heart and the pulmonary and systemic circuits.

Heart Conduction System

The heart contains specialized muscle tissue that functions somewhat like neural tissue. It spontaneously forms impulses and transmits them to the myocardium to initiate contraction. This specialized tissue forms the **conduction system** of the heart, which consists of the sinoatrial (S-A) node, atrioventricular (A-V) node, A-V bundle, and Purkinje fibers. Observe the location of the conduction system and its parts in figure 12.9.

The **sinoatrial** (sī-nō-ā′-trē-al) **node,** the pacemaker of the heart, is located in the right atrium at the junction of the superior vena cava. It rhythmically forms impulses to initiate each heartbeat. The impulses are transmitted to the myocardium of the atria, where they produce a simultaneous contraction of the atria. At the same time, the impulses are carried to the **atrioventricular node,** which is located in the right atrium near the junction with the ventricular septum.

There is a brief time delay as the impulses pass through the A-V node, which allows time for the ventricles to fill with blood. From the A-V node, the impulses pass along the **A-V bundle** (bundle of His), a group of large fibers that divide into left and right branches extending down the ventricular septum and up the lateral walls of the ventricles. The smaller **Purkinje fibers** arise from the branches of the A-V

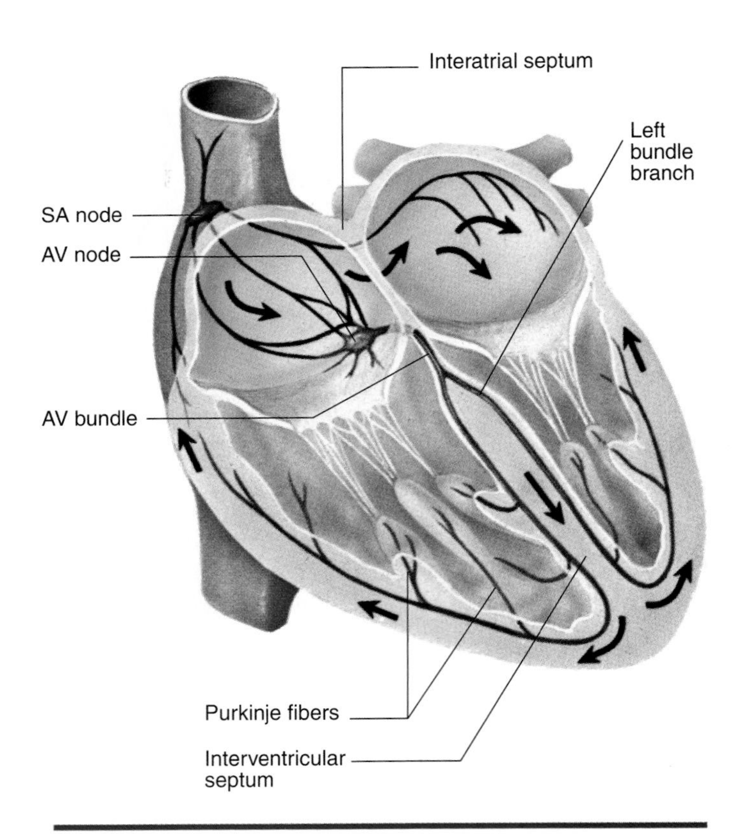

Figure 12.9 The heart conduction system. *Arrows* indicate the flow of impulses from the SA node.

Some irregularities in heart rhythms result from improper transmission of impulses by the heart conduction system. In patients where the S-A node or A-V node malfunction, a normal heartbeat may be obtained by implanting an artificial *pacemaker* in the chest wall. This battery-operated device synchronizes heart contractions and controls the heart rate by sending weak electrical pulses to the heart to initiate contraction.

bundle and carry the impulses to the myocardium of the ventricles, where they stimulate ventricular contraction. The distribution of the A-V bundle and the Purkinje fibers causes the ventricles to contract from the apex upward so that blood is forced into the pulmonary trunk and aorta.

Electrocardiogram

The origination and transmission of impulses through the conduction system of the heart generates electrical currents that may be detected by electrodes placed on the body surface. An instrument called an *electrocardiograph* is used to transform the electrical currents picked up by the electrodes into a recording called an **electrocardiogram** (**ECG** or **EKG**).

Figure 12.10 shows a normal EKG of five cardiac cycles, and figure 12.11 shows an enlargement of a normal EKG of one cardiac cycle. Note that an EKG consists of several deflections, or waves. These waves correlate with the flow of impulses during particular portions of the cardiac cycle.

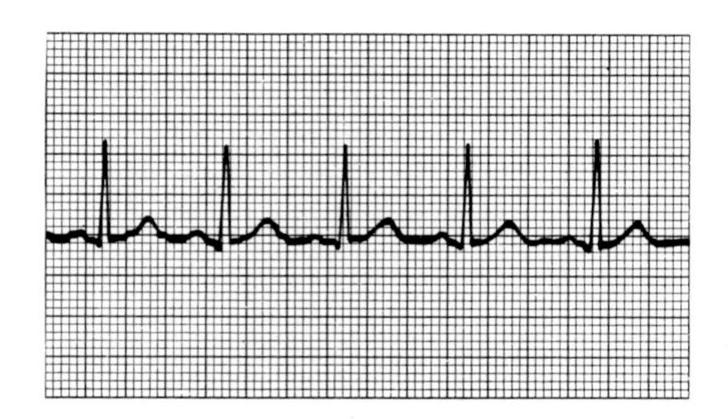

Figure 12.10 A normal electrocardiogram showing five cardiac cycles.

An electrocardiogram has three distinct waves: the P wave, QRS wave, and T wave. The *P wave* is a small upward wave. It is produced by the depolarization of the atria. The *QRS wave* is produced by the depolarization of the ventricles. The greater size of the QRS wave is due to the greater muscle mass of the ventricles. The last wave is the *T wave,* which is produced by the repolarization of the ventricular myocardium. The repolarization of the atria is not detected because it is masked by the stronger QRS wave.

What composes the cardiac conduction system?
What is the basis of an electrocardiogram?
What events produce the waves of an electrocardiogram?

Regulation of Heart Rate

Without regulating factors, the S-A node produces a rhythmic, but constant, heart rate. However, there are a number of factors that act on the S-A node to increase or decrease the heart rate in response to body needs.

Autonomic Regulation

Heart rate regulation is primarily under the control of the **cardiac center** located within the medulla

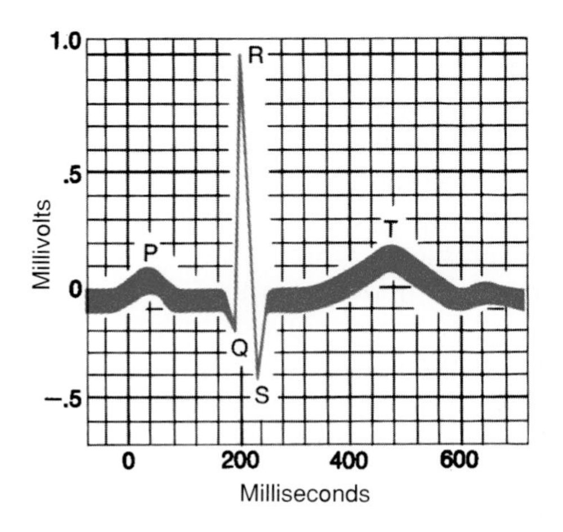

Figure 12.11 A normal electrocardiogram. The P wave results from depolarization of the atria. The QRS wave results from the depolarization of the ventricles. The T wave results from the repolarization of the ventricles.

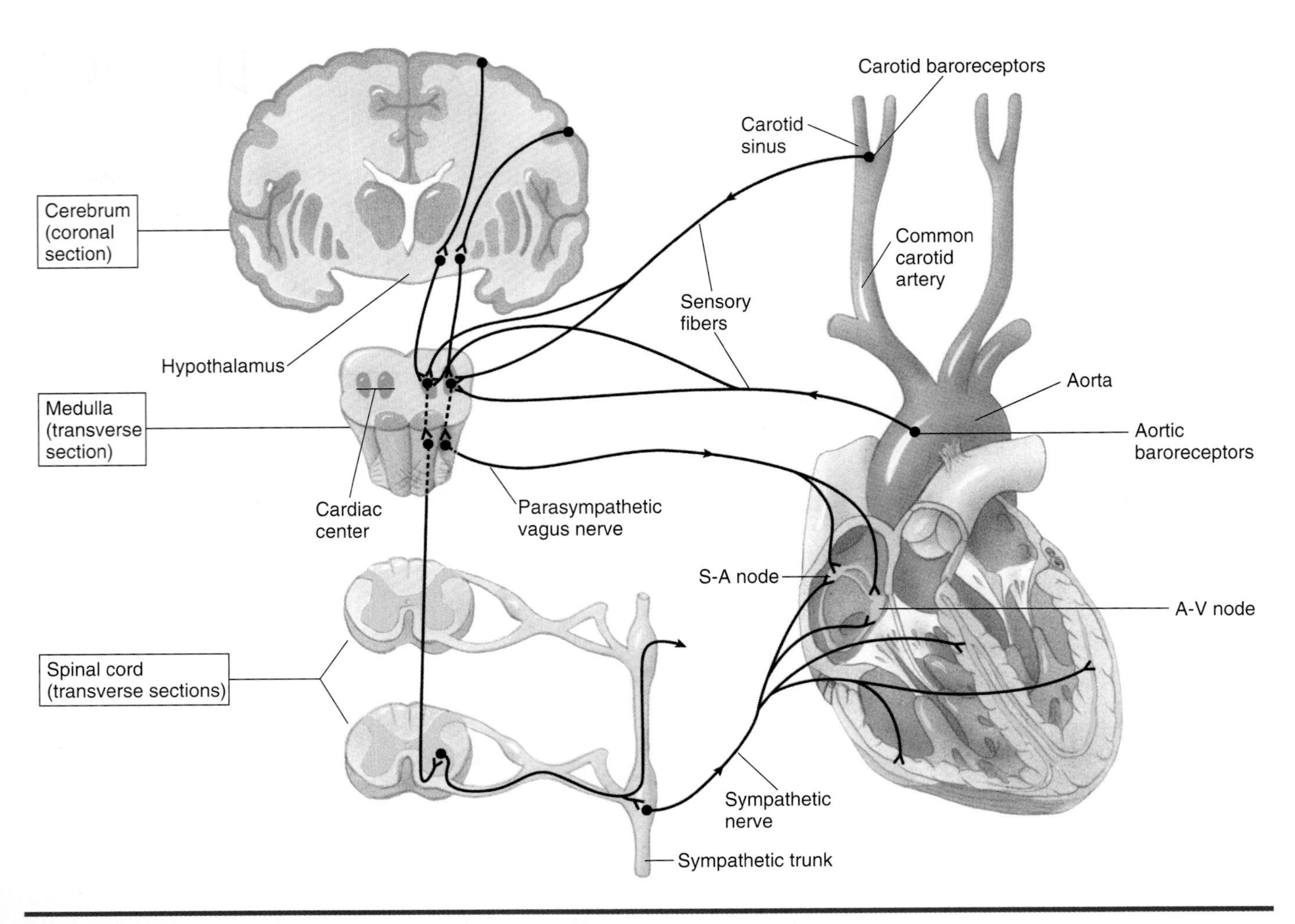

Figure 12.12 The rate and strength of heart contractions are regulated by the antagonistic actions of sympathetic and parasympathetic divisions of the autonomic nervous system.

oblongata of the brain. It receives sensory information about the level of blood pressure from pressure receptors (baroreceptors) located in the aortic arch and in the carotid sinuses of the carotid arteries. It is also affected by emotions, which generate impulses from the cerebrum and hypothalamus.

The cardiac center consists of both sympathetic and parasympathetic components. Impulses transmitted to the heart via sympathetic fibers cause an increase in heart rate, while impulses transmitted by parasympathetic fibers cause a decrease in heart rate. The cardiac center constantly adjusts the frequency of sympathetic and parasympathetic impulses to produce a heart rate that meets the changing needs of body cells (figure 12.12).

Neurons of the sympathetic division extend from the cardiac center down the spinal cord to the thoracic region. There the sympathetic fibers exit the spinal cord via thoracic nerves to innervate the S-A node, A-V node, and portions of the myocardium. The transmission of impulses causes the sympathetic fibers to secrete *norepinephrine* at synapses in the heart. Norepinephrine increases the heart rate and strengthens the force of myocardial contraction. Physical and emotional stresses, such as exercise, excitement, anxiety, and fear, stimulate sympathetic fibers to increase the heart rate.

Parasympathetic nerve fibers arise from the cardiac center and exit in the vagus nerve (X) to innervate the S-A and A-V nodes. The transmission of impulses causes the parasympathetic fibers to secrete *acetylcholine* at the heart synapses, which slows the heart rate. The greater the frequency of parasympathetic impulses sent to the heart, the slower the heart rate. Excessive blood pressure and emotional factors, such as grief and depression, stimulate the parasympathetic nerve fibers to slow the heart rate.

There is a predominance of parasympathetic impulses reaching the heart when the body is at rest, and

Table 12.2 Comparison of Arteries, Capillaries, and Veins

Type of Vessel	Function	Structure
Arteries	Carry blood from the heart to the capillaries	Relatively thick muscular layer to withstand higher blood pressure
Capillaries	Enable exchange of materials between blood and tissue cells	Microscopic vessels composed of a single layer of endothelial cells
Veins	Return blood from the capillaries to the heart	Relatively thin muscular layer; internal valves located at intervals to prevent backflow of blood

they slow the heart rate to produce the resting heart rate. As cellular needs for blood increase, a decrease in the frequency of parasympathetic impulses and an increase in sympathetic impulses cause the heart rate to increase.

Other Factors Affecting Heart Rate

Age, sex, physical condition, temperature, epinephrine, thyroxine, and the blood levels of calcium and potassium ions also affect the heart rate.

The resting heart rate gradually declines with age, and it is slightly faster in females than in males. Average resting heart rates in females are 72 to 80 beats per minute, as opposed to 64 to 72 beats per minute in males. People who are in good physical condition have a slower resting heart rate than those in poor condition. Athletes may have a resting heart rate of only 40 to 60 beats per minute. An increase in body temperature, which occurs during exercise or when feverish, increases the heart rate.

Epinephrine, which is secreted by the adrenal glands during stress or excitement, affects the heart like norepinephrine—it increases and strengthens the heart rate. An excess of thyroxine produces a lesser, but longer-lasting, increase in heart rate.

Reduced levels of Ca^{++} slow the heart rate, while increased Ca^{++} levels increase the heart rate and prolong contraction. In extreme cases, an excessively prolonged contraction may result in death. Excessive levels of K^+ decrease both heart rate and force of contraction. Abnormally low levels of K^+ may cause potentially life-threatening abnormal heart rhythms.

How is the heart rate regulated?
What other factors affect the heart rate?

An electrocardiogram provides important information in the diagnosis of heart disease and abnormalities. In reading an EKG, physicians pay close attention to the height of each wave and to the time required for each wave.

Types of Blood Vessels

There are three basic types of blood vessels: arteries, capillaries, and veins. They form a closed system of tubes that carry blood from the heart to the body cells and back to the heart. Table 12.2 compares these three types.

Structure of Arteries and Veins

The walls of arteries and veins are composed of three distinct layers. The *tunica externa*, the outer layer, is formed of fibrous connective tissue that includes both collagenous and elastic fibers. These fibers provide support and elasticity for the vessel. The *tunica media,* the middle layer, usually is the thickest layer. It consists of smooth muscle fibers that encircle the blood vessel. The smooth muscle fibers not only provide support but also produce changes in the diameter of the blood vessel by contraction or relaxation. The *tunica interna,* the inner layer, forms the inner lining of blood vessels. It consists of a single layer of squamous epithelial cells, called the *endothelium.* It is supported by thin layers of connective tissue that lie between it and the smooth muscle fibers.

The walls of arteries and veins have the same basic structure. However, arterial walls contain more

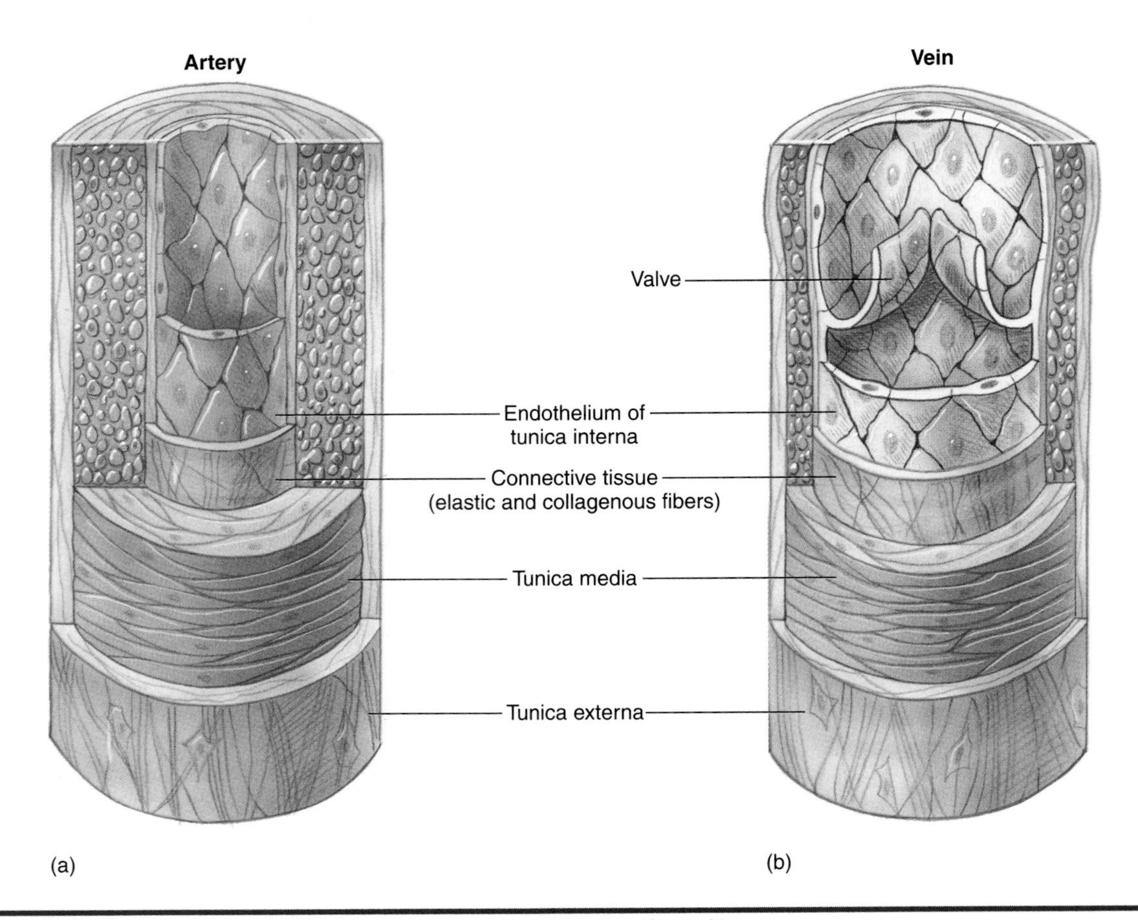

Figure 12.13 (*a*) The wall of an artery. (*b*) The wall of a vein.

smooth muscle and elastic connective tissues as an adaptation to the higher blood pressure found in them. Another difference is that large veins, but not arteries, contain valves formed of endothelium. Venous valves prevent a backflow of blood. Compare the structure of arteries and veins in figure 12.13.

Arteries

Arteries carry blood away from the heart. They branch repeatedly into smaller and smaller arteries and ultimately form microscopic arteries called **arterioles** (ar-te′-rē-ōls). As arterioles branch and form smaller arterioles, the thickness of the muscle layer decreases. The walls of the smallest arterioles consist of only endothelium and a few encircling smooth muscle fibers, as shown in figure 12.14. Arteries, especially the arterioles, play an important role in the control of blood flow and blood pressure.

Capillaries

Arterioles connect with **capillaries,** the most numerous and the smallest blood vessels. A capillary's diameter is so small that red blood cells must pass through it in single file. The wall of capillaries consists only of endothelium, which allows the exchange of materials between blood in capillaries and body cells.

The distribution of capillaries in body tissues varies with the metabolic activity of each tissue. Capillaries are especially abundant in active tissues, such

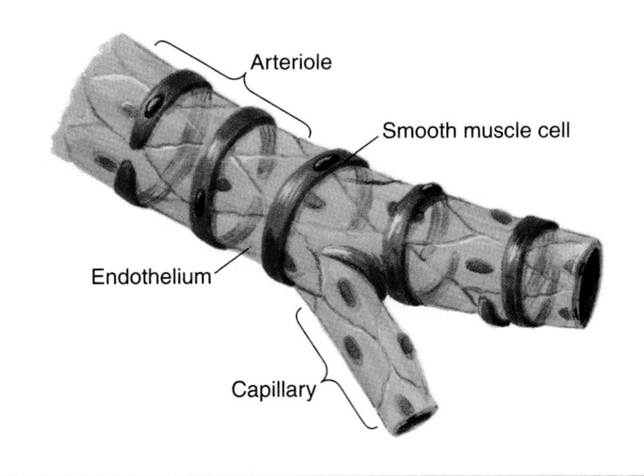

Figure 12.14 Small arterioles have smooth muscle fibers in their walls. Capillaries lack these fibers.

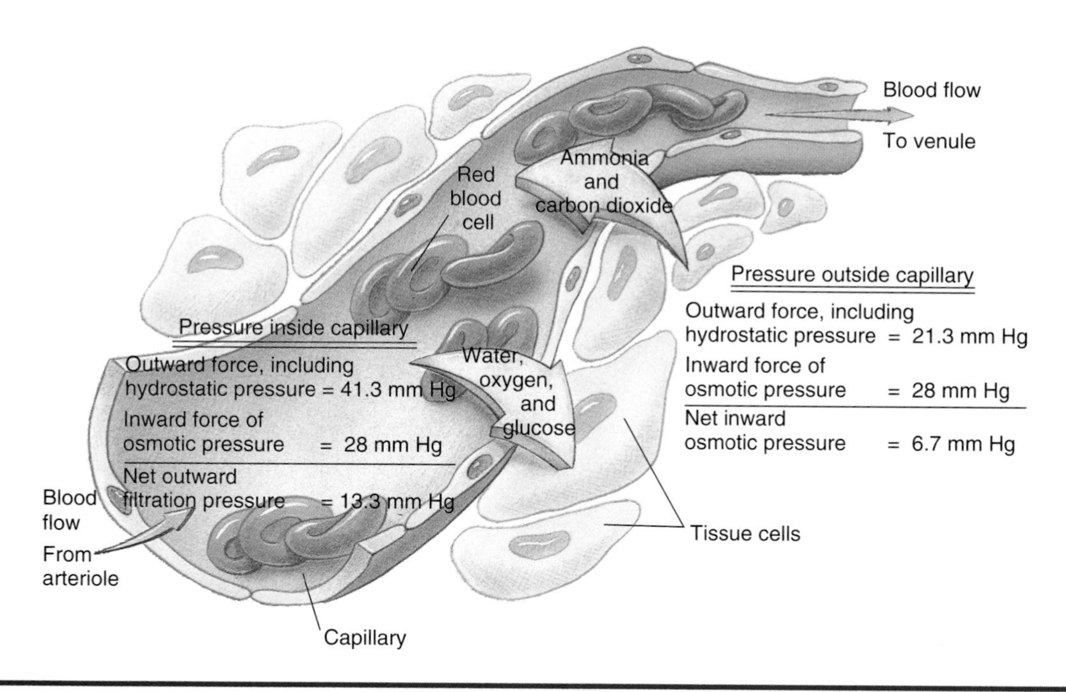

Figure 12.15 Water and other substances leave capillaries because of a net outward filtration pressure at capillaries' arteriolar ends. Water enters at capillaries' venule ends because of a net inward force of osmotic pressure. Substances move in and out along the length of capillaries according to their respective concentration gradients.

as muscle and nerve tissues, where nearly every cell is near a capillary. Capillaries are less abundant in connective tissues, and they are absent in some tissues, such as cartilage, epidermis, and the lens and cornea of the eye.

Blood flow in capillaries is controlled by **precapillary sphincters,** smooth muscle fibers encircling the bases of capillaries at the arteriole-capillary junctions. Contraction of a precapillary sphincter inhibits blood flow to its capillary network. Relaxation of the sphincter allows blood to flow into its capillary network to provide oxygen and nutrients for the tissue cells. The flow of blood in capillary networks occurs intermittently. When some capillary networks are filled with blood, others are not. Capillary networks receive blood according to the needs of the cells that they serve. For example, during physical exercise blood is diverted from capillary networks in the digestive tract to fill the capillary networks in skeletal muscles. This pattern of blood distribution is largely reversed after a meal.

Exchange of Materials

The continual exchange of materials between the blood and tissue cells is essential for life. Cells require oxygen and nutrients to perform their metabolic functions, and they produce carbon dioxide and other metabolic wastes that must be removed by the blood.

The cells of tissues are enveloped in a thin film of extracellular fluid called **interstitial fluid,** or **tissue fluid,** that fills the tissue spaces and lies between the tissue cells and the capillaries. Therefore, all materials that pass between the blood and tissue cells must pass through the interstitial fluid.

The exchange of materials occurs by *diffusion, osmosis,* and *filtration.* Dissolved substances, which can pass through capillary walls, diffuse either in or out of a capillary from the area of their higher concentration to an area of their lower concentration. Osmosis of water works the same way. Filtration is different. It is the forcing of some of the water and dissolved substances through capillary walls by blood pressure.

Filtration and osmotic pressure are major forces in the exchange of materials. At the arteriole end of a capillary, blood (hydrostatic) pressure exceeds the osmotic pressure of the blood (figure 12.15). Therefore, some of the water and dissolved substances are forced out of the capillary and into the tissue fluid by filtration. This loss of water from the capillary decreases blood pressure as blood flows through a capillary. Simultaneously, the retention of plasma proteins, which are too large to pass through a capillary wall, increases the osmotic pressure of the blood. At the venule end of the capillary, osmotic pressure exceeds blood pressure. Therefore, water is "pulled" into the capillary by osmosis, and dissolved substances follow by diffusion.

Veins

After blood flows through the capillaries, it enters the **venules,** the smallest **veins.** Several capillaries merge to form a venule. The smallest venules consist only of endothelium and connective tissue, but larger venules also contain smooth muscle tissue. Venules unite to form small veins. Small veins combine to form progressively larger veins as blood is returned to the heart. Larger veins, especially those in the legs and arms, contain valves that prevent a backflow of blood and aid the return of blood to the heart.

Since nearly 60% of the blood volume is in veins at any instant, veins may be considered as storage areas for blood that can be carried to other parts of the body in times of need. Venous sinusoids in the liver and spleen are especially important reservoirs. If blood is lost by hemorrhage, both blood volume and pressure decline. In response, the sympathetic nervous system sends impulses to constrict the muscular walls of the veins, which reduces the venous volume and compensates for the blood loss. A similar response occurs during strenuous muscular activity in order to increase the blood flow to skeletal muscles.

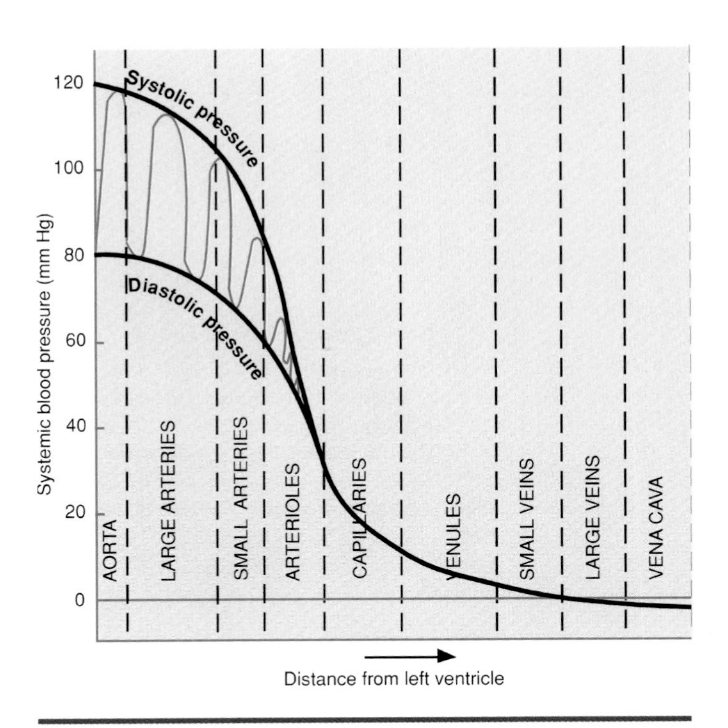

Figure 12.16 Blood pressure decreases as distance from the left ventricle increases.

> ***Compare the structure and function of arteries, capillaries, and veins.***
> ***How does the exchange of materials occur between blood in capillaries and tissue cells?***

Blood Flow

Blood circulates because of differences in blood pressure. Blood flows from areas of higher pressure to areas of lower pressure. Blood pressure is greatest in the ventricles and lowest in the atria. Figure 12.16 shows the decline of blood pressure in the systemic circuit with increased distance from the left ventricle.

Contraction of the ventricles creates the blood pressure that propels the blood through the arteries. However, the pressure declines as the arteries branch into an increasing number of smaller and smaller arteries and finally connect with the capillaries. The decline in blood pressure occurs because the overall cross-sectional area of the combined arteries is greatly increased as the arteries continue to branch. By the time blood has left the capillaries and entered the veins, there is very little blood pressure remaining to return the blood to the heart. The return of venous blood is assisted by two additional forces: *skeletal muscle contractions* and *respiratory movements.*

Contractions of skeletal muscles compress the veins, forcing blood from one valved segment to another and on toward the heart since the valves prevent a backflow of blood. This method of moving venous blood toward the heart is especially important in the return of blood from the arms and legs, and it is illustrated in figure 12.17.

Respiratory movements aid the movement of blood upwards toward the heart in the abdominal and thoracic cavities. The downward contraction of the diaphragm during inspiration decreases the pressure within the thoracic cavity and increases the pressure within the abdominal cavity. The higher pressure on abdominal veins forces blood upwards into thoracic veins, where the pressure is reduced since valves in leg veins prevent a backflow of blood.

Velocity of Blood Flow

The velocity of blood flow varies inversely with the overall cross-sectional area of the combined blood vessels. Therefore, the velocity progressively decreases as blood flows through an increasing number of smaller and smaller arteries and into the capillaries. Then, the velocity progressively increases as the blood flows into a decreasing number of larger and larger veins on its way back to the heart.

Blood flow is fastest in the aorta and slowest in the capillaries, an ideal situation providing for the rapid circulation of the blood and yet sufficient time for the exchange of materials between blood in the capillaries and tissue cells.

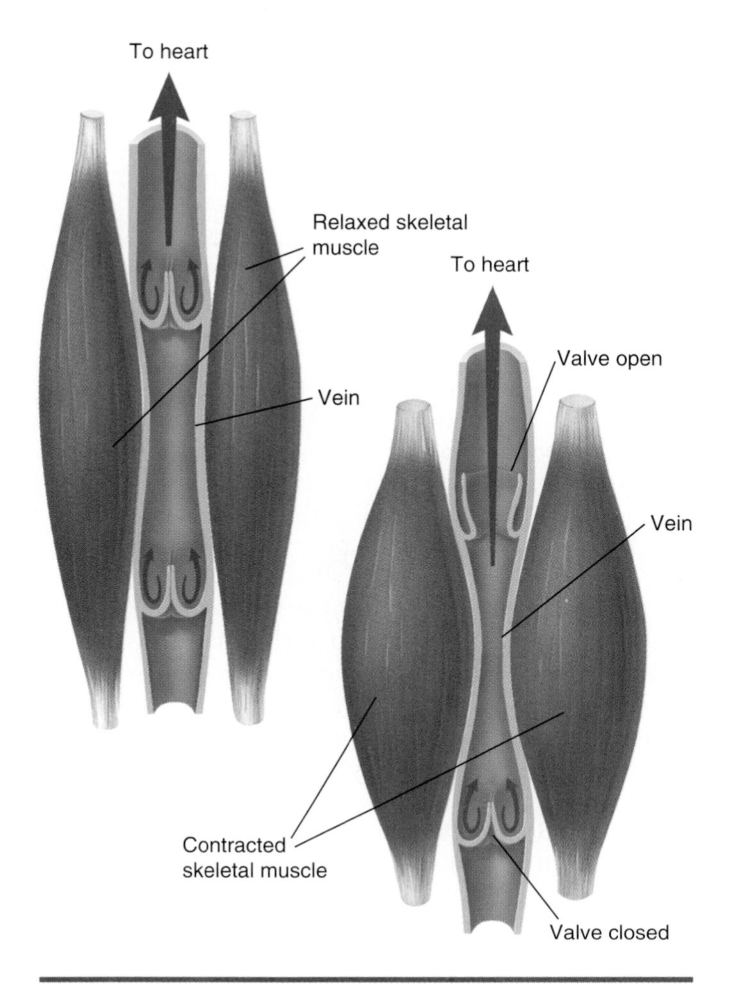

Figure 12.17 Contraction of skeletal muscles compresses veins and aids the movement of blood toward the heart.

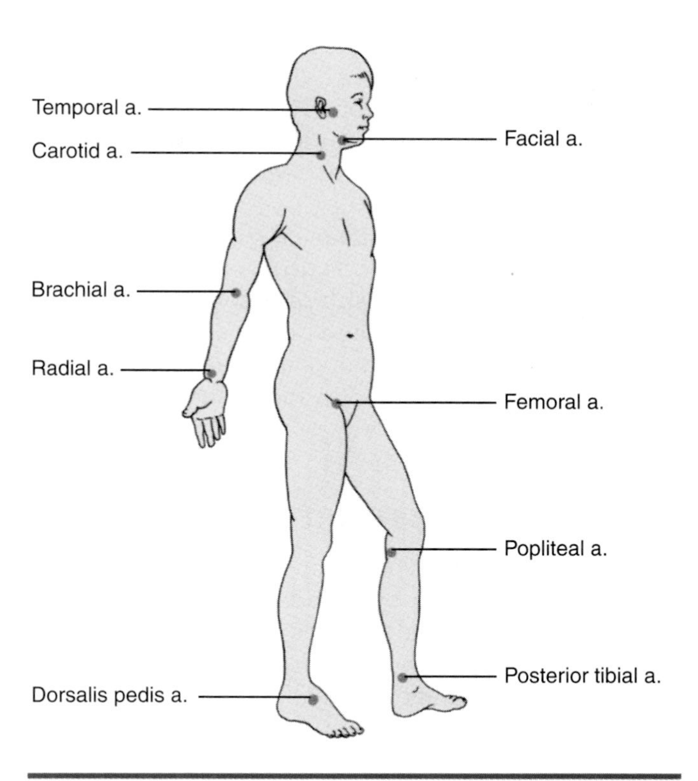

Figure 12.18 Locations and arteries where the pulse may be detected. (a. = artery)

Blood Pressure

The term *blood pressure* usually refers to arterial blood pressure in the systemic circuit—in the aorta and its branches. Arterial blood pressure is greatest during ventricular contraction (systole) as blood is pumped into the aorta and its branches. This pressure is called the **systolic blood pressure,** and it normally averages 120 ± 10 mmHg (mercury) when measured in the brachial artery. The lowest arterial pressure occurs during ventricular relaxation (diastole). This pressure is called the **diastolic blood pressure,** and it normally averages 80 ± 10 mmHg (see figure 12.16).

The difference between the systolic and diastolic blood pressures is known as the *pulse pressure.* The alternating increase and decrease in arterial blood pressure during ventricular systole and diastole causes a comparable expansion and contraction of the elastic arterial walls. This pulsating expansion of the arterial walls follows each ventricular contraction, and it may be detected as the *pulse* by placing the fingers on a surface artery. Figure 12.18 identifies the name and location of surface arteries where the pulse may be detected.

Factors Affecting Blood Pressure

Four major factors affect blood pressure: cardiac output, blood volume, peripheral resistance, and blood viscosity. An increase in any of these factors causes an increase in blood pressure.

Cardiac output is the volume of blood pumped by the heart in one minute. It is determined by the heart rate and the volume of blood pumped at each contraction. An increase or decrease in cardiac output causes a comparable change in blood pressure. Changes in cardiac output occur primarily by changes in heart rate.

Blood volume may be decreased by severe hemorrhage, vomiting, diarrhea, or reduced water intake. As soon as the lost fluid is replaced, blood pressure returns to normal. Conversely, if the body retains too much fluid, blood volume and blood pressure increase.

Peripheral resistance is the friction of blood against the walls of blood vessels. Arterioles play an important role in controlling blood pressure by changing their diameters, which alters peripheral resistance. As arterioles constrict, peripheral resistance and blood pressure increase. As arterioles dilate, peripheral resistance and blood pressure decrease.

Viscosity is the resistance of a liquid to flow. For example, water has a low viscosity, while honey has a high viscosity. The viscosity of blood is determined by the concentration of blood cells and plasma proteins, so an increase in either of these factors increases blood viscosity, which, in turn, increases blood pressure.

Control of Peripheral Resistance

The sympathetic nervous system primarily controls peripheral resistance by regulating the diameter of blood vessels, especially arterioles. The control center is the *vasomotor center* in the medulla. An increase in the frequency of sympathetic impulses to the smooth muscle of blood vessels produces *vasoconstriction,* which raises the blood pressure and increases blood velocity. This response accelerates the rate of oxygen transport to cells and the removal of carbon dioxide from blood by the lungs. A decrease in impulse frequency results in *vasodilation,* which lowers blood pressure and slows blood velocity.

Like the cardiac center, the activity of the vasomotor center is modified by epinephrine, impulses from higher brain areas, and sensory impulses from pressure receptors and chemoreceptors in the aorta and carotid arteries. For example, a decrease in pressure, pH, or oxygen concentration of the blood stimulates vasoconstriction. Conversely, an increase in these values promotes vasodilation.

In addition, arterioles and precapillary sphincters are affected by localized changes in blood concentrations of oxygen, carbon dioxide, and pH. These local effects override the control by the vasomotor center, a process called *autoregulation,* and increase the rate of exchange of materials between tissue cells and the capillaries. For example, if a particular muscle group is active for an extended period, a localized decrease in oxygen concentration and an increase in carbon dioxide concentration results. These chemical changes stimulate the dilation of local arterioles and precapillary sphincters, which increases the flow of blood into capillary networks of the affected muscles to provide more oxygen and to remove more carbon dioxide. Then, as soon as the blood leaves the localized area, it flows more rapidly to the heart and lungs in vessels that are constricted under the control of the vasomotor center.

How does blood pressure affect the flow of blood through blood vessels?
Compare systolic and diastolic blood pressure.
How do cardiac output, blood volume, and peripheral resistance affect blood pressure?

Circulation Pathways

As noted earlier, the heart is a double pump that serves two distinct circulation pathways: the pulmonary and systemic circuits. These circuits were shown earlier in figure 12.8.

Pulmonary Circuit

The **pulmonary circuit** carries deoxygenated blood to the lungs, where oxygen and carbon dioxide are exchanged between the blood and the air in the lungs. The right ventricle pumps deoxygenated blood into the **pulmonary trunk,** a short, thick artery that divides to form the left and right **pulmonary arteries.** Each pulmonary artery enters a lung and divides repeatedly to form arterioles, which continue into the capillary networks that surround the air sacs (alveoli) of the lungs. Oxygen diffuses from the air in the lungs into the capillary blood, while carbon dioxide diffuses from the blood into the air in the lungs. Blood then flows from the capillaries into venules, which merge to form small veins, which, in turn, join to form progressively larger veins. Two **pulmonary veins** emerge from each lung to carry oxygenated blood back to the left atrium of the heart.

Systemic Circuit

The **systemic circuit** carries oxygenated blood to the tissue cells of the body and returns deoxygenated blood to the heart. The left ventricle pumps the freshly oxygenated blood, received from the pulmonary circuit, into the **aorta** for circulation to all parts of the body except the lungs. The aorta branches to form many major arteries, which continually branch to form arterioles leading to capillaries, where the exchange of materials between the blood and tissue cells takes place. Oxygen diffuses from the capillary blood into the tissue cells, while carbon dioxide diffuses from the tissue cells into the blood. From the capillaries, blood enters venules, which merge to form small veins, which join to form progressively larger veins. Ultimately, veins from the arms, shoulders, head, and neck join to form the **superior vena cava,** which returns blood from these regions back to the right atrium. Similarly, veins from the legs and lower trunk enter the **inferior vena cava,** which also returns blood into the right atrium.

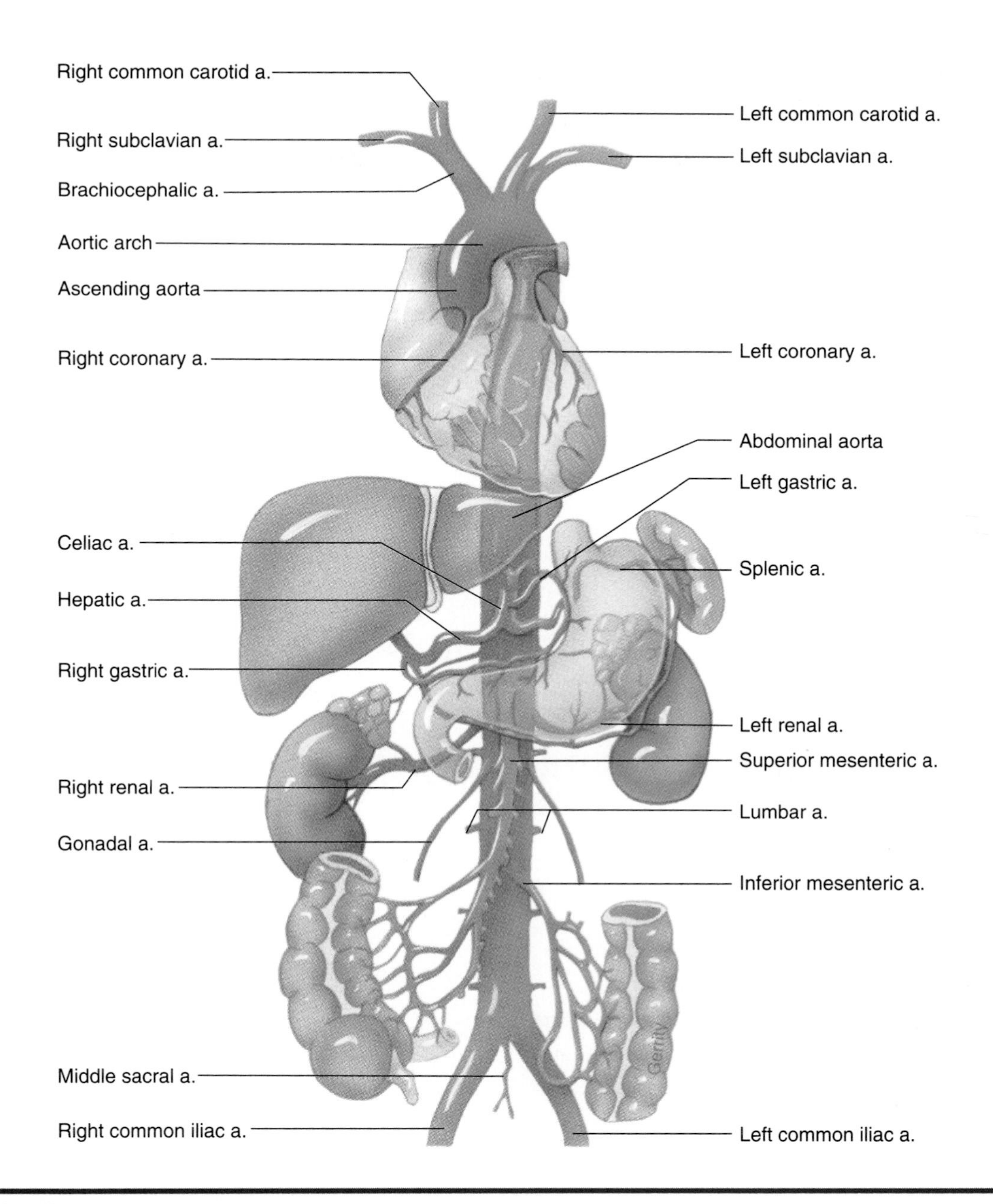

Figure 12.19 The major arteries that branch from the aorta. (a. = artery)

Systemic Arteries

All of the systemic arteries are branch arteries of the aorta, which carries oxygenated blood from the left ventricle of the heart.

Major Branches of the Aorta

The aorta ascends from the heart, arches to the left and behind the heart, and descends through the thorax and abdomen just anterior to the vertebral column. Figure 12.19 shows the major branches of the aorta and their relationships to the internal organs. Tables 12.3 and 12.4 list the major branches of the aorta and the organs and body regions that they supply.

The first arteries to branch from the aorta are the left and right **coronary arteries,** which supply blood to the heart. They branch from the aorta just distal to the aortic semilunar valve in the base of the *ascending aorta.*

Three major arteries branch from the *aortic arch.* In order of branching, they are the **brachiocephalic** (brāk-ē-ō-se-fal′-ik) **artery,** the **left common carotid** (kah-rot′-id) **artery,** and the **left subclavian** (sub-klā′-vē-an) **artery.**

Pairs of small **intercostal** (in-ter-kos′-tal) **arteries** branch from the *thoracic aorta* to supply the intercostal

Table 12.3 Major Arteries Branching from the Ascending Aorta, Aortic Arch, and Thoracic Aorta

Artery	Origin	Region Supplied
Coronary	Ascending aorta	Heart
Brachiocephalic	Aortic arch	Head and right arm
Right common carotid	Brachiocephalic	Right side of head and neck
Right subclavian	Brachiocephalic	Right shoulder and arm
Left common carotid	Aortic arch	Left side of head and neck
External carotid	Common carotid	Scalp, face, and neck
Internal carotid	Common carotid	Brain
Left subclavian	Aortic arch	Left shoulder and arm
Vertebral	Subclavian	Neck and brain
Axillary	Subclavian	Armpit
Brachial	Axillary	Upper arm
Radial	Brachial	Forearm and hand
Ulnar	Brachial	Forearm and hand
Intercostal	Thoracic aorta	Thoracic wall

Table 12.4 Major Arteries Branching from the Abdominal Aorta

Artery	Origin	Region Supplied
Celiac	Abdominal aorta	Liver, stomach, spleen
Hepatic	Celiac	Liver
Gastric	Celiac	Stomach
Splenic	Celiac	Spleen
Renal	Abdominal aorta	Kidney
Superior mesenteric	Abdominal aorta	Small intestine; part of colon
Gonadal	Abdominal aorta	Ovaries or testes
Lumbar	Abdominal aorta	Lumbar region of back
Inferior mesenteric	Abdominal aorta	Colon and rectum
Common iliac	Abdominal aorta	Pelvic region and leg
Internal iliac	Common iliac	Reproductive and urinary organs; rectum
External iliac	Common iliac	Pelvic region and leg
Femoral	External iliac	Thigh
Popliteal	Femoral	Knee
Anterior tibial	Femoral	Lower leg (anterior) and foot
Posterior tibial	Femoral	Lower leg (posterior) and foot

muscles between the ribs and other organs of the thoracic wall. A number of other small arteries supply the organs of the thorax.

Once the aorta descends through the diaphragm, it is called the *abdominal aorta,* and it gives off several branch arteries to the abdominal wall and visceral organs. The **celiac** (sē′-lē-ak) **artery** is a short artery that divides to form three branch arteries: (1) the **gastric artery** supplies the stomach, (2) the **splenic artery** supplies the spleen and pancreas, and (3) the **hepatic artery** supplies the liver.

The **superior mesenteric** (mes-en-ter′-ik) **artery** supplies most of the small intestine and the first portion of the large intestine. The left and right **renal ar-**

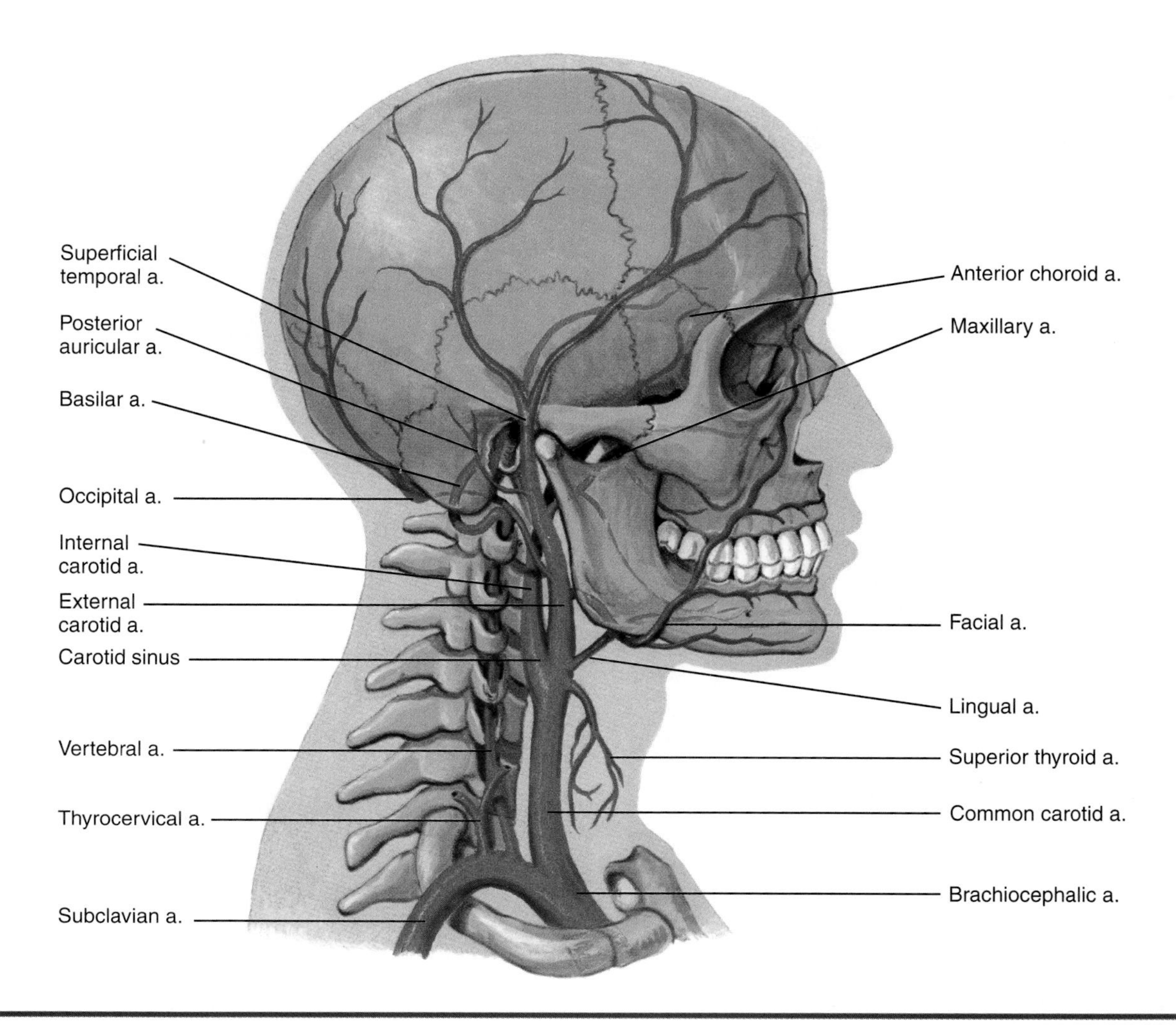

Figure 12.20 Major arteries supplying the head and neck. (a. = artery)

teries supply the kidneys. The left and right **gonadal** (gō-nad′-al) **arteries** supply the ovaries in females and the testes in males.

Several pairs of **lumbar arteries** supply the walls of the abdomen and back. The **inferior mesenteric artery** supplies the distal half of the large intestine.

At the level of the iliac crests, the aorta divides to form two large arteries, the left and right **common iliac** (il′-ē-ak) **arteries,** which carry blood to the lower portions of the trunk and to the legs.

Arteries Supplying the Head and Neck

The head and neck receive blood from several arteries that branch from the carotid and subclavian arteries. Note in figures 12.19 and 12.20 that the brachiocephalic artery branches to form the **right common carotid artery** and the **right subclavian artery.** The left common carotid and left subclavian arteries branch directly from the aorta.

Each common carotid artery divides in the neck to form an **external carotid artery** and the **internal carotid artery.** The slight enlargement at the junction of these arteries is a *carotid sinus,* the site of pressure receptors and chemoreceptors that send sensory impulses to the cardiac and vasomotor centers in the medulla. The external carotids give rise to a number of smaller arteries that carry blood to the neck, face, and scalp. The internal carotids enter the cranium through foramina in the temporal bones and provide the major supply of blood to the brain.

Two pairs of smaller arteries branch from each subclavian artery to serve the head and neck. The **vertebral arteries** pass upwards through foramina in transverse processes of the cervical vertebrae, enter the cranium, and merge at the base of the brain to form the **basilar artery.** The basilar artery branches to form two **posterior cerebral arteries,** which are part of an arterial circle known as the **circle of Willis** (figure 12.21). The **anterior cerebral arteries,** which branch from the internal carotids, form the anterior portion of this circle. The circle of Willis is an example of an

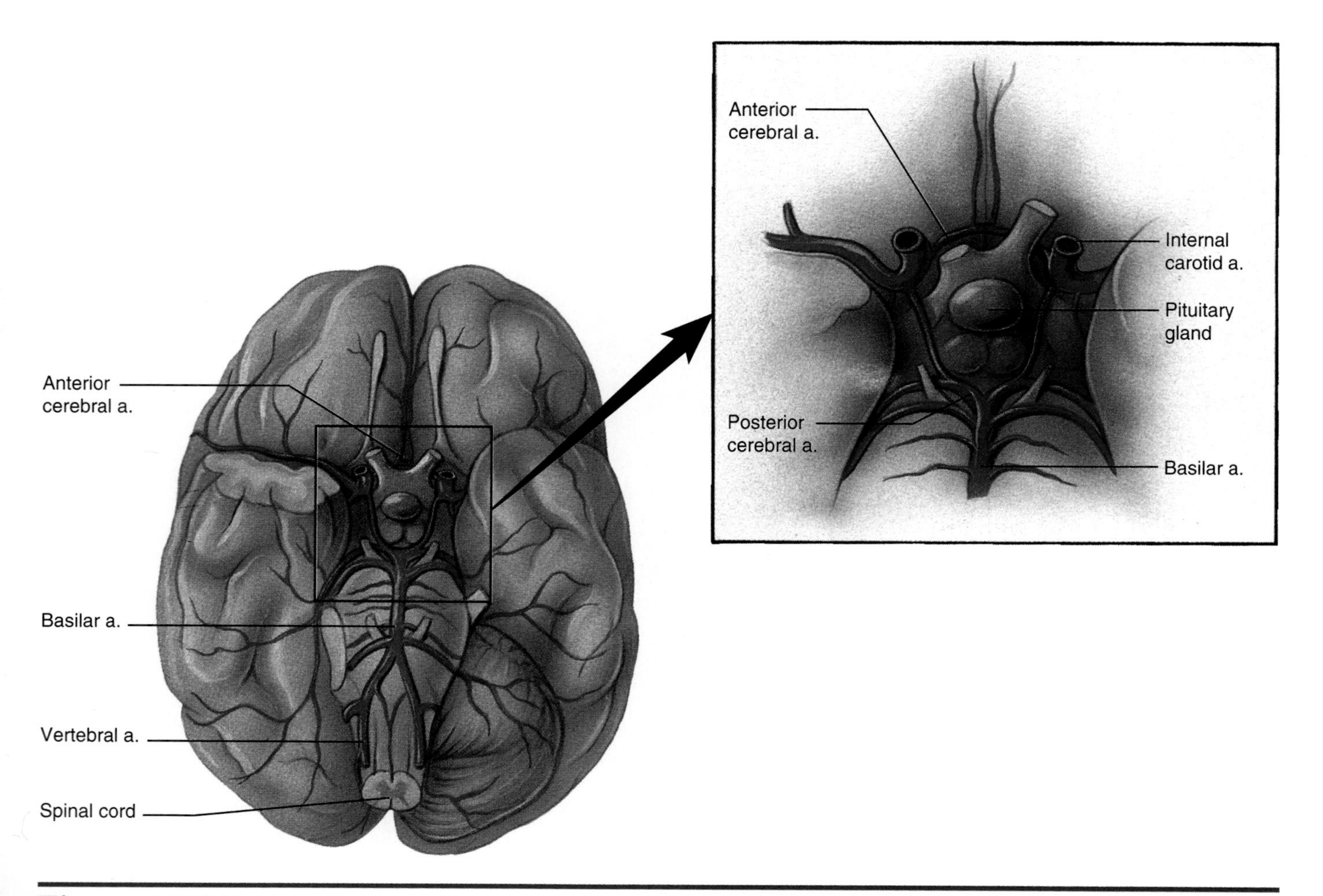

Figure 12.21 The circle of Willis forms from the anterior and posterior cerebral arteries, which join the internal carotid arteries. (a. = artery)

anastomosis, the merging of two or more arteries or two or more veins to provide alternate pathways for blood flow in case one pathway is blocked.

Many small arteries branch from the vertebral arteries to supply the muscles and vertebrae of the neck. Within the cranium, the basilar arteries supply the medulla, pons, midbrain, and cerebellum. The **thyrocervical** (thī-rō-ser′-vi-kal) **arteries** also supply organs and tissues of the neck.

Arteries Supplying the Shoulders and Arms

The subclavian artery provides branches to the shoulder and passes under the clavicle to become the **axillary artery,** which supplies branches to the thorax wall and axillary region. The axillary artery continues into the arm to become the **brachial artery,** which provides branches to serve the arm. At the elbow, the brachial artery divides to form a **radial artery** and an **ulnar artery,** which supply the forearm and wrist and merge to form a network of arteries supplying the hand (figure 12.22 and table 12.3).

Arteries Supplying the Pelvis and Legs

As noted earlier, the left and right common iliac arteries branch from the inferior end of the aorta. Each common iliac branches within the pelvis to form internal and external iliac arteries. The **internal iliac artery** is the smaller branch that supplies the pelvic wall, pelvic organs, external genitalia, and medial thigh muscles. The **external iliac artery** is the larger branch, and it supplies the anterior pelvic wall and continues into the leg, where it becomes the femoral artery.

The **femoral artery** gives off branches that supply the anterior and medial muscles of the thigh. The largest branch is the **deep femoral artery,** which serves the posterior and lateral thigh muscles. As the femoral artery descends, it passes posterior to the knee and becomes the **popliteal** (pop-li-tē′-al) **artery,**

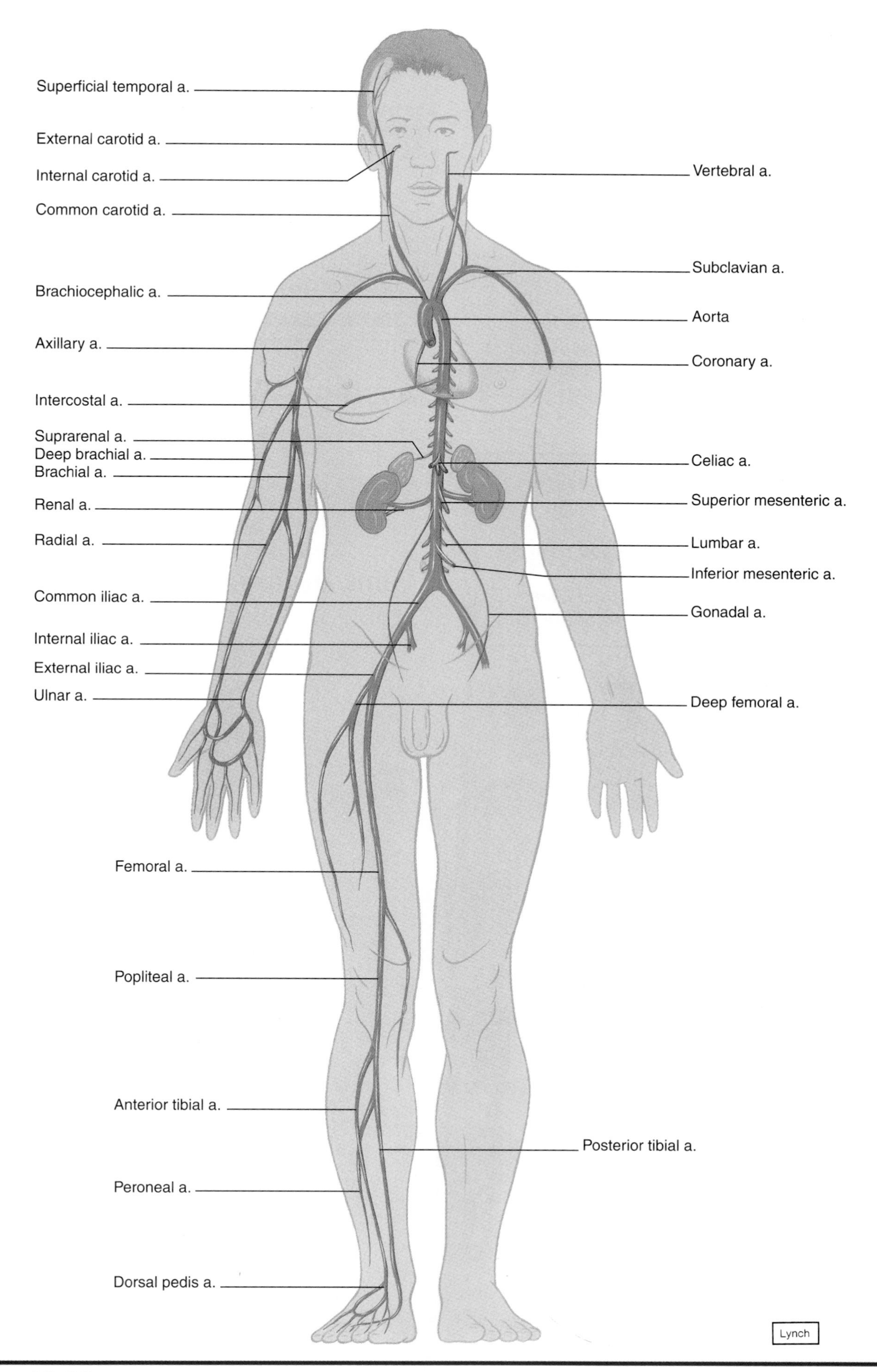

Figure 12.22 Major systemic arteries. (a. = artery)

which supplies certain muscles of the thigh and leg, as well as the knee. The popliteal artery branches just below the knee to form the anterior and posterior tibial arteries.

The **anterior tibial artery** descends between the tibia and fibula to supply the anterior and lateral portions of the leg, and it continues to become the **dorsal pedis artery,** which supplies the ankle and foot. The **posterior tibial artery** lies posterior to the tibia and supplies the posterior portion of the leg, and it continues to supply the ankle and the plantar surface of the foot. Its largest branch is the **peroneal** (per-ō-nē′-al) **artery,** which serves the lateral leg muscles (table 12.4).

What is the arterial pathway of blood from the left ventricle to the right side of the brain?
What is the arterial pathway of blood from the left ventricle to the small intestine?
What is the arterial pathway of blood from the left ventricle to the dorsal surface of the foot?

Determination of the pulse and blood pressure involves certain arteries. The pulse may be taken at any surface artery, but the radial artery at the wrist and the common carotid artery in the neck are the most commonly used sites. Blood pressure is usually taken in the brachial artery of the upper arm.

Systemic Veins

The systemic veins receive deoxygenated blood from capillaries and return the blood to the heart. Ultimately, all systemic veins merge to form two major veins, the superior and inferior venae cavae, that empty into the right atrium of the heart.

Veins Draining the Head and Neck

As shown in figure 12.23, superficial areas of the head and neck are drained by the left and right **external**

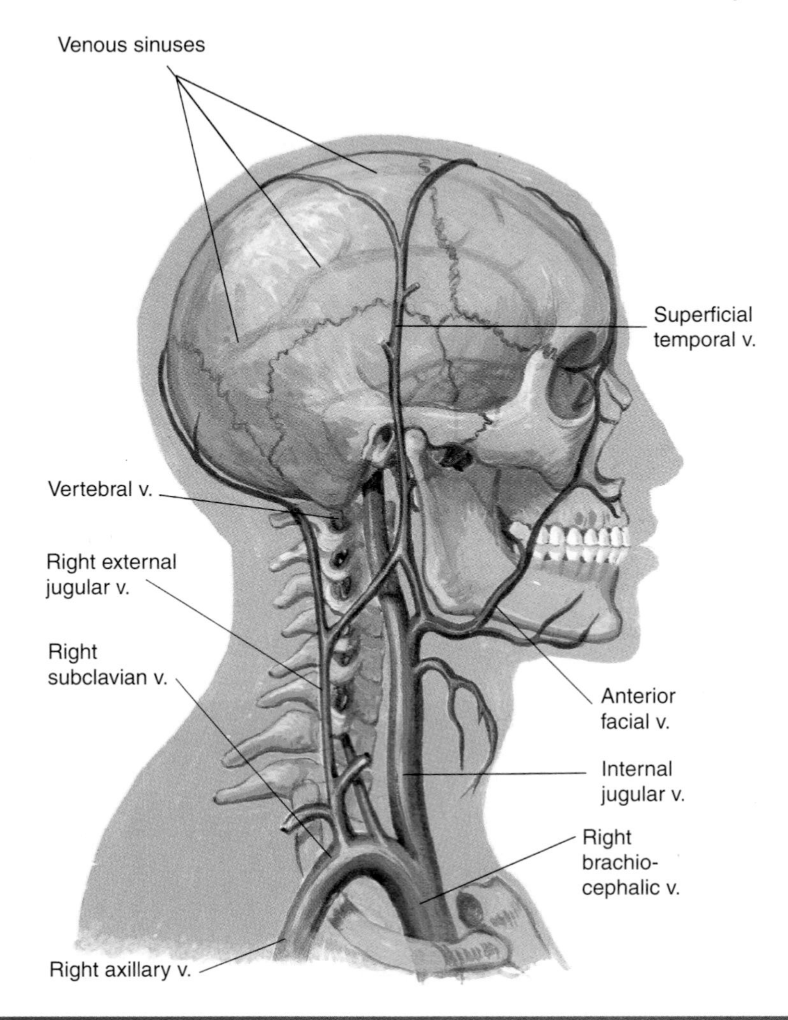

Figure 12.23 Major veins draining the head and neck. (v. = vein)

jugular (jug′-ū-lar) **veins,** which lead into the left and right **subclavian veins,** respectively. The left and right **vertebral veins** carry blood from the brain and deep neck regions into the subclavian veins as well.

Most of the blood from the brain, face, and neck is carried by the left and right **internal jugular veins.** Each internal jugular vein merges with a subclavian vein to form a **brachiocephalic vein.** The left and right brachiocephalic veins join to form the **superior vena cava,** which returns blood to the right atrium of the heart (table 12.5).

Veins Draining the Shoulders and Arms

Deep regions of the forearm are drained by the **radial** and **ulnar veins,** which join at the elbow to form the **brachial vein,** which drains the deep areas of the arm (figure 12.24).

Superficial regions of the hand, forearm, and arm are drained by the laterally located **cephalic** (se-fal′-ik) **vein** and the medially located **basilic** (bah-sil′-ik) **vein.** Note the **median cubital** (kyū′-bi-tal) **vein,** which connects the basilic and cephalic veins.

The basilic and brachial veins merge in the axilla to form the **axillary vein,** which, in turn, joins with the cephalic vein to form the **subclavian vein.** As noted earlier, the subclavian vein joins with the internal jugular vein to form the brachiocephalic vein (table 12.5).

Veins Draining the Pelvis and Legs

The **anterior** and **posterior tibial veins** drain the foot and deep regions of the leg. They join below the knee to form the **popliteal vein.** The **small saphenous** (sah-fē′-nus) **vein** drains the superficial posterior part of the leg and merges with the popliteal vein. The **peroneal vein** drains the lateral portion of the leg and joins with the popliteal vein at the knee to form the **femoral vein,** which drains the deep regions of the thigh and hip.

The median cubital and great saphenous veins have important clinical uses. The median cubital vein is the vein of choice when drawing a sample of blood for clinical tests. It is easily located just under the skin on the anterior surface of the elbow joint. In coronary bypass surgery, the great saphenous vein is removed from a leg, and a segment of this vein is grafted to the afflicted coronary artery on each side of the blockage. This allows blood to flow through the segment of vein, bypassing the blockage to reach the distal part of the coronary artery.

Table 12.5 Major Veins Draining to the Superior Vena Cava

Vein	Region Drained and Location	Receiving Vein
Head and Neck		
External jugular	Face, scalp, and neck	Subclavian
Vertebral	Head and neck (deep)	Subclavian
Internal jugular	Brain and neck (deep)	Brachiocephalic
Arm and Shoulder		
Radial	Hand and forearm	Brachial
Ulnar	Hand and forearm	Brachial
Basilic	Upper arm	Axillary
Cephalic	Forearm and upper arm (deep)	Axillary
Brachial	Upper arm	Axillary
Axillary	Armpit	Subclavian
Subclavian	Shoulder	Brachiocephalic
Upper Trunk		
Brachiocephalic	Upper trunk	Superior vena cava
Azygos	Upper trunk (deep)	Superior vena cava

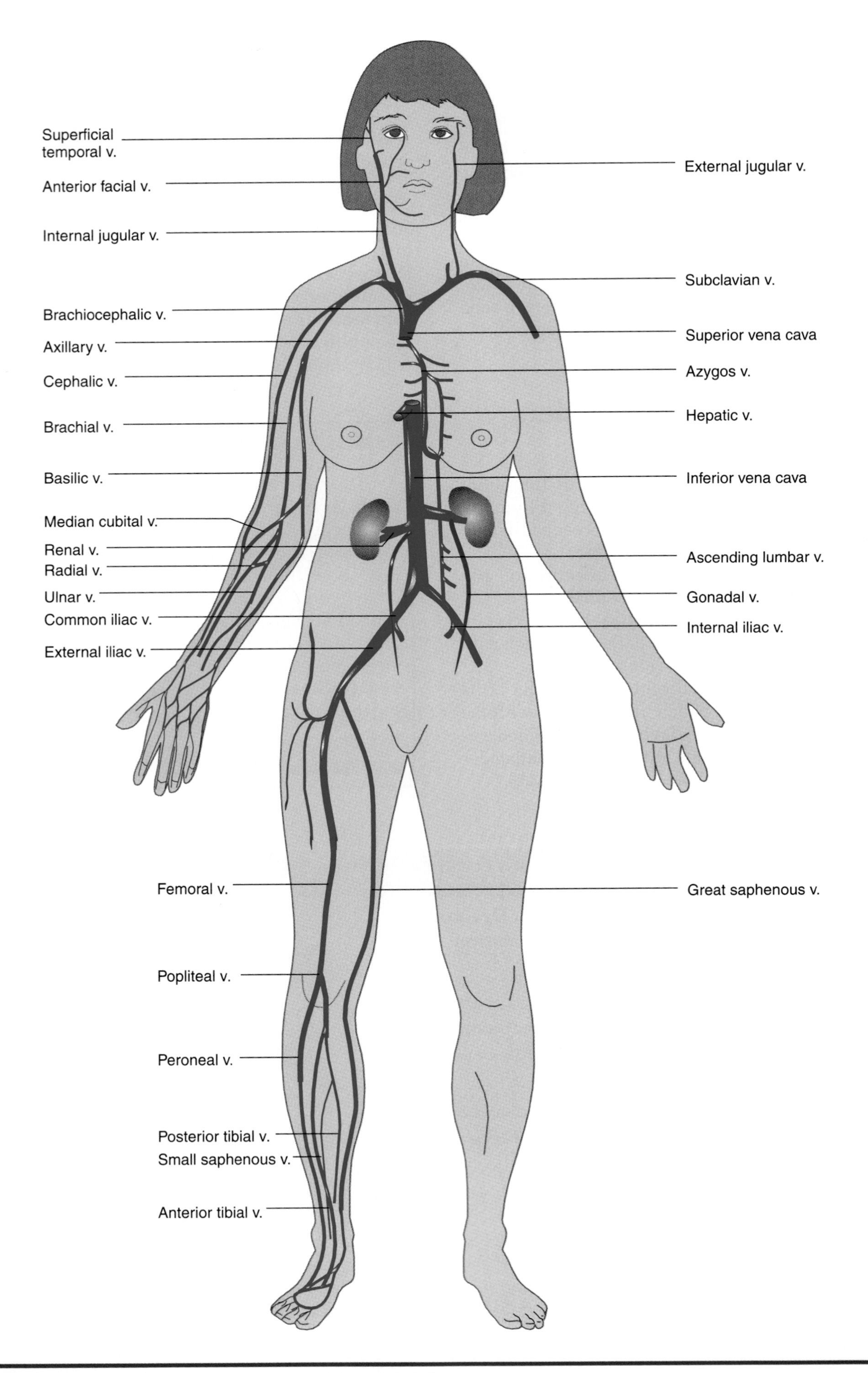

Figure 12.24 Major vessels of the venous system. (v. = vein)

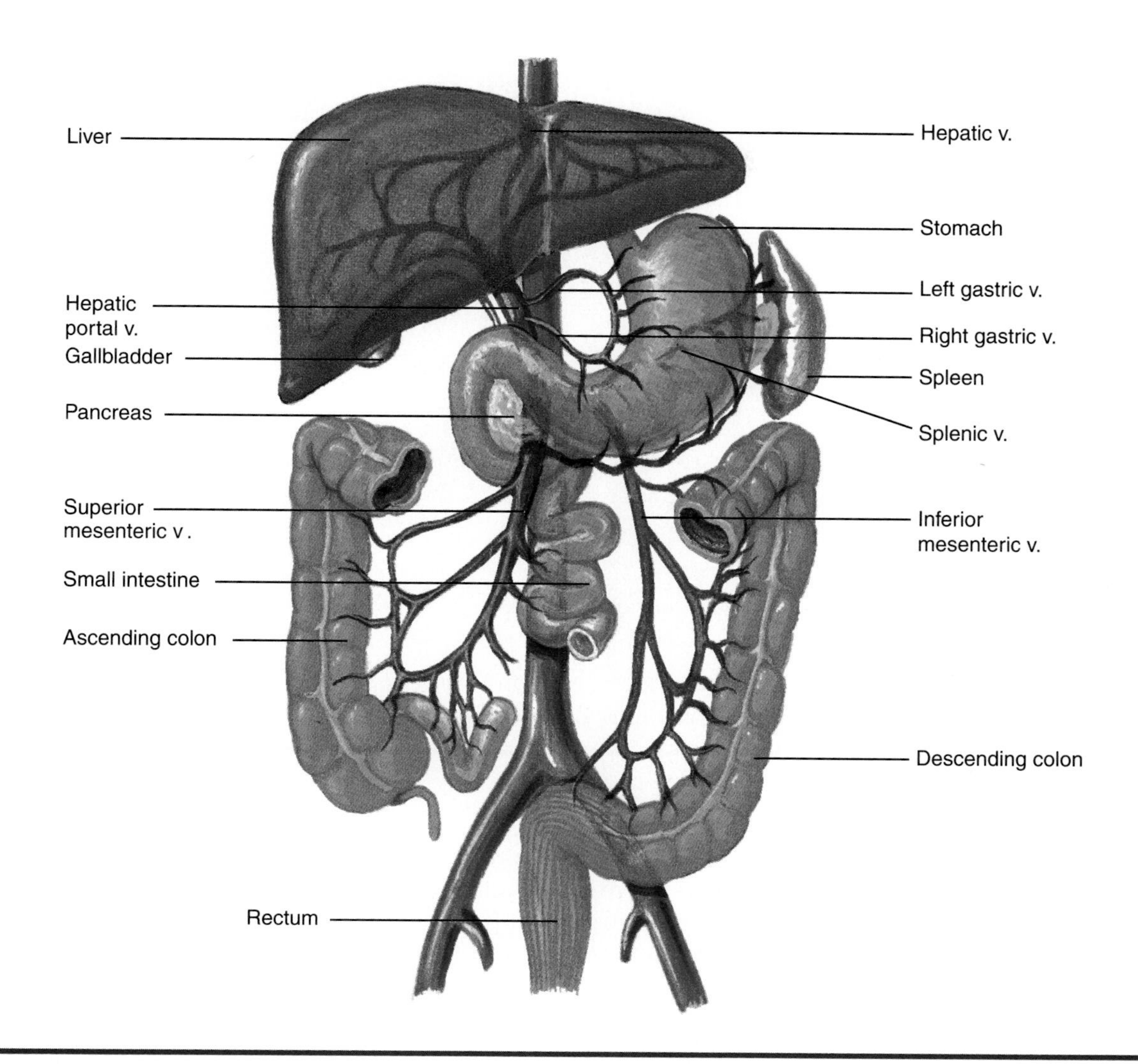

Figure 12.25 Veins that drain the abdominal viscera. (v. = vein)

The **great saphenous vein** originates from the venous arches in the foot, and it drains the medial and superficial portions of the foot, leg, and thigh. It merges with the femoral vein to form the **external iliac vein.** The external iliac vein and the **internal iliac vein** receive branches that drain the upper thigh and pelvic areas, and they merge to form the **common iliac vein.** The left and right common iliac veins merge to form the **inferior vena cava,** which returns blood to the right atrium of the heart (see figure 12.24).

Veins Draining the Abdominal and Thoracic Walls

As shown in figure 12.24, the **azygos** (az′-i-gōs) **vein** drains most of the thoracic and abdominal walls, and it empties into the superior vena cava near the right atrium. The azygos vein receives blood from a number of smaller veins, including the **intercostal veins** and the **ascending lumbar vein,** which drains the wall of the abdomen.

Veins Draining the Abdominal Viscera

The **inferior mesenteric vein** and the **superior mesenteric vein** drain the large and small intestines and merge to form the **hepatic portal vein.** The **gastric veins,** which drain the stomach, the **splenic vein,** which drains the spleen, and the **pancreatic vein,** which drains the pancreas all empty into the hepatic portal vein. All of these veins compose the **hepatic portal system.**

The hepatic portal vein carries blood from the stomach, intestines, spleen, and pancreas to the liver instead of the inferior vena cava. After entering the liver, the blood flows through the venous sinusoids, where materials are either removed or added before the blood enters the **hepatic veins,** which empty into the inferior vena cava (figure 12.25). The hepatic portal system allows the liver to monitor and adjust the concentrations of substances in blood coming from the digestive tract before it enters the general circulation.

Table 12.6 Major Veins Draining to the Inferior Vena Cava

Vein	Region Drained and Location	Receiving Vein
Leg		
Anterior and posterior tibial	Foot and lower leg	Popliteal
Popliteal	Knee	Femoral
Small saphenous	Posterior leg	Popliteal
Great saphenous	Foot, leg, and thigh	Femoral
Femoral	Thigh	External iliac
Trunk		
Hepatic portal	Intestines and stomach	Liver sinusoids
Hepatic	Liver	Inferior vena cava
Renal	Kidneys	Inferior vena cava
Gonadal	Ovaries or testes	Inferior vena cava
External iliac	Pelvic wall	Common iliac
Internal iliac	Viscera in pelvis and lower abdomen	Common iliac
Common iliac	Pelvic and lower abdominal regions	Inferior vena cava

The left and right **renal veins** carry blood from the kidneys, and the left and right **gonadal veins** return blood from the ovaries in females and the testes in males. Renal and gonadal veins empty into the inferior vena cava (table 12.6).

What is the venous pathway of blood from the left side of the head to the right atrium?
What is the venous pathway of blood from the small intestine to the right atrium?
What is the venous pathway of blood from the posterior portion of the ankle to the right atrium?

Disorders of the Heart and Blood Vessels

These disorders are grouped according to whether they affect primarily the heart or blood vessels. In some cases, the underlying cause of a heart ailment is a blood vessel disorder.

Heart Disorders

Arrhythmia (ah-rith′-mē-ah), or dysrhythmia, refers to an abnormal heartbeat. It may be caused by a number of factors, including damage to the heart conduction system, drugs, electrolyte imbalance, or a diminished supply of blood via the coronary arteries. In addition to irregular heartbeats, arrhythmia includes:

- Bradycardia—a slow heart rate of less than 60 beats per minute.
- Tachycardia—a fast heart rate of over 100 beats per minute.
- Heart flutter—a very rapid heart rate of 200 to 300 beats per minute.
- Fibrillation—a very rapid heart rate in which the contractions are uncoordinated so that blood is not pumped from the ventricles. Ventricular fibrillation is usually fatal without prompt treatment.

Congestive heart failure (CHF) is the acute or chronic inability of the heart to pump out the blood returned to it by the veins. Symptoms include fatigue; edema (accumulation of fluid) of the lungs, feet, and legs; and excess accumulation of blood in internal organs. CHF may result from atherosclerosis of the coronary arteries, which deprives the myocardium of adequate blood.

Heart murmurs are unusual heart sounds. They are usually associated with defective heart valves, which allow a backflow of blood. Unless there are complications, heart murmurs have little clinical significance.

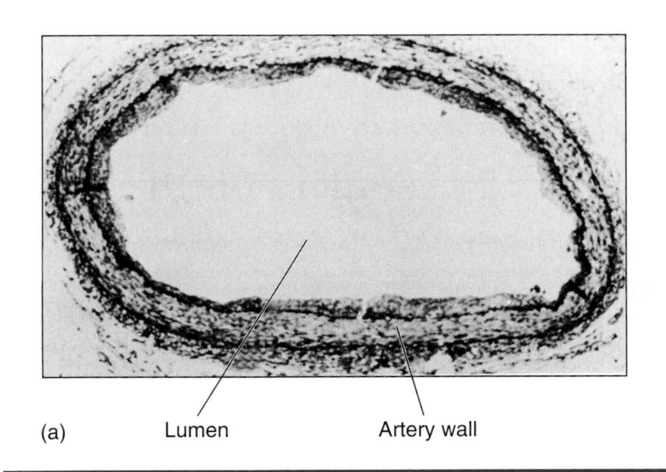

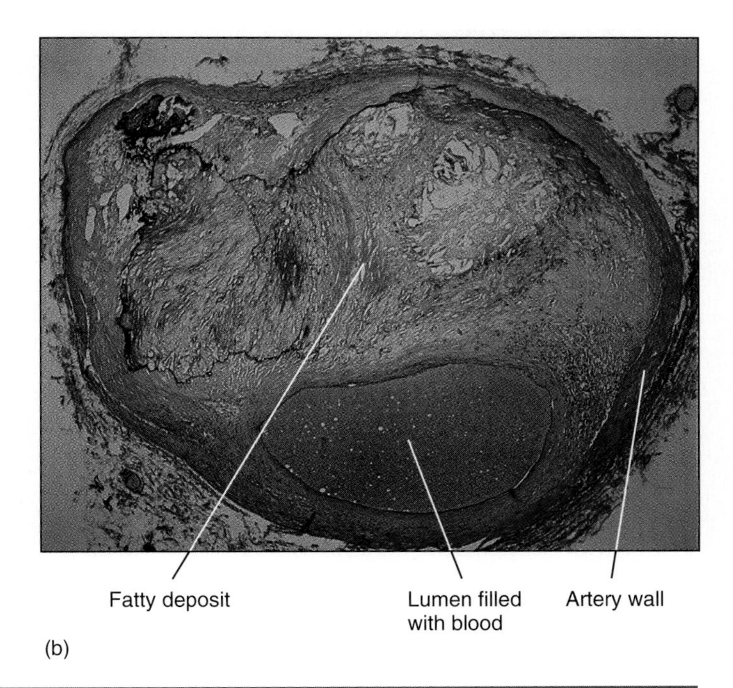

Figure 12.26 Cross sections of (*a*) a normal artery and (*b*) an atherosclerotic artery whose lumen is diminished by fatty deposits. Atherosclerosis promotes the formation of a blood clot within the artery.

Myocardial infarction (mī-ō-kar′-dē-al in-fark′-shun) is the death of a portion of the myocardium due to an obstruction in a coronary artery. The obstruction is usually a blood clot that has formed as a result of atherosclerosis. This event is commonly called a "heart attack," and it may be fatal if a large portion of the myocardium is deprived of blood.

Pericarditis is the inflammation of the pericardium and is usually caused by a viral or bacterial infection. It may be quite painful as the inflamed membranes rub together during each heart cycle.

Blood Vessel Disorders

An **aneurysm** (an′-yū-rizm) is a weakened portion of a blood vessel that bulges out, forming a balloonlike sac filled with blood. Rupture of an aneurysm in a major artery may produce a fatal hemorrhage.

Arteriosclerosis (ar-te″-rē ō-skle-rō′-sis) is hardening of the arteries. It results from calcium deposits that accumulate in the tunica media of arterial walls and is usually associated with atherosclerosis.

Atherosclerosis is the formation of fatty deposits (cholesterol and triglycerides) in the tunica interna of arterial walls. The atherosclerotic plaques reduce the lumen of the arteries and increase the probability of blood clots being formed. Such deposits in the coronary, carotid, or cerebral arteries may lead to serious circulatory problems (figure 12.26).

Hypertension refers to chronic high blood pressure. It is the most common disease affecting the heart and blood vessels. Blood pressure that exceeds 140/90 is indicative of hypertension. Hypertension may be caused by a variety of factors, but persistent stress and smoking are commonly involved.

Phlebitis (flē-bī′-tis) is inflammation of a vein, and it most often occurs in a leg. If it is complicated by the formation of a blood clot, it is called *thrombophlebitis.*

Varicose veins are veins that have become dilated and swollen because their valves are not functioning properly. Heredity seems to play a role in their occurrence. Pregnancy, standing for prolonged periods, and lack of physical activity reduce venous return and promote varicose veins in the legs. Chronic constipation promotes their occurrence in the anal canal, where they are called **hemorrhoids** (hem′o-royds).

CHAPTER SUMMARY

Structure of the Heart

1. The heart is enveloped by the pericardial membranes, which provide protection and support.
2. The heart wall consists primarily of the myocardium, a thick layer of cardiac muscle. It is lined on the inside by the thin endocardium and on the outside by the thin epicardium.
3. The heart contains four chambers. The upper chambers are the left and right atria, which receive blood returning to the heart. The lower chambers are the left and right ventricles, which pump blood from the heart. There are no openings between the atria or between the ventricles.
4. Atrioventricular valves allow blood to flow between each atrium and its corresponding ventricle but prevent a backflow of blood. The bicuspid A-V valve lies between the left atrium and left ventricle. The tricuspid A-V valve lies between the right atrium and right ventricle.
5. Semilunar valves allow blood to be pumped from the ventricles into their associated arteries. The aortic semilunar valve is located in the base of the aorta. The pulmonary semilunar valve is located in the base of the pulmonary trunk.

Cardiac Cycle

1. The cardiac cycle includes both contraction (systole) and relaxation (diastole) phases.
2. During atrial diastole, blood returns to the atria and flows on into the ventricles. Atrial systole forces more blood into the ventricles to fill them.
3. During ventricular diastole, blood flows into the ventricles. Ventricular systole pumps blood from the ventricles into their associated arteries.
4. The normal lub-dup heart sound is caused by the closure of the heart valves. The first sound results from the closure of the atrioventricular valves. The second sound results from the closure of the semilunar valves.
5. The right atrium receives deoxygenated blood from the superior and inferior venae cavae, while the left atrium receives oxygenated blood from the pulmonary veins.
6. The right ventricle pumps deoxygenated blood into the pulmonary trunk, which divides into the left and right pulmonary arteries, which lead to the lungs. At the same time, the left ventricle pumps oxygenated blood into the aorta, which leads to all parts of the body except the lungs.
7. The heart tissues receive blood from the coronary arteries, which branch from the ascending aorta. Blood is returned from the heart tissues by the cardiac veins, which open into the coronary sinus, which leads to the right atrium.

Heart Conduction System

1. The S-A node is the pacemaker, which rhythmically initiates impulses that cause the heart contractions.
2. Impulses pass through the atria, causing atrial systole and simultaneously reach the A-V node.
3. Impulses pass from the A-V node along the A-V bundle to the Purkinje fibers, which transmit the impulses to the myocardium, causing ventricular contraction.
4. An electrocardiogram is a recording of the formation and transmission of impulses through the heart conduction system.
5. An electrocardiogram consists of a P wave, a QRS wave, and a T wave, and it is used in the diagnosis of heart ailments.

Regulation of Heart Rate

1. Heart rate is controlled by the autonomic nervous system. The cardiac control center is in the medulla oblongata. It receives sensory impulses from pressure receptors in the aorta and carotid arteries.
2. Sympathetic fibers release norepinephrine at heart synapses, which causes an increase in the heart rate. Parasympathetic fibers release acetylcholine at heart synapses, which causes a decrease in heart rate.
3. The dynamic balance in the frequency of sympathetic and parasympathetic impulses reaching the heart adjusts the heart rate to meet body needs.
4. Heart rate is also affected by age, sex, physical condition, temperature, epinephrine, thyroid hormone, and the concentration of Ca^{++} and K^{+}.

Types of Blood Vessels

1. The three basic types of blood vessels are arteries, capillaries, and veins. Large arteries and veins are formed of an outer fibrous layer of connective tissue, a middle layer of smooth muscle, and an inner layer of simple squamous epithelium.
2. Arteries have thick, muscular walls and carry blood from the heart. Large arteries divide repeatedly to form the smallest arteries, arterioles, which connect with capillaries.
3. Capillaries are the smallest and most numerous blood vessels. Their thin walls allow an exchange of materials between the blood and the tissue cells. This exchange is by diffusion and filtration. Capillaries merge to form venules.
4. Veins have thinner walls than arteries and carry blood from capillaries toward the heart. The smallest veins are venules, which lead from capillaries and merge to form small veins. Veins contain valves that prevent a backflow of blood.

Blood Flow

1. Blood circulates from areas of higher pressure to areas of lower pressure. Blood pressure is highest in the ventricles and lowest in the atria.
2. Systemic blood pressure declines as blood is carried to the capillaries. Skeletal muscle contractions and respiratory movements are important forces that aid the return of venous blood.
3. Blood velocity varies inversely with the cross-sectional area of the combined blood vessels. Blood velocity is highest in the aorta and lowest in the capillaries.

Blood Pressure

1. Normal systolic blood pressure is 120 ± 10 mmHg. Normal diastolic blood pressure is 80 ± 10 mmHg.
2. The difference between systolic and diastolic pressures is the pulse pressure. The pulse may be detected by palpating surface arteries.
3. Blood pressure is determined by four factors: cardiac output, blood volume, peripheral resistance, and blood viscosity.
4. The vasomotor center in the medulla oblongata provides the autonomic control of blood vessel diameter. In this way, the autonomic nervous system controls peripheral resistance and blood pressure.
5. Local autoregulation of arterioles overrides autonomic control and regulates blood flow in capillaries according to the needs of the local tissues.

Circulation Pathways

1. The pulmonary circuit carries blood from the heart to the lungs and back again.
2. The systemic circuit carries blood from the heart to all parts of the body, except the lungs, and back again.

Systemic Arteries

1. The major branch arteries of the aorta are the coronary, brachiocephalic, left common carotid, left subclavian, intercostals, celiac, superior mesenteric, renal, gonadal, lumbar, inferior mesenteric, and common iliac arteries.
2. The major arteries supplying the head and neck are paired arteries. Each common carotid artery branches to form the external and internal carotid arteries. Each vertebral artery enters the cranium and becomes the basilar artery. The internal carotid and basilar arteries branch to form the circle of Willis. Thyrocervical arteries supply portions of the neck.
3. Each shoulder and arm is supplied by a subclavian artery, which becomes the axillary artery, which becomes the brachial artery of the arm. The brachial artery branches to form the radial and ulnar arteries of the forearm.
4. Each common iliac artery branches to form internal and external iliac arteries. The external iliac enters the leg to become the femoral artery, which becomes the popliteal artery near the knee. The popliteal branches below the knee to form the anterior and posterior tibial arteries.

Systemic Veins

1. Veins draining the head and neck are paired veins. On each side, the external jugular and vertebral veins empty into the subclavian vein. The internal jugular vein and subclavian merge to form the brachiocephalic vein. The left and right brachiocephalic veins join to form the superior vena cava.
2. The ascending lumbar vein and the intercostal veins enter the azygos vein, which opens into the superior vena cava.
3. Radial and ulnar veins of the forearm join to form the brachial vein of the arm. The basilic vein joins the brachial vein to form the axillary vein, which, in turn, receives the cephalic vein to form the subclavian vein.
4. Anterior and posterior tibial veins merge below the knee to form the popliteal vein. The popliteal vein receives the small saphenous and peroneal veins to form the femoral vein. The great saphenous vein extends from the foot to join with the femoral vein near the hip, which forms the external iliac vein. The external iliac vein joins with the internal iliac vein to form the common iliac vein. The left and right common iliac veins merge to form the inferior vena cava.
5. The inferior and superior mesenteric veins merge to form the hepatic portal vein, which then receives the gastric, splenic, and pancreatic veins. These veins form the hepatic portal system. The hepatic portal vein empties into the liver sinusoids. The hepatic vein carries blood from the liver to the inferior vena cava.
6. The paired gonadal and renal veins empty into the inferior vena cava.

Disorders of the Heart and Blood Vessels

1. Disorders of the heart include arrhythmia, congestive heart failure, heart murmurs, myocardial infarction, and pericarditis.
2. Disorders of blood vessels include aneurysm, arteriosclerosis, atherosclerosis, hypertension, phlebitis, and varicose veins.

BUILDING YOUR VOCABULARY

1. **Selected New Terms**
   anastomosis, p. 256
   atrioventricular (A-V) bundle, p. 244
   atrioventricular (A-V) node, p. 244
   bicuspid valve, p. 241
   cardiac center, p. 245
   cardiac output, p. 251
   chordae tendineae, p. 241
   electrocardiogram, p. 245
   interstitial fluid, p. 249
   pericardial cavity, p. 238
   peripheral resistance, p. 251
   sinoatrial (A-V) node, p. 244
   tricuspid valve, p. 241

2. **Related Clinical Terms**
   angiospasm (an′-jē-ō-spa-zum) Muscular spasm of a blood vessel.
   arteriography (ar-ter-ē-og′-rah-fē) An X ray of arteries after injection of a radiopaque dye.
   cardiology (kar-dē-ol′-ō-jē) Medical specialty treating disorders of the heart.
   phlebitis (flē-bī′-tis) Inflammation of a vein, most often in a leg.
   thrombophlebitis (throm-bō-flē-bī′-tis) A blood clot in a vein resulting from inflammation; usually forms in a leg.

STUDY ACTIVITIES

1. Complete the **Chapter 12 Study Guide,** which begins on page 459.
2. Complete the **Learning Objectives** listed on the first page of this chapter.

CHECK YOUR UNDERSTANDING

(Answers are located in appendix B.)

1. The membranous sac around the heart is the _____ .
2. The _____ is the heart chamber receiving oxygenated blood from the lungs.
3. The _____ valve prevents a backflow of blood from the right ventricle into the _____ .
4. During _____ diastole blood fills the atria; during ventricular _____ blood is pumped into arteries leading from the heart.
5. The heartbeat originates in the _____ node: the _____ node relays impulses along the A-V bundle and Purkinje fibers to the ventricular myocardium.
6. Impulses from _____ fibers increase the heart rate; impulses from the _____ fibers decrease the heart rate.
7. Blood is carried from the heart in _____ and returned to the heart in _____ .
8. Vessels with the thickest walls are _____ ; those with the thinnest walls are _____ .
9. The exchange of materials between capillary blood and tissue fluid occurs by _____ , osmosis, and _____ .
10. The heart chambers and vessels in the pulmonary circuit are _____ ventricle, pulmonary _____ , pulmonary _____ , capillaries in the lungs, pulmonary _____ , _____ atrium.
11. The arterial pathway of blood from the heart to the right side of the brain is ascending aorta, aortic arch, _____ , common carotid, and _____ .
12. The arterial pathway of blood from the heart to the liver is ascending aorta, aortic arch, _____ aorta, abdominal aorta, _____ , and _____ .
13. The venous pathway returning blood from the small intestine to the heart is _____ , liver, _____ , and _____ vena cava.
14. The venous pathway returning blood from the back of the knee is _____ , femoral, _____ , _____ , and _____ vena cava.
15. The venous pathway from the little finger to the heart is basilic, _____ , _____ , _____ , and _____ vena cava.

CHAPTER

The Lymphatic System and Defenses Against Disease

THIRTEEN

13

Chapter Preview & Learning Objectives

Lymph
- Explain the source of lymph.

Lymphatic Capillaries and Vessels
- Describe the pathway of lymph in its return to the blood.

Transport of Lymph
- Describe how lymph is propelled through lymphatic vessels.

Lymphatic Organs
- Describe the locations and functions of the lymph nodes, spleen, and thymus gland.

Nonspecific Resistance Against Disease
- Identify the components of nonspecific resistance against disease.

Immunity
- Compare nonspecific resistance and specific resistance against disease.
- Explain the mechanism of cell-mediated immunity.
- Explain the mechanism of antibody-mediated immunity.

Immune Responses
- Compare primary and secondary immune responses.
- Explain the basis of immunization.

Rejection of Organ Transplants
- Explain how rejection of a transplanted organ occurs.

Disorders of the Lymphatic and Immune Systems
- Describe the common disorders of the lymphatic and immune systems.

Chapter Summary
Building Your Vocabulary
Study Activities
Check Your Understanding

SELECTED KEY TERMS

Allergen A foreign substance that stimulates an allergic reaction.
Antibody (anti = against) A substance produced by B cells in response to a specific antigen.
Antigen A foreign substance capable of stimulating the production of antibodies.
Complement A group of plasma proteins that works with antibodies to destroy cells of pathogens.
Immunity (immun = free) Resistance to specific disease-producing antigens.
Inflammation (inflam = to set on fire) A localized response to damaged or infected tissues that is characterized by swelling, redness, pain, and heat.
Lymph (lymph = clear water) The fluid transported in lymphatic vessels.
Lymph node A lymph-filtering node of lymphatic tissue located on a lymphatic vessel.
Lymphatic vessel A vessel that transports lymph.
Pathogen A disease-causing organism.
Spleen The largest lymphatic organ.

During the exchange of materials between capillary blood and interstitial fluid, more fluid enters the interstitial fluid than is returned to the blood. The **lymphatic system,** which consists of lymphatic vessels and organs, returns the excess interstitial fluid to the blood for reuse. By recycling interstitial fluid, the lymphatic system helps to maintain homeostasis by conserving water and dissolved substances. This relationship of the lymphatic system to the cardiovascular system is shown diagrammatically in figure 13.1.

The lymphatic system also works with white blood cells to protect the body against **pathogens** (path′-ō-jens)—disease-causing organisms. Pathogens

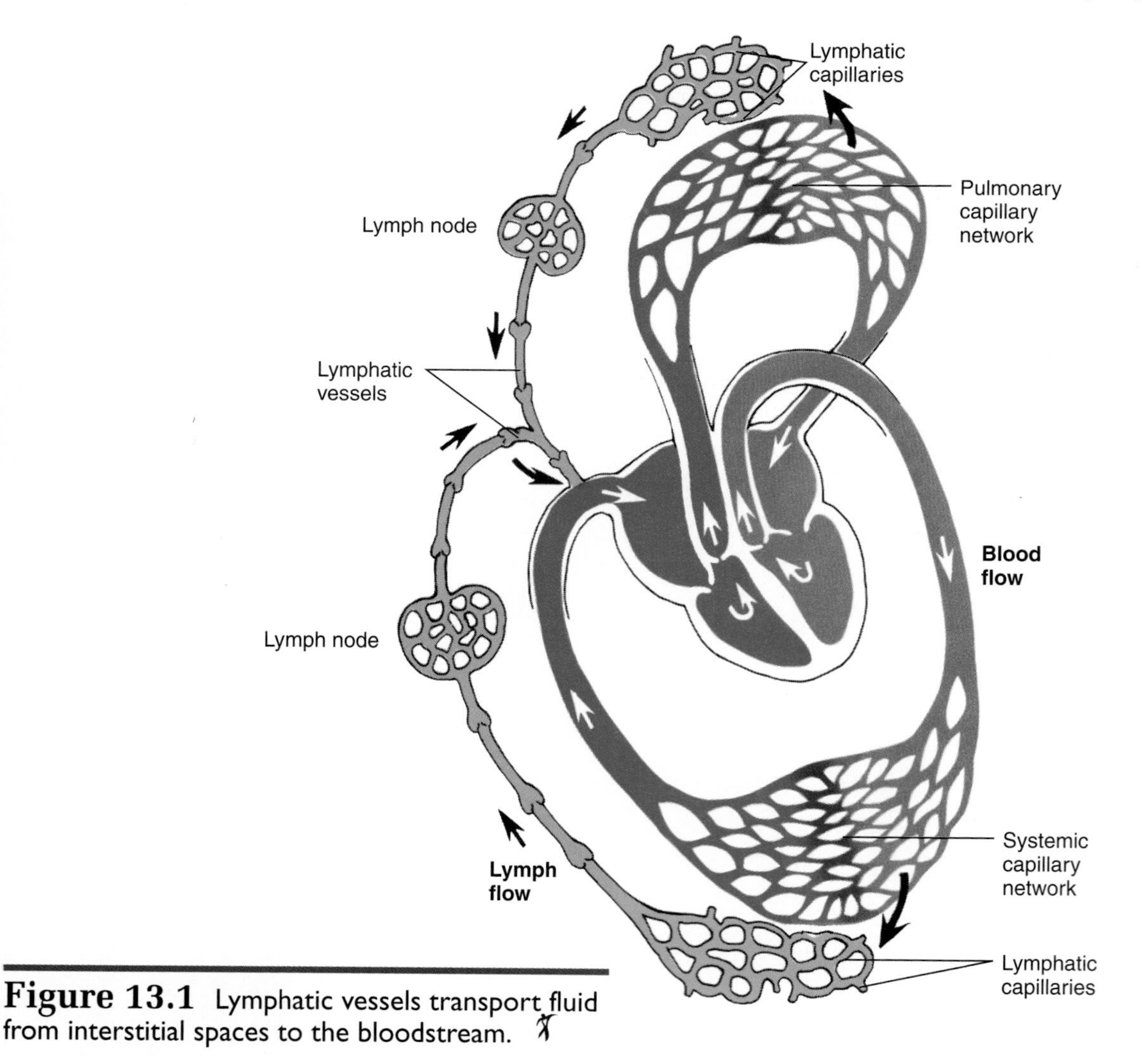

Figure 13.1 Lymphatic vessels transport fluid from interstitial spaces to the bloodstream.

include viruses, bacteria, fungi, protozoa, and other parasites that may invade the body. The body's defense mechanisms are categorized into two broad groups: nonspecific resistance and specific resistance, or immunity.

Lymph

Interstitial fluid is derived from blood plasma as materials move from the blood in capillaries to the tissue cells. There is a certain amount of leakage of small proteins from capillaries into the interstitial fluid. The accumulation of these proteins increases the osmotic pressure of interstitial fluid, promotes the movement of water into the tissue spaces, and slows the return of water into capillaries. Thus, interstitial fluid tends to accumulate in the tissue spaces, and, as it does, it is picked up and removed by lymphatic vessels. Once interstitial fluid enters a lymphatic capillary, it is called **lymph.** The pickup of interstitial fluid by lymphatic vessels and the return of lymph to the blood prevents *edema* of the tissues and helps to maintain the normal blood volume.

Lymphatic Capillaries and Vessels

Lymphatic vessels provide a pathway for the one-way movement of lymph from body tissues to the subclavian veins (figure 13.1). The smallest lymphatic vessels are the microscopic **lymphatic capillaries,** closed-ended tubes that are found in spaces within body tissues (figure 13.2). Lymphatic capillaries are immersed in interstitial fluid. Like blood capillaries, lymphatic capillaries consist of a single layer of endothelial cells (squamous epithelium). The edges of adjacent cells overlap each other slightly, which forms tiny flaplike valves between the cells. Fluid can move easily into the lymphatic capillary, but the valves inhibit movement of fluid out of the capillary.

Lymphatic capillaries merge to form **lymphatic vessels,** which are about as numerous as the veins in the body. Lymphatic vessels have a structure similar to veins, including valves that prevent a backflow of lymph. The smaller lymphatic vessels merge to form larger vessels called *lymphatic trunks* that drain large body regions. Locate these trunks in figure 13.3.

Lymphatic trunks drain into one of two lymphatic *collecting ducts.* Lymphatic trunks from the right side of the head and neck, right arm and shoulder, and right side of the thorax empty into the **right lymphatic duct.** The right lymphatic duct is very

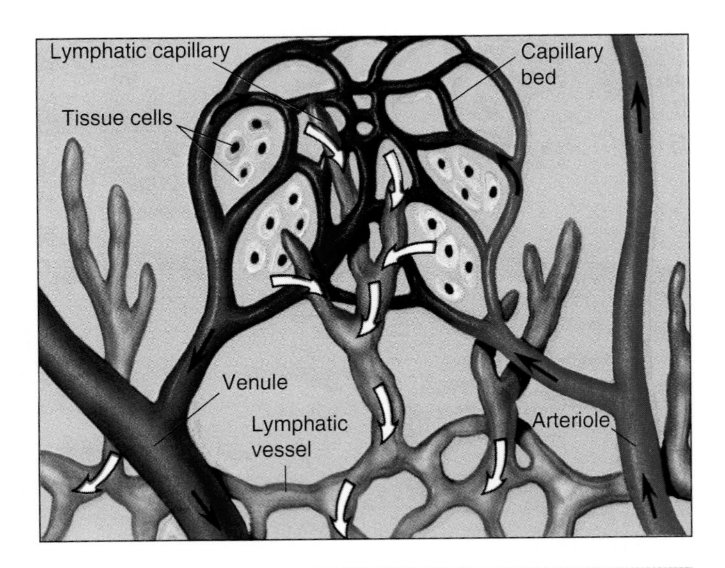

Figure 13.2 Lymphatic capillaries are microscopic, closed-ended tubes that begin in the interstitial spaces of most tissues.

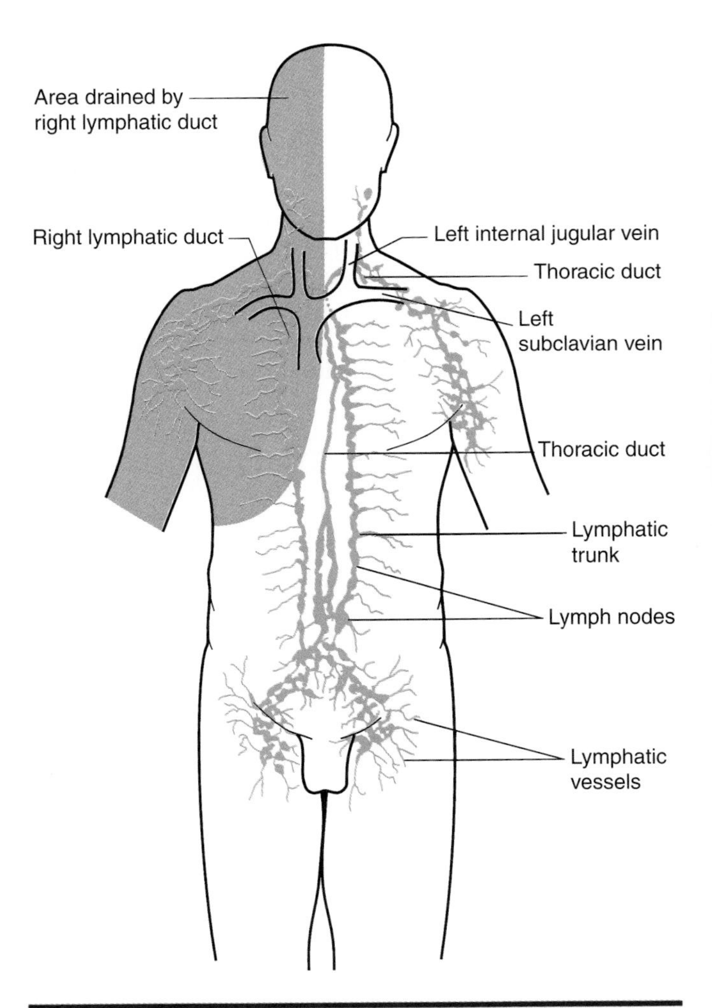

Figure 13.3 The right lymphatic duct drains lymph from the upper right side of the body, and the thoracic duct drains lymph from the rest of the body.

Table 13.1 Components of the Lymphatic System

Component	Characteristics	Function
Lymphatic capillaries	Microscopic closed-ended tubes in tissue spaces	Collect interstitial fluid from tissue spaces; once in a lymphatic capillary, interstitial fluid is called *lymph*
Lymphatic vessels	Formed by merging of lymphatic capillaries; structure similar to veins; contain valves; merge to form lymphatic trunks that drain into either the *right lymphatic duct* or the *thoracic duct*	Transport lymph and empty it into the subclavian veins
Lymphatic Organs		Sites of lymphocyte production and immune responses
Lymph nodes	Small, bean-shaped organs arranged in groups along lymphatic vessels	Filter and cleanse lymph; destroy pathogens carried by lymph
Tonsils	Masses of lymphatic tissue at entrance to throat	Intercept and destroy pathogens entering throat from nasal and oral cavities
Spleen	Large lymphatic organ containing venous sinuses	Filters and cleanses blood; destroys pathogens carried by blood; contains reservoir of blood; primary site for removal of old RBCs
Thymus	Bilobed gland located above heart; size decreases with age	Site of T cell maturation; secretes hormone *thymosin,* which stimulates maturation of T cells

short, and it empties into the right subclavian vein. Lymphatic trunks from the rest of the body drain into the **thoracic duct,** a large lymphatic duct that extends from the upper abdomen and through the thorax to empty into the left subclavian vein. In this way, lymph is returned to the blood plasma for reuse.

Transport of Lymph

The movement of lymph is dependent upon outside forces since there is no pumping mechanism within the lymphatic system. These forces are the contractions of skeletal muscles and respiratory movements—the same forces that aid the return of venous blood to the heart.

The contraction of skeletal muscles compresses lymphatic vessels. Since the backflow of lymph is prevented by the valves in the vessels, compression of a lymphatic vessel forces lymph from one valved segment to another. This mode of lymph movement is especially important in the arms and legs.

Movement of lymph in vessels within the thorax and abdomen is enabled by respiratory movements that change the pressure within these body cavities. During inspiration, the pressure within the thoracic cavity is decreased while the pressure within the abdominal cavity is increased. This difference in pressure causes lymph to flow from abdominal vessels upward into the thoracic vessels along the pressure gradient. Of course, the valves play an important role as well by preventing a backflow of lymph.

Table 13.1 outlines the components of the lymphatic system.

What is lymph and how is it returned to the blood?
Why is the return of lymph to the blood important?

Lymphatic Organs

There are several lymphatic organs in the body, including the lymph nodes, tonsils, spleen, and thymus gland. All lymphatic organs are sites of lymphocyte production. Many of the lymphocytes remain in the lymphatic organs, but some are released into the blood and are carried to other sites, especially when a pathogen invades the body.

Lymph Nodes

Lymph nodes, or lymph glands, usually occur in groups along the larger lymphatic vessels. They are widely distributed in the body, but they do occur more

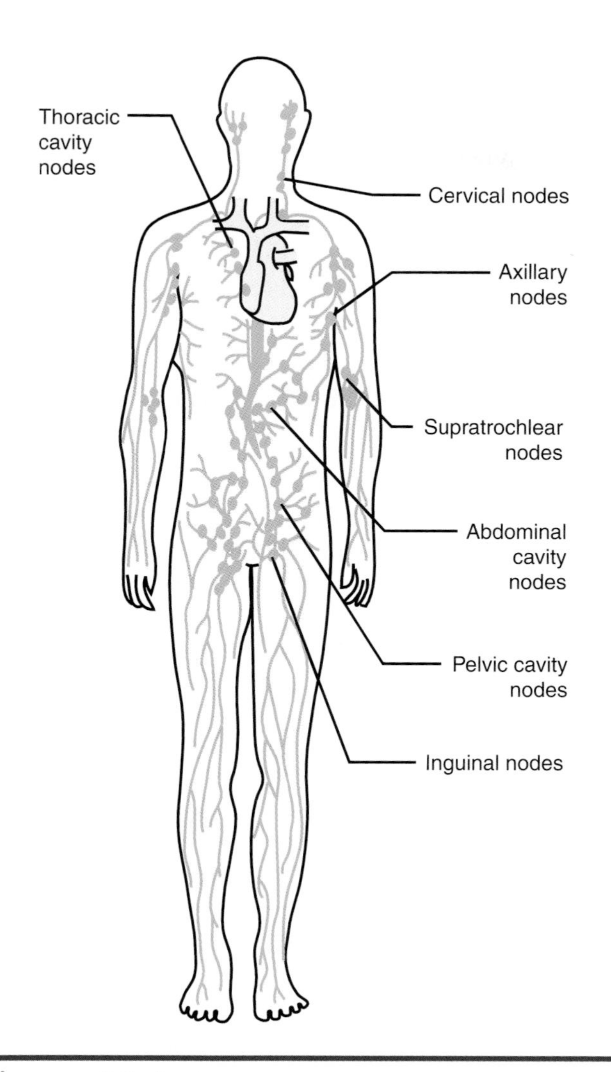

Figure 13.4 Major locations of lymph nodes.

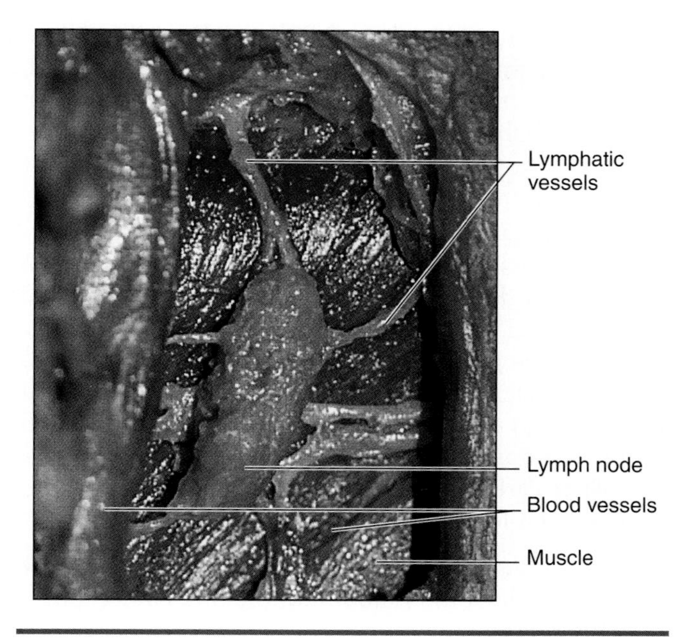

Figure 13.5 A lymph node and its associated vessels.

frequently in certain body regions, as shown in figure 13.4. Interestingly, there are no lymph nodes in the central nervous system.

Structure of Lymph Nodes

Lymph nodes are roughly bean-shaped and 1.0 to 2.5 mm in length. Figure 13.5 is a photograph of a lymph node *in situ* that shows lymphatic vessels leading to and from the node. Figure 13.6 shows the structure of a section of a lymph node. Note that the lymph node consists of a number of small subunits called **nodules.** The nodules are production sites for new lymphocytes.

Lymph enters a lymph node through several *afferent* (af′-er-ent) *lymphatic vessels* and flows through the *lymph sinuses,* which surround the nodules. Lymph is collected from the sinuses and enters *efferent* (ef′-er-ent) *lymphatic vessels,* which carry lymph away from the lymph node. The indentation of the node where efferent lymphatic vessels emerge is called the *hilum.*

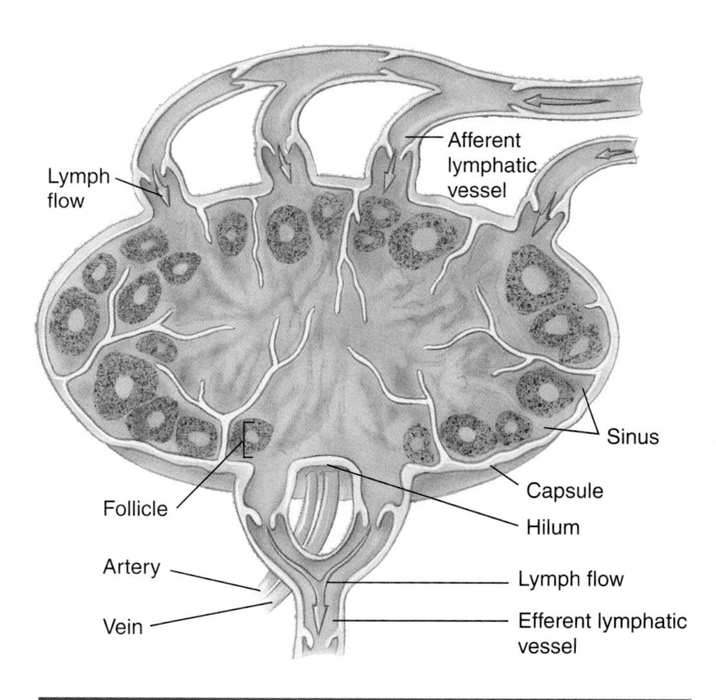

Figure 13.6 Section of a lymph node.

Function of Lymph Nodes

A major function of lymph nodes is the filtration and cleansing of the lymph as it passes through a node. They are the only lymphatic organs that filter the lymph. Damaged cells, cancerous cells, cellular debris, bacteria, and viruses become trapped in the fibrous network of the lymph node and are destroyed by the action of lymphocytes and macrophages. Lymphocytes act against cancerous cells and pathogens, such as bacteria and viruses. Macrophages engulf cellular debris, immobilized or dead bacteria, and viruses.

Tonsils

Tonsils (ton′-sils) are clusters of lymphatic tissue located just under the mucous membrane at the back of the nasal and oral cavities. Like all lymphatic tissue, they produce lymphocytes and contain both lymphocytes and macrophages to fight disease. Their function is to intercept and destroy pathogens that enter through the nose and mouth before they can reach the blood.

There are three kinds of tonsils that are strategically located to carry out this function:

1. The *palatine* (pal′-ah-tine) *tonsils* are located on each side at the junction of the oral cavity and the pharynx (throat).
2. The *pharyngeal* (fah-rin′-jē-al) *tonsil* is located on the back wall of the nasal cavity, a region known as the nasopharynx. This tonsil is commonly called the *adenoid.*
3. The *lingual* (ling′-gwal) *tonsils* are located on the back of the tongue.

Tonsils are larger in young children and play an important role in their defense against disease. Sometimes the palatine tonsils and pharyngeal tonsil become so overloaded and swollen with bacteria that they must be surgically removed by a tonsillectomy. But this is usually avoided, if possible.

Spleen

The **spleen** is the largest lymphatic organ. It is located behind the stomach near the diaphragm in the upper left quadrant of the abdominal cavity. The lower ribs provide protection against physical injury. The spleen is a soft, purplish organ 5 to 7 cm (2–3 in) wide and 13 to 16 cm (5–6 in) long. It contains numerous centers for lymphocyte production and an abundant venous sinus filled with blood.

Before birth, the spleen is a major blood-forming organ, but this function is later taken over by red bone marrow. After birth, the spleen's role is related to both lymphatic and cardiovascular functions:

- It cleanses and filters the blood much like lymph nodes cleanse lymph. Lymphocytes and macrophages destroy pathogens, and macrophages clean up the debris.
- It stores a reserve supply of red blood cells, which can be released into the blood in times of need, such as after a hemorrhage.
- It is a major site for red blood cell destruction and removal. Macrophages play a major role in this process.
- It is a major site of lymphocyte production.

In spite of these important functions, the spleen is not essential for life. But following a splenectomy, a person may be more susceptible to potent pathogens and the effects of hemorrhage.

The spread of cancerous cells occurs when cells break away from the primary tumor and are carried to other sites via the lymphatic system or the blood. This process is called *metastasis* (me-tas′-ta-sis), and it generates secondary cancerous growths called *metastases* wherever the cancerous cells lodge. Since metastatic cells are frequently carried in the lymph, secondary cancerous growths often occur in lymph nodes on lymphatic vessels draining the region of the primary cancer. Knowledge of the location of lymphatic vessels and lymph nodes, and the direction of lymph flow, is important in the detection and treatment of metastases.

Thymus

The **thymus** is a soft, bilobed gland located in the mediastinum above the heart (figure 13.7). It is large in infants and children, but after puberty it begins to atrophy and becomes quite small in adults. The thymus plays a key role in the development of the immune system before birth and during early childhood. Until the immune system matures at about two years of age, an infant is more susceptible to disease than older children. Only after the immune system is well-developed can vaccination provide protection for children against certain diseases.

The major function of the thymus is the differentiation (maturation) of a class of lymphocytes called **T lymphocytes,** or **T cells.** The hormone *thymosin* promotes the differentiation and division of T cells, which makes them *immunocompetent,* that is, capable of recognizing and attacking foreign antigens. After maturation, T cells are distributed by blood to lymphatic tissues throughout the body.

What are the functions of the lymph nodes, tonsils, spleen, and thymus gland?

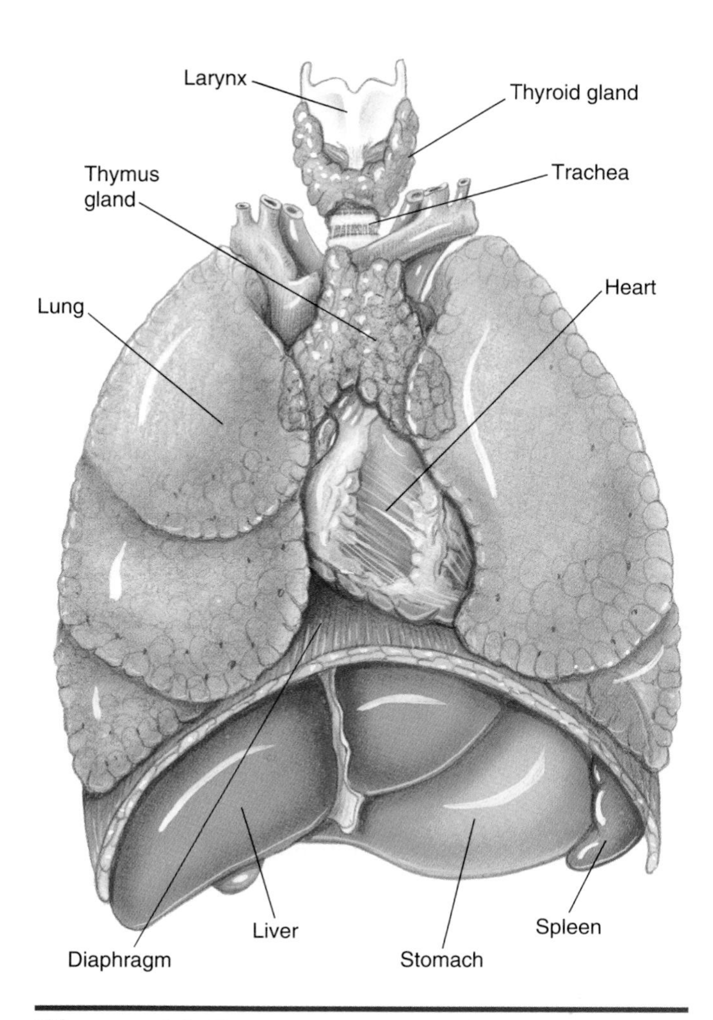

Figure 13.7 The thymus gland is located in the mediastinum above the heart. It is the site of T lymphocyte maturation.

Nonspecific Resistance Against Disease

Nonspecific resistance provides protection against all pathogens and foreign substances, but it is not directed against a specific pathogen. Nonspecific defense mechanisms include mechanical barriers, chemical actions, phagocytosis, inflammation, and fever.

Mechanical Barriers

The most obvious *mechanical barriers* against infectious agents are the skin and the mucous membranes that line the digestive, respiratory, urinary, and reproductive passages. The closely packed epidermal cells of the skin make penetration by pathogens very difficult, and the acid pH of the skin discourages bacterial growth. Mucous membranes are less effective barriers than the skin. However, they secrete mucus that entraps microorganisms and airborne particles and usually prevents their contact with the underlying membranes. The continuous flow of tears over the eyes, the production and swallowing of saliva, and the passage of urine through the urethra are examples of fluid mechanical barriers that help to flush away microbes before they can attack body tissues.

Chemical Actions

Various body chemicals, including certain enzymes, provide a nonspecific defense against pathogens. A few examples will illustrate the effect of these chemicals.

Tears, saliva, nasal secretions, and perspiration contain the enzyme *lysozyme,* which destroys bacteria and helps to protect underlying tissues.

Mucus is continuously produced by epithelium lining the respiratory and digestive passageways. Microorganisms entering the nose and mouth tend to be trapped in the mucus, which is carried to the throat and is swallowed at frequent intervals. Upon reaching the stomach, most microorganisms are destroyed by the gastric juice, either by its low pH or by the enzyme *pepsin.* Pepsin acts by digesting the proteins composing the microbes.

When infected with a virus, lymphocytes and macrophages produce *interferon,* a substance that stimulates uninfected cells to synthesize special proteins that inhibit the replication of viruses within cells. In this way, the rapid growth of viruses may be inhibited. Interferon is now produced commercially by biotechnology and may be used, via injections, to fight serious viral diseases.

Phagocytosis

Phagocytosis (fag-ō-sī-tō′-sis) is the engulfment and destruction (by digestion) of microorganisms, damaged cells, and cellular debris by granulocytic leukocytes and macrophages. When an infection occurs, neutrophils and monocytes are quickly attracted to the infected tissues. Monocytes entering the tissues become transformed into **macrophages** (mak′-rō-fāj-es), large cells that are especially active in phagocytosis.

Some macrophages wander among the tissues, searching out and phagocytizing microorganisms and cellular debris. Others become fixed in particular locations in the body, where they phagocytize microbes that are passing by. Fixed macrophages are especially abundant along the inner walls of blood and lymphatic vessels and in the spleen, lymph nodes, liver, brain, and bone marrow. The wandering and fixed

Table 13.2 Summary of Major Components of Nonspecific Resistance

Component	Function
Mechanical Barriers	
Intact skin	Closely packed cells and multiple cell layers prevent entrance of pathogens
Intact mucous membranes	Closely arranged cells retard entrance of pathogens; not as effective as intact skin
Mucus	Traps pathogens in digestive and respiratory tracts
Saliva	Washes pathogens from oral surfaces
Tears	Washes pathogens from surface of eye
Urine	Flushes pathogens from urethra
Chemical Actions	
Acid pH of skin	Retards growth of many bacteria
Gastric juice	Kills pathogens that are swallowed
Interferon	Helps to prevent viral infections
Lysozyme	Antimicrobial enzyme in nasal secretions, perspiration, saliva, and tears kills some pathogens
Fever	Speeds up body processes and inhibits growth of pathogens
Inflammation	Promotes nonspecific resistance; confines infection
Phagocytosis	Macrophages and granulocytic leukocytes engulf and destroy pathogens

macrophages compose the **tissue macrophage system** (formerly called the reticuloendothelial system), which plays a major role in the destruction of potential pathogens.

Inflammation

Inflammation is a localized response to infection or injury that promotes the destruction of pathogens and the healing process. It is characterized by redness, pain, heat, and swelling of the affected tissues.

When infection occurs, cells release several types of chemicals, such as histamine, that cause a dilation of the arterioles and increase the permeability of capillaries in the affected area. The increased blood flow to the local area produces redness and heat. Blood from deeper tissues carries heat to inflamed surface tissues. The increased movement of fluids from the capillaries produces swelling (edema) of the tissues. Pain results from irritation of nerve endings by pathogens, swelling, or chemicals released by infected cells.

White blood cells are attracted to the affected area to fight the infection. In bacterial infections, neutrophils and macrophages actively phagocytize the pathogens and damaged cells. The accumulated mass of living and dead leukocytes, tissue cells, and bacteria may form a thick, whitish fluid called **pus.**

Fluids from capillaries that enter the affected area contain both fibrinogen and fibroblasts. Fibrinogen may be converted into fibrin to form a clot that is subsequently penetrated and enveloped by fibers formed by the fibroblasts. This action tends to seal off the infected area and prevent the spread of pathogens to neighboring tissues.

The continued action of leukocytes usually brings the infection under control. Then, the dead pathogens and cells are cleaned up by phagocytes, and new cells are produced by cell division to repair any damage to the tissues.

Fever

Fever is an abnormally high body temperature that accompanies widespread infections. It serves a useful purpose as long as the body temperature does not get too high. The increased body temperature inhibits growth of certain pathogens and increases the rate of body processes, including those that fight infection.

Table 13.2 summarizes the major components of nonspecific resistance.

Identify and explain the method of action of barrier, chemical, and cellular mechanisms providing nonspecific resistance against disease.

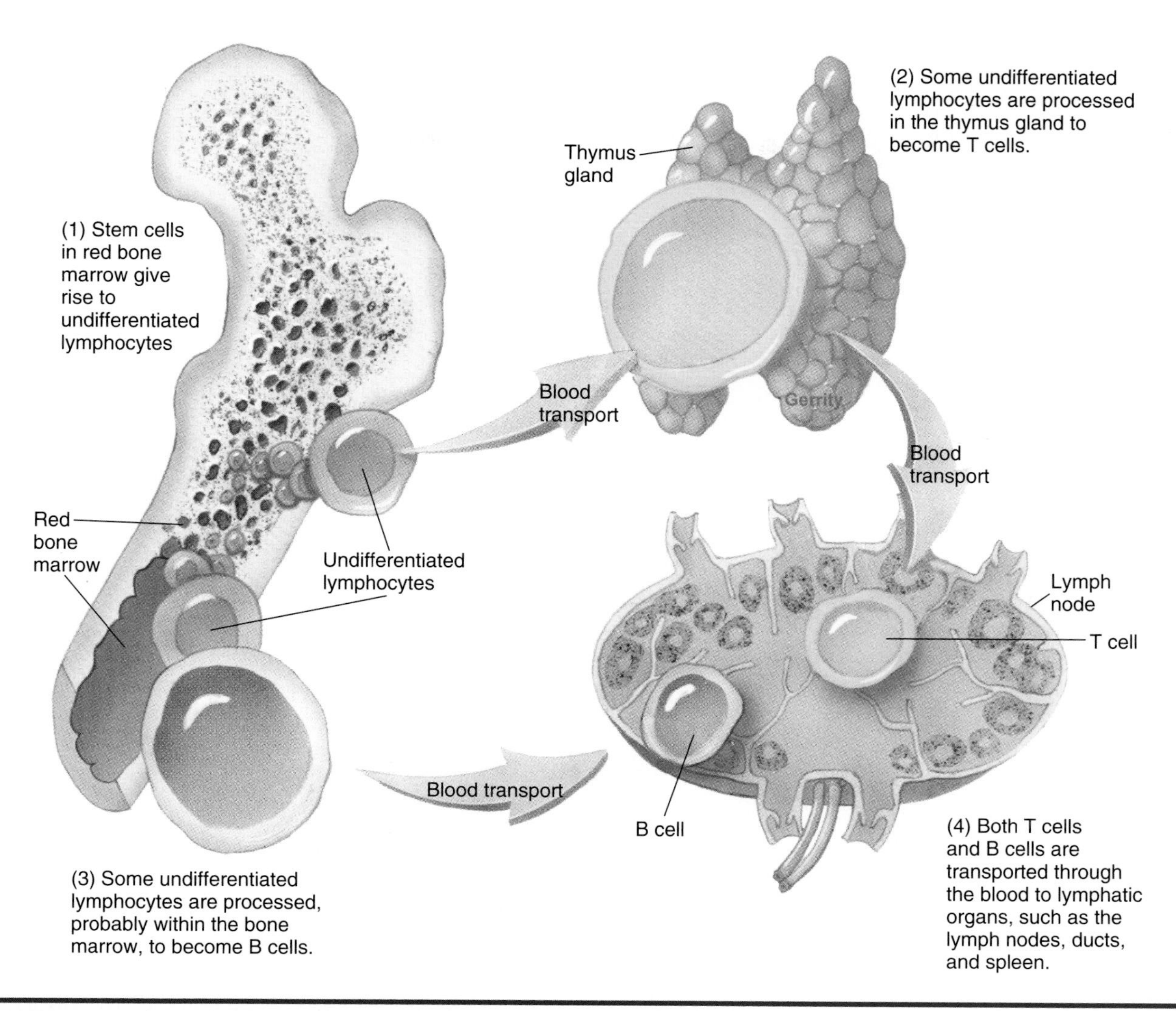

Figure 13.8 Differentiation of T lymphocytes and B lymphocytes.

Immunity

In contrast to nonspecific resistance, **immunity** (i-mū′-ni-tē), or specific resistance, is directed at specific pathogens and foreign cells. An immune response involves the production of specific cells and substances to attack a specific invader. And immunity has a "memory." If the same pathogen should reenter the body at a later date, the immune response is quicker and stronger than during the first encounter. Lymphocytes and macrophages play important roles in immunity.

Specialization of Lymphocytes

During fetal development and in newborn infants, lymphocytes are formed in red bone marrow, but they must mature and become specialized before they can participate in immunity. About half of the unspecialized (undifferentiated) lymphocytes pass to the thymus gland, where they become specialized (differentiated) to become **T lymphocytes,** or **T cells.** The other half of the lymphocytes become specialized in a presently unknown site, perhaps in the bone marrow, spleen, or liver, to become **B lymphocytes,** or **B cells.** T and B lymphocytes are carried by blood to lymphatic organs, such as the lymph nodes, spleen, and tonsils, where they divide to form large populations of T and B cells, and some are released into the blood. About 75% of circulating lymphocytes are T cells, and about 25% are B cells (figure 13.8).

Recognizing Pathogens

All cells, including human cells, have surface recognition molecules called **antigens.** (Recall that antigens are used in typing blood.) In fact, the cells of each person have a unique set of antigens. Antigens are large molecules, such as proteins and glycoproteins.

During the specialization process, lymphocytes "learn" to distinguish "self" antigens from "nonself,"

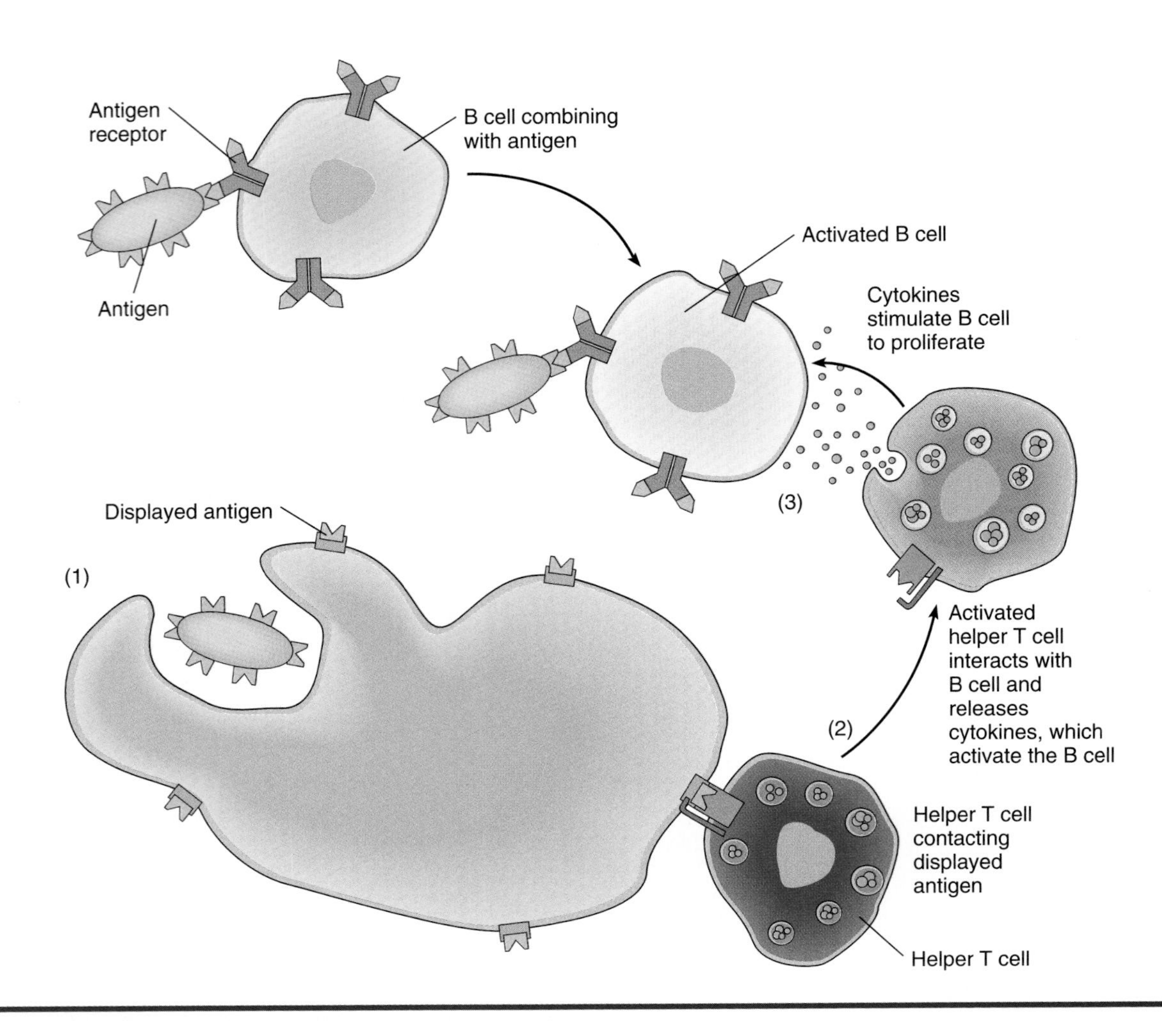

Figure 13.9 Recognition of a foreign antigen by a helper T cell starts the immune process. (*1*) An antigen is engulfed by a macrophage and part of it is presented at the cell surface. (*2*) A helper T cell binds to the displayed antigen and becomes activated. (*3*) The activated helper T cell releases cytokines, which activate B cells with receptors bound to an identical antigen.

or foreign, antigens. Thereafter, lymphocytes can recognize an invading pathogen or cells with abnormal antigens, such as cancer cells and cells infected by a virus, and launch an attack. Unfortunately, lymphocytes also recognize a transplanted organ as foreign and in some cases fail to recognize certain body tissues as "self" and attack the body's own tissues.

The maturation process produces thousands (and perhaps millions) of different kinds of T and B cells, each with specific receptors capable of binding with (recognizing) a specific antigen. Therefore, when an antigen of a pathogen or foreign cell enters the body, only the T or B cells that have a receptor that can bind with the specific antigen are involved in the immune response.

Starting an Immune Response

The first step in an immune response against a pathogen is recognizing that its antigens are foreign (nonself). This involves an **antigen-presenting cell,** usually a macrophage, and a special type of T cell called a **helper T cell.** A pathogen is first engulfed by a macrophage, and part of the pathogen's antigen is moved to the surface of the macrophage, where it is "presented" along with the macrophage's own self antigens. A helper T cell that binds with the displayed foreign antigen and the self antigen becomes activated and starts the immune response (figure 13.9).

The immune response involves one or both of two distinct mechanisms. **Antibody-mediated immunity** involves T cells, macrophages, and B cells. **Cell-mediated immunity** involves T cells and macrophages. Table 13.3 summarizes the roles of B and T cells in immunity.

What is immunity?
How is an immune reaction started?

Table 13.3 Roles of B Cells and T Cells in an Immune Response

Cell	Function
B cells	When activated, produce a clone of plasma cells and memory B cells that have the same antigen receptors as the parent-activated B cell
Plasma cells	Produce specific antibodies against the antigen that activated the parent B cell
Memory B cells	Dormant cells that are activated if an antigen that fits their antigen receptors reenters the body; launch a secondary immune response
Helper T cells	Become activated by binding to an antigen displayed by a macrophage; when activated, start an immune response by releasing cytokines, which help activate B cells and stimulate clone formation in B cells and T cells; form clone composed of helper T cells, killer T cells, memory T cells, and suppressor T cells
Killer T cells	Bind to antigens on the surface of target cells and rupture plasma membrane, killing cells
Memory T cells	Dormant cells that are activated if an antigen that fits their antigen receptors reenters the body; launch a secondary immune response
Suppressor T cells	Stop an immune response after pathogens have been destroyed

Table 13.4 Classes of Antibodies

Class	Location	Functions
IgA	Saliva, tears, mucus, breast milk	Protects mucous membranes from pathogens; provides passive immunity for breast-fed infants
IgD	B lymphocytes	Serves as a receptor on B cells
IgE	Binds to mast cells and basophils	Triggers allergic reactions by causing release of histamine when it binds with allergen
IgG	Blood plasma	Provides long-term immunity following vaccination or recovery from infection; crosses placenta to provide passive immunity for newborn infant
IgM	B lymphocytes	Released from B cells and agglutinates antigens

Antibody-Mediated Immunity

A B cell that binds to a specific antigen and is stimulated by *cytokines* from an activated helper T cell becomes activated (figure 13.9). An activated B cell immediately undergoes rapid cell division (proliferation), which produces a **clone** of B cells that have the same antigen receptor as the activated B cell. The clone consists of *plasma cells* and *memory B cells* (figure 13.10). Plasma cells produce antibodies against the specific antigen and release the antibodies into the blood and lymph for transport throughout the body. After the pathogen has been destroyed, memory B cells remain to launch an even stronger immune response if the same pathogen ever reenters the body.

Antibodies do not destroy pathogens directly. Instead, they bind to the antigens, forming *antigen-antibody complexes* that tag pathogens for destruction by other means. For example, an antigen-antibody complex involving bacteria, or other cellular pathogens, creates a site for binding **complement,** a group of plasma proteins. The binding of complement is known as **complement fixation.** The fixed complement punches holes in the pathogen's plasma membrane, causing the cell to burst and the pathogen to be destroyed. Subsequently, the resulting debris is cleaned up by phagocytes (neutrophils and macrophages).

Antibodies neutralize bacterial toxins by binding to the antigens, which prevents the toxins from attaching to receptors of body cells. Body cells escape damage because the toxins cannot enter the cells. Subsequently, the antigen-antibody complexes are engulfed and destroyed by macrophages and neutrophils.

Antibodies are proteins known as globulins, so another name for antibodies is **immunoglobulins,** which have a shorthand designation of **Ig.** There are several classes of antibodies, and each class plays a special role in antibody-mediated immunity, as noted in table 13.4.

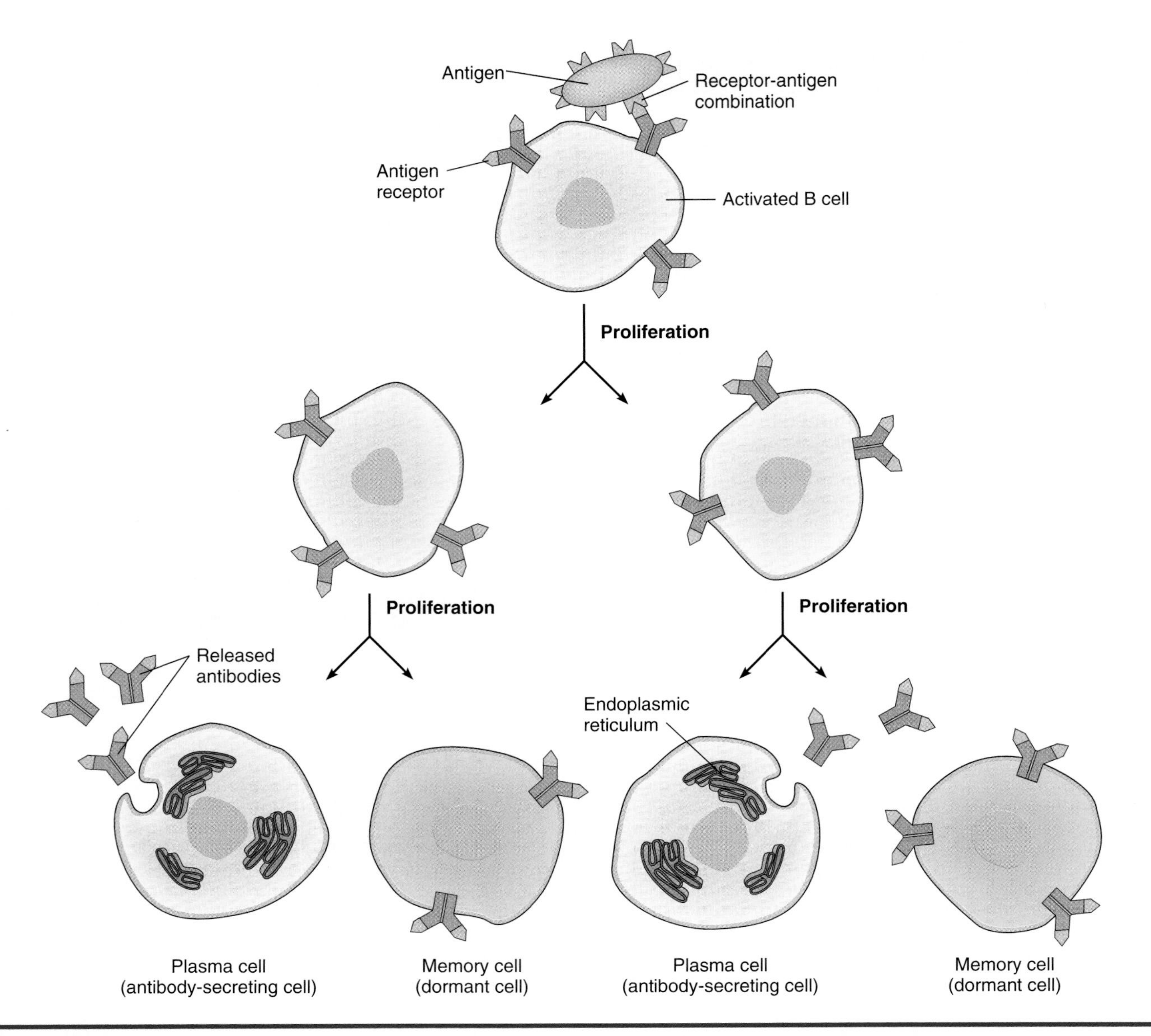

Figure 13.10 An activated B cell rapidly divides, producing a clone of B cells with the same antigen receptor. The clone consists of plasma cells and memory cells.

Cell-Mediated Immunity

An activated T cell undergoes rapid cell division to produce a clone of T cells with the same antigen receptor as the activated T cell. The clone consists of *helper T cells, killer (cytotoxic) T cells, memory T cells,* and *suppressor T cells.*

When activated, helper T cells release cytokines, which stimulate clone formation of T and B cells and stimulate killer T cells. Their most important role is to start and intensify an immune response.

Killer T cells attack cells bearing the specific antigen to which they attach. They primarily attack and destroy abnormal body cells, that is, cells infected by a virus or an intracellular parasite, cancer cells, and foreign cells, such as cells of a transplanted organ. They destroy cells by releasing a cytotoxin that ruptures the plasma membrane of the target cell. Cytokines released by killer T cells attract additional lymphocytes and macrophages to join in the battle. The cellular debris is removed by macrophages and neutrophils.

Once the target cells or pathogens have been destroyed by antibody-mediated and/or cell-mediated mechanisms, suppressor T cells slow and finally stop the immune response. Memory T cells remain to launch an even stronger immune response should the same antigen reenter the body.

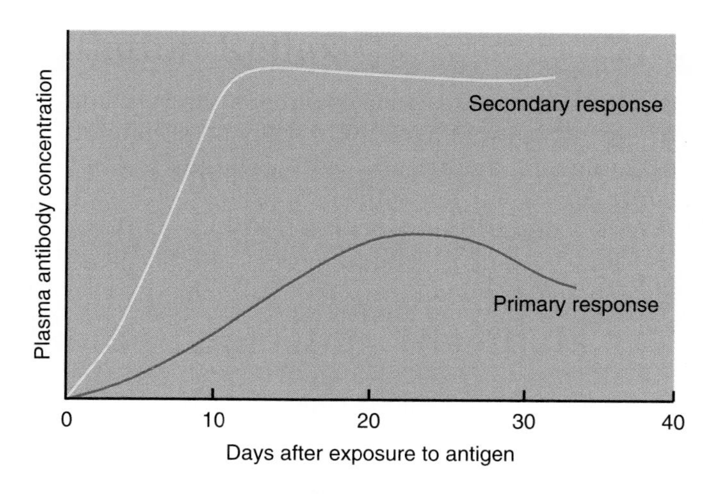

Figure 13.11 Antibodies are produced more rapidly and at a higher concentration during a secondary immune response.

Immune Responses

When an antigen is encountered for the first time, it stimulates T cells and B cells to produce clones that attack and destroy the invading antigen. This is the **primary immune response,** and it also produces memory cells that are able to recognize the same antigen if it should reenter the body. If another invasion of the same antigen occurs at a later date, the memory T cells and memory B cells recognize it and launch a **secondary immune response,** which is more rapid and intense than the primary immune response. A secondary immune response occurs each time the same antigen is detected by the memory cells. Figure 13.11 shows that the concentration of antibodies in the secondary immune response is much higher than in the primary response.

Types of Immunity

There is more than one way for a person to be immune to a particular disease, and these mechanisms may be grouped into two broad categories: active immunity and passive immunity. A person is directly involved in the development of **active immunity** but not in **passive immunity.** Also, active immunity is longer lasting than passive immunity, but the duration of active immunity varies in different diseases.

Naturally acquired active immunity results after a person contracts a disease and recovers, leaving antibodies and memory B and T cells to fight the pathogen if it reenters the body.

Artificially acquired active immunity results after a person receives a vaccine (vak-sēn′) of dead or weakened (attenuated) pathogens or inactivated toxins (toxoid), which trigger a primary immune response, leaving antibodies and memory B and T cells to fight the pathogen if it reenters the body. Booster shots may be used to trigger a secondary immune response to build up the concentration of antibodies even higher.

Naturally acquired passive immunity occurs in infants who have received maternal antibodies via the placenta and in mother's milk after birth. You can see that breast-feeding provides more benefits to the infant than nutrients.

Artificially acquired passive immunity results from receiving injections of antibodies (immunoglobulins) produced in another person or animal. Tetanus antitoxin, commercially produced in horses, is an example.

Why are helper T cells so important to immunity?
How does a vaccination provide immunity against disease?

Rejection of Organ Transplants

Organ transplants are viable treatment options for persons with terminal disease of certain organs such as the heart, kidneys, and liver. Except for the surgery, the major problem encountered is that the patient's immune system recognizes the transplanted organ as foreign (nonself) and launches an attack against it. The problem is reduced by carefully matching compatibility of the tissues of both donor and recipient. This is usually done by comparing the antigens on the surfaces of leukocytes, called the *human leukocyte antigens group A (HLA)* of donor and recipient. A match of 75% is considered minimal for making a transplant.

To overcome the normal immune reaction and organ rejection, immunosuppressive therapy is administered following transplant surgery. The immune system must be suppressed sufficiently to prevent rejection of the organ but not enough to eliminate immunity against pathogens. Achieving this delicate balance has been aided by the use of *cyclosporine,* a selective immunosuppressive drug derived from fungi. Cyclosporine inhibits T cell functions but has minimal effects on B cells. Since T cells are primarily responsible for organ rejection, rejection is minimized, and B cells are able to provide antibody-mediated immunity against pathogens. In spite of advances in immunosuppressive therapy, bacterial and viral infections are the primary causes of death among organ transplant recipients.

Immunization is dependent on recognition by memory cells of antigens to which they have previously been sensitized. The injection of vaccines produces the primary immune reaction. A secondary immune response occurs if the pathogen actually invades the body or if additional injections of the vaccine are given. Booster shots of vaccines trick the memory cells into producing a secondary immune response as if a real reinvasion of the antigen has occurred. This secondary immune response raises the concentration of T cells, B cells, and antibodies, which increases the person's immunity against that antigen.

This secondary immune response is the basis for repeated childhood immunizations. Most children in the United States receive *DPT vaccine,* which stimulates immune reactions against diphtheria, pertussis (whooping cough), and tetanus, at selected ages: 2 to 3 months, 4 to 5 months, 6 to 7 months, 15 to 19 months, and 4 to 6 years. Similarly, *oral poliomyelitis vaccine* is administered at 4 to 5 months, 6 to 7 months, 15 to 19 months, and 4 to 6 years.

Disorders of the Lymphatic and Immune Systems

These disorders may be grouped according to infectious or noninfectious origins.

Infectious Disorders

Acquired immune deficiency syndrome (AIDS) is a viral disease that is approaching epidemic proportions. It is caused by the *human immunodeficiency virus (HIV),* which attacks and kills helper T cells and invades macrophages, which serve as a reservoir for the virus. In time, the immune defenses of the victim are greatly reduced, and the patient becomes susceptible to opportunistic diseases that ultimately lead to death. These secondary diseases include pneumonia caused by *Pneumocystis carinii* and a cancer known as Kaposi's sarcoma, disorders that are rarely encountered except in AIDS patients. At present, there is no cure for AIDS, although a major research effort has provided drugs to slow its progress.

Although HIV has been found in most body fluids, it appears that sufficient HIV concentrations for transmission to other persons are not present in tears or saliva. Transmission does not occur through routine, nonintimate contact; instead, it primarily occurs through blood exchanges and sexual intercourse. Transmission occurs through exchanges of blood, most commonly by the use of contaminated hypodermic needles and by exposure of open wounds or mucous membranes to infected blood. Vaginal fluids and semen of infected persons are effective transmitting agents in sexual intercourse. Also, infected mothers may transmit HIV to infants during childbirth.

Elephantiasis (el-e-fan-tī′-ah-sis) is a tropical disease. It is a chronic condition characterized by greatly swollen (edemous) legs or other body parts, which become "elephantlike" in appearance. This occurs because the lymphatic vessels are blocked by masses of microscopic roundworms, which causes fluid to accumulate excessively in the tissues drained by the plugged lymphatic vessels. The microscopic worms are transmitted by the bites of certain species of mosquitoes.

Tonsillitis is the inflammation of the tonsils. It usually results from bacterial infections that cause the tonsils to become sore and swollen. If this condition becomes chronic and interferes with breathing or swallowing, or becomes a persistent focal point for spreading infections, the tonsils may be surgically removed. Tonsillectomies are much less common now that effective antibiotics are available and the role of tonsils in immunity is better understood.

Noninfectious Disorders

Allergy (al′-er-jē), or **hypersensitivity,** refers to an abnormally intense immune response to an antigen that is harmless to most people. Such antigens are called *allergens* to distinguish them from antigens associated with disease. Once a person is sensitized to an allergen, an allergic reaction results whenever subsequent exposure to that allergen occurs. Allergic reactions may be either immediate or delayed.

Immediate reactions result when allergens bind with IgE antibodies on the surface of basophils and mast cells. This interaction causes these cells to secrete substances such as histamine that stimulate an inflammatory response. Immediate allergic reactions may be either localized or systemic (whole body).

Localized reactions, such as hay fever, hives, allergy-based asthma, and digestive disorders, are unpleasant but rarely life threatening. In contrast, systemic allergic reactions, also known as *anaphylaxis* (an-ah-fi-lak′-sis), are often life threatening. They quickly impair breathing and may cause circulatory failure, due to a sudden drop in blood volume as blood vessels dilate and fluid moves into the tissues. Allergic reactions to penicillin and bee stings are examples of systemic allergic responses.

Delayed allergic reactions appear one to three days after exposure to the antigen. Delayed allergic reactions result from cytokines released by T cells. They are caused by poison ivy and certain chemicals.

Autoimmune diseases result when T and B cells, for unknown reasons, recognize certain body tissues as foreign antigens and produce an immune response against them. This problem may result because certain body molecules have changed slightly and are no longer recognizable as self. Among the major autoimmune diseases are rheumatoid arthritis, rheumatic fever, myasthenia gravis, multiple sclerosis, and systemic lupus erythematosus.

Lymphoma (lim-fō′-mah) is a general term referring to any tumor of lymphatic tissue. There are several types of lymphomas. One type of malignant lymphoma is *Hodgkin's disease.* It is characterized by enlarged, but painless, lymph nodes, fatigue, and sometimes fever and night sweats. Early treatment with chemotherapy or radiation yields a high cure rate.

What is so serious about the AIDS virus destroying helper T cells?
What is the cause of autoimmune diseases?

CHAPTER SUMMARY

Lymph

1. As interstitial fluid accumulates in tissue spaces, it is picked up by lymphatic capillaries. Once the fluid is within a lymphatic vessel, it is called lymph.

Lymphatic Capillaries and Vessels

1. Lymph is carried by lymphatic vessels and is ultimately returned to the blood. Lymphatic vessels contain valves that keep the lymph moving in one direction.
2. Lymphatic vessels merge to form lymphatic trunks, which drain major regions of the body. Lymphatic trunks ultimately join one of two collecting ducts.
3. The right lymphatic duct receives lymphatic trunks that drain the upper right portion of the body; it empties into the right subclavian vein.
4. The thoracic duct receives lymphatic trunks from the rest of the body; it empties into the left subclavian vein.

Transport of Lymph

1. The propulsive forces that move lymph through the vessels are skeletal muscle contractions and respiratory movements.

Lymphatic Organs

1. Lymphatic organs include lymph nodes, tonsils, the spleen, and the thymus gland. Lymphatic organs are sites of lymphocyte production and immune reactions.
2. Lymph nodes occur in groups along lymphatic vessels, where they cleanse lymph as it passes through them.
3. Tonsils are groups of lymphatic tissues located near the entrance to the pharynx. They intercept pathogens entering the pharynx from the mouth and nose.
4. The spleen filters and cleanses the blood and contains a reservoir of blood that is squeezed into circulation if more blood is needed. It destroys old and damaged red blood cells.
5. The thymus gland is located above the heart within the mediastinum. It is the site of T lymphocyte differentiation.

Nonspecific Resistance Against Disease

1. Nonspecific resistance provides general protection against all pathogens, but it is not directed at any particular pathogen.
2. Mechanical barriers include the skin, mucous membranes, mucus, tears, saliva, and urine.
3. Protective chemicals include gastric juice, interferon, enzymes such as lysozyme and pepsin, and fluids with an acid pH.
4. Phagocytosis is a major mechanism. Neutrophils, monocytes, and wandering and fixed macrophages actively engulf and destroy microorganisms. The wandering and fixed macrophages form the tissue macrophage system, which plays a major role in protection against disease.
5. Inflammation helps control infection by attracting leukocytes and macrophages and by increasing the blood supply to the affected area.

6. Fever increases the rate of defense processes and inhibits the growth of certain microorganisms.

Immunity

1. Immunity provides protective mechanisms against specific antigens. Lymphocytes and macrophages play key roles in immunity.
2. Maturing lymphocytes learn to distinguish molecules composing the body (self) from foreign (nonself) molecules. Large nonself molecules that can stimulate an immune response are called antigens.
3. Undifferentiated lymphocytes are formed in fetal bone marrow and are released into the blood. About half of the lymphocytes are carried to the thymus, where they mature to become T lymphocytes. The remainder mature in unknown tissues to become B lymphocytes. Differentiated T cells and B cells are disbursed to lymphatic organs throughout the body and also circulate in the blood.
4. B cells provide antibody-mediated immunity. B cells are activated when a specific antigen binds to their antigen receptors. The activated helper T cells stimulate B cells to divide and form a clone. Most of the B cells become plasma cells, which secrete antibodies to attack the antigen; some remain dormant as memory B cells, which can activate an immune response if the original antigen reappears.
5. Antibodies are specific proteins produced by plasma cells against specific antigens. There are five classes of antibodies: IgA, IgD, IgE, IgG, and IgM. Each performs a special role.
6. T cells provide cell-mediated immunity. Antigens are engulfed and presented on the cell surface of macrophages. Those helper T cells that can bind with the antigen are activated (sensitized) to produce a clone of T cells to attack the antigen.
7. Several subtypes of T cells compose the clone. Killer T cells directly attack and destroy the antigen. Helper T cells stimulate B cells to produce antibodies and increase the phagocytic activity of macrophages. Suppressor T cells slow or stop these actions when the antigen has been eliminated. Memory T cells remain dormant but are able to recognize the original antigen and promote an immune response if it should reappear.

Immune Responses

1. The first contact with a specific antigen produces the primary immune response. Subsequent contacts with the same antigen produce secondary immune responses that are more rapid and intense.
2. There are two basic types of immunity: active and passive. Active immunity results when the body produces its own memory cells and antibodies. Passive immunity results from introduced antibodies that have been produced by another organism. The use of monoclonal antibodies provides passive immunity.

Rejection of Organ Transplants

1. The normal response of the immune system is to attack and destroy organ transplants since they are foreign. Rejection is minimized by a good match between donor and recipient and by the administration of immunosuppressive drugs.
2. The use of cyclosporine inhibits functions of T cells but leaves functions of B cells essentially intact.

Disorders of the Lymphatic and Immune Systems

1. Infectious disorders include AIDS, elephantiasis, and tonsillitis.
2. Disorders of the immune system include allergy, autoimmune diseases, and lymphoma.

BUILDING YOUR VOCABULARY

1. **Selected New Terms**
   active immunity, p. 279
   antigen-presenting cell, p. 276
   B cell, p. 275
   clone, p. 277
   cytokine, p. 277
   fever, p. 274
   helper T cell, p. 276
   immunoglobulin, p. 277
   interferon, p. 273
   killer T cell, p. 278
   memory B cell, p. 277
   memory T cell, p. 278
   passive immunity, p. 279
   plasma cell, p. 277
   suppressor T cell, p. 278
   T cell, p. 275
   thymus, p. 272
   tonsil, p. 272
2. **Related Clinical Terms**
   immunocompetence (im-mū-nō-kom′-pe-tens) The ability of the immune system to respond to a foreign antigen.
   immunodeficiency (im-mū-nō-dē-fish′-en-sē) The inability of the immune system to respond to a foreign antigen.
   lymphadenectasis (lim-fad-in-nek′-tah-sis) Enlargement of a lymph node.
   splenectomy (splē-nek′-tō-mē) The surgical incision in the spleen.

STUDY ACTIVITIES

1. Complete the **Chapter 13 Study Guide,** which begins on page 465.
2. Complete the **Learning Objectives** listed on the first page of this chapter.

CHECK YOUR UNDERSTANDING

(Answers are located in appendix B.)

1. The lymphatic system removes excess _____ from tissue spaces.
2. Lymphatic vessels carry _____ from body tissues to the _____ veins.
3. Filtering and cleansing lymph is the function of _____ , which also are sites of _____ production.
4. Filtering and cleansing blood is a function of the _____ , which also removes worn-out _____cells.
5. Mucous membranes, tears, and gastric juice provide _____ resistance against disease.
6. The major phagocytes involved in nonspecific resistance are _____ and _____ .
7. Fighting infection is aided by _____ , which increases the blood flow and attracts phagocytes to the affected area.
8. Lymphocytes that are processed by the _____ become T lymphocytes.
9. To start an immune response, part of an antigen must be displayed by an _____ cell where a specific _____ cell can bind to it.
10. Subtype cells formed in a clone of T cells are _____ T cells, _____ T cells, _____ T cells, and _____ T cells.
11. B cells produce _____ -mediated immunity, and T cells produce _____ -mediated immunity.
12. Lymphokines released by _____ cells promote rapid cell division in activated _____ cells and _____ cells.
13. An activated B cell forms a clone consisting of _____ cells, which produce _____, and memory B cells.
14. Immunity acquired by contact with a pathogen and producing antibodies against it is known as _____ immunity.
15. _____ artificially introduces an antigen into the body and produces a primary immune response, so if the real pathogen subsequently enters the body, a faster, more intense _____ immune response destroys it.

FOURTEEN

14

C H A P T E R

The Respiratory System

Chapter Preview & Learning Objectives

Organs of the Respiratory System
- List the parts of the respiratory system and describe their functions.

Breathing
- Describe the mechanism of breathing.

Respiratory Volumes
- Describe the various respiratory volumes and explain the significance of each one.

Control of Breathing
- Describe the neural control of breathing.

Factors Influencing Breathing
- Identify the factors that influence breathing and explain how they produce their effect.

Gas Exchange
- Describe the mechanisms of gas exchange in the lungs and body tissues.

Transport of Respiratory Gases
- Describe how oxygen and carbon dioxide are transported by the blood.

Disorders of the Respiratory System
- Describe the major disorders of the respiratory system.

Chapter Summary
Building Your Vocabulary
Study Activities
Check Your Understanding

SELECTED KEY TERMS

Alveolus (alveol = small cavity) A microscopic air sac within a lung.
Bronchial tree (bronch = windpipe) The branching tubes that carry air to the alveoli of the lungs.
Expiration (ex = from; spirat = breathe) Movement of air out of the lungs.
External respiration The exchange of oxygen and carbon dioxide between the air in alveoli and the blood in alveolar capillaries.
Glottis The opening between the vocal cords within the larynx.
Inspiration Movement of air into the lungs.
Internal respiration The exchange of oxygen and carbon dioxide between the tissue cells and the blood in tissue capillaries.
Larynx (laryn = gullet) The cartilaginous box containing the vocal cords.
Pharynx (pharyn = throat) The throat; the passageway for air from the nasal cavity and mouth to the larynx.
Surfactant A chemical in alveoli that reduces the cohesion of water molecules.
Trachea (trache = windpipe) The tube carrying air from the larynx to the bronchi.

The cells of the body require a continuous supply of oxygen for cellular respiration, which produces carbon dioxide as a waste product. Both the respiratory and cardiovascular systems work together to provide oxygen to the cells and to remove carbon dioxide from them. The overall process of the exchange of respiratory gases (oxygen and carbon dioxide) between the atmosphere and the body cells is called **respiration,** and it involves four distinct events:

1. The movement of air into and out of the lungs is accomplished by breathing (pulmonary ventilation).
2. The exchange of respiratory gases between the air and the blood in the lungs occurs by diffusion. Oxygen moves from the air in the lungs into the blood, and carbon dioxide moves from the blood into the air in the lungs. This exchange is known as *external respiration.*
3. The transport of respiratory gases between the lungs and the body cells is accomplished by the cardiovascular system.
4. The exchange of gases between the blood and the body cells occurs by diffusion. Oxygen moves from the blood into the body cells, and carbon dioxide moves from the body cells into the blood. This exchange is known as *internal respiration.*

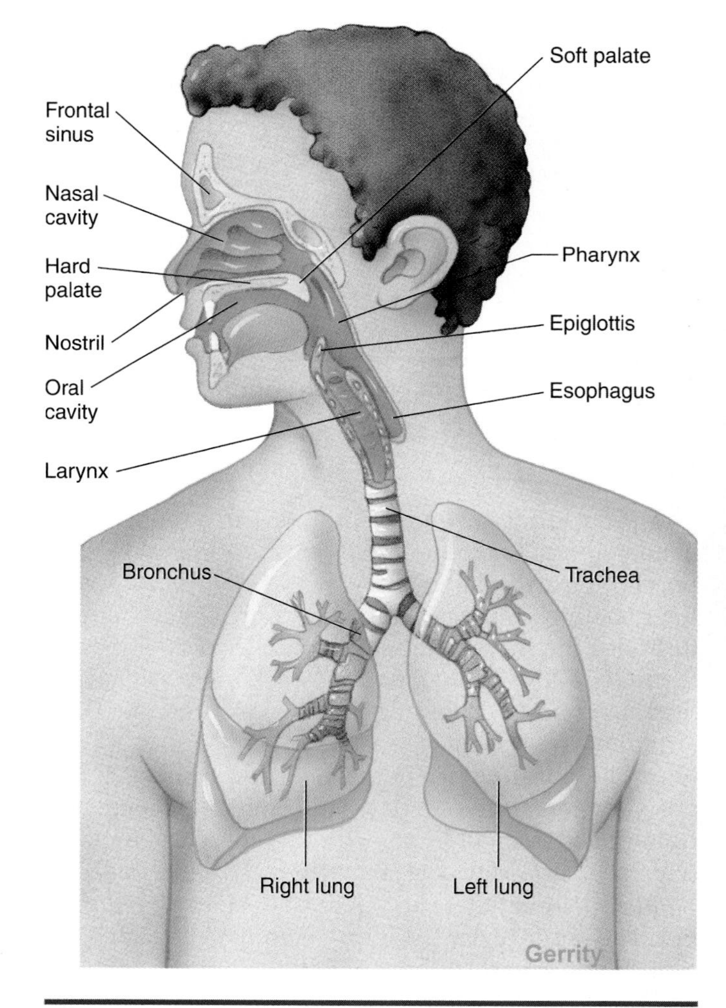

Figure 14.1 The respiratory system.

Organs of the Respiratory System

The **respiratory system** is subdivided into upper and lower respiratory tracts. The *upper respiratory tract* is that portion not located within the thorax: nose, pharynx, larynx, and part of the trachea. The *lower respiratory tract* is located within the thorax: part of the trachea, the bronchial tree, and the lungs. Figure 14.1 illustrates the major components of the respiratory system. Figure 14.2 shows the major parts of the upper respiratory tract.

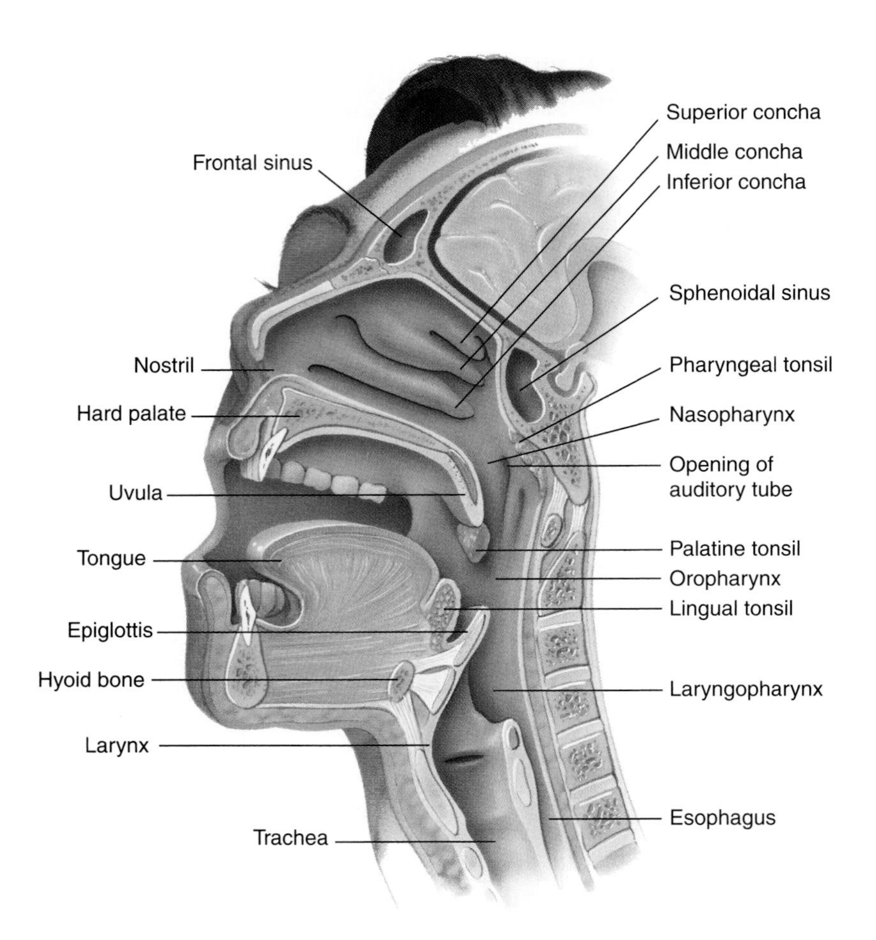

Figure 14.2 Major structures of the upper respiratory tract.

Nose

The protruding portion of the **nose** is supported by bone and cartilage. The nasal bones form a rigid support for the bridge of the nose, while cartilage supports the remaining portions and is responsible for the flexibility of the nose. The **nostrils,** the two openings in the nose, allow air to enter and leave the nasal cavity. Stiff hairs around the nostrils tend to keep out large airborne particles and insects.

The **nasal cavity** is the interior chamber of the nose, and it is surrounded by skull bones. It is separated from the oral cavity by the **palate** (roof of the mouth), which consists of two basic portions. The anterior *hard palate* is supported by bone; the posterior *soft palate* is not. The nasal cavity is divided into left and right portions by the **nasal septum,** a vertical partition of bone and cartilage that is located on the midline. Three **nasal conchae** (kong′-kē) project from each lateral wall and serve to increase the surface area of the nasal cavity.

The nasal cavity is lined with pseudostratified ciliated columnar epithelium, whose goblet cells produce mucus that coats the epithelial surface. As air flows through the nasal cavity, it is warmed by the blood-rich mucous membranes and is moistened by the mucus. In addition, airborne particles, including microorganisms, are trapped in the mucus layer. Cilia of the epithelium slowly move the layer of mucus with its entrapped particles toward the pharynx, where it is swallowed. Upon reaching the stomach, most microorganisms in the mucus are destroyed by the gastric juice (figure 14.3).

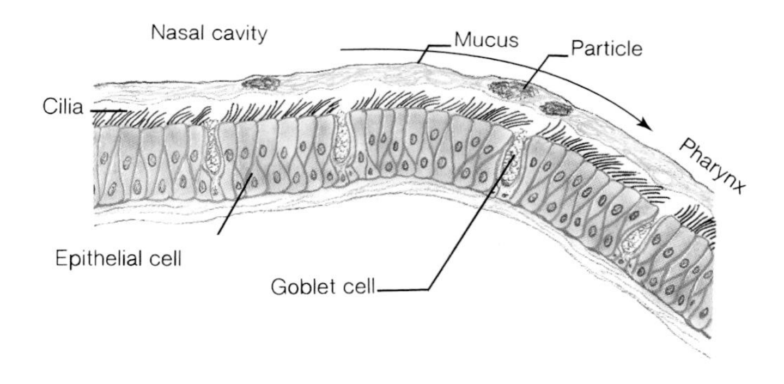

Figure 14.3 Mucus and entrapped particles are moved by cilia from the nasal cavity to the pharynx.

Several bones surrounding the nasal cavity contain **paranasal sinuses,** air-filled cavities. Sinuses are located in the ethmoid, frontal, maxillary, and sphenoid bones adjacent to the nasal cavity. The sinuses lighten the skull and serve as sound-resonating chambers during speech. The sinuses open into the nasal cavity, and they are lined with ciliated mucous membranes that are continuous with the mucous membranes of the nasal cavity.

The paranasal sinuses are lined with mucous membranes that are continuous with those of the nasal cavity. The secreted mucus drains into the nasal cavity. *Sinusitis* results from allergic reactions or infections, and it may cause the membranes to swell so that drainage is reduced or prevented. When this occurs, the increased fluid pressure may cause sinus headaches.

Pharynx

The **pharynx** (fayr′-inks), commonly called the throat, is a short passageway that lies posterior to the nasal and oral cavities and extends to the larynx and esophagus. It has a muscular wall and it is lined with mucous membranes. As shown in figure 14.2, the pharynx consists of three parts: the *nasopharynx* lies behind the nose; the *oropharynx* lies behind the mouth; and the *laryngopharynx* lies behind the larynx.

The auditory (eustachian) tubes, which extend to the middle ear, open into the nasopharynx. Air moves in or out of the auditory tubes to equalize the air pressure on each side of the tympanic membrane.

The tonsils, clumps of lymphatic tissue, occur at the openings to the pharynx. The **palatine tonsils** are located on each side of the pharynx at the junction with the oral cavity. The **pharyngeal tonsil** (adenoids) is located at the upper end of the pharynx, and the **lingual tonsils** are found on the back of the tongue. Tonsils are sites of immune reactions and may become sore and swollen when infected. Enlargement of the palatine tonsils tends to make swallowing painful and difficult. A swollen pharyngeal tonsil tends to block the flow of air from the nasal cavity into the pharynx, which promotes mouth breathing. When breathing through the mouth, air is not adequately warmed, filtered, and moistened.

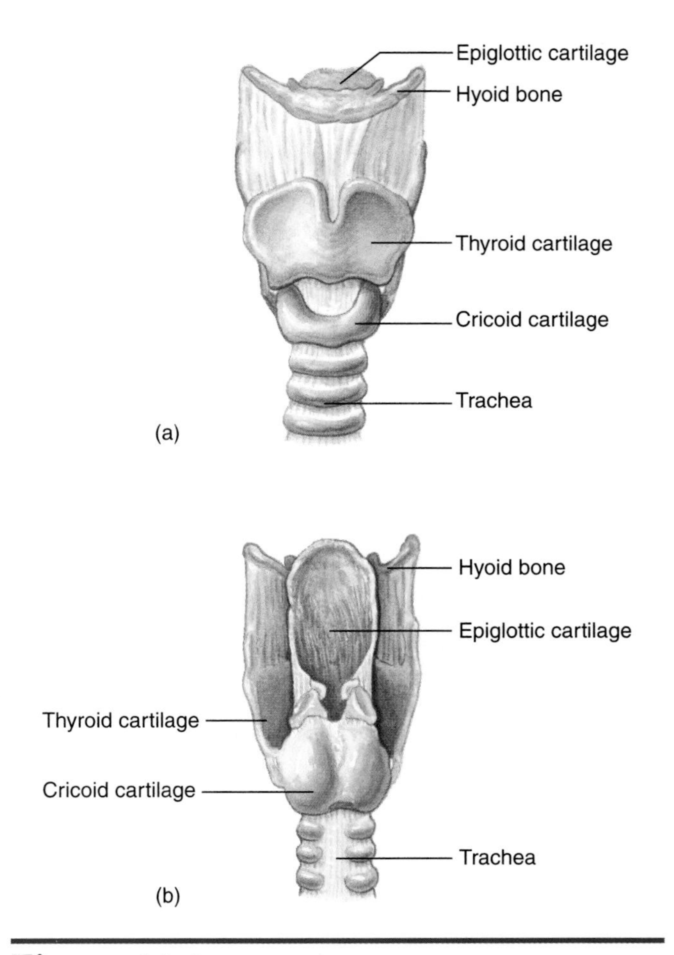

Figure 14.4 The larynx. (*a*) Anterior view. (*b*) Posterior view.

Larynx

The **larynx** (layr′-inks) is a cartilaginous, boxlike structure that provides a passageway for air between the pharynx and the trachea. The three largest cartilages are the *thyroid cartilage,* which projects anteriorly to form the Adam's apple, the *cricoid cartilage,* which forms the attachment to the trachea, and the *epiglottis,* a cartilaginous flap that helps to keep food from entering the larynx. The larynx is supported by ligaments that extend from the hyoid bone (figure 14.4).

The **vocal cords,** folds of the mucous membrane, are located within the larynx (figure 14.5). They are relaxed during normal breathing, but when contracted, they vibrate to produce vocal sounds when exhaled air passes over them. The opening between the vocal cords, the **glottis,** leads to the trachea.

Since the oropharynx and laryngopharynx are also passageways for food, a mechanism exists to prevent food from entering the larynx and to direct food into the esophagus (ē-sof′-ah-gus), the flexible tube that carries food to the stomach. When swallowing,

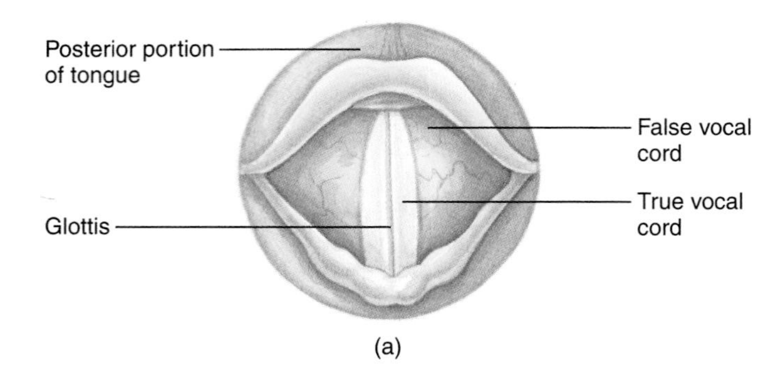

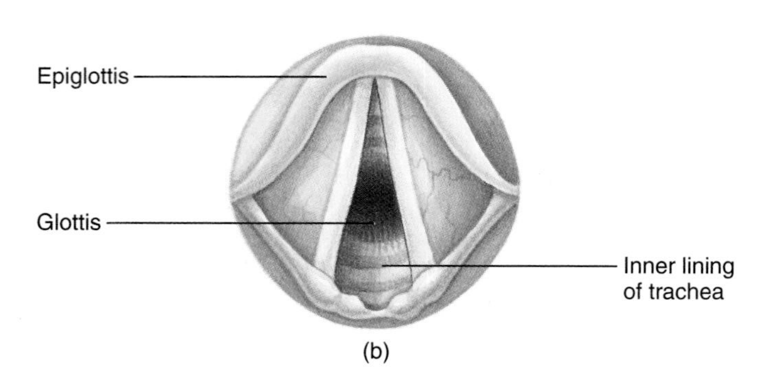

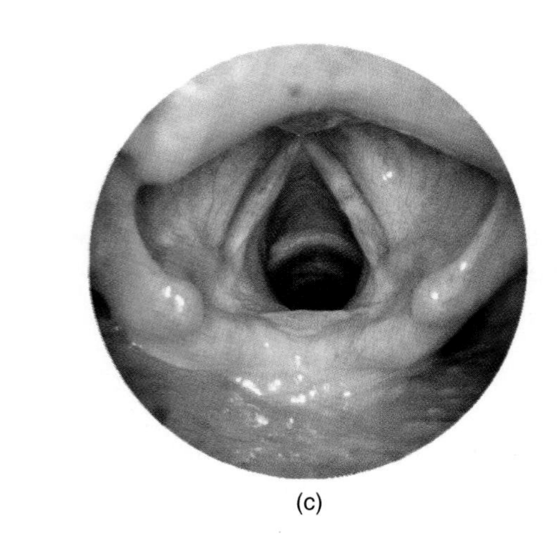

Figure 14.5 Vocal cords as viewed from above with the glottis (*a*) closed and (*b*) open. (*c*) Photograph of the glottis and vocal folds.

muscles lift the larynx upward, which causes the epiglottis to flop over and cover the opening into the larynx. This action directs food into the esophagus, whose opening is located just posterior to the larynx. Sometimes this mechanism does not work perfectly and a small amount of food or drink enters the larynx, stimulating a coughing reflex that usually expels the substance.

What are the functions of the nose?
What are the three divisions of the pharynx, and where are they located?
What are the functions of the glottis and epiglottis?

Trachea

The **trachea** (trā′-kē-ah), or windpipe, is a tube that extends from the larynx into the thoracic cavity, where it branches to form the primary bronchi. The walls of the trachea are supported by C-shaped **cartilaginous rings** that hold the passageway open in spite of the air pressure changes that occur during breathing. The open portion of the cartilaginous rings is oriented posteriorly against the esophagus. This orientation allows the esophagus to expand slightly as food passes down to the stomach.

The inner wall of the trachea is lined with the same type of ciliated mucous membrane that lines the upper respiratory passages. Its mucus functions in a similar way—to trap airborne particles and microorganisms. The beating cilia move the mucus along with entrapped particles upward to the pharynx, where it is either swallowed or spat out.

Infections of the mucous membranes lining the nasal cavity and pharynx may spread to the middle ear via the auditory (eustachian) tubes to cause *otitis media,* a middle ear infection that is common in young children.

Bronchial Tree

The trachea branches at about midchest into the left and right **primary bronchi** (brong′-kī). Each primary bronchus enters its respective lung, where it branches to form smaller **secondary bronchi,** one for each lobe of the lung. Within each lobe the bronchi continue to branch into smaller and smaller tubules. Because the trachea, bronchi, and smaller tubules resemble a tree and its branches, they are collectively called the **bronchial tree** (figure 14.6).

The walls of the bronchi contain cartilaginous rings similar to those of the trachea, but as the branches get progressively smaller, the amount of cartilage gradually decreases and finally is absent in the

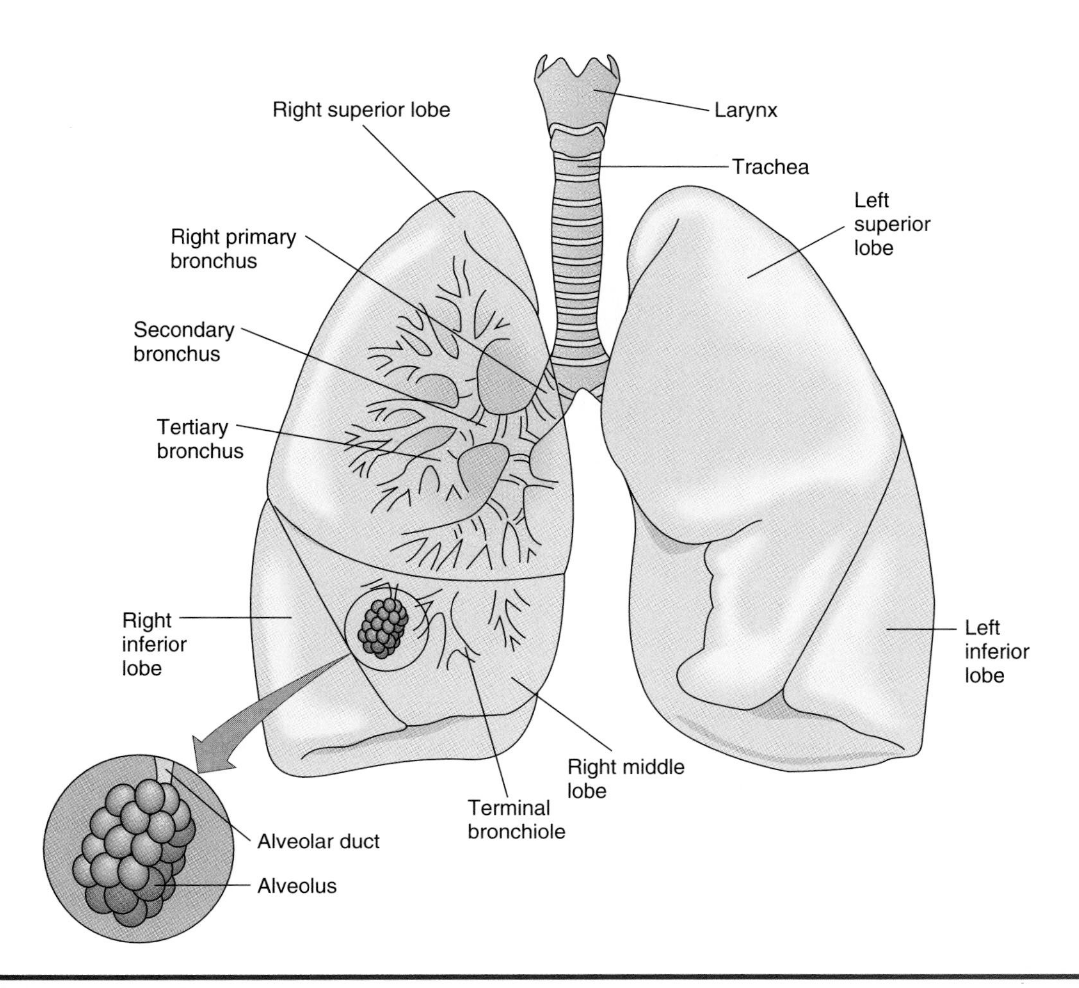

Figure 14.6 The lungs and the bronchial tree. The bronchial tree provides a passageway for air between the trachea and the alveoli.

very small tubes called the **bronchioles** (brong′-kē-ōls). As the amount of cartilage decreases, the amount of smooth muscle increases. Air passageways larger than bronchioles are lined with ciliated mucous membranes that continue to trap and remove airborne particles. Bronchioles and smaller tubules are composed of simple cuboidal epithelium, so foreign particles that reach them are not effectively removed. The terminal bronchioles branch to form microscopic **alveolar ducts,** which terminate in tiny air sacs called **alveoli** (al-vē′-ō-lī). Alveoli resemble tiny grapes clustered about an alveolar duct (figure 14.7).

The primary function of the bronchial tree is to carry air into and out of the alveoli during breathing. The exchange of respiratory gases occurs between the air in the alveoli and the blood in the capillary networks that surround the alveoli. Alveoli are extremely numerous—about 300 million in each lung. They have a combined surface area of about 75 square meters and can hold about 6,000 ml of air.

The interior of the bronchi may be viewed via the use of a *bronchoscope,* an illuminated, flexible tube that can be passed down the trachea to the bronchi. This procedure is known as a *bronchoscopy.*

Since alveoli contain very small spaces and the interior surfaces of alveoli are coated with fluid, the attraction between water molecules in the fluid tends to collapse the alveoli. Normally, this is prevented because special cells of the alveoli secrete **surfactant** (ser-fak′-tant), which reduces the attraction (cohesion) between water molecules within the alveoli. In this way, surfactant plays a vital role in keeping the alveoli

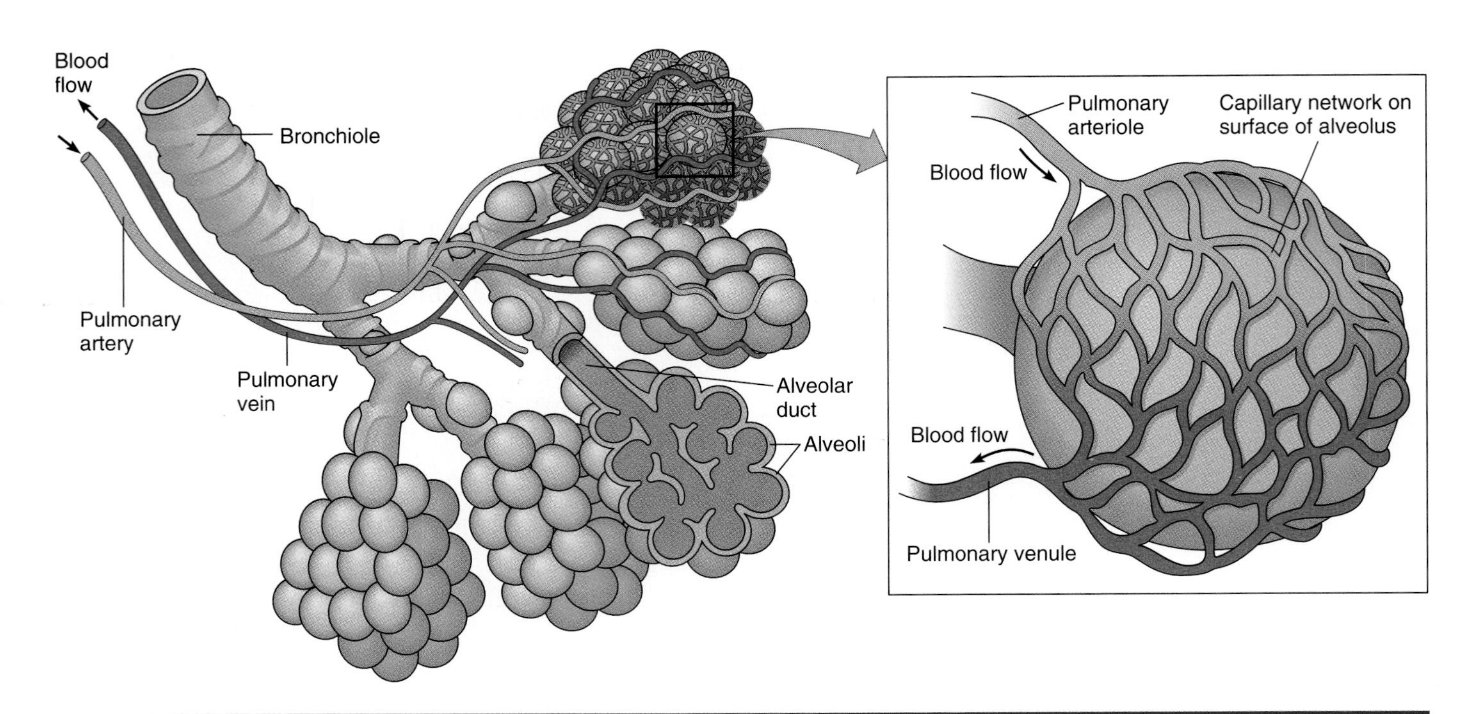

Figure 14.7 Terminal bronchioles branch to form alveolar ducts; and each alveolar duct opens into many alveoli, which cluster around it. Each alveolus is enveloped by a capillary network.

open so that they may be filled with air during inspiration. Without surfactant, alveoli would collapse and become nonfunctional.

Lungs

The paired **lungs** are large organs that occupy much of the thoracic cavity. They are roughly cone-shaped and are separated from each other by the mediastinum and the heart. Each lung is divided into lobes. The left lung has two lobes and it is somewhat smaller than the right lung, which is composed of three lobes. Each lobe is supplied by a secondary bronchus, blood and lymphatic vessels, and nerves. The lungs consist primarily of air passages, alveoli, blood and lymphatic vessels, and connective tissues giving the lungs a soft, spongy texture.

Two serous membranes enclose and protect each lung. The **visceral pleura** is firmly attached to the surface of each lung, and the **parietal pleura** lines the inner wall of the thorax and forms part of the mediastinum. The potential space between the visceral and parietal pleurae is known as the **pleural cavity.** A thin film of serous fluid occupies the pleural cavity and reduces friction between the pleural membranes as the lungs inflate and deflate during breathing. Although lungs are elastic and tend to contract, the attraction of water molecules in the serous fluid within the pleural cavity keeps the visceral and parietal pleurae pressed together.

Table 14.1 summarizes the functions of the respiratory organs.

What is the function of cartilaginous rings in the walls of the trachea and bronchi?
What is the function of the bronchial tree?
Why is surfactant important?

Breathing

Breathing (pulmonary ventilation) is the process that exchanges air between the atmosphere and the alveoli of the lungs. Air moves into and out of the lungs along an air pressure gradient—from regions of higher pressure to regions of lower pressure. There are three pressures that are important in breathing:

1. **Atmospheric pressure** is the pressure of the air that surrounds the earth. Atmospheric pressure at sea level is 760 mmHg (mercury), but at higher elevations it decreases because there is less air at higher elevations. Atmospheric pressure forces air into the lungs.
2. **Intrapulmonary pressure** is the air pressure within the lungs. As we breathe in and out, this

Table 14.1 Summary of Functions of the Respiratory Organs

Component	Function
Nose	Nostrils allow air to enter the nasal cavity; the nasal cavity filters, warms, and moistens air as it passes toward the pharynx
Pharynx	Carries air from the nasal cavity to the larynx; filters, warms, and moistens the air; also serves as passageway for food from the mouth to the esophagus
Larynx	Carries air from the pharynx to the trachea; contains vocal cords that produce sounds in vocalization; prevents objects from entering the trachea
Trachea	Receives air from the larynx and carries it to the primary bronchi; filters, warms, and moistens the air
Bronchial tree	Air from the trachea enters the primary bronchi, which enter the lungs and branch repeatedly to form the bronchial tree; passageways larger than bronchioles filter, warm, and moisten the air
Lungs	Gas exchange occurs between air in the alveoli and blood in surrounding capillaries

pressure fluctuates between being lower than atmospheric pressure and then higher than atmospheric pressure.

3. **Intrapleural pressure** is the pressure within the pleural cavity. It is normally about 756 mmHg, which is slightly less than atmospheric pressure. This lower intrapleural pressure is often described as a "negative pressure," and it keeps the lungs pressed against the inner walls of the thorax, even as the thorax expands and contracts during breathing. If the intrapleural pressure were to equal atmospheric pressure, the lungs would collapse and be nonfunctional.

Inspiration

The process of breathing air into the lungs is called **inspiration,** or inhalation. When the lungs are at rest, the air pressure in the lungs is the same as the atmospheric air pressure. In order for air to flow into the lungs, the air pressure within the lungs must be reduced to less than the atmospheric air pressure. This is accomplished by the contraction of the primary breathing muscles: the diaphragm and the external intercostals.

The dome-shaped **diaphragm** is a thin sheet of skeletal muscle separating the thoracic and abdominal cavities. When it contracts, the diaphragm pulls downward and becomes flattened, which increases the volume of the thoracic cavity. At the same time, contraction of the **external intercostal muscles** lifts the ribs upward and outward, and pushes the sternum forward, which further increases the volume of the thoracic cavity (figure 14.8).

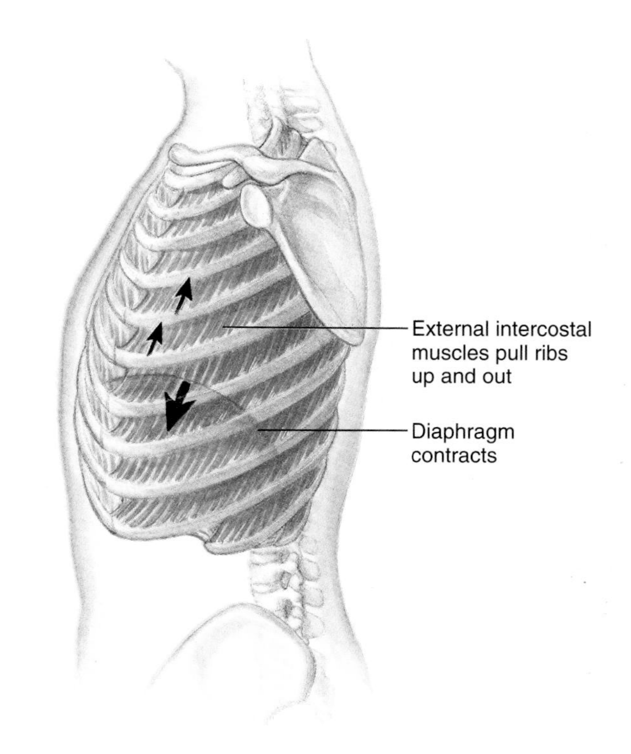

Figure 14.8 Inspiration results when the thoracic cavity is increased by contraction of the diaphragm and the external intercostal muscles. At the end of inspiration, these muscles relax, which decreases the thoracic cavity resulting in expiration.

When the thoracic cavity expands, the lungs are pulled along, causing an increase in lung volume and a decrease in the intrapulmonary pressure. The higher atmospheric pressure forces air through the air passageways into the lungs until intrapulmonary and atmospheric pressures are equal (figure 14.9).

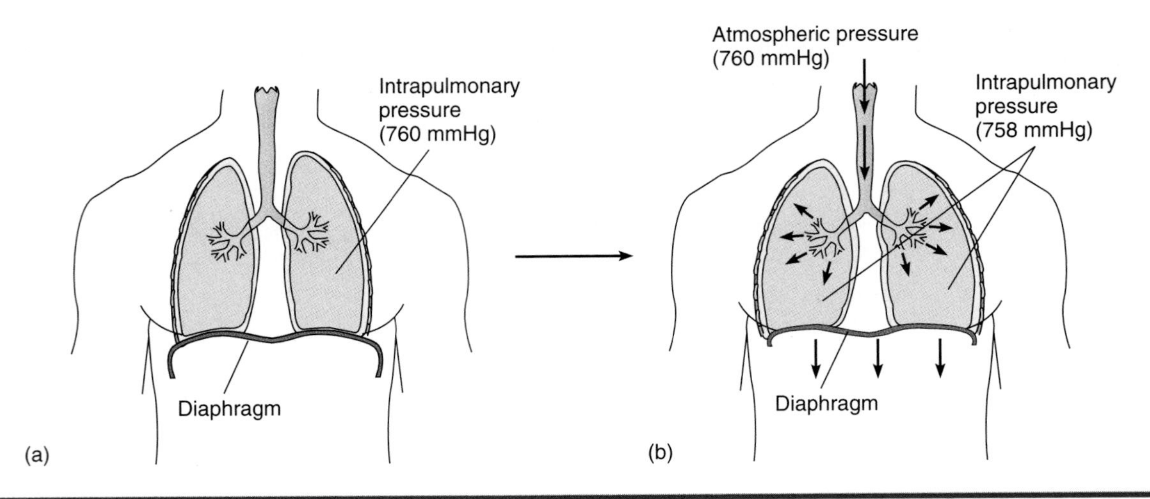

Figure 14.9 (*a*) When the lungs are at rest, air pressure in the lungs is equal to atmospheric pressure (760 mmHg). (*b*) When the thoracic cavity enlarges, air pressure in the lungs decreases and the higher atmospheric pressure forces air into the lungs.

The presence of air in the pleural cavity is called a *pneumothorax* (nū-mō-thō′-raks). This may occur due to a thoracic injury or surgery that allows air to enter the pleural cavity. It also occurs in emphysema patients when air escapes from ruptured alveoli into the pleural cavity. The result of a pneumothorax is that the affected lung collapses and is nonfunctional. Since each lung is in a separate pleural cavity, the collapse of one lung does not adversely affect the other lung. Treatment involves removing the intrapleural air to restore the normal pressure so that the lungs may inflate.

Expiration

Expiration, or exhalation, occurs when the diaphragm and external intercostal muscles relax, allowing the thoracic cage and lungs to return to their original size. This results in a decrease in the volume of the thoracic cavity and lungs. The decrease in lung volume increases intrapulmonary pressure to a level higher than atmospheric pressure. The higher intrapulmonary pressure forces air out of the lungs until intrapulmonary and atmospheric pressures are equal.

Expiration during quiet breathing is a rather passive process since the abundant elastic tissue in the lungs and thoracic wall causes them to return to their original size as soon as the muscles of inspiration relax. However, a forceful expiration is possible by contraction of the **internal intercostal muscles,** which pull the ribs down and inward, and by the muscles of the abdominal wall, which force the abdominal viscera and diaphragm upwards. These contractions further decrease the volume of the lungs, which increases the air pressure in the lungs, causing more air to flow out of the lungs.

How do intrapleural pressure and serous fluid in the pleural cavity affect breathing? Describe the mechanisms of inspiration and expiration.

Respiratory Volumes

Normal adults average 12 to 15 quiet breathing cycles per minute. A *breathing cycle* is one inspiration followed by one expiration. The volume of air inhaled and exhaled in a quiet or forceful breathing cycle varies with size, sex, age, and physical condition. The average respiratory volumes have been determined by size, age, and sex in order to enable evaluation of pulmonary functions. Respiratory volumes that are 80% or less than the normal average usually indicate some form of pulmonary disease.

An instrument called a *spirometer* is used to determine respiratory volumes. It produces a *spirogram,* a graphic record of the volume of air exchanged.

The volume of air exchanged in a quiet breathing cycle is about 500 ml, and it is known as the **tidal volume (TV).** Forceful inspirations and expirations can exchange a much greater volume of air. The maximum volume of air that can be forcefully inhaled in addition to a tidal inspiration is about 3,000 ml, and it is known as the **inspiratory reserve volume (IRV).**

The maximum volume of air that can be forcefully exhaled after a tidal expiration is about 1,100 ml, and it is known as the **expiratory reserve volume**

Table 14.2 Summary of Respiratory Volumes

Name	Definition	Average Volume
Tidal volume (TV)	Volume of air inhaled or exhaled during quiet breathing	500 ml
Inspiratory reserve volume (IRV)	Volume of air that can be forcefully inhaled after a tidal volume inhalation	3,000 ml
Expiratory reserve volume (ERV)	Volume of air that can be forcefully exhaled after a tidal volume expiration	1,100 ml
Vital capacity (VC)	Maximum volume of air that can be forcefully exhaled after a maximum forceful inhalation VC = TV + IRV + ERV	4,600 ml
Residual volume (RV)	Volume of air that always remains in the lungs	1,200 ml
Total lung capacity (TLC)	Maximum volume of air that the lungs can contain TLC = VC + RV	5,800 ml

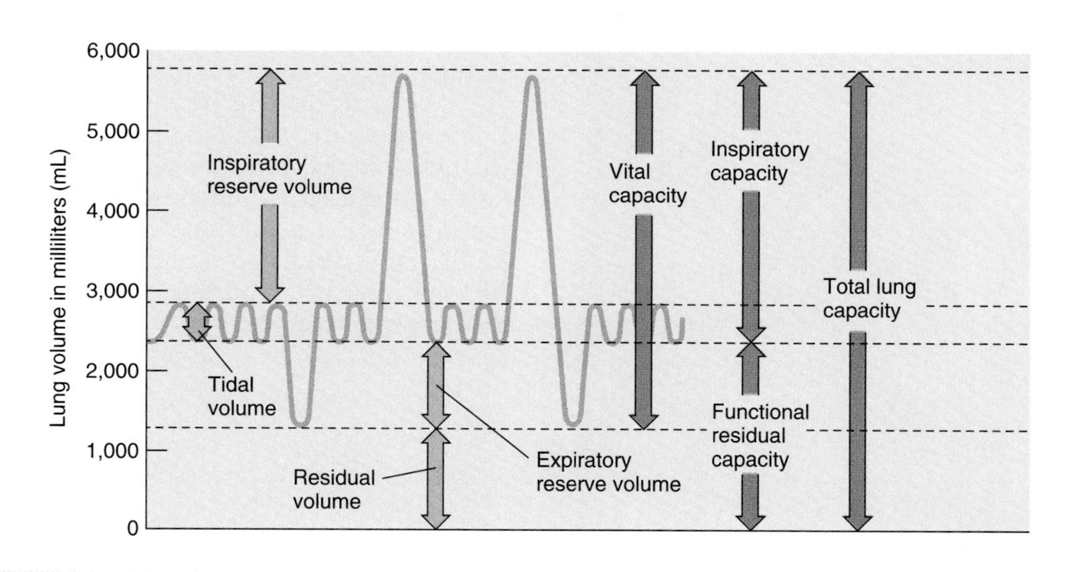

Figure 14.10 Respiratory volumes and capacities.

(ERV). About 1,200 ml of air remains in the lungs after a maximum forced expiration. This residual air is known as the **residual volume (RV).** Once an infant takes its first breath, there is always residual air in the lungs. The residual air keeps the alveoli open, preventing collapse of the lungs.

The maximum amount of air that can be forcefully exchanged is known as the **vital capacity (VC),** and it is equal to the sum of the tidal volume, inspiratory reserve volume, and the expiratory reserve volume—about 4,600 ml. The **total lung capacity (TLC)** is equal to the sum of the vital capacity and the residual volume—about 5,800 ml.

The respiratory volumes are summarized in table 14.2 and are graphically shown in figure 14.10.

What distinguishes inspiratory reserve and expiratory reserve?
What distinguishes tidal volume and vital capacity?

Control of Breathing

Breathing is controlled by groups of neurons composing the **respiratory center** in the brain stem. The respiratory center consists of groups of neurons that are located in both the pons and the medulla (figure 14.11).

The determination of the lung volumes is useful in identifying the two basic categories of pulmonary disease: obstructive and restrictive disorders. Each category has a characteristic pattern of abnormal test results. It is possible for a patient to exhibit both patterns simultaneously.

The *obstructive pattern* occurs where there is airway obstruction from any cause, such as in asthma, bronchitis, and emphysema. In this pattern, residual volume is increased and the RV/TLC ratio is increased.

The *restrictive pattern* occurs when there is a loss of lung tissue or when expansion of the lungs is limited. This pattern may result from lung tumors, weakness of respiratory muscles, pulmonary edema, or fibrosis of the lungs. In this pattern, the TLC is decreased, the RV/TLC ratio is normal or increased, and the VC is decreased.

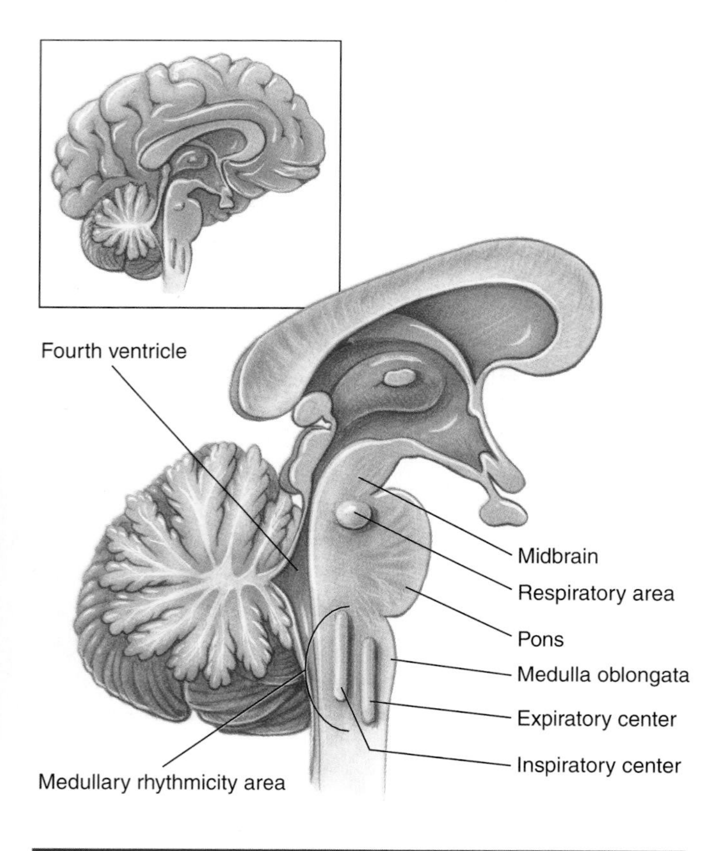

Figure 14.11 The respiratory center controls breathing. It is located in the pons and medulla oblongata of the brain stem.

Medullary Respiratory Center

The medulla contains the **respiratory center,** which controls the rhythmic nature of breathing. It is composed of two components: an inspiratory area and an expiratory area.

Neurons of the **inspiratory area** rhythmically depolarize, sending impulses via the *phrenic nerves* to the diaphragm and via the *intercostal nerves* to the external intercostal muscles. The impulses stimulate these breathing muscles to contract, producing inspiration. When the inspiratory area stops sending the impulses, the muscles relax and expiration occurs.

When deeper inspirations and forceful expirations are required, such as during strenuous exercise, the frequency of impulses sent to the breathing muscles is increased. To quickly exhale the greater volume of air, the **expiratory area** is activated, and it sends impulses that stimulate the contraction of the internal intercostal muscles and abdominal muscles, which produce forceful expirations. The expiratory area is not involved in expirations during quiet breathing.

Pons Respiratory Center

The **pons respiratory center** coordinates the actions of the medullary respiratory center to produce smooth inspirations and expirations. It contains two kinds of neurons: those whose impulses stimulate and those whose impulses inhibit the medullary respiratory rhythmicity center. These impulses either increase or decrease the depth and length of inspiration. The antagonistic actions of these impulses produce a homeostatic balance of the breathing rate and depth of inspiration to meet body needs.

Factors Influencing Breathing

The respiratory center is influenced by a number of factors that cause adjustments in the depth and rate of breathing. Some factors stimulate the respiratory center directly, while others are detected by sensors that send impulses to the respiratory center.

Inflation Reflex

Stretch receptors in the visceral pleurae are sensitive to the degree of stretching by the lungs. During inspiration, impulses from the stretch receptors are sent to the respiratory center via the vagus nerve, where they inhibit the formation of impulses causing inspiration. This promotes expiration and prevents excessively deep inspirations that may damage the lungs.

Higher Brain Centers

Impulses from higher brain centers may affect the respiratory center. These impulses may be voluntarily generated, as when a person chooses to alter the normal pattern of quiet breathing. However, these voluntary controls are limited. For example, if a little child tries to "punish" his mother by holding his breath, as soon as the level of CO_2 in his blood increases to a critical point, the impulses from higher brain centers are ignored and involuntary breathing resumes.

Involuntary impulses may be formed by higher brain centers during emotional experiences, such as anxiety, fear and excitement, and during chronic pain. At such times, the breathing rate is increased. Similarly, a sudden emotional experience, a sharp pain, or a sudden cold stimulus tend to momentarily stop breathing, a condition called **apnea** (ap′-nē-ah).

Body Temperature

An increase in body temperature, such as occurs during strenuous exercise or a fever, increases the breathing rate. Conversely, a decrease in body temperature lowers the breathing rate.

Chemicals

The most important chemical factors affecting respiration are the concentrations of CO_2, hydrogen ions (H^+), and O_2 in the blood. Sensory receptors that are sensitive to these factors are called **chemoreceptors,** and they occur in the respiratory center, the carotid bodies, and the aortic bodies. The *carotid bodies* are located in both common carotid arteries, where they divide into the internal and external carotid arteries. The *aortic bodies* are located in the aortic arch. You can see that they are strategically located, especially to monitor blood going to the brain.

You may wonder why the concentration of H^+ is involved in respiratory control. This occurs because the mechanism for transporting CO_2 in the blood releases H^+ as a by-product. Therefore, an increase in CO_2 concentration produces an increase in the H^+ concentration.

The chemoreceptors in the respiratory center are sensitive to changes in CO_2 and H^+ concentrations in the blood and cerebrospinal fluid. If the concentrations of CO_2 and H^+ increase, the respiratory center is stimulated to increase the rate and depth of breathing, which increases the rate of CO_2 and H^+ removal and returns their concentrations to normal resting levels. The result is that the rate and depth of breathing also return to normal quiet levels.

If the CO_2 and H^+ concentrations are abnormally low, brief periods of apnea may occur, and the breathing is slow and shallow until their concentrations return to normal levels.

The respiratory center is not sensitive to a decline in O_2 concentration in the blood. Chemoreceptors that detect oxygen levels are located in the carotid bodies and the aortic bodies. These chemoreceptors send impulses via sensory nerves to inform the respiratory center of changes in oxygen levels in the blood. Usually, a drop in O_2 concentration is not a strong stimulus for increasing the rate and depth of breathing, and its main effect seems to be to increase the sensitivity of chemoreceptors in the respiratory center to changes in the CO_2 concentration. The oxygen level of the blood has little effect on normal breathing since it must be very low to have a direct effect on breathing.

How is breathing controlled by the nervous system?
What factors influence the control of breathing?

Gas Exchange

Respiratory gases are exchanged between the air in the alveoli and the blood in the capillaries that surround them. Oxygen and carbon dioxide must diffuse through the *respiratory membrane,* which is composed of the cell forming an alveolar wall and the cell forming a capillary wall (figure 14.12).

Alveolar air has a higher concentration of oxygen and a lower concentration of carbon dioxide than does the capillary blood. Since molecules tend to move from an area of higher concentration to areas of lower concentration, oxygen diffuses from the alveolar air into the blood, and carbon dioxide diffuses from the blood into the alveolar air.

Blood entering a capillary network of an alveolus is oxygen-poor and carbon dioxide-rich. Following

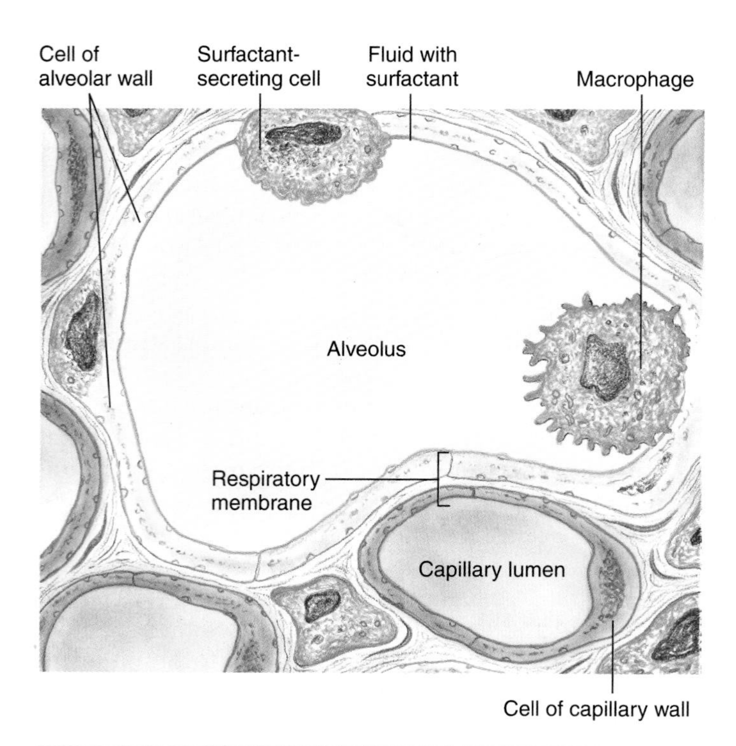

Figure 14.12 The respiratory membrane consists of alveolar and capillary walls.

the gas exchange, blood leaving the capillary is oxygen-rich and carbon dioxide-poor (figure 14.13).

After blood has been oxygenated, it returns to the heart and is pumped throughout the body to supply the tissue cells. Blood in capillaries supplying body tissues contains a higher concentration of oxygen and a lower concentration of carbon dioxide than the tissue cells. Therefore, oxygen diffuses from the blood into the tissue cells, while carbon dioxide diffuses from the tissue cells into the blood. In this way, cells are supplied with oxygen for their metabolic activities, and carbon dioxide, which is produced by cellular metabolism, is removed.

Blood entering a capillary network at the tissue level is oxygen-rich and carbon dioxide-poor. Following gas exchange, blood leaving the capillary is oxygen-poor and carbon dioxide-rich (figure 14.13).

Transport of Respiratory Gases

The red blood cells play a major role in the transport of both oxygen and carbon dioxide.

Oxygen Transport

In the lungs, oxygen diffuses from the air in alveoli into the blood of surrounding capillaries. Most of the oxygen enters red blood cells and combines with the heme (iron-containing) portions of hemoglobin (Hb) to form **oxyhemoglobin (HbO_2)**. About 97% of the oxygen is transported as oxyhemoglobin. Only 3% is dissolved in the plasma.

In body tissues, oxyhemoglobin releases oxygen, and it diffuses from capillary blood into the tissue cells. Actually, only about 25% of the oxygen is released, so oxyhemoglobin is present even in deoxygenated blood.

The reason that hemoglobin is such an effective carrier of oxygen is because the chemical bond between oxygen and hemoglobin is relatively unstable. When the surrounding oxygen concentration is high, as in the lungs, hemoglobin combines readily with oxygen, but when the surrounding oxygen concentration is low, as in body tissues, hemoglobin releases oxygen. The pickup and release of oxygen is summarized as follows:

$$\underset{\text{Hemoglobin}}{\text{Hb}} + \underset{\text{Oxygen}}{O_2} \underset{\text{In tissues}}{\overset{\text{In lungs}}{\rightleftharpoons}} \underset{\text{Oxyhemoglobin}}{\text{HbO}_2}$$

Carbon Dioxide Transport

The transport of carbon dioxide is more complex. When carbon dioxide diffuses from body cells into the capillary blood, it takes three pathways:

1. About 7% is dissolved in the plasma.
2. About 23% enters red blood cells and combines with hemoglobin to form **carbaminohemoglobin ($HbCO_2$)**. Carbon dioxide combines with the globin (protein) portion of hemoglobin, so carbon dioxide and oxygen have different combining sites on hemoglobin. Therefore, hemoglobin can transport oxygen and carbon dioxide at the same time.
3. The remaining 70% of the carbon dioxide also enters red blood cells, but it quickly combines with water to form **carbonic acid (H_2CO_3)**. This reaction is catalyzed by the enzyme **carbonic anhydrase.** Carbonic acid rapidly breaks down (dissociates) into hydrogen ions (H^+) and **bicarbonate ions (HCO_3^-)**. Carbon dioxide is now part of the bicarbonate molecule. The hydrogen ions are picked up (buffered) by hemoglobin, preventing acidosis. Most of the bicarbonate ions diffuse into the plasma, while chloride ions (Cl^-) move into the red blood cells to take their place (figure 14.14).

When the blood gets back to the lungs, all of these reactions run in reverse, releasing carbon dioxide,

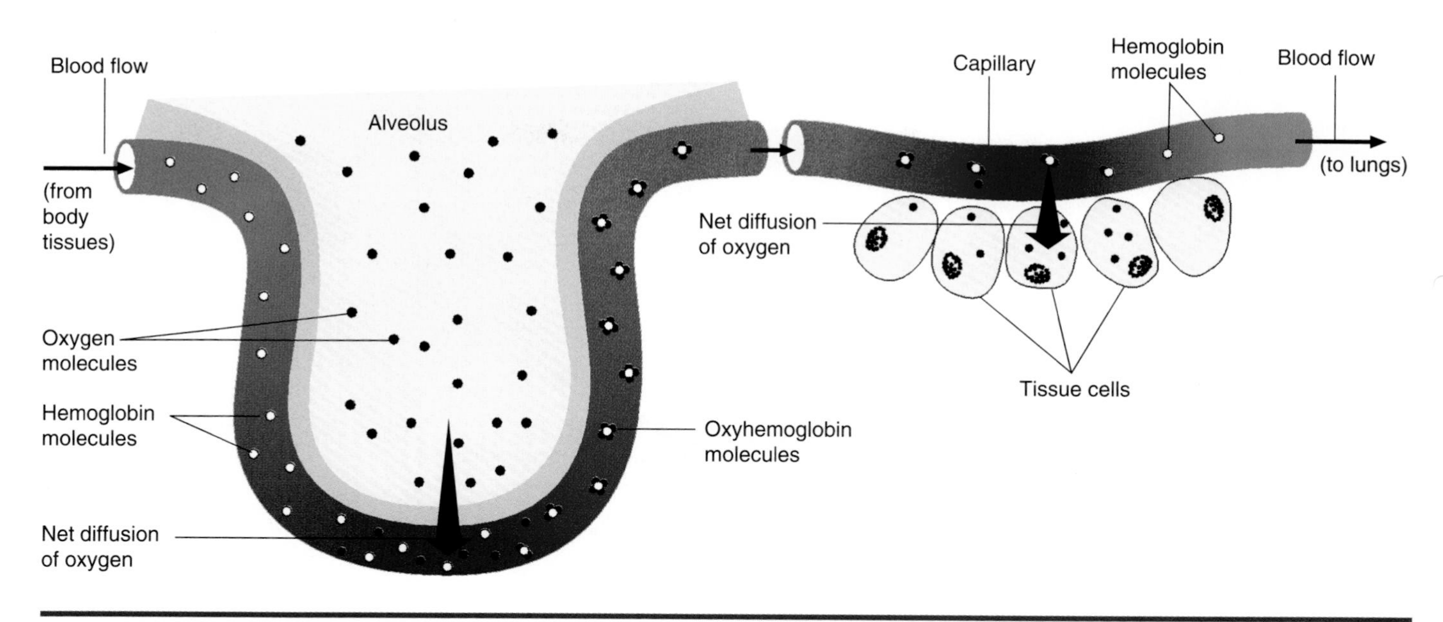

Figure 14.13 Pickup and release of oxygen. In the lungs, oxygen diffuses from alveoli into the blood and combines with hemoglobin to form oxyhemoglobin. In the tissues, oxyhemoglobin releases oxygen that diffuses from the blood to the tissue cells.

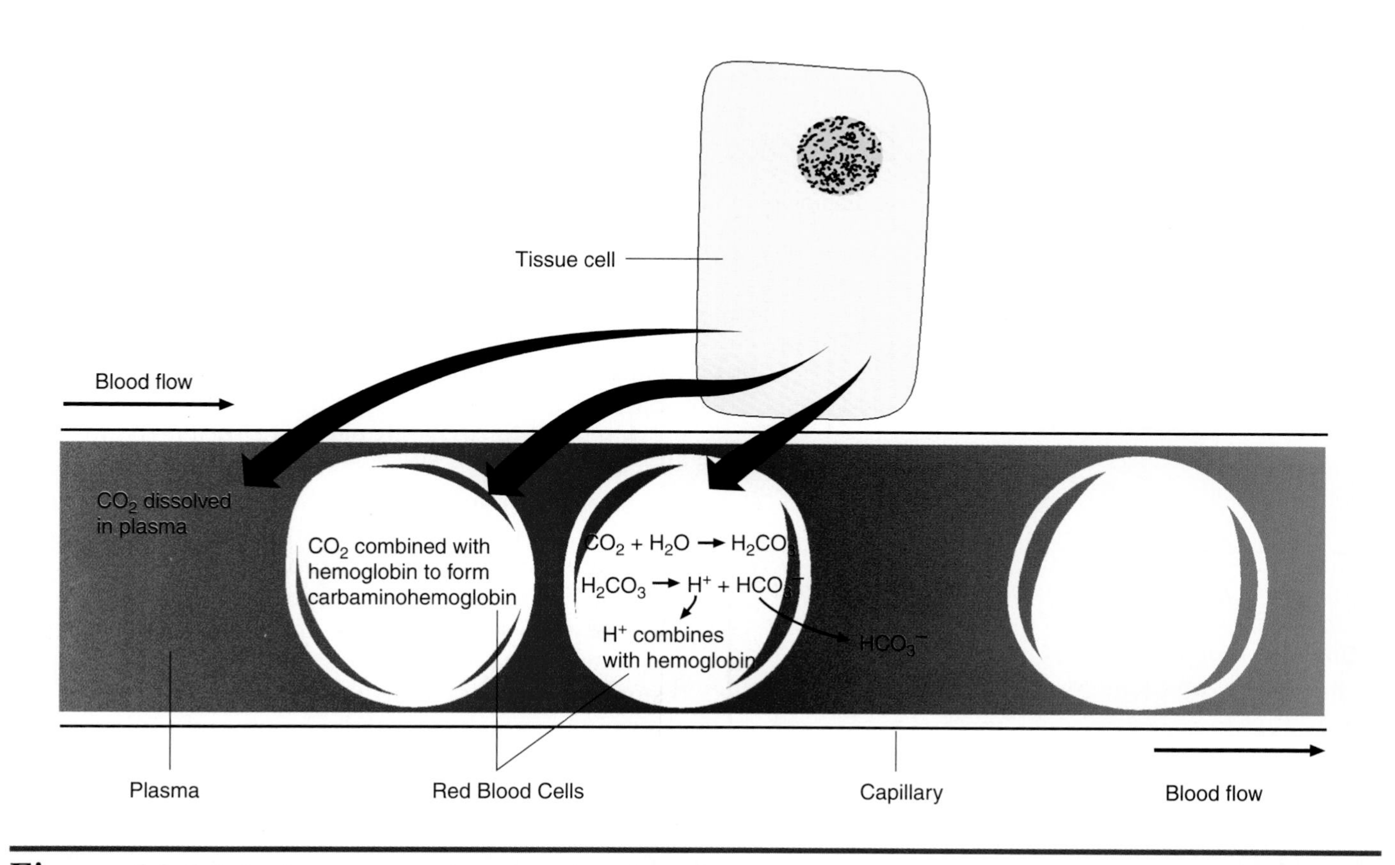

Figure 14.14 In the tissues, carbon dioxide diffuses from tissue cells into the blood for transport to the lungs. Carbon dioxide is carried in three ways: dissolved in plasma, as carbaminohemoglobin, and mostly as bicarbonate ions (HCO_3^-).

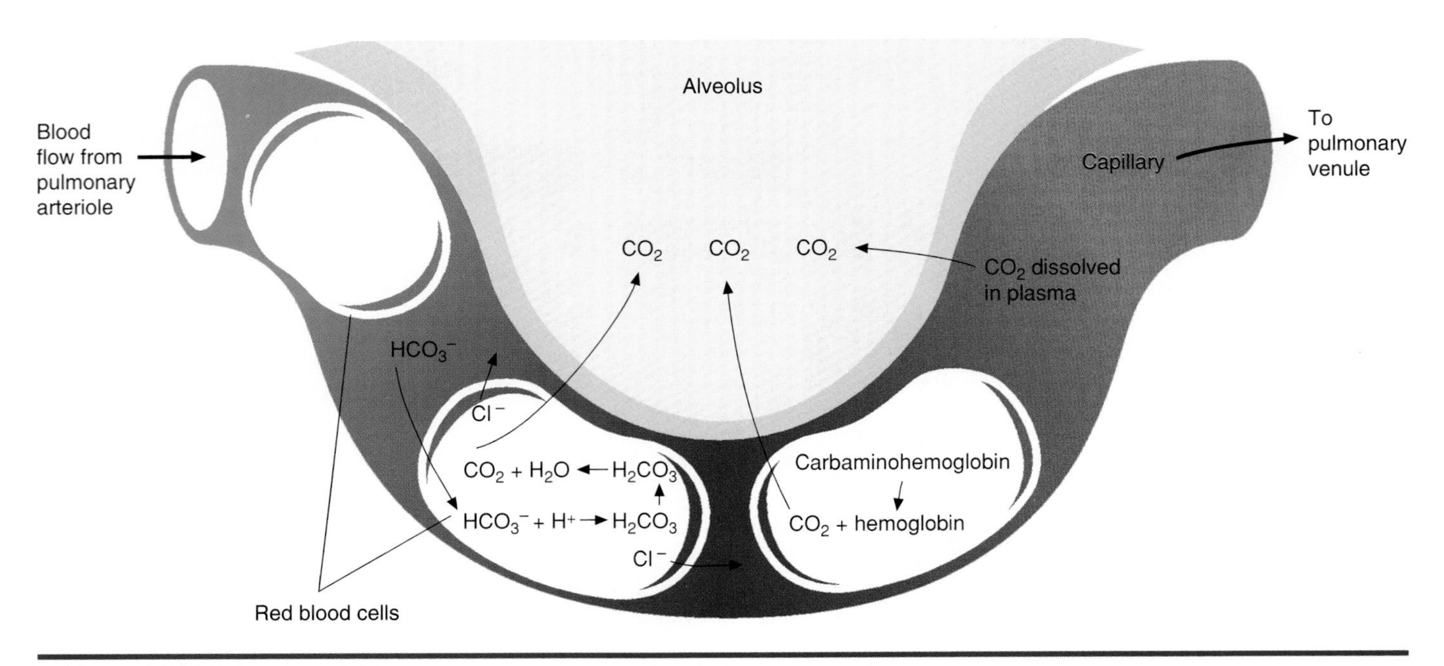

Figure 14.15 In the lungs, carbon dioxide diffuses from the CO_2-rich blood into the alveoli.

which diffuses into the alveoli (figure 14.15). The pickup and release of carbon dioxide is summarized as follows:

$$\underset{\text{Carbon dioxide}}{CO_2} + \underset{\text{Hemoglobin}}{Hb} \underset{\text{In lungs}}{\overset{\text{In tissues}}{\rightleftharpoons}} \underset{\text{Carbaminohemoglobin}}{HbCO_2}$$

$$\underset{\text{Carbon dioxide}}{CO_2} + \underset{\text{Water}}{H_2O} \underset{\text{In lungs}}{\overset{\text{In tissues}}{\rightleftharpoons}} \underset{\text{Carbonic acid}}{H_2CO_3}$$

$$\underset{\text{Carbonic acid}}{H_2CO_3} \underset{\text{In lungs}}{\overset{\text{In tissues}}{\rightleftharpoons}} \underset{\text{Hydrogen ion}}{H^+} + \underset{\text{Bicarbonate ion}}{HCO_3^-}$$

> ***How does gas exchange occur in the lungs and in body tissues?***
> ***How are oxygen and carbon dioxide transported by the blood?***

Disorders of the Respiratory System

These disorders may be categorized as inflammatory or noninflammatory.

Inflammatory Disorders

Asthma (az′-mah) is characterized by wheezing and **dyspnea** (labored breathing) that result from constriction of the bronchioles by the contraction of smooth muscles in their walls. It is often caused by an allergic reaction to airborne substances but also may result from hypersensitivity to bacteria or viruses infecting the bronchial tree.

Bronchitis is inflammation of the bronchi, and it is characterized by excessive mucus production that partially obstructs air flow. Acute bronchitis is usually caused by viral or bacterial infections. Chronic bronchitis occurs in chronic asthmatics, and it is common in smokers due to persistent exposure to irritants in tobacco smoke.

The **common cold** may be caused by a number of viruses, and it often involves rhinitis, laryngitis, and sinusitis. Excessive mucus production, sneezing, and congestion are common symptoms.

Emphysema (em-fi-sē′-mah) results from long-term exposure to airborne irritants, especially tobacco smoke. It is characterized by a rupture of the alveoli, forming larger spaces in the lungs, and excess mucus production, which plugs terminal bronchioles, trapping air in the alveoli. This reduces the respiratory surface area and impairs gas exchange.

A reduced expiratory reserve volume and an increased residual volume are symptomatic. Exhaling requires voluntary effort by the patient. The disease is uncommon except among long-term smokers. It usu-

Carbon monoxide (CO) is an odorless, colorless gas that is produced by burning carbon fuels. It competes with oxygen for the same binding sites on hemoglobin molecules, and it combines with hemoglobin about 200 times more readily than oxygen. Further, CO binds so tightly with hemoglobin that it is hard to remove. Therefore, even small concentrations of CO can displace oxygen from hemoglobin molecules and deprive tissues of needed oxygen. CO poisoning is the leading cause of death from fires. It is especially treacherous because it kills quietly without attracting attention. Treatment includes the administration of 100% oxygen to flush out the CO.

ally can be prevented and progressive deterioration can be stopped by removing the airborne irritant—usually tobacco smoke. Otherwise, there is no cure.

Influenza, or flu, is caused by one of several viruses. Symptoms are fever, chills, headache, and muscular aches followed by coldlike symptoms.

Laryngitis is the inflammation of the mucous membrane lining the larynx. It usually causes a thickening of the vocal cords, which lowers the voice. Viral or bacterial infections or allergens may cause laryngitis.

Pleurisy (pler′-i-sē) is inflammation of the pleural membranes. It often results in a decrease in secretion of serous fluid, which causes sharp pains with each breath. Pleurisy can also cause the opposite effect—an increase in serous fluid secretion. This type exerts pressure on the lungs and impairs expansion of the lungs.

Pneumonia (nū-mōn′-yah) is an acute inflammation of the alveoli that may be caused by viral or bacterial infections. The alveoli become filled with fluid, pathogens, and white blood cells, which reduce space for air exchange. Blood oxygen levels may be greatly reduced. Pneumonia is one of the common causes of death among older people.

Rhinitis (rī-nī′-tis) is inflammation of the membranes lining the nasal cavity. It is characterized by an increased mucus secretion. Causes may be viral or bacterial infections or airborne allergens.

Sinusitis is the inflammation of the sinuses, and it is characterized by an increased mucus secretion and a buildup of pressure within the sinuses. Causes may be viral or bacterial infections or airborne allergens.

Tuberculosis (tū-ber″-kū-lō′-sis) is an inflammation caused by the bacterium *Mycobacterium tuberculosis,* which is transmitted by inhalation. When it infects the lungs, the destroyed lung tissue is replaced by fibrous connective tissue that retards gas exchange and reduces lung elasticity. Fortunately, modern drugs are effective in treating this disease.

Noninflammatory Disorders

Lung cancer usually develops from long-term exposure to irritants, and the most common irritant producing this malignancy is tobacco smoke. The link between lung cancer and cigarette smoking has been firmly established. Lung cancer metastasizes rapidly and is not usually detected until it has spread to other parts of the body. Treatment includes surgical removal of the diseased lung, if detected prior to metastasis, and anticancer drug therapy. Since over 90% of lung cancers occur in smokers, the most effective prevention is the elimination of cigarette smoking.

Pulmonary edema is the accumulation of fluid in the lungs. It results from excessive fluid passing from lung capillaries into the alveoli, which may be due to congestive heart failure. Symptoms include labored breathing and a feeling of suffocation. Treatment includes administration of oxygen, diuretics, drugs that dilate the bronchioles, suctioning air passageways, and mechanical ventilation.

Pulmonary embolism refers to a blood clot or gas bubble that blocks a pulmonary artery and prevents blood from reaching a portion of a lung. Gas exchange cannot occur in the affected parts of the lung. A massive embolism affecting a large portion of a lung may cause a cardiac arrest.

Respiratory distress syndrome (RDS), or hyaline membrane disease, is a disease of newborn infants, especially premature infants. It results from an insufficient production of surfactant in the alveoli, leading to alveolar collapse.

At birth, the respiratory system of an infant goes through a transition from a nonfunctional, fluid-filled system to a functional, air-filled system. Normally, an infant's first breath is the most difficult because it must open the collapsed alveoli. Succeeding breaths are easier because surfactant keeps the alveoli open after expiration. Without adequate surfactant, alveoli tend to collapse at each expiration, and the infant must expend a great amount of energy to force them open at each inspiration. RDS causes about 20,000 infant deaths every year.

CHAPTER SUMMARY

Organs of the Respiratory System

1. The major organs of the respiratory system are the nose, pharynx, larynx, trachea, bronchial tree, and lungs.
2. The external portion of the nose is supported by bone and cartilage, while the nasal cavity is surrounded by skull bones. The nasal conchae increase the surface area of the ciliated mucous membrane that lines the nasal cavity.
3. Air enters and leaves the nasal cavity via the nostrils. Inhaled air is filtered, warmed, and moistened by the mucous membrane of the nasal cavity.
4. The pharynx is a short passageway for both air moving from the nasal cavity to the larynx and food passing from the mouth to the esophagus. The pharyngeal, lingual, and palatine tonsils are clumps of lymphoid tissue associated with the pharynx.
5. The larynx is a cartilaginous boxlike structure that conducts air from the pharynx to the trachea, and it houses the vocal cords. During swallowing, the epiglottis prevents food from entering the larynx and directs it into the esophagus.
6. The trachea extends from the larynx into the thoracic cavity, where it branches to form the primary bronchi, which enter the lungs.
7. The bronchial tree consists of the trachea, primary bronchi, secondary bronchi, tertiary bronchi, bronchioles, and alveolar ducts that lead to the alveoli. The bronchial tree carries air into and out of the alveoli.
8. Portions of the bronchial tree (except the bronchioles) are supported by cartilaginous rings and are lined with ciliated mucous membranes. These portions filter, warm, and moisten the air.
9. Lungs fill most of the thoracic cavity. They consist of air passageways of the bronchial tree, alveoli, blood and lymphatic vessels, nerves, and connective tissues. Gas exchange occurs between air in the alveoli and the blood in the alveolar capillaries.
10. Surfactant in the alveoli prevents the collapse of the alveoli.
11. The visceral pleurae cover the outer surfaces of the lungs, and the parietal pleurae line the thoracic cavity. The pleural cavity is the potential space between the pleural membranes.

Breathing

1. Breathing involves inspiration and expiration. Air moves into and out of the lungs along a pressure gradient.
2. Inspiration results from contraction of the diaphragm and external intercostal muscles, which increases the volume and decreases the pressure within the thoracic cavity and lungs. The higher atmospheric pressure causes air to flow into the lungs until the atmospheric and intrapulmonary pressures are equalized.
3. Expiration results from relaxation of these muscles, which decreases the volume and increases the pressure within the thoracic cavity and lungs. The higher intrapulmonary pressure causes air to flow out of the lungs until the intrapulmonary and atmospheric pressures are equalized. A forced expiration involves the contraction of internal intercostal and abdominal muscles.

Respiratory Volumes

1. Respiratory air volumes vary with size, sex, age, and physical condition. Variations from the norm usually indicate a pulmonary disorder.
2. The average values for respiratory volumes are tidal volume—500 ml; inspiratory reserve volume—3,000 ml; expiratory reserve volume—1,100 ml; vital capacity—4,600 ml; residual volume—1,200 ml; total lung capacity—5,800 ml.

Control of Breathing

1. Breathing is controlled by the respiratory center, which is located in the pons and medulla oblongata.
2. The respiratory center in the medulla controls the rhythmic nature of breathing. The inspiratory area generates impulses that stimulate contraction of the diaphragm and external intercostal muscles, resulting in inspiration. When impulse formation ceases, inspiration stops and expiration occurs. The expiratory area is activated only in forceful expirations.
3. The pons respiratory center coordinates the functions of the medullary respiratory center.

Factors Influencing Breathing

1. The stretching of the visceral pleurae during inspiration triggers the inflation reflex, which inhibits excessive inspiration and promotes expiration.
2. Higher brain centers can influence the respiratory center either voluntarily or involuntarily. Sudden pain or cold produce momentary apnea. Chronic pain and anxiety increase the breathing rate.
3. The breathing rate varies directly with changes in body temperature.
4. Chemoreceptors in the respiratory center are sensitive to changes in concentrations of carbon

dioxide and hydrogen ions in the blood. An increase in their concentrations is the primary stimulus for inspiration. The breathing rate and depth vary directly with changes in carbon dioxide and hydrogen ion concentrations.
5. Chemoreceptors in the carotid and aortic bodies are sensitive to the concentration of oxygen in the blood. They send impulses to the respiratory center when a drop in oxygen concentration is detected. Oxygen concentration must be very low to produce a direct effect on breathing.

Gas Exchange

1. Gas exchange between air in the alveoli and the blood in alveolar capillaries occurs by diffusion, and it is called external respiration.
2. Oxygen diffuses from air in the alveoli into the blood; carbon dioxide diffuses from the blood into air in the alveoli.
3. Gas exchange between tissue cells and blood in capillaries occurs by diffusion, and it is called internal respiration.
4. Oxygen diffuses from the blood into the tissue cells; carbon dioxide diffuses from the tissue cells into the blood.

Transport of Respiratory Gases

1. In the lungs, oxygen combines with hemoglobin to form oxyhemoglobin. In tissues, oxyhemoglobin releases oxygen to tissue cells. About 97% of the oxygen is carried as oxyhemoglobin; only about 3% is carried dissolved in blood plasma.
2. Carbon dioxide is mostly carried in bicarbonate ions. When carbon dioxide diffuses from tissues into the blood, most of it enters the red blood cells. Carbonic anhydrase in red blood cells catalyzes the combination of carbon dioxide and water to form carbonic acid, which ionizes to form hydrogen and bicarbonate ions. In the lungs, the reaction reverses to release carbon dioxide into the alveoli.
3. Some carbon dioxide is carried as carbaminohemoglobin, and a little carbon dioxide is carried dissolved in the plasma.

Disorders of the Respiratory System

1. Inflammatory disorders include asthma, bronchitis, common cold, emphysema, influenza, laryngitis, pleurisy, pneumonia, rhinitis, sinusitis, and tuberculosis.
2. Noninflammatory disorders include lung cancer, pulmonary edema, pulmonary embolism, and respiratory distress syndrome.

BUILDING YOUR VOCABULARY

1. **Selected New Terms**
   aortic body, p. 295
   bicarbonate ion, p. 296
   carbaminohemoglobin, p. 296
   carbonic anhydrase, p. 296
   carotid body, p. 295
   intrapleural pressure, p. 291
   laryngopharynx, p. 287
   larynx, p. 287
   lower respiratory tract, p. 285
   nasopharynx, p. 287
   oropharynx, p. 287
   oxyhemoglobin, p. 296
   respiratory membrane, p. 295
   upper respiratory tract, p. 285
2. **Related Clinical Terms**
   apnea (ap′-nē-ah) Cessation of breathing.
   dyspnea (disp-nē′-ah) Labored breathing.
   hyperventilation Rapid, deep breathing.
   pneumothorax (nū-mō-thor′-aks) Air in the pleural cavity causing collapse of a lung.

STUDY ACTIVITIES

1. Complete the **Chapter 14 Study Guide,** which begins on page 469.
2. Complete the **Learning Objectives** listed on the first page of this chapter.

CHECK YOUR UNDERSTANDING

(Answers are located in appendix B.)

1. Inhaled air is moistened, _____ , and warmed by the _____ lining the nasal cavity and most of the bronchial tree.
2. Vocal cords are located within the _____ , and the space between relaxed vocal cords is the _____ .
3. The walls of the trachea and bronchi are supported by _____ , which are absent in _____ and alveolar ducts.
4. The lungs consist largely of blood vessels, air passageways, and tiny air sacs called _____ where gas exchange occurs.
5. The potential space between the visceral and parietal pleurae is the _____ .
6. Alveoli do not collapse because _____ reduces the cohesion of water molecules within alveoli.
7. During inspiration, air moves into the lungs because _____ pressure is greater than _____ pressure.
8. The _____ volume, about _____ ml, is the amount of air inhaled and exhaled in quiet breathing.
9. The rhythmic cycle of breathing is controlled by a respiratory center in the _____ , and it is coordinated by a respiratory center in the _____ .
10. The rate and depth of breathing increases when the blood levels of _____ and _____ increase.
11. In the lungs, _____ moves from alveoli into capillary blood and _____ moves from capillary blood into alveoli.
12. In the blood, oxygen is primarily carried as _____ , and carbon dioxide is primarily carried as _____ .

C H A P T E R

The Digestive System

FIFTEEN

15

Chapter Preview & Learning Objectives

Digestion: An Overview
- Compare mechanical and chemical digestion.
- Describe the role of digestive enzymes.

Alimentary Canal: General Characteristics
- Identify the layers composing the wall of the alimentary canal and describe their functions.

Mouth
- Compare deciduous and permanent teeth.
- Describe the basic structure of a tooth.
- Describe digestion in the mouth.

Pharynx
- Describe the location and function of the pharynx.

Esophagus
- Describe the location and function of the esophagus.

Stomach
- Describe the structure and functions of the stomach.
- Describe the control of gastric secretions.

Pancreas
- Describe the control and functions of pancreatic secretions.

Liver
- Describe the location and functions of the liver.
- Explain how bile release is stimulated.

Small Intestine
- Describe digestion in the small intestine.
- Explain how the end products of digestion are absorbed.

Large Intestine
- Describe the structure and functions of the large intestine.

Nutrients: Sources and Uses
- Identify the sources and uses of carbohydrates, lipids, proteins, vitamins, and major minerals.

Disorders of the Digestive System
- Describe the major disorders of the digestive system.

Chapter Summary
Building Your Vocabulary
Study Activities
Check Your Understanding

SELECTED KEY TERMS

Absorption The process by which nutrients pass from the digestive tract into the blood.
Accessory organs Organs that have a digestive function but are not part of the alimentary canal.
Alimentary canal (aliment = food) The tube through which food passes from the mouth to the anus.
Chyme (chym = juice) Liquified food entering the small intestine from the stomach.
Digestion (digest = to dissolve) Mechanical and chemical processes that convert nonabsorbable foods into absorbable nutrients.
Duodenum The first part of the small intestine, which receives bile and pancreatic juice.
Nutrient (nutri = to nourish) A substance required for the functioning of body cells.
Peristalsis (peri = around; stalsis = constriction) Wavelike contractions that move food through the alimentary canal.
Sphincter (sphin = squeeze) A circular muscle that closes an opening.
Villus (vill = hairy) Tiny projections of the mucous membrane in the small intestine.
Vitamin (vita = life) An organic compound required by the body in minute amounts.

Body cells require a continuous supply of **nutrients** in order to carry out their vital functions. Nutrients include carbohydrates, proteins, lipids, vitamins, and minerals, and they come from the food we eat. However, most food molecules are too large to pass directly into the blood, so they must be digested to break them down into absorbable molecules. Digestion of food and absorption of nutrients are the major functions of the digestive system.

The **digestive system** consists of the *alimentary canal,* a long tube through which food passes, and *accessory organs.* The major parts of the alimentary canal are the mouth, pharynx, esophagus, stomach, small intestine, large intestine, and anus. The major accessory organs are the teeth, salivary glands, liver, gallbladder, and pancreas (figure 15.1).

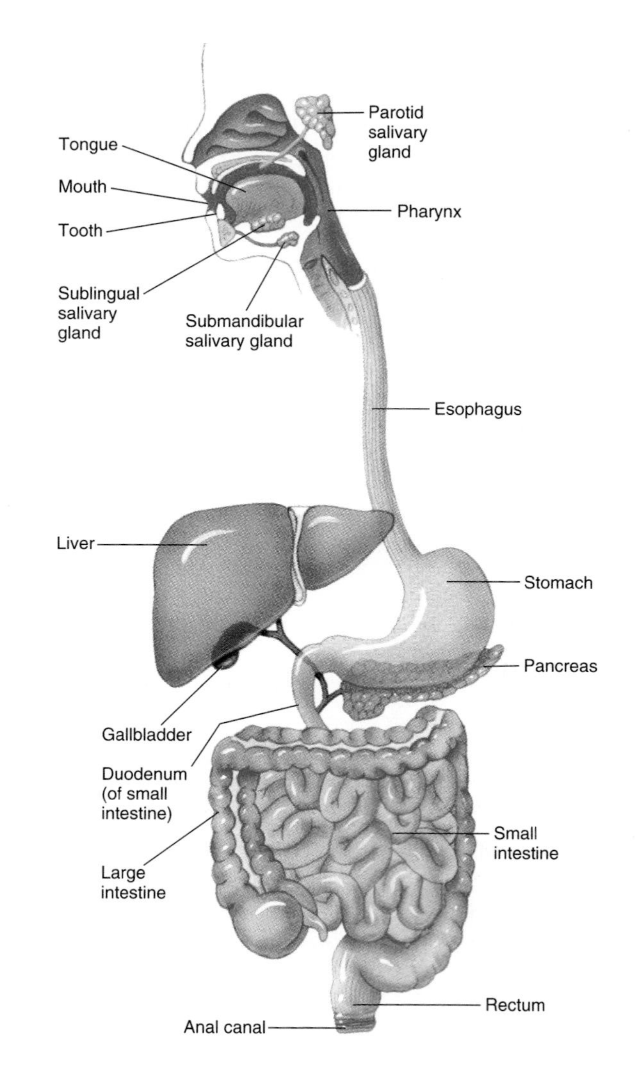

Figure 15.1 Major organs of the digestive system.

Digestion: An Overview

Digestion involves both mechanical and chemical processes. *Mechanical digestion* is the physical breakdown of food into smaller pieces, which provides a greater surface area for contact with digestive secretions. *Chemical digestion* is the splitting of complex, nonabsorbable food molecules into small, absorbable nutrient molecules by the addition of water—a process known as *hydrolysis* (hī-drol′-i-sis). Since hydrolysis is normally very slow, it is the action of digestive enzymes that speeds up digestion and enables the formation of small, absorbable nutrients within the alimentary canal. Chemical digestion may be summarized as follows.

$$\text{Nonabsorbable food molecules} + \text{Water} \xrightarrow{\textit{Digestive enzymes}} \text{Absorbable nutrient molecules}$$

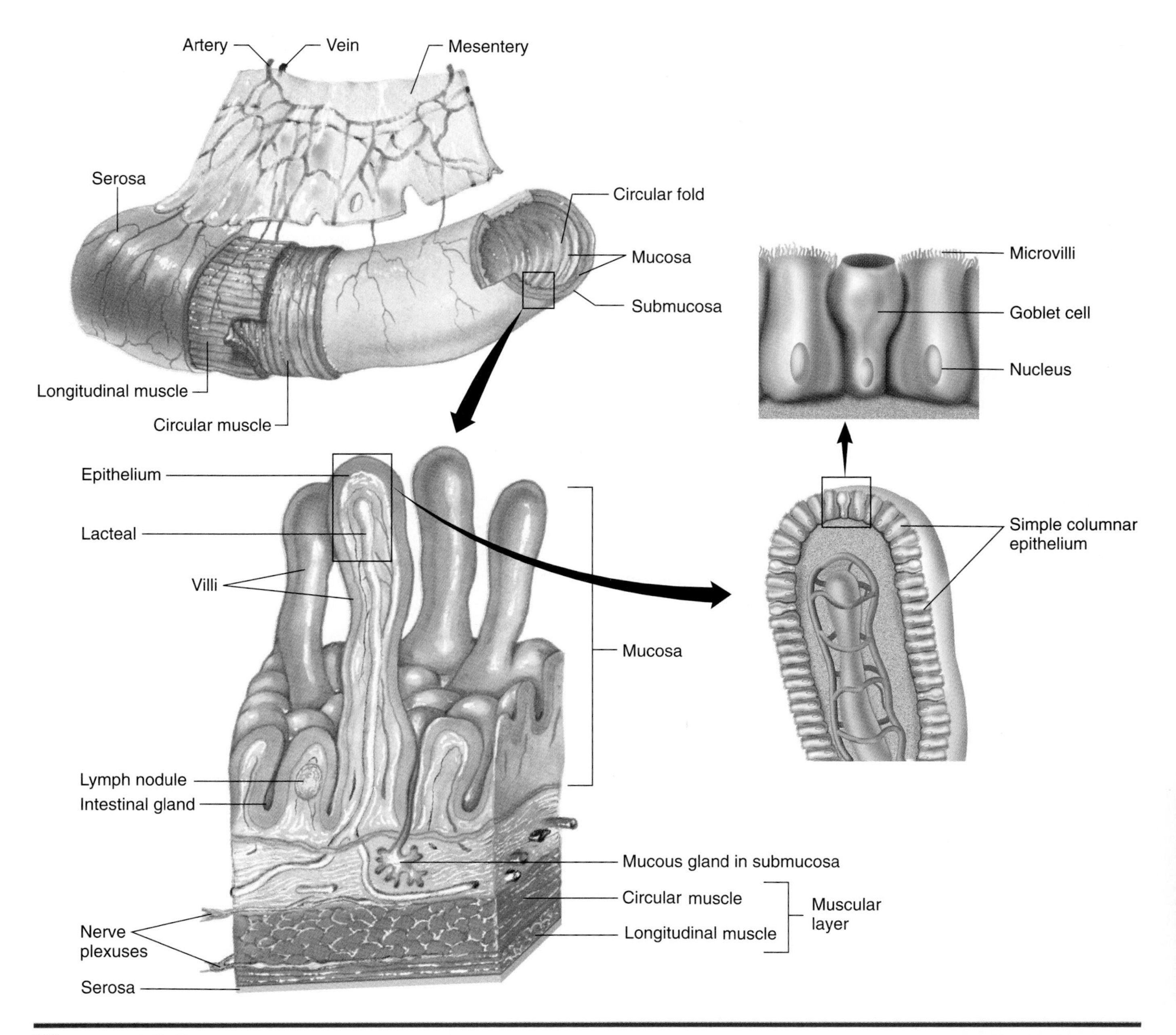

Figure 15.2 The wall of the alimentary canal consists of four tissue layers as shown here in a section of the small intestine. From outside in, they are the serous layer, muscular layer, submucosa, and mucosa.

A number of different types of enzymes are involved in digestion. Each type of digestive enzyme acts on a particular type of food molecule and speeds up its breakdown into smaller molecules. A series of digestive reactions involving several digestive enzymes from different parts of the digestive system are usually required to break down complex food molecules into absorbable nutrients.

Alimentary Canal: General Characteristics

The **alimentary canal** is a muscular tube about 9 m (29 ft) in length that extends from the mouth to the anus. Various portions of the alimentary canal are specialized to perform different digestive functions.

Structure of the Wall

The wall of the alimentary canal consists of four layers. From outside in, they are the serosa, muscular layer, submucosa, and mucosa. These layers may be modified in the various regions but remain as distinct layers. Compare figure 15.2 to the discussion that follows.

The **serosa,** or **serous layer,** is the outer layer. It is formed of the *visceral peritoneum* and is continuous with the *parietal peritoneum,* which lines the inner abdominal wall. Cells of the peritoneum secrete serous fluid, which keeps the membrane surfaces moist and reduces friction as parts of the alimentary canal rub against each other and the abdominal wall.

The **muscular layer** lies just under the serosa. It consists of two layers of smooth muscle that differ in the orientation of their muscle fibers. Muscle fibers of the outer layer are arranged longitudinally. Their contractions shorten the tube. Muscle fibers of the inner layer are arranged circularly around the tube. Their contractions constrict the tube. Contractions of these muscular layers mix food with digestive secretions and move food through the alimentary canal.

The **submucosa** lies between the muscular layer and the mucosa. It contains nerves, lymphatic vessels, and blood vessels embedded in loose connective tissue.

The innermost layer is the **mucosa.** It consists of a surface layer of columnar epithelial tissue supported by underlying connective tissue that contains a few smooth muscle cells. The epithelial cells produce digestive secretions and mucus. The epithelium is often folded to increase the surface area that is in contact with food. The mucosa has different functions in different parts of the digestive tract. In some regions, it secretes only mucus, which protects underlying cells. In others, it secretes mucus and digestive fluids containing enzymes, and it absorbs nutrients.

Movements

Contraction of the smooth muscle layers produces two different types of movement in the alimentary canal: mixing movements and propelling movements. Mixing movements involve alternating rhythmic contractions of muscle fibers in short segments of the alimentary canal. These ripplelike contractions mix food substances with digestive secretions.

The movements that propel food through the alimentary canal are called **peristalsis** (per-i-stal′-sis). In peristalsis, contraction of circular muscle fibers produces a ringlike constriction that moves along the alimentary canal in a wavelike manner, pushing food in front of it. In this way, peristaltic contractions move food from one portion of the alimentary canal to another.

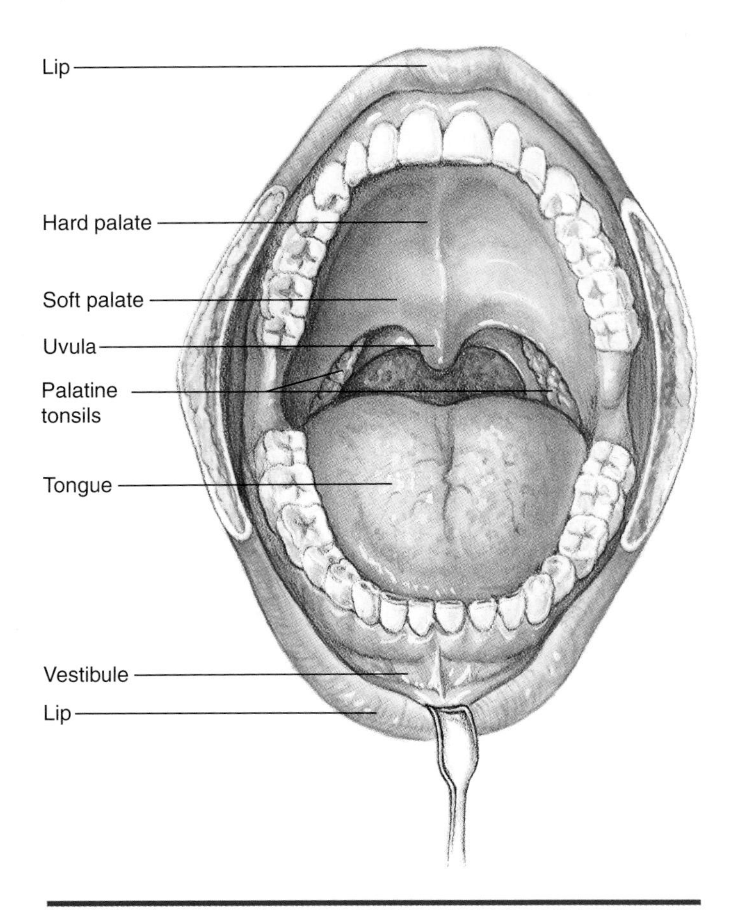

Figure 15.3 Major structures associated with the mouth.

> ***What organs compose the alimentary canal?***
> ***What is chemical digestion?***
> ***How is food moved through the alimentary canal?***

Mouth

The **mouth,** or **oral cavity,** is involved in the intake of food, mechanically breaking it into small pieces, mixing it with saliva, and swallowing it. The mouth is surrounded by the cheeks, palate, and tongue. Examine figures 15.3 and 15.4.

Cheeks

The **cheeks** form the lateral walls of the mouth. Skin covers their outer surfaces, and nonkeratinized squamous epithelium lines their inner surfaces. Contractions of muscles located in the cheeks produce facial expressions. The anterior portions of the cheeks form the *lips,* which surround the opening into the mouth.

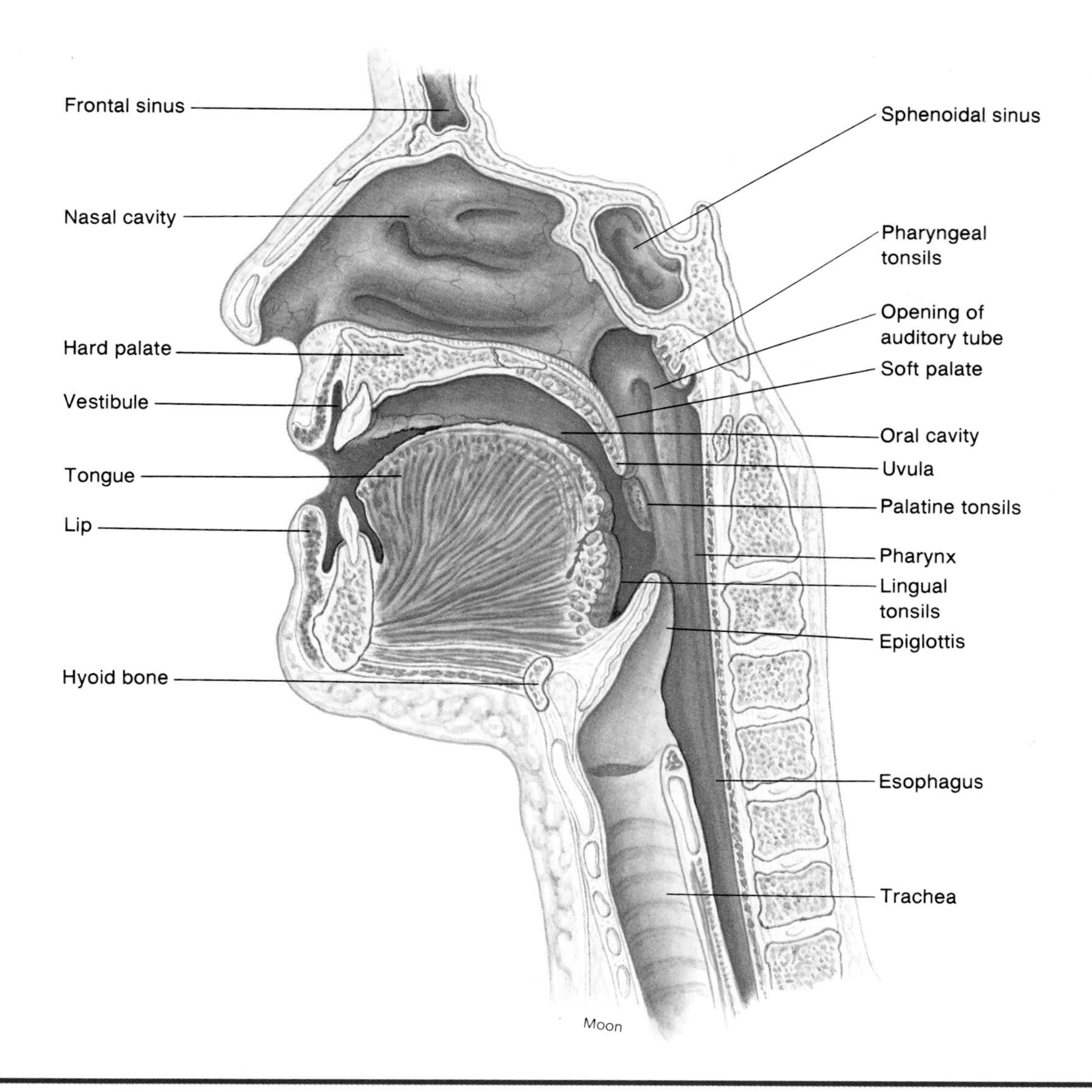

Figure 15.4 The structural relationships of the nasal cavity, mouth, pharynx, esophagus, and larynx are shown in sagittal section.

Lips are sensitive, highly mobile structures. Their pinkish color results from numerous blood vessels near their surfaces.

Palate

The **palate** forms the roof of the mouth and separates the oral cavity from the nasal cavity. The anterior portion is known as the *hard palate* since it is supported by bone. The posterior *soft palate* lacks bony support. The soft palate ends posteriorly in a cone-shaped *uvula* that hangs downward at the back of the oral cavity. The uvula is very sensitive to touch stimuli. This sensitivity causes the soft palate to move upward during swallowing, which closes off the nasal cavity and directs food downward into the pharynx.

Tongue

The **tongue** forms the floor of the oral cavity. It is composed primarily of skeletal muscle that is covered by a mucous membrane. An anterior fold of the mucous membrane on the undersurface of the tongue attaches the tongue to the floor of the mouth. This membranous attachment is known as **lingual frenulum** (ling′-gwal fren′-ū-lum), and it limits the posterior movement of the tongue.

The upper surface of the tongue contains numerous tiny projections called **papillae** (pah-pil′-ē) that

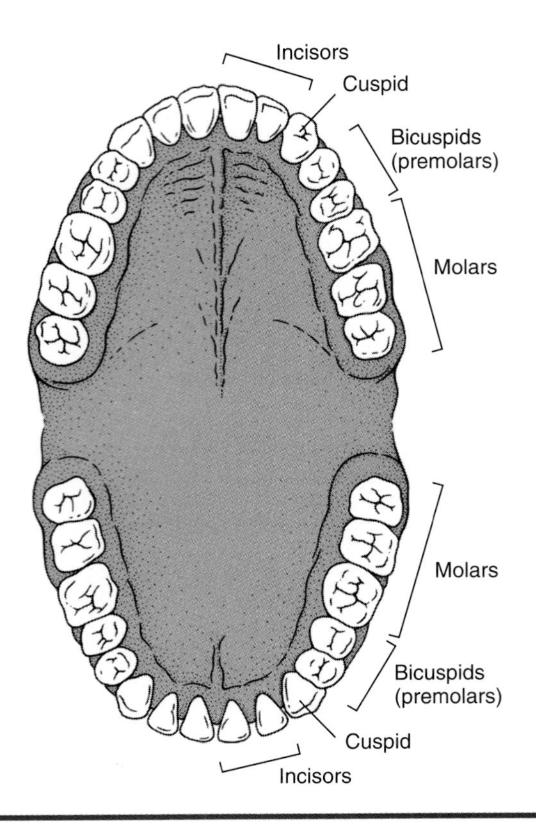

Figure 15.5 There are 32 permanent teeth; 16 in each jaw. Note the location and shape of the incisors and molars.

give a rough texture to the tongue and aid in its manipulation of food. Taste buds are located on the papillae. The tongue moves the food about during chewing and aids in mixing it with saliva. In swallowing, the tongue pushes food posteriorly into the pharynx.

> Sometimes the lingual frenulum is too short and restricts tongue movements that are required for normal speech. A person with this problem is said to be "tongue-tied." Cutting the frenulum to allow freer movement usually solves the problem.

Teeth

Teeth are important accessory digestive structures that mechanically break food into smaller pieces during **mastication** (mas-ti-kā′-shun), or chewing. Humans develop two sets of teeth: deciduous and permanent teeth.

The **deciduous teeth,** the first set, start to erupt through the gums at about six months of age. Central incisors (in-sī-zers) come in first, and second molars erupt last. There are 20 deciduous teeth, 10 in each jaw, and all of them are in place by three years of age. Deciduous teeth are gradually shed starting at about six years of age, and they are lost in the same order as their emergence. The loss of a deciduous tooth occurs because the growth of a permanent tooth underneath it causes resorption of the root. Later, the emerging permanent tooth pushes out the deciduous tooth.

Table 15.1 Deciduous and Permanent Teeth

Type	Number	
	Deciduous	Permanent
Incisors		
Central	4	4
Lateral	4	4
Cuspid	4	4
Bicuspid		
First	0	4
Second	0	4
Molar		
First	4	4
Second	4	4
Third	0	4
Totals	20	32

The **permanent teeth** begin appearing at about six years of age when the first molars (six-year molars) erupt. All of the permanent teeth, except the third molars, are in place by age 16. The third molars (wisdom teeth) erupt between 17 and 21 years of age, or they may never emerge. In many persons, there is insufficient room for the third molars, so they become impacted and often must be surgically removed. The 32 permanent teeth, 16 in each jaw, consist of four different types: incisors, cuspids, bicuspids, and molars (figure 15.5).

There are four chisel-shaped *incisors* at the anterior of each jaw. Incisors are adapted for biting off pieces of food. Lateral to the incisors are the *cuspids* (kus′-pids), or canines, two in each jaw, that are used to grasp and tear tough food morsels. Posterior to the cuspids are four *bicuspids* (premolars) and six *molars* in each jaw. Their somewhat flattened surfaces are used to crush and grind the food. Examine figure 15.5 and table 15.1.

Figure 15.6 shows the basic structure of a tooth. Each tooth consists of two major parts: a root and a crown. The **crown** is the portion that protrudes above the *gingiva* (jin-ji-vah), or gum, covering the alveolar bone. The **root** is embedded in a socket in the alveolar bone of the jaw. The junction of the crown and root is known as the *neck* of the tooth.

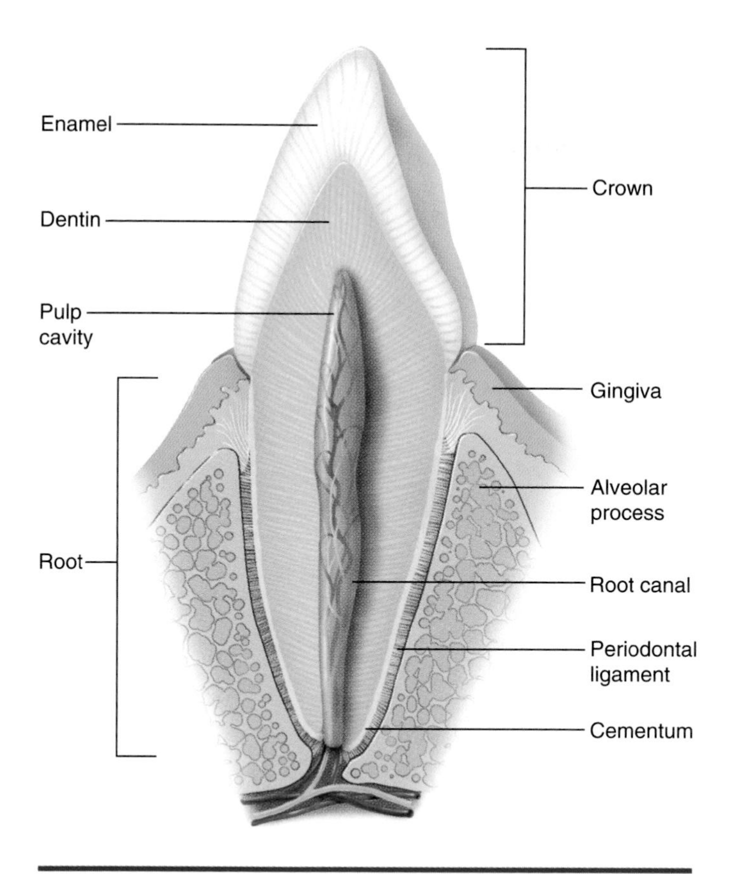

Figure 15.6 The structure of a tooth is shown in this section of a cuspid.

The root of a tooth is attached firmly to the alveolar bone of the jaw. A hard substance called *cementum* attaches the root to the tough *periodontal membrane,* whose fibers penetrate into the bone.

Most of a tooth is composed of **dentin,** a hard, bonelike substance. The crown of the tooth has a layer of **enamel** overlying the dentin. Enamel is the hardest substance in the body, and it is appropriately located to resist the abrasion caused by chewing hard foods. A *pulp cavity* occupies the central portion of a tooth. Pulp consists of loose connective tissue that supports blood vessels and nerves, which enter at the root tip and extend through the *root canal* and into the pulp cavity.

Enamel and dentin are not replaced after they have been eroded by tooth decay. The repair of a decayed tooth requires removal of the decayed material, sterilization of the area, and the insertion of a filling. A filling seals the area and provides an abrasion-resistant surface.

Salivary Glands

The **salivary glands** secrete **saliva** into the mouth, where it is mixed with food during chewing. The presence of food (or nonfood objects) in the mouth activates neural reflexes that stimulate the flow of saliva. The mere thought of appetizing food can stimulate the flow of saliva. The functions of saliva include binding food particles together, dissolving certain foods, cleansing and lubricating the mouth, and starting carbohydrate digestion. The sense of taste is dependent upon saliva because only dissolved food molecules can stimulate the taste buds. There are three pairs of major salivary glands: the parotid, submandibular, and sublingual salivary glands (figure 15.7).

The largest salivary glands are the *parotid* (pah-rot′-id) *glands.* A gland is located in front of each ear over the masseter muscle. Parotid glands secrete saliva that is rich in amylase. Parotid secretions are emptied through a duct into the vestibule of the mouth near the upper second molars.

The *submandibular* (sub-man-dib′-ū-lar) *glands* are found in the floor of the mouth. They produce a watery saliva that contains relatively little mucus. Secretions of the submandibular glands are emptied through ducts into the anterior part of the mouth at the base of the lingual frenulum.

The *sublingual* (sub-ling′-gwal) *glands* lie on the floor of the mouth under the tongue. They are the smallest of the major salivary glands. Their secretions consist mostly of mucus, and they are emptied by several ducts into the floor of the mouth under the tongue.

Saliva consists mostly of water (99.5%), which helps to dissolve food substances, plus a small amount of other substances. Mucus helps to hold food particles together during chewing and swallowing. Saliva contains two enzymes. **Salivary amylase** is a digestive enzyme that speeds up the breakdown of starch and glycogen, and *lysozyme* is an enzyme that kills certain bacteria.

Mumps is a common childhood disease. Infection by the mumps virus causes inflammation of the parotid glands. The glands become swollen and sore, making opening the mouth, chewing, and swallowing quite painful.

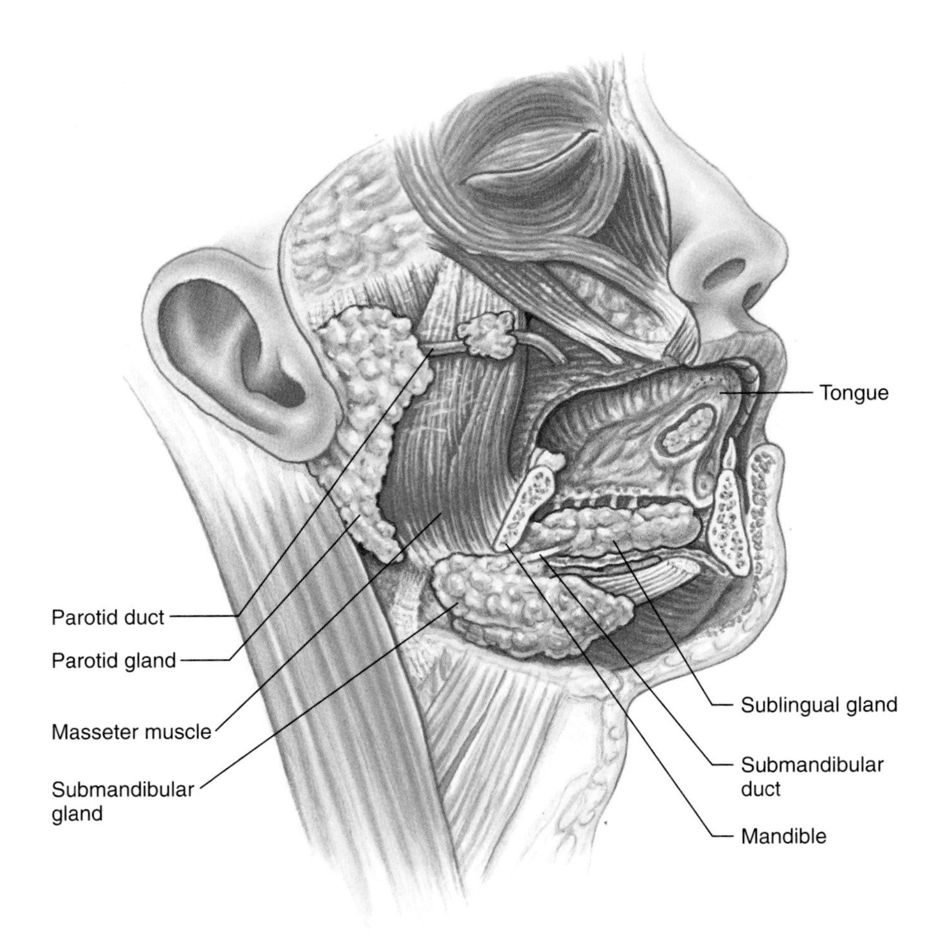

Figure 15.7 The major salivary glands.

Digestion in the Mouth

Both mechanical and chemical digestion take place in the mouth. Mechanical digestion in the mouth consists of breaking food into smaller pieces and mixing it with saliva during chewing. This improves chemical digestion because the smaller pieces have an increased surface area upon which digestive secretions may act. Chemical digestion starts in the mouth with the breakdown of certain complex carbohydrates. **Salivary amylase** acts on *starch* and *glycogen* (polysaccharides) and speeds up their breakdown into *maltose,* a disaccharide sugar. The action of salivary amylase continues for a few minutes after food enters the stomach, but it soon stops because it is inactivated by the strong acidity of gastric juice.

What is the structure of a tooth?
What digestive processes occur in the mouth?

Pharynx

The **pharynx** (fayr′-inks) is the passageway that connects the nasal and oral cavities with the larynx and esophagus. It is part of both the respiratory and digestive systems. Its digestive function is the transport of food from the mouth to the esophagus during swallowing.

The swallowing reflex is activated when food is pushed into the pharynx by the tongue. The soft palate contracts upward, preventing food from entering the nasal cavity, and directs food downward into the pharynx. At the same time, muscle contractions raise the larynx, and this causes the epiglottis to flop over and cover the opening into the larynx. This action prevents food from entering the larynx and directs it into the esophagus.

Esophagus

The **esophagus** (ē-sof′-ah-gus) is a muscular tube that extends from the pharynx down through the thoracic

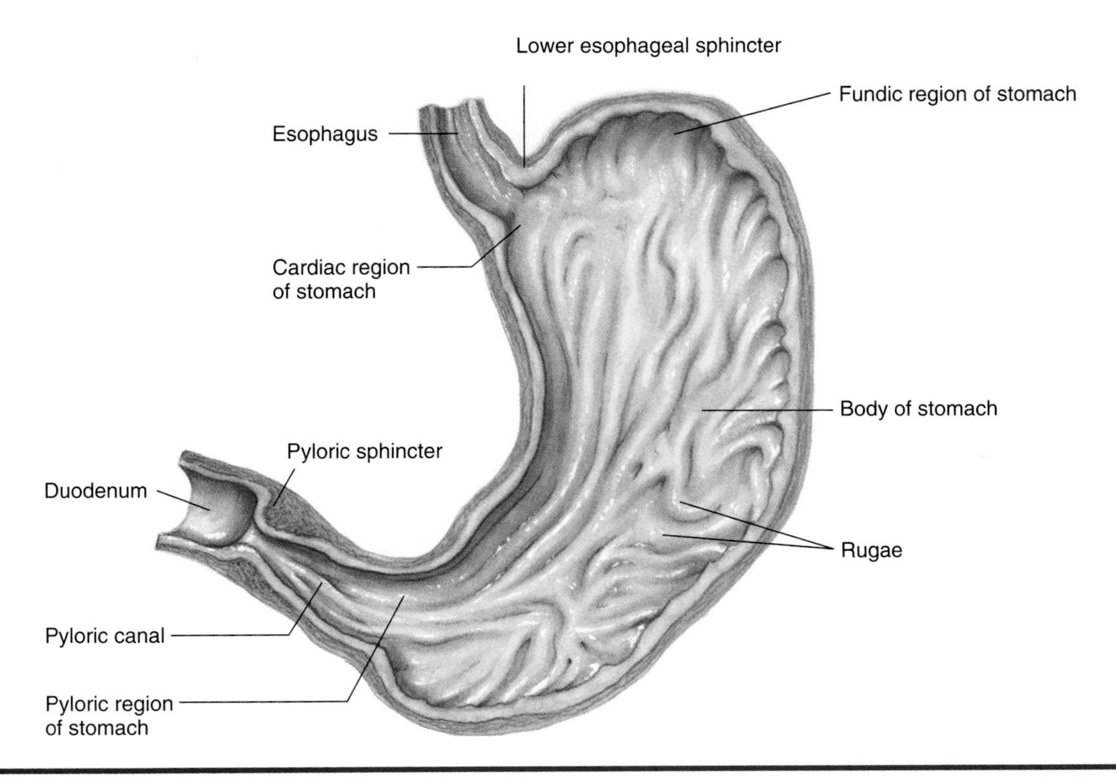

Figure 15.8 The pouchlike stomach receives food from the esophagus and releases chyme into the duodenum.

cavity and the diaphragm to join with the stomach. Food is carried to the stomach by peristalsis. The esophageal mucosa produces mucus that lubricates the esophagus and aids the passage of food.

At the junction of the esophagus and stomach there is a thickened ring of circular muscle fibers known as the **lower esophageal,** or **cardiac, sphincter** (sfink′-ter) muscle. Normally, this sphincter is constricted to prevent regurgitation of stomach contents into the esophagus, but it opens to allow food to enter the stomach from the esophagus.

Describe the swallowing reflex.
How does the esophagus carry food to the stomach?

Stomach

As shown in figure 15.8, the J-shaped **stomach** is a pouchlike portion of the alimentary canal. It lies just below the diaphragm in the upper left quadrant of the abdomen. The basic functions of the stomach are temporary storage of food, mixing food with gastric juice, and starting the digestion of proteins.

Structure

The stomach may be subdivided into four regions: the cardiac, fundic, pyloric regions, and body. The *cardiac region* (closest to the heart) is a relatively small area that receives food from the esophagus. The *fundic region* expands above the level of the cardiac region and serves as a temporary storage area. The *pyloric region* is the narrow portion located near the junction with the duodenum. The *body* is the largest region of the stomach, and it is located between the fundic and pyloric regions.

The **pyloric sphincter** is a thickened ring of circular muscle fibers that is located at the junction of the stomach and duodenum. This muscle usually is contracted, closing the stomach outlet, but it relaxes to let food pass into the small intestine.

The mucous membrane lining the stomach is quite thick and, in an empty stomach, is organized into numerous folds called **rugae** (rū-jē). These folds allow the mucosa to stretch as the stomach fills with food. The mucosa is dotted with numerous pores called *gastric pits.* Gastric pits are openings of the **gastric glands** that extend deep into the mucosa.

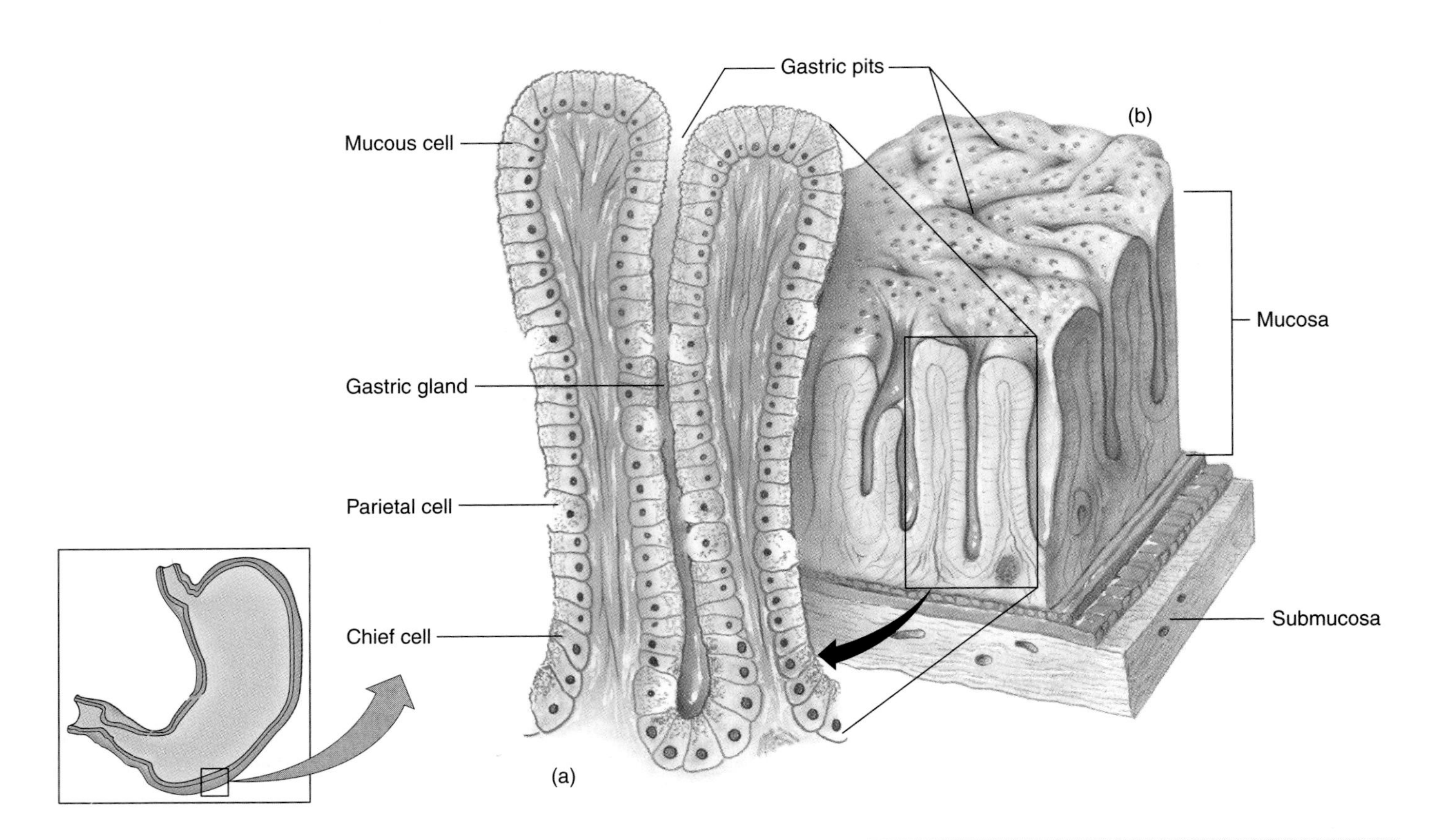

Figure 15.9 (*a*) An enlargement of the gastric glands shows the locations of mucous cells, parietal cells, and chief cells. (*b*) The thick stomach mucosa is dotted with gastric pits, the openings of the gastric glands.

Gastric Juice

The secretion of the gastric glands is known as **gastric juice.** Cells near the opening of the gastric glands secrete mucus that coats and protects the mucosa from the action of digestive secretions. *Chief cells,* located at the bottom of the glands, secrete digestive enzymes. *Parietal cells,* located in the midportion of the glands, secrete hydrochloric acid (figure 15.9).

Sometimes the lower esophageal sphincter inadvertently allows a small amount of gastric juice to regurgitate into the base of the esophagus. When this occurs, the acidic gastric juice irritates the esophageal lining and produces a burning sensation called *heartburn.*

As food is mixed with gastric juice and as digestion occurs, it is converted into a semiliquid substance called **chyme** (kīm). Small amounts of chyme are released intermittently into the duodenum by the opening of the pyloric sphincter.

Control of Gastric Secretion

The rate of gastric secretion is controlled by both neural and hormonal means. Gastric juice is produced continuously, but its secretion is greatly increased whenever food is on the way to, or already in, the stomach. The sight, smell, or thought of appetizing food, food in the mouth, or food in the stomach stimulate the transmission of parasympathetic impulses that increase the secretion of gastric juice. These impulses also stimulate certain stomach cells to secrete a hormone called **gastrin.** Gastrin is absorbed into the blood and is carried to stomach mucosal cells, increasing their secretions (figure 15.10).

As stomach contents are gradually emptied into the small intestine, there is a decrease in the frequency of parasympathetic impulses to the stomach, which reduces the secretion of gastric juice. When chyme

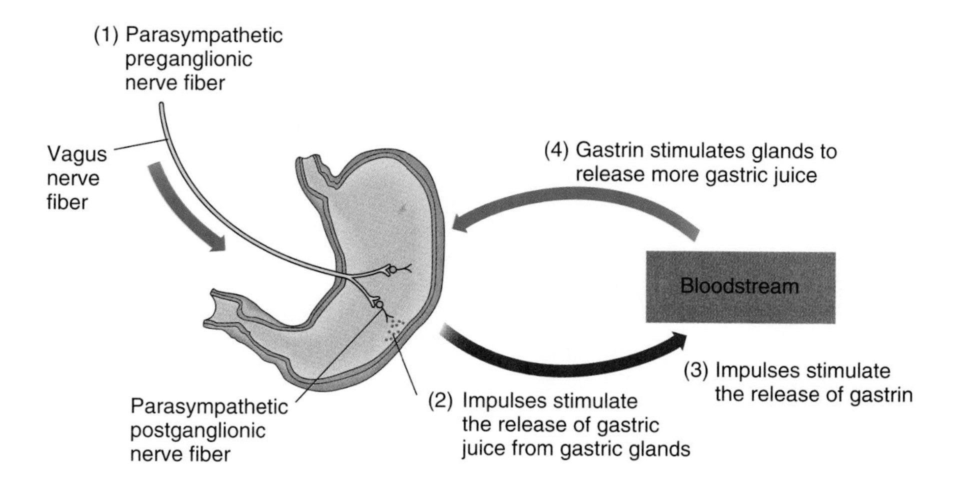

Figure 15.10 Neural and hormonal control of gastric secretions.

passes from the stomach into the small intestine, it stimulates the intestinal mucosa to release two hormones: **cholecystokinin** (kō-lē-sis-tō-kīn′-in) **(CCK)** and **secretin** (se′-krē′-tin), which reduce the secretion of gastric juice.

Digestion and Absorption

Food entering the stomach is thoroughly mixed with gastric juice by ripplelike, mixing contractions of the stomach wall. Gastric juice is very acidic (pH 2) due to an abundance of hydrochloric acid. **Pepsin** is the most important digestive enzyme in gastric juice, and it is secreted in an inactive form that prevents digestion of the cells secreting it. Once it is released into the stomach, pepsin is activated by the strong acidity of gastric juice. Pepsin acts on proteins and breaks these complex molecules into smaller molecules called *peptides.* However, peptides are still much too large to be absorbed.

Rennin is an enzyme in the gastric juice of infants. It curdles milk proteins, which keeps them in the stomach longer and makes them more easily digestible by pepsin. Gastric juice also contains a substance known as **intrinsic factor** that is essential for the absorption of vitamin B_{12} by the small intestine. Except for a few substances such as water, minerals, some drugs, and alcohol, little absorption occurs in the stomach.

What digestive processes occur in the stomach?
How are gastric secretions controlled?

Gastric ulcers produce stomach pain one to three hours after eating. Without treatment, they can perforate the stomach wall, producing internal bleeding or peritonitis. Gastric ulcers result from persistent erosion of the alkaline mucus that coats the stomach lining. Most recurring gastric ulcers are caused by an acid-resistant bacterium, *Helicobacter pylori,* which erodes the protective mucous layer, allowing gastric juice to attack the underlying cells. Other contributing factors that promote hydrochloric acid (HCl) secretion or reduce mucus production include stress, smoking, alcohol, coffee, aspirin, and nonsteroidal antiinflammatory drugs. Treatment involves antibiotics to kill the bacteria and drugs to reduce gastric secretion.

The next section of the alimentary canal is the small intestine. Digestion in the small intestine involves not only intestinal fluids but also secretions from the pancreas and liver that are emptied into the duodenum. These accessory organs will be considered before studying the small intestine.

Pancreas

The **pancreas** is a small, pennant-shaped gland located posterior to the pyloric portion of the stomach. It is connected by a duct to the inner curve of the duodenum, the first part of the small intestine. The pancreas has both endocrine and exocrine functions. The

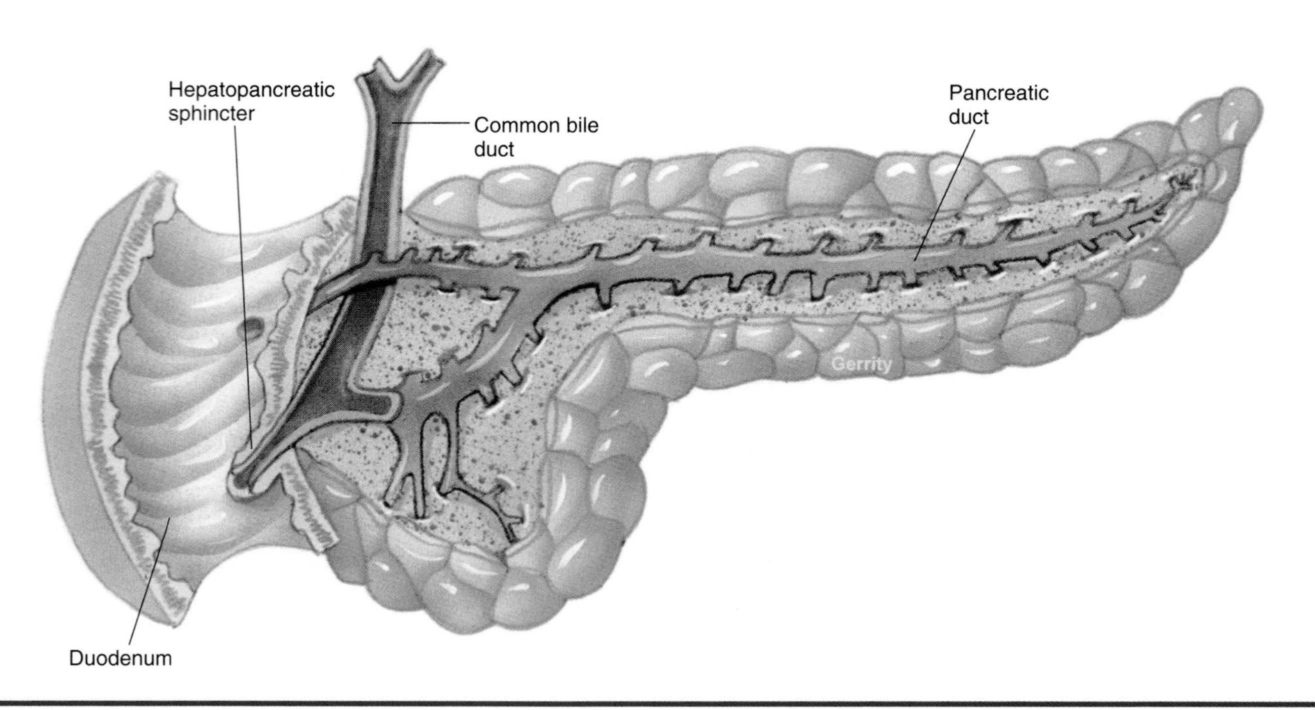

Figure 15.11 Pancreatic juice is carried by the pancreatic duct into the common bile duct, which opens into the duodenum.

digestive function of the pancreas is the secretion of **pancreatic juice,** and the cells that produce it compose most of the pancreas. Pancreatic juice is collected by tiny ducts that enter the **pancreatic duct,** which leaves the pancreas and enters the **common bile duct** just before the bile duct enters the duodenum. The *hepatopancreatic sphincter* dilates to allow pancreatic juice and bile to enter the duodenum (figure 15.11).

Control of Pancreatic Secretion

Pancreatic secretion, like gastric secretion, is controlled by both neural and hormonal mechanisms. Neural control is via parasympathetic fibers. When parasympathetic impulses activate the stomach mucosa, they also stimulate the pancreas to secrete pancreatic juice.

Hormonal control of pancreatic secretion results from two hormones that stimulate different types of pancreatic cells. Acid chyme entering the duodenum stimulates the intestinal mucosa to release the hormone secretin, which is carried by blood to the pancreas, where it stimulates secretion of pancreatic juice that is rich in carbonates (figure 15.12). Carbonates neutralize the acidity of the chyme entering the small intestine. Fat-laden chyme stimulates production of cholecystokinin (CCK) by the intestinal mucosa. CCK stimulates secretion of pancreatic juice, which is rich in digestive enzymes.

Table 15.2 summarizes the major hormones that regulate digestive secretions.

Digestion by Pancreatic Enzymes

Pancreatic juice contains enzymes that act on each of the major classes of energy foods: carbohydrates, fats, and proteins. Their digestive actions occur within the small intestine.

Pancreatic amylase, like salivary amylase, acts on *starch* and *glycogen,* splitting these polysaccharides into *maltose,* a disaccharide.

Pancreatic lipase acts on *fats* (triglycerides) and splits them into *monoglycerides* and *fatty acids* that are absorbable. A monoglyceride is a glycerol molecule with one fatty acid attached.

Trypsin is the major pancreatic enzyme in pancreatic juice. Trypsin splits *proteins* into shorter amino acid chains called *peptides.* Like pepsin in the stomach, it is secreted in an inactive form and is activated when mixed with intestinal secretions within the small intestine. This mode of secretion prevents the pancreatic cells from being digested by their own enzymatic secretions.

What composes pancreatic juice?
How is pancreatic secretion controlled?

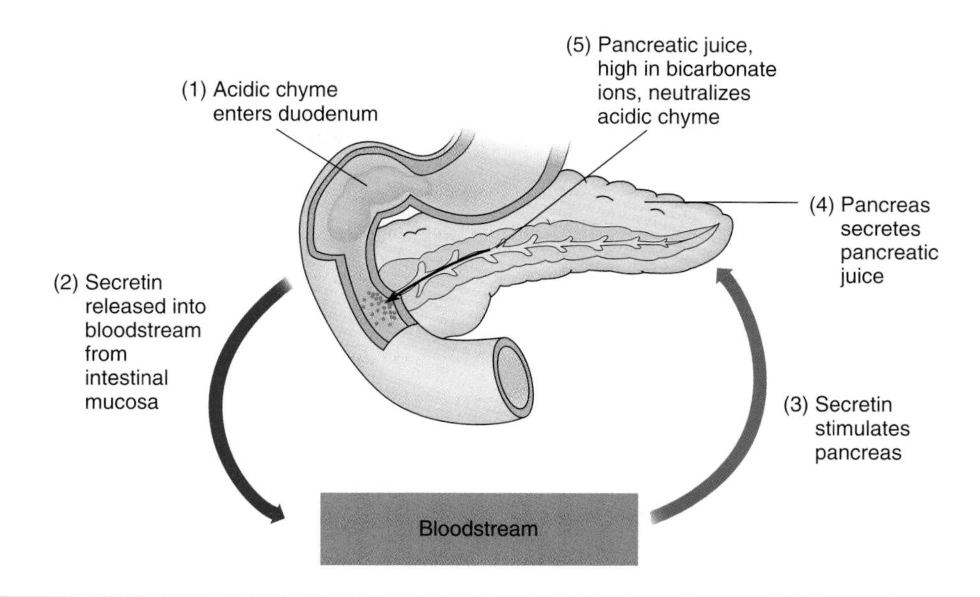

Figure 15.12 Acidic chyme entering the duodenum triggers the release of secretin, a hormone that stimulates the secretion of pancreatic juice.

Table 15.2 Major Hormones Regulating Digestive Secretions

Hormone	Action	Source
Gastrin	Stimulates gastric glands to increase their rate of secretion	Gastric mucosa; food in the stomach stimulates release of gastrin
Cholecystokinin	Reduces secretion of gastric glands; stimulates secretion of pancreatic juice that is rich in digestive enzymes; stimulates contraction of gallbladder, causing release of bile	Intestinal mucosa; fatty chyme stimulates the release of cholecystokinin
Secretin	Stimulates secretion of pancreatic juice that is rich in carbonates; inhibits gastric secretion	Intestinal mucosa; acid chyme stimulates the release of secretin

Liver

The **liver** is the largest gland in the body. It weighs about 1.4 kg (3 lb) and is dark reddish brown in color. The liver is located mostly in the upper right quadrant of the abdomen just below the diaphragm, where it is protected by the lower ribs.

The liver is a vital organ. It has many important functions, but most are not associated with digestion. Its only digestive function is the secretion of bile. Later in the chapter, you will see how the liver plays a key role in carbohydrate, lipid, and protein metabolism. In addition to these functions, the liver detoxifies poisons and harmful chemicals, such as alcohol and other drugs; removes worn-out blood cells; and stores fat, glycogen, iron, and several vitamins.

The liver is encased in a fibrous connective tissue capsule that, in turn, is covered by the peritoneum, which provides support. The liver is divided into four lobes, but the left and right lobes are most obvious (figure 15.13).

The liver receives blood from two sources. The *hepatic artery* brings oxygenated blood to the liver cells. The *hepatic portal vein* brings deoxygenated, nutrient-rich blood from the digestive tract. As blood flows through the liver, liver cells remove, modify, or add substances to the blood before it leaves the liver via the *hepatic vein.*

As noted, the production of bile is the only digestive function of the liver. Bile is collected in tiny ducts that merge to form the *hepatic duct,* which carries bile from the liver. The hepatic duct and the *cystic duct,* a short duct that extends from the gallbladder, merge to form the **common bile duct,** which carries bile to the duodenum. The cystic duct carries bile to and from the **gallbladder,** a small,

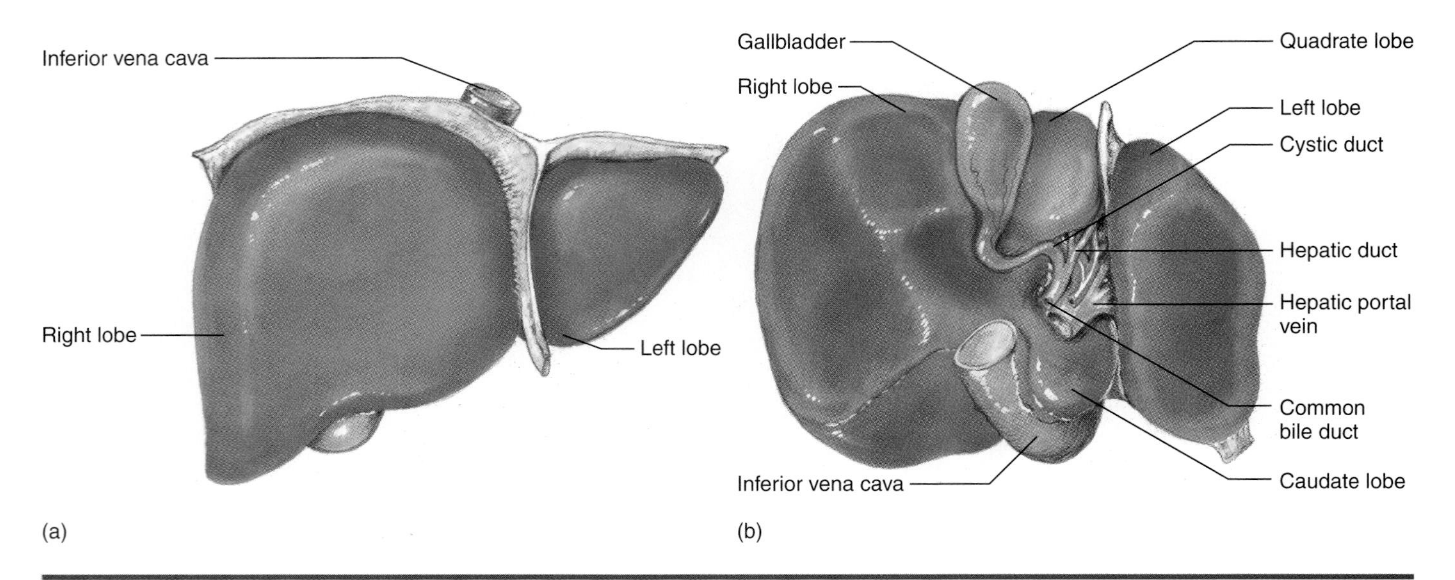

Figure 15.13 The liver and associated structures as viewed (*a*) anteriorly and (*b*) from below.

pear-shaped sac that stores bile temporarily between meals (figure 15.13).

Bile

Liver cells continuously produce **bile**, a yellowish green liquid. Bile consists of water, bile salts, bile pigments, cholesterol, and minerals. Bile pigments, mostly *bilirubin*, are waste products of hemoglobin breakdown.

Bile salts are the only bile components that play a digestive role. When in contact with fatty substances, they break up the lipid globules into very small droplets, a process called *emulsification* (ē-mul-si-fi-kā′-shun). Emulsification greatly increases the surface area of the lipid substances exposed to water and lipid-digesting enzymes. In this way, bile salts aid the digestion of lipids. Bile salts also aid the absorption of fatty acids, cholesterol, and fat-soluble vitamins by the small intestine.

Release of Bile

Bile normally enters the duodenum only when food is present. When the intestine is empty, the hepatopancreatic sphincter at the base of the common bile duct is closed, and this forces bile to enter the gallbladder, where it is stored temporarily.

When fat-laden chyme enters the duodenum, it stimulates the release of cholecystokinin from the intestinal mucosa. CCK is carried by the blood to the gallbladder, where it stimulates contraction of muscles in the gallbladder wall ejecting bile from the gallbladder into the common bile duct. Further, CCK relaxes the hepatopancreatic sphincter muscle so bile is injected into the small intestine. Note that this hormonal control releases bile only when it is needed in the small intestine (figure 15.14).

Where is the liver located?
What is the digestive function of the liver?
How is bile release controlled?

Small Intestine

The small intestine is about 2.5 cm (1 in) in diameter and 6.4 m (21 ft) in length. It begins at the pyloric sphincter of the stomach, fills much of the abdomen, and empties into the large intestine. Most of the digestion of foods and absorption of nutrients occur in the small intestine.

Structure

There are three sequential segments composing the small intestine. The **duodenum** (dū-o-dē′-num) is a very short section, about 25 cm long, that receives chyme from the stomach. The middle section is the **jejunum** (je-jū′-num), and it is about 2.5 m long. The last and longest segment is the **ileum** (il′-ē-um), which is about 3.6 m long. The ileum joins with the large intestine at the **ileocecal** (il-ē-ō-sē′-kal) **sphincter.**

The small intestine is suspended from the posterior abdominal wall by the **mesentery** (mes′-en-ter-ē), double folds of the peritoneum that provide support but allow movement. Blood vessels, lymphatic vessels, and nerves serving the small intestine are also supported by the mesentery (figure 15.15).

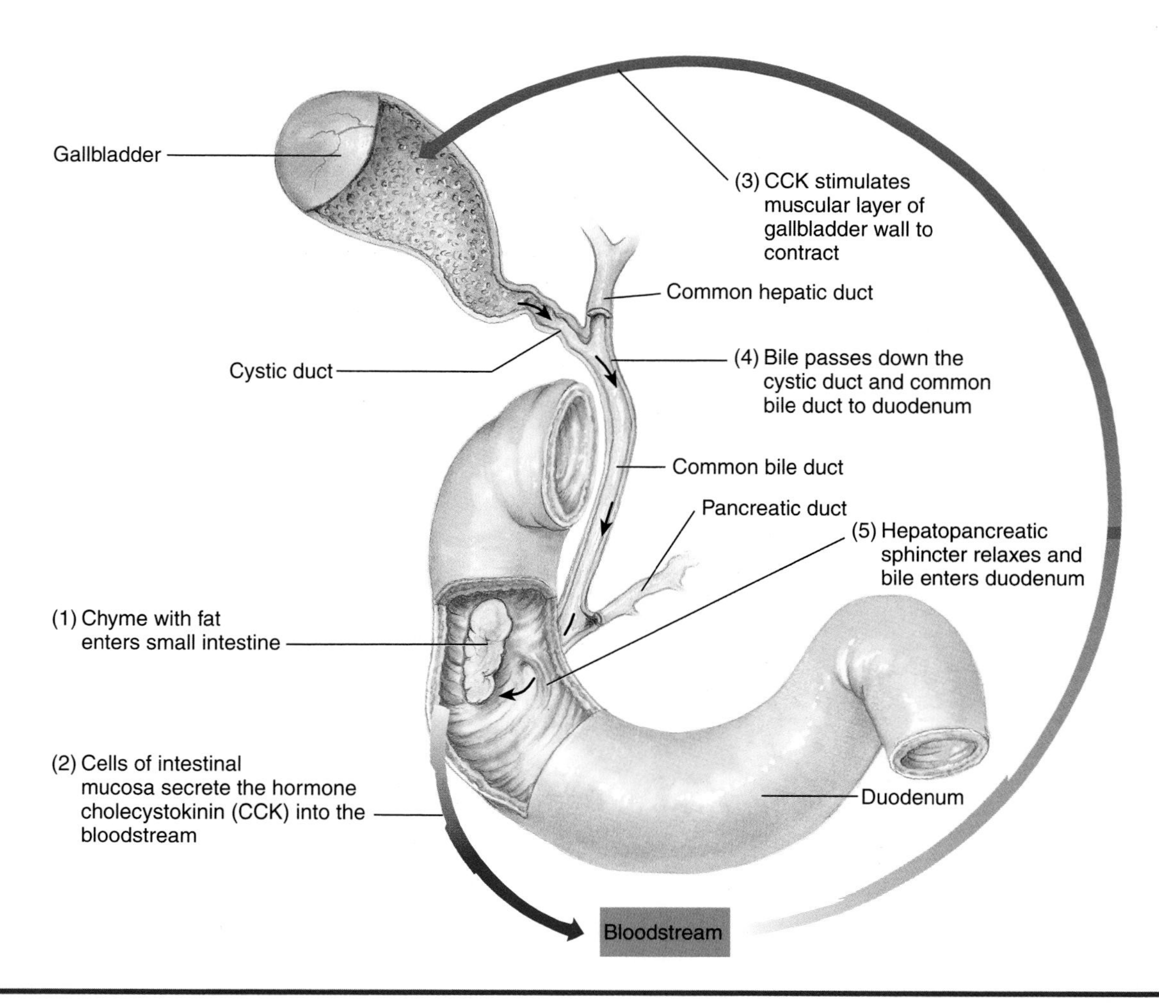

Figure 15.14 Fatty chyme entering the duodenum triggers the secretion of cholecystokinin, the hormone that stimulates release of bile from the gallbladder.

The mucosa of the small intestine is modified to provide a very large surface area. The distinctive velvety appearance of the intestinal mucosa results from the presence of **intestinal villi,** tiny projections from the mucosa that are extremely abundant (see figure 15.2). Each villus, like all of the mucosa, is covered by columnar epithelium and contains a centrally located **lacteal,** a lymphatic capillary. A blood capillary network surrounds the lacteal. At the bases of the villi are tiny pits that open into **intestinal glands,** which secrete intestinal juice and mucus (figure 15.16).

The mucosal surface area in contact with chyme and digestive fluids is further increased by the presence of numerous **microvilli.** The microscopic microvilli are formed by folds of the plasma membranes that form the exposed surfaces of the epithelial cells (figure 15.17).

Intestinal Juice

The fluid secreted by the intestinal glands is known as **intestinal juice.** It is slightly alkaline and contains abundant water and mucus. Intestinal juice provides an appropriate environment for the action of bile salts and pancreatic digestive enzymes within the small intestine. Recall that trypsin in pancreatic juice is activated only after being mixed with intestinal secretions.

Regulation of Intestinal Secretion

The presence of chyme in the small intestine provides mechanical stimulation of the mucosa that activates the secretion of intestinal juice and enzymes. Also, the chyme causes an expansion of the intestinal wall, which triggers a neural reflex, sending parasympathetic impulses to the mucosa. The impulses stimulate an increase in the rate of intestinal secretions.

Digestion and Absorption

Vigorous contractions of the small intestine mix chyme with bile, pancreatic juice, and intestinal juice. The emulsification of fats by bile and the continued digestion of carbohydrates, fats, and proteins by pancreatic and intestinal enzymes occurs within the small

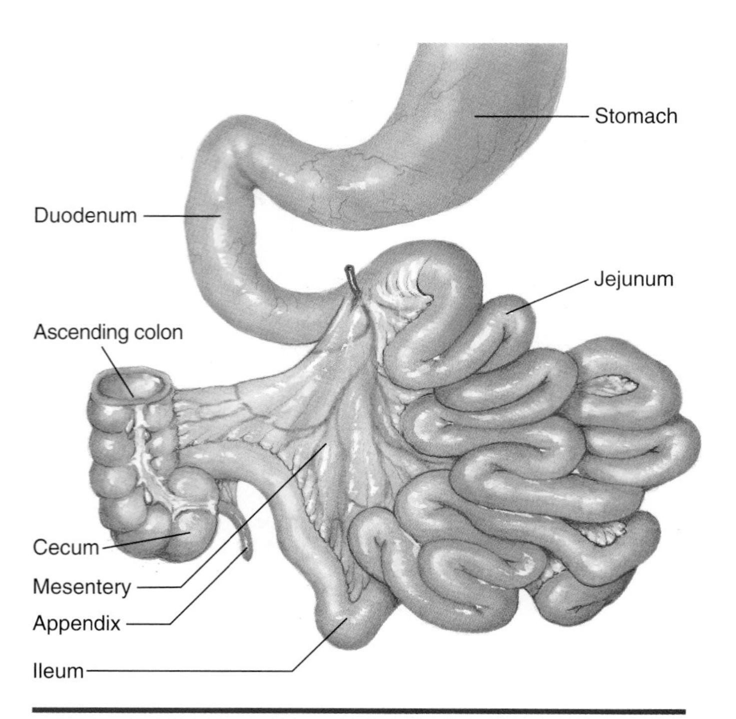

Figure 15.15 The small intestine consists of the duodenum, jejunum, and ileum. Chyme from the stomach enters the duodenum. After digestion and absorption, food residues pass from the ileum into the large intestine.

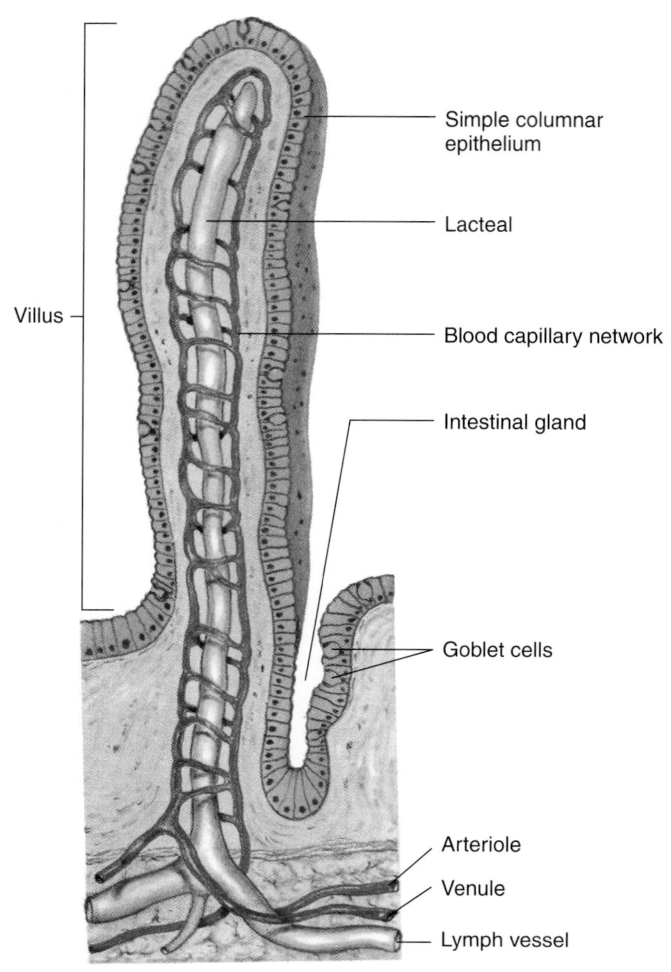

Figure 15.16 The structure of a villus.

intestine. The digestive process is completed by intestinal enzymes within the small intestine.

There are three intestinal enzymes that split disaccharides into monosaccharides. (1) **Maltase** converts *maltose* into *glucose;* (2) **sucrase** converts *sucrose* into *glucose* and *fructose;* and (3) **lactase** converts *lactose* into *glucose* and *galactose.*

Enzymes acting on fats and proteins are also present. **Intestinal lipase** splits *fats* (triglycerides) into *monoglycerides* and *fatty acids.* **Peptidase** splits *peptides* into *amino acids.* Table 15.3 summarizes the enzymes involved in the digestion of carbohydrates, fats, and proteins.

Carbohydrate digestion begins in the mouth and concludes in the small intestine. The end products of carbohydrate digestion are monosaccharides, the simple sugars *glucose, fructose,* and *galactose.* These sugars are absorbed into the capillaries of the villi primarily by active transport, although diffusion accounts for some of the absorption.

Fat (triglyceride) digestion occurs in the small intestine. The end products are **monoglycerides** and **fatty acids,** and they diffuse into the epithelial cells of the villi. Once inside the epithelial cells, they recombine to form fat molecules. Then, the fat molecules combine in small clusters, along with cholesterol and phospholipids, and become coated with protein, forming particles known as **chylomicrons** (kī-lō-mī′-krons). The protein coat makes chylomicrons water-soluble. Chylomicrons move from the epithelial cells and enter the lacteals of the villi, as shown in figure 15.17. They are carried by lymphatic vessels to the left subclavian vein, where lymph from the intestine enters the blood. Very small fatty acids seem to enter the capillary networks of villi directly without being recombined to form fat molecules.

Protein digestion begins in the stomach and concludes in the small intestine. The end products are **amino acids,** and they are actively absorbed in the capillaries of villi.

In addition to the end products of digestion, other needed substances are absorbed in the small intestine. For example, water, minerals, and vitamins are absorbed into the capillaries of villi. Materials absorbed into the blood are carried from the intestines to the liver via the hepatic portal vein. After processing by the liver, appropriate concentrations of nutrients are released into the general circulation to serve the needs of tissue cells.

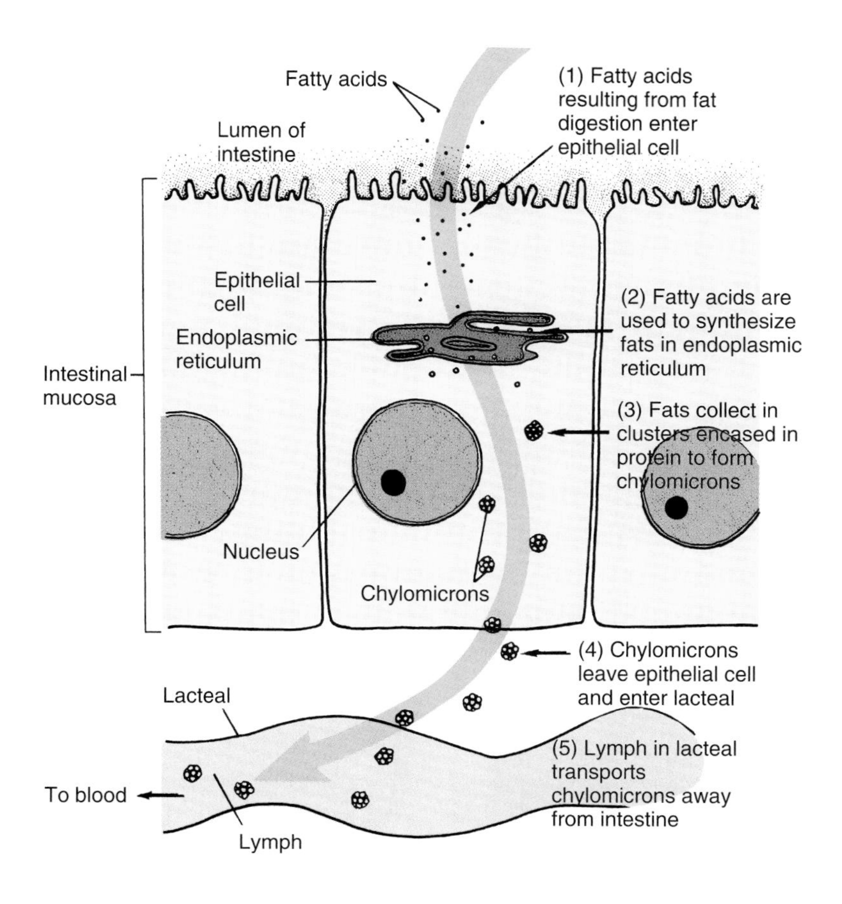

Figure 15.17 Absorption of fatty acids.

Table 15.3 Summary of the Major Digestive Enzymes and Their Actions

Enzyme	Substrate	Product
Saliva		
Amylase	Starch and glycogen	Maltose
Gastric Juice		
Pepsin	Proteins	Peptides
Pancreatic Juice		
Amylase	Starch and glycogen	Maltose
Lipase	Triglycerides	Monoglycerides* and fatty acids*
Trypsin	Proteins	Peptides
Intestinal Juice		
Maltase	Maltose	Glucose*
Sucrase	Sucrose	Glucose* and fructose*
Lactase	Lactose	Glucose* and galactose*
Lipase	Triglycerides	Monoglycerides* and fatty acids*
Peptidase	Peptides	Amino acids*

*End products of digestion

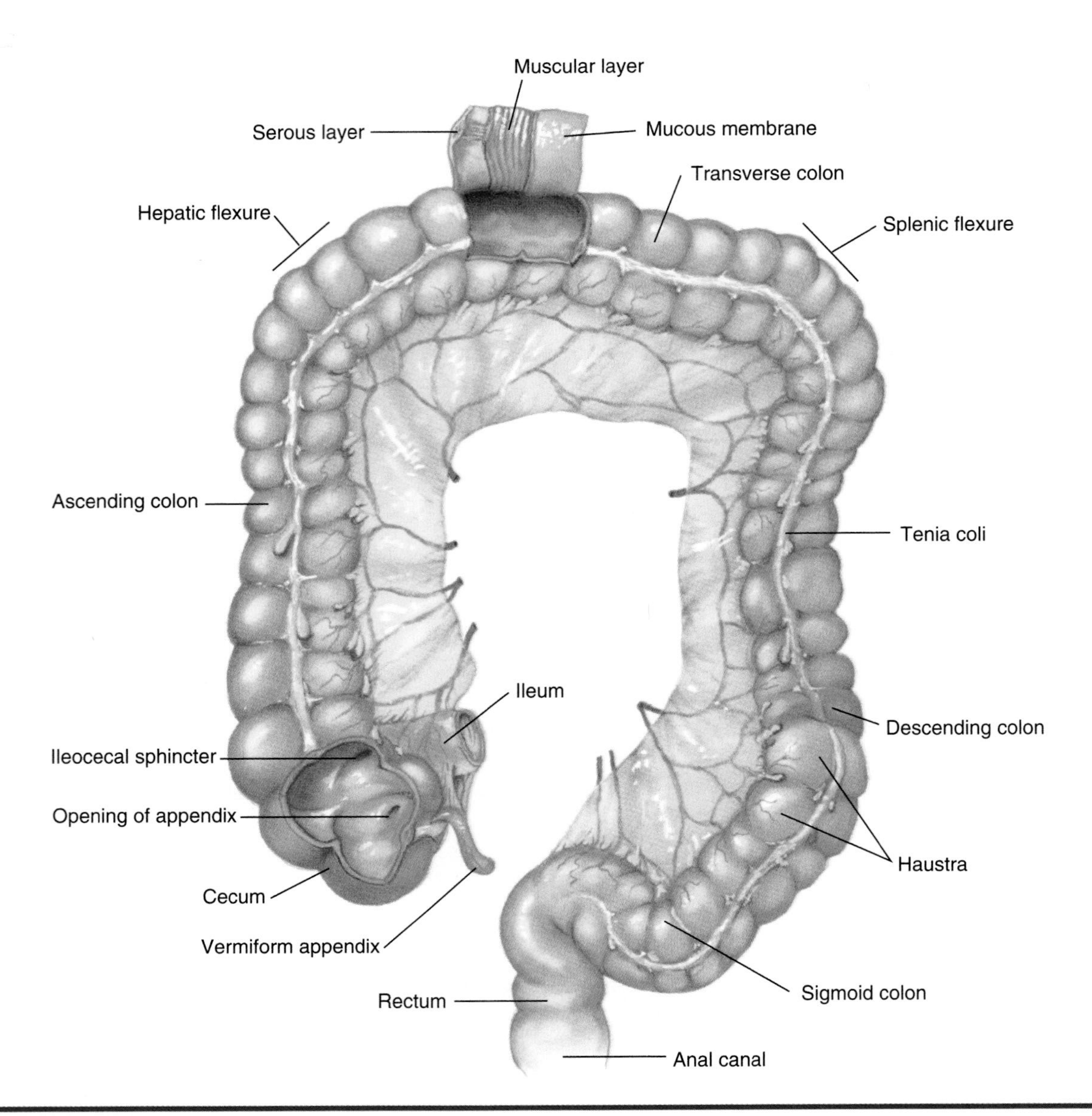

Figure 15.18 Anterior view of the large intestine.

> ***What are the three parts of the small intestine?***
> ***What digestive processes occur in the small intestine?***
> ***How are end products of digestion absorbed?***

Large Intestine

The small intestine joins with the large intestine at the ileocecal sphincter. This valve is closed most of the time, but opens to allow food residues to enter the large intestine.

Structure

The **large intestine** gets its name because its diameter (6.5 cm; 2.5 in) is larger than that of the small intestine, although its length (1.5 m; 5 ft) is much shorter. The large intestine consists of three segments: cecum, colon, and rectum.

The first portion of the large intestine is the pouchlike **cecum,** which bulges below the ileocecal sphincter. The slender wormlike **appendix** extends from the cecum and has no digestive function.

The **colon** forms most of the large intestine, and it is subdivided into four segments. The *ascending colon* extends upwards from the cecum along the right side of the abdomen. As it nears the liver, it turns left to become the *transverse colon.* Near the spleen, the transverse colon turns downward to become the *descending colon* along the left side of the abdomen. Near the pelvis, the descending colon becomes the *sigmoid colon,* which is characterized by an S-shaped curvature leading to the rectum (figure 15.18).

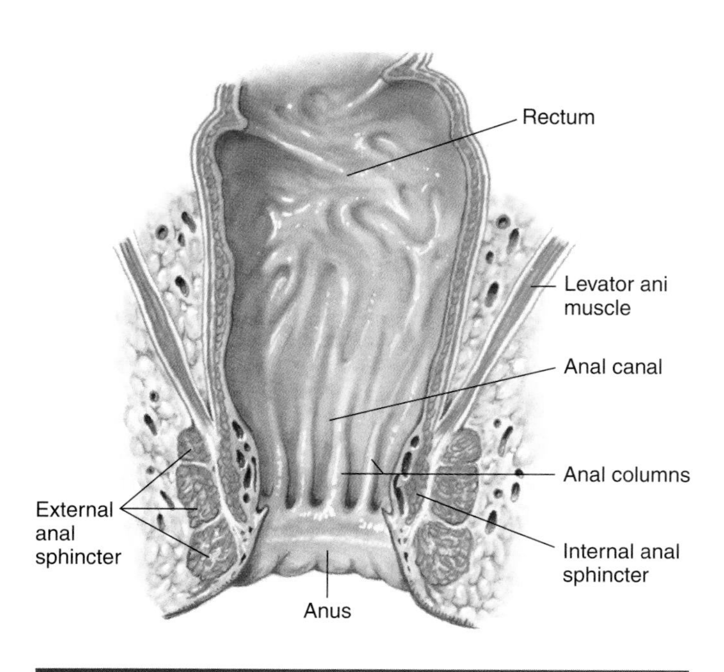

Figure 15.19 The rectum and the anal canal are located at the distal end of the alimentary canal.

The **rectum** is the short, terminal portion of the large intestine. The last portion of the rectum is known as the **anal canal** and its exterior opening is the **anus.** The mucosa of the anal canal is folded to form the *anal columns,* which contain networks of arteries and veins. The anus is kept closed except during defecation by the involuntarily controlled *internal anal sphincter* and the voluntarily controlled *external anal sphincter* (figure 15.19).

The large intestine has a puckered appearance when viewed externally. This results because the longitudinal muscles are not uniformly layered but are reduced to three longitudinal bands, the *taeniae coli,* that run the length of the colon. Contractions of these muscle fibers gather the colon into a series of pouches, the *haustra.* Like the small intestine, the large intestine is supported by the mesentery.

The mucosa of the large intestine is also different from that of the small intestine. Villi are absent, and the columnar epithelium contains numerous mucus-producing goblet cells.

Functions

Food residue entering the large intestine contains water, minerals, bacteria, and other substances that were not digested or absorbed while in the small intestine. There are no digestive enzymes formed by the large intestine. Instead, intestinal bacteria decompose the nondigested food residues. This action yields certain B vitamins and vitamin K and also produces gas (flatus). The mucosa of the large intestine secretes large quantities of mucus that lubricate the intestinal lining and reduce abrasion as materials are moved along.

A major function of the large intestine is the absorption of water, some minerals, and vitamins as the contents slowly move through the colon. Much of this absorption occurs before the contents reach the descending colon, where they are congealed to form the **feces** (fē-sēz). Feces contain large amounts of bacteria, mucus, and water as well as nondigested food residues.

Lactose intolerance is caused by a deficiency or absence of intestinal lactase. The presence of undigested lactose in the intestines produces an osmotic gradient that prevents the normal reabsorption of water into the blood and, even worse, actually causes water to be drawn into the intestines from interstitial fluid. The result is diarrhea, flatulence, bloating, and intestinal cramps. Afflicted persons can avoid this problem if they take a tablet or liquid containing lactase before meals containing milk or milk products.

Movements

The mixing and propelling movements of the large intestine are more sluggish than those of the small intestine. Vigorous peristalsis occurs only two to four times a day, usually following a meal. Such peristaltic contractions are called *mass movements* because they move the contents of the descending and sigmoid colons toward the rectum. The defecation reflex is activated when the rectum fills with feces and its wall is stretched. This reflex stimulates contractions that increase pressure within the rectum and open the internal anal sphincter. If the voluntarily controlled external anal sphincter is relaxed, the feces are expelled through the anus. If its contraction is voluntarily maintained, defecation is postponed.

What are the parts of the large intestine?
What is the function of the large intestine?

Table 15.4 Summary of Liver Functions

Function	Comment
Carbohydrate metabolism	Regulates glucose levels of the blood; converts excess glucose into glycogen or fat for storage; converts glycogen, fat, and amino acids into glucose as needed
Lipid metabolism	Forms lipoproteins for transport of fatty acids, triglycerides, and cholesterol; synthesizes cholesterol and uses it to form bile salts
Protein metabolism	Removes amine groups from amino acids, making the remainder of amino acid molecules available for cellular respiration or conversion into glucose or fat; converts ammonia into urea; forms plasma proteins, including fibrinogen and immunoglobulins
Storage	Stores fat, glycogen, iron, and vitamins A, D, E, K, and B_{12}
Detoxification	Modifies many drugs and toxic chemicals to form less toxic compounds
Phagocytosis	Removes worn-out blood cells and any bacteria that are present
Secretion	Produces bile, which is emptied into the small intestine; releases heparin, an anticoagulant, into the blood

Nutrients: Sources and Uses

Nutrients are substances present in foods that are used in the normal growth and maintenance of the body. The required nutrients are carbohydrates, lipids, proteins, vitamins, minerals, and water. The liver plays an important role in the metabolism of various nutrients. See table 15.4 for a summary of liver functions.

Carbohydrates

Nearly all carbohydrates in the diet come from plant foods. Sugars are abundant in fruits, and starch is a major component of cereals, legumes (e.g., beans, peas, peanuts), and root vegetables. Only small amounts of glycogen occur in meats, and lactose sugar is found in milk and milk products. The recommended daily allowance (RDA) for carbohydrates is 125 to 175 g, or 55% to 60% of the total daily caloric intake.

Evidence suggests that high-fiber, low-fat diets reduce the risk of certain colon disorders, such as diverticulitis and cancer.

Cellulose is a polysaccharide that is abundant in plant foods, but it cannot be digested by humans. However, it is an important dietary component, because it provides fiber (roughage) that increases the bulk of the intestinal contents, which aids the function of the large intestine.

Carbohydrates are primarily used as an energy source by the body. Glucose is the primary energy source for cellular respiration in the body cells. It is carried by blood to body cells, where it is broken down by cellular respiration to release energy. Body cells, especially nerve cells, require a continuous supply of glucose for cellular respiration to meet their energy needs.

The liver, along with the hormones insulin and glucagon, is involved in the regulation of glucose concentration in the blood. Excess glucose is converted into glycogen and is stored in the liver and skeletal muscles. If excess glucose still remains, it is converted into triglycerides (fats) and is stored in adipose tissues. When blood glucose levels decline, the liver converts glycogen into glucose. If still more glucose is needed, fats are converted into glucose (figure 15.20).

Lipids

Most **lipids** in the diet are triglycerides (neutral fats) that occur in both animal and plant foods. Saturated fats occur mostly in meats and milk products. Unsaturated fats are found primarily in vegetable oils, seeds, and nuts. Cholesterol occurs in milk products, egg yolks, and red meats.

The recommended daily allowance for lipids is 80 to 100 g, or no more than 30% of the total daily caloric intake. The American Heart Association recommends that saturated fats compose only 10% or less of the total fat consumed, because they are more

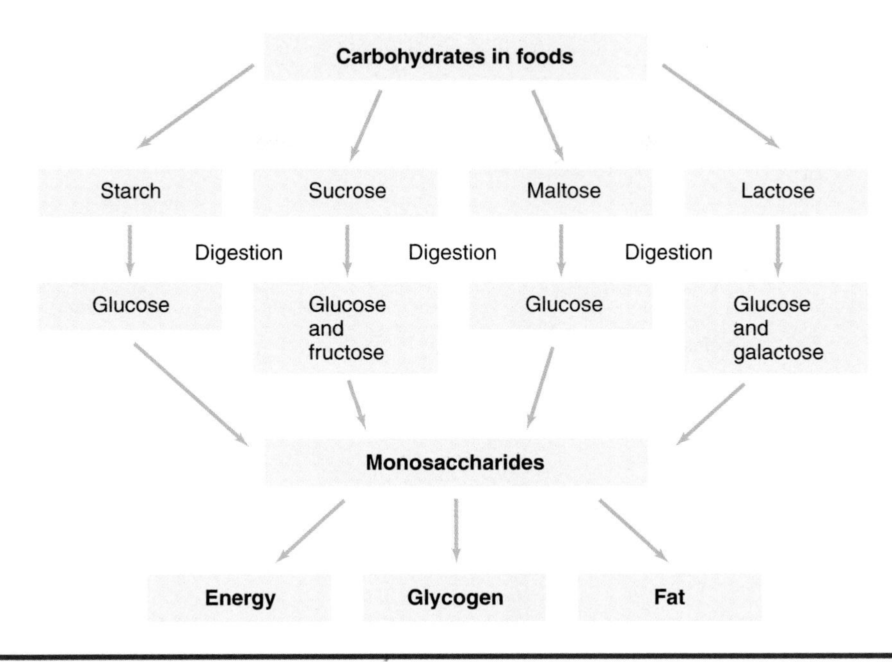

Figure 15.20 Monosaccharides are the end products of carbohydrate digestion. They are used primarily as energy sources for cells, but they may be converted into glycogen or fat.

easily converted by the liver into cholesterol than are unsaturated fats. It is recommended that cholesterol intake not exceed 250 mg (one egg yolk) per day.

High-fat diets and high blood levels of triglycerides and cholesterol are associated with an increased risk of atherosclerosis and coronary heart disease.

Lipids are essential components of the diet, although excessive amounts are not desirable. Phospholipids form the major portion of cellular membranes, and fatty acids help form the myelin sheaths of neurons. Triglycerides are important energy sources for many cells, including liver and skeletal muscle cells. Excess triglycerides are stored in adipose tissue, where they form the largest energy reserve in the body.

Cholesterol is not used as an energy source, but it forms parts of plasma membranes and is used in the synthesis of bile salts and steroid hormones. The liver helps to regulate the concentration of triglycerides and cholesterol in the blood. Fatty acids, triglycerides, and cholesterol are combined with proteins in the liver to form lipoproteins, the form in which lipids are transported in the blood (figure 15.21).

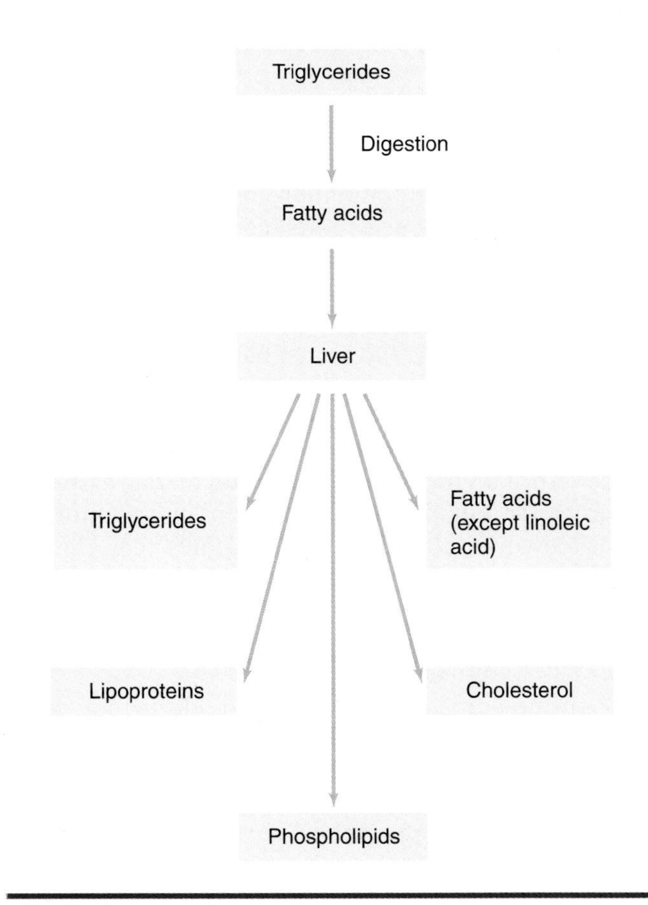

Figure 15.21 The liver uses fatty acids to synthesize different types of lipids.

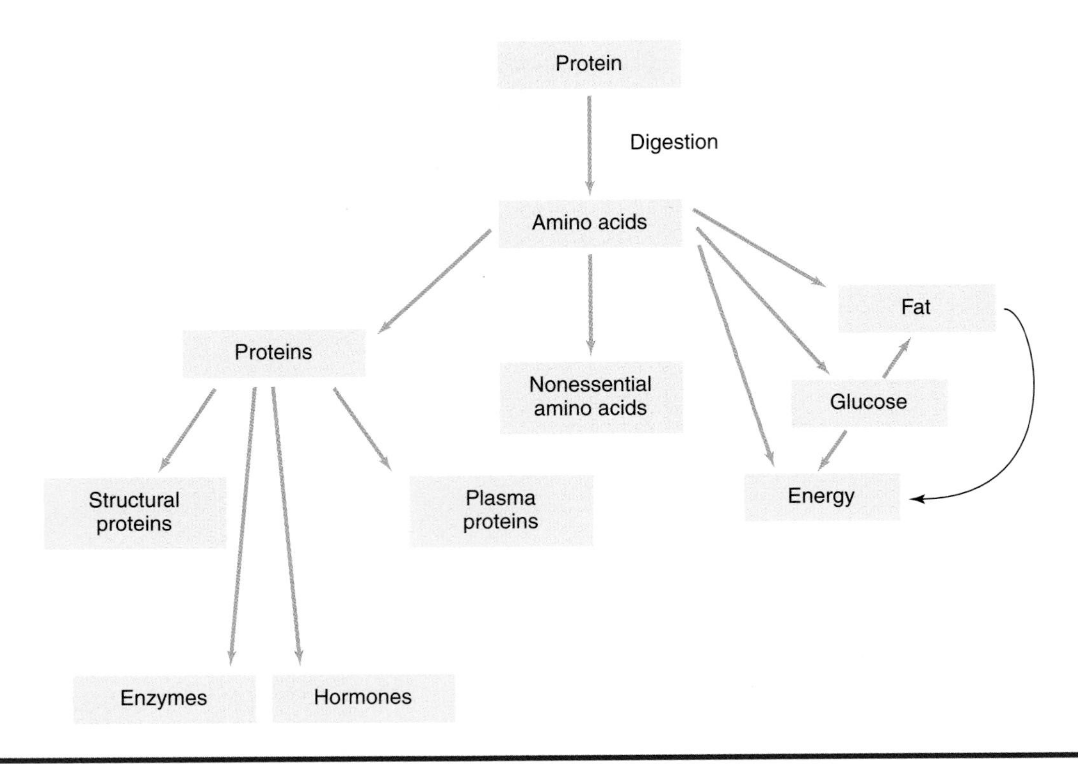

Figure 15.22 Amino acids are the end products of protein digestion. They are used primarily to synthesize proteins, but they may be used as an energy source or they may be converted into glucose or fat.

Proteins

Prime sources of proteins include red meat, poultry, fish, milk products, eggs, nuts, cereals, and legumes. Of the 20 kinds of amino acids composing proteins, eight cannot be synthesized from other amino acids by the liver. These eight amino acids are the **essential amino acids** because they must be present in the diet. All essential amino acids must be present in the body at the same time in order for the body to synthesize proteins necessary for normal growth and maintenance. Animal proteins contain all of the essential amino acids, but plant proteins lack one or more of them. However, if cereals and legumes (e.g., beans, peas) are eaten together, this combination provides all of the essential amino acids. The recommended daily allowance for proteins is at least 0.8 g per kilogram (2.2 lb) of body weight.

Amino acids are used by the body primarily to synthesize proteins: plasma proteins, certain hormones, enzymes, and proteins that form structural components of cells. However, they may be used as an energy source in cellular respiration if there is a deficient supply of glucose or fats. If there is an excess of amino acids, they may be converted to glucose or fat.

For amino acids to be used as an energy source, or converted to glucose or fat, the liver first removes the amine groups (—NH_2) so that the remainders of the amino acid molecules are available for these alternative pathways.

The amine groups react to form ammonia, a toxic substance. However, the liver converts ammonia into *urea,* a less toxic substance that is released into the blood and is excreted by the kidneys in urine. Note the uses of amino acids in figure 15.22.

Weight loss by dieting results from the intake of fewer calories than required for the normal functioning of the body. This forces the body to mobilize stored fats and use them as an energy source.

Vitamins

Vitamins are organic compounds that are required in minute amounts for the normal functioning of the body. They are not an energy source, but they are essential for the utilization of energy foods. Most vitamins act with enzymes to bring about particular metabolic reactions.

Table 15.5 Water-Soluble Vitamins

Vitamin	Sources	Importance
C (ascorbic acid)	Citrus fruits, cabbage, tomatoes, leafy green vegetables	Required for synthesis of steroid hormones, absorption of iron, and formation of connective tissues
B_1 (thiamine)	Meats, eggs, cereals, beans, peas, leafy green vegetables	Required for cellular respiration and the synthesis of ribose sugars
B_2 (riboflavin)	Meats, cereals, leafy green vegetables; formed by colon bacteria	Required for cellular respiration
Niacin	Meats, liver, beans, peas, peanuts	Required for cellular respiration and synthesis of proteins, nucleic acids, and fats
B_6	Meats, poultry, fish, cereals, beans, peas, peanuts; formed by colon bacteria	Required for protein synthesis and formation of antibodies
Pantothenic acid	Meats, fish, cereals, beans, peas, fruits, vegetables, milk; formed by colon bacteria	Required for cellular respiration, conversion of amino acids and lipids into glucose, synthesis of cholesterol and steroid hormones
B_{12}	Meats, poultry, fish, milk, eggs, cheese	Required for formation of red blood cells and cellular respiration of amino acids
Folic acid	Liver, cereals, leafy green vegetables; formed by colon bacteria	Required for synthesis of DNA, RNA, and normal blood cells
Biotin	Liver, eggs, beans, peas, peanuts; formed by colon bacteria	Required for cellular respiration and synthesis of fatty acids

Table 15.6 Fat-Soluble Vitamins

Vitamin	Sources	Importance
A	Eggs, milk, butter, liver; green, yellow, and orange vegetables and fruits	Required for healthy skin and mucous membranes; for development of visual pigments in rods and cones; for healthy bones and teeth
D	Formed by skin when exposed to ultraviolet light; fish liver oils, milk, eggs	Required for healthy bones and teeth; promotes absorption of calcium and phosphorus
E	Cereals, vegetable oils, fruits, and vegetables	Prevents oxidation of fatty acids and vitamin A; helps keep plasma membranes intact
K	Synthesized by colon bacteria; leafy green vegetables, cabbage, pork, liver, soybean oil	Required for formation of prothrombin, an enzyme involved in the formation of blood clots

Vitamins are classified according to their solubility: water-soluble or fat-soluble. Water-soluble vitamins include vitamin C and the B vitamins. They are sensitive to heat and easily destroyed by cooking. Fat-soluble vitamins are vitamins A, D, E, and K. They are more resistant to heat and are not easily destroyed by cooking.

The sources and functions of vitamins are shown in tables 15.5 and 15.6.

Minerals

Minerals are inorganic substances that plants absorb from the soil. They are present in both plant foods and animal foods since animals obtain them by eating plants. Humans need adequate amounts of the seven major minerals noted in table 15.7 but require only trace amounts of other minerals that occur in the body.

Table 15.7 Major Minerals in the Body

Mineral	Sources	Importance
Calcium (Ca)	Milk, milk products, leafy green vegetables	Forms bones and teeth; required for blood clotting, conduction of nerve impulses, and muscle contraction
Phosphorus (P)	Meats, cereals, nuts, milk, milk products, legumes	Forms bones and teeth; component of proteins, nucleic acids, ATP, and phosphate buffers
Potassium (K)	Meats, nuts, potatoes, bananas, cereals	Required for conduction of nerve impulses and muscle contractions
Sulfur (S)	Meats, milk, eggs, legumes	Component of vitamins biotin and thiamine, the hormone insulin, and some amino acids
Sodium (Na)	Table salt, cured meats, cheese	Helps maintain osmotic pressure of body fluids; required for nerve impulse transmission
Chlorine (Cl)	Table salt, cured meats, cheese	Helps maintain osmotic pressure of body fluids; required for formation of HCl in gastric juice
Magnesium (Mg)	Milk, milk products, cereals, legumes, nuts, leafy green vegetables	Required for normal nerve and muscle functions; involved in ATP-ADP conversions

About 4% of the body weight consists of minerals. Calcium and phosphorus, the most abundant minerals, account for about 75% of the total minerals in the body.

In the body, minerals may be incorporated into organic molecules, deposited as mineral salts (as in bone), or occur as ions in body fluids. Table 15.7 identifies the sources and functions of the major minerals.

How are monosaccharides, fatty acids, and amino acids used in the body?
What are the major functions of the liver?

Disorders of the Digestive System

These disorders may be grouped as inflammatory or noninflammatory.

Inflammatory Disorders

Appendicitis is an acute inflammation of the appendix. First symptoms include referred pain in the umbilical region and nausea. Later, pain is localized in the right lower quadrant of the abdomen. Surgical removal of the appendix is the standard treatment.

Colitis (kō-lī-tis) is the inflammation of the mucosa of the large intestine. The cause is unclear, but chronic stress may contribute to this condition. Diarrhea and cramps are typical symptoms.

Diverticulitis (dī-ver-tik-ū-lī′-tis) is a disorder of the large intestine. Small saclike outpockets of the colon often develop from a diet lacking sufficient fiber (bulk). Diverticulitis is the inflammation of these diverticula, and it may cause considerable pain, bloating, or diarrhea.

Hemorrhoids is a condition where one or more veins in the anal canal become enlarged and inflamed. Chronic constipation contributes to the development of hemorrhoids.

Hepatitis is inflammation of the liver, and it may be caused by several factors, including viruses, drugs, or alcohol. It is characterized by jaundice, fever, and liver enlargement. There are three basic types of hepatitis.

Hepatitis A (infectious hepatitis) is caused by the hepatitis A virus, which is spread by person-to-person contact and fecal contamination of food and water. Recovery takes four to six weeks.

Hepatitis B (serum hepatitis) is caused by the hepatitis B virus, and it is spread by transfusions, contaminated needles, saliva, and sexual contact. Most people recover completely, but some retain live viruses for years and unknowingly serve as carriers of the disease. Chronic hepatitis may produce cirrhosis or lead to liver cancer.

Hepatitis C is spread by person-to-person contact and fecal contamination of food and water. Symptoms are usually mild, with recovery in four to six weeks. This type of hepatitis is much more serious during the third trimester of pregnancy.

Periodontal disease refers to a variety of conditions characterized by inflammation, bleeding gums, and degeneration of the gingivae, cementum, periodontal ligaments, and alveolar bone, which causes loosening of the teeth. Poor dental hygiene contributes to this condition.

Peritonitis is the acute inflammation of the peritoneum that lines the abdominal cavity and covers abdominal organs. It may result from bacteria entering the peritoneal cavity due to contamination in accidents or surgery or by a ruptured intestine or appendix.

Noninflammatory Disorders

Cirrhosis of the liver is characterized by scarring, which results from connective tissue replacing destroyed liver cells. It may be caused by hepatitis, alcoholism, nutritional deficiencies, or liver parasites.

Constipation is a condition where defecation is difficult, and the feces are hard and dry. This results from feces remaining in the colon for a longer than normal period, which allows more water to be absorbed.

Dental caries, or tooth decay, results from the excess acid produced by certain microorganisms that live in the mouth and use food residues for their nutrients. Residues of carbohydrates, especially sugars, nurture microorganisms that cause decay.

Diarrhea is the production of watery feces due to the abnormally rapid movement of food residues through the colon. Increased peristalsis may result from a number of causes, including inflammation and chronic stress.

Eating disorders result from an obsessive concern about weight control, especially among young adult females. There are two major types of eating disorders: anorexia and bulimia.

Anorexia nervosa is self-imposed starvation that results in malnutrition and associated physiological changes. Patients with this disorder see themselves as overweight, although others see them as very thin. Death can occur due to the complications of prolonged starvation.

Bulimia is characterized by frequent overeating and purging by self-induced vomiting. Fears of being overweight, depression, and stress are associated factors. The exact cause is unknown. Bulimia may lead to such complications as an imbalance of electrolytes, erosion of tooth enamel by stomach acids, and constipation.

Gallstones result from crystallization of cholesterol in bile. They commonly occur in the gallbladder, but they may be carried into bile ducts, where they block the flow of bile. Severe pain, and often jaundice, accompanies such blockage. Treatment may include drugs that dissolve the gallstones, shock-wave therapy to break up the stones, or surgical removal of the gallstones and gallbladder.

CHAPTER SUMMARY

Digestion: An Overview

1. Digestion involves both mechanical and chemical processes. Mechanical digestion is the physical process of breaking food into smaller particles. Chemical digestion is an enzymatic process that converts nonabsorbable foods into absorbable nutrients.
2. A number of different digestive enzymes are required for the chemical digestion of foods.

Alimentary Canal: General Characteristics

1. The alimentary canal is the long tube through which food passes from the mouth to the anus. Digestion of food and absorption of nutrients occur in some portions of the alimentary canal.
2. The wall of the alimentary canal is composed of four layers. From outside in, they are the serosa, muscular layer, submucosa, and mucosa.
3. Food is moved through the alimentary canal by peristaltic contractions. Ripplelike contractions mix food with digestive secretions.

Mouth

1. The mouth is surrounded by the cheeks, palate, and tongue. Teeth are embedded in the dental arches of the maxilla and mandible.
2. Humans have two sets of teeth: deciduous and permanent. There are 32 permanent teeth divided into four types: incisors, cuspids, bicuspids, and molars.
3. A tooth is composed of two major parts: a crown covered with enamel and a root embedded in alveolar bone. Dentin forms most of the tooth. The centrally located pulp cavity contains blood vessels and nerves. The root is anchored to bone by cementum and by a periodontal membrane.
4. Three pairs of salivary glands (parotid, submandibular, and sublingual) secrete saliva into the mouth. Salivary secretion is under reflexive neural control. Saliva cleanses and lubricates the mouth and binds food together.
5. Saliva contains salivary amylase, which breaks down starch and glycogen into maltose.

Pharynx

1. The pharynx connects the oral and nasal cavities with the esophagus and larynx. Its digestive function is to transport food from the mouth to the esophagus.
2. The swallowing reflex causes the epiglottis to cover the laryngeal opening, directing food into the esophagus.
3. Pharyngeal, palatine, and lingual tonsils are located near entrances to the pharynx.

Esophagus

1. The esophagus carries food by peristalsis from the pharynx to the stomach.
2. The lower esophageal sphincter opens to let food enter the stomach.

Stomach

1. The stomach is a pouchlike enlargement of the alimentary canal. It is located in the upper left quadrant of the abdomen and consists of four regions: cardiac, fundic, and pyloric regions and the body.
2. The pyloric sphincter opens to allow chyme to enter the duodenum.
3. The stomach mucosa contains gastric glands that secrete gastric juice, whose components include hydrochloric acid and pepsin. Gastric juice converts food into chyme.
4. The secretion of gastric juice is regulated by neural reflexes and the hormones gastrin and cholecystokinin.
5. Pepsin acts on proteins and breaks them into peptides.
6. Rennin is a gastric enzyme secreted by infants. It curdles milk, which aids digestion of milk proteins.
7. Gastric juice contains intrinsic factor, which is necessary for absorption of vitamin B_{12}.
8. Little absorption occurs in the stomach.

Pancreas

1. The pancreas is located adjacent to the duodenum and pyloric region of the stomach. The pancreas secretes pancreatic juice, which is carried to the duodenum by the pancreatic duct. The hormones secretin and cholecystokinin stimulate the secretion of pancreatic juice.
2. Pancreatic digestive enzymes act on each type of energy food. Amylase converts starch and glycogen into maltose. Lipase converts fats into monoglycerides and fatty acids. Trypsin converts proteins into peptides.

Liver

1. The liver is located in the upper right quadrant of the abdomen. The liver removes, modifies, or adds substances to the blood as it flows through the sinusoids.
2. Bile is continuously secreted by the liver. Between meals, bile is stored in the gallbladder. It is released when the hormone cholecystokinin, formed by the intestinal mucosa, stimulates contraction of the gallbladder. Bile is carried to the duodenum by the common bile duct.
3. Bile emulsifies lipids, which aids their digestion by lipases.

Small Intestine

1. The small intestine occupies much of the abdominal cavity, and it consists of three parts: duodenum, jejunum, and ileum. Most of the digestion and absorption of food occur in the small intestine.
2. The mucosa contains numerous intestinal glands that secrete intestinal juice and villi that absorb nutrients. Secretion of intestinal juice is activated by a neural reflex. Intestinal digestive enzymes act on all three types of energy foods.
3. The enzymes maltase, sucrase, and lactase act on corresponding disaccharides to form the monosaccharides: glucose, fructose, and galactose. Lipase converts fats into monoglycerides and fatty acids. Peptidase converts peptides into amino acids.
4. End products of digestion are absorbed into the villi. Monosaccharides, amino acids, vitamins, and minerals enter the capillaries of the villi. Monoglycerides and fatty acids enter mucosal cells and are recombined to form fat molecules. Clusters of fat molecules are coated with protein to form chylomicrons, which enter the lacteals of the villi.
5. Nondigested and nonabsorbed materials exit the small intestine and enter the large intestine through the ileocecal sphincter.

Large Intestine

1. The large intestine consists of the cecum, colon, and rectum. The appendix is a nonfunctional appendage of the cecum.
2. The large intestine is gathered into a series of pouches by the taeniae coli. Its mucosa lacks villi and secretes only mucus.
3. The absorption of water and the formation and expulsion of feces are major functions of the large intestine. Bacteria decompose the nondigested materials.
4. Mass peristaltic movements propel the feces into the rectum, initiating the defecation reflex, which opens the internal anal sphincter. Voluntary relaxation of the external sphincter allows expulsion of the feces.

Nutrients: Sources and Uses

1. Nutrients include carbohydrates, lipids, proteins, vitamins, minerals, and water. The liver is involved in the metabolism of many nutrients.
2. Dietary carbohydrates come primarily from plant foods. Cellulose is a nondigestible polysaccharide that provides fiber in the diet.

3. The liver regulates the concentration of glucose in the blood. Excess glucose is converted to glycogen or fats for storage, and these reactions may be reversed to release more glucose into the blood.
4. Dietary lipids are mostly triglycerides that occur either as saturated fats, usually in animal foods, or as unsaturated fats, usually in plant foods. Cholesterol occurs in egg yolks, milk, and meats.
5. Lipids form important parts of plasma membranes and myelin sheaths of neurons. Fats are an energy source for many cells. Excess fats are stored in adipose tissue. The liver helps to regulate the concentration of triglycerides and cholesterol in the blood.
6. Dietary proteins occur in meats, milk, eggs, cereals, nuts, and legumes. The eight essential amino acids cannot be synthesized by the body. Only animal proteins contain all of the essential amino acids.
7. Amino acids are used primarily to synthesize protein in the body. These proteins form plasma proteins, enzymes, certain hormones, and structural parts of cells. Amino acids may be deaminated by the liver and used to form glucose or fat or may be used as an energy source by cells.
8. Vitamins are organic molecules required in minute amounts for normal functioning of the body. Water-soluble vitamins include the B vitamins and vitamin C. Fat-soluble vitamins include vitamins A, D, E, and K.
9. Vitamins act with enzymes to bring about essential chemical reactions, such as cellular respiration, in cells.
10. Minerals are inorganic substances that are necessary for the normal functioning of the body. Minerals are obtained by plants from the soil, and they are passed on to animals, including humans, eating the plants.
11. Many minerals are a part of organic molecules in the body. Other minerals are deposited as salts in bones and teeth, and some occur as ions in body fluids.
12. There are seven major minerals that are required in moderate amounts in the diet: calcium, phosphorus, potassium, sulfur, sodium, chlorine, and magnesium. Other minerals of the body are required in trace amounts.

Disorders of the Digestive System

1. Inflammatory disorders include appendicitis, colitis, diverticulitis, hemorrhoids, hepatitis, periodontal disease, and peritonitis.
2. Other disorders include eating disorders (anorexia nervosa and bulimia), cirrhosis, constipation, dental caries, diarrhea, and gallstones.

BUILDING YOUR VOCABULARY

1. **Selected New Terms**
   bile, p. 316
   cholecystokinin, p. 313
   chylomicrons, p. 318
   dentin, p. 309
   enamel, p. 309
   essential amino acids, p. 324
   gastrin, p. 312
   lactase, p. 318
   lipase, p. 314
   maltase, p. 318
   mastication, p. 308
   pepsin, p. 313
   peptidase, p. 318
   rennin, p. 313
   rugae, p. 311
   saliva, p. 309
   salivary amylase, p. 309
   secretin, p. 313
   sucrase, p. 318
   trypsin, p. 314
2. **Related Clinical Terms**
   dysphagia (dis-fā′-jē-ah) Difficulty in swallowing.
   colonoscopy (kō-lon-os′-kō-pē) Viewing the interior of the colon with a flexible, tubelike instrument called an endoscope.
   gastrectomy (gas-trek′-tō-mē) Removal of all or part of the stomach.
   gastroenterology (gas-trō-en-ter-ol′-ō-jē) Medical specialty treating disorders of the digestive tract.
   ileus (il′-ē-us) Paralysis of the digestive tract.

STUDY ACTIVITIES

1. Complete the **Chapter 15 Study Guide,** which begins on page 475.
2. Complete the **Learning Objectives** listed on the first page of this chapter.

CHECK YOUR UNDERSTANDING

(Answers are located in appendix B.)

1. It is _____ that speed up the hydrolysis of food molecules to yield absorbable _____ .
2. Digestive secretions are produced by the _____ , the inner lining of the alimentary canal.
3. Food is moved through the digestive tract by wavelike contractions called _____ .
4. A tooth is composed mostly of _____ , but the crown of a tooth is covered by _____ .
5. Chemical digestion begins in the mouth, where the enzyme _____ acts on starch to yield _____ .
6. When food enters the esophagus, it is carried by _____ and enters the stomach when the _____ opens.
7. Gastric juice contains HCl and the enzyme _____ , which breaks down _____ to peptides.
8. Gastric secretion is stimulated by impulses carried by _____ fibers and by the hormone _____ .
9. The hormones _____ and _____ activate the production of pancreatic juice, and the hormone _____ stimulates the release of bile from the gallbladder.
10. The three enzymes in pancreatic juice are _____ , which acts on starch, _____ , which acts on fats, and _____ , which acts on proteins.
11. Digestion is completed in the _____ , where disaccharides are converted to _____ and peptides are converted to _____ .
12. End products of digestion are absorbed into _____ of the small intestine, amino acids and monosaccharides enter the _____ , and fatty acids and monoglycerides enter the _____ .
13. Absorption of water and dissolved minerals and vitamins is the primary function of the _____ .
14. The _____ processes and modifies absorbed nutrients before they enter the general circulation.
15. Organic substances that are essential to health in minute amounts are called _____ .

CHAPTER

Urinary System

SIXTEEN

16

Chapter Preview & Learning Objectives

Kidneys
- Identify the organs of the urinary system and describe their general functions.
- Describe the structure and blood supply of the kidney.
- Describe the structure and functions of a nephron.

Urine Formation
- Compare filtration, tubular reabsorption, and tubular secretion.
- Explain how urine is formed.

Maintenance of Blood Plasma Composition
- Explain how the kidneys maintain the volume and composition of the blood plasma.

Excretion of Urine
- Describe the structure and function of the ureters, urinary bladder, and urethra.
- Describe the control of micturition.

Characteristics of Urine
- Indicate the normal components of urine.

Disorders of the Urinary System
- Describe the common disorders of the urinary system.

Chapter Summary
Building Your Vocabulary
Study Activities
Check Your Understanding

SELECTED KEY TERMS

Glomerular filtration The forcing of water and small solutes from the blood in a glomerulus into a glomerular capsule.
Glomerulus (glomus = ball) The cluster of capillaries enveloped by the glomerular capsule.
Juxtaglomerular apparatus (juxta = next to) Specialized cells next to the glomerulus that help to regulate blood pressure.
Micturition (micture = to urinate) Urination.
Nephron (nephros = kidney) The structural and functional unit of the kidneys.
Peritubular capillaries (peri = around) Capillaries surrounding a renal tubule.
Renal corpuscle (ren = kidney) The portion of a nephron composed of a glomerulus and its enveloping glomerular capsule.
Renal tubule The portion of a nephron composed of a proximal convoluted tubule, a loop of Henle, and a distal convoluted tubule.
Tubular reabsorption The reabsorption of substances from the filtrate in a renal tubule back into the blood.
Tubular secretion The secretion of substances from blood in peritubular capillaries into the filtrate in a renal tubule.

The normal metabolic activities of body cells produce a number of waste materials which tend to change the balance of water and dissolved substances in body fluids. The basic function of the **urinary system** is to maintain the volume and composition of body fluids within normal limits.

The urinary system consists of the kidneys, ureters, urinary bladder, and urethra. The paired *kidneys* maintain the composition and volume of body fluids by removing wastes and excess substances in the formation of urine. *Ureters* are slender tubes that carry urine from the kidneys to the *urinary bladder* for temporary storage. Urine is carried from the urinary bladder and is expelled from the body through the *urethra* (figure 16.1).

Kidneys

The **kidneys** are reddish brown, bean-shaped organs located on each side of the vertebral column and against the upper, posterior wall of the abdominal cavity. They lie posterior to the parietal peritoneum, which covers their anterior surfaces, that is, they are located in a retroperitoneal position. The kidneys are located between the levels of the twelfth thoracic vertebra and the third lumbar vertebra, and they are protected by the lower ribs. Each kidney is supported by connective tissue and is encased in a cushioning capsule of thick adipose tissue. The innermost layer of supporting connective tissue forms the **renal capsule,** which closely envelops each kidney and provides support for the soft internal tissues.

Structure

Each kidney is convex laterally and concave medially with a medial indentation called the *hilum.* Blood vessels, lymphatic vessels, nerves, and the ureter enter or exit at the hilum. An adult kidney is about 12 cm long, 7 cm wide, and 2.5 cm thick.

The internal macroscopic structure of a kidney is best observed in coronal section, as shown in figure 16.2. Three regions of the kidney are evident: the cortex, medulla, and pelvis. The **renal cortex** is the relatively thin, outermost layer. Interior to the cortex is the **renal medulla,** which contains the cone-shaped *renal pyramids.* The apex, or *renal papilla,* of each pyramid extends toward the renal pelvis. The lines extending from the base to the apex of each pyramid are formed by microscopic **collecting ducts.** Narrow portions of the cortex, the *renal columns,* extend into the medulla between the renal pyramids.

The work of the kidneys is performed by microscopic tubules called **nephrons** (nef′-rons), which are extremely abundant in the kidneys. Nephrons originate in the cortex, dip into the medulla, return to the cortex, and ultimately join a collecting duct, as shown in the enlargements in figure 16.2. Nephrons form urine that flows into the collecting ducts of renal pyramids.

Medial to the medulla is the **renal pelvis,** a flat, funnel-like cavity that is continuous with the ureter. The **calyces** (kā′-li-sēz) (singular, *calyx*), branches of the pelvis, extend to form cuplike receptacles enclosing the papillae of renal pyramids. Urine flows from collecting ducts of the renal pyramids into the calyces and passes through the pelvis into the ureter. Urine is carried by the ureter to the urinary bladder by peristalsis.

Nephron

Each kidney contains about 1 million nephrons, the functional units of the kidneys. A nephron consists of

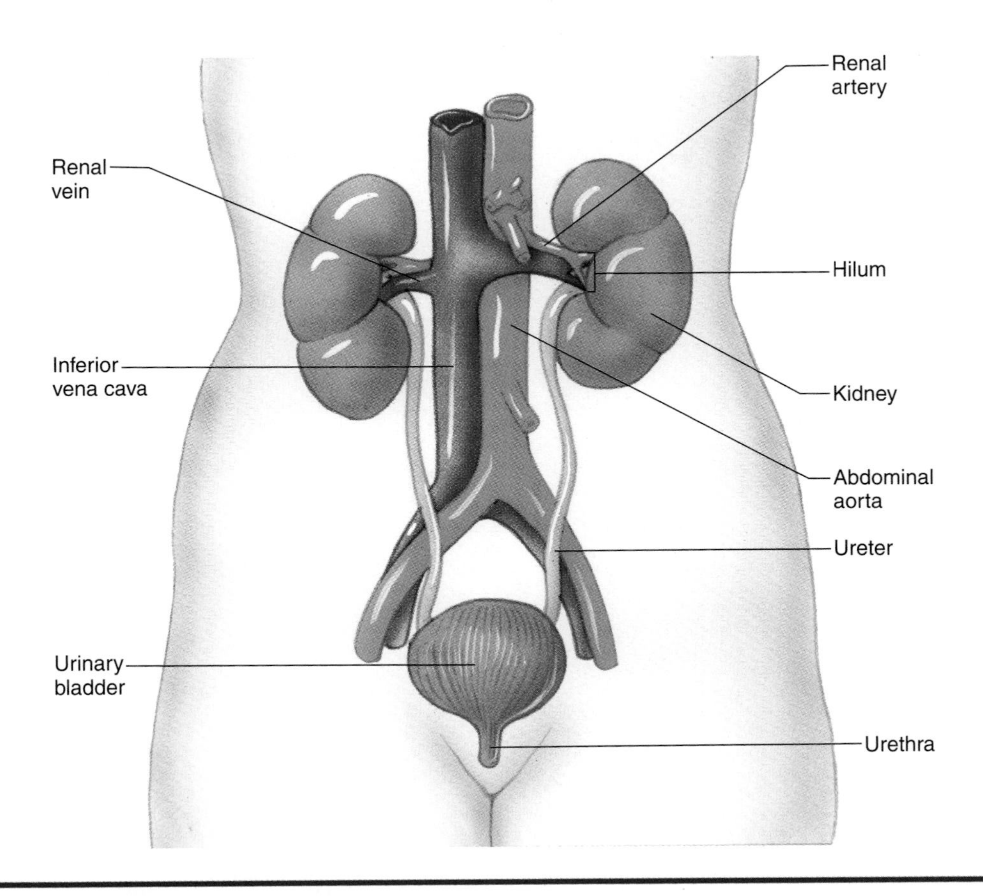

Figure 16.1 Organs of the urinary system

two major parts: a renal corpuscle and a renal tubule. Figure 16.3 shows the structure of a nephron and its associated blood vessels.

Renal corpuscles are located in the renal cortex of the kidneys. Each renal corpuscle is composed of a **glomerulus** (glō-mer′-ū-lus), a tuft of arterial capillaries, which is enclosed in a double-walled **glomerular (Bowman's) capsule.** The glomerular capsule is an expanded extension of a renal tubule.

A **renal tubule** consists of three segments: proximal convoluted tubule, loop of Henle, and distal convoluted tubule. The highly coiled **proximal convoluted tubule** lies within the renal cortex. It extends from the glomerular capsule to the U-shaped **loop of Henle,** which extends into the medulla. The *descending limb* of the loop of Henle passes deep into the medulla, where it makes a sharp U-turn and becomes the *ascending limb,* which returns into the cortex. The diameter of most of the loop of Henle is less than the diameter of the rest of the tubule. The ascending limb joins the highly coiled **distal convoluted tubule** within the cortex. After numerous coils, the distal convoluted tubule joins a collecting duct. Several nephrons unite with a single collecting duct. Collecting ducts extend from the base of a renal pyramid to the papilla, where they open into a calyx.

Blood Supply

The kidneys receive a large volume of blood—1,200 ml per minute, which is about one-fourth of the total cardiac output. Each kidney receives blood via a **renal artery,** which branches from the abdominal aorta. Within each kidney, the renal artery branches to form smaller and smaller arteries.

In the cortex, **afferent arterioles** branch from the smallest arteries, and each afferent arteriole carries blood to a glomerulus. Blood leaves a glomerulus in an **efferent arteriole.** Note that a glomerulus is an arterial capillary tuft sandwiched between two arterioles. The efferent arteriole leads to a **peritubular capillary network,** which is entwined around the renal tubule. Blood from the peritubular capillary enters a venule, progressively larger veins, and finally the renal vein. A **renal vein** carries blood from each kidney to the inferior vena cava (figures 16.3 and 16.4).

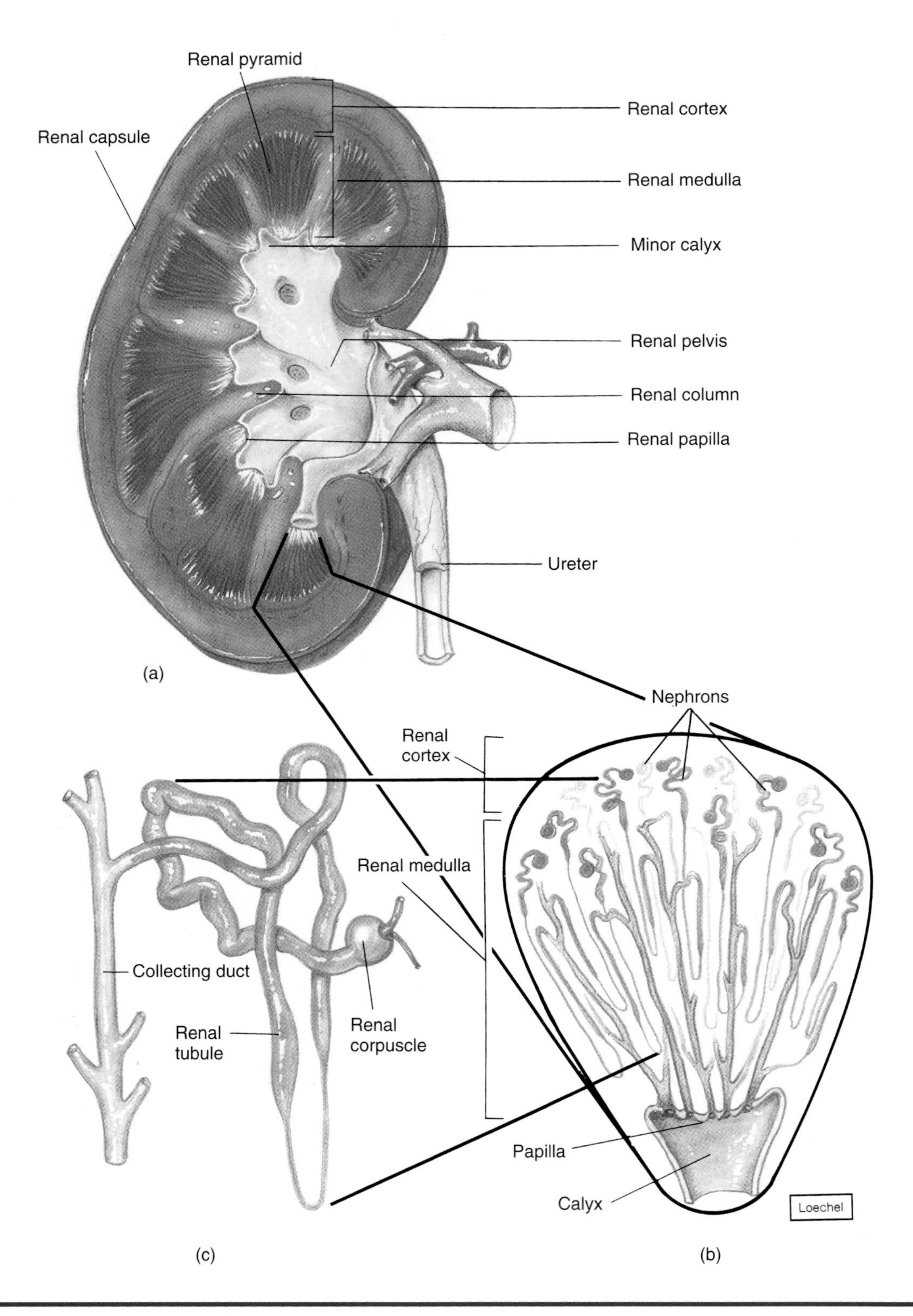

Figure 16.2 (*a*) A coronal section of a kidney; (*b*) a renal pyramid showing nephrons and collecting ducts; (*c*) a single nephron and its collecting duct.

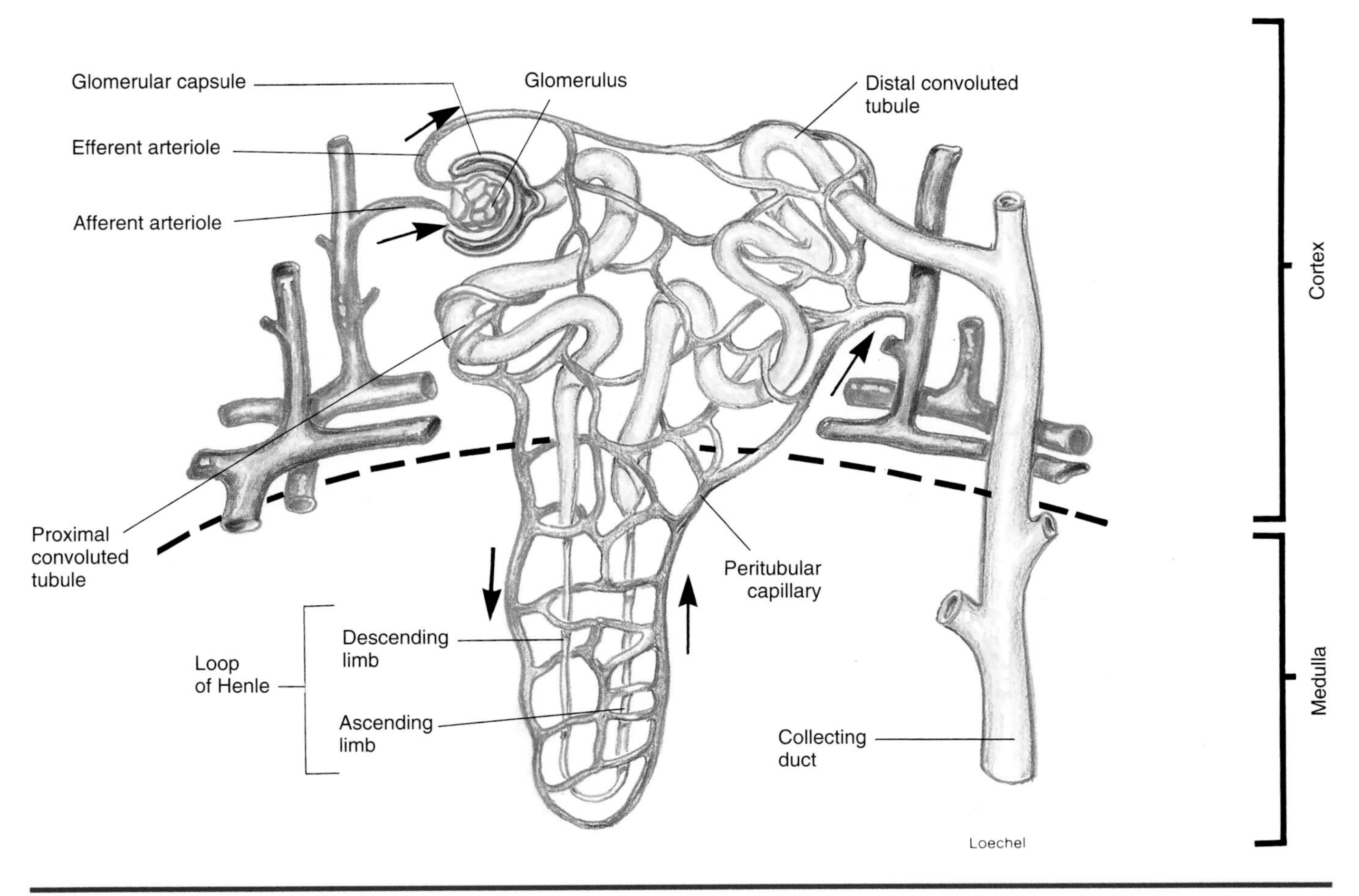

Figure 16.3 A nephron and its associated blood vessels.

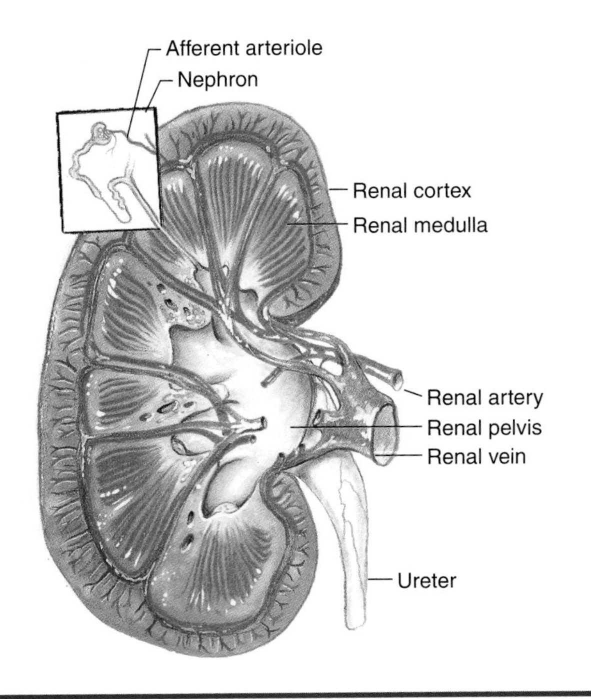

Figure 16.4 Main branches of the renal artery and renal vein.

Juxtaglomerular Apparatus

The distal convoluted tubule extends back into the region of the renal capsule and contacts the afferent arteriole near the glomerulus. The cells of both the afferent arteriole and the distal tubule are modified at the point of contact. These modified cells compose the **juxtaglomerular** (juks-tah-glō-mer′-ū-lar) **apparatus,** whose function helps to regulate blood pressure (figure 16.5).

Overview of Kidney Functions

1. **Production of urine.** The major function of the kidneys is to keep the volume and composition of blood plasma within normal limits. This is accomplished by the removal of waste products and excess substances from the blood through the formation of urine.
2. **Secretion of renin.** Whenever systemic blood pressure decreases, the kidneys secrete **renin.**

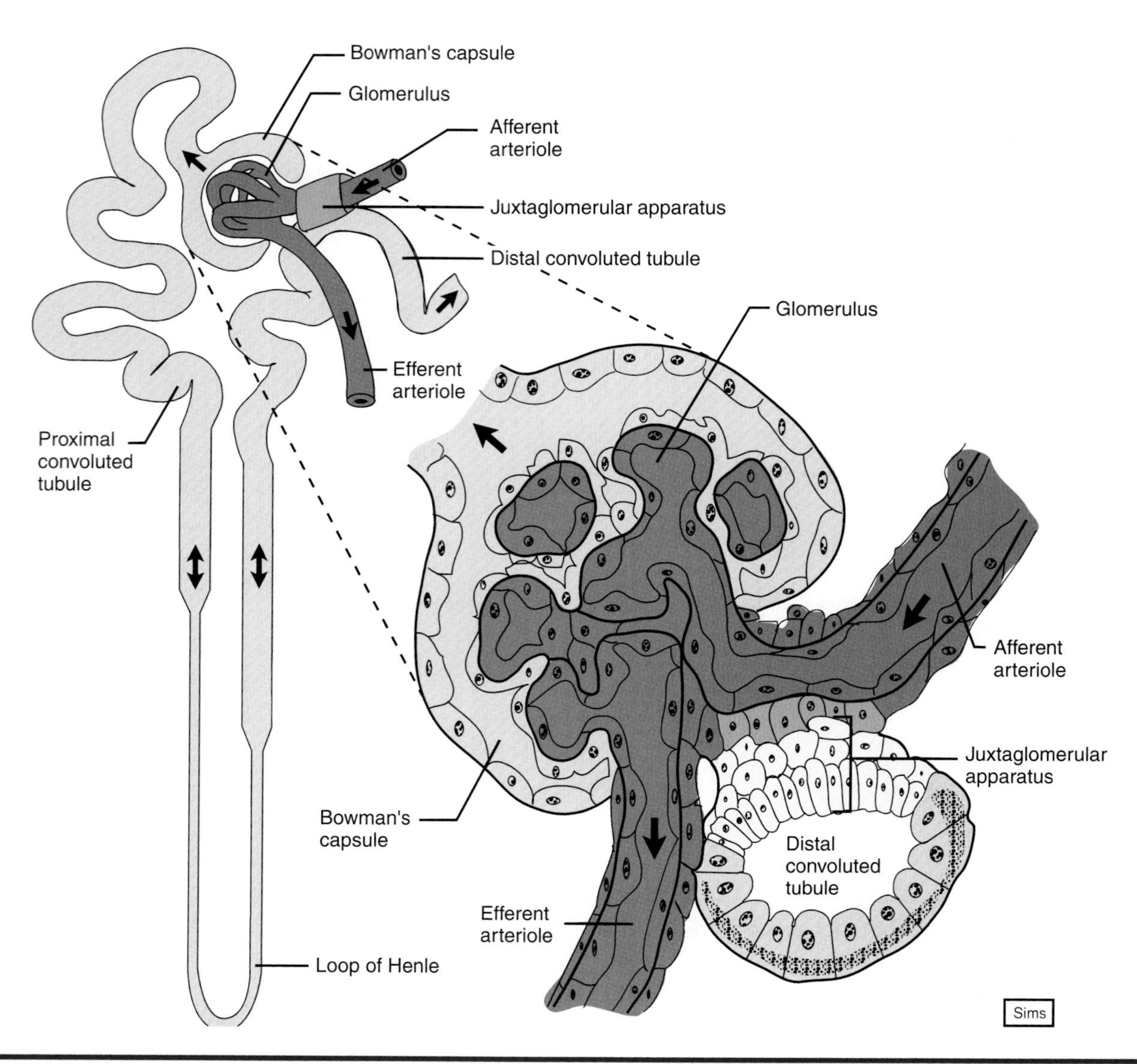

Figure 16.5 The juxtaglomerular apparatus.

Renin is an enzyme that triggers the **renin-angiotensin mechanism,** which returns blood pressure to a normal level.

3. **Secretion of erythropoietin.** When the blood oxygen level falls below normal (hypoxia), the kidneys release more **erythropoietin,** which stimulates red blood cell formation by red bone marrow. The added red blood cells help to return blood oxygen to a normal level.

What organs compose the urinary system, and what are their functions?
What is the general function of the urinary system?

Urine Formation

Nephrons perform three basic functions: (1) they regulate the volume and composition of the blood plasma by removing certain amounts of water and dissolved substances; (2) they help regulate blood pH; and (3) they remove toxic wastes and excess materials from the blood. The result of these processes is the formation of **urine.** Three processes are involved in urine formation: filtration, tubular reabsorption, and tubular secretion.

Filtration

Urine formation starts with **glomerular filtration,** a process that forces some of the water and dissolved substances in blood plasma from the glomeruli into

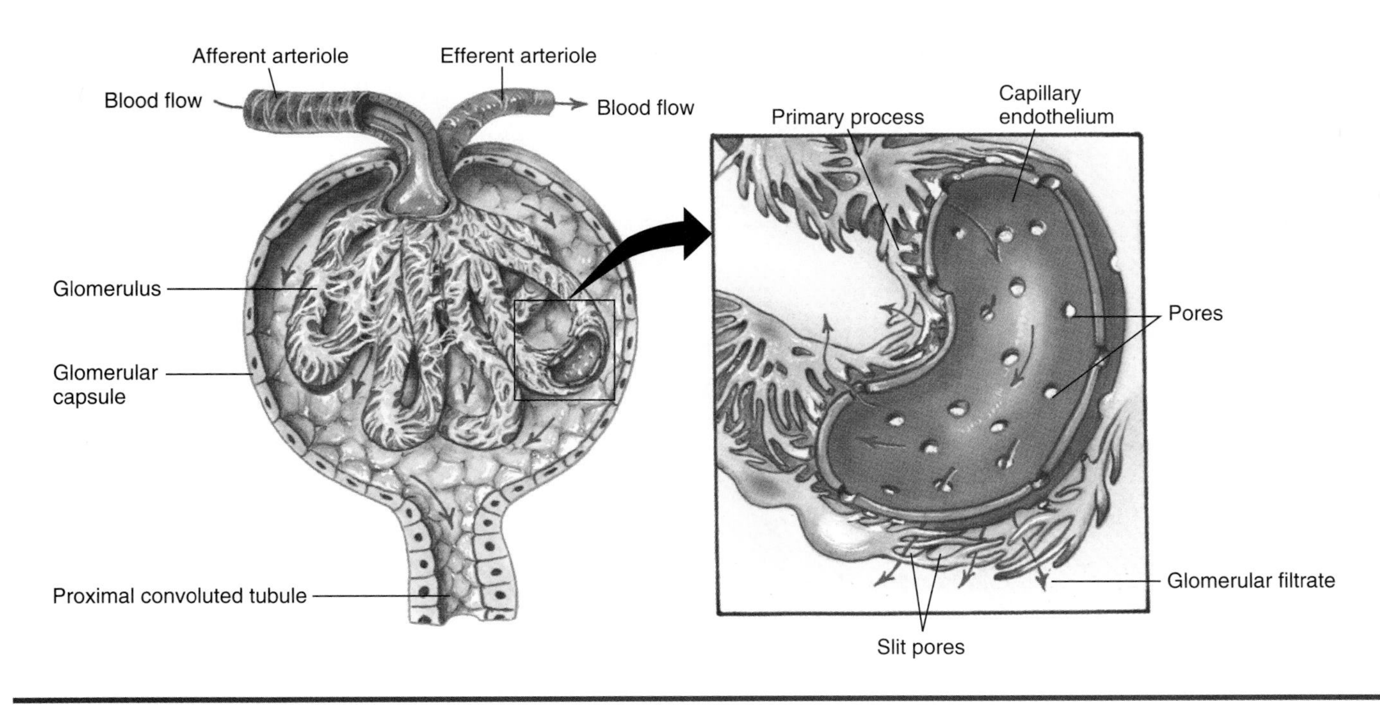

Figure 16.6 Filtration is the first step in urine formation. Water and dissolved substances in plasma are filtered through pores in the glomerular capillaries into the glomerular capsule.

the glomerular (Bowman's) capsules. Two major factors are responsible for filtration: (1) the greater permeability of glomerular capillary walls and (2) the higher blood pressure within the glomeruli.

Glomerular capillaries are much more permeable to substances in the blood than are other capillaries because their walls contain numerous pores, as shown in figure 16.6. These pores allow water and most dissolved substances to easily pass through the capillary walls into the glomerular capsules, but they are small enough to prevent the passage of plasma proteins and blood cells.

The higher glomerular blood pressure results because the diameter of the efferent arteriole is smaller than that of the afferent arteriole. Since blood can enter a glomerulus at a faster rate than it can leave it, a buildup of blood pressure within the glomerulus occurs. Blood pressure within a glomerulus is about four times higher than in systemic capillary networks in the rest of the body. It is the glomerular blood pressure that provides the force for filtration.

The fluid that enters the glomerular capsule is known as the **filtrate.** It consists of the same substances as those composing the blood plasma, except for plasma proteins that are too large to pass through the pores of the glomerular capillaries. As noted, blood cells are normally absent from the filtrate. Since filtration is a nonselective process, the concentrations of these substances are the same in both blood plasma and filtrate (table 16.1).

Filtration Rate

Glomerular filtration rate (GFR) is about 125 ml per minute, or 7.5 liters per hour. This means that the entire volume of blood is filtered every 40 minutes! In 24 hours, about 180 liters (nearly 45 gallons) of filtrate is produced. However, most of the filtrate is reabsorbed, as you will see shortly.

Maintenance of a relatively stable GFR is necessary for normal kidney function. The GFR varies directly with glomerular blood pressure, which, in turn, is primarily determined by systemic blood pressure. **Renal autoregulation** keeps the GFR rather constant in spite of normal fluctuations in systemic blood pressure. This is primarily accomplished by regulating the diameter of the afferent arterioles to keep blood flow into glomeruli within normal limits.

A severe drop in blood pressure, perhaps caused by hemorrhage, activates the sympathetic nervous system and the renin-angiotensin mechanism. The sympathetic system raises blood pressure by constricting arterioles throughout the body. The renin-angiotensin mechanism is a chemical control system that raises blood pressure by (1) constricting arterioles,

Table 16.1 Concentrations of Selected Chemicals in Plasma, Filtrate, and Urine*

Chemicals	Plasma (g/l)	Filtrate (g/l)	Urine (g/l)
Protein	44.4	0.0	0.0
Chloride (Cl^-)	3.5	3.5	6.3
Sodium (Na^+)	3.0	3.0	3.8
Bicarbonate (HCO_3^-)	1.7	1.7	0.4
Glucose	1.0	1.0	0.0
Urea	0.3	0.3	31.3
Potassium (K^+)	0.2	0.2	5.0
Uric acid	0.05	0.05	1.0
Creatinine	0.01	0.01	1.9

*Based on 180 liters of filtrate and 1.25 liters of urine produced in a 24-hour period

(2) conserving water to maintain blood volume, and (3) stimulating water intake to increase blood volume. By raising systemic blood pressure, both of these mechanisms help to maintain filtration.

The renin-angiotensin mechanism is triggered when the juxtaglomerular apparatus detects a reduced filtration rate and releases the enzyme renin. Renin converts a plasma protein (angiotensinogen), which is formed by the liver, into angiotensin I. Then an enzyme from the lungs changes angiotensin I into **angiotensin II,** which produces the actions to increase systemic blood pressure (figure 16.7). Angiotensin II augments the mechanisms regulating the secretion of aldosterone and antidiuretic hormone (ADH). These hormones play major roles in tubular reabsorption and in water-electrolyte balance.

Autoregulatory mechanisms cannot maintain glomerular pressure when the mean (average) systemic blood pressure drops to 50 mmHg or less. At this point, filtration ceases and urine is not formed.

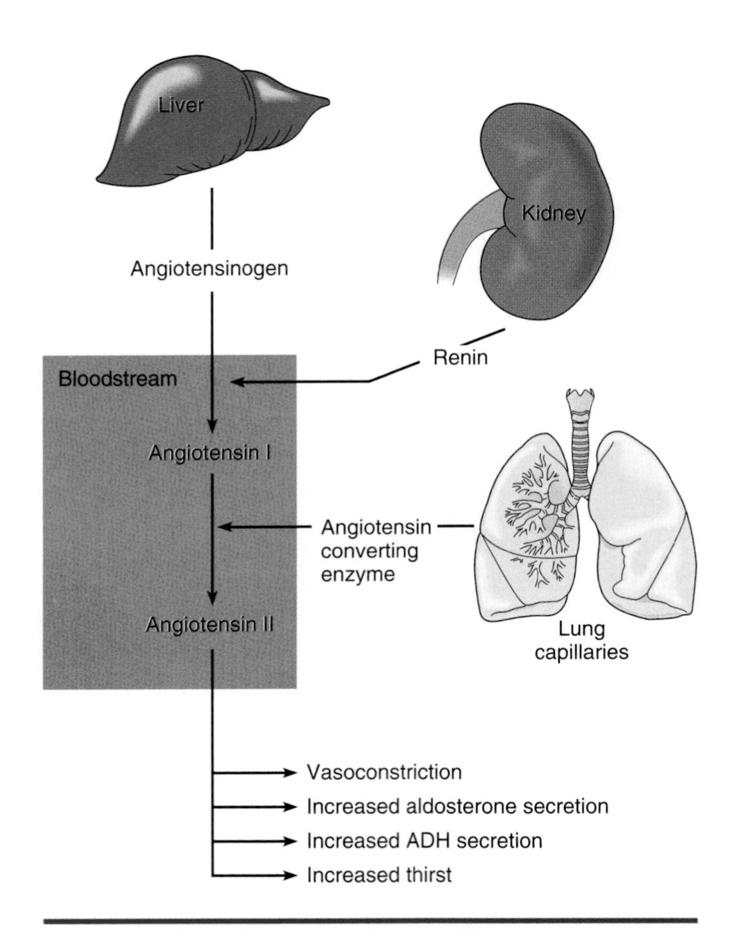

Figure 16.7 The renin-angiotensin mechanism. The multiple actions of angiotensin II help to maintain filtration by increasing systemic blood pressure.

Tubular Reabsorption

Although 180 liters of filtrate are produced daily, only 1 to 2 liters of urine are formed each day. The relatively small amount of urine results because **tubular reabsorption** passes vital substances from the filtrate in the renal tubule into the blood of the peritubular capillary. Most of the tubular reabsorption occurs in proximal convoluted tubules, but other parts of the tubule are also involved (figure 16.8).

Reabsorption involves both active and passive transport mechanisms. It is a selective process that conserves needed materials while allowing excess and waste materials to be excreted in the urine. For example, normally all of the glucose and amino acids are reabsorbed from the filtrate, and about 99% of the water is reabsorbed. However, excess mineral ions and nitrogenous wastes remain in the filtrate and are ex-

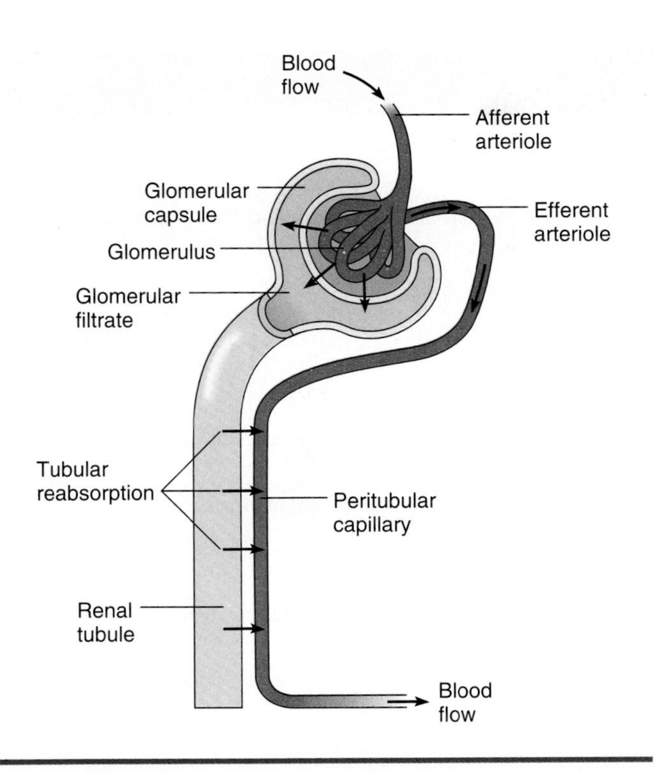

Figure 16.8 Tubular reabsorption transports needed substances from the filtrate in the renal tubule back into the blood of the peritubular capillary.

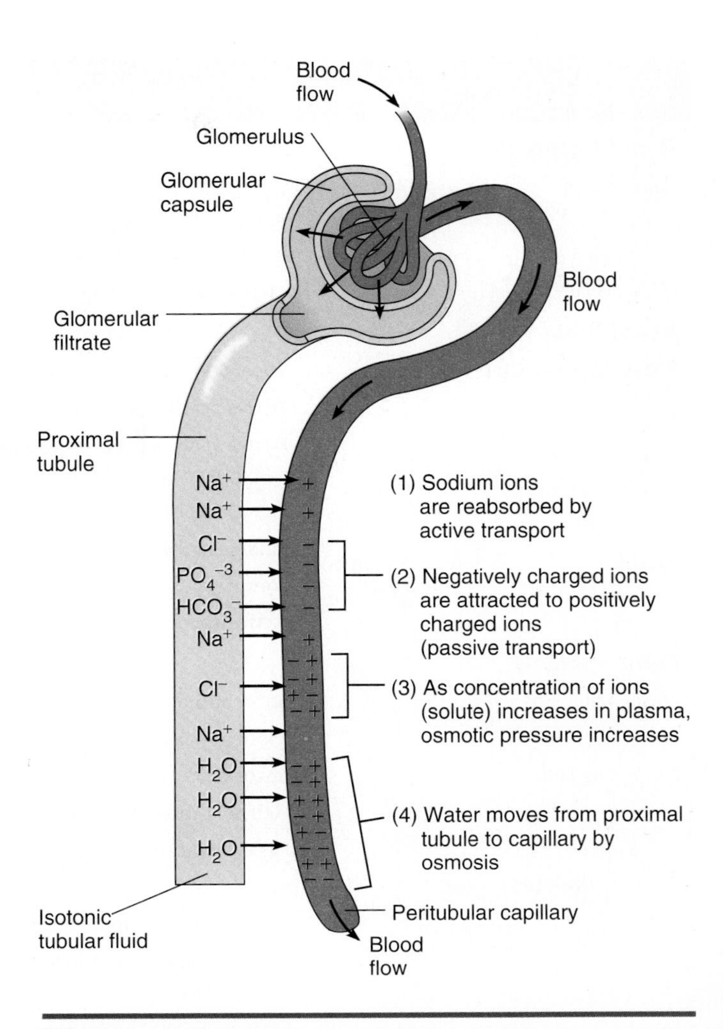

Figure 16.9 Passive reabsorption of water results from the active reabsorption of sodium and potassium ions.

creted in the urine. The reabsorption of water from the filtrate concentrates the remaining substances in the urine (table 16.1).

Active Transport

Active transport is powered by ATP and uses carrier molecules to reabsorb substances against a concentration gradient. Nutrients (e.g., glucose and amino acids), vitamins, and positively charged ions (Na^+, K^+, and Ca^{++}) are actively reabsorbed in the proximal tubule, while chloride ions (Cl^-) are actively reabsorbed in the ascending limb of the loop of Henle.

Passive Transport

Mechanisms of passive transport include electrochemical attraction, diffusion, and osmosis. The effectiveness of passive transport for any substance is determined by (1) the permeability of tubule segments to that substance and (2) the amount of water that is reabsorbed.

Negatively charged ions move passively from the proximal tubule into the peritubular capillary because of their electrochemical attraction to positively charged ions that are actively reabsorbed. For example, the active reabsorption of sodium (Na^+) passively attracts negatively charged ions like chloride (Cl^-) and bicarbonate (HCO_3^-) into the peritubular capillaries. In the descending limb of the loop of Henle, this process is reversed—chloride ions are actively reabsorbed, while sodium and potassium ions are reabsorbed passively by electrochemical attraction.

As the concentration of ions in the peritubular capillary increases, the osmotic pressure of the blood increases. This causes the passive reabsorption of water by osmosis from the proximal tubule into the peritubular capillary along the concentration gradient. About 80% of the water reabsorption occurs in the proximal tubule as a consequence of the active reabsorption of sodium. Figure 16.9 summarizes how the active absorption of sodium ions results in the passive transport of negatively charged ions and water. About 20% of the water reabsorption occurs by osmosis in the loop of Henle and the distal tubule. Reabsorption of water in the distal tubule is dependent upon antidiuretic hormone (ADH) increasing the permeability of the tubule to water.

Table 16.2 Summary of Nephron Functions

Renal Corpuscle	
Glomerulus	Filtration Glomerular blood pressure forces some of the water and dissolved substances (except proteins) from the blood plasma through the pores in the glomerular capillary walls
Glomerular capsule	Receives filtrate from glomerulus
Renal Tubule	
Proximal convoluted tubule	Reabsorption Active reabsorption of all nutrients, including glucose and amino acids Active reabsorption of positively charged ions like sodium, potassium, calcium, and magnesium in addition to organic acids like lactic, ascorbic, and uric acids Passive reabsorption by electrochemical attraction of negatively charged ions such as bicarbonate Passive reabsorption of water by osmosis Secretion Active secretion of hydrogen ions and substances such as histamine and creatinine
Loop of Henle	
Descending limb	Reabsorption Passive reabsorption of water by osmosis
Ascending limb	Reabsorption Active reabsorption of chloride ions Passive reabsorption by electrochemical attraction of sodium and potassium ions
Distal convoluted tubule	Reabsorption Active reabsorption of sodium ions under influence of aldosterone Passive reabsorption by electrochemical attraction of negatively charged ions such as chloride and bicarbonate Passive reabsorption of water by osmosis under the influence of ADH Secretion Active secretion of hydrogen ions Secretion of potassium ions by passive electrochemical attraction and under the influence of aldosterone

The continued reabsorption of water concentrates the filtrate and allows some substances to diffuse into the peritubular capillary along a concentration gradient. For example, about 40% of the urea in the filtrate is passively reabsorbed by diffusion, and the remainder is excreted in urine. Reabsorption of urea by diffusion is reduced because tubule cells are much less permeable to urea than to water. However, there is always a small amount of urea and other nitrogenous wastes in the blood; only the excess is excreted in the urine.

Each part of the renal tubule is specialized to reabsorb certain substances actively or passively. Table 16.2 summarizes the major reabsorption activities of each segment of a renal tubule.

Tubular Secretion

Tubular secretion is essentially the reverse of tubular reabsorption, and it plays an important role in urine formation (figure 16.10). In this process, substances are actively or passively secreted from blood in the peritubular capillary into the filtrate in the renal tubule. Active tubular secretion involves carrier molecules and requires ATP to power the process.

Tubular secretion allows the body to rid itself of excess substances and helps control the pH of the blood. Substances like histamine, ammonia, uric acid, creatinine, and hydrogen ions (H^+) are actively secreted into the proximal tubule. Hydrogen ions are also actively secreted into the distal tubule. Potassium

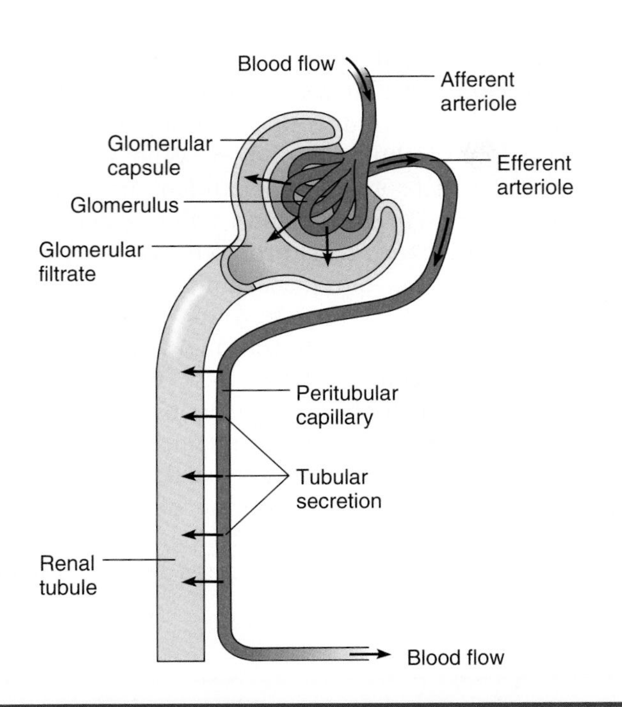

Figure 16.10 Tubular secretion moves unwanted substances from the blood in the peritubular capillary into the renal tubule.

ions (K^+) are passively secreted into the distal tubule because the active reabsorption of sodium ions and other positively charged ions leaves an excess of negatively charged ions in the filtrate. Therefore, potassium ions move passively into the filtrate due to their electrochemical attraction to the negatively charged ions. Aldosterone promotes sodium reabsorption and potassium secretion in the distal tubule.

Table 16.2 summarizes the secretion activities of each part of a nephron.

What are the mechanisms of filtration, tubular reabsorption, and tubular secretion?
What substances are reabsorbed and secreted?

Maintenance of Blood Plasma Composition

The composition and volume of blood plasma is affected by diet, cellular metabolism, and urine production. The intake of food and liquids provides the body with water and a variety of nutrients, including minerals, that are absorbed into the blood. Cellular metabolism depletes nutrients and produces waste products, including nitrogenous wastes. Urine production retains essential nutrients and minerals in the blood but removes some water along with excess substances and nitrogenous wastes. In healthy people, the kidneys are able to keep the composition and volume of the blood plasma relatively constant in spite of variations in diet and cellular activity.

Water and Electrolyte Balance

Two important components of blood plasma and other body fluids are water and electrolytes, and their concentrations in body fluids must be maintained within normal limits. Recall that water is the solvent of body fluids in which the chemical reactions of life occur. **Electrolytes** are substances that form ions when dissolved in water, and they are so named because they can conduct an electric current when dissolved in water. For example, sodium chloride is an electrolyte that forms sodium (Na^+) and chloride (Cl^-) ions when dissolved in water.

The concentrations of water and electrolytes in body fluids are interrelated because the concentration of one affects the concentration of the other. For example, the concentration of electrolytes establishes the concentration gradient that enables water to be reabsorbed by osmosis (see figure 16.9).

Substances that increase the production of urine are known as *diuretics*. Caffeine is a common diuretic. Physicians often prescribe a diuretic to reduce the volume of body fluids in patients with edema or hypertension.

Water Balance

The intake of water is largely regulated by the *thirst center* located in the hypothalamus of the brain. The thirst center is activated when it detects a decrease in water concentration in the blood. It is also activated by angiotensin II when blood pressure declines significantly. An awareness of thirst stimulates water intake to replace water lost from body fluids. Water intake must balance water loss, and this averages about 2,500 ml per day.

The body loses water in several ways, but about 60% of the total water loss occurs in urine. In addition, water is lost in the humidified air exhaled from the lungs, in feces, and in perspiration (figure 16.11). However, it is the kidneys that regulate the concentration of water in the blood plasma by controlling the volume of water lost in urine.

The volume of water lost in urine varies with both the volume of water lost by other means and the

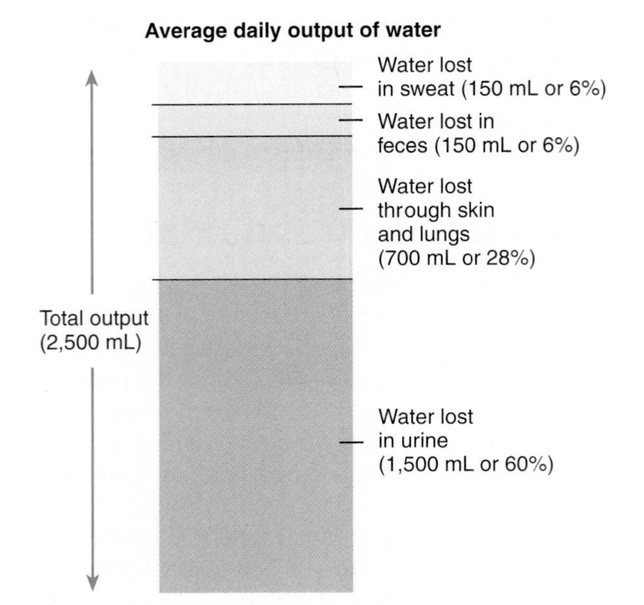

Figure 16.11 Pathways of water loss. Urine formation is the most important process that regulates water loss.

volume of water intake. These factors affect the action of the kidneys simultaneously, but we consider them separately to better understand how they influence kidney function.

In general, the more water that is lost through other means, the less water that is lost in urine. For example, if excessive water loss occurs through perspiration or diarrhea, more water is reabsorbed from the renal tubule. The result is a smaller volume of more concentrated urine. Conversely, if water loss through other means is minimal, water reabsorption is reduced, and a larger volume of more dilute urine is produced.

Similarly, the greater the intake of water, the less water is reabsorbed, and a larger volume of more dilute urine is produced. Conversely, the less the intake of water, the more water is reabsorbed, and a smaller volume of more concentrated urine is produced.

If water loss significantly exceeds water intake for several days, extracellular fluids may become more concentrated, causing water to move out of the cells by osmosis. This condition, known as *cellular dehydration,* may lead to serious complications unless the water loss is quickly restored. In serious cases, dehydration may result in fever, mental confusion, or coma.

You can see that regulating water balance is a dynamic process and that water balance is largely controlled by the amount of water reabsorbed from renal tubules into peritubular capillaries. Whether more or less water is reabsorbed is dependent upon antidiuretic hormone (ADH) secreted by the posterior lobe of the pituitary gland. ADH promotes water reabsorption by increasing the permeability of the distal tubules and collecting ducts to water.

When the water concentration of blood is excessive, ADH secretion declines, less water is reabsorbed, and a greater volume of urine is produced. Conversely, when the water concentration of blood decreases, ADH secretion is increased, more water is reabsorbed, and a smaller volume of urine is produced. ADH minimizes water loss in urine, but it cannot prevent it. Thus, water must be replenished daily by fluid intake.

If ADH secretion is deficient, a condition known as *diabetes insipidus* results. It is characterized by an excessive production of dilute urine due to the inability to reabsorb sufficient water from the filtrate. Up to 15 liters of urine may be produced in a single day, which triggers extreme thirst and the almost continuous intake of water. Without treatment, the excessive water loss may lead to severe dehydration and electrolyte imbalances. Patients with this disorder are treated with synthetic ADH to promote water reabsorption.

Electrolyte Balance

Important electrolytes in body fluids include ions of sodium, potassium, calcium, chloride, phosphate, sulfate, and bicarbonate. Electrolytes are obtained from the intake of food and fluids. A craving for salt results when electrolytes are in low concentration in body fluids.

Electrolyte balance is regulated largely by active reabsorption of positively charged ions, which, in turn, secondarily controls the passive reabsorption of negatively charged ions by electrochemical attraction. Sodium ions are the most important ions to be regulated since they compose about 90% of the positively charged ions in extracellular fluids. Certain hormones play important roles in maintaining electrolyte balance.

Aldosterone is a hormone that regulates the balance of sodium and potassium ions in the blood by

Table 16.3 Hormones Acting on the Kidneys

Hormone	Source	Action
Aldosterone	Adrenal cortex	Stimulates reabsorption of Na^+ from the filtrate into the blood; stimulates the secretion of K^+ from blood into the filtrate
Antidiuretic hormone	Posterior pituitary	Stimulates the reabsorption of water from the filtrate into the blood by making the distal tubule and collecting ducts more permeable to water; decreases the volume of urine produced
Parathyroid hormone	Parathyroid glands	Stimulates the reabsorption of Ca^{++} from the filtrate into the blood; stimulates the secretion of PO_4^-

stimulating the reabsorption of sodium ions and the secretion of potassium ions by the distal tubule. Thus, aldosterone causes an exchange of sodium and potassium ions between the filtrate and the blood until the blood concentrations of these two ions returns to normal. The adrenal cortex is stimulated to secrete aldosterone by (1) an increase of potassium ions (K^+) in the blood, (2) a decrease of sodium ions (Na^+) in the blood, and (3) angiotensin II. As long as blood concentrations of sodium and potassium ions are normal, aldosterone is not secreted.

The blood concentration of calcium ions (Ca^{++}) is regulated mainly by the antagonistic actions of parathyroid hormone and calcitonin. When the blood calcium concentration declines, the parathyroid glands are stimulated to secrete *parathyroid hormone (PTH).* PTH promotes an increase in blood calcium by stimulating three different processes: (1) the reabsorption of calcium ions from renal tubules; (2) the resorption of calcium from bones into the blood; and (3) the absorption of calcium in foods by the small intestine. When the blood calcium level returns to normal, parathyroid hormone secretion is curtailed. Table 16.3 summarizes the effect of hormones that act on the kidneys.

In contrast, *calcitonin* is secreted by the thyroid gland when the level of blood calcium is too high. Calcitonin promotes the deposition of calcium in bones, which reduces the level of blood calcium.

Nitrogenous Wastes

A major function of the kidneys is the removal of excess nitrogenous wastes. The kidneys do not remove all of these wastes but keep their concentrations in the blood within tolerable limits. The primary nitrogenous wastes produced by cellular metabolism are urea, uric acid, and creatinine (see table 16.1).

Urea is a waste product of amino acid metabolism. In order for amino acids to be used as an energy source in cellular respiration or converted into glucose or fat, the liver removes the amine ($—NH_2$) groups from them. The amine groups react to form ammonia, which is converted to the less toxic urea by the liver. About 40% of the urea in the filtrate is passively reabsorbed by diffusion, and the remainder is excreted in urine.

Uric acid is a waste product of nucleic acid metabolism. Much of the uric acid in the filtrate is reabsorbed by the renal tubules, but enough is actively secreted into the filtrate to keep the blood concentration of uric acid within normal limits.

Creatinine is a waste product of muscle metabolism. It is actively secreted by the distal tubule.

An abnormally high concentration of uric acid in the blood and the deposition of uric acid crystals in joints is characteristic of a hereditary disorder called *gout.* Joints of the hands and feet are often the sites of uric acid deposition, which produces inflammation and severe pain.

pH Balance

The pH of the blood must be maintained within rather narrow limits—pH 7.35 to pH 7.45—in order for body cells to function properly. Cellular metabolism produces products that tend to upset the pH balance. These products, such as lactic acid, phosphoric acid, and carbonic acid, tend to make the blood more acidic, as shown in figure 16.12.

Acids are substances that release hydrogen ions (H^+) when they are in water, which lowers the pH and increases the acidity of the liquid. Strong acids release more hydrogen ions than weak acids. **Bases** are substances that, when placed in water, release ions that can combine with hydrogen ions, such as hydroxyl (OH^-) or bicarbonate (HCO_3^-) ions. Body fluids contain both acids and bases.

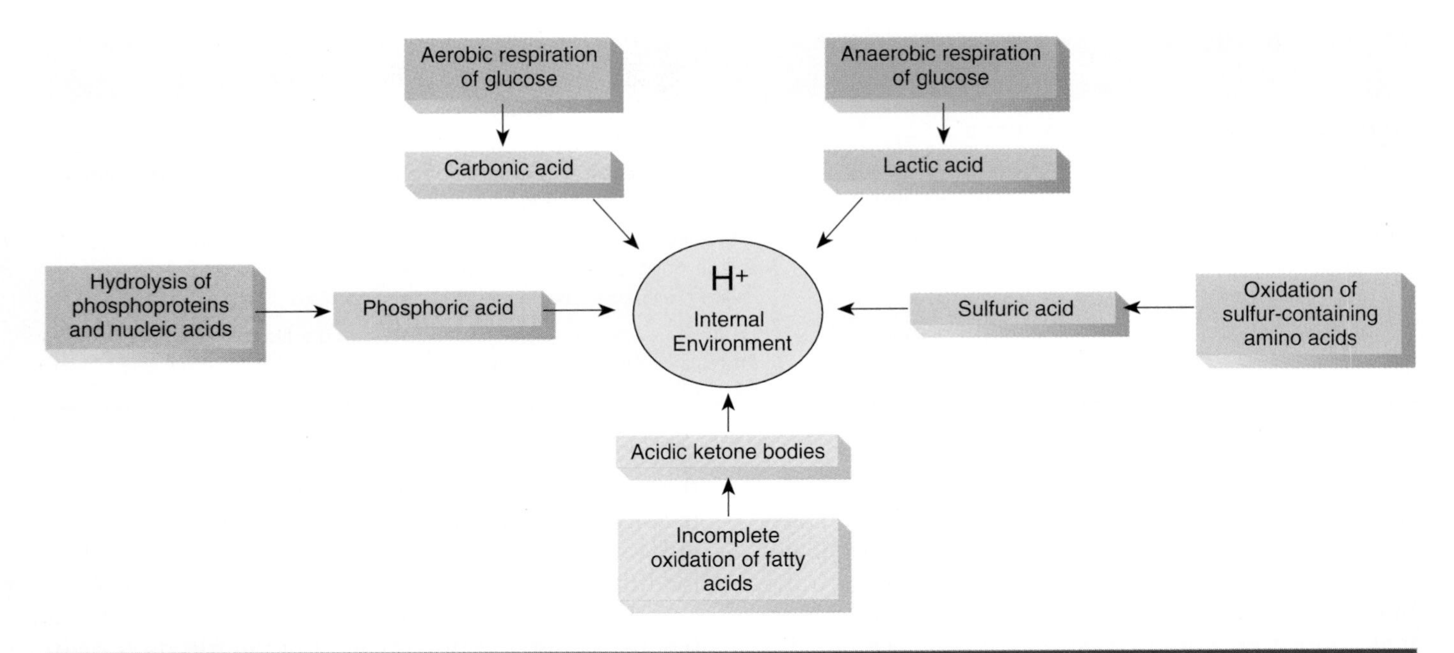

Figure 16.12 Examples of metabolic processes that increase the hydrogen ion concentration of body fluids.

The kidneys have a tremendous functional reserve. Renal insufficiency becomes evident only after about 75% of the renal functions have been lost. As the development of renal failure progresses, patients must rely on *hemodialysis* as a means of removing wastes and excessive substances from the blood. In hemodialysis, the patient's blood is pumped through selectively permeable tubes that are immersed in a dialyzing solution within a "kidney machine." Nitrogenous wastes and excessive electrolytes diffuse from the blood into the dialyzing solution, while certain needed substances, such as buffers, diffuse from the dialyzing solution into the blood. In this way, the concentration of wastes and electrolytes in the patient's blood are temporarily restored within normal limits. Hemodialysis may be required two to three times per week for patients with chronic kidney failure.

An alternative method is called *continuous ambulatory peritoneal dialysis (CAPD).* In this technique, 1 to 3 liters of dialyzing fluid are introduced into the peritoneal cavity through an opening made in the abdominal wall. Waste products and excessive substances diffuse from blood vessels in the peritoneal membrane into the dialyzing solution, which is drained after two to three hours. This technique is less costly, may be done at home, and allows the patient to move about during the procedure. However, it must be done more frequently than dialysis using a kidney machine, and there is a greater chance of serious infection.

The blood and other body fluids contain chemicals known as **buffers** that prevent significant changes in pH. Buffers are able to combine with or release hydrogen ions as needed to stabilize the pH. If the hydrogen ion concentration is excessive, buffers combine with some hydrogen ions to reduce their concentration. Conversely, if too few hydrogen ions are present, buffers release some hydrogen ions to increase their concentration to within normal limits. In this way, buffers help to keep the pH of the blood relatively constant.

$$\text{Buffer} + H^+ \rightleftharpoons \text{Buffer}{-}H^+$$

The removal of excess hydrogen ions by the kidneys prevents the buffers from being overwhelmed. Tubular secretion actively transports excess hydrogen

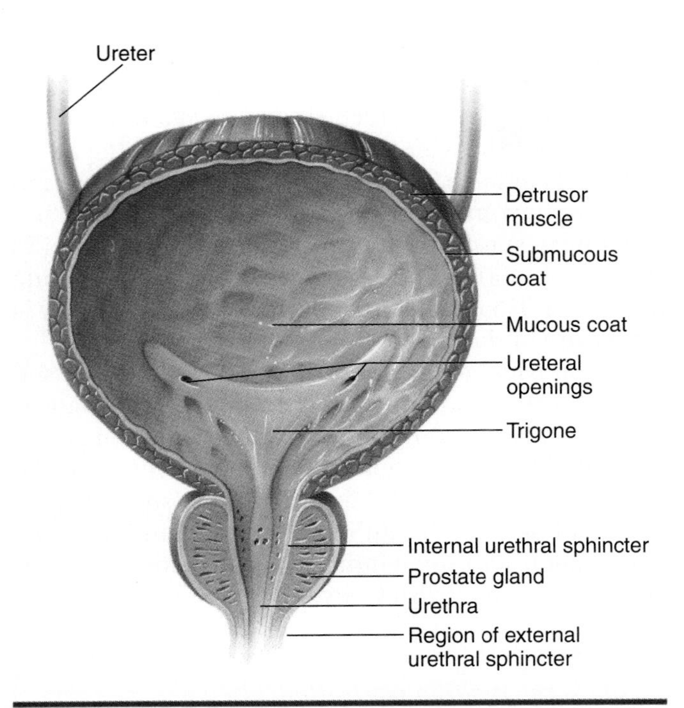

Figure 16.13 A male urinary bladder in coronal section.

ions from blood in peritubular capillaries into the filtrate in the renal tubules. By excreting excess hydrogen ions in urine, kidneys help to maintain the normal pH of body fluids.

> ***What hormones help regulate water-electrolyte balance, and what are their effects?***
> ***What are the nitrogenous wastes in urine, and how are they formed?***

Excretion of Urine

Once the filtrate passes from collecting ducts into the renal pelvis, it is called urine. Urine passes from the renal pelvis into the ureter and is carried by peristalsis to the urinary bladder. Urine is voided from the bladder through the urethra.

Ureters

Each **ureter** is a slender tube about 25 cm (10 in) long that extends from a kidney to the urinary bladder. It begins at the kidney with the funnel-shaped renal pelvis and enters the lower lateral margin of the urinary bladder (figures 16.1 and 16.13).

The wall of a ureter is formed of three layers. The outer fibrous layer is composed of supportive connective tissue. The middle layer consists of smooth muscle fibers that produce peristaltic waves for urine transport. The inner layer is a mucous membrane that is continuous with that of the kidney pelvis and the urinary bladder. A flaplike fold of mucous membrane in the urinary bladder covers the opening of the ureter, and it functions as a valve that prevents urine from backing up into the ureter.

Urinary Bladder

The **urinary bladder** is a hollow, muscular organ located posterior to the symphysis pubis within the pelvic cavity. It lies inferior to the parietal peritoneum. The urinary bladder provides temporary storage of urine, and its shape varies with the volume of urine that it contains. When distended with urine, it is almost spherical as its upper surface expands upward. When empty, its upper portion is lowered, giving a deflated appearance.

The internal floor of the bladder contains the **trigone** (tri′-gōn), a smooth, triangular area that contains an opening at each of its angles. The openings of the ureters are located at the two posteriorly located base angles, and the opening of the urethra is located at the anteriorly located apex angle (figure 16.13).

Four layers compose the wall of the bladder. The innermost layer is the *mucosa,* which is composed of transitional epithelium that is adapted to the repeated stretching of the bladder wall. The epithelium stretches, and its thickness varies as the bladder fills with urine.

The mucosa is supported by the underlying *submucosa* formed of connective tissue containing an abundance of elastic fibers. Blood vessels and nerves supplying the urinary bladder are present in the submucosa.

Smooth muscle fibers compose the third, and thickest, layer. These fibers form the *detrusor* (dē-trū′-sor) *muscle.* The detrusor muscle is relaxed as the bladder fills with urine, and it contracts as urine is expelled. Fibers of the detrusor muscle form an **internal urethral sphincter** at the junction of the bladder and urethra.

The outermost layer consists of the parietal peritoneum, but it covers only the upper portion of the urinary bladder. The remainder of the bladder surface is coated with fibrous connective tissue.

Urethra

The **urethra** is a thin-walled tube that carries urine from the bladder to the outside. The urethral wall contains smooth muscle fibers and is supported by connective tissue. The inner lining is a mucous membrane that is continuous with the mucous membrane of the urinary bladder. An **external urethral sphincter,** which is composed of skeletal muscle fibers in the floor of the pelvis, is located where the urethra penetrates the pelvic floor.

The female urethra is quite short, about 3 to 4 cm (1.5 in) in length. Its external opening lies anterior to the vagina. The male urethra is much longer, about 16 to 20 cm (6–8 in) in length, because the urethra runs the length of the penis. The external opening is at the tip of the penis.

An infant is not able to be toilet trained until neural development allows control of the external urethral sphincter muscle. Voluntary control is possible shortly after two years of age.

Micturition

Micturition (mik-tū-rish′un), or urination, is the act of expelling urine from the bladder. Although the bladder may hold up to 1,000 ml of urine, micturition usually occurs long before that volume is attained. When 200 to 400 ml of urine have accumulated in the urinary bladder, stretch receptors in the bladder wall are stimulated and they trigger the **micturition reflex.** This reflex sends parasympathetic impulses to the detrusor muscle, causing rhythmic contractions. As this reflex continues, it causes the involuntarily controlled internal urethral sphincter to open. If the voluntarily controlled external urethral sphincter is relaxed, micturition occurs; if it is not relaxed, micturition is postponed.

Micturition may be postponed by keeping the external sphincter voluntarily closed, and in a few moments the urge to urinate subsides. After more urine enters the urinary bladder, the micturition reflex is activated again, and the urge to urinate returns. Micturition cannot be postponed for long periods of time. After a while the reflex overwhelms voluntary control, and micturition occurs, ready or not.

Characteristics of Urine

The volume of urine produced in a 24-hour period usually is 1.5 to 2.0 liters. In healthy persons, fresh urine is usually clear with a pale yellow to light amber color and a characteristic odor. The color is due to the presence of **urochrome,** a substance produced by the breakdown of bile pigments in the intestine. The more concentrated the urine, the darker is its color.

Urine is usually slightly acidic (about pH 6), although the normal pH range extends from 4.7 to 8.0. Variations in pH usually result from the diet. High-protein diets increase the acidity. Conversely, vegetarian diets tend to make the urine more alkaline.

Urine is heavier than water because of the many solutes it contains. The term *specific gravity* is used to compare how much heavier urine is than water. Pure water has a specific gravity of 1.000, while the specific gravity of normal urine ranges from 1.001 (dilute urine) to 1.035 (concentrated urine). Specific gravity is a measure of the concentration of a urine sample.

The usual solutes of urine have been discussed earlier in the section on urine formation. Substances not normally found in urine are glucose, proteins, red or white blood cells, hemoglobin, and bile pigments. The presence of any of these substances suggests possible pathological conditions. Normal values of urine components and some indications of abnormal values are listed on inside back cover.

How is micturition controlled?
What substances, if present in urine, indicate a urinary disorder?

Disorders of the Urinary System

These disorders are grouped as inflammatory or noninflammatory.

Inflammatory Disorders

Cystitis (sis-tī′-tis) is the inflammation of the urinary bladder. It is often caused by bacterial infection. Females are more prone to cystitis because their shorter urethra makes it easier for bacteria to reach the urinary bladder.

Glomerulonephritis (glō-mer-ū-lō-ne-frī′-tis) is the inflammation of a kidney involving the glomeruli. It may be caused by bacteria or bacterial toxins. The inflamed glomeruli become more permeable, allowing blood cells and proteins to leak into the filtrate and remain in the urine.

Pyelonephritis (pī-e-lō-ne-frī′-tis) is the inflammation of the renal pelvis and nephrons. If only the renal pelvis is involved, the condition is called *pyelitis.* These conditions result from bacteria carried by blood from other places in the body or by migration of bacteria from lower portions of the urinary tract.

Urethritis is the inflammation of the urethra. It may be caused by several types of bacteria, but the bacterium *Escherichia coli* is the most common. Urethritis is more common in females.

Noninflammatory Disorders

Diuresis is the excessive production of urine. It results from inadequate tubular reabsorption of water and is characteristic of diabetes insipidus.

Renal calculi (kal′-kū-li), or kidney stones, result from crystallization of uric acid or of calcium or magnesium salts in the renal pelvis. They can cause extreme pain, especially when moving through a ureter by peristalsis. A new treatment using ultrasound waves to break up the stones is proving successful as an alternative to surgery.

Renal failure is characterized by a reduction in urine production and a failure to maintain the normal volume and composition of body fluids. It may occur suddenly (acute) or gradually (chronic). Renal failure leads to *uremia,* a toxic condition caused by excessive nitrogenous wastes in the blood, and ultimately to *anuria,* a cessation of urine production. Hemodialysis and/or a kidney transplant may be necessary.

A routine urinalysis is a common clinical test that provides information about kidney function and also about general health of the body. Kidney function is also assessed by two blood tests. The *blood urea nitrogen (BUN)* test evaluates how effectively the kidney removes urea from the blood and other substances as well. The normal BUN value is 5 to 25 mg per 100 ml of blood. In acute kidney failure and in later stages of chronic kidney failure, BUN may range from 50 to 200 mg per 100 ml of blood. *Blood (serum) creatinine* is another test that assesses kidney effectiveness. Creatinine levels in the blood are normally stable (0.6–1.5 mg/100 ml), so an increase indicates a decrease in kidney function.

CHAPTER SUMMARY

Kidneys

1. The paired kidneys are located against the upper posterior abdominal wall, posterior to the parietal peritoneum.
2. Internal structure of a kidney consists of three recognizable parts: an outer cortex, a medulla composed of renal pyramids, and a medially located renal pelvis, which is continuous with a ureter.
3. Nephrons are the functional units of the kidneys. Each nephron consists of a renal corpuscle and a renal tubule. A renal corpuscle is composed of a glomerulus and a glomerular (Bowman's) capsule. A renal tubule is composed of a proximal convoluted tubule, the loop of Henle, and a distal convoluted tubule.
4. Each nephron joins with a collecting duct that empties into a calyx of the renal pelvis.
5. The blood supply for each kidney is provided by a renal artery and renal vein.
6. An afferent arteriole brings blood to a glomerulus. Blood exits the glomerulus via an efferent arteriole and flows through the peritubular capillary, which entwines around the renal tubule.
7. The juxtaglomerular apparatus consists of modified cells of the afferent arteriole and the distal convoluted tubule at their point of contact.
8. The functions of the kidneys are the production of urine and the secretion of renin and erythropoietin.

Urine Formation

1. The process of urine formation regulates the composition and volume of blood plasma by removing excess nitrogenous wastes and surplus substances from the blood plasma.
2. Urine is formed by three sequential processes: filtration, tubular reabsorption, and tubular secretion.
3. In filtration, water and dissolved substances (except plasma proteins) in blood plasma are filtered from the glomerulus into the glomerular capsule. Filtration results from the greater permeability of glomerular capillaries and the greater blood pressure within the glomerulus.
4. About 180 liters of filtrate are formed in a 24-hour period.
5. Filtration rate is proportional to the glomerular blood pressure. Glomerular blood pressure is maintained by mechanisms that control the diameters of the afferent and efferent arterioles.
6. Glomerular blood pressure generally varies directly with systemic blood pressure. If a severe drop in blood pressure occurs, the parasympathetic nervous system and the renin-angiotensin mechanism help to return it to a normal level.
7. In tubular reabsorption, needed substances are reabsorbed back into the blood of the peritubular capillary by either active or passive transport.
8. Positively charged ions, amino acids, and glucose are actively reabsorbed. Negatively charged ions are

passively reabsorbed by electrochemical attraction to the positively charged ions. Water is passively reabsorbed by osmosis, and substances not actively absorbed, such as urea, may be partially reabsorbed by diffusion.
9. Most tubular reabsorption occurs in the proximal tubule, but other portions of the tubule are also involved. Substances remaining in the filtrate are concentrated by the reabsorption of water.
10. In tubular secretion, certain substances are actively or passively secreted into the filtrate from the blood. Uric acid and hydrogen ions are actively secreted. Potassium ions are secreted passively.

Maintenance of Blood Plasma Composition

1. The prime function of the kidneys is to maintain the volume and composition of the blood plasma in spite of variations in diet and metabolic processes.
2. Water intake must equal water loss. Most water is lost in urine, but other avenues include exhaled air, perspiration, and feces.
3. The volume of water lost in urine is decreased when water loss via other means is increased, and vice versa.
4. Antidiuretic hormone (ADH), which is released from the posterior pituitary gland, increases the permeability of distal tubules and collecting ducts to water and thereby promotes water reabsorption by osmosis.
5. Electrolyte intake must replace electrolyte loss. Electrolytes are conserved largely by the active reabsorption of positively charged ions that passively pull along negatively charged ions by electrochemical attraction.
6. Aldosterone regulates the plasma concentration of sodium and potassium ions by stimulating the active reabsorption of sodium ions and the secretion of potassium ions.
7. The concentration of calcium ions in the blood is regulated by the antagonistic actions of two hormones. Parathyroid hormone stimulates the reabsorption of calcium from bones into the blood, the reabsorption of calcium ions by the kidneys, and the absorption of calcium by the small intestine. In contrast, calcitonin promotes the deposition of calcium in bones.
8. A major function of the kidneys is the removal of excess nitrogenous wastes—urea, uric acid, and creatinine—in order to keep their concentrations in the blood within normal limits.
9. The maintenance of the pH of the blood between 7.35 and 7.45 includes three major mechanisms: buffers in the blood either combine with or release hydrogen ions as needed; carbon dioxide is removed by the lungs; and renal tubules secrete excess hydrogen ions into the filtrate.

Excretion of Urine

1. Urine is the fluid that flows from collecting tubules into the renal pelvis and on into the ureter. A ureter is a slender tube that carries urine by peristalsis into the urinary bladder.
2. Urine is temporarily held in the urinary bladder. The urinary bladder is located posterior to the symphysis pubis in the pelvic cavity.
3. The wall of the urinary bladder consists of the mucosa, submucosa, muscle layer, and fibrous connective tissue. The parietal peritoneum covers only its upper surface. The muscular layer consists of smooth muscle fibers forming the detrusor muscle.
4. A thickening of the detrusor muscle at the bladder-urethra junction forms the internal urethral sphincter. The external urethral sphincter is formed of skeletal muscle fibers in the floor of the pelvis.
5. Micturition is the process of voiding urine from the urinary bladder. Urine is expelled from the urinary bladder through the urethra.
6. When the bladder contains 200 to 400 ml of urine, the micturition reflex is triggered, causing the detrusor muscle to contract rhythmically. Continued contractions open the involuntarily controlled internal urethral sphincter. If the voluntarily controlled external urethral sphincter is relaxed, micturition occurs. If not, micturition is postponed.

Characteristics of Urine

1. The daily production of urine is 1.5 to 2.0 liters. Normal urine is a clear, pale yellow to amber fluid with a characteristic odor. The color is due to the presence of urochrome.
2. Urine is usually slightly acidic but the pH may range from 4.7 to 8.0.
3. Urine is heavier than water due to the dissolved substances that it contains.
4. Abnormal substances that may be in urine are glucose, proteins, blood cells, hemoglobin, and bile pigments.

Disorders of the Urinary System

1. Inflammatory disorders include cystitis, glomerulonephritis, pyelonephritis, and urethritis.
2. Noninflammatory disorders include diuresis, renal calculi, and renal failure.

BUILDING YOUR VOCABULARY

1. **Selected New Terms**
   afferent arteriole, p. 333
   angiotensin, p. 338
   calyx, p. 332
   creatinine, p. 343
   efferent arteriole, p. 333
   renal cortex, p. 332
   renal medulla, p. 332
   renal pelvis, p. 332
   renal pyramid, p. 332
   renin, p. 335
   urea, p. 343
   ureter, p. 345
   urethra, p. 345
   uric acid, p. 343
   urochrome, p. 346

2. **Related Clinical Terms**
   cystocoele (sis′-tō-sel) Herniation of the urinary bladder into the vagina; often a result of tearing pelvic muscles during childbirth.
   enuresis (en-ū-rē′-sis) Involuntary urination.
   incontinence (in-kon′-ti-nens) Inability to control urination or defecation.
   pyelogram (pī-el′-ō-gram) An X ray of the kidney and ureter, usually after intravenous injection of a contrast material.
   uremia (u-rē′-mē-ah) An excessive level of urea in the blood.
   urology (ū-rol′-ō-jē) The medical specialty dealing with disorders of the urinary system.

STUDY ACTIVITIES

1. Complete the **Chapter 16 Study Guide,** which begins on page 481.
2. Complete the **Learning Objectives** listed on the first page of this chapter.

CHECK YOUR UNDERSTANDING

(Answers are located in appendix B.)

1. The functional units of the kidneys are called _____ .
2. Renal corpuscles, proximal convoluted tubules, and distal convoluted tubules are located in the _____ of a kidney.
3. Renal pyramids are located in the _____ of a kidney, and they are composed mostly of _____ .
4. The force powering filtration is _____ blood pressure, which, in turn, is primarily determined by the _____ blood pressure.
5. In filtration, water and dissolved substances that are forced from blood in the glomerulus into the _____ compose the _____ .
6. In tubular reabsorption, certain substances pass from the _____ in the renal tubules into blood in the _____ .
7. Glucose, amino acids, and positively charged ions are _____ reabsorbed, and water is _____ reabsorbed by _____ .
8. A severe drop in blood pressure causes the juxtaglomerular apparatus to secrete _____ , which triggers a mechanism to raise blood pressure by chemical means.
9. The _____ hormone promotes the reabsorption of water by making the distal tubule and _____ more permeable to water.
10. The hormone _____ promotes the reabsorption of sodium ions and secretion of _____ ions.
11. Fluid in the renal pelvis is called _____ .
12. The _____ carry urine to the urinary bladder by _____ .
13. Urine is voided from the urinary bladder via the _____ .
14. The _____ urethral sphincter is involuntarily controlled, and the _____ sphincter is voluntarily controlled.
15. Water lost in urine composes about _____ % of the total water loss from the body.

SEVENTEEN

17

CHAPTER
The Reproductive Systems

Chapter Preview & Learning Objectives

Male Reproductive System
- Identify the organs of the male reproductive system and describe the functions of each one.
- Describe spermatogenesis.

Male Sexual Response
- Describe the male sexual response.

Hormonal Control of Reproduction in Males
- Describe the actions of testosterone.
- Describe the hormonal control of testosterone secretion.

Female Reproductive System
- Identify the organs of the female reproductive system and describe the functions of each one.
- Describe oogenesis.

Female Sexual Response
- Describe the female sexual response.

Hormonal Control of Reproduction in Females
- Describe the ovarian and menstrual cycles and their regulation by hormones.

Mammary Glands
- Describe the structure of mammary glands and breasts.

Birth Control
- Identify the various methods of birth control, describe how they work, and outline the effectiveness of each one.

Disorders of the Reproductive Systems
- Describe the common disorders of the reproductive systems.

Chapter Summary
Building Your Vocabulary
Study Activities
Check Your Understanding

SELECTED KEY TERMS

Gamete A sex cell, either a sperm or an ovum.
Gonads (gone = seed) The sex glands—the ovaries and testes—that produce sex cells.
Meiosis (mei = less) A form of cell division in which the daughter cells contain one-half the number of chromosomes as the parent cell.
Menopause (men = month; paus = stop) The cessation of monthly female reproductive cycles.
Menstrual cycle The monthly female reproductive cycle.
Oogenesis (oo = ovum, egg; genesis = origin) The process of ova formation in the ovaries.
Ovulation The release of a female sex cell from an ovary.
Puberty (puber = grown up) The age at which reproductive organs mature.
Semen (semin = seed) Fluid composed of sperm and secretions of male accessory reproductive glands.
Spermatogenesis The process of sperm formation in the testes.

eproduction is the process by which life is sustained from one generation to the next. The human male and female reproductive systems are specially adapted for their roles in reproduction. The **gonads** (gō′-nads)—the ovaries and testes—form the **gametes** (sex cells). Other reproductive organs nurture or transport male and female sex cells to sites where they may unite. The fertilized egg develops within the female reproductive system and culminates in the birth of a baby. Sexual maturation and the development of sex cells in both sexes and pregnancy in females are regulated by hormones secreted by the pituitary gland and the gonads.

Male Reproductive System

The primary functions of the male reproductive system are the production of male hormones, the formation of sperm cells (spermatozoa), and the placement of sperm cells in the female reproductive tract, where one sperm can unite with a female sex cell. The organs of the male reproductive system include (1) paired testes, which produce sperm and male hormones; (2) ducts that store and transport sperm; (3) accessory glands, whose secretions form part of the semen; and (4) external accessory structures including the scrotum and penis (figure 17.1).

Testes

The paired **testes** (tes′-tēz), or **testicles** (tes′-ti′kulz), are the male gonads, or sex glands. Each testis is protected and supported by a capsule of fibrous connective tissue. Septa (partitions) of connective tissue radiate into the testis from its posterior surface, dividing the testis into internal subdivisions called *lobules.* Each lobule contains several highly coiled **seminiferous** (se-mi-nif′-er-us) **tubules.** Seminiferous tubules are lined with **germinal** (jer′-mi-nal) **epithelium,** which is formed of spermatogenic cells and supporting cells. *Spermatogenic cells* divide to produce sperm, while *supporting cells* support and nourish the spermatogenic cells. The cells that fill the spaces between the seminiferous tubules are known as **interstitial** (in-ter-stish′-al) **cells,** and they produce the male sex hormone (figure 17.2).

Spermatogenesis

Spermatogenesis (sper-mah-tō-jen′-e-sis) is the process that produces sperm cells by the division of the spermatogenic cells in the germinal epithelium. Sperm production begins at **puberty** (pū′-ber-tē), the development of reproductive potential, and continues throughout the life of a male. Sexual maturity and sperm production are controlled by follicle-stimulating hormone (FSH) and luteinizing hormone (LH) from the anterior pituitary gland and by testosterone from the testes. LH is often called interstitial cell–stimulating hormone (ICSH) in males. Hormonal relationships are discussed later in this chapter.

The large, outermost cells of a seminiferous tubule are known as **spermatogonia** (sper-mah-to-gō′-nē-ah), and they compose the germinal epithelium. Each spermatogonium contains 46 chromosomes, the normal number of chromosomes for human body cells. Each spermatogonium divides by mitosis to produce two spermatogonia, each with 46 chromosomes. One spermatogonium remains next to the margin of the tubule. It will serve as the "stem" spermatogonium and will divide repeatedly by mitosis. The spermatogonium that is pushed inward toward the lumen of the tubule undergoes changes to become a **primary spermatocyte.**

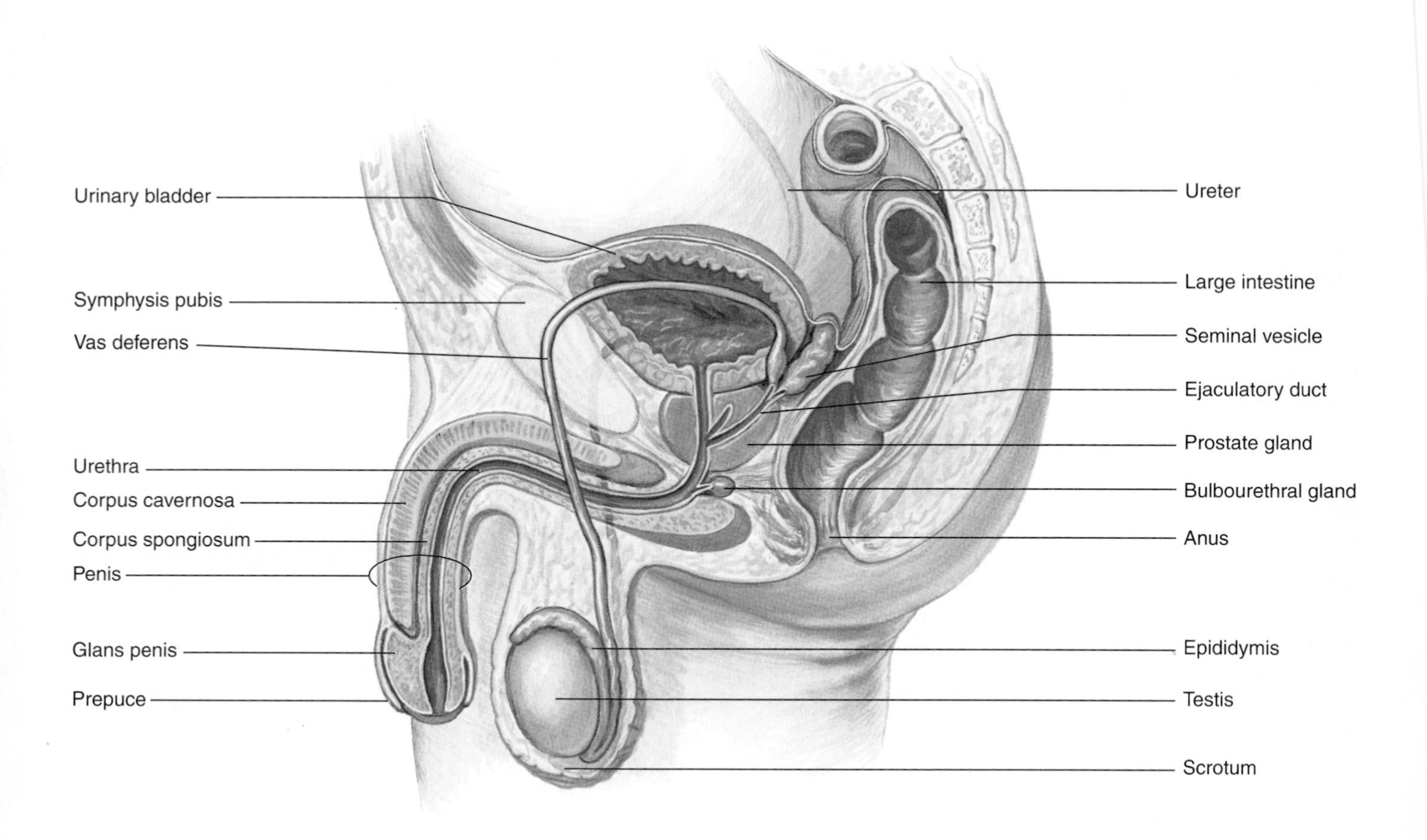

Figure 17.1 Male reproductive system in a midsagittal section.

Primary spermatocytes divide by **meiosis** (mī-ō′-sis), a special type of cell division. Meiosis requires two successive divisions and reduces the number of chromosomes in the daughter cells by one-half.

Meiotic divisions in spermatogenesis are summarized in figure 17.3, and the orientation of the cells in a seminiferous tubule is shown in figure 17.4*b*. Each primary spermatocyte, containing 46 chromosomes, divides in the first meiotic division to form two **secondary spermatocytes,** each containing 23 chromosomes. In the second meiotic division, each secondary spermatocyte divides to form two **spermatids,** each containing 23 chromosomes. Each spermatid attaches to a supporting cell, gradually loses much of its cytoplasm, and develops a flagellum to form a sperm cell containing 23 chromosomes. Examine figure 17.4 and note how cells in stages of spermatogenesis are arranged in sequence from outside inward, with sperm cells located in the lumen of the tubule. Once the sperm cells are completely formed, they are carried into the epididymis, where they are temporarily stored while they mature.

A mature **sperm cell** consists of a head, body, and flagellum. The flattened *head* is composed of a compact nucleus containing 23 chromosomes. The anterior portion of the head is covered by a caplike structure, the *acrosome.* The acrosome contains an enzyme that helps the sperm penetrate a female sex cell. The *body* contains mitochondria, where ATP is formed to power the movements of the *flagellum,* which enables movement (figure 17.5).

Accessory Ducts

Sperm cells pass through a series of accessory ducts as they are carried from the testes to outside the body. Sperm from a testis pass through the epididymis, vas deferens, ejaculatory duct, and urethra.

Epididymis

The seminiferous tubules of a testis lead to a number of small ducts that open into the epididymis. The **epididymis** (ep-i-did′-i-mis) (plural, *epididymides*) appears as a comma-shaped organ that lies along the superior and posterior margins of a testis (see figure 17.1). Upon close examination, the epididymis is shown to be a long (6 m), tightly coiled, slender tube that is continuous with the vas deferens (see figure 17.2).

Sperm cells mature as they are slowly moved (10–14 days) through the epididymis by weak peristaltic contractions.

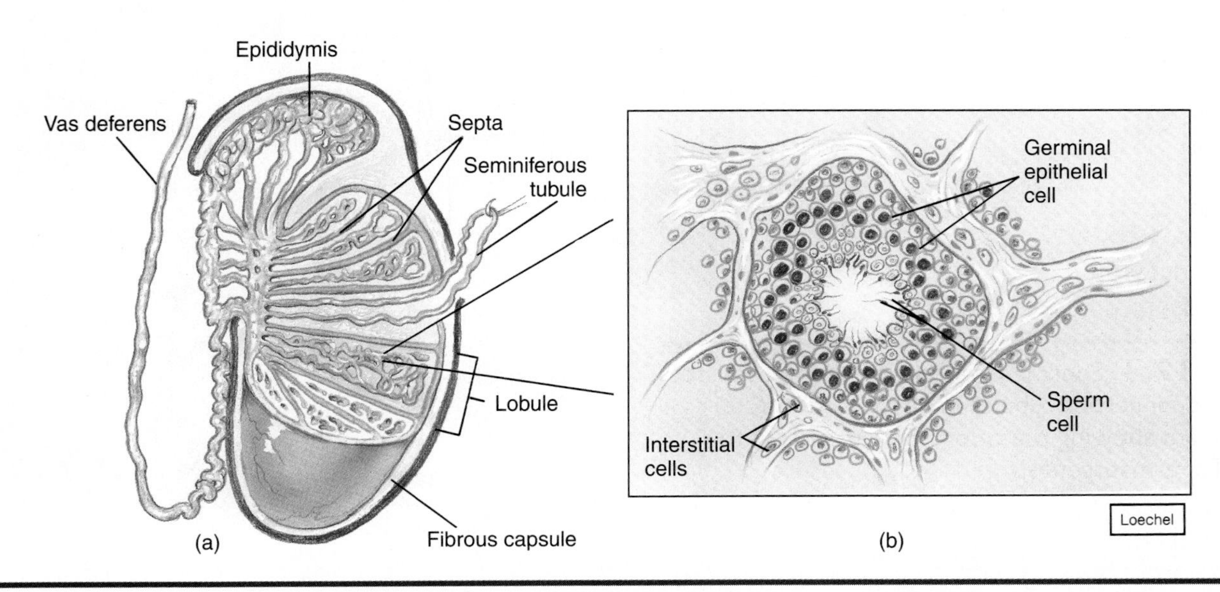

Figure 17.2 (*a*) Testis in sagittal section; (*b*) seminiferous tubule in cross section.

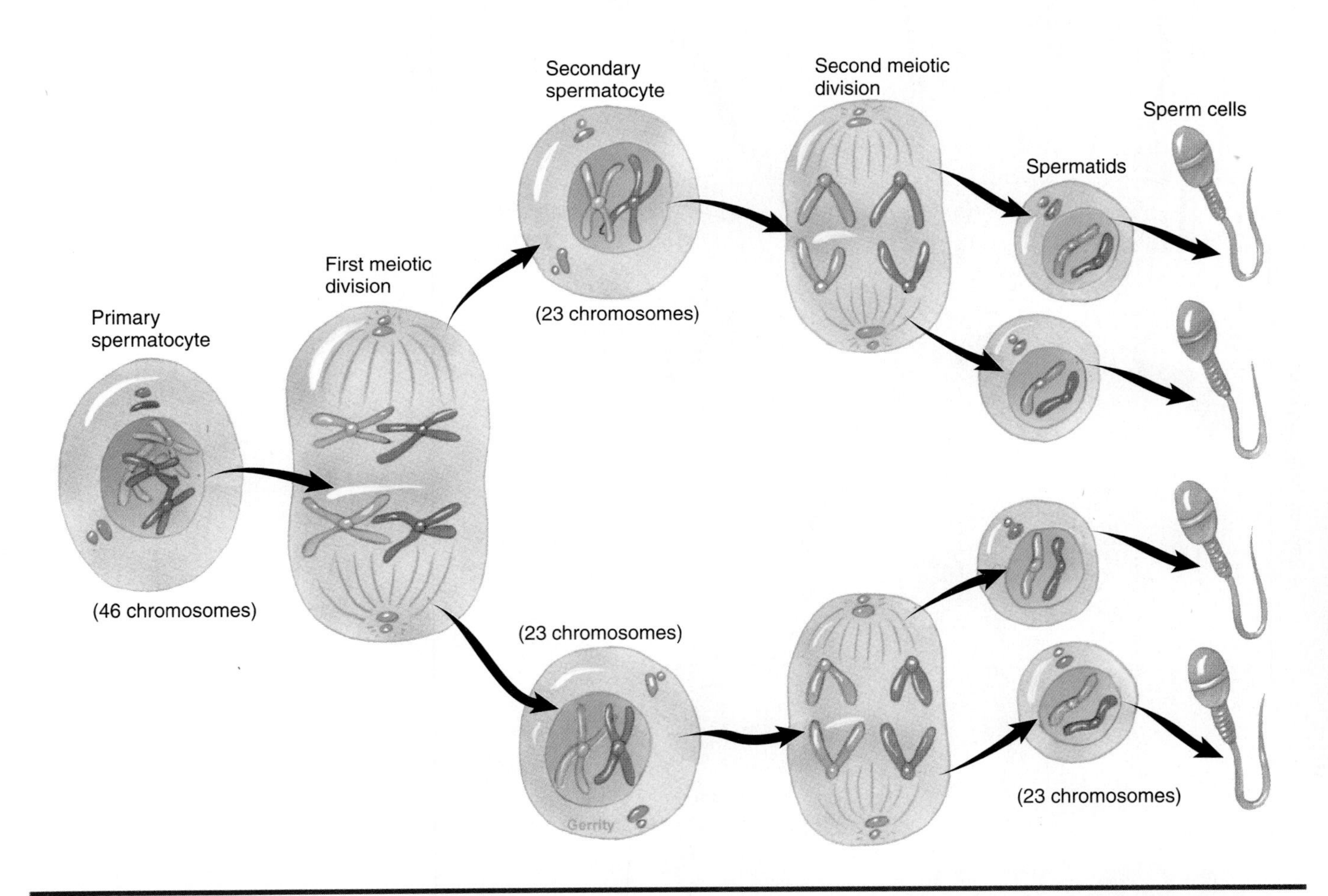

Figure 17.3 Spermatogenesis requires two successive meiotic divisions plus the development of spermatids into sperm cells.

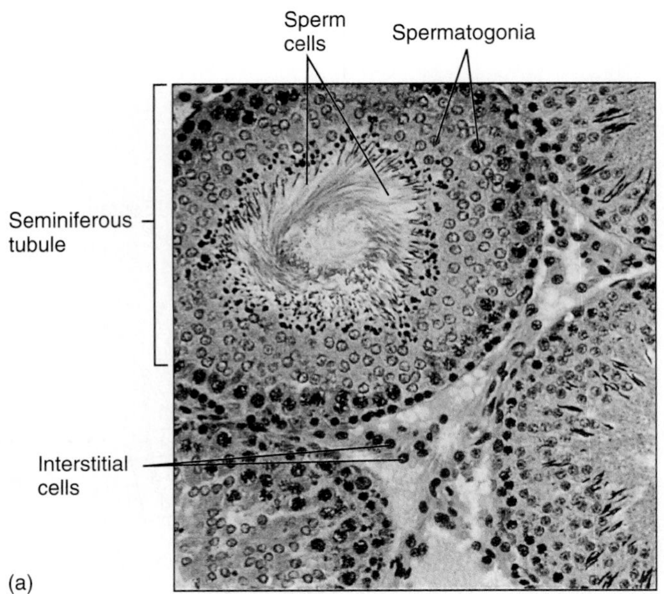

Figure 17.4 Spermatogenesis. (*a*) A photomicrograph of seminiferous tubules in cross section (200×); (*b*) a diagram showing the orientation of cells in the stages of spermatogenesis.

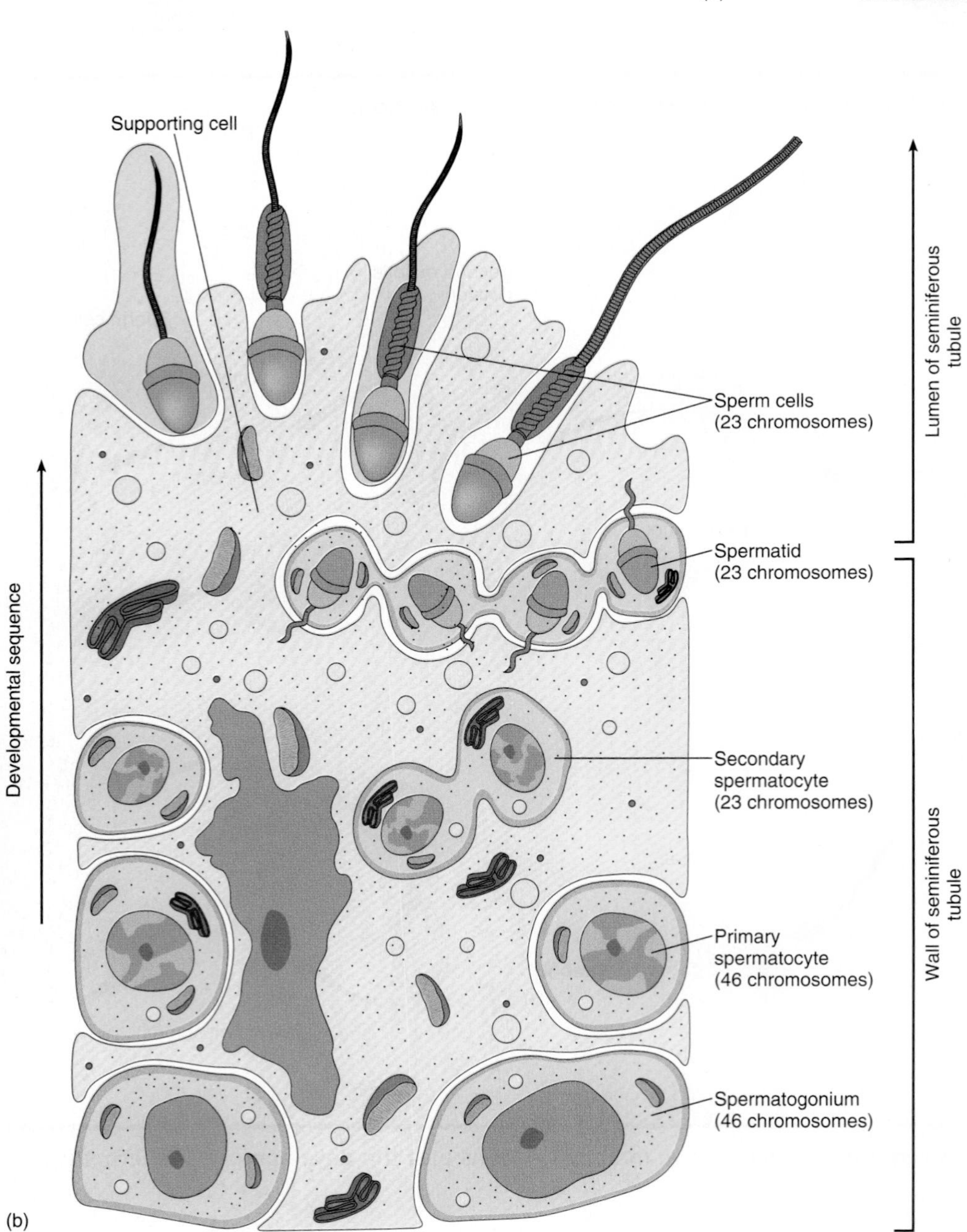

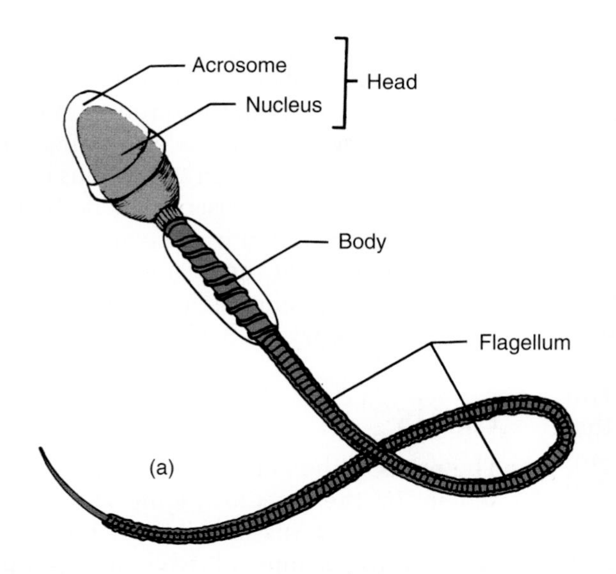

Figure 17.5 (*a*) Parts of a mature sperm cell. (*b*) Scanning electron photomicrograph of human sperm cells (1,400×).

Vas Deferens

As shown in figure 17.1, a **vas deferens** (vas def'-er-enz) (plural, *vasa deferentia*) extends from the epididymis upwards in the scrotum, passes through the inguinal canal, and enters the pelvic cavity. It runs along the lateral surface of the urinary bladder and merges with the duct from a seminal vesicle under the urinary bladder. The duct formed by this merger is an ejaculatory duct that passes through the prostate gland and opens into the urethra. The vasa deferentia have rather thick, muscular walls that move the sperm by peristalsis.

Ejaculatory Duct

A short **ejaculatory duct** is formed by the merger of a vas deferens and a duct from a seminal vesicle. Contraction of the muscular walls of the ejaculatory duct forces sperm cells and fluid from the seminal vesicles into the urethra.

Urethra

The **urethra** is a thin-walled tube that extends from the urinary bladder to the end of the penis. The urethra serves a dual role in the male. It transports urine from the urinary bladder during micturition, as noted in chapter 16, and it also carries semen, which includes sperm, during ejaculation. Control mechanisms prevent both fluids passing at the same time.

Describe how sperm cells are formed.
Trace the path of sperm from a testis to outside the body.

Accessory Glands

Three different types of glands produce secretions involved in the reproductive process. These glands are the seminal vesicles, prostate gland, and bulbourethral glands. See figure 17.1.

Seminal Vesicles

The **seminal vesicles** are paired glands. The duct of each gland merges with a vas deferens near the posterior surface of the prostate gland. The alkaline secretions of the seminal vesicles are mixed with sperm cells during ejaculation and help to regulate the pH of the semen. In addition, these secretions contain fructose, which serves as an energy source for sperm cells.

Prostate Gland

The **prostate gland** is a doughnut-shaped gland that encircles the urethra where it exits the urinary bladder. The ejaculatory ducts pass through the posterior portion of the prostate to join with the urethra within the prostate. The prostate gland has numerous small ducts that open into the urethra. The secretion of the prostate gland is an alkaline, somewhat milky fluid that keeps the pH of the semen slightly alkaline. It also activates the swimming movements of the sperm.

Bulbourethral Glands

The **bulbourethral** (bul-bō-ū-rē'-thral) **glands** are two small, spherical glands that are located below the prostate gland near the base of the penis (see figure 17.1). These glands secrete an alkaline, mucuslike fluid into the urethra in response to sexual stimulation. This secretion neutralizes the acidity of the urethra and lubricates the end of the penis in preparation for sexual intercourse.

Semen

The seminal fluid, or **semen** (sē′-men), is the fluid passed from the urethra during ejaculation. It consists of the fluids secreted by the bulbourethral glands, seminal vesicles, and prostate gland along with sperm cells from the testes. It is slightly alkaline (pH 7.5) and contains nutrients for sperm cells and substances that activate their motility. The volume of semen in a single ejaculation may vary from 3 to 6 ml, with 50 to 150 million sperm per milliliter. Although only one sperm participates in fertilization, many sperm are necessary for fertilization to occur.

> Male infertility may be caused by a number of factors, such as hormone imbalances, duct blockage, a low sperm count, abnormal sperm, or a low fructose concentration in the semen. The minimum sperm count for male fertility is considered to be at least 20 million sperm per milliliter of semen.

Scrotum

The **scrotum** (skrō′-tum) is the external sac of skin and subcutaneous tissue that contains the testes. It hangs from the trunk midline posterior to the penis. A median partition keeps each testis in a separate chamber within the scrotum. Testes develop within the pelvic cavity but descend into the saclike scrotum through the inguinal canals near the end of the seventh month of fetal development. This migration occurs under the stimulation of testosterone, the male sex hormone, and it is essential for the production of viable sperm.

> Failure of the testes to descend into the scrotum produces sterility and increases the chance of testicular cancer. Undescended testes may be stimulated to descend by testosterone administration, or surgical means may be used to move the testes into the scrotum. Both treatments are done in early childhood.

The subcutaneous layer of the scrotum contains smooth muscle fibers that reflexively contract or relax to shorten or lengthen the scrotum in response to temperature. In addition, muscles attached to the testes respond similarly. Through the actions of these muscles, the testes are raised closer to the body in cold temperatures and lowered in warm temperatures. In this way, the temperature of the testes is kept about 3°F lower than the normal body temperature. This lower temperature is necessary for the production of viable sperm.

Penis

The **penis** is the male copulatory organ that deposits semen in the female vagina during sexual intercourse. It contains specialized erectile tissues that enable it to become enlarged and rigid during sexual excitement.

Three columns of erectile tissue compose the body of the penis. The *corpora cavernosa* are two columns of erectile tissue located dorsally, while the *corpus spongiosum* is located ventrally. The urethra extends throughout the length of the corpus spongiosum (see figure 17.1). The corpora spongiosum expands at the tip to form the *glans penis,* which contains numerous sensory nerve endings and the urethral opening. A loose sheath of skin, the *prepuce,* extends forward to cover the glans.

Table 17.1 summarizes the functions of male reproductive organs.

> ***What composes semen?***
> ***What is the function of the scrotum?***

> For many American male babies, *circumcision,* the surgical removal of the prepuce, is performed in the delivery room or within three days after birth. In Jewish culture, it may be performed on the eighth day as a religious rite. It is believed that circumcision improves male hygiene and decreases the risk of penile infections, and it may reduce the risk of cervical cancer in sexual partners.

Male Sexual Response

In the absence of sexual stimulation, the vascular sinusoids in the erectile tissue of the penis contain a small amount of blood, and the penis is flaccid (flak′-

Table 17.1 Summary of Functions of Male Reproductive Organs

Organ	Function
Testis	Seminiferous tubules produce sperm; interstitial cells secrete testosterone
Epididymis	Site of sperm maturation and temporary storage; carries sperm to vas deferens
Vas deferens	Carries sperm to ejaculatory duct
Ejaculatory duct	Carries sperm and secretion from seminal vesicle to the urethra
Urethra	Carries semen to outside the body
Bulbourethral gland	Secretes watery fluid that neutralizes acidity of the urethra
Seminal vesicle	Secretes alkaline fluid containing nutrients for sperm
Prostate gland	Secretes alkaline fluid that activates motility of sperm
Scrotum	Contains and protects testes
Penis	Inserted into vagina during sexual intercourse; deposits semen in vagina; contains sensory nerve endings associated with feelings of sexual pleasure

sid), or soft. Sexual stimulation initiates parasympathetic nerve impulses that dilate arterioles and constrict venules supplying the erectile tissue. These vascular changes cause the erectile tissue to become engorged with blood, which produces **erection,** a condition in which the penis swells, lengthens, and becomes erect.

Continued sexual stimulation of the glans, as in sexual intercourse, culminates in **orgasm,** which is characterized by **ejaculation** (ē-jak-ū-lā′-shun) of semen and a feeling of intense pleasure. Just prior to ejaculation, sympathetic nerve impulses stimulate peristaltic contractions of the epididymides, vasa deferentia, and ejaculatory ducts along with contractions of the seminal vesicles and prostate gland. These contractions force semen (sperm and glandular secretions) into the urethra. Then, ejaculation occurs as certain skeletal muscles at the base of the penis contract rhythmically, forcing semen through the urethra to the outside. A feeling of general relaxation follows.

Immediately after ejaculation, the vascular changes that produced erection are reversed. Sympathetic impulses constrict the arterioles and dilate the venules supplying the erectile tissue, allowing the accumulated blood to leave the penis. The penis becomes flaccid again, as blood is carried away. After orgasm, erection is not possible for a time period that varies from less than an hour to several hours.

Hormonal Control of Reproduction in Males

The onset of male sexual development begins around the ages of 11 or 12 and is completed by ages 15 to 17. Processes initiating the onset of puberty are not well understood, but the sequence of events is known. Hormones of the hypothalamus, anterior pituitary gland, and testes are involved.

At puberty, unknown stimuli trigger the hypothalamus to secrete **gonadotropin releasing hormone (GnRH),** which is carried to the anterior pituitary gland by the blood. GnRH activates the release of **luteinizing hormone (LH)** and **follicle-stimulating hormone (FSH)** from the anterior pituitary gland. In males, LH is sometimes called *interstitial cell–stimulating hormone (ICSH).* LH (ICSH) promotes growth of the interstitial cells of the testes and stimulates their secretion of **testosterone** (tes-tos′-te-rōn), the primary male hormone. FSH and testosterone in combination act on the seminiferous tubules, stimulating spermatogenesis.

Action of Testosterone

Male hormones are collectively called **androgens** (an′-dro-jens), and testosterone is the most important one. Small quantities of androgens are secreted by the adrenal cortex during fetal development, where they promote formation of the male sex organs, but there is little androgen production between birth and the onset of puberty.

At puberty, testosterone stimulates the maturation of the male reproductive organs, the continuation of spermatogenesis, and the development of the male secondary sexual characteristics. Male **secondary sexual characteristics** include (1) growth of body hair, especially on the face, axillary, and pubic regions; (2) increased muscular development; (3) development of heavy bones, broad shoulders, and narrow pelvis; and (4) deepening of the voice due to enlargement of the larynx and thickening of the vocal cords. Less obvious effects are increases in the rate of cellular

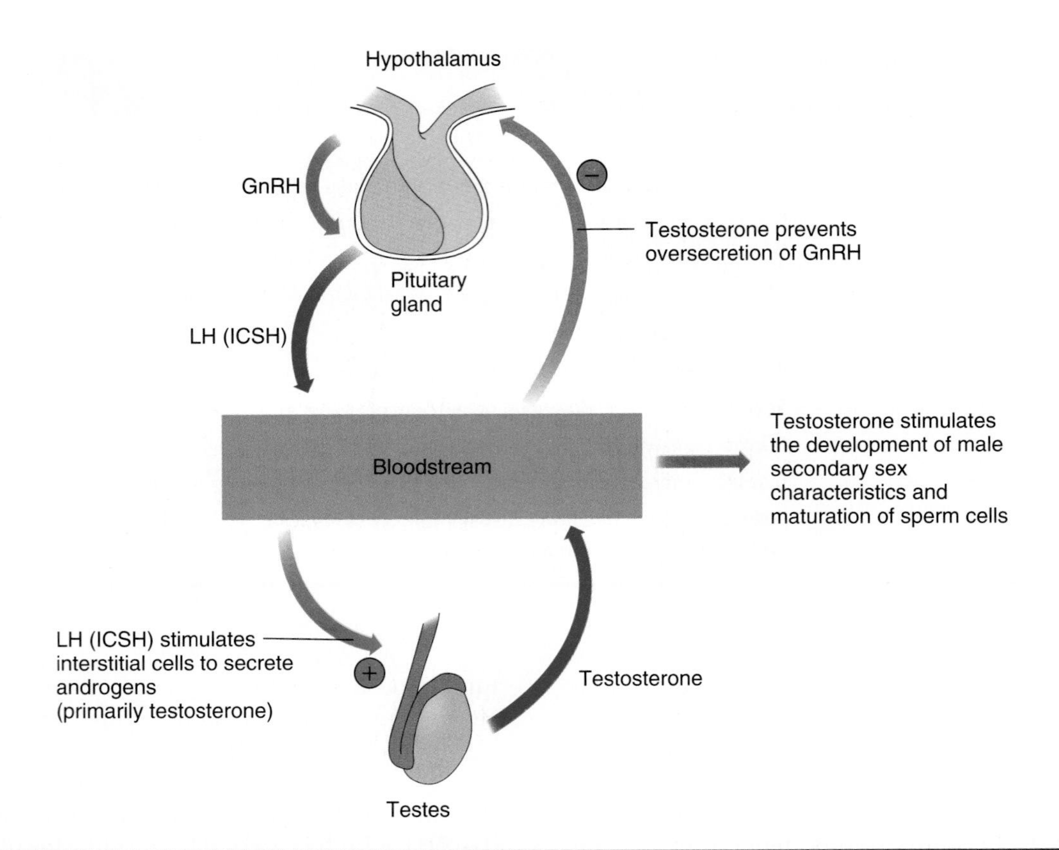

Figure 17.6 A negative feedback control system regulates testosterone production.

metabolism and red blood cell production. Both metabolic rate and the concentration of red blood cells in the blood are greater in males than in females.

Regulation of Testosterone Secretion

Like many other hormones, testosterone production is controlled by a *negative feedback system,* as shown in figure 17.6. After puberty, this mechanism maintains the testosterone level in the blood within normal limits.

As testosterone concentration increases in the blood, it inhibits the production of GnRH by the hypothalamus, which, in turn, reduces the release of LH from the anterior pituitary gland. The decrease in LH production causes a decrease in testosterone secretion by the interstitial cells of the testes, which reduces testosterone concentration in the blood.

Conversely, as testosterone concentration in the blood declines, the hypothalamus is stimulated to secrete GnRH, which, in turn, promotes the release of LH by the anterior pituitary. The increase in LH production causes an increase in testosterone secretion by the testes, which increases the testosterone concentration in the blood.

How is sperm production regulated?
What are the functions of testosterone?
How is testosterone secretion regulated?

Testosterone production starts to decline at about 40 years of age, resulting in a gradual decline in sexual activity. However, male sexual activity continues into old age.

Female Reproductive System

The female reproductive system produces female hormones and female sex cells and transports the sex cells to a site where they may unite with a sperm cell. In addition, the female reproductive system provides a suitable environment for the development of the offspring and is actively involved in the birthing process. The organs of the female reproductive system include the paired ovaries, which produce female

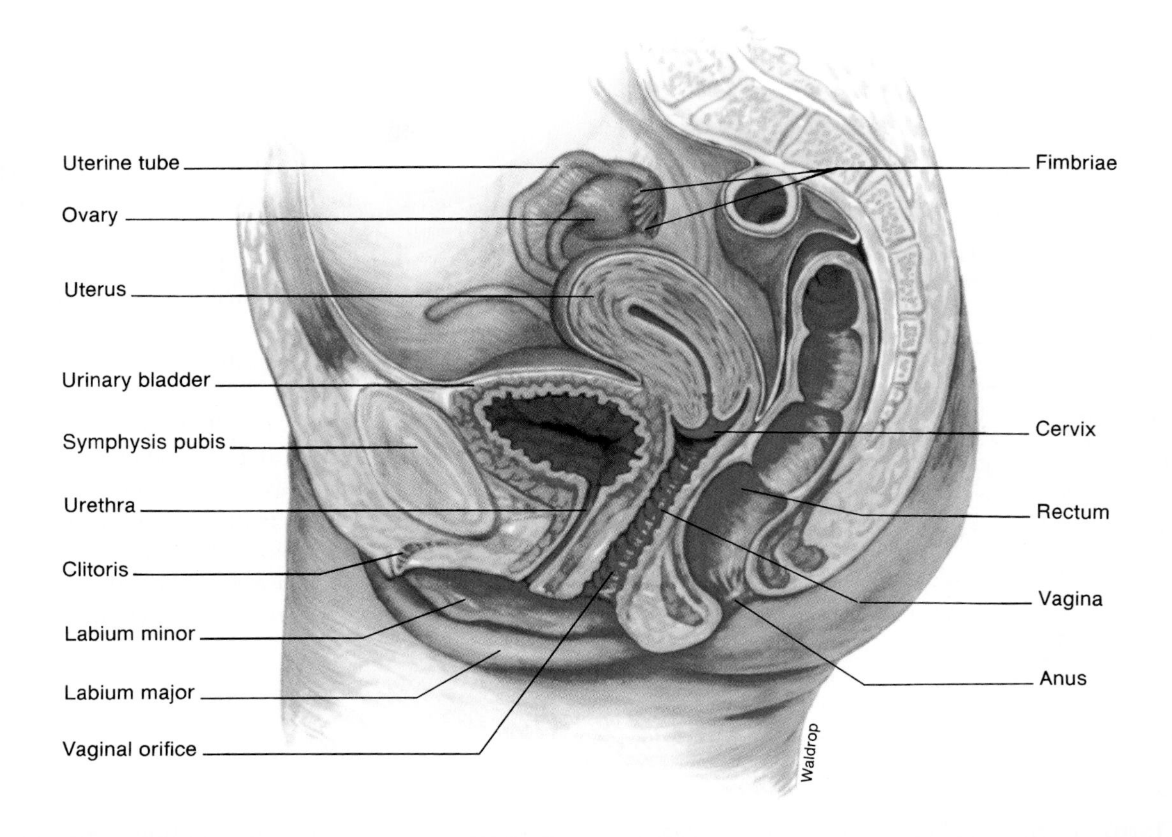

Figure 17.7 Female reproductive system in a midsagittal section.

hormones and sex cells; paired uterine tubes, which transport the female sex cells; a uterus, where internal development of the offspring occurs; a vagina, which serves as the female copulatory organ and birth canal; and accessory glands and external organs (figure 17.7).

Ovaries

The **ovaries** are located in the upper, lateral portions of the pelvic cavity, one on each side of the uterus. They are about the same size and shape as large almonds, and they are supported by several ligaments. The outer surface of an ovary is covered by **germinal epithelium,** which starts the formation of immature sex cells by cell division. Interior to the germinal epithelium is a more dense region, which contains immature ovarian follicles. The interior of an ovary consists of loose connective tissue supporting nerves and blood vessels and contains follicles in several stages of development.

Oogenesis

Oogenesis (ō-ō-jen′-e-sis) is the process producing ova (eggs), the female sex cells. Oogenesis is similar to spermatogenesis with a few notable exceptions. The process of oogenesis is summarized in figure 17.8, and the sequential stages within an ovary are shown in figure 17.9.

By the fifth month of development, the developing ovaries of a female baby contain several million **oogonia** (ō-ō-gō′-nē-ah), which are formed from the germinal epithelium by mitotic cell division. Then, oogonia formation ceases. Most of these oogonia degenerate. Later in fetal development, the remaining oogonia enlarge slightly and become **primary oocytes** (ō′-ō-sītz), which contain 46 chromosomes and are capable of meiotic cell division. However, meiotic cell division does not occur until puberty. Each tiny primary oocyte is enclosed by a single layer of epithelial cells forming the **primordial follicles.** There are about 1 million primordial follicles in the ovaries of a female baby at birth, but most degenerate. Only about 400,000 follicles remain at puberty, at which point they are called **primary follicles.**

Starting at puberty, follicle-stimulating hormone (FSH) stimulates enlargement of the ovaries and the growth of several primary follicles each month. One primary follicle develops more rapidly than the others, whose development is suppressed. As the dominant primary follicle develops, a cavity forms within it that is filled with follicular fluid. The accumulated

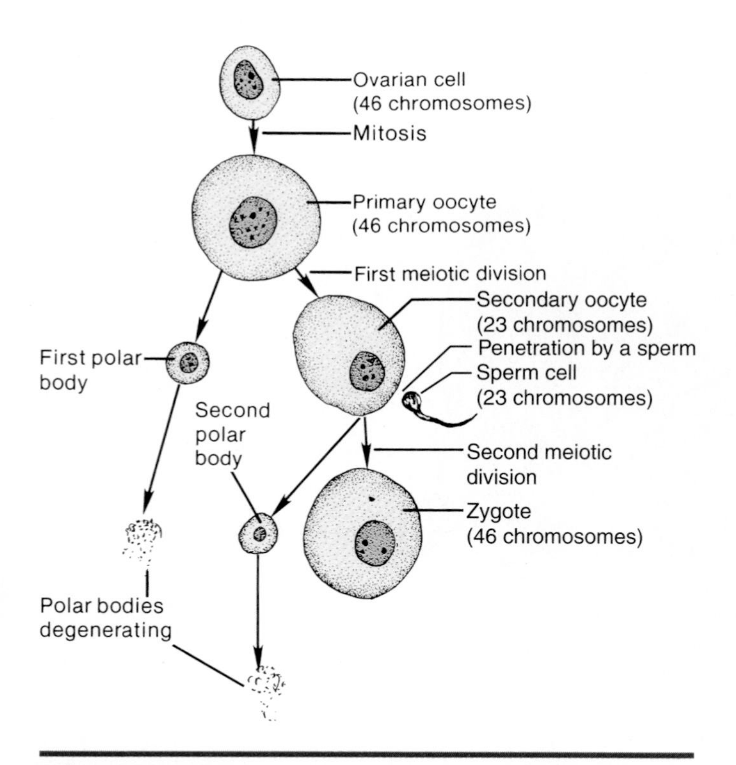

Figure 17.8 Oogenesis. Meiosis I produces a single secondary oocyte. If it is penetrated by a sperm, meiosis II forms an egg cell. Fusion of sperm and egg nuclei produce a zygote.

fluid presses the primary oocyte to one side, where it is covered by a few layers of follicular cells.

FSH stimulation causes the primary oocyte in a growing follicle to undergo the first meiotic division, forming two cells of unequal size, each containing 23 chromosomes. The large **secondary oocyte** receives most of the cytoplasm. The small first **polar body** contains little cytoplasm and has no reproductive function. Formation of a polar body is simply a convenient way to get rid of 23 chromosomes. At this stage, the follicle is considered to be a **secondary follicle.**

It is the secondary oocyte that is released from an ovary in ovulation. Further division of the oocyte stops at this point unless it is penetrated by a sperm. If the secondary oocyte is penetrated by a sperm, it undergoes the second meiotic division, which produces an **ovum** (egg cell) and a second polar body. Both of these new cells also contain 23 chromosomes (see figure 17.8).

Ovulation

As the secondary follicle continues to grow under the influence of FSH, it produces a bulge on the ovary surface. At about this time it is called a **mature follicle.** Continued enlargement of the follicle and a weaken-

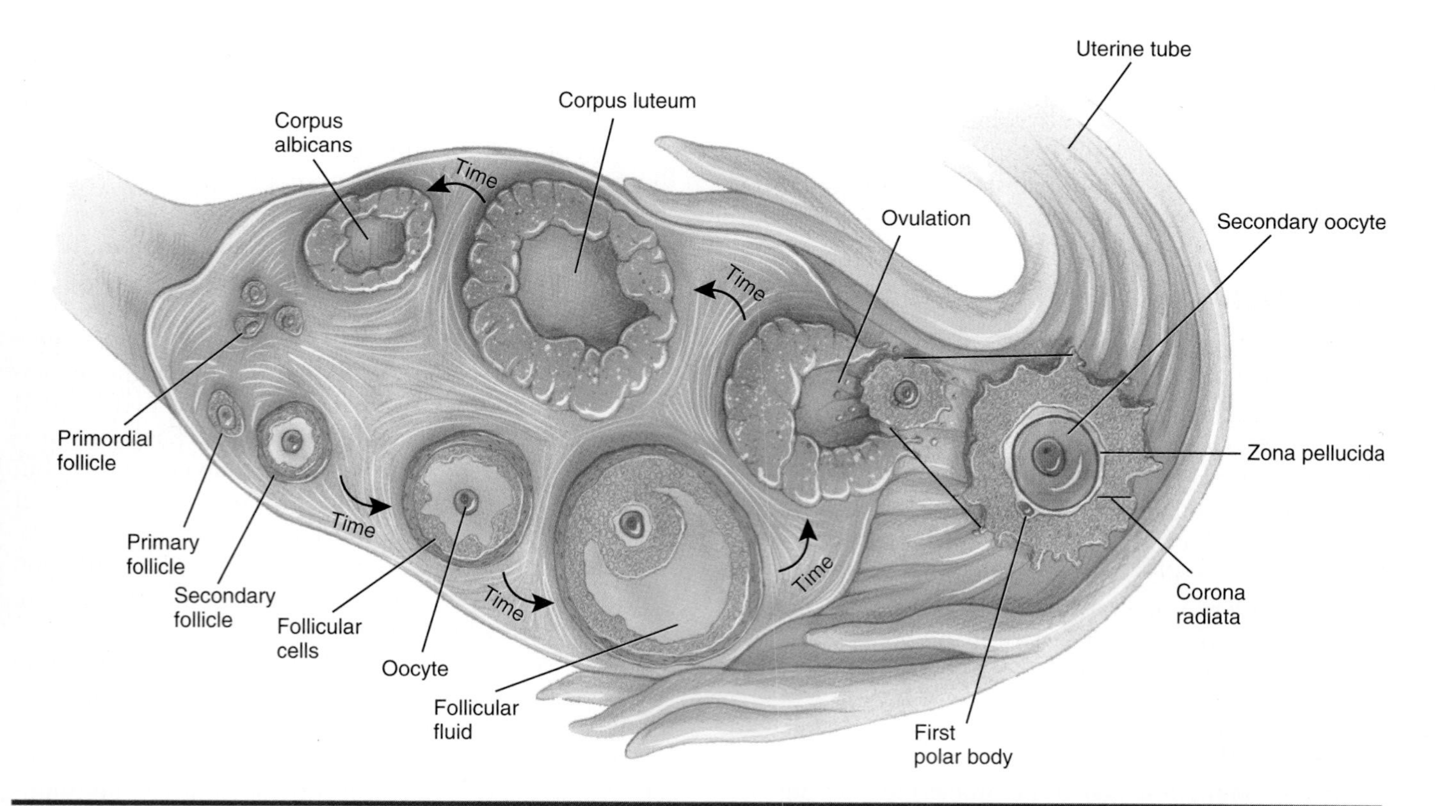

Figure 17.9 Stages of follicle development, release of a secondary oocyte at ovulation, and the formation of a corpus luteum.

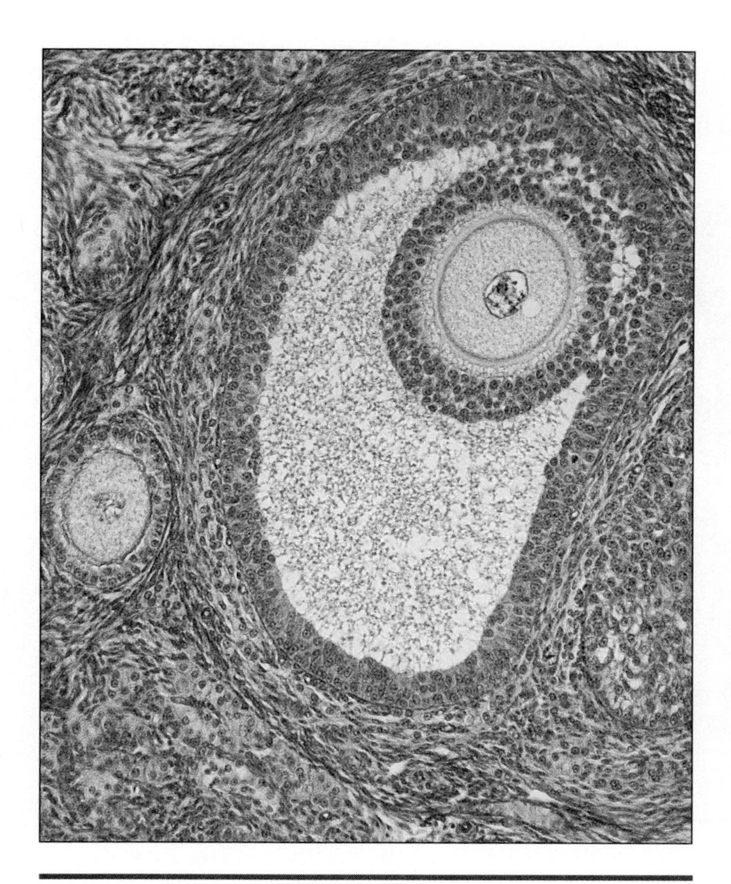

Figure 17.10 Photomicrograph of a maturing follicle showing an oocyte covered by follicular cells (250×).

ing of the ovarian wall results in the rupture of the follicle and the release of the secondary oocyte—a process known as **ovulation.** The secondary oocyte, polar body, and covering follicular cells are discharged at the opening of a uterine tube. The surrounding follicular cells are collectively called the *corona radiata* (figures 17.9 and 17.10).

Uterine Tubes

The paired **uterine tubes** are sometimes called fallopian tubes, or oviducts. Each uterine tube extends laterally from the superior lateral surface of the uterus to an ovary. The end of each tube forms a funnel-shaped expansion, the **infundibulum** (in-fun-dib′-ū-lum), that partially envelops an ovary but is not connected to it. Each infundibulum is subdivided into a number of fingerlike processes called *fimbriae* (fim′-brē-ē) that may touch the ovary (figure 17.11).

The inner lining of a uterine tube consists of simple ciliated columnar epithelium and secretory cells. The beating of the cilia creates a current that helps draw the ovulated secondary oocyte into the infundibulum. Then, the beating cilia move the oocyte slowly toward the uterus.

Infections of the female reproductive tract can easily migrate up the uterine tubes into the pelvic cavity, where the infections become far more serious.

Uterus

The **uterus** (ū′-ter-us) is located in the pelvic cavity posterior to the urinary bladder. As shown in figure 17.7, it is located superior to the vagina and is bent anteriorly over the urinary bladder. The uterus is a hollow organ with thick, muscular walls. Its primary function is to provide an appropriate internal environment for a developing baby.

The uterus has two major regions. The *cervix,* the lower tubular portion, is inserted into the upper end of the vagina. The *body,* the upper portion, is enlarged and rounded, and it joins with the uterine tubes laterally.

The walls of the uterus are composed of three layers. The **endometrium** (en-dō-mē′-trē-um) is the mucosal layer that lines the interior of the uterus. The **myometrium** is the thick layer of smooth muscle that forms most of the wall thickness. The **perimetrium** is the thin, serous layer that covers the exterior of the uterus (see figure 17.11).

Vagina

The **vagina** (vah-jī′-nah) is the collapsible tube that extends from the uterus to the exterior. It serves as the female copulatory organ and the birth canal. The vagina is located posterior to the urethra and anterior to the rectum.

The vaginal wall consists of three layers: an inner mucosa of stratified squamous epithelium and supportive connective tissue; a thin muscle layer consisting of smooth muscle; and an outer fibrous layer composed of dense connective tissue.

External Accessory Organs

The openings of the urethra and vagina are surrounded by external accessory reproductive organs

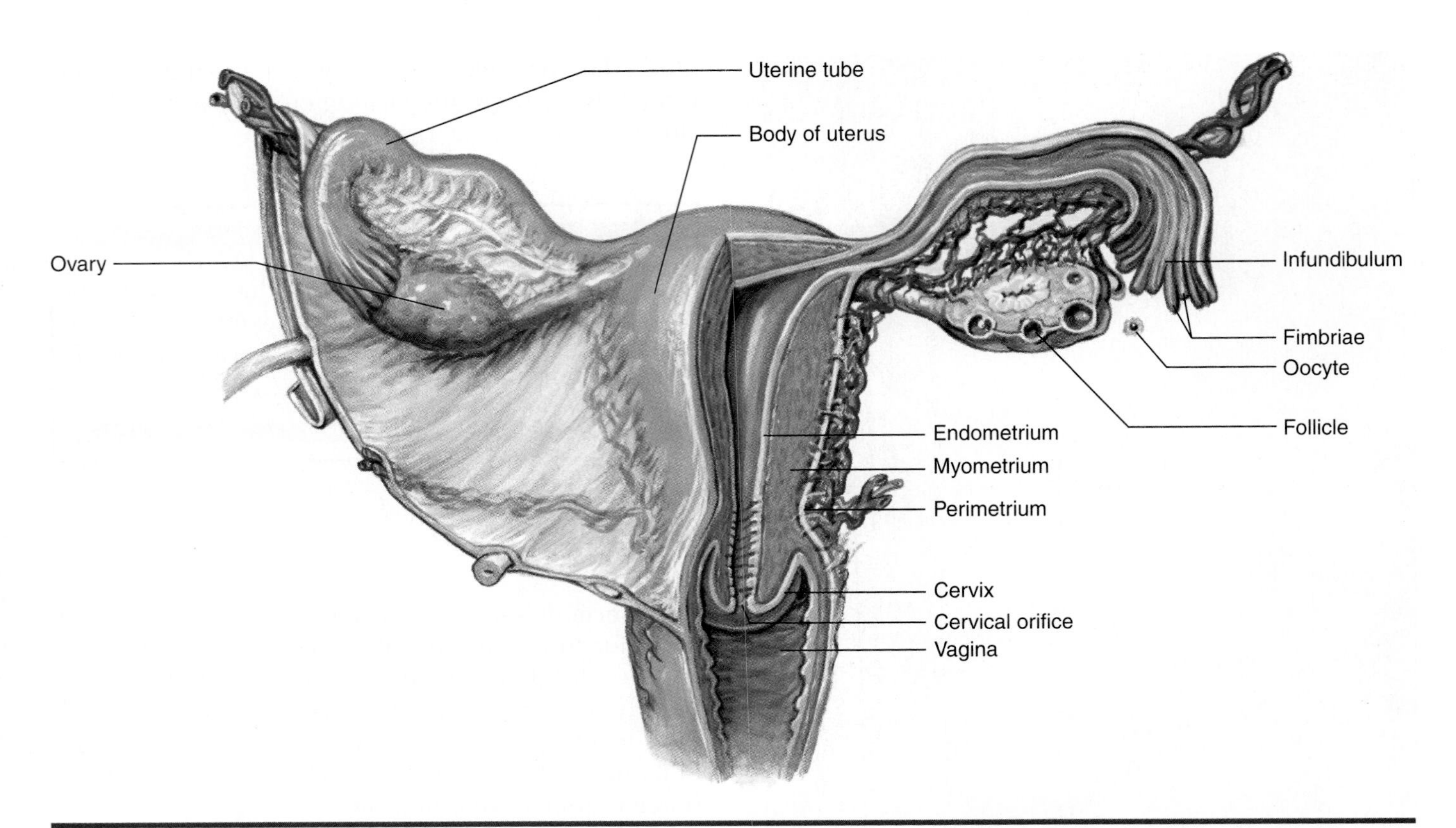

Figure 17.11 Female reproductive organs.

that compose the external genitals, or **vulva.** Figure 17.12 illustrates the female external genitals and **perineum** (per-i-nē′-um), the pelvic floor. The portion of the perineum between the vaginal opening and the anus is often called the *obstetrical perineum.*

The paired **labia majora** (lā′-bē-ah ma-jō′-rah) are rounded longitudinal folds of adipose tissue and a thin layer of smooth muscle covered by skin. The labia majora enclose the other external accessory structures, and their medial margins are separated by a narrow cleft. They join anteriorly at the *mons pubis,* a rounded cushion of adipose tissue over the anterior surface of the pubic symphysis. The labia majora are formed of the same embryonic tissues that form the scrotum in males.

The **labia minora** are paired, thinner, longitudinal folds that lie medial to the labia majora. They join anteriorly to form a hoodlike covering of the clitoris. Posteriorly, they join with the labia majora.

The narrow space between the labia minora is known as the **vestibule.** The urethra opens into its anterior region, and the vagina opens into its posterior region. A pair of **vestibular (Bartholin's) glands** lie on each side of the vaginal opening and release their secretions into the vestibule.

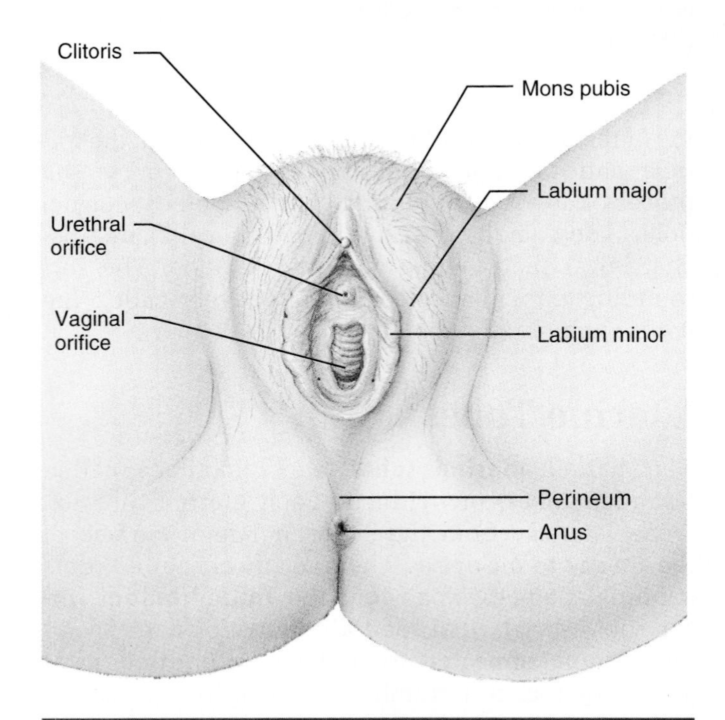

Figure 17.12 Female external genitals.

Table 17.2 Summary of Functions of Female Reproductive Organs

Organ	Function
Ovary	Produces egg cells; secretes estrogen and progesterone
Uterine tube	Receives secondary oocyte, site of fertilization; carries early embryo to uterus
Uterus	Holds embryo and fetus during pregnancy
Vagina	Receives penis in sexual intercourse; serves as birth canal
Labia majora	Protect other external reproductive organs
Labia minora	Enclose vestibule; protect vaginal and urethral openings
Clitoris	Contains sensory nerve endings associated with feelings of sexual pleasure
Vestibular glands	Secrete fluid that lubricates vaginal opening and vestibule

The **clitoris** (kli′-tōr-is) is a nodule of erectile tissue located at the anterior union of the labia minora. It consists of erectile tissue and corresponds to the penis in males. The *glans,* its anterior portion, contains abundant sensory nerve endings whose stimulation leads to the female sexual response.

Table 17.2 summarizes the functions of the female reproductive organs.

> ***Describe oogenesis and ovulation.***
> ***What are the parts of the vulva?***

Female Sexual Response

In the absence of sexual stimulation, the erectile tissue of the clitoris contains a small amount of blood. Sexual stimulation causes parasympathetic nerve impulses to dilate the arterioles and constrict the venules supplying the erectile tissue. These vascular changes cause the erectile tissue to become engorged with blood and produce erection of the clitoris.

Simultaneously, nerve impulses cause enlargement of the vaginal mucosa and breasts and erection of the nipples due to increased blood flow to these areas. Secretion of the vestibular glands is increased, lubricating the vestibule and aiding entry of the penis.

As in the male, the female sexual response culminates in orgasm, but there is no ejaculation. Instead, muscles of the pelvic floor and walls of the uterus and uterine tubes contract rhythmically. These contractions aid the movement of sperm through the uterus and toward the upper ends of the uterine tubes. Orgasm produces a sensation of intense pleasure followed by general relaxation and a feeling of warmth throughout the body.

Hormonal Control of Reproduction in Females

Reproduction in females is controlled by hormones produced by the hypothalamus, anterior pituitary gland, and ovaries.

Female Sex Hormones

There are two groups of female sex hormones—estrogen and progesterone—and both are produced by the ovaries in nonpregnant females. Ovarian follicles under the stimulation of FSH produce **estrogen** (es′-trō-jen), which stimulates the maturation of the female sex organs and the development and maintenance of the female **secondary sexual characteristics.** The female secondary sexual characteristics include development of the mammary glands and breasts, a broad pelvis, increased deposition of adipose tissue in the subcutaneous layer (especially in the breasts, buttocks, and thighs), and increased blood supply to the skin. The development of axillary and pubic hair probably is stimulated by the small amount of androgen produced by the adrenal glands.

After ovulation, the remainder of the ovarian follicle becomes a **corpus luteum** (kor′-pus lū′-tē-um) (see figure 17.9). Under stimulation by LH, the corpus luteum secretes the other female sex hormone, **progesterone** (prō-jes′-te-rōn), as well as estrogen. The major role of progesterone is the development and maintenance of the uterine lining in pregnancy, but it also inhibits uterine contractions and dilation of the cervix. Both estrogen and progesterone play major roles in the regulation of the female reproductive cycles.

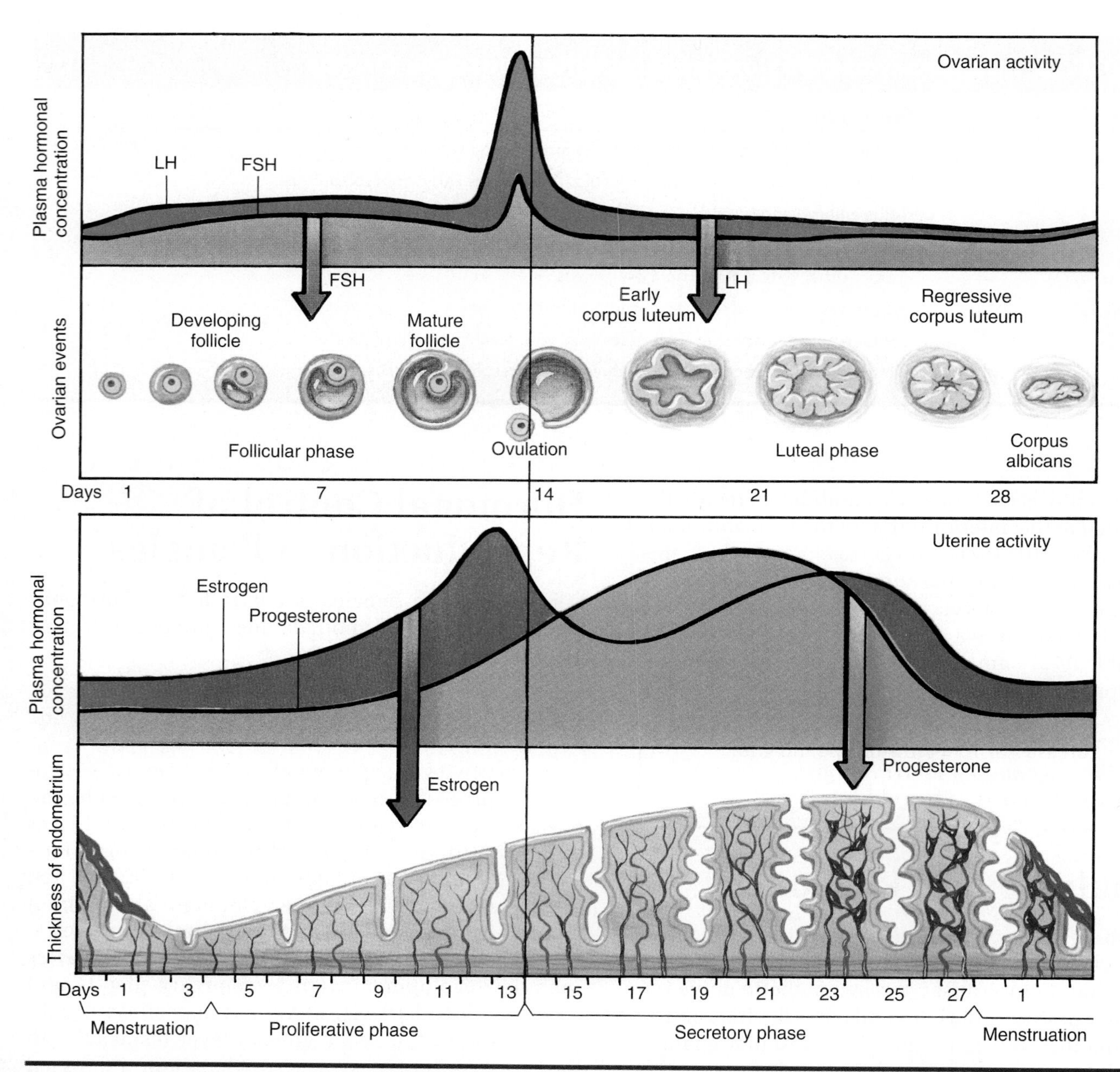

Figure 17.13 Female reproductive cycle.

Female Reproductive Cycles

The two female reproductive cycles are hormonally controlled and occur simultaneously starting at puberty: ovarian cycles and menstrual cycles. **Ovarian cycles** involve the development of ovarian follicles and ovulation. **Menstrual cycles** involve repetitive changes in the uterine lining that lead to monthly menstrual bleeding. The lengths of these cycles range from 24 to 35 days in different women, but 28 days is about average. Figure 17.13 illustrates the stages of menstrual and ovarian cycles along with the concentrations of the hormones that control them in a 28-day cycle. Table 17.3 summarizes the events of the ovarian and menstrual cycles.

Except for periods of pregnancy and nursing, women experience monthly reproductive cycles from puberty (menarche) until menopause. **Menopause,** the cessation of reproductive cycles, usually occurs between 45 and 55 years of age.

Ovarian Cycle

Puberty in females begins at about 11 years of age, with the first menses occurring at about 13 years of age. The female reproductive cycles begin when the

Table 17.3 Major Events in the Female Reproductive Cycle

1. GnRH secreted by the hypothalamus stimulates the anterior pituitary to release FSH and LH.
2. FSH stimulates enlargement of a few primary ovarian follicles and the subsequent maturation of the dominant follicle. FSH (with LH involvement) promotes estrogen secretion by follicular cells.
3. Estrogen stimulates the thickening of the uterine lining and exerts a negative feedback inhibition on GnRH secretion by the hypothalamus.
4. A rapid rise in estrogen concentration peaks on day 12 as the dominant follicle enlarges.
5. The anterior pituitary produces a sharp increase in LH and a smaller increase in FSH that peak on day 13.
6. The primary oocyte undergoes the first meiotic division, forming the secondary oocyte, and the surge of LH causes ovulation on day 14.
7. LH transforms the ruptured follicle into a corpus luteum and stimulates the corpus luteum to secrete increasing amounts of progesterone and estrogen during days 17 to 23.
   a. Estrogen promotes the continued thickening of the uterine lining.
   b. Progesterone promotes the formation of blood vessels and glands in the uterine lining, preparing it to receive an embryo.
   c. Progesterone exerts a negative feedback on GnRH secretion by the hypothalamus, inhibiting the release of FSH and LH from the anterior pituitary.
8. If fertilization does not occur, the corpus luteum degenerates, and the secretion of estrogen and progesterone declines rapidly from days 25 to 28.
9. The reproductive cycle starts again as the low levels of estrogen and progesterone
   a. result in the breakdown of the uterine lining with menstruation starting on day 1 of the next menstrual cycle;
   b. no longer inhibit the hypothalamus, allowing the secretion of GnRH, which stimulates release of FSH and LH from the anterior pituitary, starting a new ovarian cycle.

hypothalamus secretes gonadotropin releasing hormone (GnRH), which activates the anterior pituitary gland to release follicle-stimulating hormone (FSH) and luteinizing hormone (LH).

FSH promotes the development of a few primary follicles in the ovaries each month. In each monthly cycle, one of these follicles becomes the dominant follicle, while the others degenerate. FSH stimulates the secretion of estrogen by the follicular cells, but LH also is involved. As the dominant follicle enlarges, the secretion of estrogen gradually increases.

In a 28-day cycle, increased sensitivity of the follicular cells to FSH results in a rapid rise in estrogen production starting on day 7 and reaching a peak on day 12 (see figure 17.13). At this high concentration, estrogen exerts a *positive feedback control* that stimulates the hypothalamus to secrete GnRH and the release of LH from the anterior pituitary. This results in a sharp increase in LH and a lesser increase in FSH concentrations in the blood, which peak on day 13.

The abrupt increase in LH and FSH causes the rapid maturation of the dominant follicle and activates the primary oocyte to undergo the first meiotic division, forming the secondary oocyte and the first polar body. Thereafter, the high level of LH stimulates **ovulation,** the rupture of the mature follicle, which releases the secondary oocyte surrounded by follicular cells. Ovulation occurs 14 days before the onset of the next mensis, regardless of the length of the cycle.

After ovulation, the empty follicle is transformed into a corpus luteum under the stimulation of LH. Prior to ovulation, the follicular cells secrete estrogen and a small amount of progesterone, but after ovulation the corpus luteum secretes large amounts of progesterone as well as estrogen. The high concentration of progesterone exerts a *negative feedback control* that inhibits the secretion of GnRH by the hypothalamus and FSH and LH by the anterior pituitary.

If the secondary oocyte is not penetrated by a sperm leading to fertilization of the ovum, the corpus luteum degenerates, causing a rapid decline in estrogen and progesterone levels from days 25 to 28. Once the low level of progesterone no longer inhibits the hypothalamus, allowing secretion of GnRH, the next reproductive cycle begins.

Some women experiencing menopause have unpleasant symptoms called "hot flashes" that seem to affect primarily the upper body. Hot flashes result from sudden dilation of blood vessels and persist for less than a minute to several minutes. The cause is unknown but probably results from changes in GnRH secretion by the hypothalamus. These symptoms may be alleviated by hormone replacement therapy, the administration of estrogen and progesterone.

Menstrual Cycle

The menstrual, or uterine, cycle refers to the series of changes in the endometrium that occur each month unless pregnancy occurs. These changes are responses to fluctuating levels of estrogen and progesterone. The menstrual cycle has three phases: menstruation, proliferative phase, and secretory phase (see figure 17.13).

The low progesterone and estrogen levels at the end of the previous reproductive cycle cause the breakdown of the uterine lining, leading to menstruation. *Menstruation* starts a new menstrual cycle. It begins on the first day of the next menstrual cycle and lasts from three to five days.

The *proliferative phase* is characterized by a buildup of the endometrium under stimulation by estrogen, whose concentration in the blood increases as the dominant primary follicle develops. This phase begins at the end of menstruation and ends at ovulation.

Following ovulation, both progesterone and estrogen are produced by the corpus luteum, and they continue the development of the uterine lining in the *secretory phase.* Estrogen promotes the continued thickening of the endometrium. Progesterone stimulates the formation of blood vessels and glands in the uterine lining, preparing it to receive an early embryo. If fertilization of an ovum does not occur, the low levels of progesterone and estrogen drop rapidly, triggering the breakdown of the endometrium, which leads to menstruation on the first day of the next menstrual cycle. The secretory phase begins at ovulation and ends with the onset of menstruation.

How are FSH, LH, estrogen, and progesterone involved in ovarian and menstrual cycles?

Mammary Glands

Both males and females possess **mammary glands.** The mammary glands of males and immature females are similar. At puberty, estrogen and progesterone stimulate the development of female mammary glands. Estrogen starts breast and mammary gland development, and progesterone stimulates the maturation of the mammary glands so that they are capable of secreting milk. Another anterior pituitary hormone, prolactin, is required for milk production.

Mature mammary glands are female accessory reproductive structures that are specialized for milk production. They are located in the subcutaneous tissue of female breasts. The breasts are formed external to the pectoralis muscles. Breasts contain large amounts of fatty tissue that surround and cushion the mammary glands. Connective tissue provides support for both the breasts and the mammary glands. Externally, a circle of pigmented skin, the *areola* (ah-rē′-ō-lah), is located near the anterior portion of each breast. A *nipple* containing erectile tissue is located in the center of each areola.

Each mammary gland consists of 15 to 25 lobes containing *alveolar glands.* Alveolar glands produce milk under stimulation by *prolactin* after the birth of a baby. Milk is carried from the alveolar glands by *lactiferous ducts* that lead to the nipple and outside (figure 17.14).

Since breast cancer has the highest fatality rate among cancers in women, early detection is crucial. After 35 years of age, annual mammograms are advised and women are encouraged to develop the habit of monthly self-examinations.

Birth Control

Birth control methods may be categorized into several groups based on their mode of action: hormonal, chemical and behavioral contraceptive methods, anti-implantation devices, sterilization, and induced abortion. Figure 17.15 shows a few birth control devices, and table 17.4 summarizes the effectiveness of the various birth control methods.

Contraception

Contraceptive methods are designed to prevent sperm from reaching and penetrating a secondary oocyte, which leads to fertilization of an ovum. There are several types of contraceptives, including the use of hormones, barriers to sperm, spermicides (chemicals that kill sperm), and behavioral methods.

Hormonal Methods

Oral contraception, the use of "the pill," is a hormonal method of contraception. Although there are several types, a combination pill that contains a high concentration of progesterone and a low concentration of estrogen is most commonly used. These hormones provide a negative feedback control that inhibits the secretion of GnRH by the hypothalamus, which re-

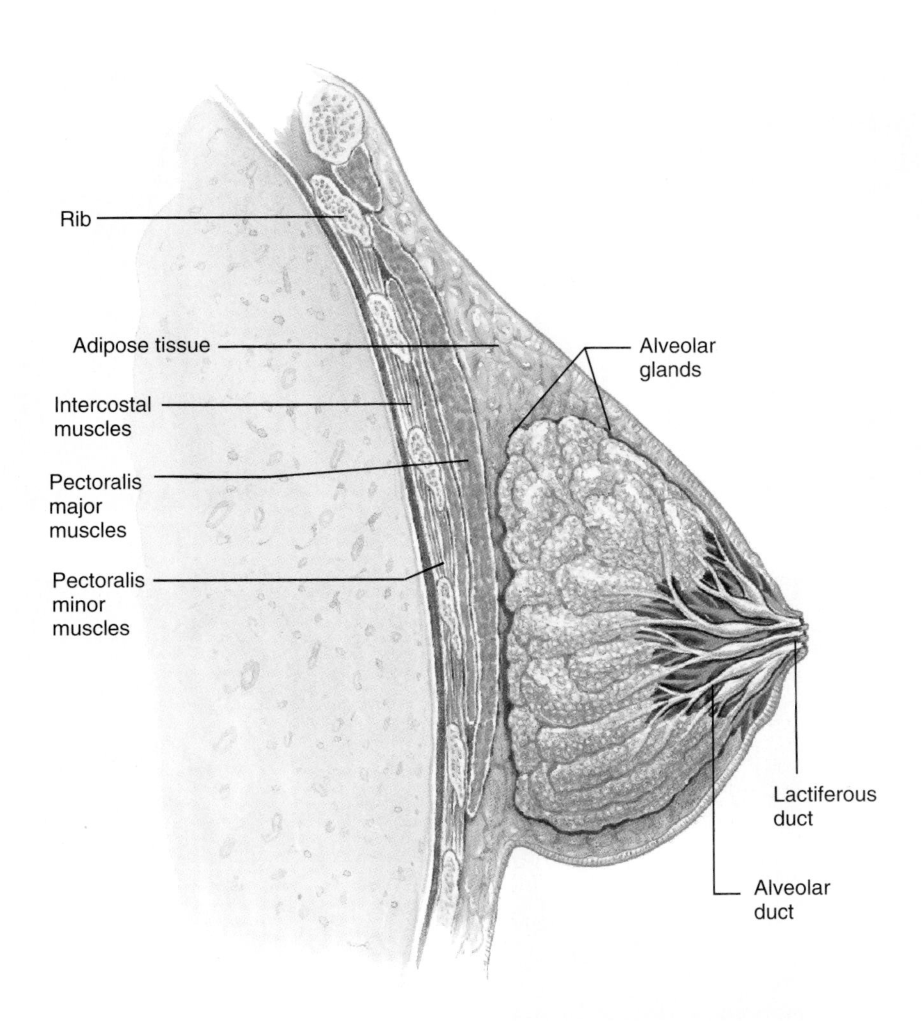

Figure 17.14 Mature breast in saggital section.

duces the secretion of FSH and LH by the anterior pituitary. The very low levels of FSH and LH usually prevent the development of mature ovarian follicles, so ovulation does not occur. Some women experience undesirable side effects, including headache, weight gain, nausea, irregular menses, and lack of menstruation. The chance of serious side effects, such as strokes, is greater among women who smoke or who have a history of thrombus formation.

Some combination birth control pills may be used in higher dosages as *morning after pills (MAPs),* for postcoital emergency contraception. When used within 72 hours of unprotected intercourse, the hormone combination disrupts normal hormonal controls and tends to prevent either fertilization or implantation.

Norplant is a set of tiny silicone rods containing synthetic progesterone that are surgically implanted under the skin of the upper arm or over the shoulder blade. The released progesterone acts similarly to an oral contraceptive by inhibiting ovulation, and the implant is effective for up to five years. The implant may be removed to terminate contraception.

Depo-Provera is a synthetic progesterone that is administered by muscular injection at three-month intervals. It protects against pregnancy by preventing ovulation and altering the uterine lining to inhibit implantation of an embryo. Its use often causes weight gain as a side effect, and it can be a considerable health risk, so it must be used under a physician's care. Women with a personal or strong family history of high blood pressure, asthma, kidney disease, migraine headaches, or breast cancer are at increased risk if using Depo-Provera.

Barriers

There are three types of **barriers** that are designed to prevent sperm from entering the uterus: condom, diaphragm, and cervical cap.

Condoms act by preventing sperm from being deposited in the vagina. A *male condom* is a thin sheath of

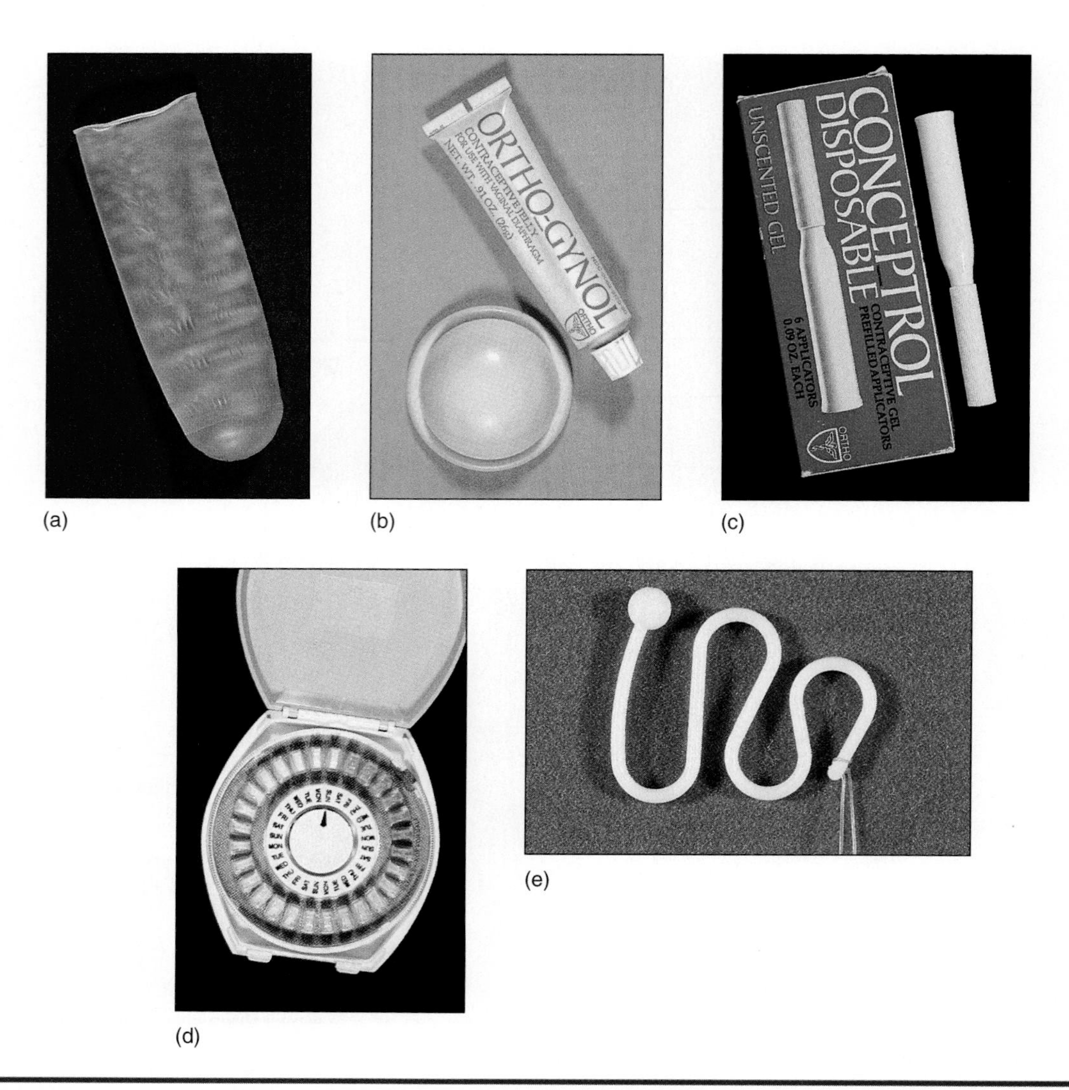

Figure 17.15 Examples of birth control substances and devices. (*a*) male condom, (*b*) diaphragm and spermicide jelly, (*c*) spermicide gel, (*d*) oral contraceptives (pills), and (e) intrauterine device (IUD).

latex rubber that is slipped over the penis prior to sexual intercourse. A *female condom* is a thin polyurethane bag with a flexible ring at each end. The woman inserts the bag into the vagina prior to sexual intercourse. Condoms reduce, but do not eliminate, the chance of infection by sexually transmitted diseases (STDs).

A **diaphragm** is a dome-shaped sheet of rubber supported by a firm, but somewhat flexible, ring. It is placed in the upper vagina over the cervix prior to intercourse. It prevents sperm from entering the uterus. Spermicidal substances are used along with a diaphragm.

A **cervical cap** is a thimble-shaped piece of latex rubber that fits snugly over the cervix. Spermicidal substances are used with a cervical cap. Females with abnormal Pap smears or cervical infections should not use a cervical cap.

Spermicides

A variety of **spermicides** are available, including creams, jellies, and suppositories. These chemicals kill sperm by destroying their plasma membranes.

Behavioral Methods

The **rhythm method** requires abstinence from sexual intercourse from three days before the day of ovulation to three days after ovulation, for a total of seven days. It is based on the fact that the secondary oocyte may be penetrated by a sperm for only about 24 hours after ovulation. One difficulty with this method is determining the time of ovulation, since few women have perfectly regular cycles. The failure rate is higher in women who have irregular cycles.

Table 17.4 Effectiveness of Birth Control Methods

Effectiveness values are approximate since they vary with how the birth control methods are used.

Method	Effectiveness*
Vasectomy	nearly 100%
Tubal ligation	nearly 100%
Depo-Provera	99%
Norplant	99%
Oral contraception	97%
Intrauterine device (IUD)	95%
Male condom	85%
Female condom	85%
Diaphragm with spermicide	80%
Cervical cap with spermicide	80%
Withdrawal	75%
Rhythm method	70%

*Effectiveness is the percentage of women who do not become pregnant in one year while using this method of birth control.

Withdrawal, or **coitus interruptus,** is the removal of the penis from the vagina just prior to ejaculation. Failures of this method result from preejaculatory emission of semen or failure to withdraw before ejaculation.

Anti-Implantation Devices

An **intrauterine device (IUD)** is a small plastic or metal object that is inserted into the uterus by a physician. The IUD causes inflammation of the uterine lining, preventing implantation of an embryo. An IUD may remain in place for long periods of time, but it should be checked at regular intervals by a physician since an IUD may be spontaneously expelled from the uterus. Undesirable side effects include excessive menstrual bleeding, painful cramps, and an increased risk of pelvic inflammatory disease and infertility.

Sterilization

Sterilization surgeries may be performed on both males and females. In males, a **vasectomy** is performed by cutting and tying the vasa deferentia within the scrotum so that sperm do not form part of the semen. The sperm disintegrate and are reabsorbed. In females, a **tubal ligation** is performed through a small abdominal incision. The uterine tubes are cut and tied to prevent the transport of a secondary oocyte towards the uterus. The ovulated oocytes disintegrate and are reabsorbed. Vasectomies and tubal ligations do not affect the production of sex hormones or the sexual response.

Induced Abortion

An **abortion** is the premature expulsion of an embryo or fetus from the uterus. Spontaneous abortions are called *miscarriages.* They usually result from hormonal disorders or serious abnormalities of the developing embryo. An **induced abortion** may be used to terminate an unwanted pregnancy. Induced abortion involves dilation of the cervix and removal of the embryo or fetus by suction or surgical means. Possible side effects of induced abortions include prolonged bleeding, perforation of the uterus, and emotional trauma.

The so-called abortion pill, *mifepristone* (RU 486), is a progesterone antagonist used in France and China, and it is in clinical trials in the United States. It causes the endometrium to break down, thereby detaching the embryo or fetus, which is passed from the uterus after administration of prostaglandins to promote uterine contractions. Its use is limited to the first five weeks of pregnancy and requires the supervision and care of a physician.

How do hormonal birth control methods work?
How does an IUD prevent pregnancy?

Disorders of the Reproductive Systems

These disorders are grouped as male disorders, female disorders, and sexually transmitted diseases.

Male Disorders

Prostatitis is the inflammation of the prostate gland. It is caused by acute or chronic infections that often are associated with tenderness and enlargement of the prostate.

An **enlarged prostate** without inflammation occurs in about 33% of males over 60 years of age. In some cases, it may restrict the flow of urine and prevent the control of micturition. Enlarged prostates are usually detected by rectal examinations. Surgical correction usually is by *transurethral resection,* a procedure in which an instrument is inserted into the urethra to remove portions of the prostate gland compressing the urethra.

Cancer of the prostate is the second leading cause of death from cancer in males. Males over 40 years of age should have annual prostate examinations.

Impotence is the inability to attain and maintain an erection long enough for sexual intercourse. It may result from organic or psychological factors. Treatment is available in most cases.

Infertility is the inability of a male to produce and deposit sufficient numbers of viable sperm in the vagina to bring about fertilization of an ovum. A low sperm count is a common cause.

Female Disorders

Amenorrhea is the absence of menstruation. *Primary amenorrhea* is the failure of a woman to begin menstruation, and it is caused by endocrine disorders or abnormal reproductive development. *Secondary amenorrhea* is the absence of one or more menstrual periods without pregnancy. This may result from excessive physical exertion or excessive weight loss.

Dysmenorrhea refers to painful menstruation that prevents a woman from doing her normal activities for one or more days during the menses. Uterine contractions are thought to be responsible for the pain.

Premenstrual syndrome is characterized by severe physical or emotional distress after ovulation and prior to menstruation. The cause is unknown, although it is probably related to ovarian hormone production.

Toxic shock syndrome is characterized by high fever, fatigue, headache, sore throat, vaginal irritation, vomiting, and diarrhea. It results from a toxin produced by a strain of bacteria (*Staphylococcus aureus*) whose growth is apparently enhanced by the use of highly absorbent tampons.

Endometriosis is the growth of endometrial tissue outside the uterus. The tissue migrates up the uterine tubes and enters the pelvic cavity. It may cause premenstrual or menstrual pain due to its breakdown during menstruation. Infertility may result from tubal obstruction.

Infertility in females is the inability to conceive. It may be caused by tubal obstruction, ovarian disease, or a lack of maintenance of the uterine lining.

Pelvic inflammatory disease (PID) is a collective term referring to any infection of the female reproductive organs and/or other pelvic tissues. The most common pathogens are those of sexually transmitted diseases that migrate into the pelvic cavity via the uterine tubes.

Cancer of the female reproductive system most often occurs as breast cancer or cervical cancer. *Breast cancer* rarely occurs before 30 years of age and is more prevalent after menopause. Any lump in a breast should be brought to a physician's attention immediately since breast cancer has a high fatality rate unless it is treated early. *Cervical cancer* is a rather slow-growing cancer and seems to be associated with penile papillomavirus infections of sexual partners. An association with multiple sex partners and uncircumcised sex partners may also be a cause. Cervical cancer usually can be detected early by a Pap smear. Annual examinations for possible breast and cervical cancers are recommended for women over 30 years of age.

Sexually Transmitted Diseases (STDs)

Acquired immunodeficiency syndrome (AIDS) results from infection with the *human immunodeficiency virus (HIV),* which attacks a group of lymphocytes known as helper T (T_4) cells. The AIDS virus is transmitted via sexual intercourse and by blood transfer, including infected needles shared by drug users. Also, it may be passed from an infected mother to her developing fetus. It is not transmitted by casual contact. AIDS is discussed in more detail in the disorders section of chapter 13.

Gonorrhea is caused by the bacterium *Neisseria gonorrhoeae.* In males, it infects the urethra, causing painful urethritis. In females, it infects the vagina and may spread to the urethra, uterus, uterine tubes, and pelvic cavity. Females may not experience symptoms until advanced stages. Gonorrhea is a major cause of female sterility due to damage to the uterine tubes, and it can cause blindness in newborn babies. Antibiotics usually provide effective treatment. Most newborn infants born in hospitals receive eyedrops of antibiotics at birth to protect against possible gonorrhea infection.

Syphilis is caused by the bacterium *Treponema pallidum.* Without treatment, syphilis progresses through several recognizable stages. The *first stage* is characterized by an open sore, a *chancre* (shang′-ker), at the site of entrance by the bacterium. It may not be readily noticed, especially in females. The chancre heals within one to five weeks. Several weeks later, the *second stage* appears as muscle and joint pain, fever, and skin rash; it persists from four to eight weeks. Then, the disease enters a latent period of variable length. When the bacterium begins to destroy organs, such as the brain and liver, the *third stage* is recognized. Treatment with antibiotics is effective prior to the third stage.

Chlamydia (klah-mid′-ē-ah) is caused by the bacterium *Chlamydia trachomatis,* and it infects nearly 5 million people each year. In males, it causes painful urethritis. In females, it may spread throughout the reproductive tract, causing damage to uterine tubes that can lead to sterility. Chlamydia may also be transmitted from an infected mother to her baby during childbirth when the bacterium enters the baby's eyes. Antibiotics are an effective treatment.

Genital herpes is caused by the herpes simplex virus type 2. This disorder is characterized by painful blisters on the reproductive organs and may be accompanied by fever or flulike symptoms. It may be transmitted from an infected mother to her baby during delivery. In the newborn baby, it may cause only mild discomfort, serious neural damage, or death. Treatment with the drug acyclovir inhibits viral replication but does not eliminate the virus. The virus remains in the body and intermittently produces the genital blisters.

Genital warts are caused by another virus, the human papillomavirus (HPV). About 1 million new cases occur each year. Patients with genital warts may have an increased risk of certain cancers of the reproductive organs. Although there is no treatment that eliminates the virus, the warts can be removed by electrocautery (burning), cryosurgery (freezing), or laser surgery.

CHAPTER SUMMARY

Male Reproductive System

1. The male reproductive system consists of the testes, accessory ducts, accessory glands, and accessory external organs.
2. Testes are the male gonads. Each testis is divided into lobules containing seminiferous tubules that produce sperm cells. Interstitial cells between the tubules produce testosterone.
3. Spermatogenesis is the process of producing sperm cells from spermatogonia. Meiotic division of primary spermatocytes produces four spermatids that mature into sperm cells, each with 23 chromosomes.
4. Sperm cells pass from the seminiferous tubules into the highly coiled epididymis, where they mature. From the epididymis, sperm pass in sequence through the vas deferens, ejaculatory duct, and the urethra to the outside.
5. Accessory glands secrete alkaline fluids. Two bulbourethral glands secrete fluids that neutralize the acidity of the urethra. Seminal vesicles release their secretions into the ejaculatory ducts. The prostate gland secretes its sperm-activating fluid directly into the urethra.
6. Semen consists of sperm cells and fluids from the accessory glands. About 3 to 6 ml of semen are produced in each ejaculation, with 50 to 150 million sperm in each milliliter.
7. Testes are contained in the saclike scrotum outside the body. Muscle fibers control the distance the testes are from the body to keep the temperature of the testes at about 95.6°F.
8. The penis contains three columns of erectile tissue that become engorged with blood during sexual excitement. The erect penis is inserted into the vagina in sexual intercourse.

Male Sexual Response

1. Sexual stimulation causes erection of the penis. In sexual intercourse, continued stimulation results in orgasm, which is characterized by ejaculation and intense pleasure.
2. After orgasm, blood leaves the erectile tissues, and the penis returns to its flaccid state.

Hormonal Control of Reproduction in Males

1. At puberty, the hypothalamus secretes GnRH, which stimulates the anterior pituitary to release FSH and LH (ICSH). LH stimulates the production of testosterone. FSH and testosterone stimulate the production of sperm.
2. Testosterone stimulates the maturation of the male sex organs, the continuation of sperm production, and the male sex drive. It also develops and maintains the male secondary sexual characteristics.
3. Testosterone secretion is regulated by a negative feedback control system.

Female Reproductive System

1. The female reproductive system consists of the ovaries, uterine tubes, uterus, vagina, accessory glands, and external accessory organs.
2. Ovaries are the female gonads, which produce female sex cells and female hormones. They are located in the pelvic cavity on each side of the uterus.
3. Oogenesis is the process of producing female sex cells from oogonia. Millions of oogonia are formed during fetal development, but most degenerate. Those that remain become primary oocytes containing 46 chromosomes and are enveloped by follicular cells, forming primordial follicles.
4. Starting at puberty, a few primary follicles are stimulated to grow each month. The dominant primary follicle enlarges, and its primary oocyte undergoes the first meiotic division, forming a secondary oocyte and the first polar body.
5. The secondary follicle grows until it ruptures, releasing the secondary oocyte in ovulation.
6. The infundibulum of a uterine tube receives the secondary oocyte, and the uterine tube carries it towards the uterus. If the secondary oocyte is penetrated by a sperm, the second meiotic division occurs, forming the ovum and a second polar body. Then, the sperm and ovum nuclei unite at fertilization.

7. The uterus is located in the pelvic cavity posterior to the urinary bladder. Its major function is to provide a suitable environment for the developing offspring. The layers composing the uterine wall are the endometrium, myometrium, and perimetrium.
8. The vagina receives the penis in sexual intercourse and serves as the birth canal in childbirth.
9. External female reproductive organs are collectively called the vulva, which includes the outer labia majora, the inner labia minora enclosing the vestibule, and the clitoris, a small nodule of erectile tissue located at the anterior juncture of the labia minora.

Female Sexual Response

1. Sexual stimulation causes erection of the clitoris and nipples and secretion by the vestibular glands.
2. Sexual response culminates in orgasm, which produces rhythmic contractions of muscles in the walls of the uterus and uterine tubes and a feeling of intense pleasure.

Hormonal Control of Reproduction in Females

1. The ovaries secrete estrogen and progesterone, the female sex hormones. Estrogen stimulates the maturation of the female reproductive organs and the development and maintenance of female secondary sexual characteristics. Progesterone maintains the uterine lining during pregnancy.
2. Female reproductive cycles involve simultaneous ovarian and menstrual cycles, which are controlled by interactions of FSH, LH, estrogen, and progesterone.
3. The start of an ovarian cycle is initiated by the secretion of GnRH by the hypothalamus, which stimulates the release of FSH and LH from the anterior pituitary gland. FSH promotes the development of ovarian follicles, which secrete increasing amounts of estrogen. Ovulation results from follicle stimulation by a sharp increase in LH and FSH.
4. After ovulation, the follicle remnants become a corpus luteum under stimulation by LH. Both progesterone and estrogen are secreted by the corpus luteum. If fertilization does not occur, secretion of progesterone and estrogen declines rapidly, which allows the hypothalamus to secrete GnRH, starting another ovarian cycle.
5. A menstrual cycle begins with the first day of menstruation, which results from the decline in progesterone and estrogen levels. Increasing levels of estrogen cause thickening of the endometrium during the proliferative phase. After ovulation, estrogen stimulates continued development of the uterine lining, and progesterone promotes the formation of blood vessels and glands in the uterine lining in the secretive phase. If fertilization does not occur, the drop in progesterone and estrogen levels triggers breakdown of the uterine lining, followed by menstruation.

Mammary Glands

1. Mammary glands are adapted to secrete milk after childbirth. They are located within the female breasts.
2. Estrogen stimulates development of the mammary glands and breasts.

Birth Control

1. Contraceptive methods include hormonal methods, barriers, spermicides, and behavioral methods.
2. Anti-implantation methods use intrauterine devices that change the uterine lining, preventing implantation.
3. Sterilization methods involve a vasectomy in males or a tubal ligation in females.
4. Induced abortion involves dilation of the cervix and evacuation of the uterine contents.

Disorders of the Reproductive Systems

1. Male disorders include an enlarged prostate, prostate cancer, impotence, and infertility.
2. Female disorders include amenorrhea, dysmenorrhea, premenstrual syndrome, toxic shock syndrome, endometriosis, infertility, pelvic inflammatory disease, and cancer.
3. Sexually transmitted diseases include AIDS, gonorrhea, syphilis, chlamydia, genital herpes, and genital warts.

BUILDING YOUR VOCABULARY

1. **Selected New Terms**
   bulbourethral gland, p. 355
   clitoris, p. 363
   contraception, p. 366
   endometrium, p. 361
   ejaculatory duct, p. 355
   epididymis, p. 352
   labia, p. 362
   menstrual cycle, p. 364
   oocyte, p. 359
   oogonium, p. 359
   ovarian cycle, p. 364
   primary follicles, p. 359
   polar body, p. 360

prostate gland, p. 355
scrotum, p. 356
seminal vesicle, p. 355
seminiferous tubules, p. 351
spermatid, p. 352
spermatocyte, p. 351
spermatogonium, p. 351
vas deferens, p. 355
vestibular gland, p. 362
vulva, p. 362

2. **Related Clinical Terms**
endometriosis (en-dō-mē-tri-ō′-sis) Spread of endometrial tissue into the pelvic cavity.
gynecology (gī-ne-kol′-o-jē) The medical specialty dealing with the female reproductive system.
hysterectomy (his-to-rek′-to-mē) The surgical removal of the uterus.
orchitis (or-kī′-tis) Inflammation of the testes.
prostatectomy (pros-tah-tek′-to-mē) Surgical removal of the prostate gland.

STUDY ACTIVITIES

1. Complete the **Study Guide for Chapter 17,** which begins on page 487.
2. Complete the **Learning Objectives** listed on the first page of this chapter.

CHECK YOUR UNDERSTANDING

(Answers are located in appendix B.)

1. Male gonads, the _____ , are formed in the pelvic cavity and migrate into the saclike _____ before birth.
2. Sperm cells are formed within _____ in the testes by a process called _____ .
3. Sperm cells mature within the _____ and pass through the vas deferens, _____ , and _____ to the outside.
4. During sexual stimulation, the erectile tissue of the penis becomes engorged with _____ .
5. The hormone _____ stimulates the secretion of _____ by interstitial cells.
6. Female sex cells are formed in the _____ by a process called _____ .
7. At ovulation, a _____ is released from an ovary and enters a _____ , which carries it toward the uterus.
8. Oogenesis is completed if a _____ is penetrated by a sperm, forming an ovum and a second _____ .
9. The _____ serves as the female copulatory organ and birth canal.
10. The onset of puberty begins with the release of _____ from the hypothalamus, which stimulates secretion of FSH and _____ by the anterior lobe of the pituitary gland.
11. FSH promotes the secretion of _____ by the ovarian follicles.
12. After ovulation, the empty follicle becomes a _____ , which secretes progesterone, which promotes buildup of the _____ of the uterus.
13. If fertilization does not occur, a rapid _____ in FSH and LH causes a breakdown of the endometrium, triggering the onset of _____ .
14. Hormonal birth control methods act by preventing _____ , while an IUD prevents _____ of an embryo.
15. Secondary sexual characteristics are produced by _____ in females and _____ in males.

EIGHTEEN

CHAPTER 18

Pregnancy, Prenatal Development, and Inheritance

Chapter Preview & Learning Objectives

SELECTED KEY TERMS

Allele An alternate form of a gene.
Embryo The developmental stage extending from the second week through the eighth week of pregnancy.
Fertilization The union of sperm and ovum nuclei.
Fetus The prenatal developmental stage extending from the ninth week of pregnancy to birth.
Gene A unit of heredity; part of a DNA molecule in a chromosome.
Genotype The genetic composition of an individual.
Germ layer One of three embryonic tissues from which all subsequent tissues develop.
Heterozygous (hetero = different) A condition in which alleles of a gene pair are different.
Homozygous (homo = the same) A condition in which alleles of a gene pair are identical.
Implantation The attachment of an early embryo to the uterine lining.
Labor The series of events composing the birth process.
Parturition The process of giving birth.
Phenotype (phen = visible) The observable characteristics of inherited traits.
Placenta (plac = flat) Temporary organ formed of fetal and maternal tissues that attaches fetus to the uterine wall.
Umbilical cord (umbil = navel) The organ connecting the fetus with the placenta.
Zygote (zygo = yolk) A fertilized egg.

Each new baby begins as a single cell formed by the fusion of a sperm cell and an ovum. At the instant of its formation, this first cell of a baby-to-be contains all of the inherited information necessary for growth and development. In addition, all inherited characteristics, such as sex and eye color, that give each baby its unique identity are established. After about 38 weeks of *prenatal* (before birth) growth and development (40 weeks from the last menses), a baby is born.

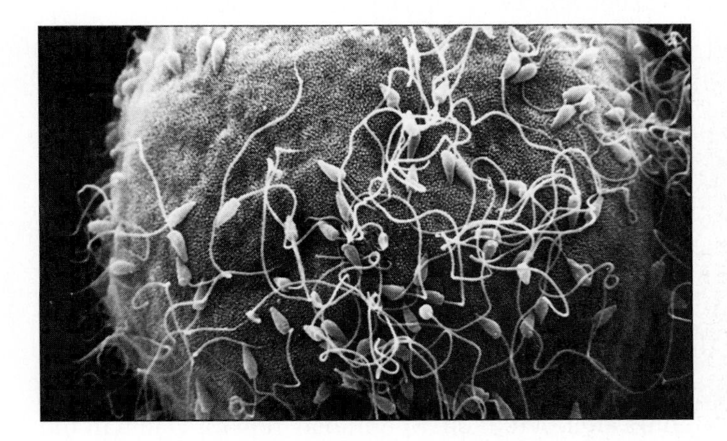

Figure 18.1 Scanning electron photomicrograph of sperm cells clustered on the surface of a secondary oocyte (1,200×). Only one sperm may enter the oocyte.

Fertilization and Early Development

Recall from chapter 17 that each **primary oocyte** undergoes the first meiotic division while still in the ovarian follicle. This division forms a **secondary oocyte** and the first **polar body,** and each cell contains 23 chromosomes. At ovulation, a secondary oocyte and the first polar body, still enclosed within a sphere of follicular cells, are released into a uterine tube. They are slowly moved toward the uterus by the beating cilia of epithelial cells lining the uterine tube.

Fertilization

After semen is deposited in the vagina, sperm begin their long journey into the uterus and on into the uterine tubes. Sperm inherently swim against the slight current of fluid that moves down the uterine tubes and uterus, and this behavior guides them into the uterus and up the tubes. Of the millions of sperm in the semen, probably only a few thousand sperm actually enter the uterine tubes. Prostaglandins in semen stimulate wavelike muscular contractions of the uterus and uterine tubes that greatly aid the migration of sperm. Sperm reach the upper portions of uterine tubes within one hour after sexual intercourse. Usually, only one uterine tube contains a secondary oocyte since only one secondary oocyte is usually released at ovulation. Sperm entering an empty uterine tube have no chance for fertilization.

When sperm reach a secondary oocyte, they are chemically attracted to it and cluster around it, trying to penetrate through the follicular cells (figure 18.1). Sperm cells release an enzyme that dissolves the "glue" holding follicular cells together so they can reach the oocyte. It takes many sperm cells to disperse the follicular cells, so that one sperm can eventually wriggle between the follicular cells and penetrate the

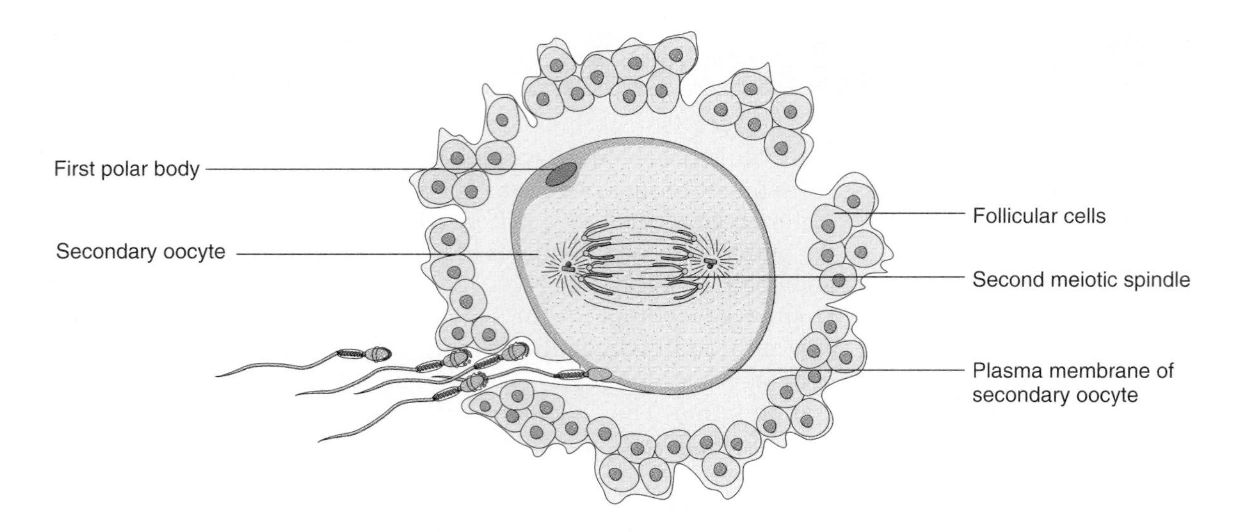

Figure 18.2 Enzymes from many sperm cells are required to separate the follicular cells, enabling one sperm to penetrate the secondary oocyte's membrane.

oocyte membrane. Once this happens, changes in the oocyte membrane prevent other sperm from entering (figure 18.2).

When a sperm enters the secondary oocyte, it triggers the second meiotic division, which forms the **ovum** and a second polar body. Then, the sperm and ovum nuclei unite in **fertilization** to form a **zygote,** the first cell of the baby-to-be. The zygote contains 46 chromosomes, 23 from the sperm and 23 from the ovum.

A secondary oocyte remains viable for about 24 hours after ovulation. Sperm may remain viable in the female reproductive tract for about 72 hours. Therefore, sexual intercourse from three days before ovulation to one day after ovulation is likely to result in fertilization.

Preembryonic Development

Immediately after fertilization, the zygote begins to divide by mitotic cell division. The early cell divisions are collectively known as *cleavage.* These divisions occur so rapidly that maximum cellular growth between divisions is not possible, so increasingly smaller cells are formed. During this time, the *preembryo* is carried down the uterine tube by the beating cilia of epithelial cells lining it. By the time the preembryo reaches the uterus, it consists of a solid ball of cells called a **morula** (mor′-u-lah) that is not much larger than the zygote.

Continued, but slightly slower, mitotic divisions form a larger hollow ball of cells called a **blastocyst** (blas′-to-sist) that soon develops within it the **inner cell mass,** a specialized group of cells from which the **embryo** later develops. The outer wall of the blastocyst is now called the **trophoblast.**

Identical, or monozygotic, twins come from a single zygote whose cells separate after the first cell division. Identical twins have a common placenta and chorion and have identical genetic characteristics. Fraternal, or dizygotic, twins are derived from two zygotes—separate ova that were fertilized by different sperm during the same female reproductive cycle. Fraternal twins do not have identical genetic traits.

Pregnancy is measured in weeks from the last menses. A full-term pregnancy lasts about 40 weeks, 10 lunar months, or 280 days from the last menses. Pregnancy is also often divided into three *trimesters,* each with a duration of three calendar months.

About the seventh day after ovulation, the blastocyst attaches to the uterine lining. Digestive enzymes, released from the trophoblast, enable the blastocyst to penetrate into the endometrium, where it is soon covered by surface layers of the uterine lining. This entire process is called **implantation,** and it is completed by the fourteenth day (figure 18.3).

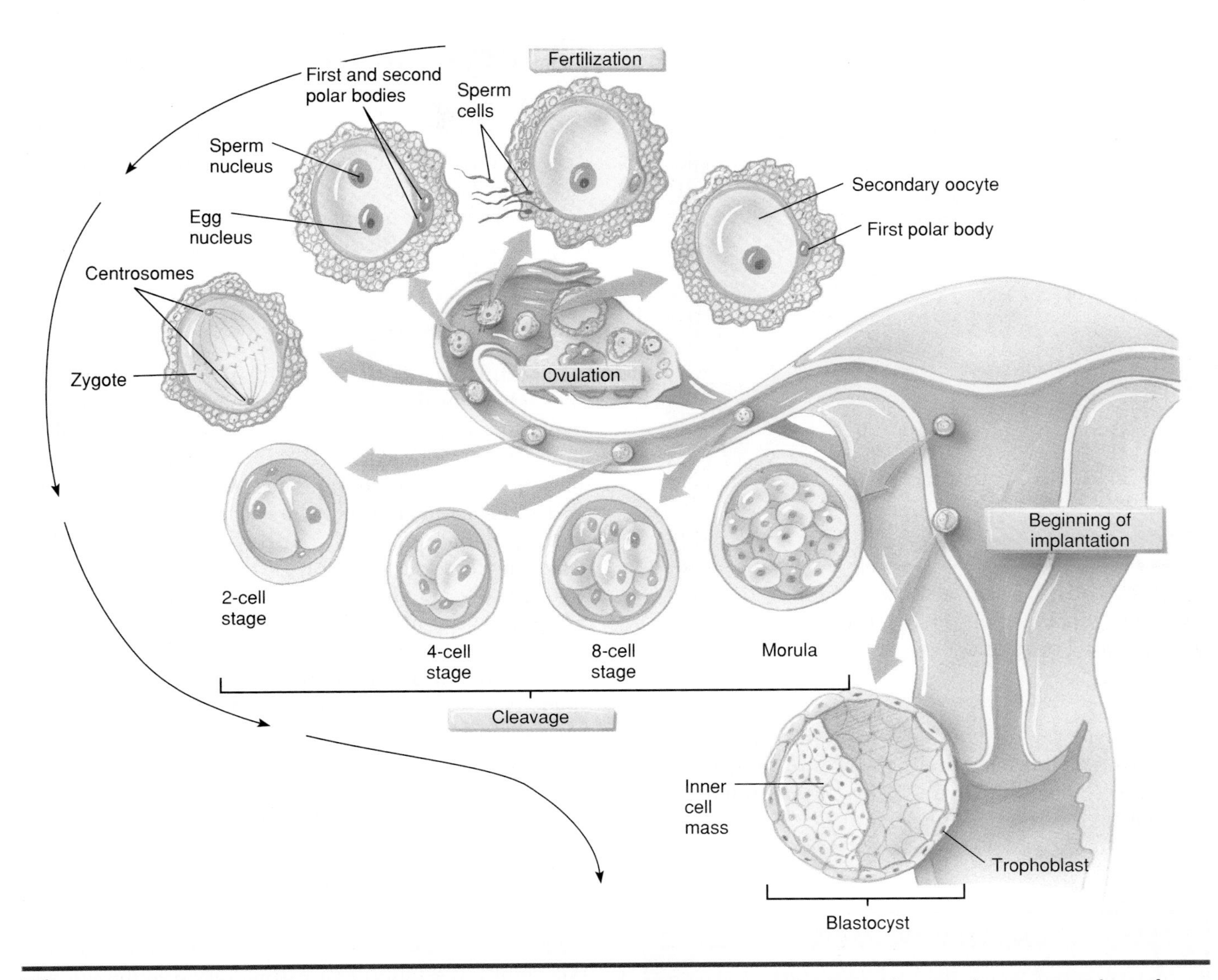

Figure 18.3 The stages of fertilization, cleavage, and implantation. Implantation begins about seven days after fertilization.

How do sperm cells reach the secondary oocyte? What are the major events from sperm penetration of a secondary oocyte to implantation?

Hormonal Control of Pregnancy

Without the formation of an early embryo, the corpus luteum disintegrates about two weeks after ovulation. The resulting decline in estrogen and progesterone levels causes the uterine lining to break down and to be shed in the menses. Pregnancy causes hormone changes that maintain the uterine lining.

The trophoblast cells of the blastocyst secrete **human chorionic gonadotropin (HCG)** (kō-rē-on′-ik gōn-ah-dō-trō′-pin), a LH-like hormone that overrides hypothalamus-pituitary controls and maintains the corpus luteum. Under stimulation of HCG, the corpus luteum continues to secrete progesterone and estrogen to maintain the uterine lining throughout pregnancy. Pregnancy tests are designed to detect the presence of HCG in a woman's blood or urine, and it is usually detectable by the end of the third week after fertilization.

The concentration of HCG in the blood rises sharply between the third and fourth week and then declines rapidly to level off at about the fourth month of pregnancy. Between the second and third month, the placenta takes over the role of producing estrogen and progesterone, and the corpus luteum degenerates.

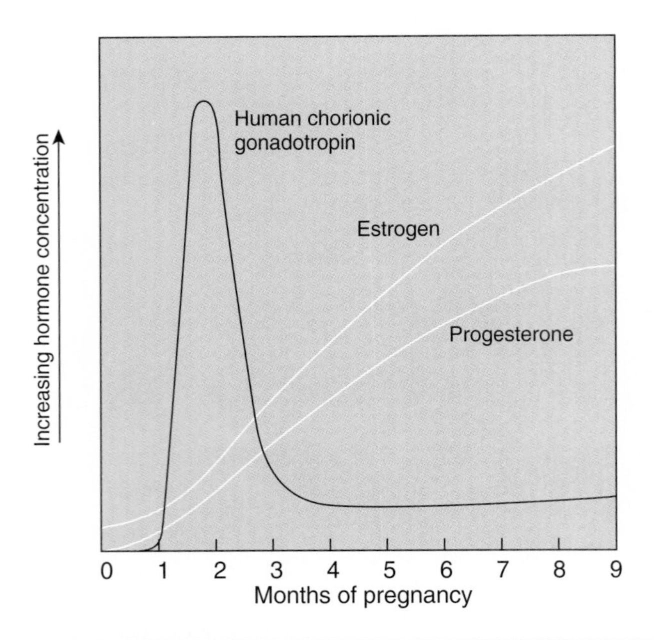

Figure 18.4 Relative concentrations of human chorionic gonadotropin (HCG), estrogen, and progesterone during pregnancy.

The ovaries remain inactive during the remainder of the pregnancy because the high level of progesterone suppresses secretion of GnRH by the hypothalamus, preventing the release of FSH and LH from the anterior pituitary.

If the corpus luteum stops secreting progesterone and estrogen too quickly or if the placenta is too slow in starting to secrete these hormones, a decline in their concentrations results. Such a decline may detach the placenta from the uterine wall, causing a miscarriage.

The concentrations of estrogen and progesterone continue to increase throughout pregnancy (figure 18.4). Note that the level of estrogen increases faster than that of progesterone. As pregnancy continues, placental estrogen and progesterone stimulate development of the mammary glands in preparation for milk secretion.

What hormonal changes are triggered by the presence of an implanted embryo?

Embryonic Development

The embryonic stage of development begins at the end of the second week and is completed at the end of the second month (eighth week). During this time, the embryo undergoes rapid development, forming the rudiments of all body organs, extraembryonic membranes, and the placenta. By the end of the eighth week, it has a distinct human appearance.

Germ Layers

After implantation, the inner cell mass grows to become the **embryonic disk,** which is supported by a short stalk extending from the wall of the blastocyst. The embryonic disk consists of three embryonic tissues: ectoderm, mesoderm, and endoderm. These embryonic tissues are called the **primary germ layers** because all body tissues and organs are formed from them. Figure 18.5 illustrates the formation of the germ layers.

Briefly, **ectoderm** forms all of the nervous system and the epidermis of the skin. **Mesoderm** forms muscles, bones, blood, and other forms of connective tissues. **Endoderm** forms the epithelial lining of the digestive, respiratory, and urinary tracts. Table 18.1 provides a more detailed listing of the major structures formed by each primary germ layer.

Extraembryonic Membranes

While the embryonic disk is forming, slender extensions from the trophoblast grow into the surrounding endometrium, firmly anchoring the blastocyst. The trophoblast of the blastocyst is now called the **chorion** (kō′-rē-on), the outermost extraembryonic membrane, and the slender extensions known as **chorionic** (kō-rē-on′-ik) **villi.**

At about the same time, two other extraembryonic membranes separate from the embryonic disk. The **amnion** (am′-nē-on) is formed dorsal (posterior) to the embryo, and the **yolk sac** is formed ventral (anterior) to the embryo (figures 18.5 and 18.6). Amniotic fluid fills the space between the embryonic disk and the amnion. As the embryo develops, the amnion margins move toward the ventral surface of the embryo. In a short time, the embryo is enveloped by the amnion.

The space between the embryo and the amnion becomes filled with **amniotic fluid.** It serves as a shock absorber for the developing embryo, and it also prevents adhesions from developing between various parts of the embryo. Later in development, the fetus swallows and inhales amniotic fluid and discharges dilute urine into it.

While the amnion is forming, the yolk sac branches, giving rise to the **allantois** (al-lan′-to-is), the last of the extraembryonic membranes. Both the allan-

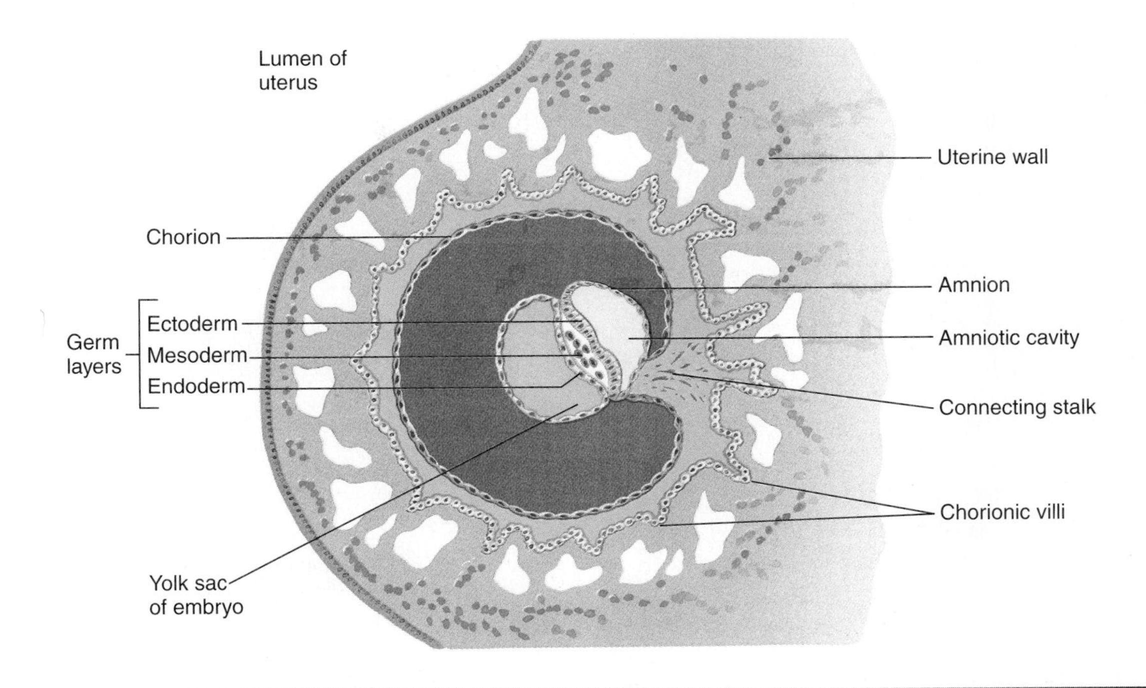

Figure 18.5 Development of the germ layers and three extraembryonic membranes—chorion, amnion, and yolk sac—in an early embryo. Note the chorionic villi.

Table 18.1 Structures Formed by the Primary Germ Layers

Ectoderm	Mesoderm	Endoderm
All nervous tissue	All connective tissues including bone and cartilage	Epithelial lining of digestive system (except oral cavity and anal canal)
Sensory epithelium of sense organs	Skeletal, cardiac, and smooth muscles	Liver and pancreas
Hypophysis, pineal gland, and adrenal medulla	Blood, bone marrow, and lymphatic tissue	Epithelial lining of respiratory system
Epidermis of the skin including nails, hair follicles, sebaceous glands, and sweat glands	Blood and lymphatic vessels	Epithelial lining of urethra and bladder
Cornea and lens of eye	Dermis of the skin	Thymus, thyroid, and parathyroid glands
Epithelial lining of oral and nasal cavities, paranasal sinuses, and anal canal	Kidneys, ureters, gonads, and reproductive ducts	
Salivary glands	Adrenal cortex	
	Epithelial lining of joints and the ventral body cavity	

tois and the yolk sac subsequently become part of the **umbilical** (um-bil′-i-kal) **cord,** which attaches the embryo to the placenta. The yolk sac forms the early blood cells for the embryo. The allantois also forms blood cells and brings embryonic blood vessels to the placenta.

Placenta

As the embryo continues to grow and the chorion enlarges, the layer of the endometrium covering the blastocyst becomes increasingly thinner. The chorionic villi in this region disintegrate, and only those in contact with the thick, spongy endometrium persist. This leads to the formation of the **placenta** (plah-sen′-tah), a disk-shaped structure formed of both embryonic and maternal tissues. The embryonic portion is formed of the chorion and chorionic villi, while the maternal portion is formed of the associated endometrium (figure 18.6).

The developing embryo is attached to the placenta by the umbilical cord. Two umbilical arteries

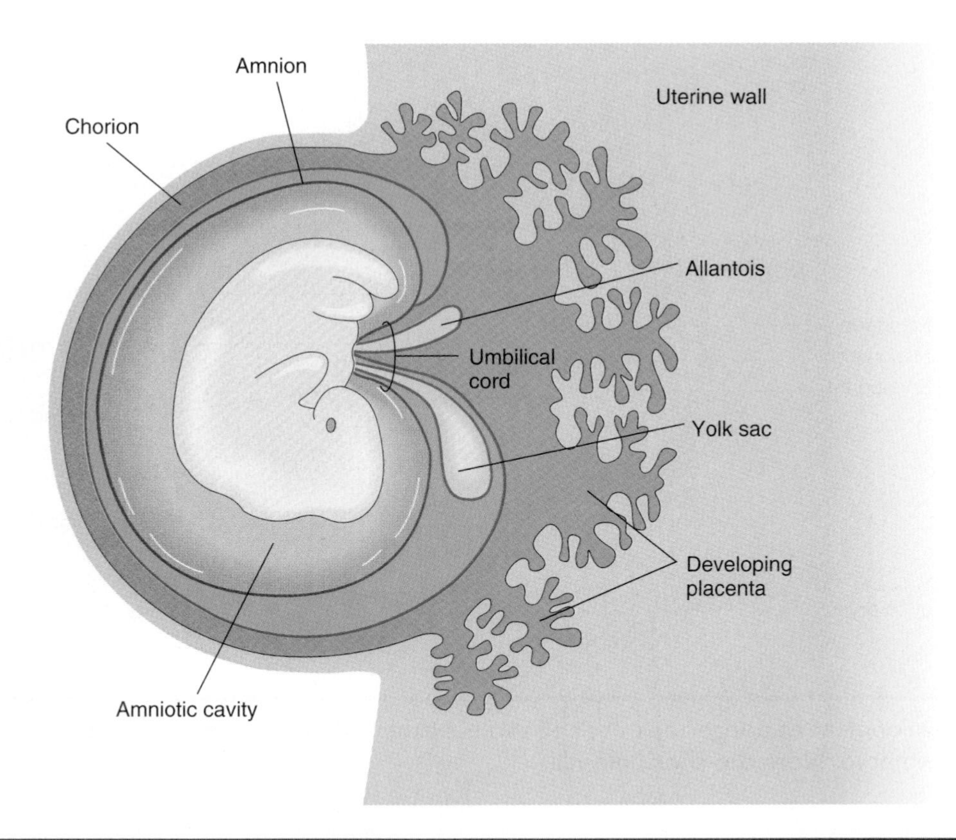

Figure 18.6 Extraembryonic membranes. Chorionic villi continue to develop where the placenta will form, the amnion envelops the developing embryo, and the yolk sac and allantois are incorporated into the umbilical cord.

bring embryonic blood to the placenta, and a single umbilical vein returns the blood to the embryo. There are no nerves in the umbilical cord.

The placenta provides an interface between embryonic and maternal bloods for the exchange of water, respiratory gases, nutrients, wastes, hormones, and antibodies. The embryonic blood must receive all required substances from the mother's blood, and it must pass metabolic wastes into the mother's blood. The placenta is usually fully functional by the end of the eighth week.

The embryonic and maternal bloods do not mix in the placenta. The maternal blood vessels open into blood-filled spaces called *lacunae* into which the embryonic blood vessels extend and branch repeatedly. This arrangement provides a large surface area for the exchange of substances between embryonic and maternal bloods. Observe this relationship in figure 18.7.

What are the germ layers, and what happens to them?
What are the extraembryonic membranes, and what are their roles?
What composes the placenta, and what is its function?

Many substances that are harmful to the embryo, such as alcohol, cocaine, heroin, nicotine, caffeine, and many therapeutic drugs, can pass across the placenta into the embryonic blood. Physicians recommend that no drugs be taken during pregnancy except those that are crucial for the mother's health.

In addition, pathogens of certain maternal infections can also pass across the placenta and infect the fetus. These infections include AIDS, German measles, syphilis, and toxoplasmosis (transmitted from cat feces). All of these diseases may cause serious fetal disorders.

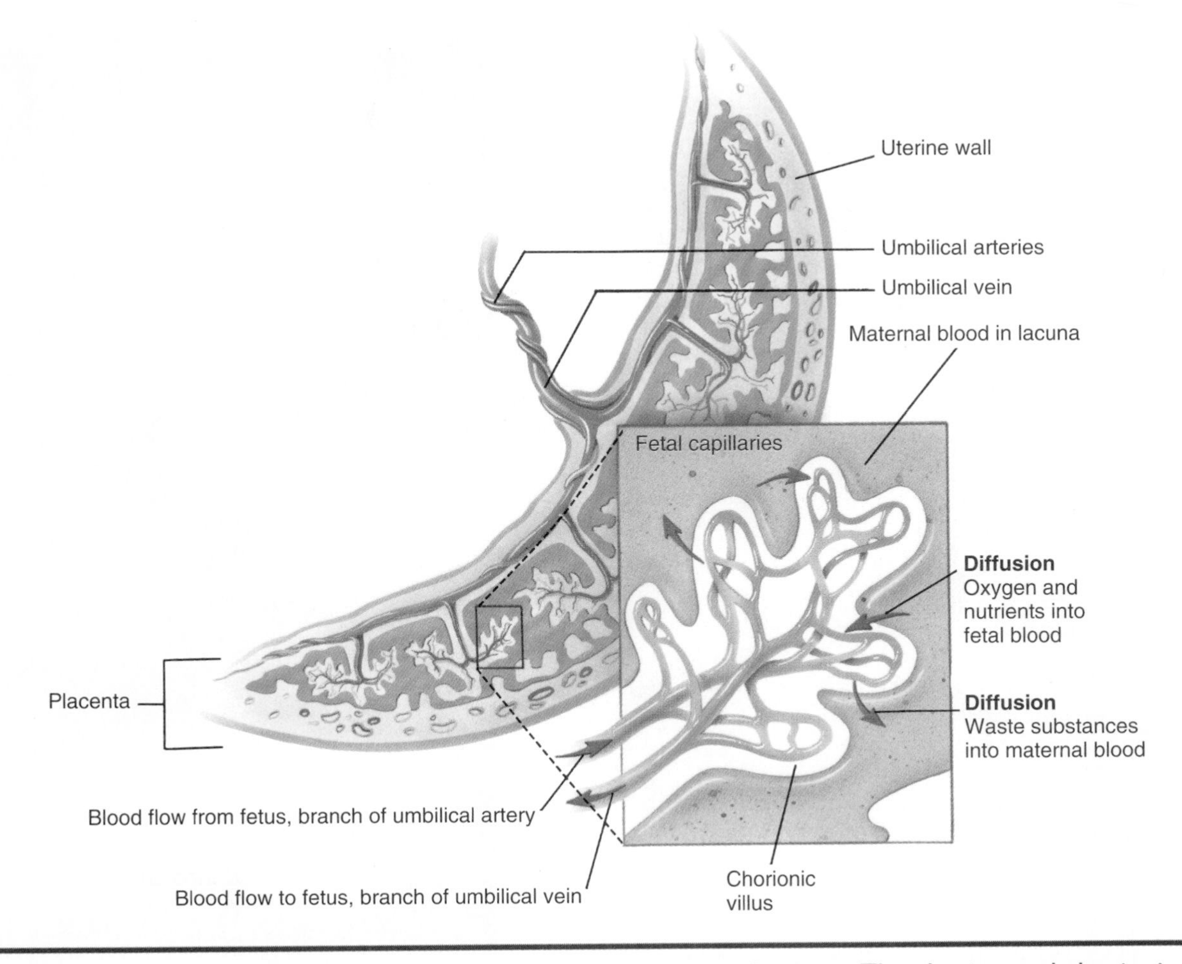

Figure 18.7 The placenta is formed of both embryonic and maternal tissues. The chorion and chorionic villi form the embryonic portion, and the endometrium of the uterine wall forms the maternal portion.

External Appearance

By the fourth week, the head and limb buds of the embryo are recognizable. By the seventh week, the rudiments of body organs are present, and the eyes and ears are visible. Figure 18.8 illustrates the visible changes in the embryo from the fourth to the seventh week.

> Some pregnant women develop hemorrhoids or varicose veins in the latter stages of pregnancy. These conditions result from inadequate venous return in veins serving the legs and anal canal due to pressure of the fetus on blood vessels in the pelvic area.

Fetal Development

When the embryo has reached the end of the eighth week of development, it has a distinctly human appearance. At this time it is called a **fetus** (figure 18.9). Table 18.2 highlights the changes in physical appearance that take place during fetal development.

Birth

During the latter stages of pregnancy, the concentration of estrogen becomes increasingly greater than that of progesterone, as noted earlier. While progesterone inhibits uterine contractions, estrogen promotes them. Therefore, there is an increasing tendency toward the onset of uterine contractions as the pregnancy approaches full term. Late in the pregnancy, the hormone **relaxin** is secreted by the placenta. It helps to relax the ligaments of the symphysis

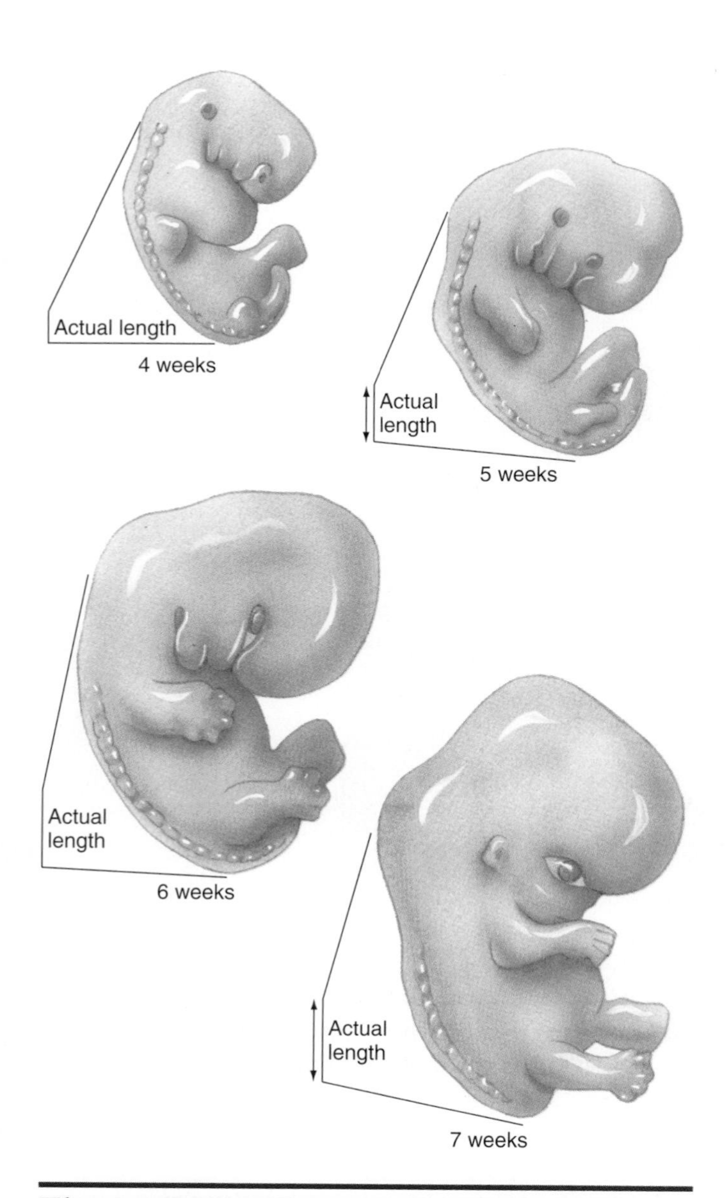

Figure 18.8 External appearance of the embryo from the fourth to the seventh week of development.

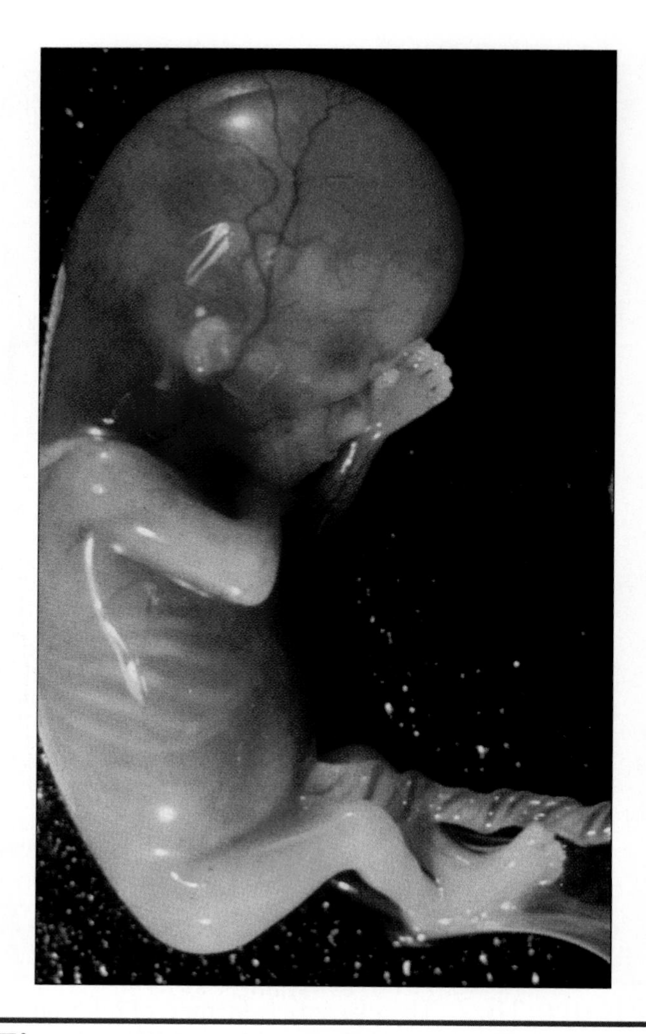

Figure 18.9 Photograph of an embryo at eight weeks, when it becomes recognizable as human.

pubis and sacroiliac joints and softens the cervix as the time of birth approaches.

Birth usually occurs within two weeks of the calculated due date, which is 280 days from the last menses. The birth process is called **parturition** (par-tū-rish′-un), and the events associated with parturition are collectively called **labor.** The onset of labor involves both hormonal and neural aspects. As the fetus reaches full term, the high estrogen level overrides progesterone's inhibition of uterine contractions, allowing uterine contractions to occur. Pressure of the fetus on the cervix stretches the cervix, stimulating neural activity that results in uterine contractions. In fact, physicians sometimes initiate labor by breaking the amnion so that increased pressure is placed on the cervix. Usually, the fetus is positioned headfirst, as shown in figure 18.10.

The **first stage of labor** is the dilation of the cervix. The stretching of the cervix triggers the formation of nerve impulses that are sent to the hypothalamus. When these reach a critical frequency, the hypothalamus activates the posterior pituitary gland to secrete *oxytocin,* increasing its concentration in the blood. Oxytocin stimulates the characteristic rhythmic contractions that begin at the upper end of the uterus and move towards the cervix, pushing the fetus towards the vagina, or **birth canal.**

The continued stretching of the cervix by uterine contractions forcing the baby's head against the cervix sends more impulses to the hypothalamus, resulting in a greater release of oxytocin from the posterior pituitary, which produces stronger and more frequent uterine contractions. This *positive feedback mechanism* produces increasingly stronger contractions until birth occurs.

Table 18.2 Changes During Fetal Development

Month	Changes
2	Recognizable human shape; head as large as the body; eyes far apart; arms and legs present with digits present; body systems present in rudimentary form
3	Head nearly as large as the body; body is growing in length; eyes nearly fully developed but eyelids still fused; brain enlarging; arms and legs well formed; nails forming on digits; ossification beginning; organ systems continuing to develop
4	Head still dominant but body getting longer; facial features well developed; eyes and ears reach characteristic positions; bones distinct; heartbeat detectable by stethoscope
5	Head still dominant; fine hair (lanugo) covers body; hair appears on head; sebaceous gland secretions (vernix caseosa) cover body; eyelashes and eyebrows develop; limb movements detectable by mother; fetus assumes fetal position due to limited space
6	Head less dominant as body grows; substantial weight gain; limbs attain final proportions
7	Body is lean; head and body are proportionate; skin is wrinkled and pinkish; eyelids separate; body systems continue development
8	Skin smoother due to subcutaneous fat deposition; skull bones ossifying but cranial bones are soft; testes descend into scrotum; sex determination possible by ultrasound; distal limb bones ossifying
9	More subcutaneous fat deposited; skin smoother and whitish pink; skull bones continue to ossify
10	More subcutaneous fat deposited; nails extend to end of toes and fingers; skull bones largely ossified but cranial bones are not firmly joined; lanugo has been shed; vernix caseosa still present; skin pinkish even in fetuses of dark-skinned parents since melanin is not produced until skin is exposed to light

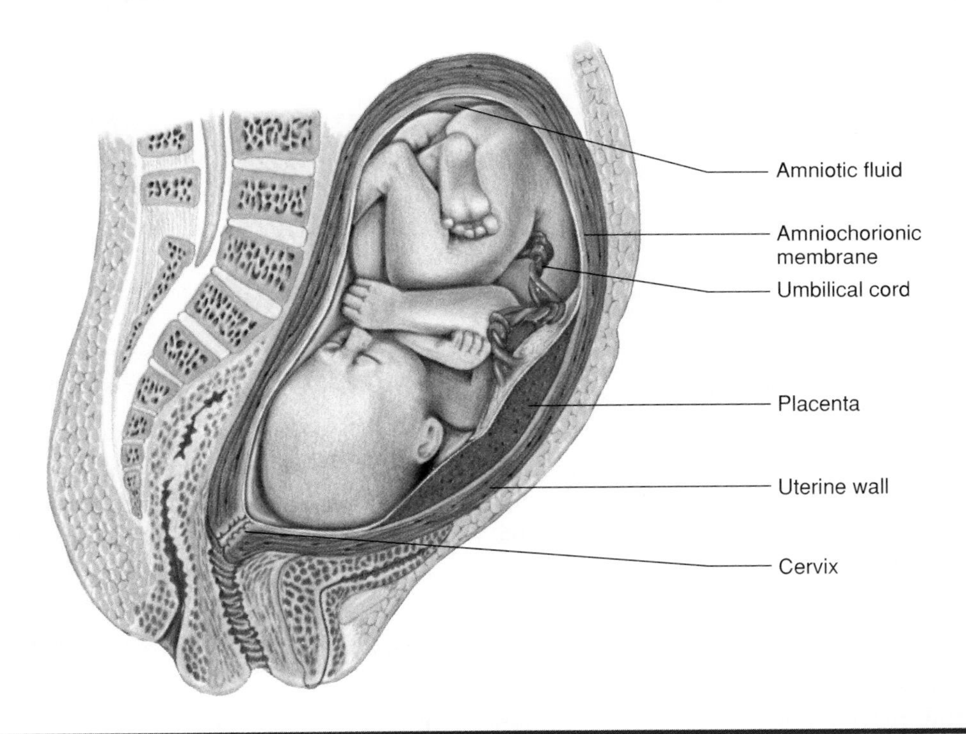

Figure 18.10 A full-term fetus positioned with the head against the cervix.

Dilation of the cervix is the longest stage of labor. It may last from 6 to 12 hours, depending on the size of the baby and whether the mother has had other children. During this time, the amnion ruptures, and the cervix dilates to the size of the baby's head (figure 18.11).

The **second stage of labor** is the delivery (expulsion) of the infant. It usually lasts less than an hour, with contractions occurring every two to three minutes and lasting about one minute. Once the head is expelled, the rest of the body exits rather quickly.

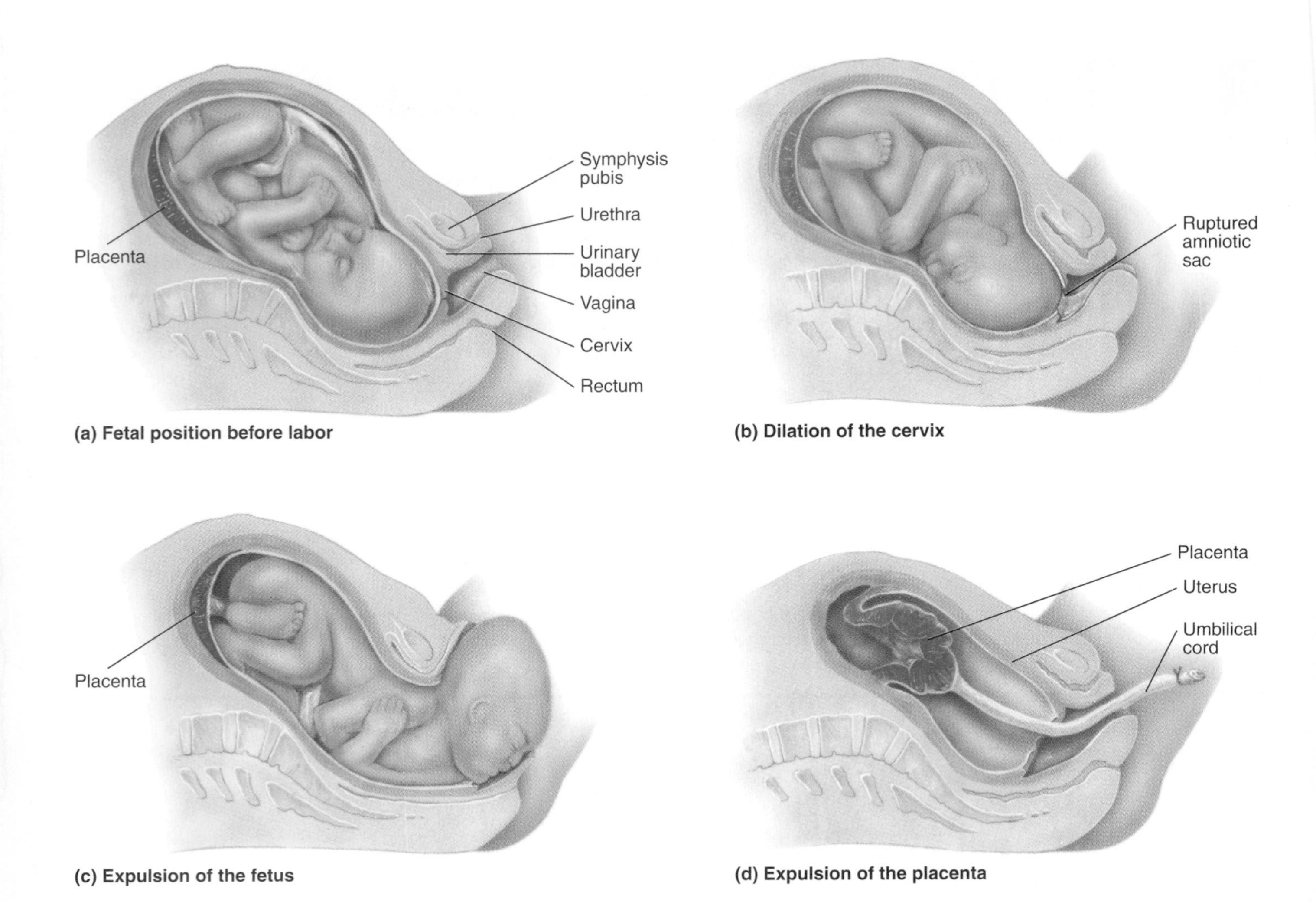

Figure 18.11 Stages of labor.

The **third stage of labor** is the delivery of the placenta. Within 15 minutes after birth of the baby, the placenta detaches. Continued contractions expel the placenta (the afterbirth). The placenta is checked carefully to see that all of it has been removed from the uterus because any residue may cause a serious uterine infection. Detachment of the placenta produces some bleeding because endometrial blood vessels at the placental site are ruptured. However, uterine contractions compress the broken blood vessels so that serious bleeding is usually avoided. Subsequently, the uterus decreases in size rather quickly.

How does an increased estrogen level contribute to the onset of labor?
What is the neural-hormonal mechanism that increases uterine contractions?

About 5% of births are *breech births* in which the fetus is presented buttocks first. This complicates the delivery and may require delivery by *cesarean section*. In a cesarean section, a low, transverse incision is made through the walls of the abdomen and uterus, through which the baby is extracted.

First Breath

Immediately after birth, the infant's nose and mouth are aspirated to remove mucus or fluid that would impair breathing. The umbilical cord is clamped and cut, separating the infant from the placenta, which has served as its prenatal respiratory organ. As carbon dioxide increases in the infant's blood, the respiratory center in the medulla is activated, and it stimulates the infant's first inspiration.

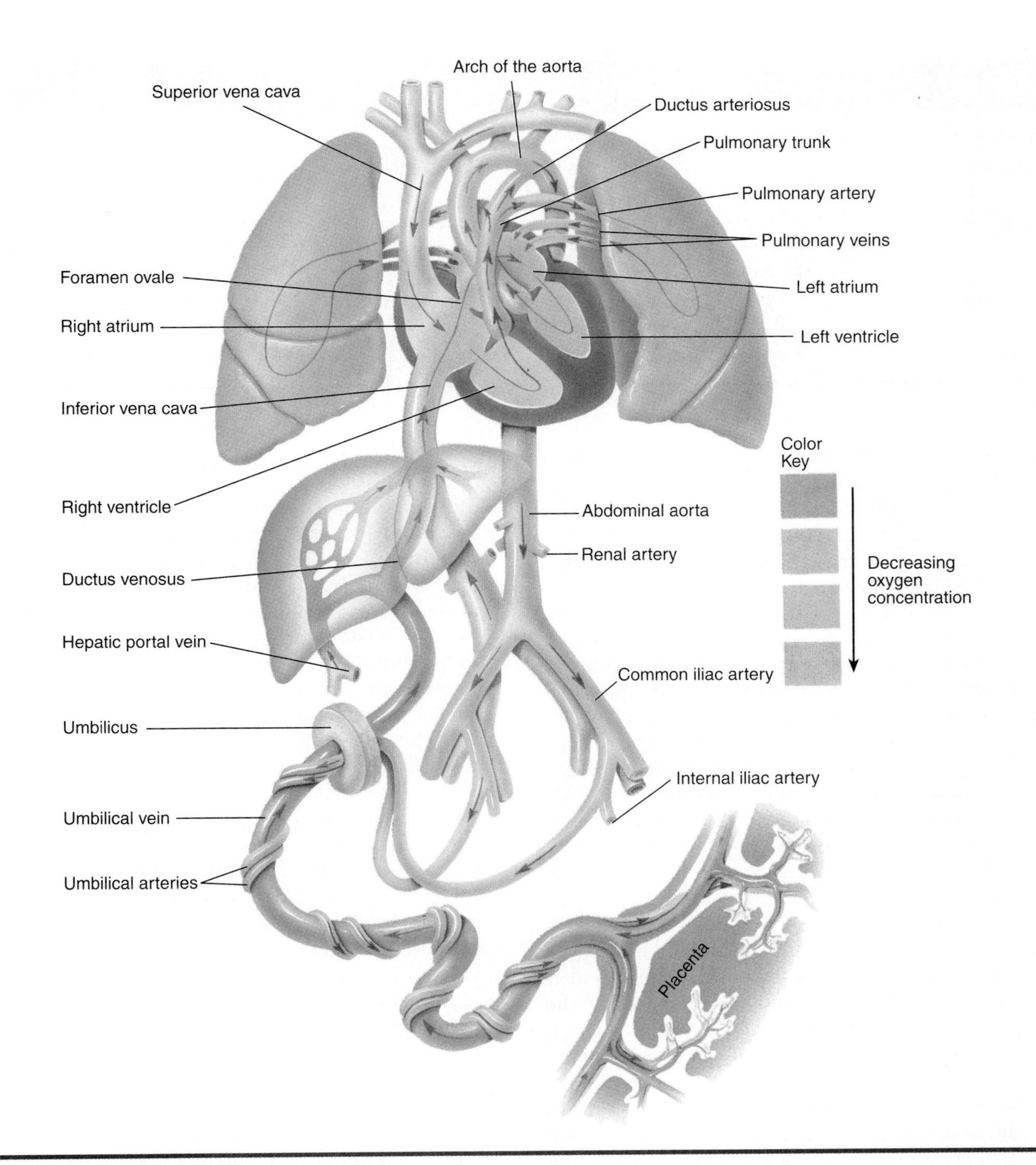

Figure 18.12 The pattern of circulation in a fetus.

The first breath is difficult because the lungs are collapsed. In a full-term fetus, surfactant in alveolar fluids reduces surface tension, making the first breath and subsequent breathing easier.

Circulatory Adaptations

Fetal circulation is quite different than adult circulation because the digestive tract, lungs, and kidneys are not functioning. Oxygen and nutrients are obtained from the maternal blood in the placenta, and carbon dioxide and other metabolic wastes are removed via the maternal blood. The pattern of fetal circulation is an adaptation to these conditions. Birth immediately separates the baby from its supply of nutrients and oxygen and stimulates circulatory changes to accommodate independent living as an air-breathing human.

Fetal Circulation

Figure 18.12 illustrates the pattern of fetal circulation. Oxygenated and nutrient-rich blood is carried from

Table 18.3 Fetal Circulatory Adaptations

Umbilical vein	Carries oxygen-rich and nutrient-rich blood from the placenta to the fetus
Ductus venosus	Carries about half of the blood in the umbilical vein into the inferior vena cava, bypassing the liver and mixing oxygenated and deoxygenated blood
Foramen ovale	Allows a large portion of the blood entering the right atrium to pass through the atrial septum directly into the left atrium, bypassing the pulmonary circuit and providing as much oxygen and nutrients as possible for body cells via the systemic circuit
Ductus arteriosus	Carries most blood from the pulmonary trunk directly into the aorta, bypassing the nonfunctional lungs and providing more blood with available oxygen and nutrients for the systemic circuit
Umbilical arteries	Carry blood from the internal iliac arteries back to the placenta

the placenta to the fetus by the **umbilical vein**, which enters the fetus at the umbilicus (navel). Within the fetus, the umbilical vein passes toward the liver, where it divides into two branches. About half of the blood enters the liver, and the other half passes through the **ductus venosus** (duk′-tus ven-ō′-sus) to enter the inferior vena cava. The oxygenated blood from the umbilical vein is mixed with deoxygenated blood in the inferior vena cava. The addition of blood from the ductus venosus increases the blood pressure within the inferior vena cava and the right atrium, which keeps the foramen ovale open.

Most of the blood entering the right atrium of the fetal heart passes directly through the **foramen ovale** (ō-vah′-lē), an opening, into the left atrium. The blood in the left atrium flows into the left ventricle, is pumped into the aorta, and is carried throughout the body. The blood entering the right ventricle is pumped through the pulmonary trunk, but most of it bypasses the lungs by flowing through the **ductus arteriosus** (duk′-tus ar-te-rē-ō′-sus) into the aortic arch. This bypass provides additional blood for transport to the body. However, sufficient blood flows through the pulmonary circuit to maintain the nonfunctional lungs.

Blood is returned to the placenta by two **umbilical arteries** that branch from the internal iliac arteries. Trace the flow of blood through the fetal cardiovascular system shown in figure 18.12. Table 18.3 summarizes these fetal circulatory adaptations.

Fetal hemoglobin can carry up to 30% more oxygen than can adult hemoglobin. Since fetal arterial blood is almost always a mixture of oxygenated and deoxygenated bloods, this adaptation ensures adequate oxygen for the fetal cells.

Postnatal Circulatory Changes

After the baby is breathing, changes are made to convert the pattern of fetal circulation into that of an air-breathing infant. This involves closure of the foramen ovale and the constriction of all of the vessels used to get blood quickly from the umbilical vein into the aorta.

The distal portions of umbilical arteries constrict, inhibiting the flow of blood to the placenta. Subsequently, they become *umbilical ligaments*. Blood continues to flow through the umbilical vein from the placenta to the newborn for about one minute, and then it constricts. It will become the *ligamentum teres*. At the same time, the ductus venosus constricts. It will subsequently become the *ligamentum venosum* in the wall of the liver.

As the umbilical vein constricts and the pulmonary circulation becomes functional, blood pressure in the right atrium decreases, while blood pressure in the left atrium increases due to increased blood flow to and from the lungs. The higher blood pressure in the left atrium closes a tissue flap in the left atrium over the foramen ovale, separating the pulmonary and systemic circuits. About the same time, the ductus arteriosus constricts and ultimately becomes the *ligamentum arteriosum*. These circulatory changes are functionally complete within 30 minutes after birth, but it takes about one year for tissue growth to make them permanent (figure 18.13).

What are the circulatory adaptations in a fetus? What circulatory changes occur in a newborn infant?

Lactation

High levels of estrogen and progesterone during pregnancy stimulate the development of the mammary glands and enlargement of the breasts in preparation

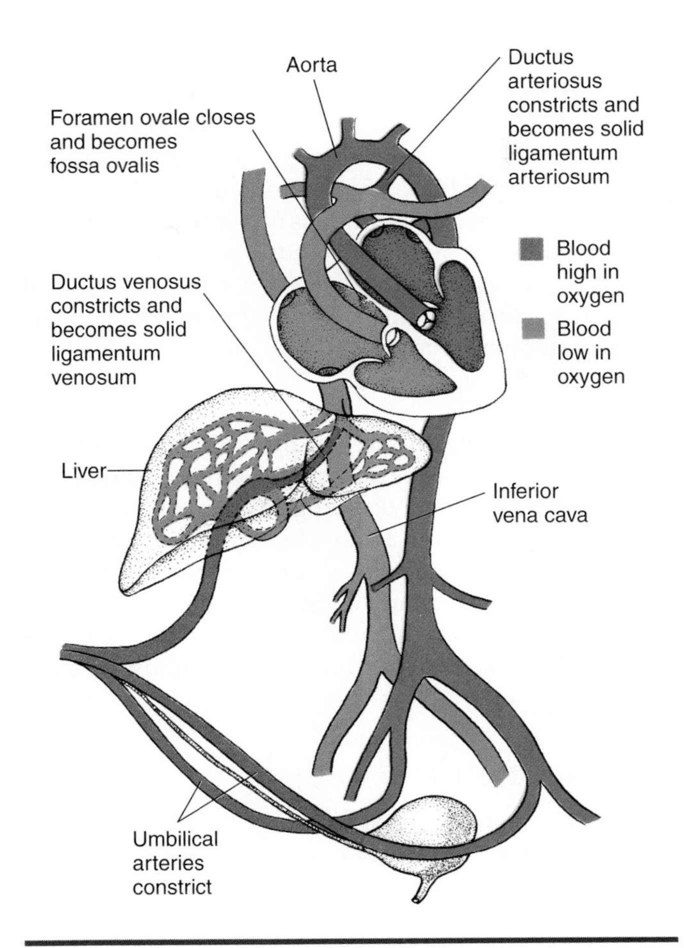

Figure 18.13 Circulatory changes in a newborn infant.

for milk production, or **lactation** (lak-tā′-shun). Although the mammary glands are capable of secreting milk, the high levels of estrogen and progesterone inhibit the hypothalamus so that milk production does not occur during pregnancy.

After birth, the levels of estrogen and progesterone drop dramatically, removing the inhibitory effect and enabling the hypothalamus to secrete **prolactin-releasing hormone (PRH).** PRH activates the anterior pituitary to secrete **prolactin,** which stimulates milk production.

Prolactin stimulates the mammary glands to secrete milk, but its effects are not evident for two to three days. In the meantime, the mammary glands produce **colostrum** (kō-los′-trum), which differs from true milk by containing higher concentrations of protein and essentially no fat. The high protein content provides an added boost of essential nutrients for protein synthesis, which is needed for the continued development and growth of the baby.

The continued secretion of prolactin and milk production are maintained by mechanical stimulation of the nipples by the suckling infant. Suckling forms

Breast-feeding seems to provide advantages for the baby, including (1) better nutrition because nutrients are easier to absorb, (2) rapid bonding due to prolonged contact with the mother, (3) antibodies that prevent digestive inflammation, and (4) enhanced cognitive development.

neural impulses that stimulate the hypothalamus to secrete PRH, which, in turn, causes continued prolactin secretion by the anterior pituitary. Thus, lactation is maintained by a *positive feedback system.*

Milk does not simply flow from the breasts. Instead, it is ejected after about 30 seconds of suckling by the infant. Stimulation of the nipple by suckling sends impulses to the hypothalamus, which triggers the release of oxytocin by the posterior pituitary. *Oxytocin* stimulates contraction of smooth muscles in the lactiferous ducts, resulting in milk ejection, or "let-down."

Milk production may continue for as long as the infant is suckled, but the volume of milk produced gradually declines. If suckling is stopped, milk accumulates, the secretion of prolactin is inhibited, and milk production ceases in about a week.

How do hormones control lactation?

The release of oxytocin by breast-feeding also results in stimulating the contraction of the uterus. This speeds up the return of the uterus to near its nonpregnant size.

Disorders of Pregnancy and Prenatal Development

These disorders are grouped as pregnancy disorders of the mother and prenatal disorders of the fetus.

Pregnancy Disorders

Eclampsia (ē-klamp′-sē-ah), or **toxemia of pregnancy,** is a disorder that occurs in two forms. *Preeclampsia* of late pregnancy is characterized by increased blood pressure, edema, and proteinuria. The cause is unknown. If unsuccessfully treated, it may develop into *eclampsia,* a far more serious disorder that may lead to convulsions and coma. Both infant and maternal mortality are high in eclampsia. Rapid termination of the pregnancy may be indicated.

An **ectopic pregnancy** is the implantation of an embryo anywhere other than in the uterus. A common site is in a uterine tube. Treatment involves surgical removal of the embryo.

A **miscarriage** is a spontaneous abortion. Most miscarriages occur early in pregnancy as a result of gross abnormalities of the embryo or placenta. Another cause is the untimely transfer of estrogen and progesterone production from the corpus luteum to the placenta during the second and third month of pregnancy.

Morning sickness is characterized by nausea and vomiting upon getting up in the morning. It lasts from one to six weeks, and the cause is unknown. About 60% of pregnant women experience this discomfort.

Prenatal Disorders

Birth defects may be inherited or may be caused by a variety of *teratogens* (ter-ah′-to-jens), substances or influences that produce physical abnormalities. Teratogens include alcohol, illegal drugs, some therapeutic drugs, X rays, and certain diseases such as German measles (Rubella). Generally, the earlier the embryo is exposed, the greater is the defect produced. Alcohol is the most common teratogen. It produces *fetal alcohol syndrome,* which is characterized by a small head, mental retardation, facial deformities, and abnormalities of the heart, genitals, and limbs.

Physiological jaundice sometimes occurs in a newborn infant due to the destruction of fetal red blood cells faster than the liver can process the bilirubin. This results in excess bilirubin in the blood. Phototherapy is a common treatment to speed up bilirubin breakdown. Jaundice may be a more serious problem in premature babies.

Infant respiratory distress syndrome (IRDS), or infant crib death, is the major cause of death in newborn infants. It occurs more often in premature babies. One cause of IRDS is insufficient surfactant production by the alveoli. Breathing is labored, and gas exchange is inadequate.

Inheritance

Inheritance is the passing of inherited traits from one generation to the next. The determiners of hereditary traits are located on **chromosomes.** A chromosome consists of DNA and protein, and it is the DNA that controls inheritance and directs cellular functions.

Each human body cell contains 46 chromosomes, 23 pairs. Chromosome pairs 1 through 22 are called *autosomes* because they control most inherited traits except gender. Gender is determined by chromosome pair 23, the *sex chromosomes.* There are two types of sex chromosomes, a large X chromosome and a small Y chromosome. The Y chromosome determines maleness, so males possess one X chromosome and one Y chromosome (XY). Females possess two X chromosomes (XX).

A person's chromosomes, including the sex chromosomes, may be examined by making a *karyotype.* The chromosomes in a dividing cell are photographed during metaphase, and the photograph is enlarged. Then the chromosomes are cut out, matched in pairs and arranged by size and location of the centromere. Figure 18.14 is a karyotype of a normal male. Note the X and Y chromosomes and that the chromosomes are arranged in pairs.

Sex Determination

Recall that gametes are formed by meiotic cell division, a process that places one member of each chromosome pair in each gamete. Each human gamete contains 23 chromosomes—22 autosomes and 1 sex chromosome. We will consider only the sex chromosomes here.

Since a female has two X chromosomes in her cells, all of her gametes contain an X chromosome. A male has both an X chromosome and a Y chromosome in his cells. Therefore, half of his gametes are X-bearing, and half are Y-bearing. If an egg is fertilized by an X-bearing sperm, the child will be a girl. If an egg is fertilized by a Y-bearing sperm, the child will be a boy. Obviously, the probability of any zygote being a girl (or a boy) is one-half or 50%. Figure 18.15 illustrates the determination of sex.

Genes

A specific part of a chromosome's DNA molecule that codes for the synthesis of a specific protein is called a **gene.** A gene is the unit of inheritance. Chemically, a gene is a specific sequence of nucleotides, so genes occur in a linear sequence in a chromosome. The DNA of a chromosome may contain hundreds of genes.

Since chromosomes occur in pairs, genes also occur in pairs. An inherited trait is determined by at least one pair of genes. There may be two or more alternate forms of a gene controlling the expression of a particular trait. These alternate forms are called **alleles**

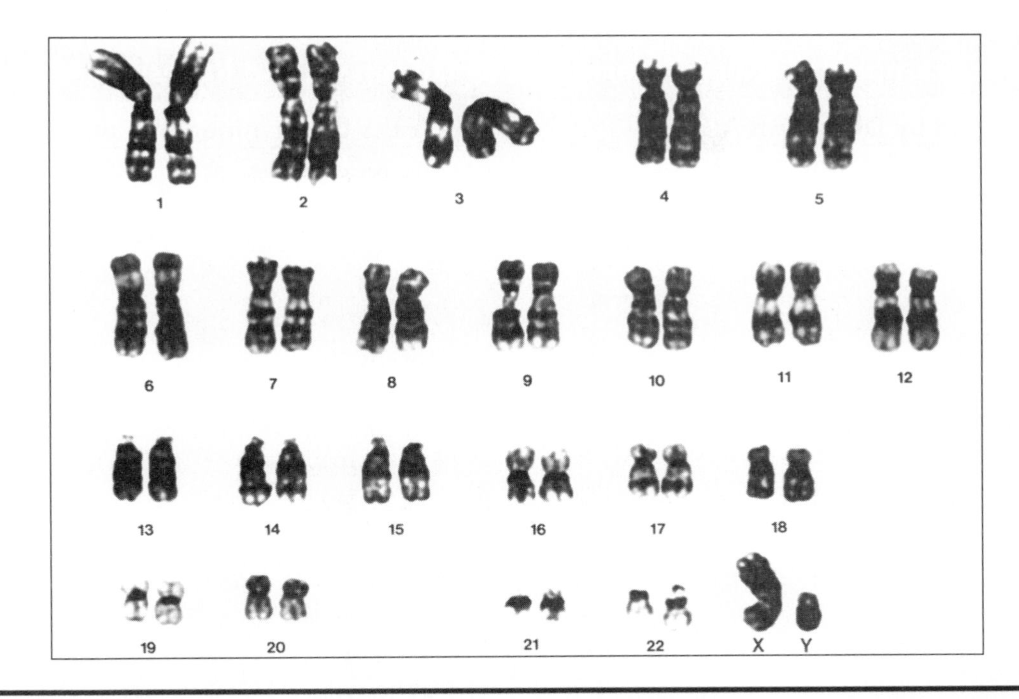

Figure 18.14 Karyotype of a normal human male. The only difference in a female karyotype would be the presence of a pair of X chromosomes in place of the XY pair of the male.

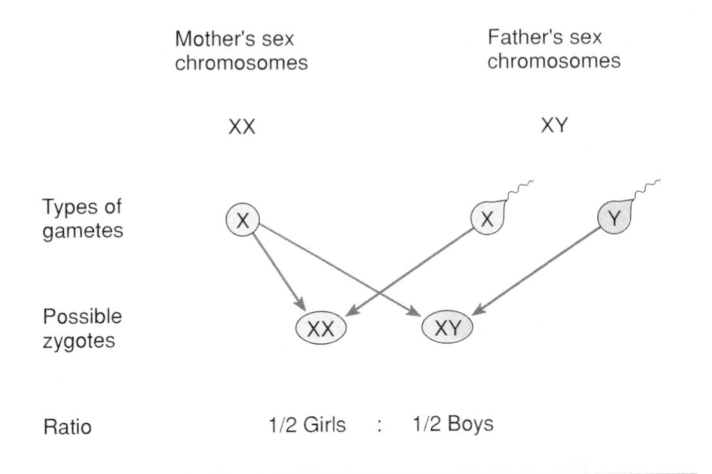

Figure 18.15 The inheritance of sex.

(ah-lēls′), and each allele affects the expression of a trait differently. So, in the simplest case, a trait is determined by one pair of alleles present in a person's cells. If the two alleles for a trait are identical, the person is *homozygous* for that trait; if they are different, the person is *heterozygous* for that trait.

Gene Expression

Each person's chromosomes contain a unique catalog of genes, the **genotype** for that person. The expression of those genes yields the observable characteristics known as the **phenotype.** What we see is the phenotype, but it is the genotype that is involved in the transmission of traits.

Dominant and Recessive Inheritance

Some alleles are dominant, and some are recessive. A *dominant allele* is always expressed, while a *recessive allele* is expressed only when both recessive alleles are present.

Consider the example of skin pigmentation. Normal skin pigmentation is controlled by a dominant allele (*A*). The absence of pigment (albinism) is controlled by a recessive allele (*a*). The possible genotypes and phenotypes are shown below. Note that it requires only one dominant allele to express the dominant trait, but that both recessive alleles must be present for the recessive trait to be expressed.

Genotype	Phenotype
AA	Normal
Aa	Normal
aa	Albino

Table 18.4 indicates a few human traits that are determined by dominant and recessive alleles.

Codominance

In some traits, both alleles are equally expressed, so they both affect the phenotype. An example is sickle-cell anemia, a condition characterized by defective

Table 18.4 Examples of Traits Determined by Dominant and Recessive Alleles

Traits Determined by Dominant Alleles	Traits Determined by Recessive Alleles
Freckles	Absence of freckles
Dimples in cheeks	Absence of dimples
Dark hair	Light hair or red hair
Full lips	Thin lips
Free earlobes	Attached ear lobes
Normal	Thalassemia
Feet with arches	Flat feet
Huntington's disease	Normal
Astigmatism	No astigmatism
Farsightedness	Not farsighted
Panic attacks	Normal
Extra fingers or toes	Normal number of digits
Normal	Cystic fibrosis
Normal	Hemophilia*
A or B blood types	O blood type
Rh+ blood type	Rh^- blood type
Color vision	Red-green color blindness*
Normal	Gout*

*Sex-linked trait

hemoglobin that cannot carry adequate oxygen. Erythrocytes with the defective hemoglobin assume a characteristic sickled or crescent shape. Sickle-cell anemia occurs among people whose ancestors lived in central Africa. About 8% of black Americans possess the allele for sickle-cell anemia.

A person who inherits both alleles for sickle-cell anemia (H^SH^S) produces abnormal hemoglobin, leading to the formation of sickled cells that cannot carry sufficient oxygen, and the sickled cells tend to plug capillaries. Symptoms include pain in joints and the abdomen and chronic kidney disease.

In the heterozygous state (HH^S), some hemoglobin molecules are normal, but others are abnormal. Fortunately, few red blood cells become sickled under normal conditions, and clinical symptoms are absent at such times. However, more red blood cells become sickled during times of oxygen deficit, a characteristic that allows detection of carriers of the sickle-cell allele. The heterozygote state affords some protective advantage against the pathogen causing malaria.

Multiple Alleles

Some traits are determined by genes that have more than two alleles. An example is the ABO blood groups. There are three alleles involved: I^A causes the production of the A antigen; I^B causes the production of the B antigen; and i^o produces neither antigen. Both I^A and I^B are dominant over the recessive i^o. If both I^A and I^B are present, both are expressed. The possible genotypes and phenotypes are:

Genotypes	Phenotypes
I^AI^A; I^Ai^o	Type A blood
I^BI^B; I^Bi^o	Type B blood
I^AI^B	Type AB blood
i^oi^o	Type O blood

Polygenes

To this point, we have considered either-or traits, but some traits do not fit this mode of inheritance. For example, people are not either tall or short. Instead, the human population exhibits a gradation of heights that fits a standard bell curve. Such traits are controlled by *polygenes,* a number of different genes that may be located on different chromosomes. Examples of such traits are height, intelligence, and skin pigment concentration. It is difficult to predict the inheritance of polygenic traits.

Let's consider height. Four pairs of genes seem to be involved. The two alleles are *T* and *t*. Each *T* allele exerts a small increase in height, while each *t* allele exerts no increase in height. If all eight alleles are *T* alleles, the person will be very tall; if all eight alleles are *t* alleles, the person will be very short. Various combinations of the two alleles produce a variety of intermediate heights.

Predicting Inheritance

Parents often wonder about the chances of their child developing certain inherited traits. This can be predicted for some traits where the inheritance pattern has been determined and if the genotypes of the parents are known. Such predictions indicate the *probability*, rather than absolute certainty, that a trait will be inherited.

Let's consider freckles. Freckles are determined by a dominant allele (*F*), while normal pigment distribution is determined by a recessive allele (*f*). The possible genotypes and phenotypes are:

Genotypes	Phenotypes
FF	Freckled
Ff	Freckled
ff	Normal

Figure 18.16 shows how to determine the probability of the freckled or nonfreckled trait in the next generation if the genotypes of the parents are known. In this example, the parents are known to be heterozygous for freckles. What is the probability that their children will be freckled?

Since each parent is heterozygous, meiotic division during gamete formation causes half of the gametes of each parent to contain an allele for freckles (*F*), and half to carry an allele for normal pigmentation (*f*). Since the union of sperm and egg occurs at random (by chance), we must allow for all possible combinations of gametes. This is accomplished by using a *Punnett square* (a chart named after Reginald Punnett, a geneticist).

The alleles in eggs are placed along the horizontal axis, and the alleles in sperm are placed along the vertical axis. Then, the allele of each egg is written in the squares below each egg, and the allele of each sperm is written in the squares to the right of each sperm. Each square now contains a possible genotype of an individual in the next generation. The Punnett square now shows all possible genotypes that may occur in the next generation.

From this information, the predicted genotype ratio may be determined. Then, knowing that the trait for freckles is dominant and that the presence of a single dominant allele (*F*) produces freckles, the predicted phenotype ratio may be determined. Note in figure 18.16 that it is possible for two heterozygous freckled parents to have a child with normal pigmentation. However, if one parent is homozygous for freckles and the other is heterozygous for freckles, all children would be freckled.

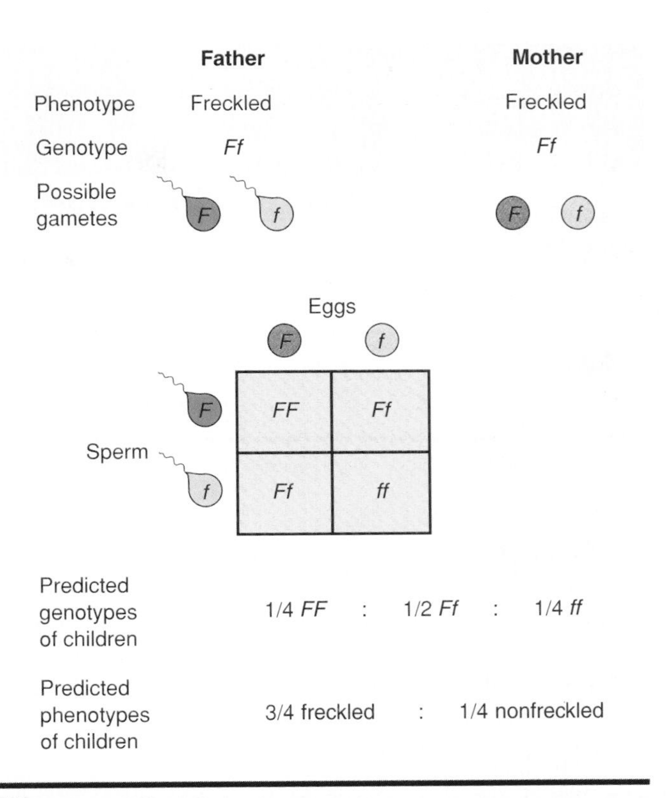

Figure 18.16 The probability of freckles being inherited by children of parents who are heterozygous for freckles.

The inheritance of any dominant/recessive trait may be determined in a similar manner.

X-Linked Traits

A few traits are determined by genes on the X chromosome. These are **X-linked,** or **sex-linked,** traits. They affect males much more frequently than females. Recessive X-linked traits are expressed in males because males possess only one X chromosome, so alleles on the X chromosome are always expressed. Since females have two X chromosomes, they usually possess a dominant normal allele, which prevents expression of a recessive allele.

Red-green color blindness is a common X-linked recessive trait. A color-blind male inherits the allele for color blindness from his mother, who provides his X chromosome. The mother may have either normal vision or be red-green color blind (table 18.5).

What are the relationships among chromosomes, DNA, genes, and alleles?
What distinguishes the dominant/recessive pattern of inheritance?
Why are recessive X-linked traits expressed more often in males than in females?

Table 18.5 Possible Genotypes and Phenotypes for Red-Green Color Blindness, an X-Linked Trait

	Genotype	Phenotype
Females	X^CX^C	Normal
	X^CX^c	Normal carrier
	X^cX^c	Color blind
Males	X^CY	Normal
	X^cY	Color blind

C = Allele for normal vision; c = Allele for color blindness.

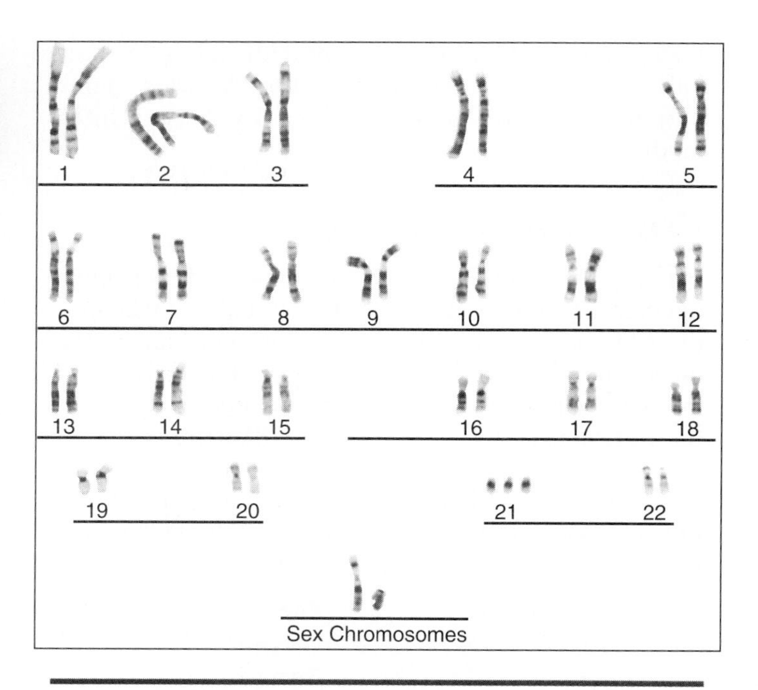

Figure 18.17 A karyotype of a male with Down syndrome caused by trisomy 21.

Inherited Diseases

Inherited, or genetic, diseases are caused by either chromosome abnormalities or specific alleles. The development of advanced techniques and new knowledge makes an understanding of genetic disease increasingly important.

Chromosome Abnormalities

Some genetic diseases are related to the presence of an additional chromosome or to the absence of a chromosome. These disorders result from errors that occur during meiotic cell division, causing some gametes to receive both members of a chromosome pair while other gametes receive neither member. If such gametes are involved in zygote formation, a genetic disorder occurs. The genetic damage usually is so severe that it causes a spontaneous abortion. In some cases, the effect is not lethal but disabling.

An example of a disabling genetic disorder is Down syndrome, one of the more common genetic disorders. It is caused by the presence of an extra chromosome 21, as shown in figure 18.17. Down syndrome is characterized by mental retardation, short stature, short digits, slanted eyes, and a protruding tongue (figure 18.18). Babies with Down syndrome are born more often to mothers over 40 years of age.

Single Gene Disorders

These disorders usually affect the baby's metabolism after birth, when it must depend on its own life processes, or they may appear later in life. Consider a few examples of single gene disorders.

Cystic fibrosis, a recessive disorder, is the most common genetic disorder among Caucasians. It is caused by a missing chloride channel on mucus-secreting cells. This causes production of thick mucus that blocks respiratory airways, and it leads to an early death from respiratory infections.

Phenylketonuria (PKU), a recessive disorder, is due to a missing enzyme needed to metabolize phenylalanine (an amino acid). Without treatment, mental and physical retardation results. A special diet that limits phenylalanine can prevent these effects if it is started at birth and continued to adulthood.

Tay-Sachs disease, a recessive disorder, primarily affects Jews of central European ancestry. An enzyme needed to metabolize a fatty substance associated with neurons is missing. The result is mental retardation, muscle weakness, seizures, and finally death, usually by two years of age.

Huntington's disease, a dominant disorder, results from one or more missing enzymes needed in cellular respiration. This causes a buildup of lactic acid in neurons in the brain. Uncontrollable muscle contractions, memory loss, and personality changes begin between 30 and 50 years of age. Death occurs within 15 years after the appearance of symptoms.

Hemophilia A and *hemophilia B,* X-linked recessive disorders, result from missing clotting factors. Prolonged bleeding can be life threatening, and joints may be painfully disabled. Patients are dependent upon frequent transfusions of normal plasma or intravenous injections of the missing clotting factor.

Genetic Counseling

Prospective parents who have genetic disorders in one or both of their families may benefit from genetic counseling. By collecting genetic information, a genetic counselor can inform prospective parents of the prob-

Figure 18.18 A girl with Down syndrome. Persons with Down syndrome have warm personalities and enjoy art and music.

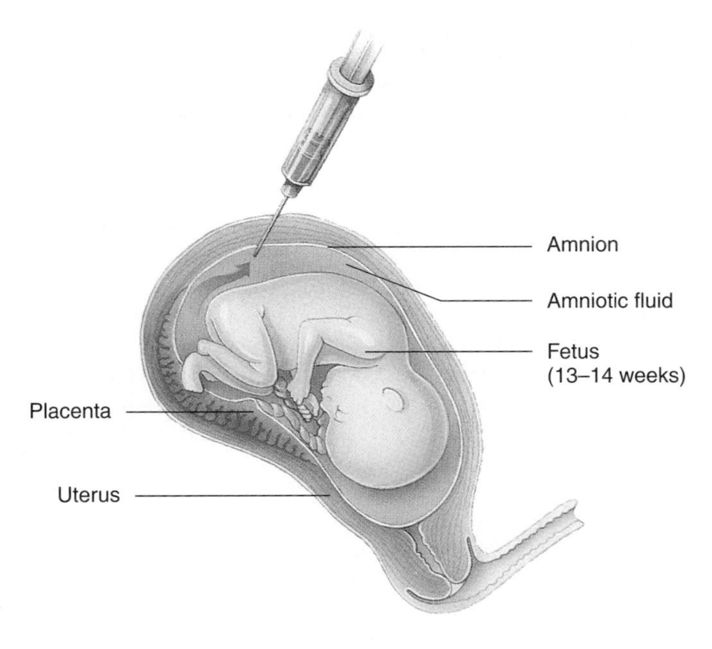

Figure 18.19 In amniocentesis, a sample of amniotic fluid is withdrawn, and the suspended fetal cells are examined for genetic abnormalities.

ability of a genetic disorder appearing in their children. Genetic information may be collected from family histories and blood tests of the prospective parents and family members. If the woman is pregnant, ultrasound may be used to detect gross fetal abnormalities, and fetal cells may be obtained for examination. Fetal cells are obtained in two ways: amniocentesis and chorionic villi sampling.

In *amniocentesis,* a hollow needle is inserted through the abdominal and uterine walls of the mother and into the amnion to draw out a sample of amniotic fluid (figure 18.19). Fetal cells in the sample are grown by tissue culture and are analyzed to see if there are chromosomal abnormalities. Also, the amniotic fluid is analyzed for the presence of specific proteins that indicate serious neural defects. Amniocentesis is not done earlier than the fourteenth week of pregnancy, when ample amniotic fluid is available for sampling without injury to the fetus.

In *chorionic villi sampling,* a narrow tube is inserted through the cervix and fetal tissue from chorionic villi is suctioned out. Chromosomes in the fetal cells are examined for abnormalities. Chorionic villi sampling may be performed at 10 weeks, and chromosome examination can be done immediately.

Both amniocentesis and chorionic villi sampling have inherent risks for mother and fetus. Fetal risks seem to be greater in chorionic villi sampling.

CHAPTER SUMMARY

Fertilization and Early Development

1. Sperm usually encounter the secondary oocyte in the upper third of a uterine tube. Only one sperm can enter an oocyte.
2. When a secondary oocyte is penetrated by a sperm, it undergoes the second meiotic division, forming an ovum and a second polar body. Fertilization occurs with the fusion of the sperm and ovum nuclei, forming a zygote.
3. The zygote undergoes cleavage divisions, forming a morula. Continued mitotic divisions form a blastocyst containing an inner cell mass.
4. About the seventh day, the blastocyst becomes implanted in the endometrium.

Hormonal Control of Pregnancy

1. The trophoblast and chorion secrete HCG, taking over the role of maintaining the corpus luteum whose progesterone maintains the uterine lining.
2. In the second or third month, the placenta takes over the production of progesterone and estrogen, and the corpus luteum disintegrates. Ovaries remain inactive for the remainder of the pregnancy.

Embryonic Development

1. The embryonic disk separates into the three germ layers: ectoderm, mesoderm, and endoderm.
2. The trophoblast becomes the chorion, the first and outermost extraembryonic membrane. The chorionic villi will become part of the placenta.
3. The amnion develops dorsal to the embryo, envelops the embryo, and becomes filled with amniotic fluid that serves as a shock absorber.
4. The yolk sac develops ventral to the embryo. It subsequently branches to form the allantois. Both of these extraembryonic membranes form blood cells prior to their production by the liver and bone marrow. The allantois brings fetal blood vessels to the placenta.
5. The placenta is formed of both embryonic and maternal tissues, and it is functional by the eighth week. It allows the exchange of materials between embryonic and maternal bloods.
6. By the seventh week, the embryo exhibits a head, body, limbs with digits, eyes, and ears.

Fetal Development

1. The embryo is called a fetus after the eighth week, and it clearly resembles a human. Organ systems are in rudimentary form, and the cardiovascular system is functioning.
2. Fetal development includes development of the organ systems to functional levels, an increase in size and weight, ossification of bones, and development of hair.

Birth

1. Placental relaxin softens the cervix and relaxes the symphysis pubis and sacroiliac joints in the late stage of pregnancy.
2. The high level of estrogen in the late stage of pregnancy counteracts the inhibitory action of progesterone against uterine contractions.
3. The onset of labor involves nerve impulses and oxytocin. Pressure of the fetus on the cervix forms nerve impulses that are sent to the hypothalamus, which stimulates oxytocin secretion by the posterior pituitary. Oxytocin stimulates uterine contractions, dilating the cervix and forming more nerve impulses. These interactions set up a positive feedback control that strengthens contractions until birth.
4. Labor involves three stages: (a) dilation of the cervix, (b) delivery of the infant, and (c) delivery of the placenta.
5. After birth, an accumulation of carbon dioxide stimulates the respiratory center to trigger the first breath. Surfactant in alveolar fluids makes breathing easier.

Circulatory Adaptations

1. The pattern of fetal circulation is adapted to life in which the lungs and digestive system are nonfunctional and nutrients and oxygen are derived from the mother's blood via the placenta. Embryonic blood is carried to the placenta by umbilical arteries and is returned by an umbilical vein.
2. Circulatory adaptations pass oxygen-rich and nutrient-rich blood as quickly as possible from the umbilical vein to the aorta to meet the needs of body cells. These adaptations include the ductus venosus, foramen ovale, and ductus arteriosus.
3. When the baby starts breathing, fetal circulatory adaptations are eliminated to enable an efficient separation of the pulmonary and systemic circuits.

Lactation

1. The high levels of estrogen and progesterone during pregnancy prepare the mammary glands for lactation but prevent hormone stimulation of lactation until birth.
2. After birth, the low levels of progesterone and estrogen allow the hypothalamus to secrete PRH, which stimulates the anterior pituitary to secrete prolactin. Prolactin stimulates the mammary glands to produce milk.
3. Colostrum is secreted first, followed by true milk after two to three days.
4. Suckling stimulates the formation of impulses that are sent to the hypothalamus, which (a) secretes PRH, ensuring the continued release of prolactin and milk production, and (b) stimulates the posterior pituitary to secrete oxytocin, which produces ejection of the milk.

Disorders of Pregnancy and Prenatal Development

1. Pregnancy disorders include eclampsia, ectopic pregnancy, miscarriage, and morning sickness.
2. Prenatal disorders include birth defects, physiological jaundice, and respiratory distress syndrome.

Inheritance

1. The determiners (genes) of hereditary traits are located on chromosomes. There are 46 chromosomes in human body cells: 22 pairs of autosomes and 1 pair of sex chromosomes.
2. Sex chromosomes for females are XX; for males, XY.
3. A gene is the unit of heredity. It is a portion of a DNA molecule that codes for a specific protein. Alternate forms of a gene are called alleles. A person with identical alleles for a trait is homozygous for that trait. If the alleles are different, the person is heterozygous.
4. The genotype is all of the genes determining a trait. The phenotype is the observable characteristics expressed for a trait.
5. Many genes have only two alleles whose expression is either by dominant/recessive inheritance or by incomplete dominance. Some traits are determined

by multiple alleles, and others are determined by polygenes.
6. The probability of transmitting traits to the next generation may be predicted using a Punnett square.
7. X-linked traits are determined by genes located on the X chromosome.

Inherited Diseases

1. Genetic diseases may result from either chromosomal abnormalities or defective alleles.
2. Prospective parents who have genetic diseases in their family histories may benefit from genetic counseling.

BUILDING YOUR VOCABULARY

1. **Selected New Terms**
   allantois, p. 378
   amnion, p. 378
   chorion, p. 378
   codominance, p. 389
   ductus arteriosus, p. 386
   ductus venosus, p. 386
   ectoderm, p. 386
   endoderm, p. 386
   foramen ovale, p. 386
   human chorionic gonadotropin (HCG), p. 377
   lactation, p. 387
   mesoderm, p. 386
   oxytocin, p. 382
   preembryo, p. 376
   prolactin, p. 387
   prolactin-releasing hormone (PRH), p. 387
   relaxin, p. 381
   trophoblast, p. 376
   umbilical arteries, p. 386
   umbilical vein, p. 386
   X-linked traits, p. 391
   yolk sac, p. 378
2. **Related Clinical Terms**
   monozygotic (mon-o-zi-got′-ik) One zygote.
   obstetrics (ob-stet′-riks) Medical specialty dealing with pregnancy and childbirth.
   placenta abruptio (ah-brup′-she-o) Premature separation of the placenta from the uterus.
   placenta previa (pre′-ve-ah) Placenta located over or adjacent to the opening into the uterus.
   postpartum (post-par′-tum) Occurring after birth.
   teratogen (te-rah′-to-jen) A substance or condition causing birth defects.

STUDY ACTIVITIES

1. Complete the **Chapter 18 Study Guide,** which begins on page 495.
2. Complete the **Learning Objectives** listed on the first page of this chapter.

CHECK YOUR UNDERSTANDING

(Answers are located in appendix B.)

1. Usually a sperm meets a secondary oocyte in the _____ .
2. Fusion of sperm and ovum nuclei is called _____ .
3. A _____ is implanted in the endometrium on about the _____ day of development.
4. Secretion of _____ by the trophoblast and chorion maintain the secretion of estrogen and _____ , which maintains the endometrium.
5. The embryonic disk forms the three _____ that subsequently form all other tissues of the embryo and fetus.
6. The _____ contains fluid in which the embryo develops, and the chorionic villi become the embryonic part of the _____ .
7. After the _____ week, the embryo is called a _____ , and it clearly resembles a human.
8. The first and longest stage of _____ is the dilation of the cervix, which results from uterine contractions stimulated by the hormone _____ .
9. In a fetus, the _____ enables oxygenated blood to pass from the right atrium into the left atrium, and the _____ passes blood from the pulmonary trunk into the aorta.
10. Lactation begins after _____ from the hypothalamus stimulates secretion of _____ by the anterior pituitary gland.
11. The units of inheritance are _____ , which are small segments of _____ that make up chromosomes.
12. Humans possess _____ chromosomes in their cells, and a person possessing two X chromosomes is a _____ .
13. X-linked recessive traits occur more often in _____ .
14. In dominant/recessive inheritance, a recessive allele is expressed only when the person is _____ for the recessive trait.
15. It is possible to predict the _____ of a trait appearing in children if the genotypes of parents are known.

1. Anatomy and Physiology

Write the terms that match the phrases in the spaces at the right.

1) The study of cells. ______
2) The study of body organization and structure. ______
3) The study of body functions. ______

2. Levels of Organization

a. List the levels of organization from the most complex to the simplest.

1) ______ 4) ______
2) ______ 5) ______
3) ______ 6) ______

b. Write the terms that match the phrases in the spaces at the right.

1) A coordinated group of organs. ______
2) Structural and functional units of the body. ______
3) An aggregation of similar cells. ______

c. Match the names of the organ systems with the phrases.

Cardiovascular	Lymphatic	Reproductive, male
Digestive	Muscle	Respiratory
Endocrine	Nervous	Skeletal
Integumentary	Reproductive, female	Urinary

1) Stomach, liver, intestines. digestive
2) Brain, spinal cord, nerves. ______
3) Provides chemical coordination. respiratory
4) Skin, hair, nails. ______
5) Returns lymph to blood; provides immunity. ______
6) Bones, ligaments, cartilages. ______
7) Contraction enables movement. muscle
8) Transports materials to and from cells. ______
9) Kidneys, ureters, urinary bladder. ______
10) Testes, penis, prostate gland. male reproductive
11) Ovaries, oviducts, uterus, vagina. ______
12) Blood, heart, arteries, veins. Respiratory
13) Supports the body. Skeletal
14) Secretes hormones that regulate functions. ______
15) Regulates volume of body fluids. ______
16) Protects underlying tissues. ______
17) Rapid coordination of body functions. ______
18) Digests food and absorbs nutrients. ______

19) Gas exchange between air and blood. ______________

20) Larynx, trachea, bronchi, and lungs. ______________

3. Directional Terms

Provide the term that correctly completes each statement.

1) The head is _____ to the neck. superior
2) The hand is _____ to the wrist. distal
3) The skin is _____ to the muscles. external
4) The mouth is _____ to the nose. inferior
5) The elbow is _____ to the wrist. proximal
6) The ear is on the _____ surface of the head. lateral
7) The umbilicus is on the _____ body surface. anterior
8) The hip is on the _____ body surface. lateral
9) The buttocks are on the _____ body surface. posterior

4. Body Regions

Label the body regions by placing the number of the label line in the space by the correct label.

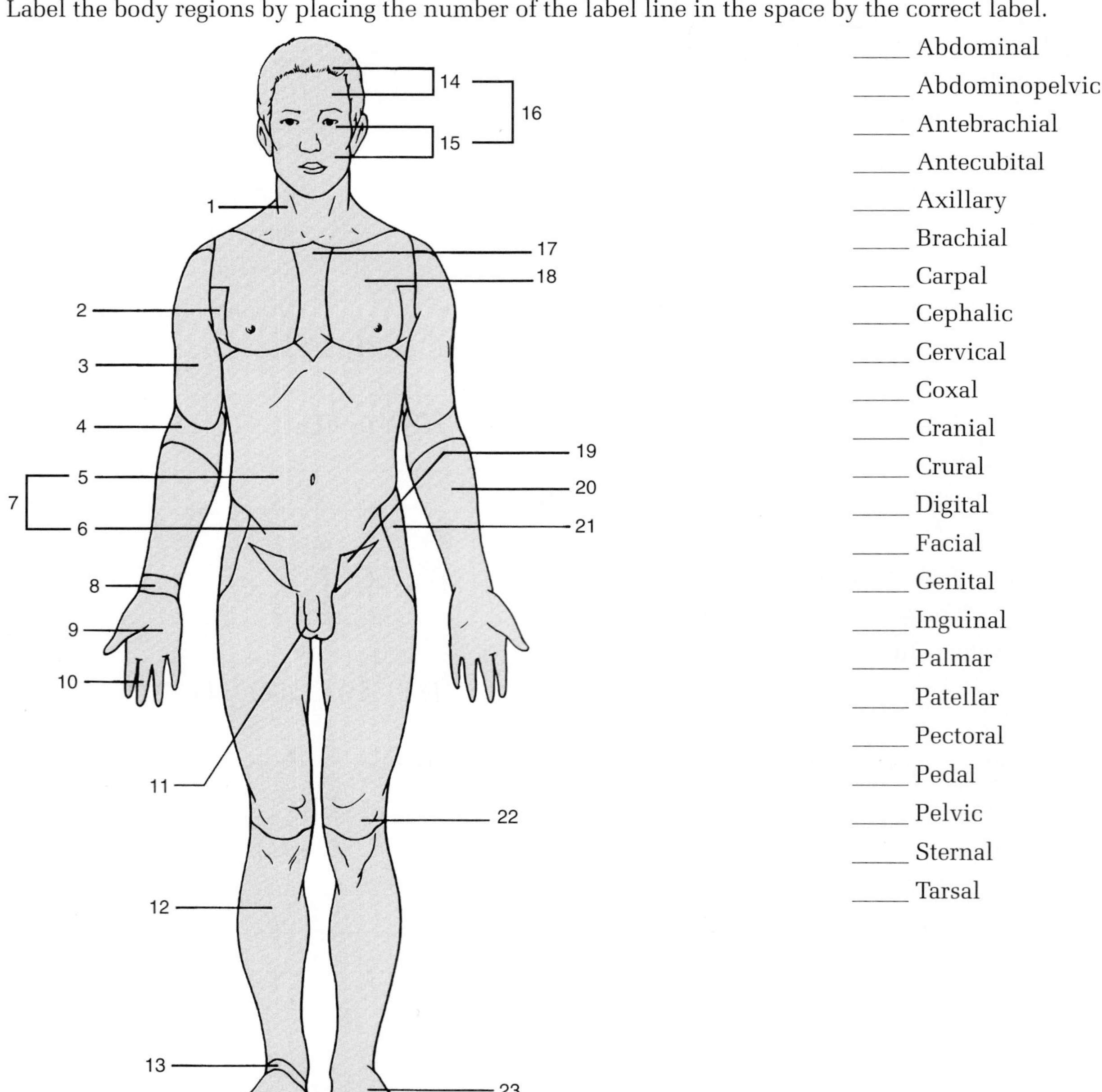

_____ Abdominal
_____ Abdominopelvic
_____ Antebrachial
_____ Antecubital
_____ Axillary
_____ Brachial
_____ Carpal
_____ Cephalic
_____ Cervical
_____ Coxal
_____ Cranial
_____ Crural
_____ Digital
_____ Facial
_____ Genital
_____ Inguinal
_____ Palmar
_____ Patellar
_____ Pectoral
_____ Pedal
_____ Pelvic
_____ Sternal
_____ Tarsal

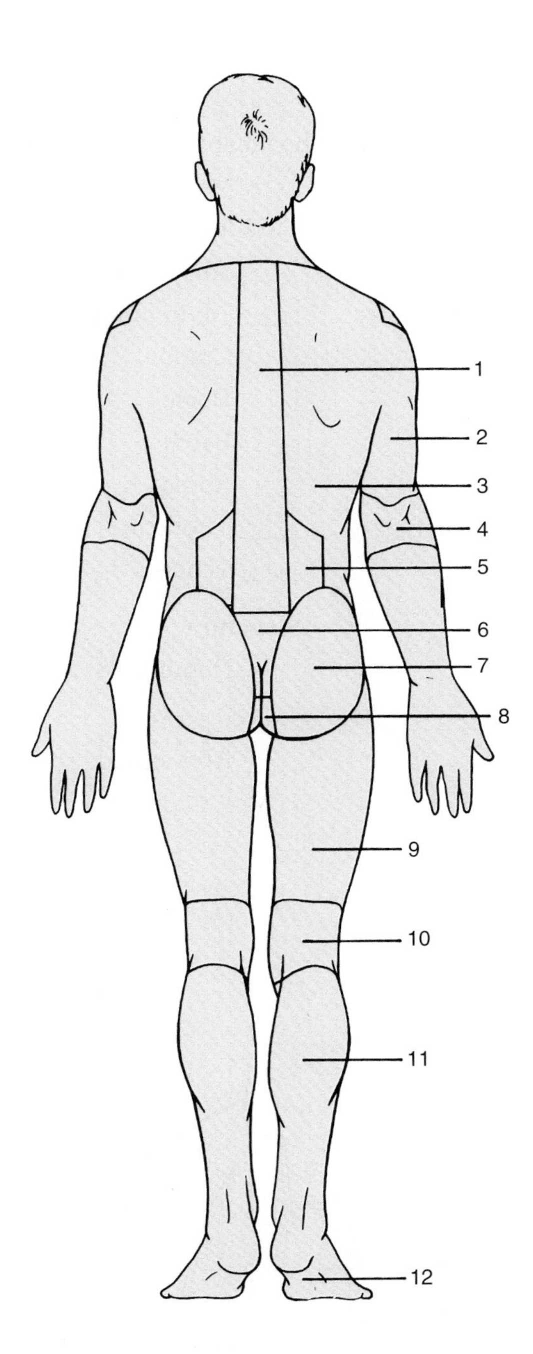

_____ Brachial
_____ Cubital
_____ Crural
_____ Dorsum
_____ Femoral
_____ Gluteal
_____ Lumbar
_____ Plantar
_____ Perineal
_____ Popliteal
_____ Sacral
_____ Vertebral

5. Body Planes and Sections

Name the planes that match the statements.

1) Divides the body into equal left and right halves. _______________

2) Divides the body into superior and inferior portions. _______________

3) Divides the body into left and right portions. _______________

4) Divides the body into anterior and posterior portions. _______________

6. Body Cavities

a. Label the body cavities and related structures by placing the number of the label line in the space by the correct label.

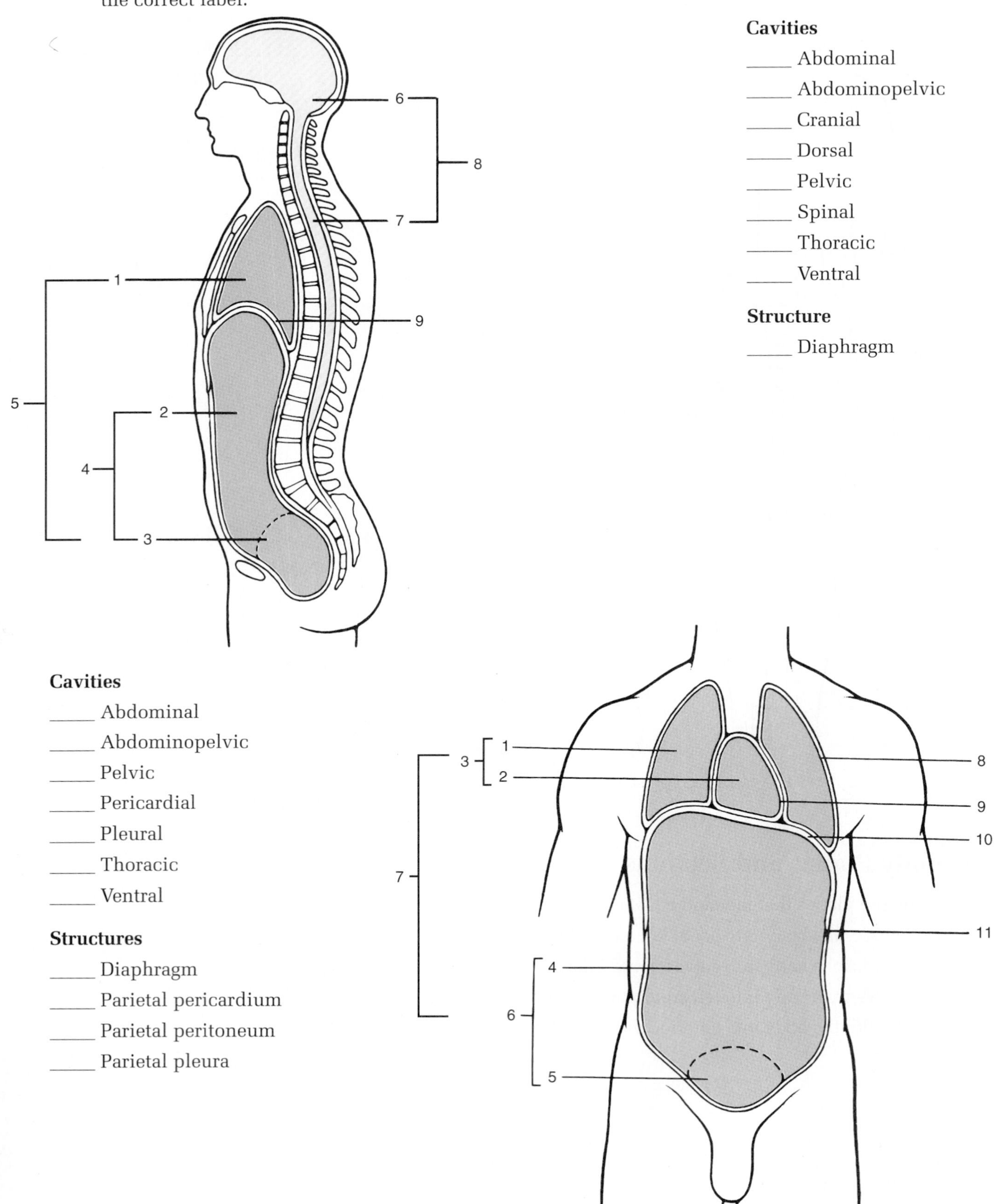

Cavities

_____ Abdominal
_____ Abdominopelvic
_____ Cranial
_____ Dorsal
_____ Pelvic
_____ Spinal
_____ Thoracic
_____ Ventral

Structure

_____ Diaphragm

Cavities

_____ Abdominal
_____ Abdominopelvic
_____ Pelvic
_____ Pericardial
_____ Pleural
_____ Thoracic
_____ Ventral

Structures

_____ Diaphragm
_____ Parietal pericardium
_____ Parietal peritoneum
_____ Parietal pleura

b. Place the number of the cavity in which the organ occurs in the space by the organ.

1) Abdominal
2) Cranial
3) Pelvic
4) Spinal
5) Thoracic, lateral parts
6) Thoracic, mediastinum

_____ Brain
_____ Gallbladder
_____ Heart
_____ Intestine, small
_____ Kidneys
_____ Liver
_____ Lungs
_____ Pancreas
_____ Spinal cord
_____ Thymus
_____ Urinary bladder
_____ Rectum

c. Write the names of the membranes that match the statements in the spaces at the right.

1) Covers the heart. _____________________
2) Covers the stomach. _____________________
3) Lines the abdominal cavity. _____________________
4) Covers the brain. _____________________
5) Lines the thoracic cavity. _____________________
6) Lines the spinal cavity. _____________________
7) Cover the lungs. _____________________
8) Forms double-membrane sac around heart. _____________________
9) Double-layered membranes supporting abdominal organs. _____________________

7. Abdominopelvic Subdivisions

Select the abdominopelvic quadrant and abdominopelvic region in which the following structures are located.

Quadrants

1. Right upper
2. Left upper
3. Right lower
4. Left lower

Regions

5. Epigastric
6. Hypogastric
7. Left hypochondriac
8. Left iliac
9. Left lumbar
10. Right hypochondriac
11. Right iliac
12. Right lumbar
13. Umbilical

_____ Gallbladder
_____ Spleen
_____ Rectum
_____ Right kidney
_____ Appendix
_____ Stomach
_____ Ascending colon
_____ Urinary bladder
_____ Left kidney
_____ Pancreas

8. Maintenance of Life

Write the terms that match the statements in the spaces at the right.

1) Maintenance of a dynamic balance of substances in body fluids. _____________________
2) The sum of all life processes. _____________________
3) Breakdown of complex substances. _____________________
4) Synthesis of complex substances. _____________________
5) Maintenance of a relatively stable internal environment. _____________________

6) Sum of the chemical reactions that occur in the body. ______________________

7) List the five basic needs essential for human life. ______________________

9. Clinical Applications

a. A patient complains of pain in the epigastric region. What organs may be involved?

____

____

b. A patient complains of pain in the right lower quadrant. What organs may be involved?

____

____

STUDY GUIDE 2

1. Atoms and Elements

a. Write the terms that match the phrases in the spaces at the right.

1) Smallest unit of an element. ____________________
2) Positively charged subatomic particle. ____________________
3) Negatively charged subatomic particle. ____________________
4) Subatomic particle with no charge. ____________________
5) Substance that cannot be broken down into any simpler substance. ____________________
6) Atoms of the same element, with different numbers of neutrons. ____________________
7) Most abundant element in the body. ____________________

b. Label the atom shown by placing the number of the component in the space by the label, then, provide the responses to the phrases below.

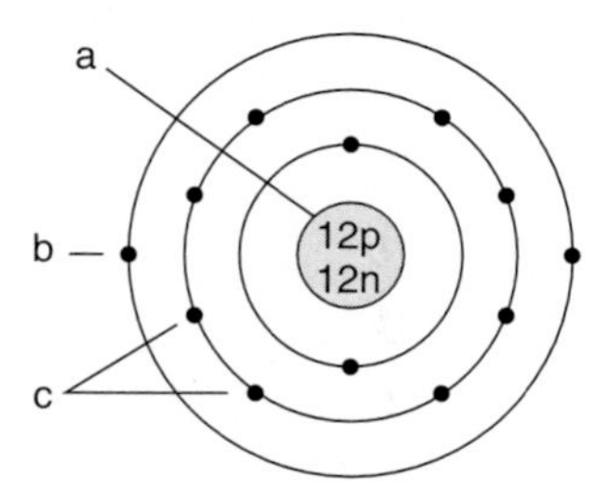

____ 1) Nonvalence electrons
____ 2) Nucleus
____ 3) Valence electron(s)

4) Atomic number of this atom. ____________________
5) Atomic weight of this atom. ____________________
6) Number of electrons needed to complete its outer shell. ____________________
7) Type of chemical bond that is likely to join this atom to another atom. ____________________
8) Symbol of this atom. ____________________

c. Diagram an atom of these elements.

Oxygen

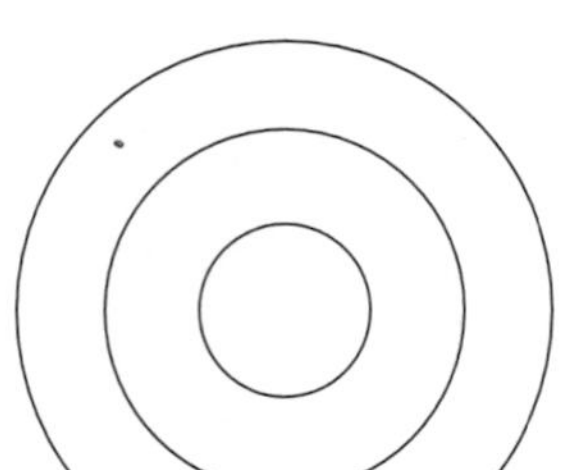

Nitrogen

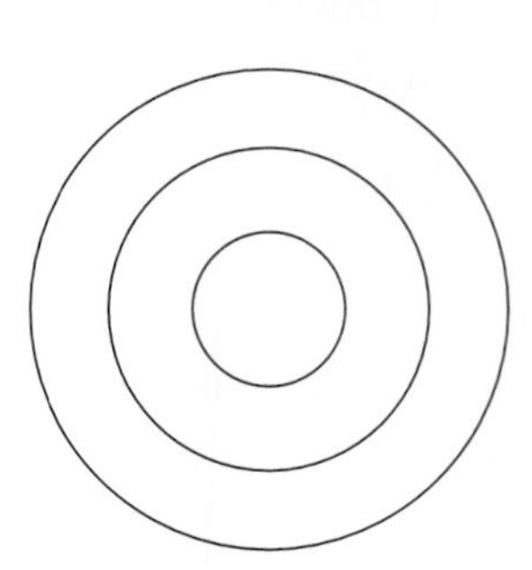

2. Molecules and Compounds

a. Write the terms that match the phrases in the spaces at the right.

1) Composed of two elements combined in a fixed ratio. ______________
2) Smallest unit of a compound. ______________
3) Number of chlorine atoms in $CaCl_2$. ______________
4) Chemical bond resulting from the donation of electron(s) from one atom to another. ______________
5) Chemical bond resulting from the sharing of valence electrons by two atoms. ______________
6) An atom with a net electrical charge. ______________
7) The attractive force between a slightly positive H atom and a slightly negative O or N atom. ______________
8) Chemical bonds forming organic molecules. ______________
9) Electrons in the outer shell. ______________

b. Indicate the kinds and numbers of atoms in a glucose molecule ($C_6H_{12}O_6$).

Kinds of Atoms	Numbers of Atoms
______________	______________
______________	______________
______________	______________

c. Identify the pH values as acid (A) or base (B). Circle the pH with the highest concentration of H^+.

_____pH 2.8 _____pH 6.8 _____pH 7.4 _____pH 9.5 _____pH 3.7

3. Compounds Composing the Human Body

a. Identify the following compounds as either organic (O) or inorganic (I).

_____ NaCl	_____ Lipids	_____ $CaPO_4$
_____ Nucleic acids	_____ Salts	_____ $C_6H_{12}O_6$
_____ Proteins	_____ Most acids	_____ CH_4
_____ Most bases	_____ Carbohydrates	_____ CO_2
_____ Amino acids	_____ Steroids	_____ Monosaccharides
_____ Fatty acids	_____ Glycerol	_____ Nucleotides

b. Write the terms that match the phrases in the spaces at the right.

1) Most abundant compound in the body. ______________
2) Substances dissolved in a liquid. ______________
3) A compound that releases H^+. ______________
4) Splitting of ionic compounds into ions. ______________
5) A measure of the H^+ concentration in a solution. ______________
6) Chemicals that keep the pH of a solution relatively constant. ______________
7) Class of compounds formed of many simple sugars joined together. ______________

8) Type of reaction that joins two glucose molecules to form maltose. ____________________

9) Storage form of carbohydrates in the body. ____________________

10) Composed of three fatty acids and one glycerol. ____________________

11) Composed of two fatty acids and a phosphate group joined to one glycerol. ____________________

12) Type of fat whose fatty acids contain no carbon–carbon double bonds. ____________________

13) Compound used to store excess energy reserves. ____________________

14) Class of lipids that includes sex hormones. ____________________

15) Class of compounds formed of 50 to thousands of amino acids. ____________________

16) Chemical bonds that determine the three-dimensional shape of proteins. ____________________

17) Bonds joining amino acids together in proteins. ____________________

18) A single-stranded nucleic acid that is involved in protein synthesis. ____________________

19) Building units of nucleic acids. ____________________

20) Steroid that tends to plug arteries when in excess. ____________________

21) Sugar in DNA molecules. ____________________

22) Primary carbohydrate fuel for cells. ____________________

23) Building units of proteins. ____________________

24) Water compartment containing 65% of water in the body. ____________________

25) Molecule releasing energy to power chemical reactions within cells. ____________________

26) Double-stranded nucleic acid. ____________________

27) Molecules catalyzing chemical reactions in cells. ____________________

28) Type of reaction breaking a large molecule into smaller molecules. ____________________

29) Molecule controlling protein synthesis in cells. ____________________

30) Element whose atoms form the backbone of organic molecules. ____________________

c. Match the four classes of organic compounds with the listed substances.

1) Carbohydrates 2) Lipids 3) Proteins 4) Nucleic acids

_____ Amino acids

_____ Steroids

_____ Glycogen

_____ Cholesterol

_____ Nucleotides

_____ Monosaccharides

_____ Triglycerides

_____ Starch

_____ Enzymes

_____ RNA

_____ DNA

_____ Fatty acids

d. Label the parts of the small portion of an RNA molecule shown and draw a line around one nucleotide.

_____ Nitrogen bases _____ Ribose sugars _____ Phosphate groups

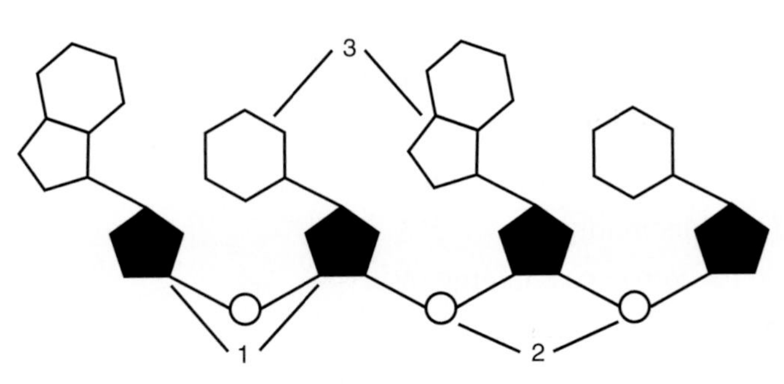

e. Show the interaction of ADP, ATP, P, and energy in the formation and breakdown of ATP by placing the numbers of the responses in the correct spaces provided.

1) ADP
2) ATP
3) Energy from cellular respiration + Ⓟ
4) Energy released for cellular work + Ⓟ

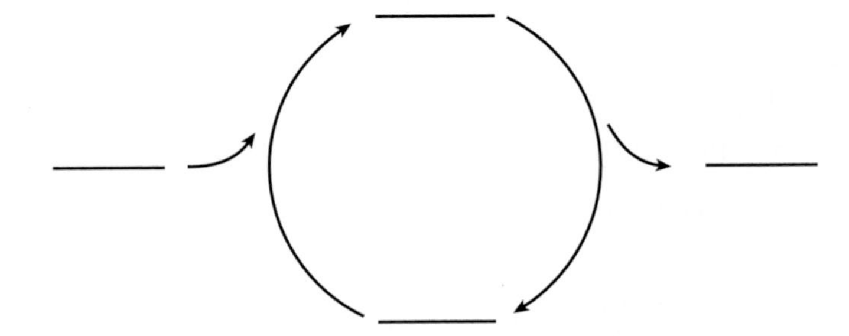

f. Explain the importance of the shape of an enzyme. ____________________

g. How does a change in pH change the shape of and inactivate an enzyme? ____________________

4. Clinical Applications

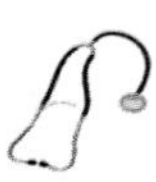

a. Why does a diet high in saturated fats increase the risk of coronary heart disease? ____________________

b. A patient in a coma is brought to the emergency room. A blood test shows that he has severe hypoglycemia (abnormally low blood glucose) and acidosis. Treatment is begun immediately to increase both blood sugar and pH.

1) Why is a normal level of blood glucose important? ____________________

2) Why is severe acidosis a problem? ____________________

STUDY GUIDE 3

1. Cell Structure

a. Label the diagram of the cell by placing the numbers of the structures by the labels listed.

- 12 Centrioles
- 4 Chromatin
- 7 Cytoplasm
- 11 Golgi complex
- 10 Microtubule
- 6 Mitochondrion
- 3 Nuclear envelope
- 2 Nucleolus
- 5 Nucleus
- 1 Plasma membrane
- 9 RER
- 8 SER

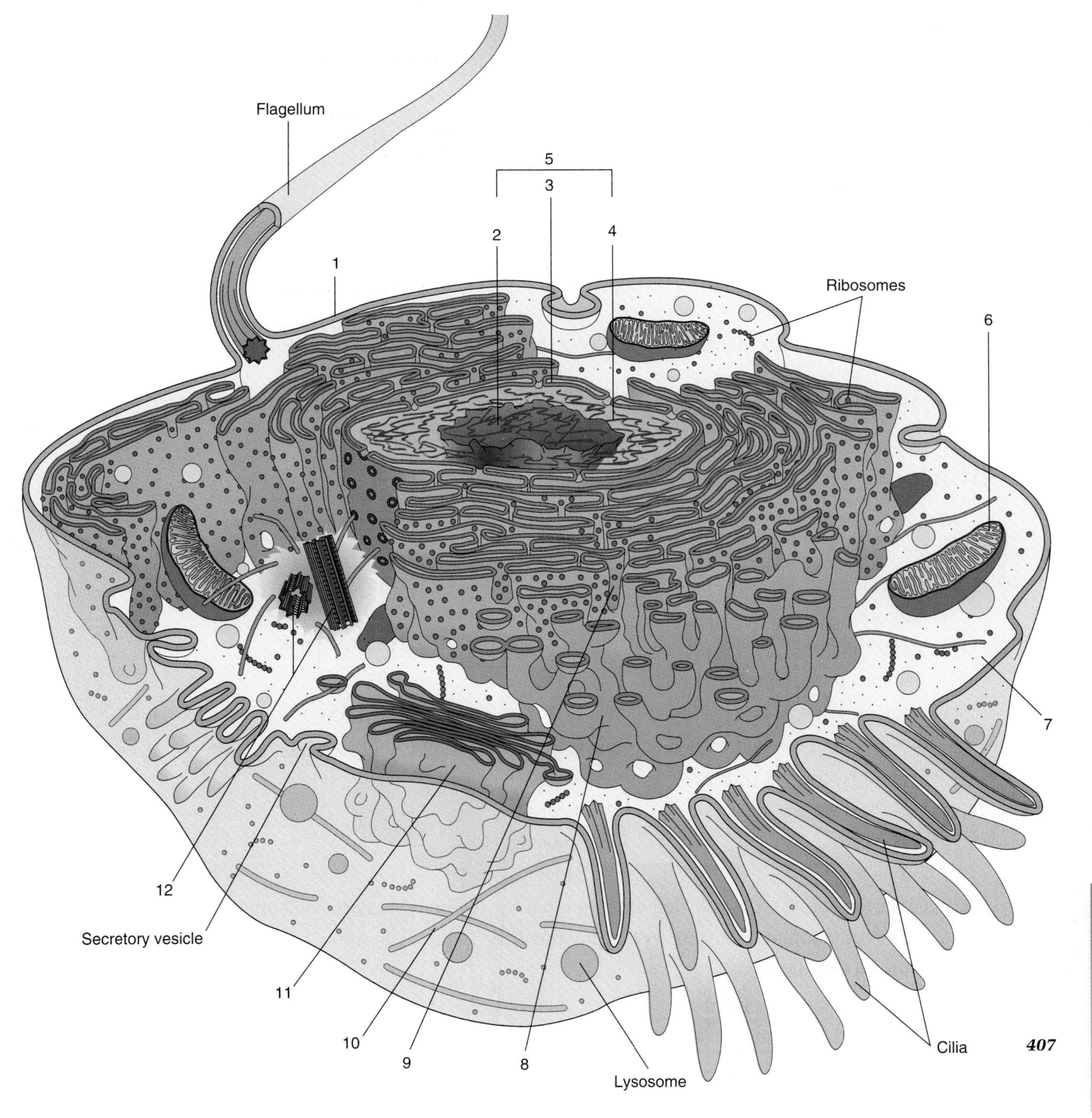

b. Write the terms that match the phrases in the spaces at the right.

1) Endoplasmic reticulum with ribosomes. rer
2) Forms cytoskeleton (two answers). microfilament
3) Packages materials for export from cell. golgi apparatus
4) Sites of protein synthesis. ribosomes
5) Composed of DNA and protein. chromosomes
6) Intranuclear site of rRNA synthesis. nucleus
7) Sites of cellular respiration. mitochondria
8) Controls movement of materials between nucleus and cytoplasm. nuclear envelope
9) Endoplasmic reticulum without ribosomes. Ser
10) Vesicles of digestive enzymes. Lysosomes
11) Provides motility for sperm. Flagellum
12) Short cylinders formed of microtubules. Centrioles
13) Semi-liquid around organelles. cytoplasm
14) Short, hairlike projections that move substances across cell surfaces. ______
15) Controls movement of materials into and out of the cell. Cell mem
16) Sites of aerobic cellular respiration. aerobic resp
17) Contains chromosomes. nucleus
18) Forms channels for material transport in the cytoplasm. endoplasm reticulum
19) Molecule determining inheritance. DNA
20) Organelle controlling cell functions. nucleus.

2. Transport Through Plasma Membranes

a. Match the terms and phrases.

Diffusion Osmosis Phagocytosis Pinocytosis

1) Diffusion of water. Osmosis
2) Engulfment of small particles. Pinocytosis
3) Engulfment of liquid droplets. Phagocytosis
4) Movement of molecules from an area of higher concentration to areas of lower concentration. diffusion
5) Results from random molecular movement. " + Osmosis

b. Identify the transport processes as either active (A) or passive (P).

A By carrier proteins
P Diffusion
A Phagocytosis
P Osmosis
A Pinocytosis
A Exocytosis

c. Consider the solutions below that are separated by a selectively permeable membrane. Use arrows to show the direction of diffusion. In 1, show the direction of *water* movement. In 2, show the direction of *solute* movement.

1) A ← B

10% protein solution	5% protein solution

2) A → B

10% salt solution	5% salt solution

3) In 1, which solution is hypotonic? B

4) In 2, which solution is hypertonic? A

5) Describe what happens when a human cell is placed in a hypotonic solution. causes it to swell then burst.

3. Cellular Respiration

a. Write the summary equation for the cellular respiration of glucose. Words may be used instead of chemical formulas. Glucose + O_2 → CO_2 + H_2O + Energy.

b. Completion

1) List the products of cellular respiration. CO_2 H_2O energy

2) The source of energy captured in ATP. glucose

c. Explain why cellular respiration is a continuous process. A continuous supply of ATP is required to provide energy

4. Protein Synthesis

Completion

1) The genetic code consists of the sequence of bases in ________ molecules. ~~protein~~ dna

2) The genetic code is transcribed to the sequence of bases in _____ molecules. MRNA

3) Molecule that carries instructions for protein synthesis to ribosomes. MRNA

4) Molecule that carries amino acids to ribosome for addition to amino acid chain. TRNA

5) Small molecules that join to form a protein during translation. ~~RRNA~~ amino acids

5. Cell Division

a. Indicate the type of cell division described by the statements.

1) Provides new cells for growth and repair. ____________

2) Forms sperm and ova. ____________

3) Daughter cells have same chromosome number and composition as parent cell. ____________

4) Daughter cells have half the number of chromosomes as the parent cell. ____________

b. Select the phase of the cell cycle described by the statements.

Interphase Prophase Metaphase Anaphase Telophase

1) Division of the cytoplasm. telephase

2) Replication of chromosomes. Interphase

3) Chromosomes appear as threadlike bodies. Prophase

4) Chromatids move toward ends of spindle. Anaphase

5) New nuclei start to form. telephase
6) Occupies most of cell cycle. Interphase
7) Chromosomes line up at equator of spindle. metaphrase
8) Cell performs its normal functions. Intphrase

c. Human body cells have 46 chromosomes. How many chromosomes are in daughter cells formed by mitotic cell division? ____________

d. Write the names of the mitotic phases in the spaces provided and place the numbers of the cell parts in the spaces by the correct label.

1 Centrioles
4 Chromosome
2 Spindle fiber
3 Centromere
5 Chromatid

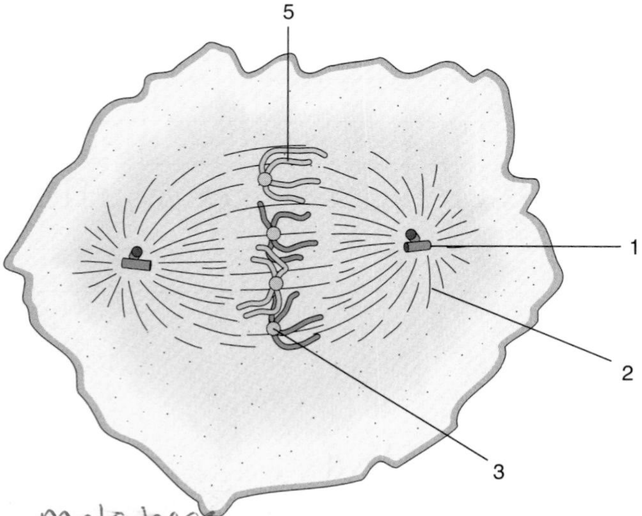

6) metaphase

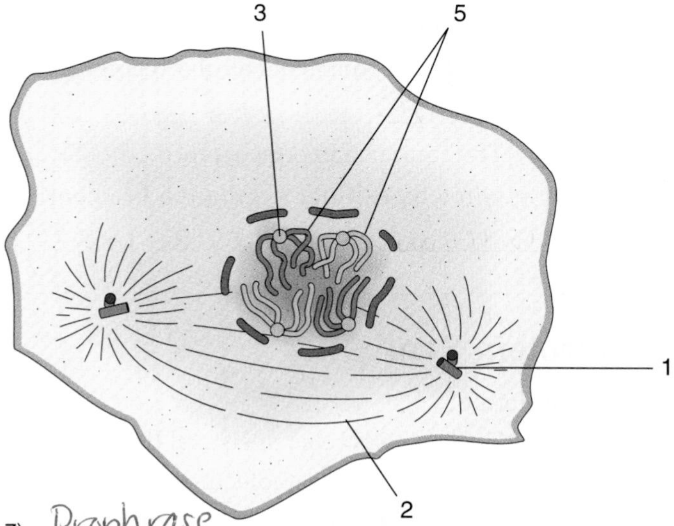

7) Prophrase

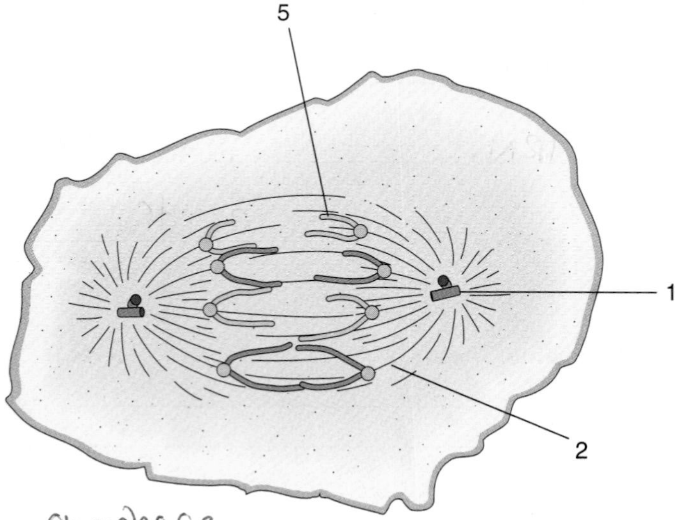

8) anaphase

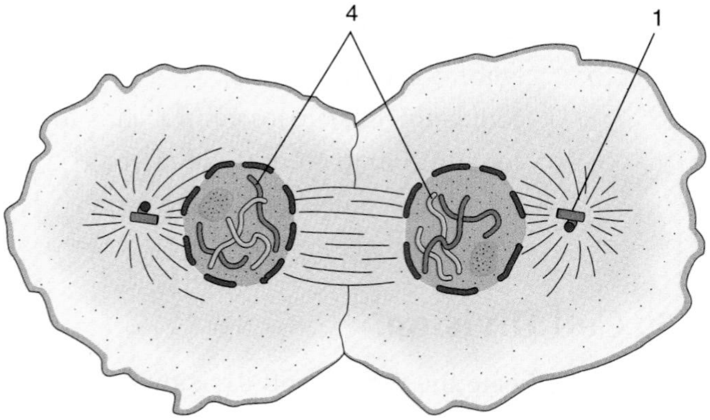

9) telephase.

6. Clinical Applications

a. When you drink a glass of water, how does the water enter the blood? ____________
Why does this occur? ____________

b. Explain why a chemical therapy drug that disrupts formation of spindle fibers kills cancerous cells.
____
____

STUDY GUIDE 4

1. Basic Tissues

Select the tissues described by the statements.

Epithelial Connective Muscle Nerve

1) Adapted for contraction. ____________________
2) Contains scattered cells in a matrix. ____________________
3) Sheets of closely packed cells. ____________________
4) Composed of neurons and supporting cells. ____________________
5) Lacks blood vessels. ____________________
6) Supports and protects organs. ____________________
7) Lines body cavities and covers organs. ____________________
8) Forms and conducts impulses. ____________________
9) Functions in absorption and secretion. ____________________

2. Epithelial Tissues

a. Write the names of the tissues in the spaces provided and place the number of each structure in the space by the correct label.

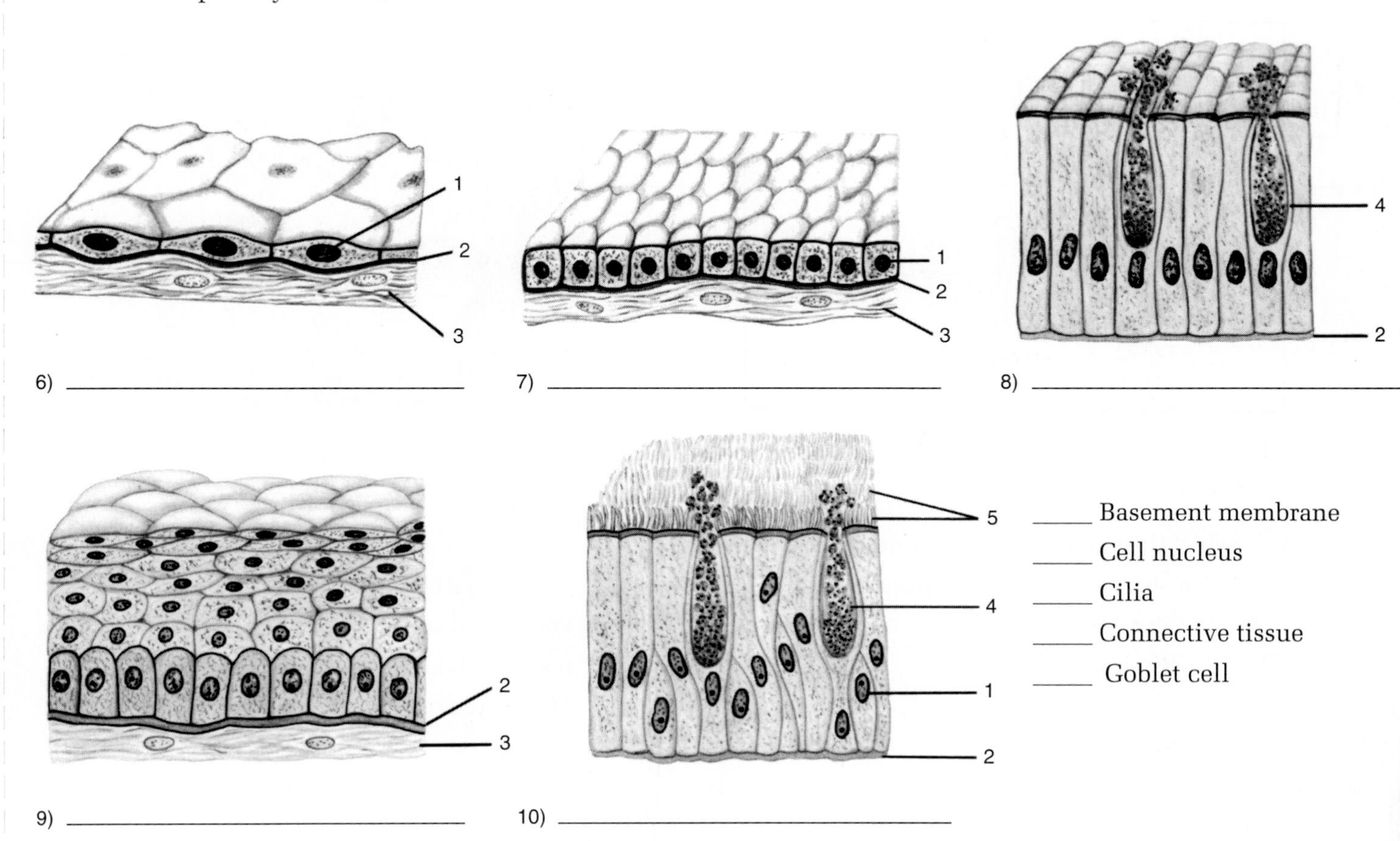

b. Write the number of the appropriate tissue described in the space provided.

1) Simple squamous
2) Simple cuboidal
3) Simple columnar
4) Pseudostratified ciliated columnar
5) Stratified keratinized squamous
6) Stratified nonkeratinized squamous
7) Transitional

_____ Forms secretory cells of glands.
_____ Lines interior of blood vessels.
_____ Lines interior of stomach and intestines.
_____ Lines upper respiratory passages.
_____ Lines ventral body cavity.
_____ Forms outer layer of skin.
_____ Lines interior of urinary bladder.
_____ Lines mouth and vagina.
_____ Contain goblet cells.
_____ Forms air sacs of lungs.

3. Connective Tissue Proper

a. Write the names of the tissues in the spaces provided and place the number of each structure in the space by the correct label.

_____ Collagenous fiber
_____ Elastic fiber
_____ Fat droplet
_____ Fibroblast
_____ Ground substance
_____ Nucleus

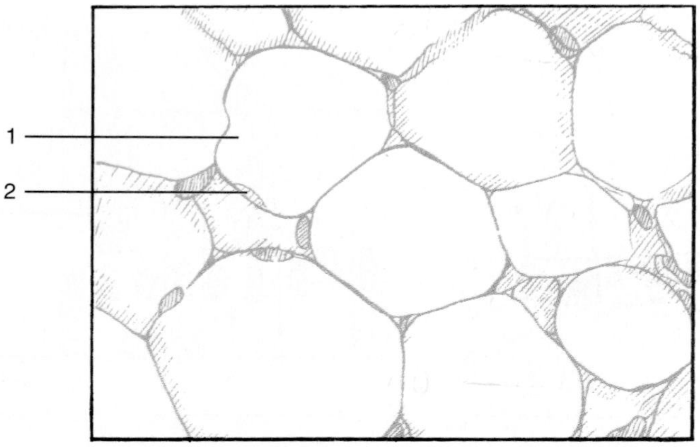

7) ______________________________

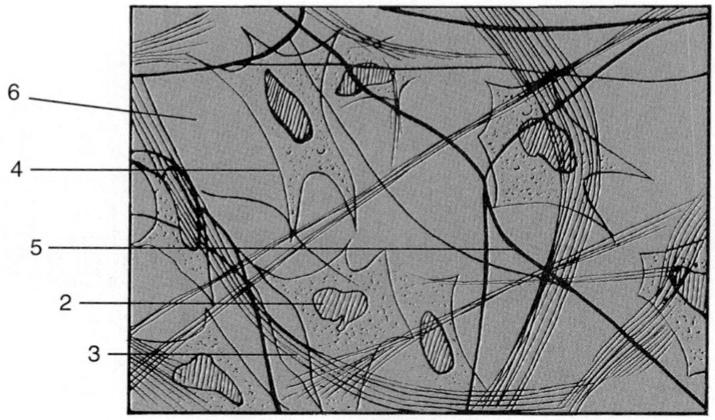

8) ______________________________

b. Select the connective tissues described by the statements.

1) Loose 2) Adipose 3) Fibrous 4) Elastic

_____ Storage area for fat.
_____ Binds skin to muscles.
_____ Forms ligaments and tendons.
_____ In walls of arteries.
_____ Supporting framework for internal organs.
_____ Very strong but pliable.
_____ Poor blood supply.
_____ Insulates body.
_____ Tightly packed collagenous fibers.
_____ Enables expansion and contraction of lungs.
_____ Protective cushion for internal organs.
_____ Inner layer of skin.

4. Special Connective Tissues

a. Write the names of the tissues in the spaces provided and place the number of each structure in the space by the correct label.

_____ Canaliculi
_____ Chondrocyte in a lacuna
_____ Elastic fibers
_____ Osteonic canal
_____ Lamellae
_____ Nonfibrous matrix
_____ Osteocyte in a lacuna

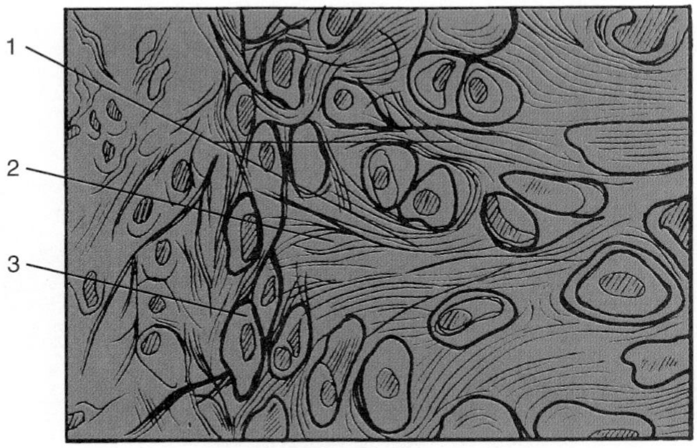

8) ______________________________

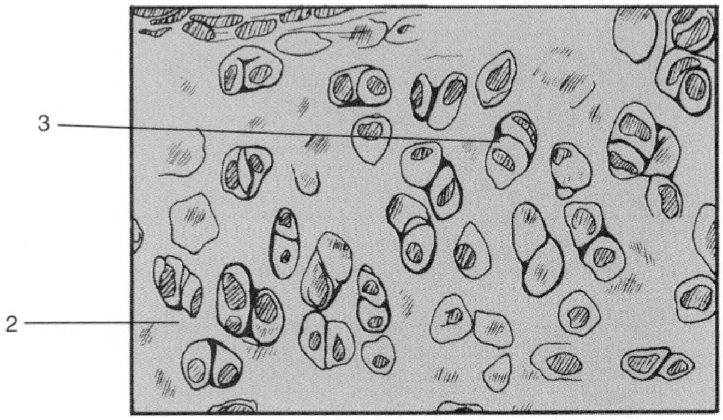

9) ______________________________

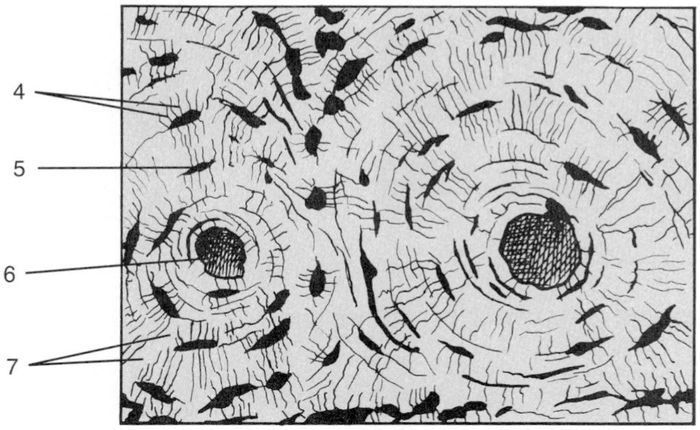

10) ______________________________

b. Select the connective tissues described by the statements.

1) Elastic cartilage
2) Fibrocartilage
3) Hyaline cartilage
4) Bone
5) Blood

_____ Intervertebral disks.
_____ Pinna of outer ear.
_____ Forms embryonic bones.
_____ Liquid matrix.
_____ Smooth, glassy matrix.
_____ Imparts resiliency.
_____ Hard, rigid matrix.
_____ Adapted to absorb shocks.

5. Muscle Tissues

a. Write the names of the muscle tissues in the spaces provided and place the number of each structure in the space by the correct label.

_____ Intercalated disk
_____ Nucleus
_____ Striations

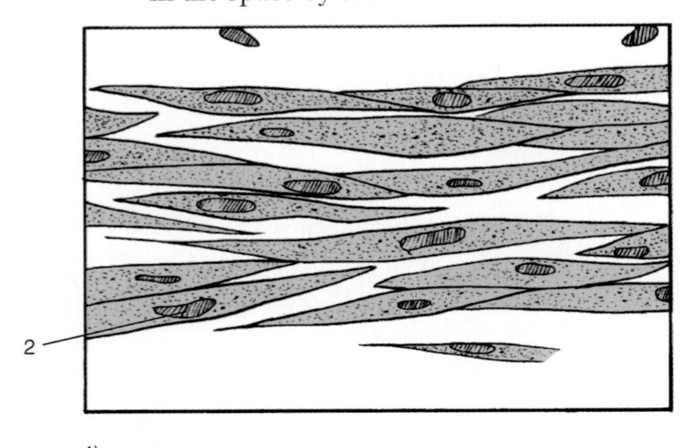

4) ______________________________

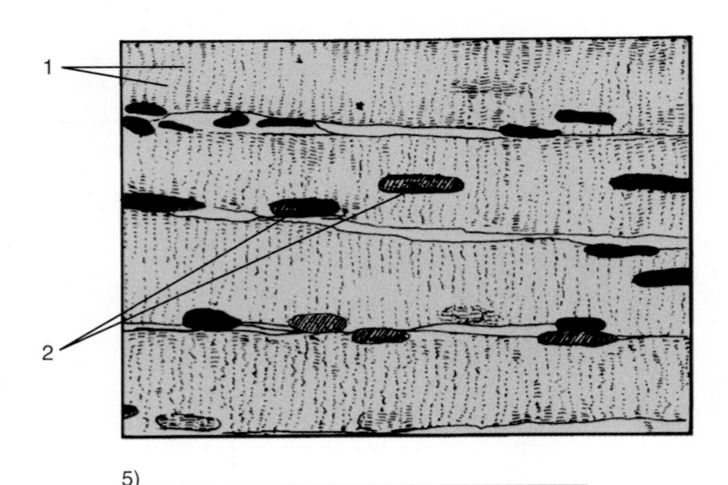

5) ______________________________

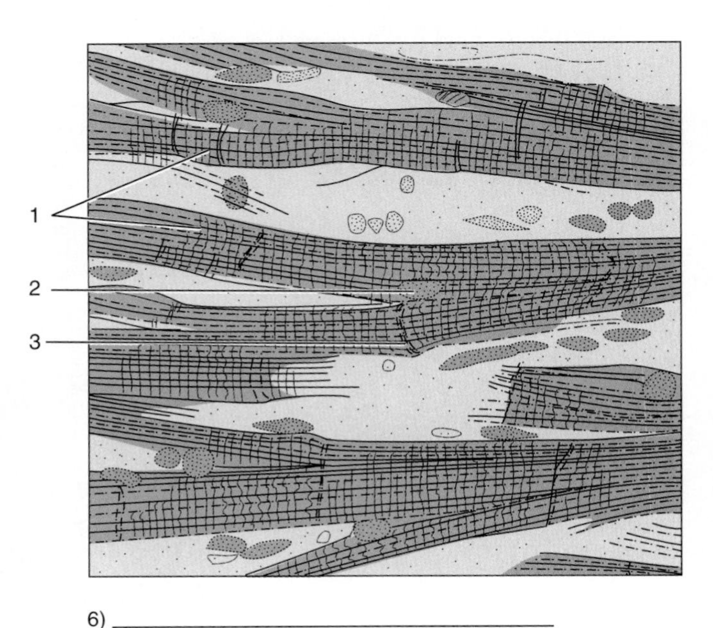

6) ______________________________

b. Select the muscle tissues described by the statements.

1) Cardiac 2) Skeletal 3) Smooth

_____ Voluntary
_____ Involuntary
_____ In walls of intestine
_____ Slow contractions
_____ Rapid contractions
_____ Rhythmic contractions

6. Nerve Tissue

Indicate whether each statement is true (T) or false (F).

_____ Nerve cells are called neurons.
_____ The nucleus of a nerve cell is located in the cell body.
_____ Nerve cells form and transmit neural impulses.
_____ Supporting cells in nerve tissue are fibroblasts.

7. Clinical Applications 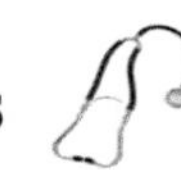

a. Most cancers are carcinomas. How do you explain this? ______________________________

____

b. Judy tore a knee cartilage on a skiing vacation. Can she expect a rapid recovery? _____ Explain. ______

____

STUDY GUIDE 5

1. Functions of the Skin

List five functions of the integumentary system.

1) ______
2) ______
3) ______
4) ______
5) ______

2. Structure of the Skin

Select the structure described by each statement.

Dermis Epidermis Stratum basale Stratum corneum Subcutaneous layer

1) Contains abundant adipose tissue. Sub layer
2) Innermost layer of the epidermis. Stratum
3) Composed of stratified squamous epithelium. Epidermis
4) Contains collagen and elastic fibers. dermis
5) Attaches skin to underlying tissues. Sub
6) Lacks blood vessels. epid
7) Outermost layer of the epidermis. Strat
8) Contains sensory receptors of the skin. dermis
9) Forms new epidermal cells. Strat b
10) Formed of dead keratinized cells. strat c
11) Provides insulation for the body. Sub
12) Provides strength and elasticity of skin. Dermis
13) Inner layer of the skin. Dermis
14) Outer cells are continuously sloughed off. Stratum cor
15) Formed of fibrous connective tissue. Dermis

3. Skin Color

a. Provide the term described by each statement.

1) Three pigments that determine skin color. Indicate the color of each pigment. ______ ______ ______
2) Protects against ultraviolet radiation. ______
3) Cells producing melanin. ______
4) Stimulates melanin production. ______
5) Ultimate determiner of skin color. ______

b. Explain why a summer tan is only temporary. ______________________________

____

____

4. Diagram of the Skin

Label the diagram

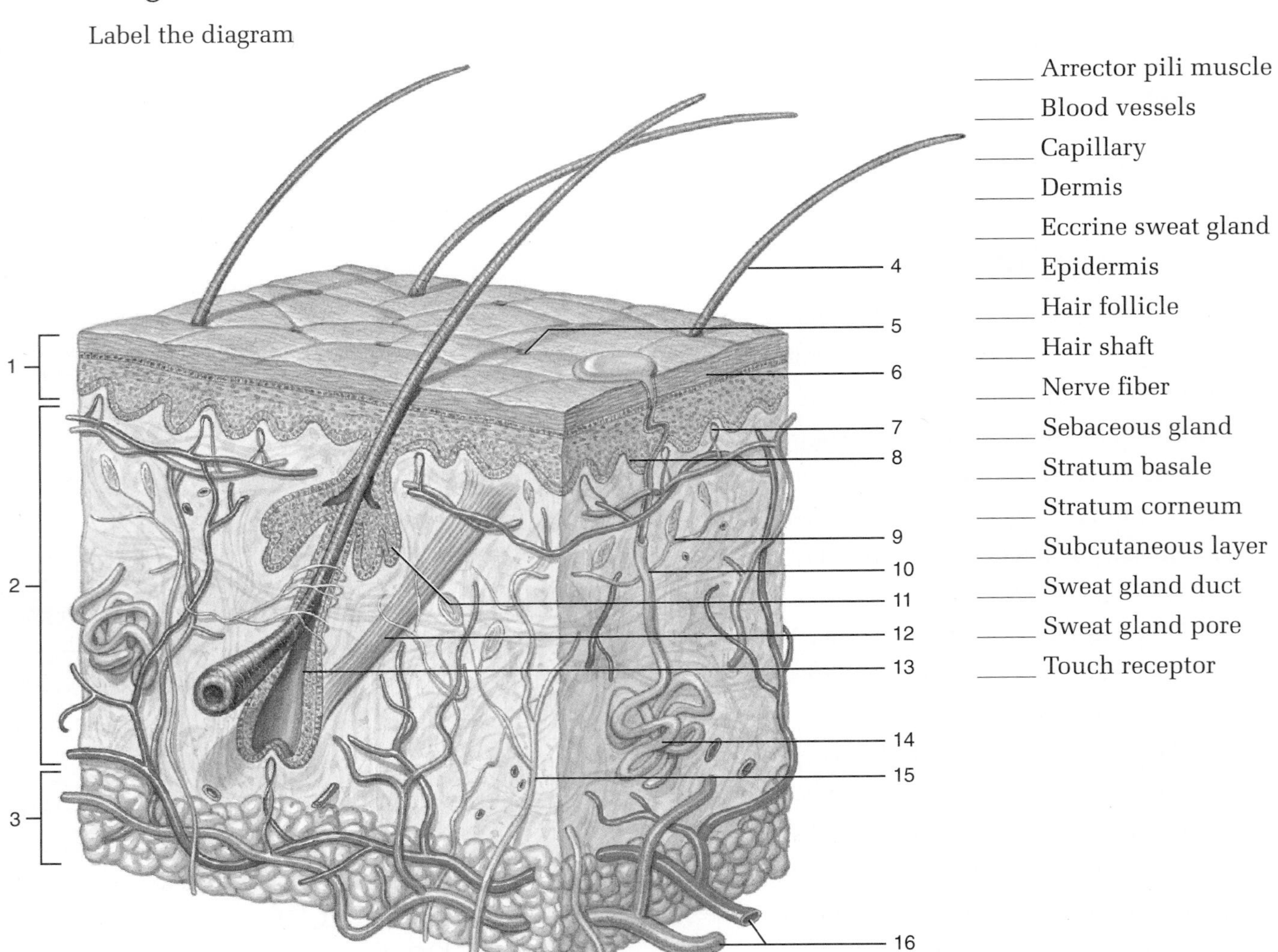

____ Arrector pili muscle
____ Blood vessels
____ Capillary
____ Dermis
____ Eccrine sweat gland
____ Epidermis
____ Hair follicle
____ Hair shaft
____ Nerve fiber
____ Sebaceous gland
____ Stratum basale
____ Stratum corneum
____ Subcutaneous layer
____ Sweat gland duct
____ Sweat gland pore
____ Touch receptor

5. Accessory Structures

Provide the term that matches each statement.

1) Tubular sheath surrounding hair root. ______________________
2) Gland producing sebum. ______________________
3) Muscle raising hair more erect. ______________________
4) Gland producing perspiration. ______________________
5) Gland producing cerumen. ______________________
6) Sweat gland opening into hair follicle. ______________________

7) Sweat gland producing watery perspiration. ______
8) Protein in cells forming hair and nails. ______
9) Basic function of hair and nails. ______
10) Secretion containing salts and urea. ______
11) Oily secretion that helps keep skin soft. ______
12) Waxy secretion found in external ear canal. ______
13) Normal color of nail beds. ______

6. Temperature Regulation

Provide the missing words in the paragraph below. Humans have a normal body temperature of __1__ °C, or __2__ °F. The heat that maintains the body temperature is generated as a by-product of cellular __3__ , especially in active organs like the liver and skeletal __4__ . Overall regulation of body temperature is controlled by the __5__ , while the __6__ serves as an important regulatory organ.

1) 37°C
2) ______
3) respiration
4) muscle
5) brain
6) skin

When the body temperature falls below normal, the flow of __7__ to the skin is decreased, which reduces secretion of __8__ by sweat glands and minimizes heat __9__ by radiation. Shivering increases cellular respiration in muscles, which generates more __10__ .

7) blood
8) perspiration
9) loss
10) heat

When the body temperature rises above normal, blood flow to the skin is __11__ , which increases heat loss by __12__ and activates __13__ glands to produce perspiration. The __14__ of perspiration from the surface of the skin increases __15__ loss and cools the body surface.

11) increases
12) radiation
13) sweat
14) evaporates
15) heat.

7. Aging of the Skin

Indicate whether each statement is true (T) or false (F).

_____ A baby's skin is thinner than an adult's.

_____ A decrease in melanin production often occurs in the elderly.

_____ Wrinkled skin results from an excess of active elastic fibers.

_____ Ultraviolet radiation accelerates the aging of the skin.

_____ Excess subcutaneous fat increases sensitivity to temperature changes.

8. Disorders of the Skin

Write the name of the disorder described by each statement.

1) Results from a chronic deficiency of circulation to a portion of skin. ______
2) Cancer of the melanocytes. ______
3) Numerous red, itchy bumps resulting from an allergic reaction. ______
4) Skin-colored tumors caused by a virus. ______
5) Slow-growing, pigmented tumors. ______
6) A burn that destroys all of the dermis. ______
7) Loss of hair as in male pattern baldness. ______
8) Thickened areas of skin on hands and feet. ______
9) Inflammation causing red, itching, scaling skin; may involve sebaceous glands. ______
10) Contagious infection in which pustules rupture and form a yellow crust. ______
11) Reddish, raised scaly patches on scalp, knees, or elbows. ______
12) Condition caused by excessive shedding of epidermal cells of the scalp. ______
13) Blisters on lips caused by *Herpes simplex.* ______
14) Itching, flaking skin between toes due to a fungus infection. ______
15) Bacterial infection of a hair follicle, sebaceous gland, and surrounding tissues. ______
16) The general term for any inflammation of the skin. ______

9. Clinical Applications

a. Name two major clinical problems expected in a patient with third-degree burns. ______

______

b. Subcutaneous injections of medications are frequently used. Why is the subcutaneous layer especially good for rapid absorption of medications? ______

______

6. The Axial Skeleton

a. Label the diagram of the skull, anterior view, by placing the number of each structure in the space by the correct label.

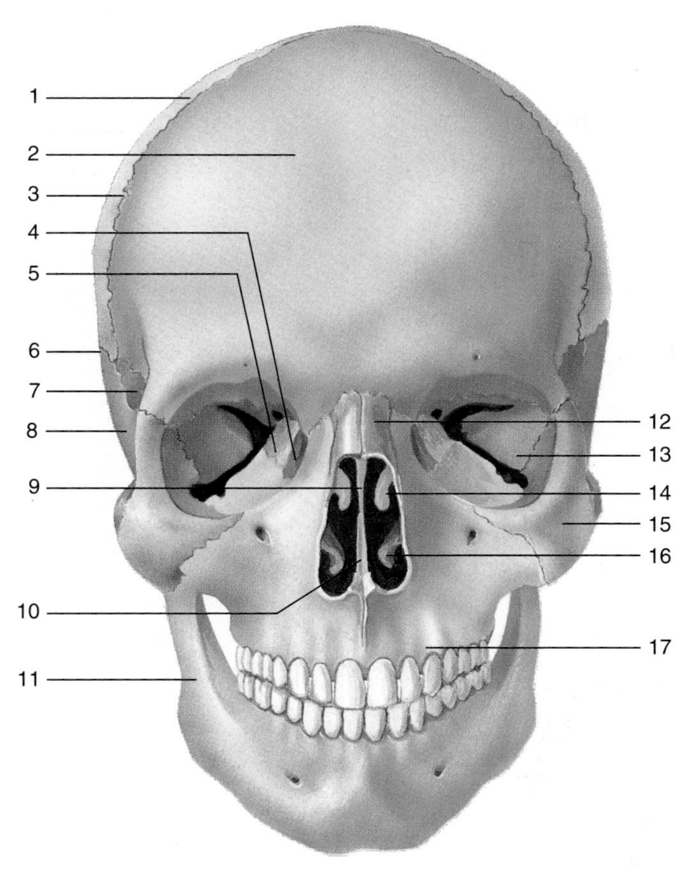

_____ Coronal suture
_____ Ethmoid (eye orbit)
_____ Ethmoid, perpendicular plate
_____ Frontal
_____ Lacrimal
_____ Mandible
_____ Maxilla
_____ Nasal
_____ Nasal concha, inferior
_____ Nasal concha, middle
_____ Parietal
_____ Squamosal suture
_____ Sphenoid (2 places)
_____ Temporal
_____ Vomer
_____ Zygomatic

b. List the skull bones that contain sinuses. ____

____

c. Label the diagram of the skull, lateral view, by placing the number of each structure in the space by the correct label.

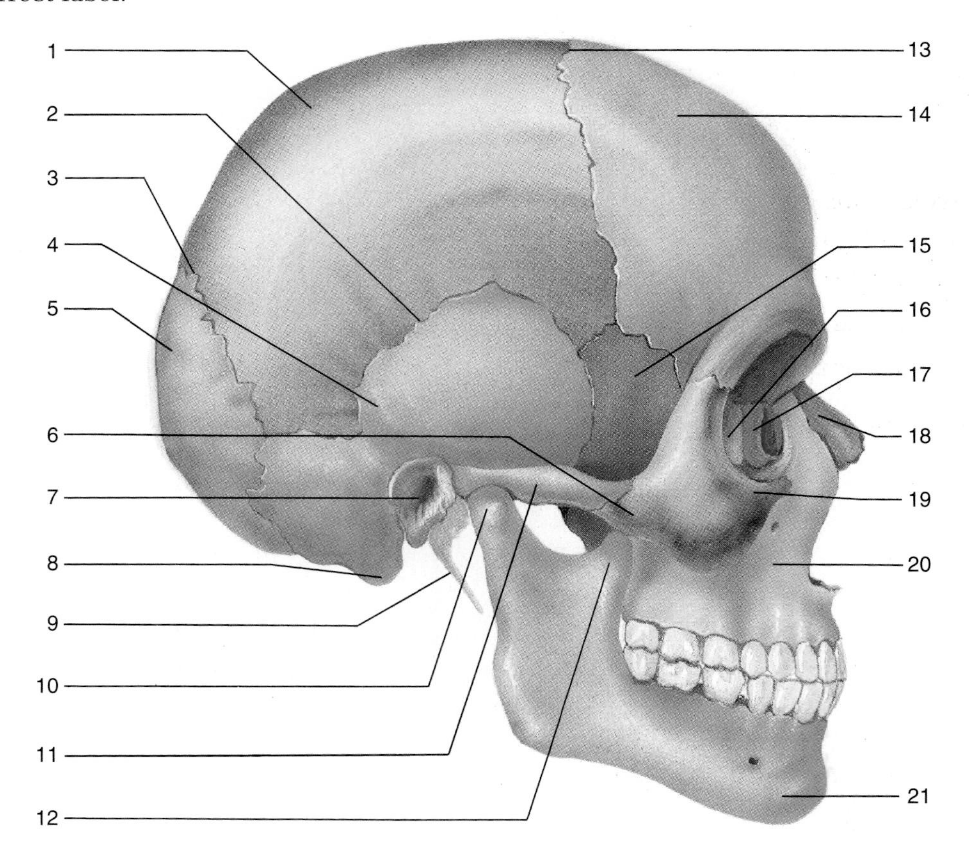

_____ Coronal suture	_____ Nasal
_____ Coronoid process	_____ Occipital
_____ Ethmoid	_____ Parietal
_____ External auditory canal	_____ Sphenoid
_____ Frontal	_____ Squamosal suture
_____ Lacrimal	_____ Styloid process
_____ Lambdoidal suture	_____ Temporal
_____ Mandible	_____ Temporal, zygomatic process
_____ Mandibular condyle	_____ Zygomatic
_____ Mastoid process	_____ Zygomatic, temporal process
_____ Maxilla	

d. Write the terms that match the statements in the spaces provided.

1) Contains the foramen magnum. ______________________
2) Forms anterior portion of hard palate. ______________________
3) Contains external auditory canal. ______________________
4) The seven vertebrae of the neck. ______________________
5) Weight-bearing portion of a vertebra. ______________________
6) Foramen through which spinal cord passes. ______________________
7) Vertebrae-bearing ribs. ______________________
8) Number of pairs of true ribs. ______________________
9) Attaches true ribs to sternum. ______________________
10) First cervical vertebra. ______________________
11) Cartilaginous pads between vertebrae. ______________________
12) Forms posterior wall of pelvic girdle. ______________________
13) Vertebrae with heaviest bodies. ______________________
14) The breastbone. ______________________

e. Name the group of bones that provides protection for the

1) Brain ______________________ 2) Heart and lungs ______________________

f. Label the vertebra by placing the number of the structure in the space by the correct label.

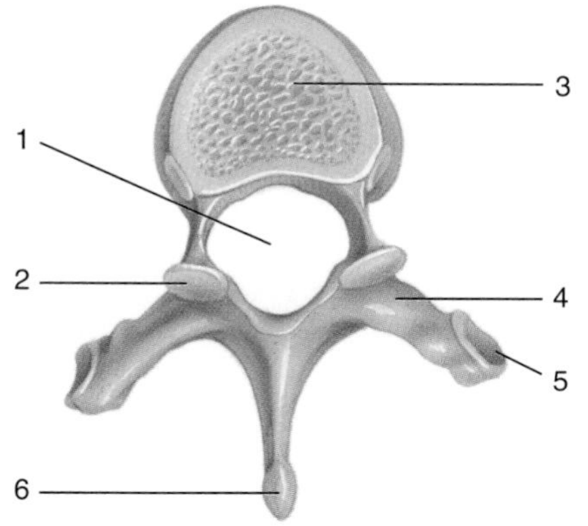

1
2
3
4
5
6

_____ Body
_____ Facet for rib
_____ Spinous process
_____ Superior articular process
_____ Transverse process
_____ Vertebral foramen

7. The Appendicular Skeleton

a. Write the missing words in the spaces at the right.

The pectoral girdle is formed of two ___1___ and two ___2___ . Its function is to support the upper ___3___ . Each ___4___ articulates with the scapula at one end and the ___5___ at the other. The scapulae are attached to the axial skeleton by ___6___ instead of ligaments.

1) ______________________
2) ______________________
3) ______________________
4) ______________________
5) ______________________
6) ______________________

b. Label these diagrams by placing the number of each structure in the space by the correct label.

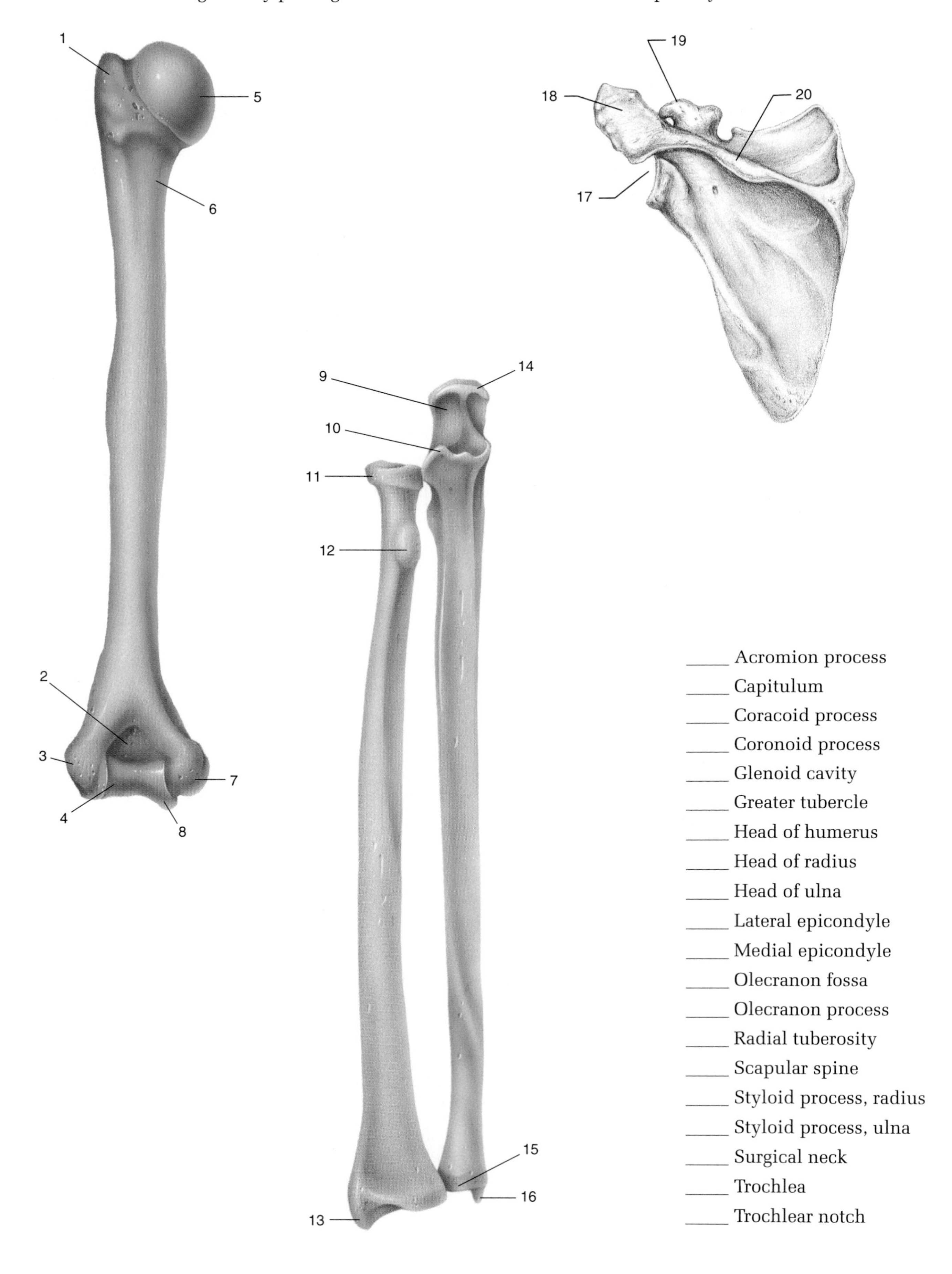

_____ Acromion process
_____ Capitulum
_____ Coracoid process
_____ Coronoid process
_____ Glenoid cavity
_____ Greater tubercle
_____ Head of humerus
_____ Head of radius
_____ Head of ulna
_____ Lateral epicondyle
_____ Medial epicondyle
_____ Olecranon fossa
_____ Olecranon process
_____ Radial tuberosity
_____ Scapular spine
_____ Styloid process, radius
_____ Styloid process, ulna
_____ Surgical neck
_____ Trochlea
_____ Trochlear notch

c. Label the diagrams by placing the number of each structure in the space by the correct label.

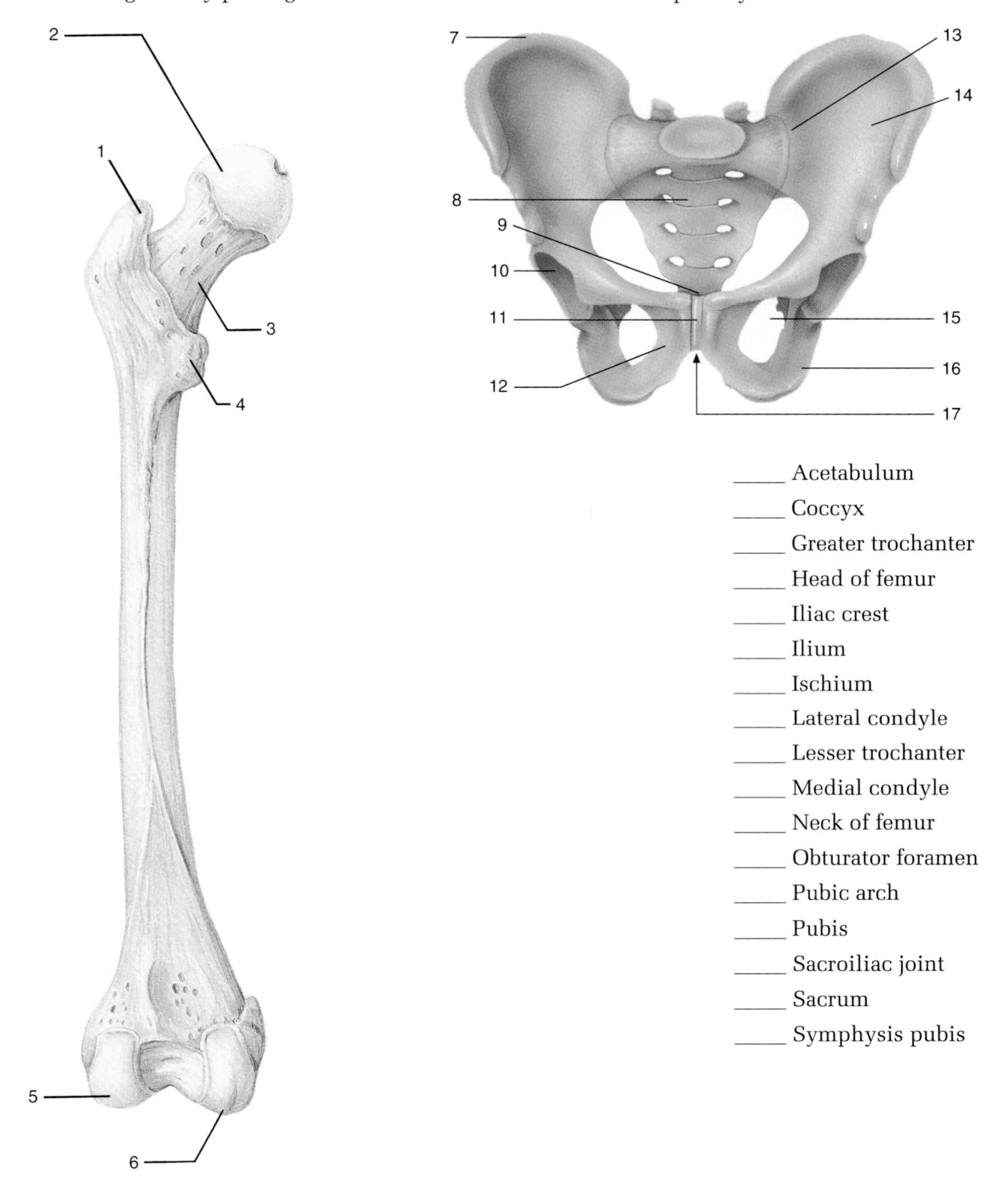

_____ Acetabulum
_____ Coccyx
_____ Greater trochanter
_____ Head of femur
_____ Iliac crest
_____ Ilium
_____ Ischium
_____ Lateral condyle
_____ Lesser trochanter
_____ Medial condyle
_____ Neck of femur
_____ Obturator foramen
_____ Pubic arch
_____ Pubis
_____ Sacroiliac joint
_____ Sacrum
_____ Symphysis pubis

d. Indicate whether each statement is associated with the fibula (F) or tibia (T).

_____ Lateral malleolus
_____ Lateral condyle
_____ Articulates with femur
_____ Medial malleolus
_____ Medial condyle
_____ Articulates with talus

8. Articulations

a. Match the type of joint with the articulation formed by the bones.

1) Immovable
2) Slightly movable
3) Ball-and-socket
4) Condyloid
5) Gliding
6) Hinge
7) Pivot
8) Saddle

6 Femur—tibia
_____ Frontal—parietal
3 Humerus—scapula
2 Vertebra—vertebra
7 Atlas—axis
_____ Carpal—carpal
_____ Trapezium—metacarpal 1
_____ Coxa—femur
_____ Maxilla—zygomatic
_____ Metacarpal—phalanx

b. Match the terms with the correct definitions.

1) Articular cartilage
2) Bursa
3) Cartilage pads
4) Joint capsule
5) Sesamoid bone
6) Synovial fluid

6 Lubricates joints.
4 Sacs of synovial fluid.
3 Support knee joint and cushion bones.
1 Protects articular surfaces of bones.
_____ Bone embedded in a tendon.
_____ Formed of ligaments.

c. Match the movements with the descriptions.

1) Abduction
2) Adduction
3) Depression
4) Elevation
5) Extension
6) Flexion
7) Dorsiflexion
8) Plantar flexion
9) Eversion
10) Inversion
11) Circumduction
12) Rotation
13) Pronation
14) Supination
15) Protraction
16) Retraction

_____ Extension of foot.
_____ Pushing mandible anteriorly.
_____ Movement of arm toward midline.
_____ Straightening arm at elbow.
_____ Turning palm of hand upward.
_____ Turning head from side to side.
_____ Decrease in angle of joint.
_____ Raising shoulders.
_____ Drawing circle on chalkboard.
_____ Turning sole of foot inward.

9. Disorders of the Skeletal System

Write the name of each disorder described in the space provided.

1) Displacement of bones forming a joint. _____
2) A lateral curvature of vertebral column. _____
3) Protrusion of intervertebral disk. _____
4) Bone broken into several pieces. _____
5) Broken bone pierces through skin. _____
6) Arthritis with invasion of fibrous tissue that calcifies, making joint immovable. _____
7) Tearing of ligaments of joint capsule. _____
8) Severe loss of calcium salts from bones. _____

10. Clinical Applications

a. A member of the soccer team is diagnosed with a torn knee cartilage. Would you expect rapid or slow recovery? Explain your answer. _____

b. Specifically, what is a broken hip? _____
Why is it more common among older persons? _____

STUDY GUIDE 7

1. Types of Muscle Tissues

Match the types of muscle tissues with the words and phrases.

1) Skeletal 2) Smooth 3) Cardiac

_____ Striated

_____ Single nucleus

_____ Involuntary

_____ Intercalated disks

_____ Branching network

_____ Walls of blood vessels

_____ Heart muscle

_____ Walls of digestive tract

_____ Skeletal muscles

_____ Easily fatigued

2. Structure of Skeletal Muscle

Write the terms that match the statements in the spaces at the right.

1) A bundle of fibers enveloped by connective tissue. Fasiculus
2) Binds all fasciculi together. Fibrous connective tissue
3) Connective tissue covering entire muscle. Deep fascia
4) Cordlike attachment of a muscle. tendon
5) Sheetlike attachment of a muscle. Aponeurosis
6) Plasma membrane of muscle cell. sarcolemma
7) Cytoplasm of a muscle cell. Sarcoplasm
8) Threadlike contractile elements. Myofibrils
9) Thinner protein filaments in myofibrils. actin
10) Thicker protein filaments in myofibrils. Myosin
11) Portion of a myofibril between Z lines. Sacromere
12) Light and dark bands on myofibrils. Striation
13) Attachment of motor axon to sarcolemma. Neuromuscular junction
14) Depression in sarcolemma receiving axon tip. Synapatic cleft
15) Motor neuron and its attached muscle fibers. motor unit

3. Physiology of Muscle Contraction

a. Write the words that complete the sentences in the spaces at the right.

The axon tip of an activated motor neuron releases ___1___ into the ___2___, where it combines with ___3___ on the sarcolemma. This stimulates the release of ___4___ from storage areas, which exposes the active sites on ___5___ filaments. Cross-bridges of ___6___ attach to the exposed active sites and exert a power stroke, which pulls the ___7___ filaments and the Z lines toward the center of the A band. This process is rapidly repeated until ___8___ is complete.

1) Acetylchonine
2) Syptic cleft
3) Reception
4) Calcium
5) actin
6) Myosin
7) actin
8) contraction

b. Write the terms that match the statements in the spaces provided.
   1) Decomposes acetylcholine.
   2) Combines with oxygen to store small amounts of oxygen in muscle cells.
   3) Phase of cellular respiration that requires oxygen.
   4) Products of pyruvic acid breakdown when adequate oxygen is present.
   5) Acid formed from pyruvic acid when adequate oxygen is not available.
   6) Provides direct energy for muscle contraction.
   7) Process releasing energy from nutrients in cells.
   8) Chemical whose accumulation produces an oxygen debt.
   9) Released from creatine phosphate to quickly re-form ATP.

c. Write the terms that match the statements in the spaces at the right.
   1) Smallest stimulus causing a contraction.
   2) Activation of a muscle fiber causes a (all-or-none, graded) contraction.
   3) Primary cause of fatigue.
   4) Type of contractions observed in whole muscles (all-or-none, graded).
   5) Smallest stimulus that activates all motor units of a muscle.
   6) Activation of an increasing number of motor units in a series of contractions.
   7) Controls the number of motor units that are activated.
   8) State of constant, partial contraction.
   9) State of constant, complete contraction.

4. Actions of Skeletal Muscles

a. Write the terms that match the statements in the spaces provided.
   1) Fixed end of a muscle.
   2) Movable end of a muscle.
   3) Muscles opposing agonists.

b. Write the names of the muscles that match the actions.
   1) Closes and puckers lips.
   2) Pulls angle of mouth upwards.
   3) Helps masseter raise the mandible.
   4) Compresses cheeks.
   5) Pair of neck muscles that flex head.

6) Pair of neck muscles that extend head. __________
7) Innermost muscle of abdominal wall. __________
8) Raises ribs during inspiration. __________
9) Elevates clavicle and scapula. __________
10) Draws scapula downward and anteriorly. __________
11) Adducts and draws humerus across chest. __________
12) Sheetlike muscle of lower back that adducts and extends humerus. __________
13) Abducts, flexes, and extends humerus. __________
14) Rotates humerus laterally. __________
15) Assists deltoid in abducting humerus. __________
16) Assists latissimus dorsi. __________
17) Assists biceps brachii (two muscles). __________ __________
18) Extends forearm. __________
19) Flexes and rotates forearm laterally. __________
20) Flexes and abducts wrist. __________
21) Flexes and adducts wrist. __________
22) Extends fingers. __________
23) Extends and adducts wrist. __________
24) Extends and abducts wrist. __________
25) Adducts, flexes, and rotates thigh laterally (two muscles). __________ __________
26) Abducts and rotates thigh medially. __________
27) Extends and rotates thigh laterally. __________
28) Flexes and abducts thigh. __________
29) Flexes thigh only (two muscles). __________ __________
30) Flexes leg and thigh. __________
31) Flexes leg and adducts thigh. __________
32) Group of four muscles that extend leg. __________
33) Three muscles that flex the leg and extend the thigh. __________ __________ __________
34) Dorsiflexes and inverts foot. __________
35) Flexes leg and plantar flexes foot. __________
36) Extends toes and dorsiflexes and everts foot. __________
37) Plantar flexes and everts foot. __________

5. Disorders of the Muscle System

Write the names of the disorders in the spaces provided.

1) Inflammation of connective tissues of muscles. ______________________
2) Involuntary, tetanic contraction of a muscle. ______________________
3) Antibodies attach to acetylcholine receptors, preventing normal stimulation of muscles. ______________________
4) Inflammation of muscle tissue. ______________________
5) A pulled muscle. ______________________
6) Abnormal increase of fibrous connective tissue in a muscle. ______________________
7) Viral disease that destroys motor neurons and paralyzes skeletal muscles. ______________________
8) Group of diseases characterized by the progressive degeneration of muscles. ______________________
9) A bacterial disease that prevents the release of acetylcholine from axon tips. ______________________
10) A bacterial disease commonly called "lockjaw." ______________________
11) Sudden, involuntary weak contractions of a muscle or group of muscles. ______________________

6. Major Skeletal Muscles

Label the muscles and associated structures in the following diagrams by writing the names of the labeled parts in the spaces provided. After labeling, color-code the muscles to help you to distinguish them.

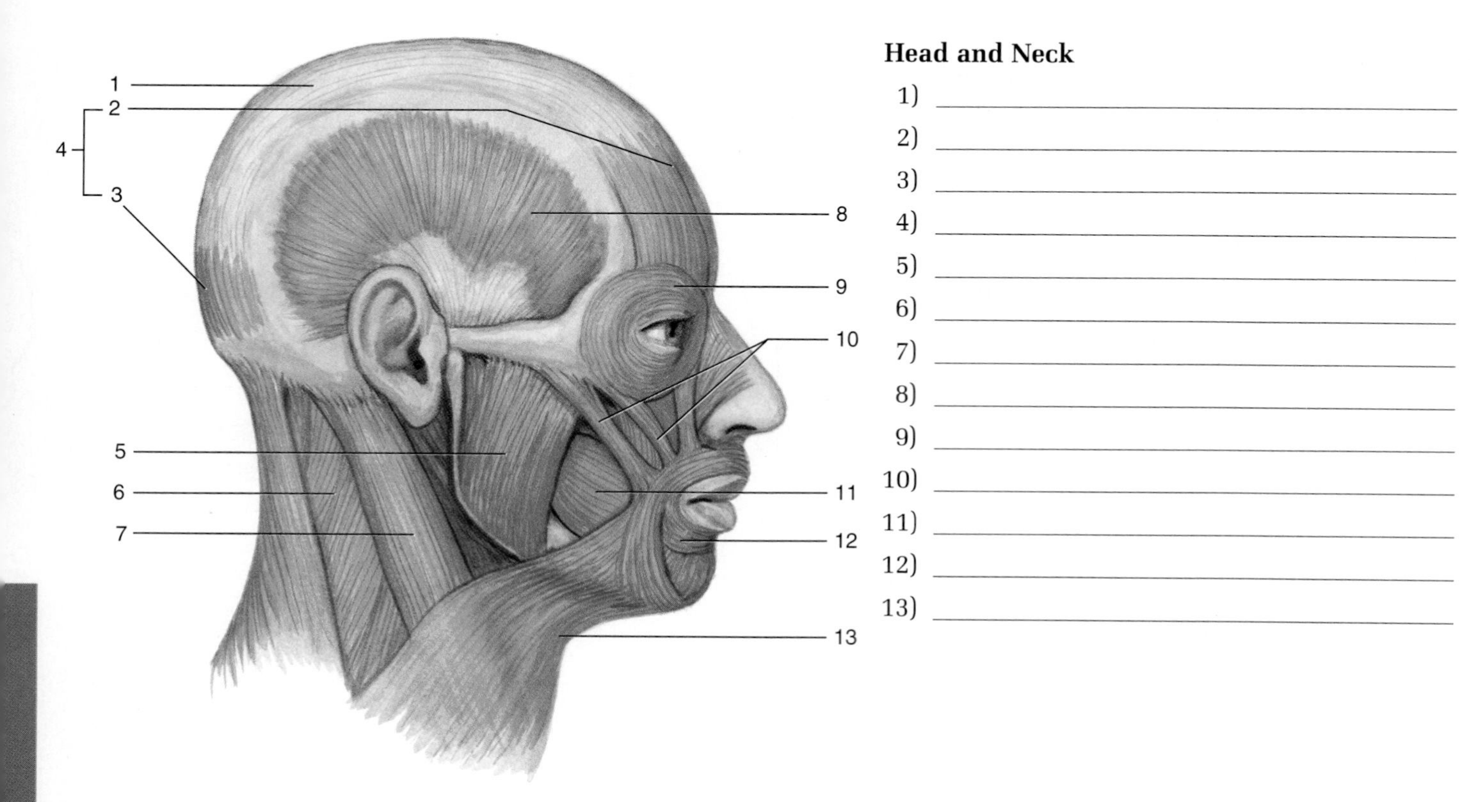

Head and Neck

1) ______________________
2) ______________________
3) ______________________
4) ______________________
5) ______________________
6) ______________________
7) ______________________
8) ______________________
9) ______________________
10) ______________________
11) ______________________
12) ______________________
13) ______________________

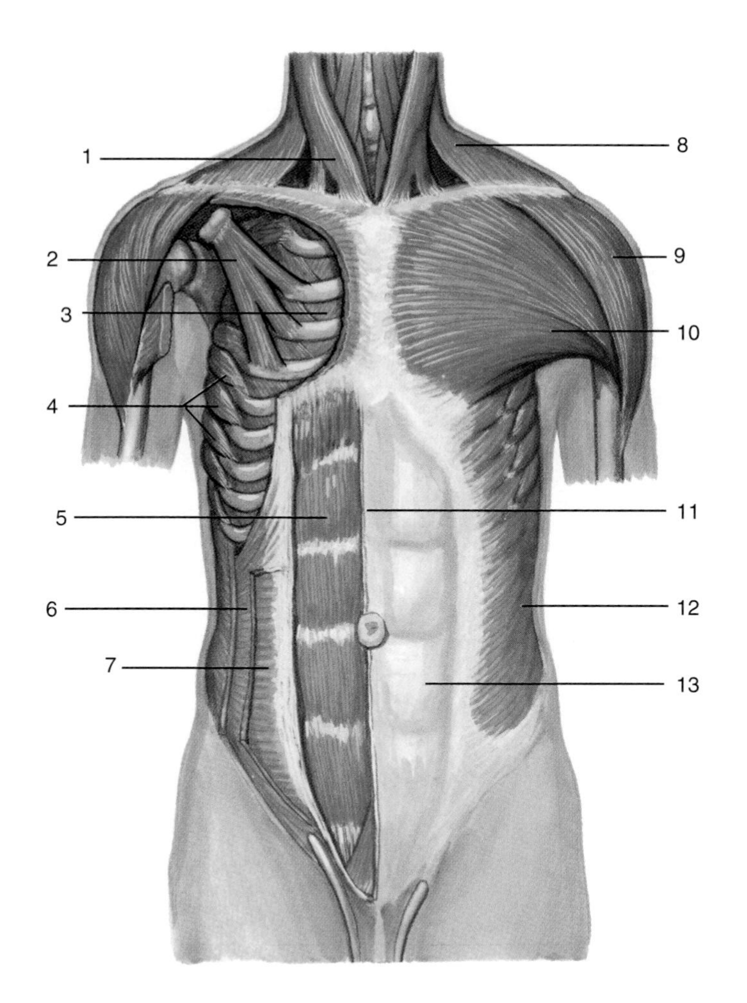

Anterior Trunk

1) ______
2) ______
3) ______
4) ______
5) ______
6) ______
7) ______
8) ______
9) ______
10) ______
11) ______
12) ______
13) ______

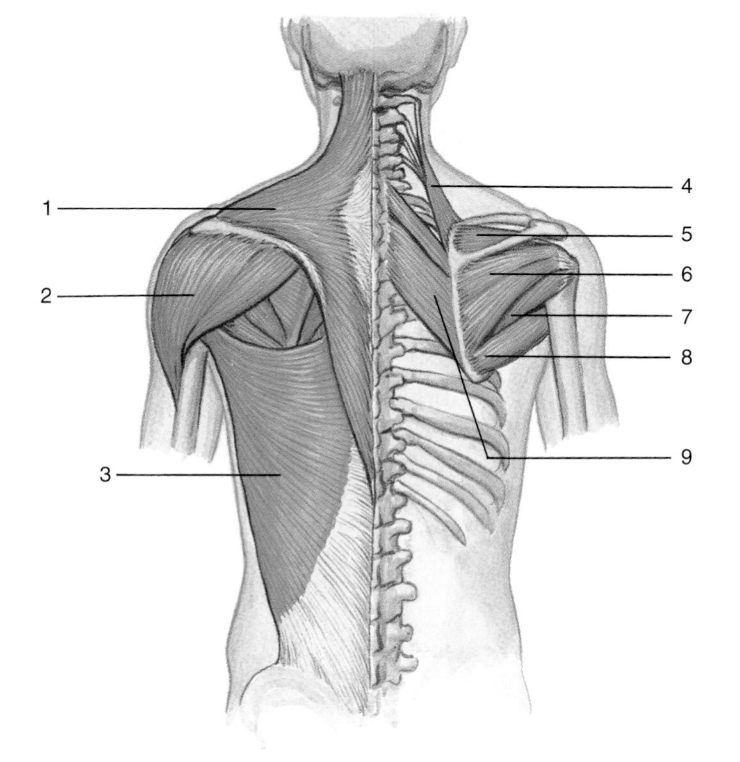

Posterior Trunk

1) ______
2) ______
3) ______
4) ______
5) ______
6) ______
7) ______
8) ______
9) ______

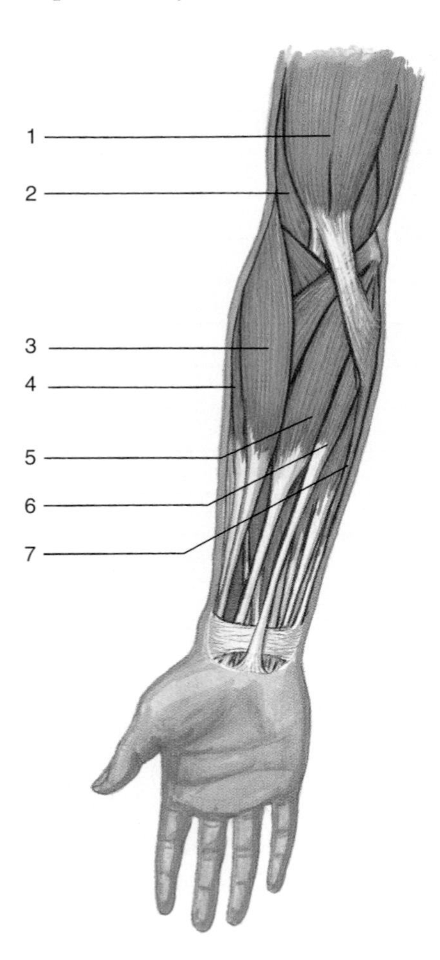

Anterior Forearm

1) ______________________
2) ______________________
3) ______________________
4) ______________________
5) ______________________
6) ______________________
7) ______________________

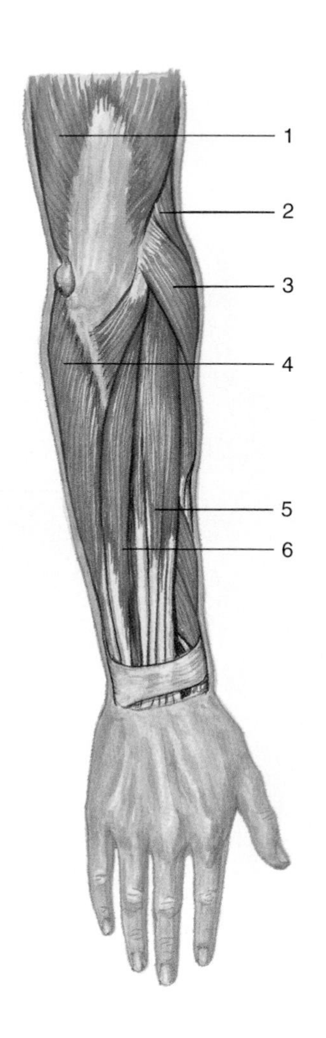

Posterior Forearm

1) ______________________
2) ______________________
3) ______________________
4) ______________________
5) ______________________
6) ______________________

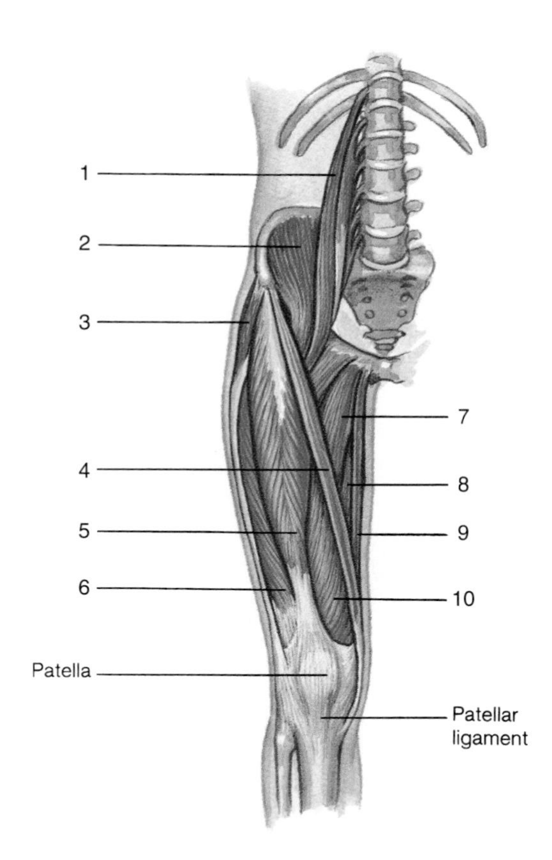

Anterior Thigh

1) ______
2) ______
3) ______
4) ______
5) ______
6) ______
7) ______
8) ______
9) ______
10) ______

Posterior Thigh

1) ______
2) ______
3) ______
4) ______
5) ______
6) ______
7) ______
8) ______

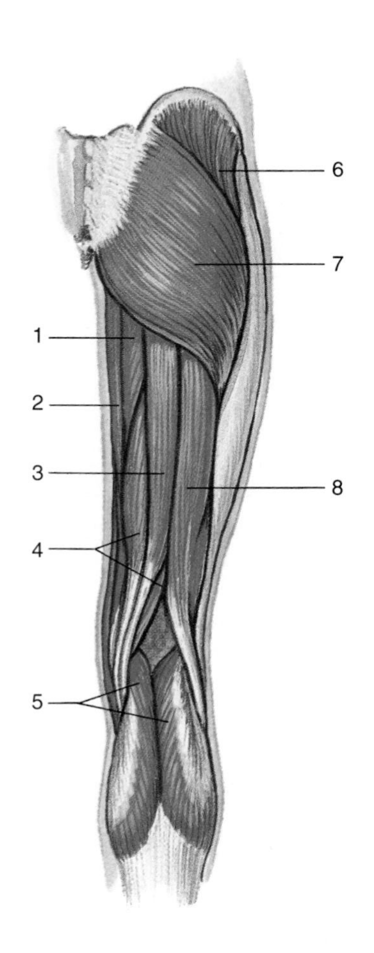

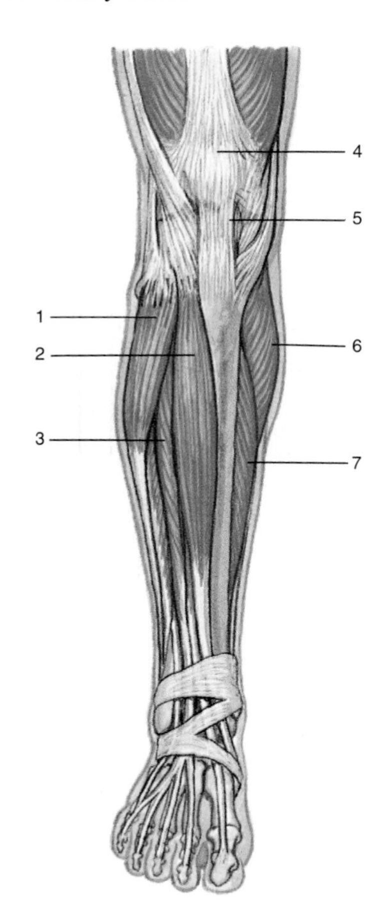

Anterior Leg

1) ______
2) ______
3) ______
4) ______
5) ______
6) ______
7) ______

Lateral Leg

1) ______
2) ______
3) ______
4) ______
5) ______
6) ______
7) ______
8) ______

7. Clinical Applications

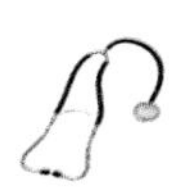

a. The accumulation of lactic acid can make muscles sore. Would heat or cold applications be best to alleviate the soreness? _____ Explain. ______

b. While playing tennis, Jim had a sudden pain on the back of his left thigh. Was this a sprain or a strain? ______ What muscles were probably involved? ______

c. Tom has been working out to build up his muscles. At the microscopic level, how does a muscle increase in size and strength? ______

STUDY GUIDE 8

1. Divisions of the Nervous System

1) List the anatomical subdivisions. ______________________

____

2) List the functional subdivisions. ______________________

____

2. Nerve Tissue

a. Write the terms that match the statements in the spaces at the right.

1) Process conducting impulses toward the cell body. ______________________
2) Process conducting impulses from the cell body. ______________________
3) Fatty insulation on some neuron processes. ______________________
4) Required for neuron process regeneration. ______________________
5) Neuron type carrying impulses toward the CNS. ______________________
6) Neuron type carrying impulses from the CNS. ______________________
7) Neuron type carrying impulses within the CNS. ______________________
8) Small spaces between Schwann cells. ______________________

b. Write the names of the neuroglial cells that match the statements.

1) Line the ventricles of the brain. ______________________
2) Form myelin of neurons in the CNS. ______________________
3) Engulf bacteria and debris. ______________________
4) Form myelin of neurons in the PNS. ______________________
5) Bind neurons with blood vessels. ______________________

c. Label the figure by placing the number of the structure by the correct label. Draw arrows to show direction of impulse transport from dendrite tip to axon tip.

_____ Axon
_____ Cell body
_____ Dendrite
_____ Myelin sheath
_____ Nodes of Ranvier
_____ Nucleus
_____ Nucleus of Schwann cell

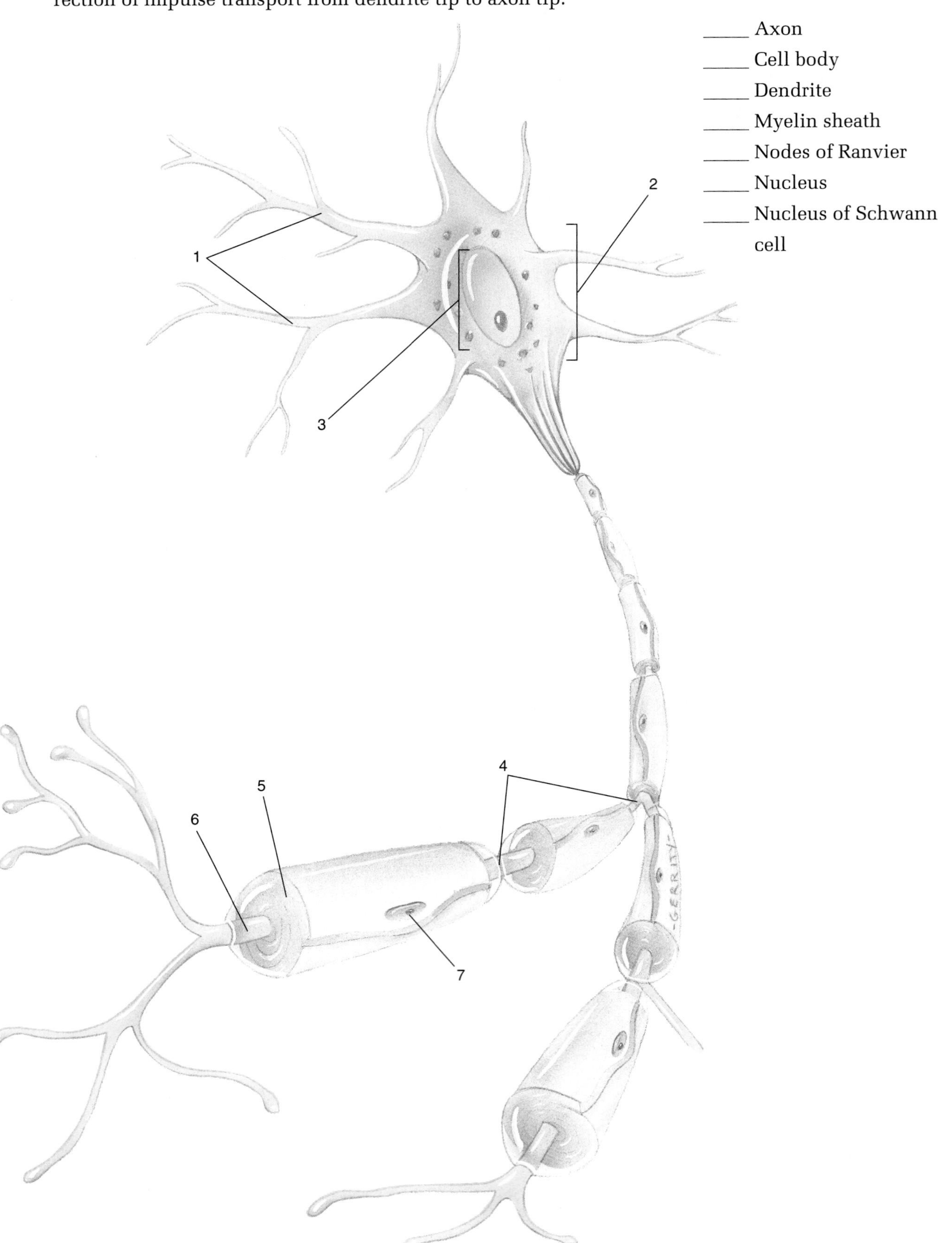

3. Neuron Physiology

a. Write the terms that complete the sentences in the spaces at the right.

In a resting neuron, __1__ ions are actively pumped out, which causes the membrane to be __2__ with an excess of positive ions __3__ the membrane and an excess of negative ions __4__ the neuron. A threshold stimulus makes the membrane permeable to __5__ ions that rapidly diffuse into the neuron, which depolarizes the membrane forming a nerve __6__. An impulse is conducted as a __7__ wave passes along a neuron. A depolarized membrane is repolarized when __8__ ions diffuse out of the neuron and replace the __9__ ions that entered the neuron. Impulse conduction is more rapid in __10__ nerve fibers.

In synaptic transmission, an impulse reaching a synaptic knob causes the release of a __11__ into the __12__. The __13__ binds with __14__ on the postsynaptic neuron, causing an __15__ to be formed. An enzyme quickly breaks down the __16__ and restores the synapse to its resting state.

1) ________________
2) ________________
3) ________________
4) ________________
5) ________________
6) ________________
7) ________________
8) ________________
9) ________________
10) ________________
11) ________________
12) ________________
13) ________________
14) ________________
15) ________________
16) ________________

b. Indicate the excitatory (+) and inhibitory (−) transmitters.

____ Acetylcholine   ____ Dopamine

____ GABA   ____ Norepinephrine

c. If a drug prevents an excitatory neurotransmitter from binding to receptors of the postsynaptic neuron, will synaptic transmission occur? ____ How might such a drug be useful in a clinical situation?

____

____

____

____

4. Protection for the Central Nervous System

Write the terms that match the phrases in the spaces at the right.

1) Outermost membrane of meninges. ________________
2) Intermediate membrane of meninges. ________________
3) Innermost membrane of meninges. ________________
4) Portion of skull protecting brain. ________________
5) Meningeal space with cerebrospinal fluid. ________________
6) Space filled with fatty connective tissue in vertebral canal. ________________
7) Provides bony protection for spinal cord. ________________

5. The Brain

a. Label the figure by placing the number of the structure by the correct label.

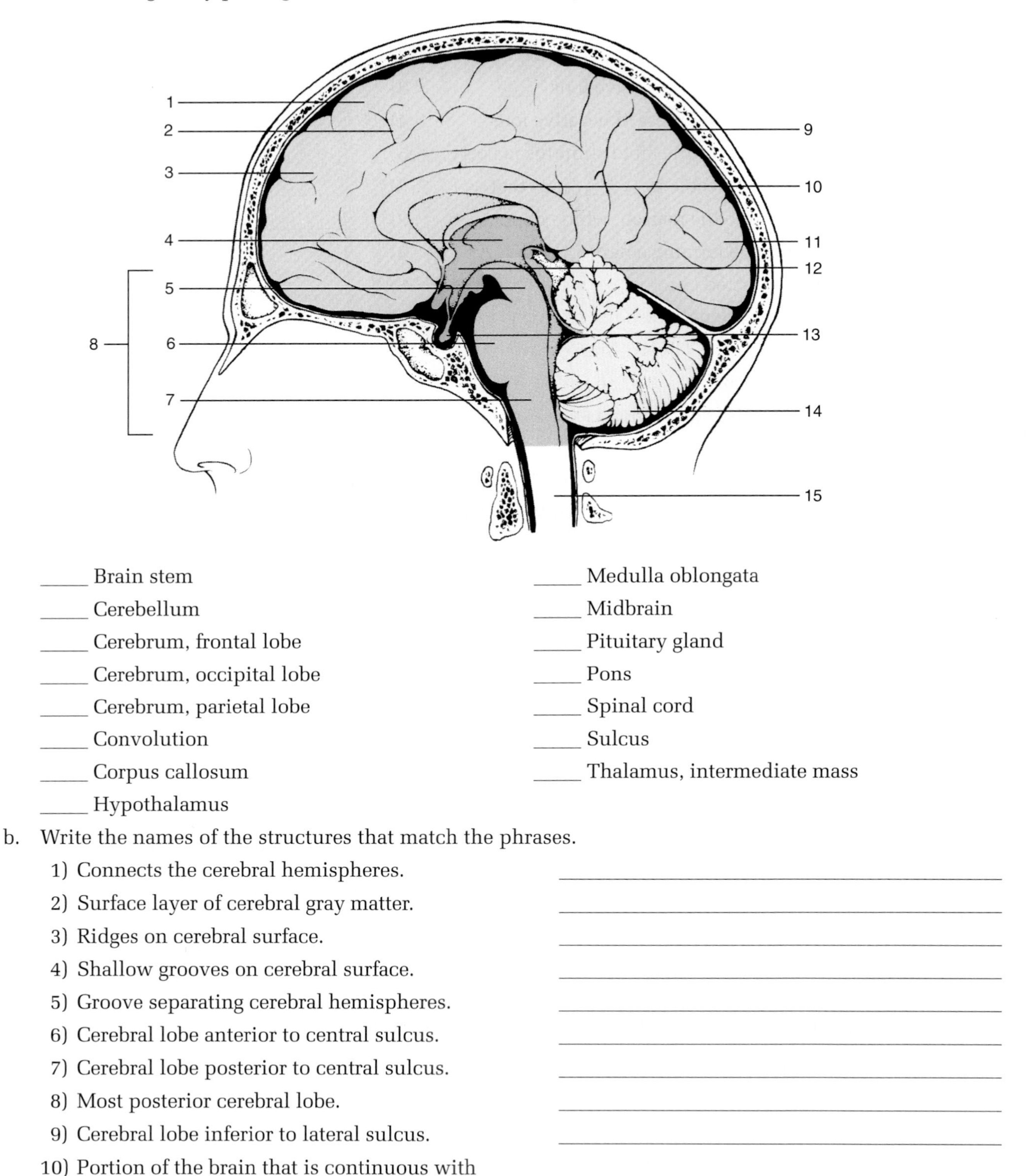

_____ Brain stem
_____ Cerebellum
_____ Cerebrum, frontal lobe
_____ Cerebrum, occipital lobe
_____ Cerebrum, parietal lobe
_____ Convolution
_____ Corpus callosum
_____ Hypothalamus
_____ Medulla oblongata
_____ Midbrain
_____ Pituitary gland
_____ Pons
_____ Spinal cord
_____ Sulcus
_____ Thalamus, intermediate mass

b. Write the names of the structures that match the phrases.

1) Connects the cerebral hemispheres. ____________
2) Surface layer of cerebral gray matter. ____________
3) Ridges on cerebral surface. ____________
4) Shallow grooves on cerebral surface. ____________
5) Groove separating cerebral hemispheres. ____________
6) Cerebral lobe anterior to central sulcus. ____________
7) Cerebral lobe posterior to central sulcus. ____________
8) Most posterior cerebral lobe. ____________
9) Cerebral lobe inferior to lateral sulcus. ____________
10) Portion of the brain that is continuous with the spinal cord. ____________
11) Two lateral masses of gray matter connected by the intermediate mass. ____________
12) Forms the floor of third ventricle. ____________

13) Formed of pons, midbrain, and medulla. ______
14) Lowest portion of the brain. ______
15) Superior portion of brain stem. ______
16) Consists of two lateral hemispheres separated by the vermis. ______
17) Cavities in the cerebral hemispheres. ______
18) Cavity continuous with the central canal of the spinal cord. ______
19) Cavity between lateral masses of the thalamus and above the hypothalamus. ______
20) Second largest portion of the brain. ______

c. Match the parts of the brain with the functions described.

1) Cerebrum
2) Cerebellum
3) Hypothalamus
4) Medulla oblongata
5) Limbic system
6) Midbrain
7) Pons
8) Reticular formation
9) Thalamus

_____ Controls body temperature and water balance.
_____ Coordinates body movements, posture, and equilibrium.
_____ Controls moods and emotional behavior.
_____ Controls voluntary actions, will, and intellect.
_____ Provides general, nonspecific awareness of sensations.
_____ Provides specific interpretation of sensations.
_____ Controls wakefulness and arouses cerebrum.
_____ Controls heart rate and blood pressure.
_____ Assists medulla oblongata in control of breathing.
_____ Major center controlling homeostasis.

d. Match parts of the cerebrum with the functions described.

1) Association areas
2) Broca's area
3) Corpus callosum
4) Left hemisphere
5) Occipital lobe
6) Precentral gyrus
7) Premotor area
8) Postcentral gyrus
9) Temporal lobe
10) Right hemisphere

_____ Interprets sensations from the skin.
_____ Center for visual sensations.
_____ Center for sound sensations.
_____ Controls ability to speak.
_____ Primary control area for contractions of skeletal muscles.
_____ Hemisphere controlling verbal and computational skills in most people.
_____ Hemisphere controlling artistic and spatial skills in most people.
_____ Conducts impulses between cerebral hemispheres.
_____ Controls complex and sequential learned motor activities.
_____ Controls will, memory, and intellectual processes.

e. Write the terms that complete the sentences in the spaces at the right.

Cerebrospinal fluid (CSF) is produced by a ___1___ in each ventricle. Most of the CSF is formed within the ___2___ ventricles and flows through two small openings into the median ___3___ ventricle. From here, it passes into the fourth ventricle in the posterior portion of the brain stem. From the ___4___ ventricle, some CSF flows into the ___5___ of the spinal cord, but most of it enters the ___6___ space and flows around the spinal cord and brain. CSF is absorbed into the blood of the ___7___ located in the ___8___ above the longitudinal fissure of the cerebrum.

1) ______________________
2) ______________________
3) ______________________
4) ______________________
5) ______________________
6) ______________________
7) ______________________
8) ______________________

6. The Spinal Cord

a. Write the terms that match the phrases in the spaces at the right.

1) Contain cell bodies of motor neurons. ______________________
2) Composed of myelinated nerve fibers. ______________________
3) Contain cell bodies of interneurons. ______________________
4) Carry impulses toward brain. ______________________
5) Carry impulses from brain. ______________________
6) Vertebra where spinal cord ends. ______________________
7) Part of brain continuous with spinal cord. ______________________
8) Nerves arising from spinal cord. ______________________

b. Label the figure by placing the number of the structure in the space by the correct label.

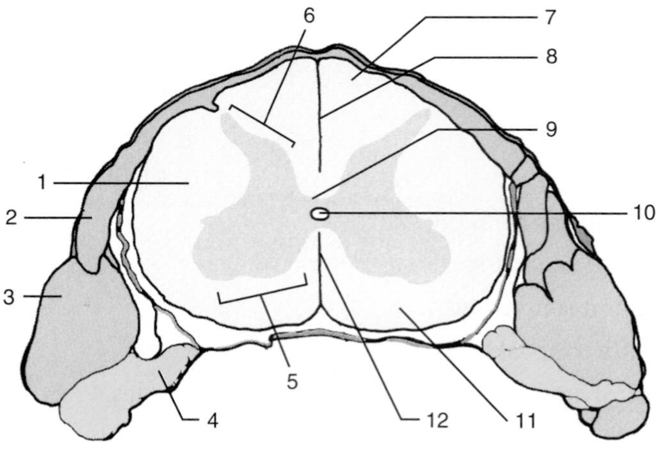

____ Anterior median fissure
____ Central canal
____ Gray matter, commissure
____ Gray matter, anterior horn
____ Gray matter, posterior horn
____ Posterior median fissure
____ Spinal nerve, anterior (ventral) root
____ Spinal nerve, posterior (dorsal) root
____ Spinal nerve, posterior root gangelion
____ White matter, anterior column
____ White matter, lateral column
____ White matter, posterior column

c. Write the terms that match the phrases in the spaces at the right.
   1) Tissue binding nerve fibers together in a nerve. ____________________
   2) Nerves arising from the spinal cord. ____________________
   3) Nerves arising from the brain. ____________________
   4) Nerves composed of both axons and dendrites. ____________________
   5) Sensory neuron processes in spinal nerves. ____________________
   6) Neurons carrying impulses from the CNS. ____________________
   7) Neurons carrying impulses to the CNS. ____________________
   8) Nerves composed of both sensory and motor neuron processes. ____________________

7. The Peripheral Nervous System

a. Write the numerals and names of the cranial nerves described by the phrases.
   1) Transmits motor impulses to facial muscles. ____________________
   2) Transmits sensory impulses from retina. ____________________
   3) Transmits motor impulses to muscles that move the eyes (three nerves). ____________________
      ____
      ____
   4) Transmits sensory impulses from thoracic and abdominal viscera. ____________________
   5) Transmits motor impulses to tongue muscles. ____________________
   6) Transmits sensory impulses from the ear. ____________________
   7) Transmits sensory impulses from "smell" receptors. ____________________
   8) Transmits motor impulses to chewing muscles. ____________________
   9) Transmits sensory impulses from the face. ____________________
   10) Transmits motor impulses to muscles of the larynx and to the trapezius muscle. ____________________
   11) Transmits motor impulses to salivary glands. ____________________
   12) Transmits motor impulses to thoracic and abdominal viscera. ____________________
   13) Transmits sensory impulses from the teeth. ____________________
   14) Transmits motor impulses to the eye that adjust pupil size and lens shape. ____________________

b. Write the terms that match the phrases in the spaces at the right.
   1) Number of pairs of spinal nerves. ____________________
   2) Root containing axons of motor nerves. ____________________
   3) Root containing sensory neurons. ____________________
   4) Spinal nerves that do *not* form plexuses. ____________________
   5) Plexuses from which nerves to the legs emerge. ____________________
   6) Rapid, involuntary, predictable responses. ____________________
   7) Where spinal nerve fibers are sorted and recombined. ____________________

c. Write the terms that complete the sentences in the spaces at the right.

In a somatic spinal reflex, a painful stimulus causes a receptor to form impulses that are carried by a __1__ neuron to the spinal cord. This neuron enters the spinal cord via the __2__ root and synapses with an __3__ in the __4__ horn. This second neuron extends to the __5__ horn, where it synapses with a __6__ neuron that exits the spinal cord via the __7__ root and carries impulses to an __8__ .

1) ____________
2) ____________
3) ____________
4) ____________
5) ____________
6) ____________
7) ____________
8) ____________

8. The Autonomic Nervous System

a. Write the terms that match the phrases in the spaces at the right.

1) Division arising from brain stem and sacral region of spinal cord. ____________
2) Division arising from thorax and lumbar regions of spinal cord. ____________
3) Autonomic ganglia forming a chain on each side of the vertebral column. ____________
4) Neurotransmitter secreted by postganglionic sympathetic neurons. ____________
5) Neurotransmitter secreted by postganglionic parasympathetic neurons. ____________
6) Division preparing body for emergencies. ____________
7) Division promoting digestion. ____________
8) Division increasing blood pressure. ____________

b. Write the terms that complete the sentences in the spaces at the right.

In a spinal autonomic visceral reflex, a __1__ neuron exits the spinal cord via an __2__ root and extends to an autonomic __3__ , where it synapses with a __4__ neuron that continues to an effector.

1) ____________
2) ____________
3) ____________
4) ____________

9. Disorders of the Nervous System

Write the disorders described by the phrases.

1) Infection of the meninges. ____________
2) Caused by a deficiency of dopamine. ____________
3) Severe memory loss caused by a loss of certain cholinergic neurons in the brain. ____________
4) Blood clot, aneurysm, or hemorrhage in brain. ____________
5) Destruction of myelin sheath in CNS. ____________

6) Prenatal brain damage, often by viral infections like German measles. ____________________
7) Convulsive seizures. ____________________
8) A severe jarring of the brain. ____________________
9) Neuritis by reactivation of chickenpox virus. ____________________
10) Results from inactive reticular formation. ____________________

10. Clinical Applications

a. A stroke patient is paralyzed on the right side of the body and cannot speak. What part of the brain is affected by the stroke? ____________________

b. A patient complains of severe pain in the left lower back that extends down the thigh. A nurse describes the pain as neuralgia. What is neuralgia? ____________________

What nerve is the likely source of the pain? ____________________

c. The poliomyelitis virus destroys neuron cell bodies in the anterior horn of the spinal cord. If the spinal nerve affected innervates the left leg, what will be the effect on the patient? ____________________

Will the patient fully recover? __________ Explain. ____________________

STUDY GUIDE 9

1. Sensations

a. Match the structures with the statements that follow.

1) Cerebral cortex 2) Nerve fiber 3) Receptor

_____ Carries impulses.
_____ Forms sensory impulses.
_____ Interprets impulses as sensations.
_____ Decreases impulse formation when repeatedly stimulated.
_____ Projects sensation back to region where impulses seem to originate.
_____ Sensitive to a particular type of stimulus.
_____ Exhibits adaptation.

b. Define adaptation. ____

____

2. General Senses

a. Match the responses with the statements that follow.

1) Cold receptors
2) Free nerve endings
3) Heat receptors
4) Meissner's corpuscles
5) Pressure receptors

_____ Pain receptors.
_____ Touch receptors.
_____ Most sensitive to temperatures over 25°C.
_____ Most sensitive to temperatures under 10°C.
_____ May be located in epidermis.
_____ Located in visceral organs.
_____ Located in superficial portion of dermis.
_____ Located in dermis and joints.
_____ Temperature receptors closest to epidermis.
_____ Touch receptor abundant in hairless skin.
_____ Only in the skin.

b. What is referred pain? ____

____

____

3. Taste and Smell

Write the terms described by the statements in the spaces at the right.

1) Organs containing taste receptors. ____________________
2) Receptors located in nasal epithelium. ____________________
3) Type of taste receptors at back of tongue. ____________________
4) Type of taste receptors at sides and tip of tongue. ____________________
5) Type of taste receptors at tip of tongue only. ____________________
6) Type of taste receptors at sides of tongue only. ____________________

4. Ear Structure

a. Label the figure by placing the number of the structure in the space by the correct label.

_____ Auditory tube	_____ Malleus	_____ Stapes
_____ Auricle	_____ Oval window	_____ Tympanic cavity
_____ Cochlea	_____ Round window	_____ Tympanic membrane
_____ External auditory canal	_____ Semicircular canals	_____ Vestibulocochlear nerve
_____ Incus		

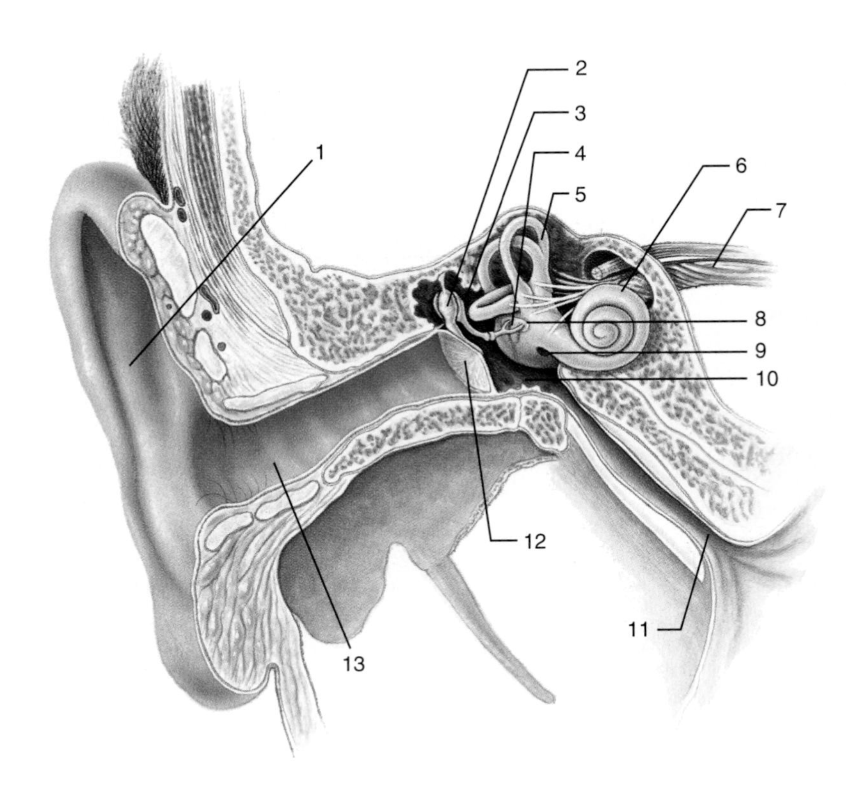

b. Write the terms that match the statements in the spaces at the right.

1) Part of the bony labyrinth that contains receptors for
   a) hearing; _______________
   b) static balance; _______________
   c) dynamic balance. _______________
2) Fills membranous labyrinth. _______________
3) Fills bony labyrinth. _______________
4) Fills tympanic cavity. _______________

c. Label the figure by placing the number of the structure in the space by the correct label.

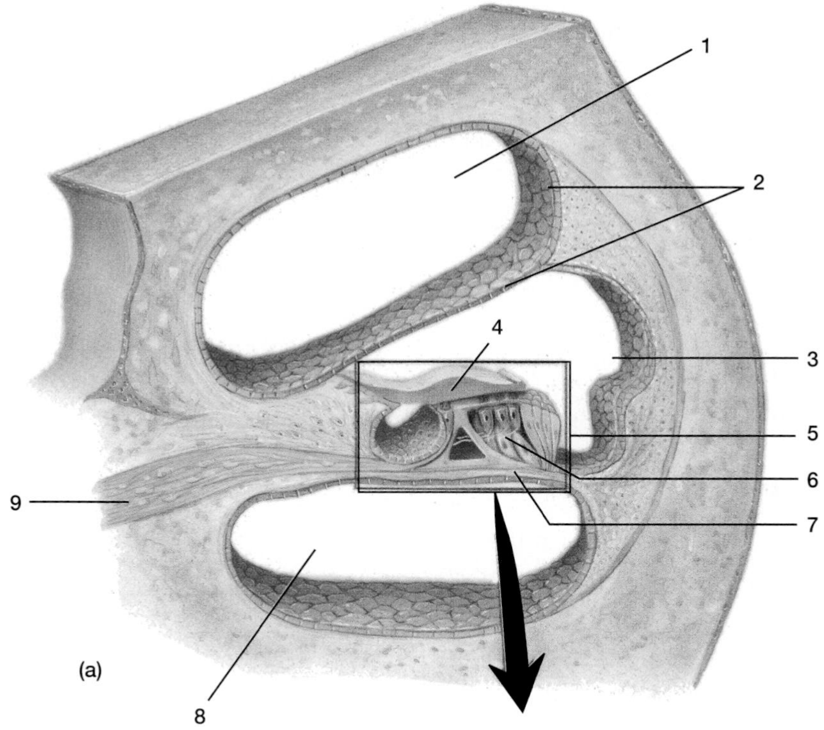

____ Basilar membrane
____ Cochlear duct
____ Cochlear portion of the vestibulocochlear nerve
____ Hair cells
____ Organ of Corti
____ Scala tympani
____ Scala vestibuli
____ Tectorial membrane
____ Vestibular membrane

5. Hearing

a. Write the terms that match the statements in the spaces at the right.

1) Receptor organ for hearing. ____________________
2) Receptor cells for hearing. ____________________
3) Carry vibrations from eardrum to perilymph. ____________________
4) Directs sound waves to tympanic membrane. ____________________
5) Contains fibers of increasing length. ____________________
6) Membrane-covered opening into scala tympani. ____________________
7) Membrane struck by sound waves. ____________________
8) Membrane that determines pitch of sound. ____________________
9) Membrane contacting hairs of receptor cells. ____________________
10) Allows air to enter tympanic cavity. ____________________

b. Write the terms that complete the sentences in the spaces at the right.

Sound waves enter the __1__ and strike the __2__, causing it to vibrate. This vibration is transmitted by the __3__ to the __4__ that fills the scala vestibuli and scala tympani. Oscillating movements of this fluid cause comparable vibrations of portions of the __5__ membrane and the __6__ that rests upon it. This causes the hair cells to contact the __7__, which stimulates them to form __8__ that are carried to the brain by the __9__ nerve. The hearing centers in the __10__ lobes interpret these impulses as sound sensations.

1) ____________________
2) ____________________
3) ____________________
4) ____________________
5) ____________________
6) ____________________
7) ____________________
8) ____________________
9) ____________________
10) ____________________

6. Equilibrium

Write the terms that match the statements in the spaces at the right.

1) Receptor organ for static equilibrium. ______________________
2) Chambers containing receptors for static equilibrium. ______________________
3) Force stimulating hair cells of macula. ______________________
4) Receptor organ for dynamic equilibrium. ______________________
5) Locations of receptor organs for dynamic equilibrium. ______________________
6) Fluid moving cupula when head is turned. ______________________
7) Part of brain controlling equilibrium. ______________________

7. Accessory Structures of the Eye

Write the terms that match the statements in the spaces at the right.

1) Lines eyelids and covers anterior sclera. ______________________
2) Group of muscles that move the eye. ______________________
3) Secretes tears. ______________________
4) Collect tears at inner corner of eye. ______________________
5) Carries collected tears into nasal cavity. ______________________

8. Eye Structure

a. Write the terms that match the statements in the spaces at the right.

   1) Fills anterior cavity. ______________________
   2) Pigmented layer of the eyeball. ______________________
   3) Substance holding retina against choroid. ______________________
   4) Opening in center of iris. ______________________
   5) Fluid mostly responsible for internal pressure in eye. ______________________
   6) Protective outer fibrous coat of eye. ______________________

b. Label the figure by placing the number of the structure in the space by the correct label.

_____ Aqueous humor in anterior cavity	_____ Optic disk
_____ Choroid coat	_____ Optic nerve
_____ Ciliary body	_____ Pupil
_____ Conjunctiva	_____ Retina
_____ Cornea	_____ Sclera
_____ Fovea centralis	_____ Suspensory ligaments
_____ Iris	_____ Vitreous humor in posterior cavity
_____ Lens	

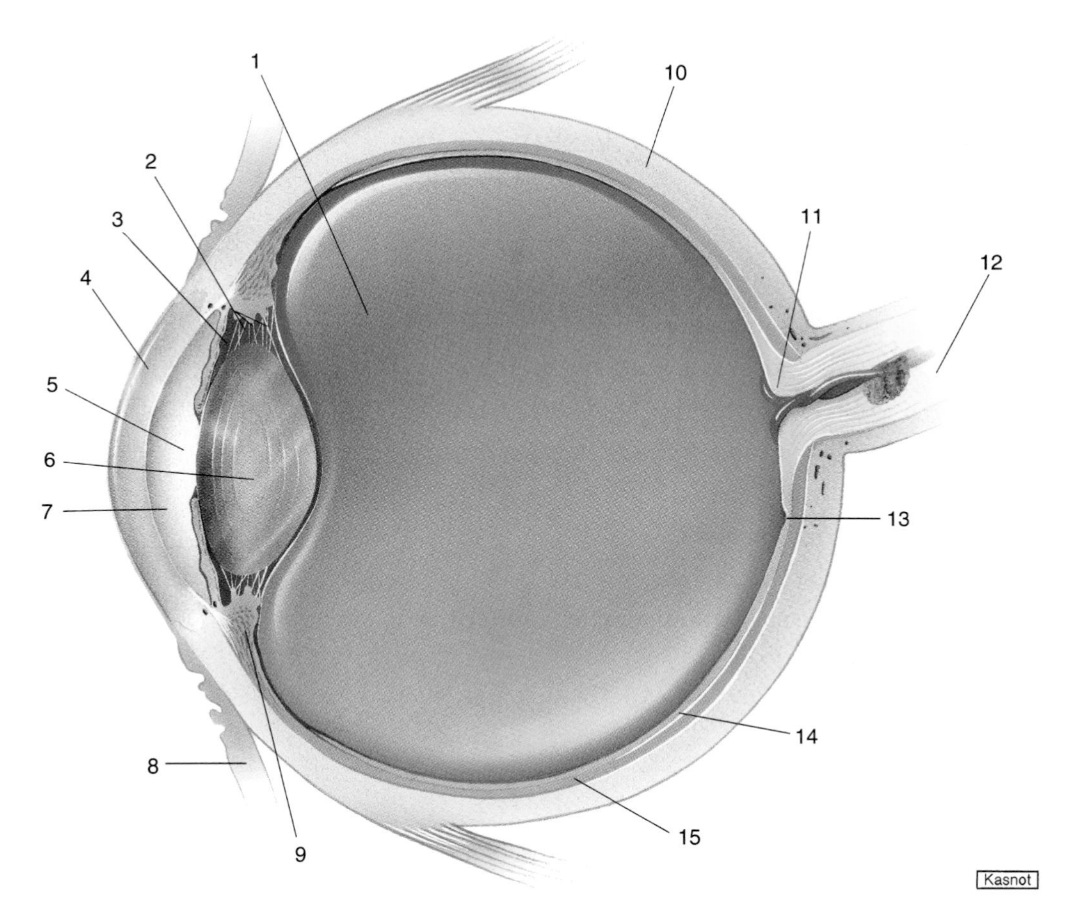

9. Vision

Write the terms that match the statements in the spaces at the right.

1) Contains photoreceptors. ______________________
2) Site of direct vision. ______________________
3) Clear window through which light enters eye. ______________________
4) Controls amount of light entering eye. ______________________
5) Focuses light rays on retina. ______________________
6) Layer containing blood vessels. ______________________
7) Changes shape of the lens. ______________________
8) Retinal area lacking photoreceptors. ______________________
9) Causes greatest bending of light rays. ______________________

10) Absorbs excessive light in eye. ____________
11) Receptors for dim light vision. ____________
12) Receptors for color vision. ____________
13) Receptors absent in fovea. ____________
14) Carries impulses from retina to brain. ____________
15) Where medial nerve fibers cross over. ____________
16) Light-sensitive pigment in rods. ____________
17) Colors of light absorbed by three types of cones. ____________
____
____
18) Vitamin required for rhodopsin synthesis. ____________

10. Disorders of Hearing and Vision

Write the disorders described in the spaces at the right.

1) An infection called "pink eye." ____________
2) Cloudiness of the lens. ____________
3) Acute infection of the middle ear. ____________
4) Deafness due to exposure to loud noises. ____________
5) Deafness correctible by hearing aids. ____________
6) Results from unequal curvatures of lens or cornea. ____________
7) Corrected by convex lenses. ____________
8) Corrected by concave lenses. ____________
9) Decreased elasticity of the lens. ____________
10) Nausea due to repeated stimulation of equilibrium receptors. ____________
11) Group of disorders producing nausea, dizziness, and tinnitis. ____________
12) Cancer of immature retinal cells. ____________
13) Caused by excessive intraocular pressure. ____________
14) Crossed eyes. ____________

11. Clinical Applications

a. An older patient calls the office and complains of pain at the base of the neck, left shoulder, and left arm. What is the probable cause of the pain? ____________ Why was the pain localized in these areas?
____
____
What would you advise the patient to do? ____________

b. An audiometry test verifies that a college student has a decreased sense of hearing. Discussion brings out that he has been working as an audio engineer at a local disco for three years. What relationship probably exists between his job and his hearing loss? ____________
____
____

STUDY GUIDE 10

1. Endocrine Glands

a. Label the endocrine glands in the figure by placing the number of the gland in the space beside the correct label.

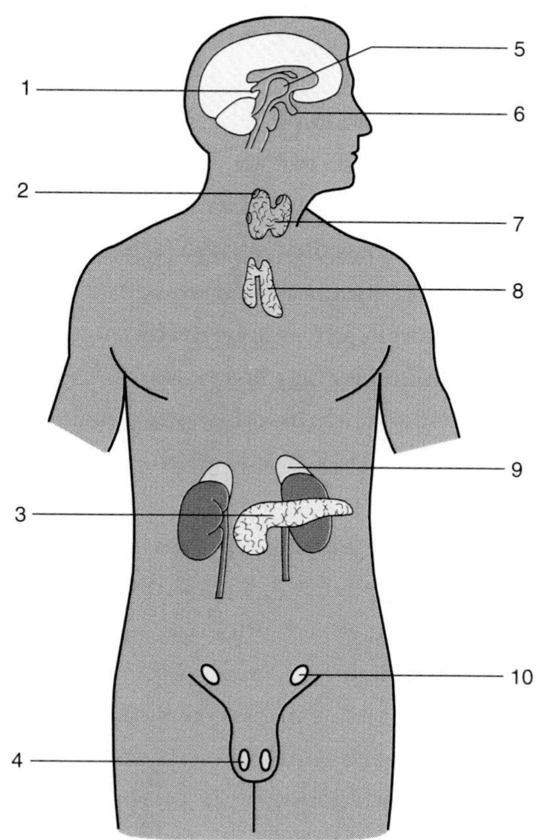

_____ Adrenal
_____ Hypothalamus
_____ Ovary
_____ Pancreas
_____ Parathyroid
_____ Pineal
_____ Pituitary
_____ Testis
_____ Thymus
_____ Thyroid

b. Contrast exocrine and endocrine glands.

1) Secretions of exocrine glands are carried by ______________________________.

2) Secretions of endocrine glands are carried by ______________________________.

2. The Nature of Hormones

a. Match the hormone with the correct statement.

1) Steroid hormone

2) Nonsteroid hormone

_____ Binds to a plasma membrane receptor.

_____ Requires a second messenger.

_____ Fat-soluble hormone.

_____ Receptor-hormone complex causes DNA to initiate synthesis of new proteins (enzymes).

_____ Binds to receptor within the target cells.

b. Write the terms that match the statements in the spaces at the right.
   1) Chemical messengers. ______
   2) Carries hormones throughout the body. ______
   3) Glands producing hormones. ______
   4) Cells containing hormone receptors. ______
   5) Excessive production of a hormone. ______
   6) Deficient production of a hormone. ______
   7) Usual regulatory mechanism for hormone production. ______
   8) Local "hormones" produced by nonendocrine cells. ______

3. Pituitary Gland

a. Write the names of the pituitary hormones that match the statements in the spaces at the right.
   1) Stimulates secretion of thyroid hormone. ______
   2) Stimulates cell growth and division. ______
   3) Stimulates secretion of estrogens. ______
   4) Stimulates secretion of testosterone. ______
   5) Stimulates secretion of cortisol. ______
   6) Stimulates sperm production. ______
   7) Stimulates water retention by kidneys. ______
   8) Stimulates contraction of uterus. ______
   9) Stimulates secretion of progesterone. ______
   10) Causes the onset of puberty. ______

b. Match the lobe with the hormone it produces.
   1) Anterior lobe 2) Posterior lobe

   1) ____ ACTH ____ 3) Prolactin ____ 5) TSH ____ 7) Growth hormone
   2) ____ Oxytocin ____ 4) FSH and LH ____ 6) ADH

4. Thyroid and Parathyroid Glands

Write the terms that match the statements in the spaces at the right.

1) Element essential for activity of thyroxine. ______
2) Hormone that increases metabolic rate. ______
3) Hormone that increases blood calcium. ______
4) Hormone whose secretion is controlled by TSH. ______
5) Hormone that decreases blood calcium. ______
6) Gland that secretes calcitonin. ______
7) Controls secretion of parathyroid hormone. ______

5. Adrenal Glands

Write the terms that match the statements in the spaces at the right.

1) Converts glycogen into glucose. ____________
2) Controls secretion of adrenal medulla. ____________
3) Two related hormones secreted by the adrenal medulla. ____________ ____________
4) Three groups of hormones secreted by adrenal cortex. ____________ ____________ ____________
5) Controls levels of electrolytes in blood. ____________
6) Inhibits inflammation; depresses immunity. ____________
7) Secretion controlled by blood levels of sodium and potassium. ____________
8) Prepares body to meet emergencies. ____________
9) Increases blood levels of sodium and water. ____________
10) Converts noncarbohydrates into glucose. ____________
11) Increases heart rate and blood pressure. ____________
12) Secretion controlled by ACTH. ____________

6. Pancreas

Write the terms that match the statements in the spaces at the right.

1) Portion of gland secreting hormones. ____________
2) Hormone decreasing blood glucose. ____________
3) Hormone aiding movement of glucose into cells. ____________
4) Hormone increasing blood glucose. ____________
5) Controls secretion of pancreatic hormones. ____________
6) Secretion stimulated by high glucose levels. ____________

7. Gonads, Pineal and Thymus Glands

Write the terms that match the statements in the spaces at the right.

1) Hormones formed by ovaries. ____________ ____________
2) Hormone secreted by testes. ____________
3) Hormone of the pineal gland. ____________
4) Hormone of the thymus gland. ____________
5) Seems to influence biorhythms. ____________
6) Stimulates development of male sex organs and secondary sexual characteristics. ____________
7) Stimulates development of female sex organs and secondary sexual characteristics. ____________
8) Involved in maturation of T lymphocytes. ____________

8. Disorders of the Endocrine System

Write the names of the disorders described below in the spaces at the right.

1) Hypersecretion of GH in adults. ______
2) Production of large amounts of dilute urine. ______
3) Enlarged thyroid due to lack of iodine. ______
4) Excessive metabolic rate and bulging eyes. ______
5) Hyposecretion of thyroid hormone in adults. ______
6) Hyposecretion of GH in growing years. ______
7) Hyposecretion of aldosterone and cortisol. ______
8) Hypersecretion of glucocorticoids. ______
9) Hyposecretion of ADH. ______
10) Continued growth of bones of face and hands. ______
11) Inability of glucose to enter body cells. ______
12) Hypersecretion of thyroxine. ______
13) Mental retardation, sluggishness, and stunted growth in an infant. ______
14) Coarse, dry skin and hair; edema; and sluggishness in adult. ______
15) Round, full face; high blood pressure; high blood glucose; and decreased immunity. ______

9. Clinical Applications

a. A patient is taken to the emergency room by her husband. She is sweating and breathing rapidly. A blood test reveals acidosis and hyperglycemia. What hormone should be administered immediately? ______ Explain. ______

b. A new mother is informed that her baby has severe hypothyroidism. How would you explain the importance of thyroxine medication for her infant? ______

c. A patient with high blood pressure and edema (water-logged tissues) is given a drug that counteracts the action of ADH. Explain why this drug was administered and how it will work. ______

STUDY GUIDE 11

1. General Characteristics of Blood

Write the answers that match the statements in the spaces at the right.

1) pH range of the blood. ______________________
2) Liquid portion of blood. ______________________
3) Blood cells and platelets. ______________________
4) Percentage of blood formed by liquid portion. ______________________
5) Percentage of blood formed by RBCs, WCBs, and platelets. ______________________
6) Average range of blood volume in males. ______________________
7) Average range of blood volume in females. ______________________
8) Basic function of blood. ______________________

2. Red Blood Cells

Write the answers that match the statements in the spaces at the right.

1) Shape of erythrocytes. ______________________
2) Red pigment in erythrocytes. ______________________
3) Primary function of RBCs is transport of ______ . ______________________
4) Iron-containing portion of hemoglobin. ______________________
5) Normal range of RBCs/mm^3 in males. ______________________
6) Normal range of RBCs/mm^3 in females. ______________________
7) Tissue-forming RBCs in children and adults. ______________________
8) Hormone-stimulating RBC production. ______________________
9) Organs releasing the hormone that stimulates RBC production. ______________________
10) Organ producing intrinsic factor. ______________________
11) Vitamins required for RBC production. ______________________
12) Intrinsic factor enables absorption of ______ . ______________________
13) Cells from which RBCs originate. ______________________
14) Average life span of RBCs. ______________________
15) Organs where old RBCs are destroyed. ______________________
16) Phagocytic cells destroying RBCs. ______________________
17) Portion of heme that is recycled. ______________________
18) Portion of heme that is excreted. ______________________
19) Organ where heme breakdown occurs. ______________________

3. White Blood Cells

a. Write the answers that match the statements in the spaces at the right.

1) Cell from which WBCs originate. ______
2) Normal range of WBCs/mm^3 of blood. ______
3) Basic function of WBCs is defense against ______ . ______
4) Where most functions of WBCs occur. ______
5) Group of WBCs with cytoplasmic granules. ______
6) Group of WBCs lacking these granules. ______
7) WBCs with lavender-staining granules. ______
8) WBCs with blue-staining granules. ______
9) Largest leukocytes. ______
10) Smallest leukocytes. ______
11) WBCs with red-staining granules. ______
12) Form 20% to 25% of WBCs. ______
13) Migrate into tissues to become macrophages. ______
14) First WBCs attracted from blood into damaged tissues. ______
15) WBCs that move into tissues to complete clean-up of tissue damage. ______
16) Form 60% to 70% of leukocytes. ______
17) Release histamine in allergic reactions. ______
18) WBCs that neutralize histamine. ______
19) Become mast cells after entering tissues. ______
20) Destroy parasitic worms. ______
21) Produce antibodies. ______
22) Compose 3% to 8% of leukocytes. ______
23) Compose 0.5% to 1.0% of leukocytes. ______
24) Compose 2% to 4% of leukocytes. ______
25) Two major phagocytic WBCs. ______

______

b. Use colored pencils to draw these white blood cells as they appear after staining.

Neutrophil **Eosinophil** **Basophil**

Monocyte **Lymphocyte**

4. Platelets

Write the answers that match the statements in the spaces at the right.

1) Alternate name for platelets. ______________________
2) Size compared to size of RBCs. ______________________
3) Number of platelets per mm^3 of blood. ______________________
4) Cells that fragment to form platelets. ______________________
5) Two functions of platelets. ______________________

5. Plasma

Write the answers that match the statements in the spaces at the right.

1) Constitutes over 90% of plasma. ______________________
2) General term for dissolved substances. ______________________
3) Most abundant plasma proteins. ______________________
4) Plasma proteins that are antibodies. ______________________
5) Plasma protein converted into fibrin. ______________________
6) Plasma proteins transporting lipids. ______________________
7) Plasma proteins helping to regulate pH and osmotic pressure of the blood. ______________________
8) Organ forming most plasma proteins. ______________________
9) Nitrogenous wastes of protein breakdown. ______________________
10) Collective term for inorganic ions in the blood plasma. ______________________

6. Hemostasis

Write the answers that match the statements in the spaces at the right.

1) Three processes of hemostasis in order of occurrence. ______________________
____
____
2) Constriction of damaged blood vessel. ______________________
3) Formed elements that temporarily plug break in damaged blood vessel. ______________________
4) Substance released by platelets and damaged tissues that starts clotting process. ______________________
5) Electrolyte required for clotting to occur. ______________________
6) Threadlike strands forming a blood clot. ______________________
7) Cells that enter clot to form new connective tissue and repair damage. ______________________
8) Enzyme converting fibrinogen into fibrin. ______________________

7. Human Blood Types

Write the answers that match the statements in the spaces at the right.

1) Location of antigens used in blood typing. ____________
2) Location of antibodies against blood typing antigens. ____________
3) Antigen(s) in type A blood. ____________
4) Antigen(s) in type AB blood. ____________
5) Antigen(s) in type O blood. ____________
6) Antibodies in type B blood. ____________
7) Antibodies in type AB blood. ____________
8) Antibodies in type O blood. ____________
9) Antibodies in Rh^- blood of person sensitized to the Rh antigen. ____________
10) Caused by maternal anti-Rh antibodies binding with Rh antigens on fetal RBCs. ____________

8. Disorders of the Blood

Write the answers that match the statements in the spaces at the right.

1) Reduced ability to form blood clots. ____________
2) Reduced capacity to carry oxygen. ____________
3) An excessive concentration of erythrocytes. ____________
4) Infection of lymphocytes by Epstein-Barr virus. ____________
5) Anemia due to a deficiency of iron. ____________
6) Cancer producing excess of leukocytes. ____________
7) Anemia due to inability to absorb vitamin B_{12}. ____________
8) Anemia due to excessive bleeding. ____________
9) Anemia due to sickling of erythrocytes. ____________
10) Anemia due to premature rupture of RBCs. ____________
11) Anemia due to loss of red bone marrow. ____________
12) Fetal blood contains erythroblasts. ____________

9. Clinical Applications

a. A person can receive platelets from anyone, no matter the blood type. How is this possible? ____________

____

b. Chemotherapy is often used to destroy the rapidly dividing cells of a cancer. What impact would chemotherapy have on the production of blood cells? ____________
Explain. ____________

____

c. Mary's blood type is A, Rh^-. She is at the hospital for delivery of her second child, and her first child is Rh^+. The attending physician wants blood available in case the baby exhibits erythroblastosis fetalis upon delivery. What blood type should he order? ____________
Explain. ____________

____

6. Blood Flow and Blood Pressure

a. Write the answers that match the statements in the spaces at the right.

1) Systemic vessel with fastest blood flow. ____________________
2) Vessels with slowest blood flow. ____________________
3) Systemic vessel with greatest blood pressure. ____________________
4) Primary force moving blood. ____________________
5) Two additional forces that help return venous blood to the heart. ____________________
 ____
6) Normal systolic blood pressure. ____________________
7) Normal diastolic blood pressure. ____________________
8) Autonomic center controlling diameter of blood vessels. ____________________
9) Systolic pressure minus diastolic pressure. ____________________
10) Effect on precapillary sphincters by a local decrease in oxygen and pH. ____________________
11) Effect on precapillary sphincters by sympathetic impulses. ____________________

b. Indicate whether the following conditions cause an *increase* (+) or *decrease* (−) in blood pressure.

_____ An increase in peripheral resistance.
_____ A marked decrease in blood volume.
_____ A decrease in cardiac output.
_____ Dilation of a great many arterioles.
_____ A significant increase in plasma proteins.
_____ Sympathetic impulses to arterioles.
_____ Constriction of most arterioles.
_____ An increase in heart rate.

7. Circulation Pathways

Trace the pathway of blood from a ventricle of the heart to the organ indicated and back to an atrium of the heart. Write the names of the correct heart chambers, arteries, and veins in the blanks.

1) Right little finger.

Left ventricle → ____________________ → ____________________ →
____________________ → ____________________ →
____________________ → ____________________ → right little finger
→ ____________________ → ____________________ →
____________________ → ____________________ → ____________________
→ ____________________ atrium.

2) Small intestine.

____________ ventricle → ____________________ → ____________________
→ small intestine → ____________________ →
____________________ → liver → ____________________ →
____________________ → right atrium.

8. Systemic Arteries

Label the figure by writing the names of the numbered arteries in the spaces.

1) ____________________
2) ____________________
3) ____________________
4) ____________________
5) ____________________
6) ____________________
7) ____________________
8) ____________________
9) ____________________
10) ____________________
11) ____________________
12) ____________________
13) ____________________
14) ____________________
15) ____________________
16) ____________________
17) ____________________
18) ____________________
19) ____________________
20) ____________________
21) ____________________
22) ____________________
23) ____________________
24) ____________________
25) ____________________
26) ____________________
27) ____________________
28) ____________________
29) ____________________
30) ____________________

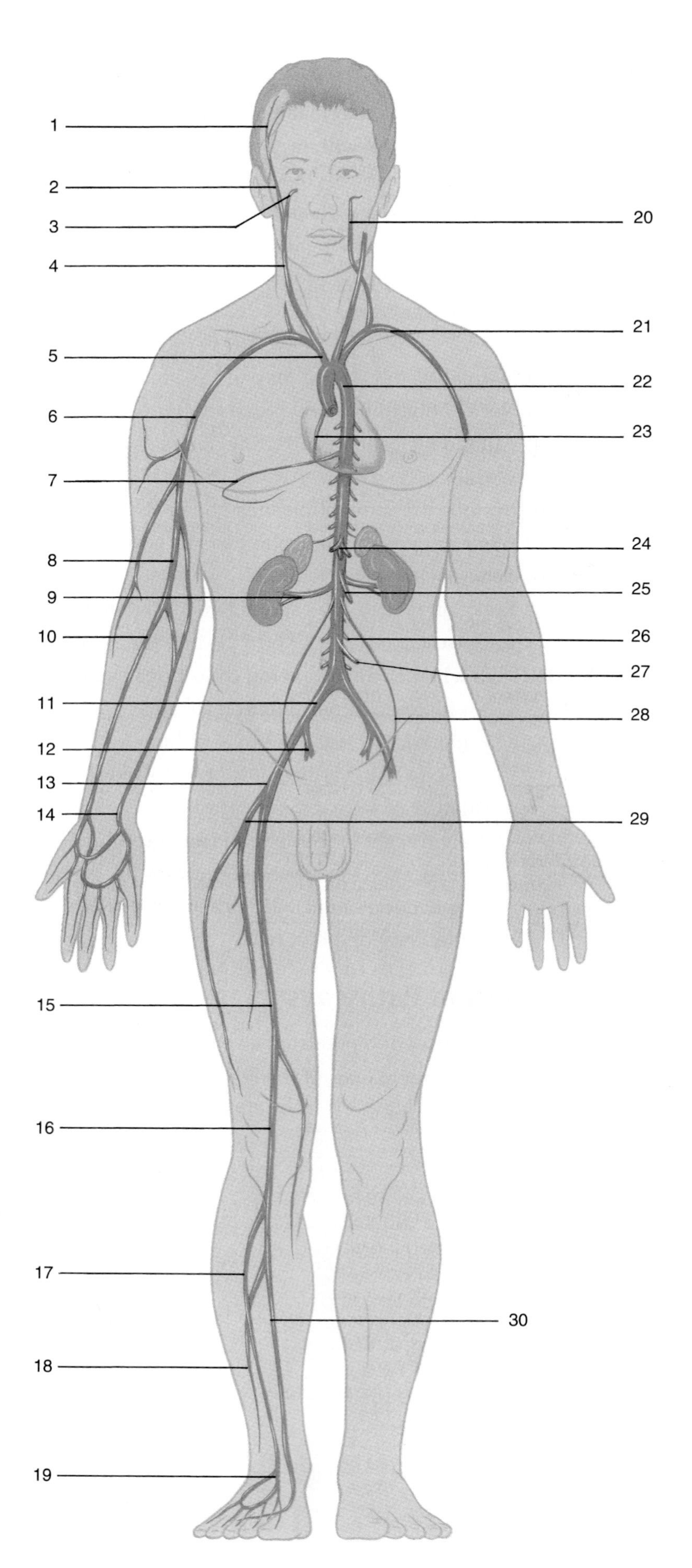

9. Systemic Veins

Label the figure by writing the names of the numbered veins in the spaces.

1) ______________________
2) ______________________
3) ______________________
4) ______________________
5) ______________________
6) ______________________
7) ______________________
8) ______________________
9) ______________________
10) ______________________
11) ______________________
12) ______________________
13) ______________________
14) ______________________
15) ______________________
16) ______________________
17) ______________________
18) ______________________
19) ______________________
20) ______________________
21) ______________________
22) ______________________
23) ______________________
24) ______________________
25) ______________________
26) ______________________
27) ______________________
28) ______________________

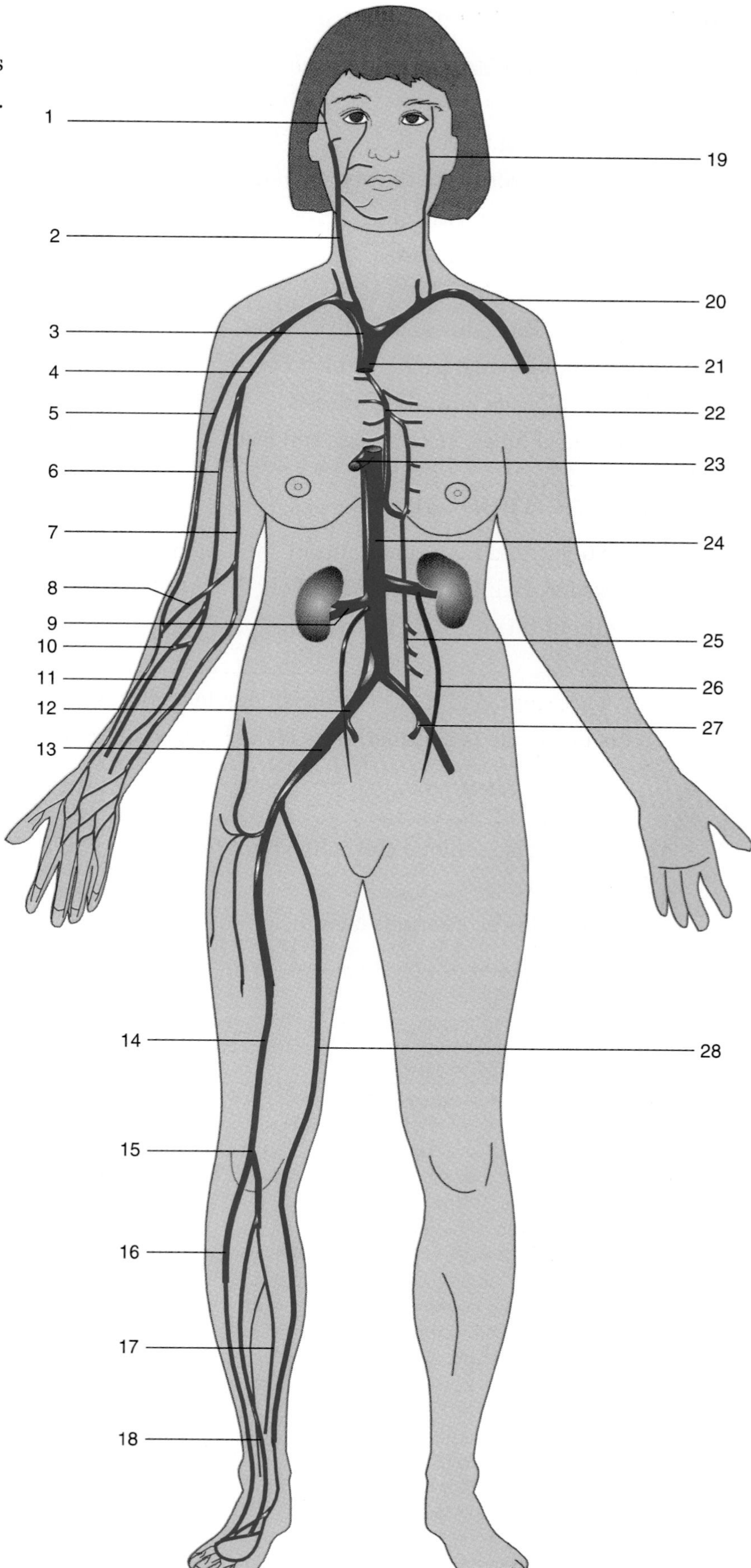

10. Disorders of the Heart and Blood Vessels

Write the disorders described by the statements in the spaces at the right.

1) Unusual heart sounds. ____________________
2) Hardening of the arteries. ____________________
3) Death of a portion of the myocardium. ____________________
4) Abnormal heart rhythm. ____________________
5) Inflammation of a vein. ____________________
6) Chronic high blood pressure. ____________________
7) Swollen veins due to defective valves. ____________________
8) Balloonlike enlargement of blood vessel. ____________________
9) Fatty deposits in walls of arteries. ____________________
10) Edema of lungs, viscera, legs, and feet. ____________________

11. Clinical Applications

a. A 60-year-old man complains of chest pain during moderate exercise. The pain goes away after he rests for a while. What is the likely cause of the pain? ____________________
Without treatment, what complications may arise? ____________________
____

b. An accident victim has lost considerable blood. His blood pressure is only slightly below normal, and his pulse rate is elevated. How is the body compensating for the loss of blood? ____________________
____
____

c. A patient has a blood clot in the right femoral vein. If a part of the clot should break loose, where is it likely to lodge? ____________________
Would this be a serious complication? ____________________ Explain.
____
____
____
____

STUDY GUIDE 13

1. Lymph and Lymphatic Vessels

a. Write the answers that match the statements in the spaces at the right.

1) Fluid within tissue spaces. ____________________
2) Fluid within lymphatic vessels. ____________________
3) Smallest lymphatic vessels. ____________________
4) Lymphatic vessels draining large body regions. ____________________
5) Forms wall of lymphatic capillaries. ____________________
6) Lymphatic duct draining upper right portion of the body. ____________________
7) Lymphatic duct draining rest of the body. ____________________
8) Prevent backflow of lymph. ____________________
9) Provide forces that move lymph. ____________________
____
10) Receives lymph from thoracic duct. ____________________
11) Empties into right subclavian vein. ____________________
12) Vessels collecting interstitial fluid. ____________________
13) Source of interstitial fluid. ____________________

b. Explain the value of the lymphatic system collecting interstitial fluid and returning it to the blood.

____
____
____
____

2. Lymphatic Organs

Write the answers that match the statements in the spaces at the right.

1) Grouped along larger lymphatic vessels. ____________________
2) Bilobed gland located above the heart. ____________________
3) Large lymphatic organ located near stomach. ____________________
4) Clustered at entrance to pharynx. ____________________
5) Organs that filter lymph. ____________________
6) Lymphatic organ that filters blood. ____________________
7) Site of T cell differentiation. ____________________
8) Vessels carrying lymph to lymph node. ____________________
9) Vessels carrying lymph from lymph node. ____________________
10) Intercept pathogens entering pharynx. ____________________
11) Contains a reserve supply of blood. ____________________
12) Hormone that promotes T cell maturation. ____________________

3. Nonspecific Resistance Against Disease

Match the type of nonspecific resistance with the statements.

1) Mechanical barriers
2) Chemicals
3) Phagocytosis
4) Inflammation
5) Fever

_____ Skin
_____ Mucus
_____ Lysozyme
_____ Mucous membranes
_____ Low pH
_____ Gastric juice
_____ Interferon
_____ Produces edema
_____ Pus formation
_____ Flow of tears
_____ Release of histamine
_____ Abnormally high body temperature
_____ Attracts neutrophils and monocytes
_____ Tissue macrophage system
_____ Flow of saliva
_____ Granulocytes and macrophages
_____ Increases local blood supply
_____ Clot seals off pathogens
_____ Speeds up body processes
_____ Pathogens are engulfed and digested

4. Immunity

a. Indicate whether the following statements are true (T) or false (F).

1) Immunity is resistance against specific pathogens. _____
2) Nonspecific resistance is directed against all pathogens. _____
3) Immunity involves granulocytes and macrophages. _____
4) Immunity requires lymphocytes to distinguish between self and nonself molecules. _____
5) Antigens are foreign molecules that cause an immune response. _____
6) Undifferentiated lymphocytes are produced in the spleen. _____
7) All lymphocytes differentiate in the thymus gland. _____
8) The majority of lymphocytes in the blood are T cells. _____
9) Differentiation of lymphocytes occurs throughout life. _____
10) T cells provide cell-mediated immunity. _____
11) B cells provide antibody-mediated immunity. _____
12) Lymphocyte receptors for specific antigens are inherited. _____
13) Lymphocyte receptors are formed by contact with specific antigens. _____
14) There are thousands of different types of B and T cells, and each type responds to a different specific antigen. _____
15) Immunity depends upon lymphocytes whose receptors fit with a specific antigen. _____
16) Immunity involves the interaction of lymphocytes, antigens, and macrophages. _____
17) At any one time, either cell-mediated immunity or antibody-mediated immunity is at work; never both at the same time. _____
18) Reproduction of differentiated lymphocytes occurs in lymphatic organs. _____

b. Write the words that complete the sentences describing cell-mediated immunity in the spaces at the right.

When a macrophage engulfs an antigen, part of it is carried to the cell surface and displayed. If a __1__ cell's __2__ can bind with the presented antigen, it does so and becomes activated. Activated __3__ cells divide, rapidly forming a __4__ of T cell subtypes that have the same antigen receptor. __5__ secrete cytotoxins that rupture antigen-bearing plasma membranes and substances that recruit additional lymphocytes and __6__. __7__ secrete chemicals that help activate B cells and stimulate __8__ by macrophages. When the pathogens have been destroyed, __9__ secrete chemicals to slow and stop the immune response. The dormant __10__ remain to recognize and start an immune response if the same __11__ should ever reenter the body.

1) __________
2) __________
3) __________
4) __________
5) __________
6) __________
7) __________
8) __________
9) __________
10) __________
11) __________

c. Write the words that complete the sentences describing antibody-related immunity in the spaces at the right.

B cells are activated when their antigen receptors bind to an __1__. Activated B cells are stimulated to divide rapidly by __2__, chemicals released from activated __3__ cells that have receptors that can bind to the same antigen. The expanding B cell population is called a __4__, which consists of __5__ cells that produce antibodies and __6__ cells that remain dormant. Once the pathogen has been eliminated, __7__ slow and stop the immune response. If the same antigen later reenters the body, __8__ start a rapid and intense __9__ response.

1) __________
2) __________
3) __________
4) __________
5) __________
6) __________
7) __________
8) __________
9) __________

d. Match the antibodies with the statements. More than one answer may apply.

IgA IgD IgE IgG IgM

1) Most abundant antibody in the blood. __________
2) Fixes complement to antigens. __________
3) Serves as receptors on B cells. __________
4) Involved in allergic reactions. __________
5) Transferred to child via mother's milk. __________
6) Transferred to fetus via placenta. __________
7) Binds with antigens. __________
8) Protects mucous membranes. __________
9) Neutralizes toxins. __________

5. Immune Responses

a. Match the immune responses with the statements.

1) Primary immune response
2) Secondary immune response

_____ Occurs when an antigen is encountered for the first time.
_____ Occurs in subsequent encounters with same antigen.
_____ Results from activation of memory cells.
_____ The more rapid and intense response.

b. Match the types of immunity with the statements.

1) Naturally acquired active
2) Artificially acquired active
3) Naturally acquired passive
4) Artificially acquired passive

_____ Immunity from antibodies received in mother's milk.
_____ Immunity from a vaccine of dead pathogens.
_____ Immunity after having the disease and recovering.
_____ Immunity from injected antibodies.
_____ Immunity from DPT injections.
_____ Immunity from monoclonal antibodies.

6. Disorders of the Lymphatic and Immune Systems

Write the answers that match the statements in the spaces at the right.

1) Microscopic worms plug lymphatic vessels. _____________________
2) An abnormally intense immune reaction. _____________________
3) HIV destroys helper T cells. _____________________
4) A tumor of lymphatic tissue. _____________________
5) Inflammation of the tonsils. _____________________
6) Lymphocytes attack own body tissues. _____________________
7) Allergy attack that involves entire body. _____________________
8) Transmitted via blood exchanges and sexual intercourse. _____________________

7. Clinical Applications

a. Mary is a grocery checker with no evidence of heart disease. She complains that when she comes home from work, her feet and legs are swollen and sometimes painful. In the morning, the swelling is gone. How do you explain this? _____________________

____

____

b. The AIDS virus attacks helper T cells. Explain how this, in time, causes immunodeficiency. _____________________

____

____

c. Infants typically receive a series of three DPT injections (vaccinations) followed by a booster shot at four to six years of age. Explain the value of the booster shot. _____________________

____

____

4. Pharynx and Esophagus

Write the terms that match the statements in the spaces at the right.

1) Tube carrying food to the stomach. ______________________
2) Relaxes to let food enter stomach. ______________________
3) Carries food from mouth to esophagus. ______________________
4) Covers laryngeal opening in swallowing. ______________________

5. Stomach

Write the terms that match the statements in the spaces at the right.

1) Region of stomach joining esophagus. ______________________
2) Region of stomach joining duodenum. ______________________
3) Glands of mucosa secreting gastric juice. ______________________
4) Hormone stimulating gastric secretion. ______________________
5) Hormones inhibiting gastric secretion. ______________________
6) Autonomic impulses stimulating gastric secretion. ______________________
7) Hormone secreted by gastric mucosa. ______________________
8) Acid in gastric juice. ______________________
9) Gastric enzyme acting on proteins. ______________________
10) Gastric enzyme curdling milk. ______________________
11) Products of gastric protein digestion. ______________________
12) Gastric substance enabling absorption of vitamin B_{12} by small intestine. ______________________

6. Pancreas

Write the terms that match the statements in the spaces at the right.

1) Carries pancreatic juice from pancreatic duct to duodenum. ______________________
2) Two hormones stimulating secretion of pancreatic juice. ______________________
3) Source of these hormones. ______________________
4) Pancreatic enzyme acting on starch. ______________________
5) Product of pancreatic starch digestion. ______________________
6) Pancreatic enzyme acting on fats. ______________________
7) Products of pancreatic fat digestion. ______________________
8) Pancreatic enzyme acting on proteins. ______________________
9) Products of pancreatic protein digestion. ______________________

7. Liver

Write the terms that match the statements in the spaces at the right.

1) Removed from amino acids and converted to urea. ____________________
2) Vessels carrying blood to liver:
   a) carries oxygen-rich blood. ____________________
   b) carries nutrient-rich blood. ____________________
3) Vessel carrying blood from liver. ____________________
4) Carbohydrate stored in liver. ____________________
5) Secretion formed by liver. ____________________
6) Stores excess bile. ____________________
7) Carries bile to duodenum. ____________________
8) Hormone contracting gallbladder. ____________________
9) Bile component emulsifying lipids. ____________________
10) Bile component from hemoglobin breakdown. ____________________

8. Small Intestine

a. Write the terms that match the statements in the spaces at the right.

1) Segment continuous with the stomach. ____________________
2) Segment continuous with the cecum. ____________________
3) Membranes supporting small intestine. ____________________
4) Relaxes to allow chyme to enter the small intestine. ____________________
5) Secretion of intestinal glands. ____________________
6) Fingerlike projections of the mucosa. ____________________
7) Microscopic folds of exposed epithelial cell membranes. ____________________
8) Hormone released by mucosa due to presence of fat-laden chyme. ____________________
9) Hormone released by mucosa due to presence of acid chyme. ____________________
10) Mechanism (neural or hormonal) that stimulates secretion of intestinal juice. ____________________
11) Enzyme acting on sucrose. ____________________
12) End products of sucrose digestion. ____________________
13) Enzyme acting on lactose. ____________________
14) Enzyme acting on maltose. ____________________
15) End product of maltose digestion. ____________________
16) End products of lactose digestion. ____________________
17) Enzyme acting on fats. ____________________
18) End products of fat digestion. ____________________
____
19) Enzyme acting on peptides. ____________________
20) End products of peptide digestion. ____________________

b. Write the terms that complete the sentences in the spaces at the right.

Monosaccharides and amino acids are absorbed into the ___1___ networks of ___2___ . Monoglycerides and fatty acids are absorbed into ___3___ cells, where they reunite to form ___4___ . Clusters of triglycerides are coated with protein, forming ___5___ that enter the ___6___ of the ___7___ .

1) ______________________
2) ______________________
3) ______________________
4) ______________________
5) ______________________
6) ______________________
7) ______________________

9. Large Intestine

Write the terms that match the statements in the spaces at the right.

1) Pouchlike first part of large intestine. ______________________
2) External opening of large intestine. ______________________
3) Colon segment along left side of abdomen. ______________________
4) Colon segment along right side of abdomen. ______________________
5) Colon segment continuous with rectum. ______________________
6) Wormlike extension of cecum. ______________________
7) Involuntarily controlled anal sphincter. ______________________
8) Voluntarily controlled anal sphincter. ______________________
9) Decompose undigested materials. ______________________
10) Fluid absorbed by large intestine. ______________________
11) Relaxes, allowing chyme to enter cecum. ______________________
12) Reflex activated by filling of rectum with feces. ______________________

10. Nutrients: Sources and Uses

Write the terms that match the statements in the spaces at the right.

1) Dietary source of most carbohydrates. ______________________
2) Plant polysaccharide providing fiber. ______________________
3) Preferred energy source for body cells. ______________________
4) Organs regulating blood glucose levels. ______________________
5) Most common lipids in the diet. ______________________
6) Type of fats common in animal foods. ______________________
7) Type of fats common in plant foods. ______________________
8) Lipid abundant in egg yolks. ______________________
9) Lipid used to form steroid hormones. ______________________
10) Lipid forming much of plasma membranes. ______________________
11) Molecules transporting lipids in blood. ______________________
12) Organ helping to regulate blood levels of triglycerides and cholesterol. ______________________
13) Amino acids that cannot be made by liver. ______________________

11. Disorders of the Digestive System

Write the names of the disorders that match the statements.

1) Inflammation of the large intestine. ____________
2) Self-induced starvation due to an abnormal concern about weight-control. ____________
3) Decay of the teeth due to acids formed by certain oral microorganisms. ____________
4) Dry, hard feces making defecation difficult. ____________
5) Crystallization of cholesterol in bile within the gallbladder. ____________
6) Replacement of destroyed liver cells by connective tissue. ____________
7) Repeated overeating and purging. ____________
8) Inflammation of the liver. ____________
9) Digestion of stomach mucosa by gastric juice. ____________
10) Inflammation, bleeding, and degeneration of the gingivae and alveolar bone. ____________
11) Watery feces due to excessive peristalsis. ____________
12) Enlarged and inflamed veins in anal canal. ____________
13) Inflammation of the appendix. ____________
14) Inflammation of the peritoneum. ____________
15) Inflammation of colon diverticula. ____________

12. Clinical Applications

a. Severe diarrhea in infants or small children can be a life-threatening event. Explain why. ____________

b. A patient is found to have a gastric ulcer. Antibiotics and a drug to reduce the secretion of gastric juice are prescribed. Explain the basis for the prescriptions. ____________

What serious results may occur with an untreated ulcer? ____________

c. A patient is admitted to the emergency room complaining of severe and spasmodic pain in the epigastric region, and the whites of his eyes are yellowish. He informs the physician that he has had similar, but milder, pains after meals for four to six weeks. What is the likely problem and the likely solution? ____________

STUDY GUIDE 16

1. Urinary System, General

a. Label the figure by placing the numbers of the structures in the spaces by the correct labels.

_____ Aorta	_____ Kidney	_____ Ureter
_____ Inferior vena cava	_____ Renal artery	_____ Urethra
_____ Hilum	_____ Renal vein	_____ Urinary bladder

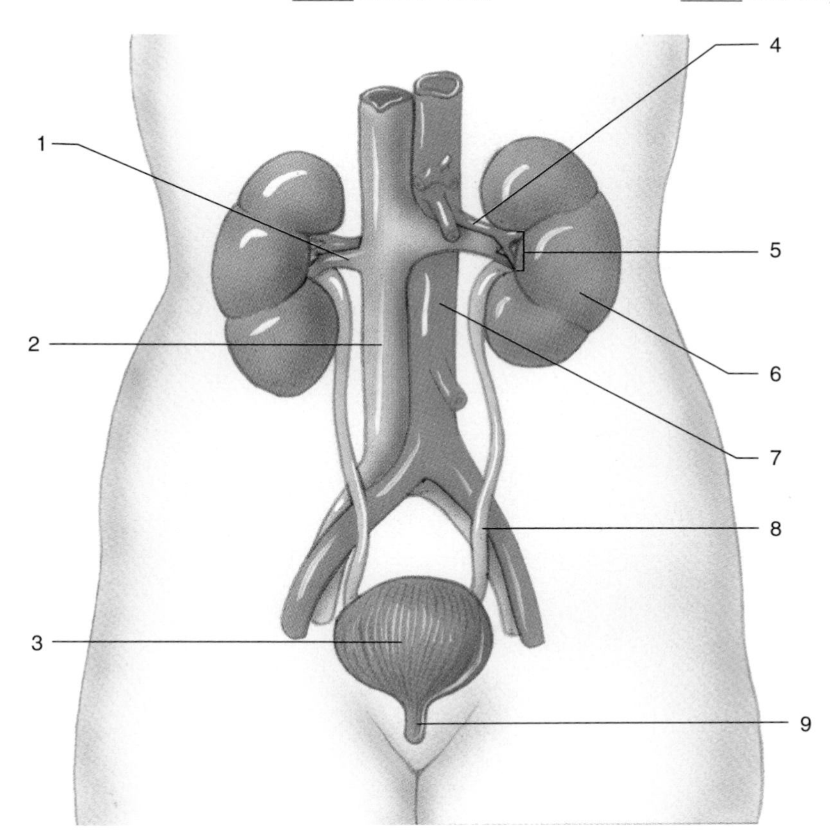

b. Write the names of the organs that match the functions in the spaces at the right.

1) Stores urine temporarily. ______________________

2) Produces urine. ______________________

3) Carries urine from the body. ______________________

4) Carries urine from the kidneys. ______________________

5) Maintains composition and volume of body fluids. ______________________

2. Kidneys

a. Write the names of the structures that match the statements in the spaces at the right.

1) Outer layer containing renal corpuscles. ____________________
2) Region containing renal pyramids. ____________________
3) Flattened cavity continuous with ureter. ____________________
4) Receptacles surrounding renal papillae. ____________________
5) Thin layer of connective tissue enveloping kidney. ____________________
6) Arterial capillaries in renal corpuscle. ____________________
7) Functional units of the kidneys. ____________________
8) U-shaped portion of renal tubule. ____________________
9) Part of renal tubule joined to glomerular capsule. ____________________
10) Part of renal tubule joined to a collecting duct. ____________________
11) Formed of modified cells at point of contact of distal tubule and afferent arteriole. ____________________

b. Label the figure by placing the numbers of the structures in the spaces by the correct labels.

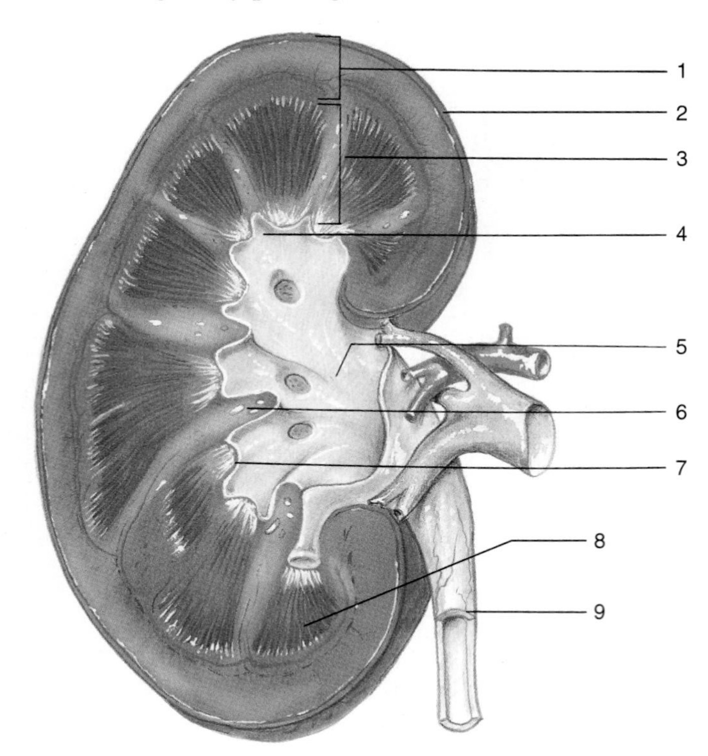

_____ Calyx
_____ Renal capsule
_____ Renal column
_____ Renal cortex
_____ Renal medulla
_____ Renal papilla
_____ Renal pelvis
_____ Renal pyramid
_____ Ureter

c. Label the figure by placing the numbers of the structures in the spaces by the correct labels.

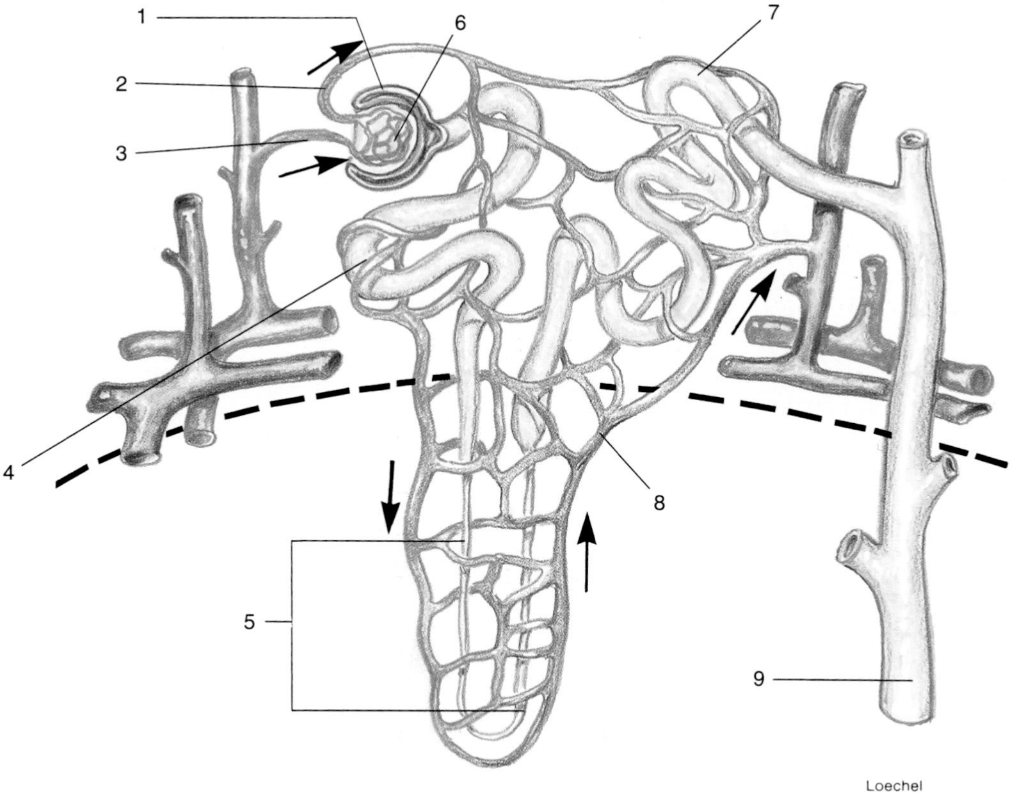

____ Afferent arteriole
____ Collecting duct
____ Distal convoluted tubule
____ Efferent arteriole
____ Glomerular capsule
____ Glomerulus
____ Loop of Henle
____ Peritubular capillary
____ Proximal convoluted tubule

3. Urine Formation

a. Write the words that complete the sentences in the spaces at the right.

A decrease in the glomerular filtration rate causes the __1__ apparatus to secrete __2__, which triggers the __3__-__4__ mechanism. The end product of these reactions is __5__, which increases systemic blood pressure by __6__ arterioles, stimulating __7__ secretion by the posterior pituitary, and stimulating __8__ secretion by the adrenal cortex.

1) ______________
2) ______________
3) ______________
4) ______________
5) ______________
6) ______________
7) ______________
8) ______________

b. Write the terms that match the statements in the spaces at the right.

1) Passage of water and solutes from glomerulus into glomerular capsule. ______________
2) Plasma component that cannot pass through glomerular pores. ______________
3) Force producing filtration. ______________
4) Fluid in glomerular capsule. ______________
5) Recovery of needed materials from filtrate into the blood. ______________
6) Volume of filtrate formed per day. ______________

7) Method of transport of sodium ions. ______________________

8) Method of transport of water. ______________________

9) Method of transport of glucose. ______________________

10) Method of transport of chloride ions. ______________________

11) Substance reabsorbed that concentrates the urine. ______________________

12) Passage of substances from blood into the filtrate. ______________________

13) Actively secreted ion. ______________________

14) Passively secreted ion. ______________________

c. Indicate whether each statement is true (T) or false (F).

_____ Urine contains waste and excessive materials removed from the blood.

_____ Urine formation depends upon maintenance of the blood pressure within the glomeruli.

_____ Most of the filtrate volume is reabsorbed.

_____ Negatively charged ions and positively charged ions are electrochemically attracted to each other.

_____ The active reabsorption of sodium ions increases the rate of water reabsorption by osmosis.

4. Maintenance of Blood Plasma Composition

a. Write the terms that match the statements in the spaces at the right.

1) Hormone promoting water reabsorption. ______________________

2) Hormone promoting reabsorption of Na^+. ______________________

3) Hormone promoting secretion of K^+. ______________________

4) Hormone decreasing blood level of Ca^{++}. ______________________

5) Hormone increasing blood level of Ca^{++}. ______________________

6) Three nitrogenous wastes in urine. ______________________

____

____

b. Indicate whether each statement is true (T) or false (F).

_____ Cellular activity does not affect plasma composition.

_____ Plasma composition is changed by the work of kidneys.

_____ Electrolytes are totally reabsorbed into the blood.

_____ About 99% of water in the filtrate is reabsorbed.

_____ Water is lost from the body only in urine.

_____ Urine volume is reduced when water intake is curtailed.

_____ Perspiring heavily may reduce the volume of urine.

_____ Electrolyte balance is largely maintained by the active reabsorption of positively charged ions.

_____ Nephrons remove all nitrogenous wastes from the blood.

_____ Urea is the most abundant nitrogenous waste in urine.

_____ Urea is formed by the kidneys from amine groups.

_____ Buffers are chemicals in body fluids that either combine with or release hydrogen ions.

_____ The production of carbon dioxide by metabolizing cells tends to make the blood more alkaline.

_____ Kidneys help to regulate the pH of body fluids by secreting excess hydrogen ions into the filtrate.

_____ Water and electrolyte balance in body fluids is essential for normal cell functioning.

_____ ADH is released by the posterior pituitary gland when the water concentration of the blood is reduced.

_____ Aldosterone is secreted by the adrenal cortex when the concentration of K^+ in the blood is reduced.

_____ ADH increases the permeability of the distal tubules and collecting ducts to water.

_____ Electrolyte concentrations in the blood affect the movement of water into cells by osmosis.

5. Excretion of Urine

a. Write the terms that match the statements in the spaces at the right.

1) Carry urine to urinary bladder. _____
2) Provides temporary storage of urine. _____
3) Muscle in wall of urinary bladder. _____
4) Carries urine from urinary bladder. _____
5) Method of urine transport by ureters. _____
6) Type of muscle composing the internal urethral sphincter. _____
7) Type of muscle composing the external urethral sphincter. _____
8) Type of muscle in walls of ureters. _____

b. Write the words that complete the sentences in the spaces at the right.

The accumulation of __1__ stretches the urinary bladder wall, which triggers the __2__ reflex. This reflex causes rhythmic involuntary contractions of the __3__ muscle and opens the involuntarily controlled __4__ urethral sphincter. If the voluntarily controlled __5__ urethral sphincter is relaxed, __6__ occurs.

1) _____
2) _____
3) _____
4) _____
5) _____
6) _____

6. Characteristics of Urine

Indicate whether each statement is true (T) or false (F).

_____ The color of urine is due to the presence of urochrome.

_____ The pH of urine is usually slightly acidic.

_____ Normal urine is never alkaline.

_____ Urine has a specific gravity greater than 1.000.

_____ Normal urine does not contain proteins or hemoglobin.

7. Disorders of the Urinary System

Write the names of the disorders matching the statements in the spaces at the right.

1) Inflammation of the glomeruli. ______________________
2) Inflammation of the urinary bladder. ______________________
3) Inflammation of nephrons and renal pelvis. ______________________
4) Excessive urine production. ______________________
5) Kidney stones. ______________________
6) Inflammation of the urethra. ______________________
7) Deposits of uric acid in the renal pelvis. ______________________
8) Characterized by uremia. ______________________
9) Characterized by protein in the urine. ______________________

8. Clinical Applications

a. A 60-year-old woman comes to the clinic with severe edema of her lower legs and feet. A diuretic is prescribed, and she is placed on a salt-free diet. She is also advised to take a 30-minute walk each morning and afternoon and to elevate her feet higher than her head for 20-minute periods morning and afternoon. Explain how the diuretic will reduce her edema. ______________________

____

____

____

Explain how the salt-free diet will help reduce her edema. ______________________

____

____

____

Explain how elevating her feet and walking will help reduce her edema. ______________________

____

____

b. Explain why women develop cystitis more frequently than men. ______________________

____

____

1. Male Reproductive System

a. Label the figure by placing the numbers of the structures by the correct labels.

_____ Bulbourethral gland
_____ Corpus cavernosum
_____ Corpus spongiosum
_____ Ejaculatory duct
_____ Epididymis
_____ Glans penis
_____ Penis
_____ Prepuce
_____ Prostate gland
_____ Scrotum
_____ Seminal vesicle
_____ Testis
_____ Urethra
_____ Urinary bladder
_____ Vas deferens

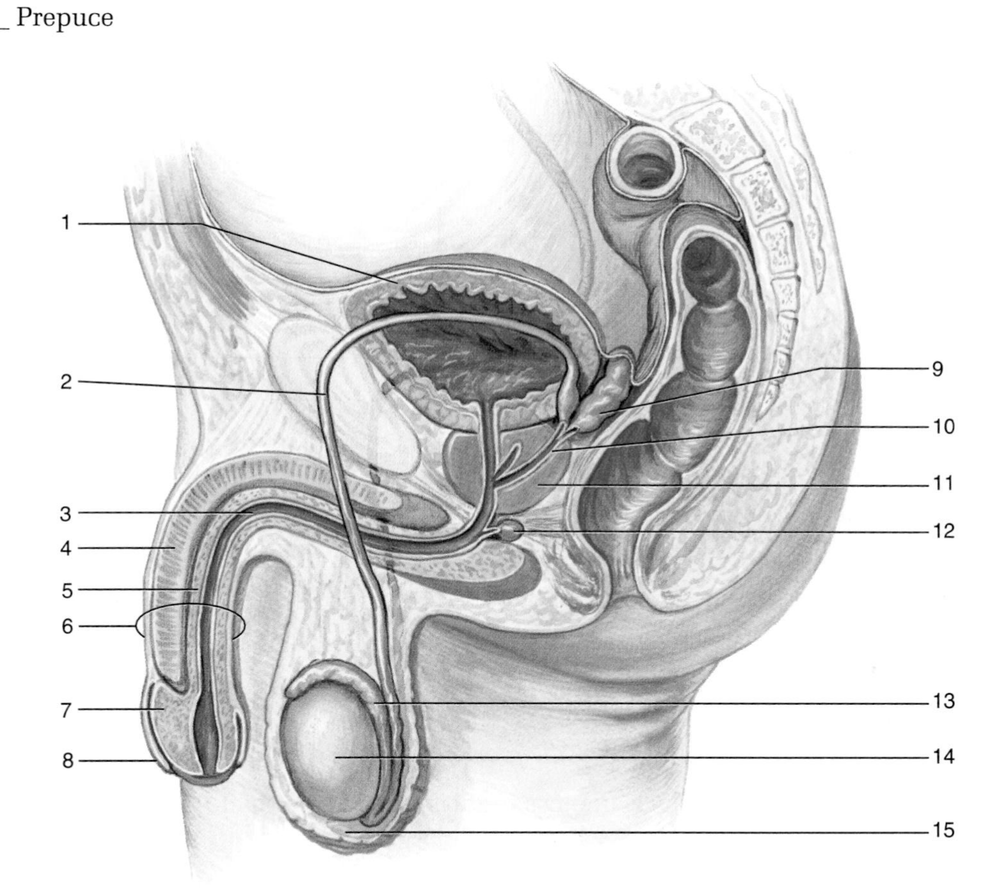

b. Trace the path of sperm cells from a testis to the outside by placing the numbers of the ducts in the spaces below.

1) Ejaculatory duct
2) Epididymis
3) Urethra
4) Vas deferens

Testis → __________ → __________ → __________ → __________ → outside.

c. Write the terms that match the statements in the spaces at the right.

1) Male gonads, or sex glands. ____________
2) Tubules producing sperm cells. ____________
3) Cells producing testosterone. ____________
4) Process of sperm formation. ____________
5) Number of chromosomes in spermatogonia and primary spermatocytes. ____________
6) Type of cell division forming primary spermatocytes. ____________
7) Type of cell division forming spermatids from primary spermatocytes. ____________
8) Number of chromosomes in each spermatid. ____________
9) Cell formed from each spermatid. ____________
10) Location of nucleus in a sperm cell. ____________
11) Provides motility for a sperm cell. ____________
12) Tubule where sperm cells mature. ____________
13) Secrete fluid, neutralizing acidity of urethra prior to ejaculation. ____________
14) Secrete fluid containing fructose. ____________
15) Secretes fluid, activating swimming movements of sperm cells. ____________
16) Mixture of glandular secretions and sperm cells. ____________
17) Contains testes outside the body. ____________
18) Approximate temperature of testes required for production of viable sperm. ____________
19) Male copulatory organ inserted into vagina during sexual intercourse. ____________
20) Erectile tissue in penis surrounding urethra. ____________
21) Two dorsal columns of erectile tissue in penis. ____________
22) Sheath of skin covering the glans penis. ____________

2. Male Sexual Response

Write the words that complete the sentences in the spaces at the right.

Sexual stimulation causes ___1___ nerve impulses to ___2___ the arterioles and ___3___ the venules serving the erectile tissue in the penis. Engorgement of the erectile tissue results in ___4___ of the penis. At the same time, the ___5___ glands secrete an alkaline fluid that neutralizes the ___6___ of the urethra. Continued sexual stimulation results in ___7___ , which is characterized by a sensation of sexual pleasure and ___8___ , the forcing of ___9___ out the urethra.

1) ____________
2) ____________
3) ____________
4) ____________
5) ____________
6) ____________
7) ____________
8) ____________
9) ____________

3. Hormonal Control of Reproduction in Males

a. Write the terms that match the statements in the spaces at the right.

1) The male sex hormone. ____________

2) Secretes gonadotropin-releasing hormone. ____________

3) Releases FSH and LH. ____________

4) Hormone stimulating testosterone secretion by testes. ____________

5) Two hormones that act together to stimulate spermatogenesis. ____________

6) Stimulates maturation of male sex organs. ____________

7) Stimulates development and maintenance of secondary sexual characteristics. ____________

8) Produces androgens in male fetus. ____________

9) Hormone whose first secretion triggers onset of puberty. ____________

10) Hormone stimulating release of FSH and LH. ____________

b. Record the numbers of the male secondary sexual characteristics listed in the space provided.

1) Maturation of the testes.
2) Enlargement of the larynx.
3) Broad shoulders.
4) Growth of body hair.
5) Production of sperm cells.
6) Increased metabolic rate.
7) Increased muscle development.
8) Increased RBC production.
9) Maturation of the penis.
10) Deepening of the voice.

____

c. Write the words that complete the sentences in the spaces at the right.

The production of testosterone by __1__ in the testes is regulated by a __2__ feedback system. When the level of testosterone in the blood declines, secretion of __3__ by the hypothalamus is __4__, causing an __5__ in the release of __6__ from the anterior pituitary, which, in turn, __7__ testosterone production. As the level of testosterone increases, it inhibits __8__ production, which decreases the release of __9__, resulting in a __10__ in testosterone production.

1) ____________
2) ____________
3) ____________
4) ____________
5) ____________
6) ____________
7) ____________
8) ____________
9) ____________
10) ____________

4. Female Reproductive System

a. Label the figure by placing the numbers of the structures by the correct labels.

_______ Anus	_______ Labium major	_______ Urinary bladder
_______ Cervix	_______ Labium minor	_______ Uterine tube
_______ Clitoris	_______ Ovary	_______ Uterus
_______ Fimbriae	_______ Urethra	_______ Vagina

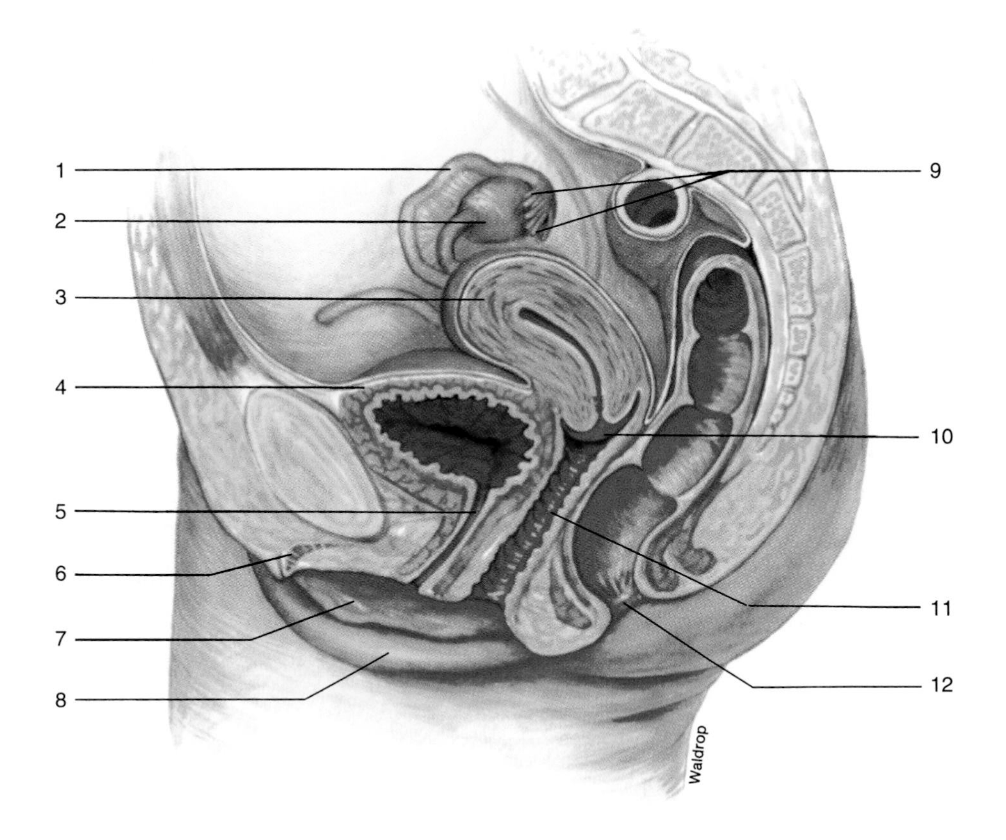

b. Write the terms that match the statements in the spaces at the right.

1) Produces female sex cells. ______________________
2) Receives penis in sexual intercourse. ______________________
3) Holds embryo/fetus during pregnancy. ______________________
4) Carries secondary oocyte toward uterus. ______________________
5) Narrow space between labia minora. ______________________
6) Birth canal during childbirth. ______________________

c. Write the words that complete the sentences in the spaces at the right.

During fetal development, __1__ cell division of germinal epithelial cells forms millions of __2__, but most of them degenerate. Survivors become __3__ oocytes containing __4__ chromosomes. Single layers of follicular cells envelop each oocyte forming __5__ follicles, but most of them degenerate; the survivors become __6__ follicles. After puberty, usually __7__ dominant follicle develops each month, and its oocyte undergoes the first __8__ division, producing a __9__ oocyte and a first __10__. Both of these cells contain __11__ chromosomes. Continued growth results in rupture of the follicle at __12__, and the released oocyte enters a __13__ tube. If the secondary oocyte is penetrated by a __14__, the second __15__ division forms an __16__ and a second __17__, each with __18__ chromosomes.

1) ____________
2) ____________
3) ____________
4) ____________
5) ____________
6) ____________
7) ____________
8) ____________
9) ____________
10) ____________
11) ____________
12) ____________
13) ____________
14) ____________
15) ____________
16) ____________
17) ____________
18) ____________

d. Write the terms that match the statements in the spaces at the right.

1) Inner lining of the uterus. ____________
2) Moves oocyte through uterine tube. ____________
3) Expanded end of uterine tube. ____________
4) Type of muscle in uterine wall. ____________
5) Collective term for the external female reproductive organs. ____________
6) Nodule of erectile tissue corresponding to the penis in males. ____________
7) Glands lubricating the vaginal opening. ____________
8) Culmination of female sexual response. ____________
9) Two organs whose walls rhythmically contract during orgasm. ____________

____

10) Tube opening into posterior portion of the vestibule. ____________
11) Larger, outer folds of vulva. ____________
12) Portion of uterus projecting into vagina. ____________

5. Hormonal Control of Reproduction in Females

a. Match the hormones listed with the following statements.

Estrogen	GnRH	Progesterone
FSH	LH	

1) Secreted by the follicular cells. ____
2) Stimulates maturation of female sex organs. ____
3) Maintains uterine lining in pregnancy. ____
4) Develops female secondary sexual characteristics. ____
5) Secreted by corpus luteum. ____
6) Secreted by the hypothalamus. ____
7) Stimulates development and function of corpus luteum. ____
8) First secretion starts onset of puberty. ____
9) Stimulates development of ovarian follicles. ____
10) Prepare endometrium for pregnancy. ____
11) High concentrations inhibit GnRH secretion. ____
12) Promotes thickening of endometrium. ____
13) Promotes formation of blood vessels in endometrium. ____

b. Write the words that complete the sentences in the spaces at the right.

Between puberty and __1__, a woman experiences one reproductive cycle per month consisting of an ovarian cycle and a __2__ cycle. These cycles are controlled by __3__ and average about __4__ days in length. A cycle is started by the secretion of __5__ by the hypothalamus, which activates the release of __6__ and __7__ by the anterior pituitary. FSH stimulates the development of a primary __8__ and the secretion of __9__ by the follicular cells. __10__ promotes the thickening of the endometrium. The increasing estrogen production triggers a sharp increase in __11__ secretion and a lesser increase in FSH production, leading to __12__ on day 14. Under stimulation by LH, the follicle remnants become a __13__ that secretes a high level of __14__ and estrogen, which together prepare the __15__ of the uterus to receive an early embryo. The high level of __16__ inhibits secretion of GnRH, preventing development of additional ovarian follicles. If pregnancy does not occur, the corpus luteum degenerates and the levels of estrogen and __17__ rapidly decline, resulting in breakdown of the endometrium leading to __18__ and secretion of GnRH starting a new reproductive cycle.

1) ____
2) ____
3) ____
4) ____
5) ____
6) ____
7) ____
8) ____
9) ____
10) ____
11) ____
12) ____
13) ____
14) ____
15) ____
16) ____
17) ____
18) ____

6. Mammary Glands

Indicate whether each statement is true (T) or false (F).

_______ Mammary glands are specialized for milk production.

_______ Breasts contain connective tissue but little fat.

_______ Alveolar glands occur in lobes of mammary glands.

_______ Estrogen stimulates the development of mammary glands.

_______ A pigmented areola surrounds a protruding nipple.

_______ Mammary glands are present, but nonfunctional, in males.

_______ Breasts are formed internal to the pectoralis muscles.

7. Birth Control

Indicate whether each statement is true (T) or false (F).

_______ Contraceptives are designed to prevent union of sperm and egg.

_______ Contraceptives and birth control are synonymous.

_______ Spermicides act by killing the sperm cells.

_______ Progesterone in the "pill" inhibits GnRH secretion.

_______ Spermicides are usually used with barrier methods.

_______ It is impossible to catch STDs when using a condom.

_______ Diaphragms and cervical caps are about equally effective.

_______ The rhythm method relies on knowing when ovulation occurs.

_______ A condom is a barrier contraceptive.

_______ Pregnancy is not possible when using a condom.

_______ The "pill" is the most effective contraceptive.

_______ There are no undesirable side effects of the "pill."

_______ An IUD prevents implantation of an embryo.

_______ Induced abortion is a contraceptive procedure.

_______ A tubal ligation prevents ovulation.

_______ Diaphragms and cervical caps are barrier contraceptives.

_______ A vasectomy prevents the ejaculation of sperm.

_______ Use of an IUD may cause pelvic inflammatory disease.

_______ Withdrawal is less effective than a condom.

_______ Undesirable side effects may result from induced abortion.

8. Disorders of the Reproductive Systems

Write the disorders that match the statements in the spaces provided.

Male Disorders

1) Inability to maintain an erection. _______________

2) Common reproductive cancer in males. _______________

3) Inflammation of the prostate glands. _______________

4) Inability to produce sufficient viable sperm. _______________

5) Causes constriction of urethra in about one-third of older males. _______________

Female Disorders

1) Physical pain during menstruation. ____________
2) Growth of endometrial tissue outside the uterus. ____________
3) Absence of menstruation without pregnancy. ____________
4) Associated with highly absorbent tampons. ____________
5) Inability to become pregnant. ____________
6) Infection in reproductive organs and/or pelvic cavity. ____________
7) Physical and emotional distress just prior to menstruation. ____________
8) Caused by toxin formed by *S. aureus.* ____________

Sexually Transmitted Diseases

1) Results from infection with herpes simplex virus type 2. ____________
2) Fatal disease resulting from HIV infection. ____________
3) Two bacterial diseases that may lead to sterility in females. ____________ ____________
4) Characterized by chancre in first stage. ____________
5) Caused by human papillomavirus (HPV). ____________
6) Bacterial diseases curable with antibiotics. ____________ ____________
7) Viral diseases for which there are no cures. ____________ ____________

9. Clinical Applications

a. Failure of the testes to descend into the scrotum (cryptochordism) causes sterility in males. Explain why. ____________

b. Secondary amenorrhea in female athletes results from strenuous activity, which blocks the hypothalmic regulation of reproduction. How does this stop the reproductive cycles? ____________

Women with amenorrhea produce little, if any, estrogen, which causes osteoporosis (bone loss). Why are such women deficient in estrogen? ____________

c. Without effective treatment, a bacterial sexually transmitted disease usually leads to pelvic inflammatory disease (PID) in females. How does this happen? ____________

STUDY GUIDE 18

1. Fertilization and Early Development

a. Write the terms that complete the sentences in the spaces at the right.

A __1__ oocyte containing __2__ chromosomes is released at ovulation, and it is enveloped by several layers of __3__ cells. After entering a __4__ tube, it is slowly carried toward the __5__ by beating __6__ of cells lining the tube. The oocyte remains viable for about __7__ hours after ovulation. Sperm deposited in the __8__ swim into the uterus and up the __9__ tubes. They usually encounter the secondary oocyte in the upper __10__ of a uterine tube. Sperm remain viable in the female reproductive tract for about __11__ hours.

Many sperm are required to separate the __12__ cells enveloping the secondary oocyte. Once a __13__ enters the secondary oocyte, chemical changes in the __14__ prevent other sperm from entering. The secondary oocyte immediately completes the __15__ meiotic division, forming an __16__ and another polar body, each containing __17__ chromosomes. Union of __18__ and __19__ nuclei complete fertilization, forming a __20__ containing __21__ chromosomes.

1) ____________________
2) ____________________
3) ____________________
4) ____________________
5) ____________________
6) ____________________
7) ____________________
8) ____________________
9) ____________________
10) ____________________
11) ____________________
12) ____________________
13) ____________________
14) ____________________
15) ____________________
16) ____________________
17) ____________________
18) ____________________
19) ____________________
20) ____________________
21) ____________________

b. Write the terms that match the statements in the spaces at the right.

1) Type of cell division in the zygote. ____________________
2) Solid ball of cells formed by cleavage. ____________________
3) Hollow ball of cells. ____________________
4) Mass of cells within the blastocyst. ____________________
5) Outer wall of the blastocyst. ____________________
6) Embedding of blastocyst in endometrium. ____________________
7) Length of preembryonic stage. ____________________
8) Length of full-term pregnancy. ____________________

2. Hormonal Control of Pregnancy

Write the terms that match the statements in the spaces at the right.

1) Hormone secreted by trophoblast. ______________________
2) Maintains corpus luteum for two to three months. ______________________
3) Hormone maintaining the uterine lining. ______________________
4) Hormone detected by pregnancy tests. ______________________
5) Takes over secretion of estrogen and progesterone from second or third month to birth. ______________________
6) Prevents GnRH secretion by hypothalamus during pregnancy. ______________________
7) Secretes estrogen and progesterone for the first two to three months of pregnancy. ______________________
8) Two hormones that prepare mammary glands for milk secretion after birth. ______________________ ______________________

3. Embryonic Development

a. Matching (more than one answer may apply).

1) Ectoderm 2) Mesoderm 3) Endoderm

_____ Primary germ layers
_____ Connective tissue
_____ Nervous system
_____ Muscles
_____ Lining of digestive tract
_____ Liver and pancreas
_____ Epidermis
_____ Dermis
_____ Kidneys and gonads
_____ Lining of respiratory passages

b. Write the terms that match the statements in the spaces at the right.

1) Becomes the chorion. ______________________
2) Connects embryo to placenta. ______________________
3) Form early embryonic blood cells. ______________________ ______________________
4) Serves as shock absorber for fetus. ______________________
5) Membrane surrounding embryo/fetus. ______________________
6) Fingerlike projections from chorion that penetrate endometrium. ______________________
7) Source of oxygen and nutrients for embryo or fetus. ______________________
8) Site of exchange of materials between embryonic and maternal bloods. ______________________
9) Name given embryo after eighth week. ______________________
10) Attaches embryo to the uterine wall. ______________________
11) Fluid in which the embryo develops. ______________________
12) Developmental stage between second and eighth weeks. ______________________

c. Label the figure by placing the numbers of the structures in the spaces by the correct labels.

_______ Allantois
_______ Amnion
_______ Amniotic cavity
_______ Chorion
_______ Developing placenta
_______ Umbilical cord
_______ Uterine wall
_______ Yolk sac

4. Birth

a. Label the figure by placing the numbers of the structures in the spaces by the correct labels.

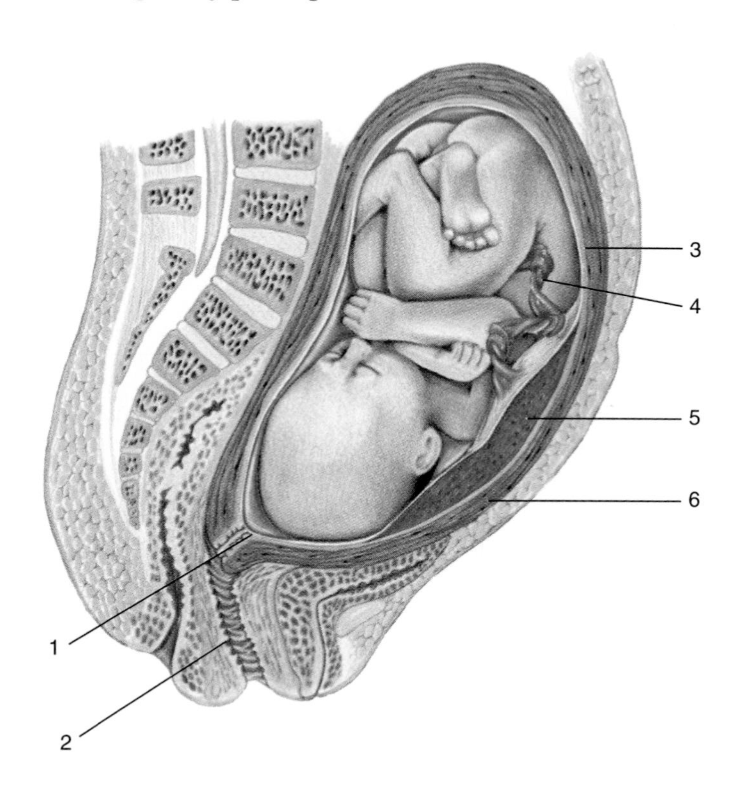

_______ Amniochorion
_______ Cervix of uterus
_______ Placenta
_______ Umbilical cord
_______ Uterine wall
_______ Vagina

b. Write the terms that match the statements in the spaces at the right.

1) Relaxes symphysis pubis as birth nears. ____________
2) Softens cervix as birth nears. ____________
3) Hormone that inhibits uterine contractions during pregnancy. ____________
4) Hormone that sensitizes uterine muscles for starting contractions as birth nears. ____________
5) Term for physical and physiological processes associated with birth. ____________
6) Hormone starting and maintaining uterine contractions. ____________
7) Receives neural impulses formed by stretching of the cervix. ____________
8) Secretes oxytocin. ____________
9) Longest stage of labor. ____________
10) Stage of labor when baby is born. ____________
11) Stage when the afterbirth is expelled. ____________
12) Name for the birth process. ____________

c. Write the words that complete the sentences in the spaces at the right.

As the time of birth approaches, the high concentration of __1__ overrides the inhibitory effect of __2__ on uterine contractions so that such contractions are possible. The __3__ feedback mechanism controlling labor seems to be started by pressure of the fetus on the __4__, which forms __5__ that are carried to the hypothalamus. The __6__ stimulates the posterior pituitary to secrete __7__ that stimulates uterine __8__. Dilation of the __9__ increases the frequency of __10__ sent to the hypothalamus, which, in turn, stimulates the posterior pituitary to release more __11__, which increases the strength and frequency of uterine __12__. This pattern of positive feedback produces increasingly stronger contractions until the baby is __13__. Shortly after birth, uterine contractions cause the detachment and expulsion of the __14__.

When the __15__ is cut, the level of __16__ increases in the infant's blood, stimulating the __17__ control center to trigger the first breath. After the first breath, breathing becomes easier because __18__ in the alveolar fluid keeps the __19__ open.

1) ____________
2) ____________
3) ____________
4) ____________
5) ____________
6) ____________
7) ____________
8) ____________
9) ____________
10) ____________
11) ____________
12) ____________
13) ____________
14) ____________
15) ____________
16) ____________
17) ____________
18) ____________
19) ____________

5. Circulatory Adaptations

a. Write the terms that match the statements relating to fetal circulation in the spaces at the right.

1) Carries blood from placenta to fetus. ____________
2) Opening between left and right atria. ____________
3) Return blood from fetus to placenta. ____________
4) Carries blood from umbilical vein to inferior vena cava, bypassing the liver. ____________
5) Carries blood from pulmonary trunk to aortic arch. ____________
6) Vein carrying oxygen-rich blood from the placenta. ____________

b. Write the words that complete the sentences regarding fetal circulation in the spaces at the right.

The fetal blood receives oxygen and nutrients from __1__ blood in the placenta. Oxygen-rich blood is carried from the placenta by the __2__ vein that enters the fetus at the __3__ . This vessel divides near the liver, and about half of the oxygenated blood passes through the __4__ , bypassing the liver, to mix with deoxygenated blood in the inferior __5__ . When this mixed blood enters the __6__ atrium, most of it passes through the __7__ into the __8__ atrium and flows into the __9__ ventricle. Contraction of the ventricle pumps blood into the __10__ to the body cells. Blood entering the __11__ ventricle is pumped into the pulmonary trunk, but some of it bypasses the lungs by flowing through the __12__ into the aorta, increasing the blood supply to body cells. A small amount of blood is carried by __13__ arteries to the nonfunctional lungs and returned to the left __14__ . Blood is returned to the placenta by two __15__ arteries.

1) ____________
2) ____________
3) ____________
4) ____________
5) ____________
6) ____________
7) ____________
8) ____________
9) ____________
10) ____________
11) ____________
12) ____________
13) ____________
14) ____________
15) ____________

c. Write the terms that match the statements relating to postnatal circulatory changes in the spaces at the right.

1) Remnant of the umbilical vein. ____________
2) Remnants of the umbilical arteries. ____________
3) Remnant of the ductus venosus. ____________
4) Remnant of the ductus arteriosus. ____________

6. Lactation

a. Write the terms that match the statements in the spaces at the right.

1) Two hormones preparing mammary glands for lactation. ____________________ ____________________
2) Hormone stimulating lactation. ____________________
3) Secretes prolactin-releasing hormone. ____________________
4) Secretes prolactin. ____________________
5) First secretion of mammary glands. ____________________
6) Two hormones whose high levels inhibit secretion of PRH. ____________________ ____________________
7) Hormone stimulating milk ejection. ____________________

b. Write the words that complete the sentences in the spaces at the right.

After birth, the drop in ___1___ and ___2___ levels allows the hypothalamus to secrete ___3___, which stimulates release of ___4___ by the anterior pituitary, promoting lactation. ___5___, the first secretion of the mammary glands, is rich in ___6___ and contains no ___7___. True ___8___ secretion starts within two to three days.

Suckling stimulates formation of ___9___ that are carried to the hypothalamus, causing it to secrete ___10___, which continues production of prolactin, which maintains ___11___, and stimulate the posterior pituitary to secrete ___12___, which stimulates contraction of lactiferous ducts, causing milk ___13___.

1) ____________________
2) ____________________
3) ____________________
4) ____________________
5) ____________________
6) ____________________
7) ____________________
8) ____________________
9) ____________________
10) ____________________
11) ____________________
12) ____________________
13) ____________________

7. Disorders of Pregnancy and Prenatal Development

Write the terms that match the statements in the spaces at the right.

1) Implantation of embryo at a site other than the uterus. ____________________
2) Spontaneous abortion. ____________________
3) Increased blood pressure, edema, and convulsions or coma in late pregnancy. ____________________
4) Nausea and vomiting in early pregnancy. ____________________
5) Major cause of death in newborn infants. ____________________
6) Substances or influences causing birth defects. ____________________
7) Results from too rapid destruction of fetal red blood cells after birth. ____________________
8) Caused by insufficient surfactant in alveoli. ____________________
9) May result from fetal exposure to X rays, alcohol, and illegal or legal drugs. ____________________
10) Most common teratogen causing birth defects. ____________________

8. Inheritance

a. Write the terms that match the statements in the spaces at the right.

1) Number of chromosomes in human body cells. ____________
2) Number of chromosomes in human gametes. ____________
3) Sex chromosomes in a female. ____________
4) Sex chromosomes in a male. ____________
5) A unit of inheritance. ____________
6) Alternate forms of a gene. ____________
7) Condition in which both alleles for a trait are identical. ____________
8) Condition in which the alleles for a trait are different. ____________
9) An allele that is always expressed. ____________
10) An allele that is expressed only when a dominant allele is absent. ____________
11) A type of gene expression in which unlike alleles are both expressed. ____________
12) A type of inheritance involving more than two dominant or recessive alleles. ____________
13) A type of inheritance involving many genes that produce a gradation of expression in the human population. ____________
14) The observable characteristics of a trait. ____________
15) All of the alleles controlling the expression of a trait. ____________
16) Traits whose alleles occur on the X chromosome. ____________
17) Type of cell division that separates chromosome pairs into gametes. ____________

b. Indicate the genotypes for the following traits.

1) Heterozygous freckled. ____________
2) Homozygous freckled. ____________
3) Homozygous nonfreckled. ____________
4) Color-blind male. ____________
5) Normal female carrying allele for color blindness. ____________
6) Color-blind female. ____________
7) Homozygous type A blood. ____________
8) Type AB blood. ____________
9) Type O blood. ____________
10) Heterozygous type B blood. ____________

c. Indicate the possible genotypes of gametes that can be formed by parents with these genotypes.
   1) Homozygous freckled. ____________
   2) Heterozygous freckled. ____________
   3) Homozygous nonfreckled. ____________
   4) Color-blind male. ____________
   5) Normal vision, carrier female. ____________
   6) Color-blind female. ____________
   7) Heterozygous type A blood. ____________
   8) Type AB blood. ____________

d. Indicate the predicted phenotype ratios for the following matings.
   1) Homozygous freckled × homozygous nonfreckled ____________
   2) Heterozygous freckled × homozygous nonfreckled ____________
   3) Type AB blood × type O blood ____________
   4) Heterozygous type A blood × type O blood ____________
   5) Normal vision, color-blind carrier mother × normal vision father ____________

e. Indicate whether each statement is true (T) or false (F).
   ______ Genetic disease may be caused by the presence of an extra chromosome.
   ______ Recessive sex-linked traits appear more frequently in females since they have two X chromosomes.
   ______ Traits that show a gradation of expression in the population are determined by polygenes.
   ______ It is possible to examine fetal cells for chromosome abnormalities.
   ______ Some genetic diseases caused by specific alleles do not show up until adulthood.
   ______ Down syndrome is caused by trisomy 21.
   ______ Amniocentesis is used to obtain a sample of amniotic fluid for examination.
   ______ Genetic counseling may be helpful for prospective parents with genetic disease in their family histories.

9. Clinical Applications

a. When the sperm count in semen falls below 20 million/ml, male infertility results. How do you explain this? ____________

b. Physicians advise women to avoid all drugs (legal and illegal) during pregnancy. What is the basis for this advice? ____________

c. What problems would occur if a newborn infant's foramen ovale failed to close? ____________

d. Why can monozygotic twins receive blood transfusions from each other without difficulty, but dizygotic twins often cannot? ____________

e. Mary and Joe have discovered that they are both heterozygous for sickle-cell anemia. They want to know what the chance is that their children will inherit sickle-cell anemia. What would you advise them? ____________

APPENDIX A

Keys to Medical Terminology

Medical terms may consist of three basic parts: a prefix, a root word, and a suffix. All terms have a root word, but some terms may lack either a prefix or a suffix. A **prefix** is the first portion of a term and comes before the root word. A **suffix** comes after the root word and is the last portion of a term. Both the prefix and suffix modify the meaning of a root word, and they may be used with many different root words. The **root word** is the main portion of the term. Root words often occur at the beginning of a term, but they also may end a term or may be sandwiched between a prefix and suffix. When determining the meaning of a term, you start with the suffix, then move to the prefix, and finally consider the root word. Consider these examples.

1. **laryngitis** becomes **laryng/itis** when broken into its component parts.

   laryng- = the root word meaning *larynx*
   -itis = the suffix meaning *inflammation*

   Thus, the meaning of laryngitis is *inflammation of the larynx.*

2. **endogastric** becomes **endo/gastr/ic** when broken into its component parts.

   endo- = the prefix meaning *within*
   gastr- = the root word meaning *stomach*
   -ic = the suffix meaning *pertaining to*

   Thus, the meaning of endogastric is *pertaining to within the stomach.*

The parts of a term are linked together in a way that aids pronunciation. This often requires the use of *combining vowels.* For example, when linking *gastr-* and *-pathy* to form a term meaning disease of the stomach, the vowel *o* is inserted to form *gastropathy.*

Some terms consist of more than one root word. For example, *gastr/o/enter/o/col/itis* means inflammation of the stomach, intestine, and colon.

You can see that once you know the meaning of common prefixes, root words, and suffixes, understanding medical terminology becomes much easier.

In the sections that follow, some common prefixes, suffixes, and root words are listed along with examples to help you understand medical terminology.

Singular and Plural Endings

Most medical terms are derived from Greek and Latin words. Therefore, changing from singular to plural is done by changing the ending of the term rather than by adding an *s* or *es* or changing a *y* to *ies* as in English terms. Examples of singular and plural endings are:

Singular Ending	Plural Ending	Example
-a	-ae	pleura; pleurae
-en	-ena	lumen; lumena
-is	-es	testis; testes
-ma	-mata	carcinoma; carcinomata
-um	-a	epicardium; epicardia
-ur	-ora	femur; femora
-us	-i	glomerulus; glomeruli
-x	-ces	appendix; appendices

Common Prefixes

Prefix	Meaning	Example
a-	without, not	a/sepsis: sterile; without germs
ab-	away from, from	ab/duct: carry away from
ad-	to, toward	ad/duct: carry toward
an-	without, not	an/ergia: without energy
ante-	before	ante/cibum: before meals
anti-	against	anti/histamine: against histamine
bi-	two	bi/lobed: having two lobes
bio-	life	bio/logy: study of life
brachy-	short	brachy/gnathia: shortness of the lower jaw
brady-	slow	brady/cardia: slow heart rate
cent-	hundred	centi/meter: 1/100 of a meter
circum-	around	circum/oral: around the mouth
co-, com-, con-,	with, together	com/press: squeeze together
de-	from, down	de/congest: reduce congestion
dia-	through	dia/rrhea: flow through
dis-	apart	dis/infect: free from infection
dys-	bad, difficult	dys/pepsia: difficult digestion
ect-	external, outer	ecto/derm: outer skin
en-	in, on	en/cranial: in the cranium
end-	within	endo/crine: secrete within
epi-	upon	epi/dermis: upon the skin
ex-	out, away from	ex/halation: to breathe out
extra-	outside of, in addition to	extra/ocular: outside the eye
hemi-	half	hemi/plegia: paralysis of one-half of the body
hyper-	above, over	hyper/trophy: excessive growth
hypo-	below, under	hypo/dermic: under the skin
infra-	below, beneath	infra/orbital: below the orbit
inter-	between	inter/cellular: between cells
intra-	within	intra/cellular: within cells
kil-	thousand	kilo/gram: 1,000 grams
macr-	large	macro/cyst: large cyst
mal-	bad, ill, poor	mal/ady: disease, disorder
mes-	middle	meso/nasal: middle of the nose
meta-	after, beyond	meta/tarsals: beyond the tarsals
micr-	small	micro/colon: abnormally small colon
milli-	one-thousandth	milli/gram: 1/1,000 of a gram
multi-	many	multi/cellular: having many cells
neo-	new	neo/plasm: new growth
ob-	against, in the way of	ob/scure: indistinct, hidden
olig-	few	oligo/spermia: few sperm
onc-	tumor	onco/genic: tumor-causing
per-	through	per/forate: to make holes
peri-	around	peri/osteum: around a bone
poly-	many	poly/morphous: many forms
post-	after	post/ocular: behind the eye
pre-	before, in front of	pre/mature: before maturation
presby-	old	presby/cardia: old heart
pro-	before, in front of	pro/chondrial: before cartilage
re-	again, back	re/flex: bend back
retr-	backward, behind	retro/nasal: back part of nose
semi-	half	semi/lunar: half moon
sub-	under	sub/cutaneous: under the skin
super-	above, superior	super/acute: strongly acute
supra-	above, superior	supra/nasal: above the nose
sym-	together, with	sym/physis: growing together
syn-	together, with	syn/dactyly: fusion of fingers
tachy-	fast	tachy/cardia: rapid heart rate

Common Suffixes

Suffix	Meaning	Example
-algia	pain	neur/algia: pain in a nerve
-centesis	puncture to aspirate fluid	amnio/centesis: puncture amnion to obtain a sample of amniotic fluid
-cide	kill	bacterio/cide: substance killing bacteria
-cis	cut	in/cision: a cut into
-cyte	cell	erythro/cyte: red blood cell
-dynia	pain	entero/dynia: intestinal pain
-ectomy	cut out	append/ectomy: procedure to cut out the appendix
-emesis	vomiting	poly/emesis: much vomiting
-emia	blood	an/emia: without blood
-gnosis	knowledge	pro/gnosis: foreknowledge
-gram	record	myo/gram: muscle record

Common Suffixes *continued*

Suffix	Meaning	Example
-graphy	making a record	myo/graphy: making a record of muscle action
-iasis	abnormal condition	Candid/iasis: *Candida* infection
-itis	inflammation	sinus/itis: inflammation of sinuses
-lepsy	seizures	narco/lepsy: seizures of numbness
-logy	study of	bio/logy: study of life
-lysis, -lytic	breakdown, dissolve	myo/lysis: breakdown of muscles
-megaly	enlargement	nephro/megaly: kidney enlargement
-oid	resembling	ov/oid: resembling an egg
-oma	tumor	carcin/oma: cancerous tumor
-osis	abnormal condition	nephr/osis: abnormal kidney condition
-ostomy	make an opening	ile/ostomy: opening into the small intestine
-pathy	disease	neuro/pathy: disease of nerves
-penia	deficiency, poor	leuko/penia: deficiency of white blood cells
-pepsia	digestion	dys/pepsia: poor digestion
-philia	attraction, love	acido/philic: attracted to acid
-phobia	abnormal fear	acro/phobia: fear of heights
-plasia	formation	hypo/plasia: deficient formation
-plasty	make, shape	angio/plasty: shaping a blood vessel
-plegia	paralysis	para/plegia: paralysis of lower body and both legs
-pnea	breath	brady/pnea: slow breathing
-rrhea	discharge, flow	pyo/rrhea: pus discharge
-soma, -some	body	chromo/some: colored body
-stasis	control, stop	hemo/stasis: stop bleeding
-therapy	treatment	thermo/therapy: heat therapy
-tomy	to cut	laparo/tomy: to cut into the abdomen
-trophy	development	hyper/trophy: excessive development
-uria	urine	glucos/uria: glucose in the urine

Common Root Words

Root Word	Meaning	Example
acr-	extremity, peak	acro/phobia: fear of heights
aden-	gland	aden/oma: tumor of a gland
angi-	blood vessel	angio/pathy: diseased vessel
arthr-	joint	arthr/itis: inflammation of joints
brachi-	arm	brachi/al: pertaining to the arm
carcin-	cancer	carcin/oma: cancerous tumor
card-	heart	cardio/logy: study of the heart
carp-	wrist	carp/al: pertaining to the wrist
cephal-	head	cephal/ic: pertaining to the head
cervic-	neck	cervic/al: pertaining to the neck
chole-	bile	chole/cyst/itis: inflammation of the gallbladder
chondr-	cartilage	chondro/cyte: cartilage cell
colp-	vagina	colpo/dynia: vaginal pain
cost-	rib	cost/algia: rib pain
crani-	skull	cranio/malacia: softening of the skull
cutan-	skin	sub/cutan/eous: under the skin
cyan-	blue	cyan/osis: bluish skin color
cyst-	bladder	cyst/algia: pain in bladder
cyt-	cell	cyto/logy: study of cells
dactyl-	fingers, toes	dactylo/megaly: abnormally large fingers or toes
dent-	tooth	denti/form: toothlike
derm-, dermato-	skin	hypo/derm/al: under the skin
dors-	back	dors/al: pertaining to the back
duct-	carry	ad/duct: carry toward
edema	swelling	edema/tous: swollen
encephal-	brain	encephal/itis: inflammation of the brain
enter-	intestine	enter/dynia: intestinal pain
erythr-	red	erythro/cyte: red blood cell
esthe-	sensation, feeling	an/esthe/tic: substance causing an absence of sensation
esthen-	weakness	my/esthenia: muscle weakness
febr-	fever	febr/ile: feverish
gastr-	stomach	gastro/spasm: stomach spasm
gen-	to produce	patho/gen: disease-causing agent
gingiv-	gum	gingiv/itis: inflammation of the gums
glu-, glyc-	sugar	hyper/glyc/emia: excessive blood sugar
gynec-, gyno-	female	gyneco/logy: study of female disorders

Common Root Words *continued*

Root Word	Meaning	Example
hem-, hemato-	blood	hemo/genesis: blood formation
hepat-	liver	hepat/ectomy: removal of the liver
hist-	tissue	histo/logy: study of tissues
hom-, home-	same	homo/sexual: attracted to the same sex
hydr-, hydra-	water	hydra/tion: gaining water
kerat-	horny, cornea	kerat/osis: condition of abnormal horny growths
lacrim-	tear	lacrim/al: pertaining to tears
lact-	milk	lacta/tion: producing milk
lapar-	abdomen	laparo/tomy: to cut into the abdomen
laryng-	larynx	laryngo/pathy: disease of the larynx
later-	side	uni/lateral: one-sided
leuc-, leuk-	white	leuko/cyte: white blood cell
lingu-	tongue	lingu/iform: tongue-shaped
lip-	lipids, fat	lip/oid: similar to fat
lith-	stone	oto/lith: ear stone
mamm-	breast	mammo/gram: X ray of breast
melan-	black	melan/in: black skin pigment
men-	monthly, mensis	meno/pause: cessation of menses
metr-	uterus	myo/metr/ium: muscle layer of uterus
morph-	shape, form	morpho/logy: study of shape
my-	muscle	myo/card/itis: inflammation of the heart muscle
myel-	marrow, spinal cord	myel/algia: pain of the spinal cord or its membranes
nas-	nose	nas/al: pertaining to the nose
nephr-	kidney	nephr/itis: inflammation of a kidney
neur-	nerve	neur/ectomy: excision of a nerve
odont-	tooth	odonto/pathy: disease of the teeth
oo-	egg, ovum	oo/genesis: formation of ova
orchid-	testis	orchid/itis: inflammation of a testis
oss-, oste-	bone	osteo/malacia: softening of bones
ot-, aur-	ear	oto/lith: ear stone aur/icular: pertaining to the ear
path-	disease	patho/logy: study of disease
pect-	chest	pecto/ral: pertaining to the chest
ped-	child	ped/iatrics: medical specialty dealing with children's disorders
pep-, peps-	digest	pep/tic: pertaining to digestion
phag-	eat	phago/cyt/osis: engulfment of particles by cells
pharyng-	throat, pharynx	pharnygo/rrhea: discharge from the pharynx
phleb-	vein	phleb/itis: inflammation of a vein
pneum-	air	pneumo/thorax: air in the chest
pneumon-	lung	pneumono/pathy: disease of a lung
proct-	rectum	procto/col/itis: inflammation of the rectum and colon
pseud-	false	pseudo/hernia: false rupture
psych-	mind	psycho/genic: originating in the mind
pulmo-, pulmon-	lung	pulmon/ary: pertaining to a lung
py-	pus	pyo/cele: cavity containing pus
pyel-	kidney pelvis	pyelo/gram: X ray of kidney pelvis
quadr-	four	quadri/plegia: paralysis of both arms and both legs
rhin-	nose	rhin/itis: inflammation of the nose
salping-	uterine tube	salping/ectomy: removal of a uterine tube
scler-	hard	scler/osis: hardening
sect-	cut	sect/ion: process of cutting
sept-	presence of microbes	septic/emia: infection of blood
sten-	narrow	sten/osis: narrowed condition
strict-	draw tight	con/strict/ion: draw tightly together
therm-	heat	hypo/thermia: low body temperature
thorac-	chest, thorax	thoraco/dynia: chest pain
thromb-	clot	thromb/us: a blood clot
tox-	poison	tox/in: poisonous substance
vas-	vessel	vaso/dilation: expansion of a vessel
viscer-	internal organ	viscero/genic: originating in the internal organs
vita-	life	vita/l: essential for life

APPENDIX B

Answers to Check Your Understanding Questions

Chapter 1

1. physiology
2. cardiovascular
3. nervous
4. distal
5. appendicular
6. popliteal
7. femoral
8. dorsal; spinal
9. right upper; epigastric
10. mediastinum
11. peritoneum
12. homeostasis

Chapter 2

1. protons
2. ion
3. molecule
4. covalent
5. water
6. less
7. synthesis
8. monosaccharides
9. amino acids
10. nucleotides
11. fat
12. ATP

Chapter 3

1. plasma membrane
2. DNA
3. mitochondria
4. ribosomes
5. nucleolus
6. endoplasmic reticulum
7. diffusion
8. filtration
9. active transport
10. cellular respiration
11. messenger RNA
12. mitosis

Chapter 4

1. one
2. little
3. pseudostratified ciliated columnar
4. simple columnar
5. epithelial
6. matrix
7. loose (areolar)
8. fibrocartilage
9. cardiac
10. smooth
11. neuroglia
12. mucous

Chapter 5

1. basale
2. keratin
3. dermis
4. subcutaneous
5. papillae
6. sebaceous
7. eccrine
8. arrector pili
9. reduces
10. wrinkles and sagging skin
11. athlete's foot
12. calluses; corns

Chapter 6

1. support; protection
2. epiphyses; spongy
3. foramen
4. intramembranous
5. periosteum
6. axial
7. mandible; temporal
8. atlas; occipital
9. costal cartilages
10. pectoral girdle
11. humerus; ulna; radius
12. ilium; ischium; pubis
13. femur; tibia
14. hinge; ball-and-socket
15. osteoporosis

Chapter 7

1. fibers (muscle cells)
2. tendons
3. sarcomere
4. acetylcholine
5. actin
6. lactic acid
7. insertion
8. masseter
9. rectus abdominis
10. latissimus dorsi
11. deltoid
12. triceps brachii
13. gluteus maximus
14. quadriceps femoris
15. gastrocnemius

Chapter 8

1. axon
2. interneurons
3. sodium
4. neurotransmitter
5. corpus callosum
6. frontal
7. cerebrum
8. hypothalamus
9. medulla oblongata
10. cerebellum
11. subarachnoid
12. spinal cord
13. anterior
14. autonomic
15. sympathetic

Chapter 9

1. Meissner's corpuscles; free nerve endings
2. referred pain
3. taste buds
4. olfactory
5. accommodation
6. hearing; equilibrium
7. endolymph
8. Corti; basilar
9. ear ossicles
10. saccule; utricle
11. cornea; lens
12. cones; color
13. ciliary body
14. vitreous humor
15. optic chiasma

Chapter 10

1. endocrine; hormones
2. receptors
3. nonsteroid
4. negative feedback
5. hypothalamus
6. TSH; adrenal cortex; FSH; LH
7. thyroxine; thyroid

8. calcitonin; parathyroid hormone
9. medulla
10. aldosterone
11. glucagon
12. estrogen; testosterone

Chapter 11
1. 45
2. hemoglobin; erythrocytes
3. hemacytoblasts
4. oxygen; erythropoietin
5. liver
6. leukocytes
7. neutrophils; monocytes
8. basophils
9. eosinophils
10. lymphocytes
11. plasma
12. thrombocytes (platelets)
13. fibrinogen; fibrin
14. antigen(s)
15. AB

Chapter 12
1. pericardium
2. left atrium
3. tricuspid; right atrium
4. atrial; systole
5. S-A; A-V
6. sympathetic; parasympathetic
7. arteries; veins
8. arteries; capillaries
9. diffusion; filtration
10. right; trunk; arteries; veins; left
11. brachiocephalic; internal carotid
12. thoracic; cephalic; hepatic
13. hepatic-portal; hepatic; inferior
14. popliteal; external iliac; common iliac; inferior
15. axillary; subclavian; brachiocephalic; superior

Chapter 13
1. interstitial fluid
2. lymph; subclavian
3. lymph nodes; lymphocyte
4. spleen; red blood
5. nonspecific
6. neutrophils; macrophages (monocytes)
7. inflammation
8. thymus
9. antigen-presenting; helper T
10. killer; helper; suppressor; memory
11. antibody; cell
12. helper T; B; T
13. plasma; antibodies
14. active natural
15. vaccination; secondary

Chapter 14
1. filtered; mucous membrane
2. larynx; glottis
3. cartilaginous rings; bronchioles
4. alveoli
5. pleural cavity
6. surfactant
7. atmospheric; intrapulmonary
8. tidal; 500
9. medulla oblongata; pons
10. carbon dioxide; hydrogen ions
11. oxygen; carbon dioxide
12. oxyhemoglobin; bicarbonate ions

Chapter 15
1. digestive enzymes; nutrient molecules
2. mucosa
3. peristalsis
4. dentin; enamel
5. salivary amylase; maltose
6. peristalsis; lower esophageal sphincter
7. pepsin; proteins
8. parasympathetic; gastrin
9. secretin; cholecystokinin; cholecystokinin
10. amylase; lipase; trypsin
11. small intestine; monosaccharides; amino acids
12. villi; capillaries; lacteal
13. large intestine
14. liver
15. vitamins

Chapter 16
1. nephrons
2. cortex
3. medulla; collecting ducts
4. glomerular; systemic
5. glomerular capsule; filtrate
6. filtrate; peritubular capillaries
7. actively; passively; osmosis
8. renin
9. antidiuretic; collecting ducts
10. aldosterone; potassium
11. urine
12. ureters; peristalsis
13. urethra
14. internal; external
15. 60

Chapter 17
1. testes; scrotum
2. seminiferous tubules; spermatogenesis
3. epididymis; ejaculatory duct; urethra
4. blood
5. FSH; testosterone
6. ovaries; oogenesis
7. secondary oocyte; uterine tube
8. secondary oocyte; polar body
9. vagina
10. GnRH; LH
11. estrogen
12. corpus luteum; endometrium
13. decrease; menstruation
14. ovulation; implantation
15. estrogen; testosterone

Chapter 18
1. uterine tube
2. fertilization
3. blastocyst; seventh day
4. human chorionic gonadotropin; progesterone
5. germ layers
6. amnion; placenta
7. eighth; fetus
8. labor; oxytocin
9. foramen ovale; ductus arteriosus
10. prolactin-releasing hormone; prolactin
11. genes; DNA
12. 46; female
13. males
14. homozygous
15. probability

GLOSSARY

A

abdomen The anterior portion of the trunk located between the diaphragm and pelvis.

abdominal cavity The portion of the abdominopelvic cavity between the diaphragm and the pelvis.

abdominopelvic cavity The portion of the anterior body cavity inferior to the diaphragm.

abdominopelvic quadrants The four divisions of the abdominopelvic surface formed by perpendicular longitudinal and transverse lines through the umbilicus.

abdominopelvic regions The nine divisions of the abdominopelvic surface formed by two equidistant longitudinal lines and two equidistant transverse lines.

abduction The movement of a body part away from the midline.

abortion The removal of an embryo or fetus from the uterus prior to birth.

absorption The taking up of substances by cells.

accessory organs Organs that assist the functions of primary organs.

accommodation Adjustment by the lens to focus an image on the retina of the eye.

acetylcholine A neurotransmitter secreted from the axon tip of many neurons.

acid A substance that ionizes in water, releasing hydrogen ions.

acne Inflammation of a hair follicle.

acquired immunodeficiency syndrome (AIDS) A progressive decrease in immune capability caused by infection of T lymphocytes and macrophages with HIV.

acromegaly A disorder caused by excessive secretion of growth hormone after epiphyseal lines have fused.

ACTH Adrenocorticotropic hormone.

actin A thin protein filament that interacts with myosin to produce contraction of a myofilament.

active immunity Immunity derived from activation of B cells and T cells by an invasion of a pathogen.

active transport A process that requires the use of energy to move substances across plasma membranes.

Addison's disease An endocrine disorder caused by a deficient secretion of hormones by the adrenal cortex.

adduction The movement of a body part toward the midline.

adenine A nitrogen base of nucleic acids that pairs with thymine in DNA and uracil in RNA.

adenosine diphosphate (ADP) A molecule used to form adenosine triphosphate.

adenosine triphosphate (ATP) A molecule that temporarily holds energy in its high-energy phosphate bonds and releases it to power cellular processes.

adipose tissue Loose connective tissue containing large numbers of fat-storing cells.

adrenal cortex The outer portion of an adrenal gland.

adrenal gland An endocrine gland located on the top of each kidney.

adrenal medulla The inner portion of an adrenal gland.

adrenocorticotropic hormone (ACTH) A hormone secreted by the anterior pituitary that stimulates the adrenal cortex to secrete hormones.

aerobic respiration The part of cellular respiration that requires oxygen.

afferent arteriole The arteriole carrying blood to the glomerulus of a nephron.

agglutination The clumping of red blood cells in an antigen-antibody reaction.

agonist A muscle whose contraction moves a body part.

agranulocyte A type of white blood cell that lacks cytoplasmic granules.

albumin An abundant plasma protein that helps maintain the osmotic pressure of blood.

aldosterone A mineralocorticoid hormone produced by the adrenal cortex that regulates potassium and sodium concentrations in the blood.

alimentary canal. The digestive tube through which food passes from the mouth to anus.

alkaline Pertaining to a base.

allantois An extraembryonic membrane that branches from the yolk sac.

allele An alternate form of a gene.

allergen A foreign substance capable of stimulating an allergic reaction.

allergy An abnormally intense immune reaction.

all-or-none response The type of response by muscle and nerve cells to simulation; total response or no response.

alopecia Excessive hair loss.

alveolar ducts Tiny air passages that open into alveoli.

alveolus An air sac in a lung.

Alzheimer's disease A disorder caused by a loss of cholinergic neurons in the brain and characterized by loss of memory.

amenorrhea The absence of menstrual cycles.

amino acid The structural unit of proteins.

amnion The extraembryonic membrane that envelops the embryo and fetus.

amphiarthrosis A slightly movable joint.

ampulla The expanded portion of a semicircular canal.

amylase An enzyme that catalyzes the digestion of starch.

anabolism Metabolic processes forming more complex molecules from simpler molecules.
anaerobic respiration The portion of cellular respiration that does not require oxygen.
anaphase The stage of mitosis in which chromatids of replicated chromosomes separate and move to opposite poles of the cell.
anatomy The study of body organization and structure.
androgen Collective term for male sex hormones.
anemia The decreased ability of the blood to carry oxygen.
aneurysm A bulging, weakened portion of a blood vessel.
antagonist A muscle whose contraction opposes that of the agonist or prime mover.
antebrachium The forearm.
antecubital The anterior surface of the elbow.
anterior Pertaining to the front of the body.
anterior cavity The body cavity composed of thoracic and abdominopelvic cavities; the ventral cavity.
anterior root The anterior attachment of a spinal nerve to the spinal cord that carries motor nerve fibers.
antibody A substance produced by B lymphocytes in response to the presence of an antigen.
antibody-mediated immunity Immunity resulting from the production of antibodies.
anticodon A group of three nucleotides of a transfer RNA molecule that pairs with a codon of a messenger RNA molecule.
antidiuretic hormone The hormone secreted by the posterior lobe of the pituitary that promotes the reabsorption of water by the kidneys.
antigen A substance that stimulates production of antibodies.
anus The terminal opening of the alimentary canal.
aorta The systemic artery receiving blood directly from the left ventricle.
aortic semilunar valve A valve at the base of the aorta preventing the backflow of blood into the left ventricle.
apocrine sweat gland A sweat gland that opens into a hair follicle.
aponeurosis A sheet of fibrous connective tissue that attaches a muscle to a bone or other connective tissues.
appendicitis Inflammation of the appendix.
appendicular Pertaining to the upper and lower extremities.
appendicular skeleton The bones of the pectoral girdle, pelvic girdle, and upper and lower extremities.
appendix A functionless tubular extension of the cecum.
aqueous humor A watery fluid filling the anterior cavity of the eye.
arrhythmia An abnormal heartbeat such as bradycardia and fibrillation.
arteriole A small artery that leads to a capillary network.
arteriosclerosis Hardening of an arterial wall due to the deposition of calcium salts.
artery A blood vessel that carries blood away from the heart and to body tissues.
arthritis An autoimmune disorder of joints.
articular cartilage The cartilage that covers the ends of bones in freely movable joints.
articulation A joint; the joining of bones to form a joint.
association area A region of the cerebrum involved in memory and reasoning.
asthma Chronic inflammation and constriction of the respiratory passages.
astigmatism A visual disorder caused by an unequal curvature of the cornea or lens.
astrocyte A type of neuroglial cell joining blood vessels to neurons.
atherosclerosis A decrease in the lumen of an artery caused by fatty deposits in the vessel wall.
athlete's foot A fungal infection of the skin.
atmospheric pressure The pressure produced by the weight of the atmosphere.
atom The smallest unit of an element.
atomic number The number of protons in an atom of an element.
atomic weight The sum of the neutrons and protons in an atom of an element.
atrioventricular bundle Specialized muscle fibers carrying impulses from the A-V node to the Purkinje fibers; the bundle of His.
atrioventricular node (A-V node) A node of specialized tissue that receives impulses from the S-A node and transmits them to the A-V bundle.
atrioventricular valve (A-V valve) A valve that prevents a backflow of blood from a ventricle into an atrium.
atrium A heart chamber that receives blood returned to the heart by veins.
auditory Pertaining to the ear.
auditory tube A tube connecting the middle ear to the pharynx; the eustachian tube.
autoimmune disease A variety of diseases caused by the immune system attacking the body's own tissues.
autonomic ganglion A cluster of autonomic neuron cell bodies outside the central nervous system.
autonomic nervous system (ANS) The portion of the nervous system that functions involuntarily in the control of internal organs.
axial Pertaining to the longitudinal axis of the body.
axial skeleton The portion of the skeleton that supports and protects the head, neck, and trunk.
axon A neuron process that carries impulses away from the neuron cell body.

B

B lymphocyte A lymphocyte that produces antibodies; a B cell.
base A substance that ionizes in water, releasing hydroxyl (OH^-) ions.
basement membrane A thin layer of noncellular material attaching epithelial tissue to underlying connective tissue.
basilar membrane The membrane supporting the organ of Corti.
basophil A granulocytic leukocyte with large blue cytoplasmic granules.
bedsores Skin ulcers caused by a deficient blood supply to localized areas.
benign tumor A tumor that does not spread to another site.

bicuspid valve The atrioventricular valve between the left atrium and left ventricle; the mitral valve.
bile Fluid secreted by the liver and stored in the gallbladder.
bilirubin A bile pigment formed by the breakdown of hemoglobin.
blastocyst A preembryonic stage of development consisting of a hollow ball of cells.
blood The specialized connective tissue that transports substances to and from cells.
boil A bacterial infection of a hair follicle and its sebaceous gland.
bone The hardest and most rigid connective tissue whose matrix consists of calcium salts.
botulism Food poisoning caused by the bacterium *Clostridium botulinum.*
Bowman's capsule A glomerular capsule enclosing a glomerulus.
brachial Pertaining to the arm.
brain stem The portion of the brain including the midbrain, pons, and medulla oblongata.
Broca's area The portion of the frontal lobe of the cerebrum that coordinates muscles involved in speech.
bronchial tree The bronchi and their branches.
bronchiole A small branch of a bronchus.
bronchitis Inflammation of the bronchi.
bronchus A large branch from the trachea that carries air to and from a lung.
buffer A substance that stabilizes the pH of body fluids by combining with or releasing hydrogen ions.
bulbourethral glands Male accessory reproductive glands that secrete an alkaline secretion into the urethra.
bursa A fluid-filled sac in or near a joint that reduces friction during movement.
bursitis Inflammation of a bursa.

C

calcitonin The thyroid hormone that helps regulate the calcium level of the blood.
callus Thickened layers of epidermis caused by chronic friction.
calyx A funnel-like receptacle receiving urine from a renal pyramid's papilla.
canaliculi Microscopic canals between lacunae in bone.
cancellous bone Spongy bone; bone formed of a network of bony plates leaving numerous spaces between them.
cancer A malignant tumor; a tumor that can spread to other sites.
capillary A microscopic blood vessel extending between an arteriole and a venule.
carbaminohemoglobin A compound formed by the combination of carbon dioxide and hemoglobin.
carbohydrate An organic compound composed of carbon, hydrogen, and oxygen with a 2:1 ratio of hydrogen atoms to oxygen atoms.
carbonic anhydrase An enzyme in RBCs that catalyzes the combination of carbon dioxide and water to form carbonic acid.
cardiac Pertaining to the heart.
cardiac conduction system Specialized fibers that transmit impulses from the S-A node to the myocardium.
cardiac control center The center in the medulla oblongata controlling the heart rate.
cardiac cycle A single heartbeat composed of systole and diastole.
cardiac muscle The muscle in the wall of the heart.
carotene A yellowish pigment that occurs in the skin; a precursor of vitamin A in plant foods.
carpus The wrist; collective term for the wrist bones.
carrier molecule A molecule that carries a specific molecule through a plasma membrane in active transport.
cartilage A type of connective tissue characterized by a semisolid matrix and cartilage cells located within lacunae.
catabolism A metabolic process that breaks down large molecules into smaller ones.
cataract A vision disorder caused by cloudiness of the lens.
cecum The pouchlike first portion of the large intestine.
cell The structural and functional unit of the body.
cell body The portion of a neuron that contains the nucleus.
cell cycle The stages through which a cell passes from formation through cell division.
cell division The division of a parent cell to form new daughter cells.
cell-mediated immunity Immunity that is characterized by a direct attack on pathogens by T lymphocytes.
cellular respiration The process that releases energy from nutrients to form ATP.
cementum The hard substance that attaches a tooth to the periodontal membrane.
central nervous system (CNS) The portion of the nervous system formed of the brain and spinal cord.
centriole A cytoplasmic organelle involved in the formation of the spindle during cell division.
centromere The portion of a chromosome that is attached to a spindle fiber during cell division.
cephalic Pertaining to the head.
cerebellum The part of the brain that coordinates body movements.
cerebral cortex The outer layer of the cerebrum that is composed of gray matter.
cerebral hemispheres The two major parts of the cerebrum that are separated by the longitudinal fissure.
cerebral palsy A motor disorder caused by brain damage, resulting in uncontrolled muscular contractions.
cerebrospinal fluid The fluid filling the ventricles of the brain and the subarachnoid space of the meninges.
cerebrovascular accident (CVA) A stroke; brain impairment caused by aneurysm, hemorrhage, embolism, or thrombus.
cerebrum The portion of the brain involved in conscious action, sensory perception, memory, and intelligence.
ceruminous gland A gland that produces cerumin (wax).
cervical Pertaining to the neck or to the cervix of the uterus.
cervix The narrow end of the uterus that opens into the vagina.
chemical bond A bond that joins two atoms together.
chemical compound A substance composed of two or more elements chemically combined.

chemical element A substance that cannot be broken down into a simpler substance by chemical means.
chemical formula A shorthand designation of the kinds and numbers of atoms in a molecule.
chemoreceptor A receptor that forms impulses when stimulated by certain substances.
chief cell A cell of a gastric gland that secretes digestive enzymes.
cholecystokinin A hormone secreted by the mucosa of the small intestine that stimulates secretion of pancreatic juice and contraction of the gallbladder.
cholesterol A steroid lipid found in meats and milk products that is a component of plasma membranes.
cholinergic fiber A neuron axon that releases the acetylcholine at its tip.
cholinesterase An enzyme promoting the breakdown of acetylcholine in synapses.
chondrocyte A cartilage cell.
chorion The outermost extraembryonic membrane derived from the blastocyst wall and forming part of the placenta.
chorionic gonadotropin The placental hormone that maintains the secretion of estrogen and progesterone during pregnancy.
chorionic villi Fingerlike projections of the chorion that penetrate into the endometrium.
choroid coat The middle, pigmented vascular layer of the eyeball.
choroid plexus A specialized mass of capillaries in the ventricles of the brain that secrete cerebrospinal fluid.
chromatid One half of a replicated chromosome.
chromatin Portions of extended, uncoiled chromosomes appearing as dark granules in a cell nucleus.
chromosome A rodlike, dark body appearing in the nucleus during cell division; composed of DNA and protein.
chylomicron A microscopic droplet of fat coated with protein.
chyme The semiliquid mass that exits the stomach into the small intestine.
cilia Microscopic, hairlike projections from the free surface of ciliated epithelial cells.
ciliary body An enlarged ring of the choroid coat containing the ciliary muscles in the eye.
circumduction The movement of a body part in a circular path.
cirrhosis The destruction of liver tissue and its replacement with connective tissue.
citric acid cycle A series of chemical reactions in the aerobic phase of cellular respiration.
cleavage The early divisions of a zygote leading to the formation of a morula.
clitoris A small nodule of erectile tissue at the anterior junction of the labia minora.
clone A population of identical cells derived from a common ancestral cell.
coagulation The formation of a blood clot.
cochlea The coiled portion of the inner ear containing the receptors for hearing.
codon Three nucleotides of messenger RNA that code for a specific amino acid and that are complementary to both the three nucleotides of DNA and an anticodon of transfer RNA.
colitis Inflammation of the colon mucosa.
collagen A fibrous protein that provides strength to fibrous connective tissues.
collagenous fibers Fibers formed of collagen.
collecting ducts Microscopic tubules composing renal pyramids.
colon The major portion of the large intestine.
color blindness The inability to see certain colors or all colors.
coma A state of unconsciousness.
common bile duct A small tube that carries bile from the cystic duct into the duodenum.
common cold A virus-caused inflammation of air passages and associated structures.
compact bone Bone tissue formed of tightly packed osteons.
concentration gradient The difference in the concentration of a substance at two different locations.
concussion A jarring of the brain caused by a blow to the head.
conductivity The ability of neurons to conduct impulses.
condyle A rounded process of a bone.
cones Receptors for color vision that are located in the retina.
congestive heart failure A disorder in which the heart is unable to pump out all of the blood it receives.
conjunctiva The mucous membrane lining the eyelids and covering the anterior surface of the eye.
conjunctivitis Inflammation of the conjunctiva.
connective tissue One of a group of supporting and protective tissues that includes cartilage, bone, fibrous tissues, and blood.
constipation Difficult defecation of hard, dry feces.
contraception Devices or procedures that prevent contact of sperm and egg in sexual intercourse.
contraction The shortening of a muscle in response to stimulation.
cornea The anterior transparent window in the outer layer of the eyeball.
coronary Pertaining to the heart.
corpus callosum A mass of neuron processes connecting the two cerebral hemispheres.
corpus luteum A ruptured ovarian follicle after ovulation.
cortex The outer layer of an organ.
cortisol A glucocorticoid hormone secreted by the adrenal cortex.
costal Pertaining to the ribs.
costal cartilages Cartilages attaching the ribs to the sternum or to other costal cartilages.
covalent bond A chemical bond between two atoms that is formed by the sharing of valence electrons.
coxal Pertaining to the hip.
cramp A sudden, continuous tetanic contraction.
cranial Pertaining to the cranium.
cranial cavity The cavity in which the brain resides.
cranial nerve A nerve that originates from the brain.
cranium The skull bones forming the cranial cavity and enveloping the brain.
creatine phosphate A compound in muscle tissue that temporarily holds additional energy for forming ATP.
creatinine A nitrogenous waste produced by muscle metabolism.
cretinism A congenital disorder due to a lack of thyroid hormone.

crista ampullaris A sense organ in the ampullae of semicircular canals that contains receptors for dynamic equilibrium.
cubital Pertaining to the elbow.
Cushing's syndrome A disorder due to the hypersecretion of glucocorticoids by the adrenal cortex.
cutaneous Pertaining to the skin.
cystic duct The duct leading from the gallbladder to the common bile duct.
cystitis Inflammation of the urinary bladder.
cytokinesis The division of the cytoplasm during telophase of cell division.
cytoplasm The semiliquid material located between the nucleus and plasma membrane of a cell.
cytosine A nitrogen base of nucleic acids that pairs with guanine.

D

deamination The removal of amine groups (—NH_2) from amino acids.
deciduous teeth The first set of teeth, which are lost and replaced by permanent teeth.
decomposition The breakdown of complex molecules into simpler molecules.
defecation The expulsion of feces through the anus.
dehydration synthesis The combining of two molecules by the removal of a water molecule.
dendrite A neuron process that carries impulses toward the cell body.
dense connective tissue A fibrous connective tissue whose matrix is composed mainly of collagen fibers.
dental caries Tooth decay.
dentin The hard substance that forms most of a tooth.
deoxyhemoglobin The molecule remaining after oxyhemoglobin has given up its oxygen.
deoxyribonucleic acid (DNA) The double-stranded nucleic acid forming the hereditary component of chromosomes.
depolarization The loss of unlike charges on each side of a membrane.
dermal papillae Conelike projections of the dermis at the dermis-epidermis border.
dermis The true skin located under the epidermis.
detrusor muscle The muscle in the wall of the urinary bladder.
diabetes insipidus A disorder caused by a deficiency of antidiuretic hormone.
diabetes mellitus A disorder caused by a deficiency of insulin.
diaphragm The sheetlike skeletal muscle separating the thoracic and abdominopelvic cavities.
diaphysis The shaft of a long bone.
diarrhea Production of watery feces.
diarthrosis A type of joint that is freely movable.
diastole The relaxation phase of the cardiac cycle.
diastolic pressure The blood pressure in systemic arteries during heart diastole.
diencephalon The portion of the brain containing the thalamus, hypothalamus, and pineal gland.
differentiation The process by which cells become specialized for various functions.
diffusion The passive movement of molecules from an area of higher concentration to an area of lower concentration.
digestion The mechanical and chemical process that breaks down food molecules into absorbable nutrient molecules.
dipeptide A molecule composed of two amino acids chemically combined.
disaccharide A molecule composed of two monosaccharides chemically combined.
dislocation Displacement of a bone forming a joint.
distal Farther from the origin of an extremity.
diuresis The excessive production of urine.
diverticulitis Inflammation of diverticula of the colon mucosa.
dopamine A neurotransmitter secreted by certain neurons in the brain.
dorsum Pertaining to the back; the posterior surface of the thorax.
ductus arteriosus A short artery that carries blood from the pulmonary trunk to the aorta in a fetal heart.
ductus venosus A short vein that carries blood from the umbilical vein to the inferior vena cava during fetal development.
duodenum The first portion of the small intestine.
dura mater The outermost membrane of the meninges.
dynamic equilibrium The maintenance of balance when the head and body are in motion.
dyslexia A reading disorder in which letters and words are reversed.
dysmenorrhea Painful menstrual periods.

E

ear ossicles Three tiny bones in the middle ear that transmit vibrations from the tympanic membrane to the inner ear.
eccrine sweat glands Sweat glands that open on the surface of the epidermis.
eclampsia A disorder of pregnancy characterized by high blood pressure, edema, and possibly convulsions and coma.
ectoderm The outermost tissue of the blastodisk.
ectopic pregnancy A condition in which the embryo is implanted at a site other than the uterus.
eczema A skin disorder characterized by red, dry, scaling skin.
edema The swelling of tissues due to the excessive accumulation of interstitial fluid.
effector A muscle or gland that responds to neural stimulation.
efferent arteriole An arteriole carrying blood from a glomerulus.
ejaculation The discharge of semen from the male urethra.
ejaculatory duct A short duct formed by the merger of the duct from a seminal vesicle and a vas deferens.
elastic fibers Protein fibers providing elasticity in certain connective tissues.
electrocardiogram A record of the electrical activity of the heart during a cardiac cycle.
electrolyte A substance that ionizes when dissolved in water.
electron A negatively charged particle that revolves around an atomic nucleus.

elephantiasis A disorder in which lymphatic vessels are plugged by an infestation of microscopic worms.

embolism The blockage of a blood vessel by a clot or gas bubble formed elsewhere and carried by blood to the site.

embryo The developmental stage from weeks 2 through 8 after fertilization.

emphysema A lung disorder in which alveolar walls rupture, reducing the respiratory surface.

enamel The hard outer layer of a tooth crown.

endocardium The inner lining of the heart chambers.

endochondral bone Bone that is first formed of hyaline cartilage, which is replaced by bone.

endocrine gland A gland that secretes hormones.

endocytosis The process by which a cell engulfs particles and liquids.

endoderm The innermost embryonic tissue of the blastodisk.

endolymph The fluid in the membranous labyrinth of the inner ear.

endometriosis The growth of endometrial tissue at sites other than in the uterus.

endometrium The inner lining of the uterus.

endoplasmic reticulum A network of membranous channels used to transport substances within a cell.

endosteum The epithelial lining of a medullary cavity.

endothelium The inner lining of the heart chambers and blood vessels.

enzyme A protein that aids and speeds up (catalyzes) a specific chemical reaction.

eosinophil A granulocytic leukocyte with red cytoplasmic granules.

ependymal cell A type of neuroglial cell that lines the ventricles of the brain.

epicardium The visceral pericardium.

epicondyle A bony projection located above a condyle.

epidermis The stratified squamous epithelium covering the dermis; the outer layer of the skin.

epididymis The highly coiled tube that carries sperm from the seminiferous tubules to the vas deferens.

epidural space The space between the dura mater and the vertebrae or cranial bones.

epigastric region The upper middle portion of the abdomen.

epiglottis A cartilaginous flap that closes over the larynx during swallowing.

epilepsy A neural disorder characterized by sudden lapses of consciousness and possible convulsions.

epinephrine A hormone secreted by the adrenal medulla in response to stress.

epiphyseal line The line of fusion between an epiphysis and diaphysis when bone growth in length is complete.

epiphyseal plate The plate or disk of hyaline cartilage located between an epiphysis and the diaphysis of a long bone.

epiphysis The enlarged end of a long bone.

epithelial tissue The body tissue that covers the free surfaces of organs and the body, and forms secretory portions of glands; epithelium.

erection The process by which erectile tissue engorges with blood.

erythroblastosis fetalis A fetal disorder caused by maternal antibodies attacking fetal Rh^+ erythrocytes.

erythrocyte A red blood cell; RBC.

erythropoietin The hormone that stimulates erythrocyte production.

esophagus The portion of the alimentary canal carrying food from the pharynx to the stomach.

estrogen A female hormone secreted by the ovaries that promotes the development of primary and secondary sexual characteristics.

eversion The movement that turns the sole of the foot outward.

excretion The removal of metabolic wastes and excessive substances from the body.

exocrine gland A gland whose secretion is carried to a specific site by a duct.

exocytosis The process by which a cell rids itself of unwanted particles and liquids.

exophthalmic goiter An endocrine disorder caused by an excessive secretion of thyroid hormone.

expiration The exhalation of air from the lungs.

extension A movement that increases the angle between two body parts forming the joint.

external Pertaining to the surface of the body.

external auditory canal A canal in the temporal bone that leads from the exterior to the middle ear.

external respiration The exchange of oxygen and carbon dioxide between air in the lungs and the blood.

extracellular Pertaining to regions external to cells.

extremity An arm or leg.

extrinsic eye muscles Muscles that move the eyeball.

F

facet A small, flat surface of a bone that is an articulation site for another bone.

farsightedness A visual disorder in which the image is focused posterior to the retina.

fascia Fibrous connective tissue supporting, covering, and separating muscles.

fat A triglyceride; adipose tissue.

fatigue The lack of response to stimulation by muscle or nerve cells.

fatty acid An organic molecule that forms part of a triglyceride.

feces Material discharged from the anus in defecation.

fertilization Union of sperm and egg nuclei.

fetus The human developmental stage from the ninth week to birth.

fever A higher-than-normal body temperature.

fever blister A cold sore; small vesicles on the lips caused by a *Herpes* virus infection.

fibril A microscopic filament.

fibrin The insoluble protein filaments that form a blood clot.

fibrinogen A soluble plasma protein converted into insoluble fibrin during blood clot formation.

fibroblast A connective tissue cell that forms intercellular protein fibers.

fibrosis The replacement of muscle tissue with fibrous connective tissue.

fibrositis Inflammation of connective tissue.

filtrate The substances removed by filtration.

filtration The forcing of water and solutes through a plasma membrane by hydrostatic pressure.

fissure A long, narrow cleft separating two parts.

flagellum A long, hairlike extension from a cell.

flexion Movement at a joint causing the angle of the bones forming the joint to be decreased.

follicle A cavity or saclike depression.

follicle-stimulating hormone (FSH) A hormone secreted by the anterior pituitary gland that stimulates development of ovarian follicles in females and sperm production in males.

follicular cells Cells that surround a developing oocyte.

fontanel Membranous regions between cranial bones of an infant.

foramen A small canal or passageway in a bone or membrane.

foramen ovale The opening between the atria in a fetal heart.

formed elements The cellular components of blood; blood cells and platelets.

fossa A depression in a bone or organ.

fovea centralis A small depression in the retina that contains only densely packed cones.

fracture A broken bone.

free nerve endings Unattached axon tips that serve as pain receptors.

frontal plane A plane dividing the body or organ into anterior and posterior portions; a coronal plane.

G

gallbladder A small sac that temporarily stores bile.

gallstones Concretions of cholesterol and salts in the gallbladder.

gamete A sex cell; a sperm or ovum.

ganglion A mass of neuron cell bodies outside the central nervous system.

gastric gland A gland of the stomach mucosa that secretes gastric juice.

gastric juice The secretion of gastric glands containing HCl and digestive enzymes.

gastrin A hormone secreted by the stomach mucosa that stimulates the secretion of gastric juice.

gene A determiner of a hereditary trait.

genotype The genetic composition of an individual.

germinal epithelium Specialized epithelium in ovaries and testes that produces sex cells.

germ layers Three cell layers of the blastodisk that form all tissues during embryonic development.

gigantism An endocrine disorder caused by excessive secretion of growth hormone prior to the formation of epiphyseal lines.

gland A cell or group of cells that produces a secretion.

glaucoma An eye disorder caused by excessive intraocular pressure that damages the retina.

globulin A type of plasma protein.

glomerulonephritis Inflammation of the kidney involving the glomeruli.

glomerulus A ball-like mass of capillaries located within a glomerular capsule of a nephron.

glottis The narrow opening between the true vocal cords.

glucagon The pancreatic hormone that promotes the formation of glucose from glycogen.

glucocorticoid A group of hormones secreted by the adrenal medulla that influences glucose metabolism.

glucose The monosaccharide that is the primary energy source for cells.

glycerol An organic compound that is the backbone of triglyceride and phospholipid molecules.

glycogen The polysaccharide that is the storage form for carbohydrates in the body.

goblet cell A mucus-producing epithelial cell.

Golgi complex A cellular organelle that packages substances for secretion from the cell.

gonad A primary sex gland; an ovary or testis.

gonadotropin A hormone from the anterior pituitary that stimulates activity of the gonads.

granulocyte A leukocyte containing cytoplasmic granules.

gray matter The portion of the central nervous system lacking myelin.

groin The anterior body region at the junction of a thigh and pelvis.

growth The enlargement of the body or a body part.

growth hormone (GH) The hormone of the anterior pituitary that promotes cell division and cell enlargement.

guanine A nitrogen base of nucleic acids that pairs with cytosine.

H

hair cells Sensory receptors for hearing and equilibrium.

hair follicle The tubelike structure in which a hair develops.

head An enlargement on the end of a bone.

heart murmur An abnormal heart sound usually caused by a leaking valve.

hematopoiesis The formation of blood cells.

hemoglobin The red pigment of erythrocytes that transports oxygen.

hemophilia A disorder in which clot formation is impaired.

hemorrhage Bleeding; excessive blood loss.

hemorrhoids Swollen, inflamed veins of the anal canal.

hemostasis The stoppage of bleeding.

hepatic Pertaining to the liver.

hepatitis Inflammation of the liver.

herniated disk A bulging intervertebral disk.

heterozygous A condition in which the paired alleles controlling a trait are different.

hives An itching rash resulting from an allergic reaction.

homeostasis The maintenance of a dynamically stable internal environment.

homozygous A condition in which the paired alleles controlling a trait are identical.

hormone A chemical secreted by an endocrine gland that affects the functions of target tissues.

hydrogen bond A weak bond between a positively charged hydrogen atom and a negatively charged atom in the same or different molecule.

hydrogen ion H^+; a single proton.

hydrolysis The breakdown of complex molecules into simpler molecules by the addition of water molecules.

hydroxyl ion OH^-.

hymen A membrane that partially covers the vaginal opening.

hyperglycemia An excessive concentration of glucose in the blood.
hypertension Chronic high blood pressure.
hypertonic solution A solution with a greater concentration of solutes than the solution with which it is compared.
hypochondriac region Either the left or right upper lateral portion of the abdomen lying lateral to the epigastric region.
hypogastric The lower middle region of the abdomen.
hypoglycemia A deficient concentration of glucose in the blood.
hypophysis The pituitary gland.
hypopituitary dwarfism Abnormal growth caused by a deficiency of growth hormone from the anterior pituitary.
hypothalamus The part of the brain lying below the thalamus and to which the hypophysis (pituitary) is attached.
hypotonic solution A solution with a lesser concentration of solutes than the solution with which it is compared.

I

iliac region Either the left or right lower abdominal portion lying lateral to the hypogastric region.
immunoglobulin Plasma proteins consisting of antibodies.
impetigo A highly contagious bacterial infection of the skin characterized by pustules that rupture and become crusted.
implantation The embedding of an embryo in the endometrium.
impotence The inability of a male to obtain and maintain an erection for sexual intercourse.
impulse A depolarization wave passing along a neuron process.
infectious mononucleosis A viral infection of lymphocytes that causes them to resemble monocytes.
infertility In females, the inability to conceive; in males, the inability to produce sperm to achieve fertilization of the ovum.
inflammation A tissue response characterized by increased blood flow, redness, and fluid accumulation.
influenza A virus-caused disorder with coldlike symptoms.
inorganic Pertaining to compounds whose molecules do not contain both carbon and hydrogen.
insertion The movable attachment of a muscle.
inspiration Inhalation; breathing air into the lungs.
insulin The pancreatic hormone that helps glucose to enter body cells.
integumentary system The skin and its associated structures.
intercalated disk A dark-staining membrane at the junction of adjoining cardiac muscle cells.
intercellular Pertaining to the spaces between cells.
internal Pertaining to the interior of the body or organ.
internal respiration The exchange of oxygen and carbon dioxide between the blood and tissue cells.
interneuron A neuron of a neural pathway that is located between sensory and motor neurons.
interphase The nondividing phase between two cell divisions.
interstitial cell A cell located between seminiferous tubules that secretes testosterone.
interstitial fluid Intercellular fluid; tissue fluid.
intervertebral disk The fibrocartilaginous pad separating two adjacent vertebrae.
intestinal gland A gland of the mucosa of the small intestine that secretes intestinal juice.
intestinal juice The secretion of intestinal glands.
intracellular Pertaining to within cells.
intramembranous bone Bone that develops within layers of membranes.
intrinsic factor A substance secreted by gastric glands that is essential for adequate absorption of vitamin B_{12}.
inversion The movement that turns the sole of the foot inward.
involuntary Without conscious control.
ion An atom or group of atoms with an electrical charge.
ionic bond A chemical bond formed between two ions with opposing electrical charges.
ionization The dissociation of molecules forming ions.
iris The colored circular muscle that controls the amount of light entering the lens of the eye.
irritability The ability of a neuron to respond to a stimulus by forming an impulse.
islets of Langerhans Clumps of pancreatic cells that secrete insulin.
isotonic solution A solution with the same concentration of solutes as a solution with which it is compared.
isotope An atom with a different number of neutrons than most atoms of the element.

J

joint An articulation; the junction of two or more bones.
juxtaglomerular apparatus Specialized cells of the afferent arteriole and distal tubule that are involved in controlling glomerular blood pressure.

K

keratin A waterproofing protein formed in the epidermis, nails, and hair.
keratinization The process by which cells form large amounts of keratin.

L

labia majora The outer folds of the vulva.
labia minora The inner folds of the vulva.
labor The series of events associated with childbirth.
labyrinth The interconnecting tubes and chambers of the inner ear.
labyrinthine disease A group of inner ear disorders that produce symptoms of dizziness and nausea.
lacrimal apparatus Structures involved in the production and removal of tears.
lacrimal gland A gland that secretes tears.
lactation Milk secretion.
lacteal A lymphatic capillary within a villus of the small intestine.

lactic acid An organic by-product formed from pyruvic acid during anaerobic cellular respiration.
lacuna A cavity in bone or cartilage.
lamellae Layers of solid bone matrix in compact bone.
laryngitis Inflammation of the larynx.
larynx The cartilaginous boxlike structure located between the pharynx and trachea that contains the vocal folds.
lateral Pertaining to the side.
lens The structure that focuses images on the retina of the eye.
leukemia A cancerous blood disorder characterized by an excess production of certain leukocytes.
leukocyte A white blood cell; WBC.
ligament A fibrous cord or sheet attaching bones to each other.
limbic system The portion of the cerebrum and diencephalon involved in emotions and moods.
lingual Pertaining to the tongue.
lipase The enzyme that catalyzes the digestion of lipids.
lipid A class of organic compounds that includes steroids, triglycerides, and phospholipids.
liver The digestive gland that secretes bile and processes absorbed nutrients prior to their entrance into the general circulation.
longitudinal fissure The deep fissure that separates the left and right cerebral hemispheres.
lumbar Pertaining to the posterior trunk region between the ribs and hips.
lumen The space within a tubular structure.
lungs The respiratory organs for gas exchange.
luteinizing hormone The anterior pituitary hormone that controls the functions of the corpus luteum in females and testosterone secretion in males.
lymph Interstitial fluid that has entered a lymphatic vessel.
lymph node A small mass of lymphoid tissue along a lymphatic vessel that filters lymph as it passes through.
lymphatic vessel A vessel transporting lymph.
lymphocyte A type of white blood cell that is involved in immune reactions.
lymphokine A chemical secreted by helper T cells that stimulates the division of B cells to form a clone.
lymphoma A cancer of lymphatic tissue.
lysis Rupture of a cell due to a rapid uptake of water.
lysosome A cellular organelle consisting of a sac of digestive enzymes.

M

macrophage A modified monocyte that has entered tissue spaces and is involved in phagocytosis and immune reactions.
macula Sense organs of the inner ear that contain receptors for static equilibrium.
macula lutea The yellow spot on the retina containing the fovea centralis.
malignant tumor A tumor with the capability of spreading (metastasizing) to other sites; a cancer.
mammary gland A milk-producing gland located within a female breast.
marrow Connective tissue filling spaces in bone.
mast cell A modified basophil in tissue spaces that releases histamine and serotonin in allergic reactions.
mastication The chewing of food.
matrix The intercellular substance in connective tissues.
matter Anything that has weight (mass) and occupies space.
maximal stimulus A stimulus producing a maximum response by a muscle.
mechanoreceptor A sensory receptor that is sensitive to mechanical stimuli.
medial Toward the midline.
mediastinum Connective tissue separating the thoracic cavity into left and right portions.
medulla The inner or central portion of an organ.
medulla oblongata The part of the brain stem that is continuous with the spinal cord.
medullary cavity The central cavity in the shaft of a long bone.
meiosis The process in meiotic cell division that reduces the number of chromosomes in daughter cells to one-half that of the parent cell.
meiotic cell division The type of cell division involved in the formation of sperm and eggs.
melanin The brown-black pigment found in the epidermis.
melanocyte A cell of the epidermis that produces melanin.
melatonin A hormone secreted by the pineal gland that influences reproductive cycles.
memory cell A dormant B or T lymphocyte produced in an initial immune response that can respond quickly if the same antigen reappears.
meninges A group of three membranes that envelops the brain and spinal cord.
meningitis Inflammation of the meninges.
menopause The cessation of menstrual cycles.
menstrual cycle The monthly female reproductive cycle characterized by the buildup, breakdown, and discharge of the uterine lining.
menstruation The breakdown and discharge of the uterine lining.
mental illness A group of disorders characterized by abnormal behaviors.
mesentery A fold of the peritoneum that supports internal abdominal organs.
mesoderm The middle embryonic tissue of the blastodisk.
messenger RNA (mRNA) The type of RNA that carries information for protein synthesis from DNA to the ribosomes.
metabolism The sum of the chemical reactions of life.
metaphase The phase of cell division characterized by the chromosomes arranged along the equator of the cell.
microfilament A microscopic protein strand within cells that is part of the cytoskeleton.
microglial cell A type of neuroglial cell that supports neurons and is involved in phagocytosis.
microtubule A microscopic tubule of protein within cells that is part of the cytoskeleton.
microvilli Microscopic projections of the plasma membrane on the free surfaces of certain epithelial cells.
micturition Urination.
midbrain The small region of the brain that is located between the thalamus and pons.
midsagittal plane A plane that divides the body or an organ into equal left and right halves.

mineralocorticoids A group of hormones secreted by the adrenal cortex that influence the concentration of electrolytes in body fluids.
miscarriage A spontaneous abortion.
mitochondrion A cellular organelle that is the site of aerobic cellular respiration.
mitosis The process in mitotic cell division that distributes replicated chromosomes so that daughter cells receive the same number of chromosomes as the parent cell.
mitotic cell division The process of cell division that produces daughter cells identical to the parent cell; is involved in growth and repair of the body tissues.
mixed nerve A nerve consisting of both sensory and motor fibers.
mole A slow-growing, pigmented skin tumor.
molecule Two or more atoms chemically combined; the smallest unit of a compound.
monocyte A large agranulocytic leukocyte that functions in phagocytosis.
monosaccharide A simple sugar; a structural unit of carbohydrates.
morning sickness A temporary disorder of pregnancy characterized by nausea upon arising.
motor area The region of the cerebrum that initiates actions of muscles and glands.
motor nerve A nerve that consists only of axons of motor neurons.
motor neuron A neuron that activates a muscle or gland.
motor unit A single motor neuron and the muscle fibers that it controls.
mucosa A mucous membrane such as the lining of the alimentary canal.
mucus Thick fluid produced by goblet cells.
multiple sclerosis A disorder characterized by the degeneration of the myelin sheath of neurons in the central nervous system.
muscle fiber A muscle cell.
muscle tissue The type of body tissue that is specialized for contraction.
muscle tone A state of partial contraction.
muscular dystrophy A disorder characterized by the atrophy of muscles.
mutation A spontaneous change in a gene.
myasthenia gravis A disorder characterized by severe muscular weakness.
myelin sheath The fatty, insulating sheaths formed around some nerve fibers.
myocardial infarction Death of a portion of the myocardium due to blockage of a coronary artery; a heart attack.
myocardium The muscle layer of the heart wall.
myofibril A thin contractile element within a muscle cell.
myoglobin A compound in muscles that temporarily stores a small amount of oxygen.
myosin A protein that interacts with actin to produce contraction of a myofibril.
myositis Inflammation of a muscle.
myxedema An adult disorder caused by a lack of thyroid hormone.

N

nasal cavity The interior cavity of the nose.
nasal septum The partition of bone separating the nasal cavity into left and right portions.
nearsightedness A vision disorder in which the image is focused in front of the retina.
negative feedback A control mechanism in which an increase in the concentration of a substance inhibits the production of that substance.
neonatal Pertaining to a newborn infant; the period from birth to the end of the first month.
nephron The functional unit of a kidney; composed of a renal corpuscle and tubule.
nerve A bundle of nerve fibers (neuron processes).
nerve impulse A depolarization wave passing along a nerve fiber.
nerve tissue Tissue specialized for the formation and conduction of nerve impulses.
nerve tract Myelinated ascending and descending bundles of nerve fibers in the central nervous system.
neuralgia Pain in a nerve.
neurilemma The outer sheath of a Schwann cell enveloping a nerve fiber.
neuritis Inflammation of a nerve.
neuroglia Tissue in the central nervous system that supports the neurons.
neuromuscular junction The junction of the axon tip of a motor neuron with a muscle fiber.
neuron A nerve cell.
neuropeptide A naturally occurring substance in the brain that modifies neural responses to neurotransmitters; enkephalins and endorphins.
neurotransmitter A chemical secreted by an axon tip that stimulates action of a muscle or gland, or the formation of an impulse in an adjoining neuron.
neutron A noncharged particle in an atomic nucleus.
neutrophil A phagocytic leukocyte with pale lavender cytoplasmic granules.
Nissl bodies Dark-staining membranous sacs in the cytoplasm of neurons comparable to the endoplasmic reticulum.
node of Ranvier A space between adjacent Schwann cells wrapped around a neuron process.
nonspecific resistance Resistance mechanisms that act against all types of pathogens.
norepinephrine A neurotransmitter released by axon tips of sympathetic neurons and secreted by the adrenal medulla.
nuclear envelope The double membrane surrounding the nucleus of a cell.
nucleic acid A compound whose molecules are composed of a series of nucleotides; either DNA or RNA.
nucleolus A dark-staining spherical structure within a cell nucleus that is composed of protein and rRNA.
nucleotide The building unit of nucleic acids; consists of a simple sugar, a phosphate group, and a nitrogen base.
nucleus The spherical cellular organelle containing the chromosomes; the core of an atom; or a mass of neuron cell bodies in the brain.

nutrient A substance in foods that is required for the normal nutrition of the body.

O

olfactory Pertaining to the sense of smell.

oligodendrocyte A type of neuroglial cell that supports neurons and forms myelin around some nerve fibers within the central nervous system.

oocyte An immature ovum (egg cell).

oogenesis The process of ovum formation.

ophthalmic Pertaining to the eye.

optic Pertaining to the eye.

optic chiasma The X-shaped site on the interior surface of the brain formed by optic nerve fibers that partially cross over to the opposite side.

optic disk The blind spot; the site of retinal nerve fibers exiting the eye to form the optic nerve.

oral Pertaining to the mouth.

organ A structure formed of two or more tissues that performs specific functions.

organ of Corti The sense organ in the inner ear that contains receptors for hearing.

organ system A group of organs that work in a coordinated fashion to carry out specialized functions.

organelle A distinct structure within a cell that performs specific functions.

organic Pertaining to compounds containing both carbon and hydrogen in their molecules.

orgasm The culmination of sexual stimulation.

origin The nonmovable attachment of a muscle.

osmosis The diffusion of water through a selectively permeable membrane.

ossification The process of bone formation.

osteoblast A cell that deposits bone matrix.

osteoclast A cell that removes bone matrix.

osteocyte A bone cell.

osteon The structural unit of compact bone consisting of lamellae and osteocytes around an osteonic canal.

osteonic canal A microscopic canal in bone that contains blood vessels and nerves, and that is encompassed by lamellae.

osteoporosis A disorder in which bone matrix is reabsorbed, producing weakened bones.

otitis media A middle ear infection.

otolith A granule of calcium carbonate associated with receptors for equilibrium.

oval window The opening in the labyrinth of the inner ear into which the stapes is inserted.

ovarian Pertaining to an ovary.

ovary The female gonad producing ova and female reproductive hormones.

ovulation The rupture of a mature ovarian follicle, releasing a secondary oocyte.

ovum A female reproductive cell; a female gamete.

oxygen debt The amount of oxygen necessary to metabolize lactic acid accumulated during anaerobic respiration.

oxyhemoglobin The compound carrying oxygen to body cells.

oxytocin The posterior pituitary hormone that stimulates uterine contractions and milk ejection.

P

pain receptor A free nerve ending.

palate The roof of the mouth formed of the hard and soft palates.

palmar Pertaining to the palm of the hand.

pancreas An abdominal organ secreting both digestive secretions and hormones.

pancreatic duct The tube carrying pancreatic juice to the common bile duct.

papilla A small, nipplelike projection.

papillary muscle A projection of the myocardium to which chordae tendineae are attached.

paralysis A disorder in which a muscle is unable to contract.

parasympathetic division The part of the autonomic nervous system arising from the brain and sacral region of the spinal cord.

parathyroid glands Small endocrine glands embedded in the posterior surface of the thyroid gland.

parathyroid hormone The secretion of the parathyroid glands that helps regulate the concentrations of calcium and phosphate in the blood.

parietal Pertaining to the wall of a cavity.

parietal cell A type of cell in gastric glands that secretes hydrochloric acid and intrinsic factor.

parietal pericardium A membranous sac enclosing the heart.

parietal peritoneum The membrane lining the wall of the abdominal cavity.

parietal pleura A pleural membrane lining the interior of the thoracic cavity.

Parkinson's disease A neural disorder characterized by muscular weakness, tremor, and rigidity.

parotid glands The largest salivary glands located just anterior and inferior to the ears.

parturition The birth process.

passive immunity Immunity provided by an injection of antibodies that have been formed in another organism.

passive transport Movement of substances through plasma membranes without the expenditure of energy.

pathogen A disease-causing virus or organism.

pectoral Pertaining to the chest.

pectoral girdle The bones that attach the upper extremities to the trunk at the shoulder.

pelvic Pertaining to the pelvis.

pelvic cavity The inferior portion of the abdominopelvic cavity.

pelvic girdle The bones that attach the lower extremities to the trunk at the hips.

pelvic inflammatory disease Inflammation of the female reproductive organs and pelvic tissues.

pelvis The ring of bones formed by the os coxae and sacrum.

penis The external male reproductive organ containing the urethra.

pepsin An enzyme secreted by gastric glands that catalyzes the splitting of proteins.

peptic ulcers Digestion of part of the stomach mucosa.

peptide A compound composed of two or more amino acids.

peptide bond A bond that joins two amino acids.
pericardial Pertaining to the pericardium.
pericardial cavity The potential space between the epicardium and parietal pericardium.
pericarditis Inflammation of the pericardium.
pericardium The saclike membrane enveloping the heart.
perichondrium The fibrous membrane surrounding a cartilage.
perilymph The fluid in the bony labyrinth of the inner ear.
perineum In males, the region between the anus and the scrotum; in females, the region between the anus and the vulva.
periodontal disease A disorder in which gums are inflamed and associated periodontal ligaments and bone degenerate.
periodontal ligament The fibrous membrane attaching a tooth to alveolar bone.
periosteum The fibrous membrane covering the surface of a bone.
peripheral nervous system (PNS) The part of the nervous system outside the central nervous system; the cranial and spinal nerves.
peripheral resistance Resistance of the flow of blood; friction between the blood and the walls of blood vessels.
peristalsis The wavelike contractions that move materials through tubular organs.
peritoneal cavity The potential space between the visceral and parietal peritoneal membranes.
peritoneum The serous membrane lining the abdominal cavity and covering internal abdominal organs.
peritonitis Inflammation of the peritoneum.
peritubular capillary The capillary network that surrounds a nephron tubule.
permeable Allowing the passage of materials.
pH A measure of the hydrogen ion concentration of a solution.
pH scale A scale that establishes the values of pH from 0 to 14.
phagocytosis The process by which cells engulf particles.
pharynx The throat; the cavity between the mouth and the esophagus or larynx.
phenotype The observable characteristics of an inherited trait.
phlebitis Inflammation of a vein.
phospholipid A molecule containing two fatty acids and a phosphate group attached to glycerol.
photoreceptor A receptor sensitive to light stimuli; a rod or cone.
physiological jaundice A disorder of newborn infants caused by the rapid destruction of fetal erythrocytes.
physiology The study of body functions.
pia mater The delicate, innermost membrane of the meninges.
pineal gland A small endocrine gland within the brain that is involved in biorhythms.
pinocytosis The process by which cells engulf liquids.
pituitary gland The endocrine gland attached to the hypothalamus; the hypophysis.
placenta The temporary organ attaching an embryo or fetus to the uterine wall.
plantar Pertaining to the sole of the foot.
plasma The liquid portion of the blood.
plasma membrane The membrane that separates a cell from its environment.
plasma protein One of several proteins dissolved in the plasma.
platelet A thrombocyte; a cell fragment formed in red bone marrow that contains enzymes initiating blood clot formation.
pleural Pertaining to the pleural membranes.
pleural cavity The potential space between the visceral and parietal pleurae.
pleural membranes The serous membranes that line the thoracic cavity and cover the surfaces of the lungs.
pleurisy Inflammation of the pleural membranes.
plexus A network of nerves or blood vessels.
pneumonia An acute inflammation of the alveoli caused by bacterial or viral infection.
polar body A small, nonfunctional cell formed during oogenesis.
polar molecule A molecule with positive or negative charges on its surface.
polarization The formation of an electrical charge on a plasma membrane due to unequal concentrations of ions on each side of the membrane.
poliomyelitis A viral disease in which motor neurons are destroyed, causing paralysis.
polycythemia A disorder in which there is an excessive number of erythrocytes.
polypeptide An organic compound formed of many amino acids.
polysaccharide An organic compound formed of many monosaccharide units.
pons The part of the brain stem located between the midbrain and the medulla oblongata.
popliteal The body region composed of the posterior portion of the knee.
positive feedback A control mechanism in which an increase in the concentration of a substance stimulates an increase in the production of that substance.
posterior Towards the back; dorsal.
posterior cavity The cavity composed of cranial and spinal cavities; the dorsal cavity.
posterior root The posterior attachment of a spinal nerve to the spinal cord that contains sensory nerve fibers; a dorsal root.
postganglionic fiber Autonomic nerve fibers leading from an autonomic ganglion.
postnatal The period after birth; opposite prenatal.
preganglionic fiber An autonomic nerve fiber leading to an autonomic ganglion.
pregnancy The female condition in which a developing baby is in the uterus.
premenstrual syndrome (PMS) A female disorder occurring just prior to menstruation characterized by pain and emotional stress.
prenatal The period from conception to birth; opposite postnatal.
presbyopia The visual condition in which the near point distance becomes greater with age.

pressure receptor A sensory receptor that is sensitive to pressure changes.
prime mover A muscle whose contraction is primarily responsible for a particular movement.
process A projection on a bone.
progesterone A female reproductive hormone secreted by the corpus luteum and the placenta that maintains the uterine lining.
projection The brain mechanism that makes a sensation seem to come from the body part being stimulated.
prolactin The anterior pituitary hormone that stimulates the production of milk by mammary glands.
prophase The phase of mitosis in which the chromosomes coil, appearing as rodlike structures.
prostaglandins Chemicals produced by cells that affect the functions of nearby cells; "localized hormones."
prostate gland An accessory male reproductive gland that surrounds the base of the urethra.
protein A complex, nitrogen-containing organic compound whose molecules consist of many amino acids.
prothrombin An inactive plasma protein that is converted to thrombin in blood-clot formation.
prothrombin activator Substance released by platelets that converts prothrombin into thrombin.
proton A positively charged particle in an atomic nucleus.
protraction The movement of a body part anteriorly.
proximal Pertaining to nearer the origin of an extremity.
pseudostratified Arrangement of cells that appears layered but is not.
psoriasis A chronic skin disorder characterized by redness, itching, and excessive scaling.
puberty The developmental stage in which the reproductive organs mature and become functional.
pulmonary Pertaining to the lungs.
pulmonary circuit Blood vessels that carry blood from the heart to the lungs and back to the heart.
pulmonary semilunar valve The valve preventing a backflow of blood from the pulmonary trunk into the right ventricle.
pulse The expansion of an artery resulting from a surge of blood generated by ventricular contraction.
pupil The opening in the iris that allows light to pass through the lens to the retina.
Purkinje fibers Specialized cardiac muscle fibers that transmit the cardiac impulse from the A-V bundle to the myocardium.
pus A thick fluid composed of leukocytes and bacteria.
pyelonephritis Inflammation of the kidney involving the renal pelvis.
pyruvic acid An organic molecule formed by the breakdown of glucose in the anaerobic phase of cellular respiration.

R

receptor molecule A specialized protein on a plasma membrane to which a specific chemical attaches.
rectum The terminal portion of the large intestine.
red marrow Blood-forming tissue in the spaces of spongy bone.
referred pain Pain that seems to originate from a site that is different from the site being stimulated.
reflex A rapid response to a stimulus that occurs without voluntary action; an automatic response.
reflex arc The neuron pathway of a reflex; usually involves sensory neuron, interneuron, and motor neuron.
refraction The bending of light rays.
relaxin The hormone secreted by the placenta that softens the cervix and loosens the pubic symphysis.
renal Pertaining to a kidney.
renal calculi Kidney stones.
renal corpuscle The first portion of a nephron consisting of a glomerulus and a glomerular capsule.
renal cortex The outer layer of a kidney.
renal medulla The interior portion of a kidney.
renal pelvis A cavity within a kidney that is continuous with a ureter.
renal tubule The tubular portion of a nephron.
renin The enzyme released by the juxtaglomerular apparatus that converts a plasma protein into angiotensin II.
renin-angiotensin pathway A chemical control mechanism that regulates the blood pressure in glomeruli.
rennin A gastric enzyme that curdles milk proteins.
reproduction The process leading to the formation of offspring.
respiratory center The areas in the pons and medulla oblongata that control breathing.
respiratory distress syndrome A disorder of newborn infants whose alveoli produce insufficient surfactant.
resting potential The polarized state of a muscle or neuron membrane.
reticular formation A neural network in the diencephalon, brain stem, and spinal cord that is connected to higher brain centers and arouses the cerebrum to wakefulness.
retina The inner layer of the eyeball that contains the photoreceptors.
retinoblastoma A cancer of immature retinal cells.
retraction The movement of a body part posteriorly.
rhinitis Inflammation of the nasal cavity.
ribonucleic acid (RNA) A single-stranded nucleic acid whose nucleotides contain ribose sugar.
ribosomal RNA (rRNA) The RNA composing ribosomes.
ribosome A tiny cellular organelle composed of protein and rRNA and serving as the site of protein synthesis.
rickets A disorder caused by a deficiency of vitamin D resulting in weakened bones due to insufficient deposition of calcium salts.
rod A photoreceptor associated with black-and-white vision.
rotation The turning of a body part on its longitudinal axis.
round window A membrane-covered opening between the middle ear and the labyrinth of the inner ear.

S

saccule A portion of the vestibule of the inner ear containing sensory receptors for static equilibrium.
sagittal A plane dividing the body or an organ into left and right portions.
saliva The secretion of salivary glands.

salivary glands Glands secreting saliva into the mouth.
sarcolemma The plasma membrane of a muscle cell.
sarcomere The contractional unit of a myofibril.
sarcoplasm The cytoplasm of a muscle cell.
sarcoplasmic reticulum A series of tubules and channels within a muscle fiber.
saturated fatty acid A fatty acid without double carbon–carbon bonds.
Schwann cell A neuroglial cell that forms the myelin sheath and neurilemma around fibers of peripheral nerves.
sciatica Neuralgia of a sciatic nerve.
sclera The fibrous outer layer of the eyeball; the white of the eye.
scrotum The external pouch containing the testes in males.
sebaceous gland An epithelial gland secreting sebum into a hair follicle.
sebum The oily secretion of a sebaceous gland.
second messenger An intracellular messenger activated by a nonsteroid hormone.
secretin A hormone secreted by the intestinal mucosa that stimulates the pancreas to secrete pancreatic juice.
selectively permeable membrane A membrane that allows some substances to pass through it while keeping out others.
semen The seminal fluid consisting of secretions from accessory reproductive glands and sperm.
semicircular canal A loop of the membranous labyrinth of the inner ear that contains receptors for dynamic equilibrium.
semilunar valve A valve located at the base of the aorta or pulmonary trunk.
seminal vesicles Accessory male reproductive organs whose secretion contribute to semen.
seminiferous tubule. A tubule within which sperm are formed in a testis.
sensation Awareness as a result of the brain interpreting sensory impulses.
sensory adaptation The decrease in impulse formation as a receptor is subjected to continuous stimulation.
sensory area The portion of the cerebrum that interprets sensory impulses.
sensory nerve A nerve composed of sensory neuron processes.
sensory neuron A neuron that carries sensory impulses from a receptor to the central nervous system.
septum A partition such as the interventricular septum.
serotonin A substance released by thrombocytes that constricts blood vessels.
serous fluid Fluid secreted by serous membranes.
serous membrane A membrane that lines enclosed body cavities.
serum The fluid remaining after blood has clotted.
sesamoid bone A small bone embedded in a tendon.
sex hormones Estrogen and testosterone.
sexually transmitted diseases Various infections caused by bacteria or viruses transmitted via sexual intercourse.
shingles Inflammation of a peripheral nerve due to reactivation of the chickenpox virus.
simple epithelium Epithelium composed of a single layer of cells.
simple goiter Enlargement of the thyroid gland caused by a deficiency of iodine in the diet.
simple sugar A monosaccharide.
sinoatrial node (S-A node) Specialized tissue in the right atrium that initiates heart contractions; the pacemaker of the heart.
sinus An air-filled cavity in a bone.
sinusitis Inflammation of a sinus.
skeletal muscle The type of muscle tissue that is attached to bones.
smooth muscle The type of muscle tissue in the walls of hollow organs except the heart.
solute A substance dissolved in a solvent.
solution A fluid composed of solutes dissolved in a solvent.
solvent A fluid that can dissolve solutes.
somatic nervous system (SNS) The portion of the nervous system involved in voluntary responses.
spasm A sudden, involuntary contraction of a muscle.
sperm A male reproductive cell; a male gamete.
spermatid A haploid cell formed in spermatogenesis that matures to become a sperm cell.
spermatocyte A cell that is involved in the meiotic cell divisions of spermatogenesis.
spermatogenesis The process of sperm production.
spermatogonium A diploid cell that divides by mitosis to form primary spermatocytes.
sphincter A circular muscle that closes an opening of a tubular structure.
spinal Pertaining to the spinal cord.
spinal cavity The portion of the posterior cavity containing the spinal cord.
spinal cord The portion of the central nervous system that occupies the vertebral canal.
spinal ganglion A small mass of sensory neuron cell bodies in the posterior root of a spinal nerve.
spinal nerve A nerve that branches from the spinal cord.
spinal plexus A network of spinal nerves in which nerve fibers are resorted before continuing on.
spindle A spool-shaped arrangement of spindle fibers formed during prophase of cell division.
spleen A large lymphatic organ located in the upper left quadrant of the abdomen.
spongy bone Cancellous bone; bone formed of thin interconnected plates with numerous spaces between them.
sprain An injury caused by the tearing of a ligament.
starch A common polysaccharide in foods derived from plants.
static equilibrium Maintenance of balance when the head and body are motionless.
steroid A group of lipids that includes sex hormones and cholesterol.
stimulus A change in the environment.
stomach The expanded portion of the alimentary canal that receives food from the esophagus.
strabismus The inability to focus both eyes simultaneously on the same object.
strain An injury caused by overuse or overstretching of a muscle.
stratified Layered.
stratum basale The innermost layer of the epidermis whose cells are involved in cell division.

stratum corneum The outermost layer of the epidermis composed of keratinized cells.
subarachnoid space The space under the arachnoid mater that is filled with cerebrospinal fluid.
subcutaneous Pertaining to under the skin.
submucosa Connective tissue under a mucous membrane.
substrate A substance acted upon by an enzyme.
sucrose Table sugar; a disaccharide composed of glucose and fructose.
sudoriferous gland A sweat gland.
sulcus A shallow groove such as between two gyri of the cerebrum.
superficial Near the surface of the body or organ.
superior Towards the head.
suppressor cell A type of T lymphocyte that inhibits the formation of antibodies.
surfactant A substance formed by alveoli that decreases the attraction between water molecules and makes it easier for alveoli to fill with air.
suture An immovable joint joining skull bones.
sympathetic nervous system The portion of the autonomic nervous system arising from the thoracic and lumbar regions of the spinal cord.
symphysis A slightly movable joint formed by bones joined by fibrocartilage.
synapse The junction between an axon tip of one neuron and a dendrite or cell body of another neuron.
synaptic cleft The space between an axon tip and another neuron within a synapse.
synaptic knob The enlarged portion of an axon tip.
synaptic transmission The passage of an impulse from one neuron to another across a synapse.
synergist A muscle whose contraction helps the prime mover perform its action.
synovial fluid Fluid secreted by synovial membranes within a joint.
synovial membrane The membrane that lines the interior of freely movable joints.
synthesis The combining of smaller molecules to form more complex molecules.
systemic circuit The blood vessels that carry blood from the heart to all parts of the body except the lungs and back to the heart.
systole The contraction phase of the cardiac cycle.
systolic pressure Arterial blood pressure during ventricular systole.

T

T lymphocyte A type of lymphocyte involved in cell-mediated immunity; a T cell.
target tissue A specific tissue on which a hormone acts.
tarsus The ankle; bones forming the ankle.
taste bud An organ on the tongue containing receptors for taste.
telophase The final phase of cell division in which daughter nuclei and cells are formed.
tendon A band or cord of fibrous connective tissue that attaches a muscle to a bone.
testis A male gonad; the primary male reproductive gland that produces sperm and testosterone.
testosterone The male sex hormone.
tetanus A potentially fatal disorder caused by a bacterial toxin; lockjaw.
tetany A continuous, maximal muscle contraction.
thalamus The portion of the brain forming the walls of the third ventricle inferior to the cerebrum.
thermoreceptor A sensory receptor sensitive to temperature changes.
thoracic Pertaining to the thorax.
thoracic cavity The portion of the anterior cavity located superior to the diaphragm.
threshold stimulus The minimal stimulus that produces formation of a nerve impulse.
thrombin The enzyme that converts fibrinogen into fibrin.
thrombocyte A cell fragment containing an enzyme that starts the formation of a blood clot; a platelet.
thrombosis A condition in which a blood clot blocks a blood vessel.
thrombus A blood clot.
thymine A nitrogen base found only in DNA that pairs with adenine.
thymosin The hormone secreted by the thymus gland.
thymus gland An endocrine gland located in the mediastinum above the heart.
thyroid gland A bilobed endocrine gland located just inferior to the larynx and anterior to the trachea.
thyroid hormone A hormone secreted by the thyroid that increases the rate of cellular metabolism.
thyroid-stimulating hormone (TSH) The anterior pituitary hormone that stimulates the secretion of thyroid hormone.
tissue A group of similar cells performing a similar function.
tissue macrophage system Wandering and fixed macrophages; formerly called the reticuloendothelial system.
tonsillitis Inflammation of the tonsils.
tonsils Masses of lymphoid tissue located around the opening to the pharynx.
toxic shock syndrome A female disorder caused by toxins released by bacteria growing in a tampon.
trachea The air passageway leading from the larynx to the bronchi; the windpipe.
transcription The rewriting of the DNA code into the codon sequence of messenger RNA.
transfer RNA (tRNA) A type of RNA that carries amino acids to a ribosome during protein synthesis.
translation The process in which the codons of mRNA place amino acids of a forming protein in a specific sequence.
transverse plane A plane that divides the body or organ into superior and inferior portions.
tremor Involuntary, repetitive weak contractions.
tricuspid valve The atrioventricular valve between the right atrium and right ventricle.
triglyceride A lipid molecule composed of three fatty acids attached to glycerol; a fat molecule.
trochanter A large, broad process of a bone.
tropic hormone A hormone that stimulates an endocrine gland.
trypsin A pancreatic enzyme that catalyzes the breakdown of protein molecules.
tubal ligation A sterilization procedure in which the uterine tubes are cut and tied.

tubercle A small, rounded process of a bone.
tuberculosis Inflammation of the lungs caused by the bacterium *Mycobacterium tuberculosis.*
tuberosity A raised portion of a bone.
tympanic membrane The thin membrane between the outer and inner ear; the eardrum.

U

umbilical cord The connecting link between the placenta and the developing embryo or fetus.
umbilical region The central abdominopelvic region surrounding the umbilicus.
unsaturated fatty acid A fatty acid with one or more double carbon–carbon bonds.
uracil A nitrogen base in RNA that pairs with adenine.
urea The nitrogenous waste formed by the liver as a result of protein metabolism.
ureter A narrow tube that carries urine from a kidney to the urinary bladder.
urethra A tube carrying urine from the urinary bladder to outside the body.
urethritis Inflammation of the urethra.
uric acid A nitrogenous waste produced from nucleic acid metabolism.
urine Waste and excessive substances removed from the blood by the kidneys and excreted from the body.
uterine Pertaining to the uterus.
uterine tube A tube extending from the uterus to an ovary; it carries female sex cells toward the uterus.
uterus The female organ that contains the embryo and fetus until birth; the womb.
utricle An enlarged portion of the membranous labyrinth of the inner ear containing receptors associated with static balance.
uvula The fleshy appendage that hangs down from the posterior margin of the soft palate.

V

vagina A tubular organ that extends from the vestibule of the vulva to the uterus.
varicose veins Dilated, swollen veins whose valves are nonfunctional.
vas deferens A slender tube leading from an epididymis to the urethra in males.
vascular Pertaining to blood vessels.
vasectomy A sterilization procedure in which the vasa deferentia are cut and tied.
vasoconstriction Narrowing of a blood vessel.
vasodilation Expansion of the lumen of a blood vessel.
vasomotor center The portion of the medulla oblongata that controls the diameter of blood vessels.
vein A blood vessel that carries blood toward the heart.
vena cava A large vein that empties blood into the right atrium of the heart.
ventricle A cavity, such as a pumping chamber of the heart or a fluid-filled space in the brain.
venule A small blood vessel that carries blood from a capillary to a vein.
vesicle A fluid-filled, membranous sac in the cytoplasm of a cell.
vestibule The space between the labia minora.
villus A small, fingerlike projection of the intestinal mucosa.
visceral Pertaining to organs in the abdominal cavity.
visceral pericardium The epicardium.
visceral peritoneum The portion of the peritoneum that covers the surfaces of organs in the abdominal cavity.
visceral pleura The portion of a pleural membrane that covers the surface of a lung.
viscosity The resistance of a fluid to flow.
vitamin An organic nutrient other than a protein, lipid, or carbohydrate that is needed in small amounts from food for normal metabolism.
vitreous humor The gel-like substance filling the posterior cavity of an eye.
vocal cords Folds of tissue within the larynx that vibrate to produce sounds as air passes over them.
voluntary Consciously controlled.
vulva The external female reproductive structures surrounding the vaginal opening.

W

wart A skin tumor caused by a localized virus infection.
water balance The state of equilibrium between water intake and water output.

Y

yellow marrow Marrow filling the medullary cavity of a long bone.
yolk sac An extraembryonic membrane that forms embryonic blood cells.

Z

zygote The cell formed by the union of sperm and ovum; a fertilized egg.

CREDITS

Chapter 1

Page 10: © SPL/Photo Researchers, Inc.

Chapter 2

Page 28: © Hank Morgan/Photo Researchers, Inc.

Chapter 3

3.3: © Stephen L. Wolfe; 3.4a: © D.W. Fawcett/Photo Researchers, Inc.; 3.6a: © Bill Longcore/Photo Researchers, Inc.; 3.7: © G. Murti/Photo Researchers, Inc.; 3.8b: American Lung Association; 3.9: © Manfred Kage/Peter Arnold, Inc.; 3.22b, 3.23b, 3.24b, 3.25b: © Edwin Reschke.

Chapter 4

4.1b, 4.2b: © Edwin Reschke; 4.3b: © Manfred Kage/Peter Arnold, Inc.; 4.4b: © Edwin Reschke; 4.5b: © Fred Hossler/Visuals Unlimited; 4.6a, 4.7b, 4.8b, 4.9b, 4.10b, 4.11b: © Edwin Reschke; 4.12b: © McGraw-Hill Higher Education/Carol D. Jacobson, Ph.D., Dept. of Veterinary Anatomy, Iowa State University; 4.13b: © Victor B. Eichler; 4.14b: © Edwin Reschke/Peter Arnold, Inc.; 4.15b: © Edwin Reschke; 4.16b: © Manfred Kage/Peter Arnold, Inc.; 4.17b, 4.18b: © Edwin Reschke.

Chapter 5

5.1: © SPL/Photo Researchers, Inc.; 5.2: © Edwin Reschke.

Chapter 6

6.3: © BioPhoto Associates/Photo Researchers, Inc.

Chapter 8

8.2: © Edwin Reschke; 8.11: © Martin Rotker/Photo Researchers, Inc.; 8.15b: © Per H. Kjeldsen, University of Michigan.

Chapter 9

9.18a: © Carroll Weiss/Camera M.D.

Chapter 11

11.5: © Warren Rosenberg/BPS; 11.6, 11.7: © Edwin Reschke; 11.8: © Edwin Reschke/Peter Arnold; 11.9: © Edwin Reschke; 11.12: © David M. Phillips/VU.

Chapter 12

12.26a-b: © Carroll Weiss/Camera M.D.

Chapter 13

13.5: © Kent Van De Graaff.

Chapter 17

17.4a: © Biophoto Associates/Photo Researchers, Inc.; 17.5b: © David M. Phillips/Photo Researchers, Inc.; 17.10: © Edwin Reschke/Peter Arnold; 17.15a-e: © McGraw-Hill Higher Education/Bob Coyle, photographer.

Chapter 18

18.1: from M. Tegner and D. Epel. Sea urchin sperm: Eggs interaction studied with the scanning electron microscope. /it/Science/xit/ 179 (16 Feb 1973) pp. 685-88, © AAAS; 18.9: © Donald Yaeger/Camera M.D.; 18.15: © Biophoto Associates/Photo Researchers, Inc.; 18.17: courtesy Integrated Genetics; 18.18: courtesy Colleen Weisz.

INDEX

A

B

C

H

I

J

K

L

M

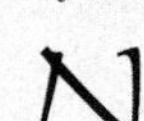

O

P

R

T

U

V

W

T

U

V